AF591397

FUNDAMENTALS OF SEISMIC WAVE PROPAGATION

Fundamentals of Seismic Wave Propagation presents a comprehensive introduction to the propagation of high-frequency, body-waves in elastodynamics. The theory of seismic wave propagation in acoustic, elastic and anisotropic media is developed to allow seismic waves to be modelled in complex, realistic three-dimensional Earth models. This book provides a consistent and thorough development of modelling methods widely used in elastic wave propagation ranging from the whole Earth, through regional and crustal seismology, exploration seismics to borehole seismics, sonics and ultrasonics. Methods developed include ray theory for acoustic, isotropic and anisotropic media, transform techniques including spectral and slowness methods such as the Cagniard and WKBJ seismogram methods, and extensions such as the Maslov seismogram method, quasi-isotropic ray theory, Born scattering theory and the Kirchhoff surface integral method. Particular emphasis is placed on developing a consistent notation and approach throughout, which highlights similarities and allows more complicated methods and extensions to be developed without difficulty. Although this book does not cover seismic interpretation, the types of signals caused by different model features are comprehensively described. Where possible these canonical signals are described by simple, standard time-domain functions as well as by the classical spectral results. These results will be invaluable to seismologists interpreting seismic data and even understanding numerical modelling results.

Fundamentals of Seismic Wave Propagation is intended as a text for graduate courses in theoretical seismology, and a reference for all seismologists using numerical modelling methods. It will also be valuable to researchers in academic and industrial seismology. Exercises and suggestions for further reading are included in each chapter and solutions to the exercises and computer programs are available on the Internet at www.cambridge.org/9780521894548.

CHRIS CHAPMAN is a Scientific Advisor at Schlumberger Cambridge Research, Cambridge, England. Professor Chapman's research interests are in theoretical seismology with applications ranging from exploration to earthquake seismology. He is interested in all aspects of seismic modelling but in particular extensions of ray theory, and anisotropy and scattering with applications in high-frequency seismology. He has developed new methods for efficiently modelling seismic body-waves and used them in interpretation and inverse problems. He held academic positions as an Associate Professor of Physics at the University of Alberta, Professor of Physics at the University of Toronto, and Professor of Geophysics at Cambridge University before joining Schlumberger in 1991. He was a Killam Research Fellow at Toronto, a Cecil H. and Ida Green Scholar at the University of California, San Diego (twice). He is a Fellow of the American Geophysical Union and the Royal Astronomical Society, and an Active Member of the Society of Exploration Geophysicists. Professor Chapman has been an (associate) editor of various journals – *Geophysical Journal of the Royal Astronomical Society, Journal of Computational Physics, Inverse Problems, Annales Geophysics* and *Wave Motion* – and is author of more than 100 research papers.

FUNDAMENTALS OF SEISMIC WAVE PROPAGATION

CHRIS H. CHAPMAN
Schlumberger Cambridge Research

CAMBRIDGE UNIVERSITY PRESS
Cambridge, New York, Melbourne, Madrid, Cape Town, Singapore,
São Paulo, Delhi, Dubai, Tokyo

Cambridge University Press
The Edinburgh Building, Cambridge CB2 8RU, UK

Published in the United States of America by Cambridge University Press, New York

www.cambridge.org
Information on this title: www.cambridge.org/9780521894548

© C. H. Chapman 2004

This publication is in copyright. Subject to statutory exception
and to the provisions of relevant collective licensing agreements,
no reproduction of any part may take place without the written
permission of Cambridge University Press.

First published 2004
Reprinted 2006
This digitally printed version 2010

A catalogue record for this publication is available from the British Library

Library of Congress Cataloguing in Publication data

Chapman, Chris H., 1945–
Fundamentals of seismic wave propagation / Chris H. Chapman.
p. cm.
Includes bibliographical references and indexes.
ISBN 0 521 81538 X
1. Seismic waves. 2. Wave-motion, Theory of. 3. Seismology – Mathematics. I. Title.
QE538.5.C48 2004
551.22 – dc22 2003062528

ISBN 978-0-521-81538-3 Hardback
ISBN 978-0-521-89454-8 Paperback

Cambridge University Press has no responsibility for the persistence or
accuracy of URLs for external or third-party internet websites referred to in
this publication, and does not guarantee that any content on such websites is,
or will remain, accurate or appropriate.

In memory of my parents
Jack (J. H.) and Peggy (M. J.) Chapman

Contents

	Preface	*page* ix
	Preliminaries	xi
	0.1 Nomenclature	xi
	0.2 Symbols	xviii
	0.3 Special functions	xxii
	0.4 Canonical signals	xxii
1	Introduction	1
2	Basic wave propagation	6
	2.1 Plane waves	6
	2.2 A point source	12
	2.3 Travel-time function in layered media	16
	2.4 Types of ray and travel-time results	26
	2.5 Calculation of travel-time functions	45
3	Transforms	58
	3.1 Temporal Fourier transform	58
	3.2 Spatial Fourier transform	65
	3.3 Fourier–Bessel transform	68
	3.4 Tau-p transform	69
4	Review of continuum mechanics and elastic waves	76
	4.1 Infinitesimal stress tensor and traction	78
	4.2 Infinitesimal strain tensor	84
	4.3 Boundary conditions	86
	4.4 Constitutive relations	89
	4.5 Navier wave equation and Green functions	100
	4.6 Stress glut source	118
5	Asymptotic ray theory	134
	5.1 Acoustic kinematic ray theory	134
	5.2 Acoustic dynamic ray theory	145

5.3 Anisotropic kinematic ray theory 163
5.4 Anisotropic dynamic ray theory 170
5.5 Isotropic kinematic ray theory 178
5.6 Isotropic dynamic ray theory 180
5.7 One and two-dimensional media 182
6 Rays at an interface 198
6.1 Boundary conditions 200
6.2 Continuity of the ray equations 201
6.3 Reflection/transmission coefficients 207
6.4 Free surface reflection coefficients 224
6.5 Fluid–solid reflection/transmission coefficients 225
6.6 Interface polarization conversions 228
6.7 Linearized coefficients 231
6.8 Geometrical Green dyadic with interfaces 237
7 Differential systems for stratified media 247
7.1 One-dimensional differential systems 247
7.2 Solutions of one-dimensional systems 253
8 Inverse transforms for stratified media 310
8.1 Cagniard method in two dimensions 313
8.2 Cagniard method in three dimensions 323
8.3 Cagniard method in stratified media 340
8.4 Real slowness methods 346
8.5 Spectral methods 356
9 Canonical signals 378
9.1 First-motion approximations using the Cagniard method 379
9.2 First-motion approximations for WKBJ seismograms 415
9.3 Spectral methods 433
10 Generalizations of ray theory 459
10.1 Maslov asymptotic ray theory 460
10.2 Quasi-isotropic ray theory 487
10.3 Born scattering theory 504
10.4 Kirchhoff surface integral method 532
Appendices
A Useful integrals 555
B Useful Fourier transforms 560
C Ordinary differential equations 564
D Saddle-point methods 569
Bibliography 587
Author index 599
Subject index 602

Preface

The propagation of high-frequency, body waves or 'rays' in elastic media is crucial to our understanding of the interior of the Earth at all scales. Although considerable progress has been made in modelling and interpreting the complete seismic spectrum, much seismic interpretation still relies on ray theory and its extensions. The aim of this book is to provide a comprehensive and consistent development of the modelling methods used to describe high-frequency, body waves in elastodynamics.

Seismology has now developed into a mature science and it would be impossible to describe all aspects of elastic wave propagation in realistic Earth models in one book of reasonable length, let alone the corresponding interpretation techniques. This book makes no pretense at being a comprehensive text. Many important topics are not even mentioned – surface waves apart from interface waves, normal modes, source functions apart from impulsive point sources, attenuation, etc. – and no real data or interpretation methods are included. Many excellent texts, some recent, already cover these subjects comprehensively, e.g. Aki and Richards (1980, 2002), Dahlen and Tromp (1998) and Kennett (2001). This book also assumes a basic understanding of seismology and wave propagation, although these are briefly reviewed. Again many recent excellent undergraduate texts exist, e.g. Shearer (1999), Pujol (2003), etc. The book concentrates on the theoretical development of methods used to model high-frequency, body waves in realistic, three-dimensional, elastic Earth models, and the description of the types of signals generated. Even so in the interests of brevity, some theoretical results that could easily have been included in the text have been omitted, e.g. body-wave theory in a sphere. Further reading is suggested at the end of each chapter, often in the form of exercises.

This book is intended as a text for a graduate or research level course, or reference book for seismologists. It has developed from material I have presented in graduate courses over the years at a number of universities (Alberta, Toronto,

California and Cambridge) and to research seismologists at Schlumberger Cambridge Research. All the material has never been presented in one course – there is probably too much – and some of the recent developments, particularly in Chapter 10, have never been in my own courses. The book has been written so that each theoretical technique is introduced using the simplest feasible model, and these are then generalized to more realistic situations often using the same basic notation as the introductory development. When only a limited amount of material can be covered in a course, these generalizations in the later part of each chapter can be omitted, allowing the important and powerful techniques developed later in the book to be included. Thus the mathematical techniques used in ray theory, reflection and transmission coefficients, transform methods and generalizations of ray theory are all first developed for acoustic waves. Although isotropic and anisotropic elastic waves introduce extra algebraic complications, the basic techniques remain the same. Only a few types of signals, particularly interface waves, specifically require the complications of elasticity.

The material in this text belongs to theoretical seismology but the results should be useful to all seismologists. Some knowledge of physics, wave propagation and applied mathematics is assumed. Most of the mathematics used can be found in one of the many undergraduate texts for physical sciences and engineering – an excellent example is Riley, Hobson and Bence (2002) – and references are not given to the 'standard' results that are in such a book. Where results are less well known or non-standard, some details or references are given in the text. Particular emphasis is placed on developing a consistent notation and approach throughout, which highlights similarities and allows more complicated methods and extensions to be developed without difficulty. Although this book does not cover seismic interpretation, the types of signals caused by different model features are comprehensively described. Where possible these canonical signals are described by simple, standard time-domain functions as well as by the classical spectral results.

Many of the diagrams were drawn using Matlab. Programming exercises suggest using Matlab and solutions have been written using Matlab. Matlab is a trademark of MathWorks, Inc.

I would like to thank various people at Schlumberger Cambridge Research for providing the time and facilities for me to write this book, in particular my department heads – Phil Christie, Tony Booer, Dave Nichols and James Martin – and manager – Mike Sheppard – who introduced personal research time which I have used to complete the manuscript. I am also indebted to Schlumberger for permission to publish. Finally, I would like to thank my family particularly Heather, my daughter, who helped me with some of the diagrams and Lillian, my wife, for her infinite patience and support.

Preliminaries

Unfortunately, the nomenclature, symbols and terms used in theoretical seismology have not been standardized in the literature, as they have in some other subjects. It would be a vain hope to rectify this situation now, but at least we can attempt to use consistent conventions, adequate for the task, throughout this book. While it would be nice to use the most sophisticated notation to allow for complete generality, rigour and developments in the future, one has to be a realist. Most seismologists have to use and understand the results of theoretical seismology, without being mathematicians. Thus the phrase *adequate* or *fit for the task* is adhered to. Unfortunately, some mathematicians will find the methods and notation naive, and some seismologists will still not be able to follow the mathematics, but hopefully the middle ground of an audience of typical seismologists and physical scientists will find this book useful and at an appropriate level.

0.1 Nomenclature

0.1.1 Homogeneous and inhomogeneous

The words *homogeneous* and *inhomogeneous* are used with various meanings in physics and mathematics. They are overused in wave propagation with at least four meanings: *inhomogeneous medium* indicating a medium where the physical parameters, e.g. density, vary with position; *inhomogeneous wave* when the amplitude varies on a wavefront; *inhomogeneous differential equation* for an equation with a source term independent of the field variable; and, *homogeneous boundary conditions* where either displacement or traction is zero. To avoid confusion, as the four usages could occur in the same problem, we will use it in the first sense and avoid the others except in some limited circumstances.

0.1.2 Order, dimensions and units

The term *dimension* is used with various meanings in physics and mathematics. In dimensional analysis, it is used to distinguish the dimensions of mass, length and time. In vector-matrix algebra, the dimension counts the number of components, e.g. the vector (x_1, x_2, x_3) has dimension 3. We have found dimensional analysis extremely useful to check complicated algebraic expressions which may include vectors. In order to avoid confusion, we use the term *units* to describe this usage, e.g. the velocity has units $[\mathrm{LT}^{-1}]$ (we appreciate that this usage is less than rigorous – in dimensional analysis, the dimensions of velocity are $[\mathrm{LT}^{-1}]$ while the units are km/s or m/s, etc. – but have been unable to find a simple alternative). We also use the term *order* to describe the order of a tensor, i.e. the number of indices. The velocity is a first-*order* tensor, of *dimension* 3 with *units* $[\mathrm{LT}^{-1}]$. The elastic parameters are a fourth-*order* tensor, of *dimension* $3 \times 3 \times 3 \times 3$ with *units* $[\mathrm{ML}^{-1}\mathrm{T}^{-2}]$.

0.1.3 Vectors and matrices

Mathematicians would rightly argue that a good notation is an important part of any problem. Generality and abstractness become a virtue. We will take a somewhat more pragmatic approach and would also argue that the notation should suite the intended audience. Most seismologists are happy with vector and matrix algebra, but become less comfortable with higher-order tensors, regarding a second-order tensor as synonymous with a matrix. The learning curve to understand fully, higher-order tensors and their notation is probably not justified by the elegance or intellectual satisfaction achieved. The ability to 'read' or 'visualize' the notation of an equation outweighs the compactness and generality that can be achieved with more sophisticated mathematics and notation. Naturally this is a very subjective choice, but we have tried to achieve a compact notation while only assuming vector and matrix algebra. We use **bold** symbols to denote vectors, matrices and tensors, without any over-arrow or under-score. The dimension of the object should be obvious from the context, e.g. we use $\mathbf{I}$ for the identity matrix whatever its dimension. A vector is normally equivalent to a column matrix, e.g.

$$\mathbf{x} = \begin{pmatrix} x \\ y \\ z \end{pmatrix}. \tag{0.1.1}$$

In the text we sometimes write a vector as, for instance, $\mathbf{x} = (x, y, z)$ but this is still understood as a column matrix. A row matrix would be written $\mathbf{x} = \begin{pmatrix} x & y & z \end{pmatrix}$. Unit or normalized vectors are denoted by a hat, e.g. $\hat{\mathbf{x}}$, and we generalize the sign function (sgn) of a scalar so

$$\boxed{\hat{\mathbf{x}} = \operatorname{sgn}(\mathbf{x}) = \mathbf{x}/|\mathbf{x}|\ .} \tag{0.1.2}$$

For a matrix, sgn is the signature of the matrix, i.e. the number of positive eigenvalues minus the number of negative eigenvalues.

An advantage of vector-matrix algebra is that it is straightforward to check that the dimensions in an equation agree. Thus if $\mathbf{a}$ is an $(l \times m)$ matrix, and $\mathbf{b}$ is $(m \times n)$, the matrix $\mathbf{c} = \mathbf{a}\,\mathbf{b}$ is $(l \times m) \times (m \times n) = (l \times n)$, i.e. the repeated dimension m cancels in the product. If $\mathbf{a}$ and $\mathbf{b}$ are (3×1) vectors, then $\mathbf{a}^{\mathrm{T}}\mathbf{b}$ is (1×1), i.e. this is the scalar product, and $\mathbf{a}\mathbf{b}^{\mathrm{T}}$ is a (3×3) matrix. It is probably worth commenting that this useful chain-dimension rule is widely broken with scalars. Thus we write $\mathbf{c} = a\mathbf{b}$, where a is a scalar and $\mathbf{b}$ and $\mathbf{c}$ vectors or matrices with the same dimension. Some care is sometimes needed to maintain the chain-dimension rule if a scalar is obtained from a scalar product. We do not use the notation $\mathbf{a}\,\mathbf{b}$ to represent a tensor (in vector-matrix algebra, the dimensions would be inconsistent).

We use an underline to indicate a Green function, i.e. solutions for elementary, point sources. Thus $\mathbf{u}$ is the particle displacement while $\underline{\mathbf{u}}$ is the particle displacement Green function and typically contains solutions for three unit-component, body-force sources. The underline indicates an extra dimension, i.e. $\underline{\mathbf{u}}$ is 3×3.

For cartesian vectors we use the notations x, y and z, and x_i, $i = 1$ to 3 interchangeably. The former is physically more descriptive, whereas the latter is mathematically more useful as we can exploit the *Einstein summation convention*, i.e. $a_i b_i$ with a repeated index means $a_1 b_1 + a_2 b_2 + a_3 b_3$. Similarly, we use $\hat{\mathbf{i}}$, $\hat{\mathbf{j}}$ and $\hat{\mathbf{k}}$ for unit cartesian vectors or $\hat{\mathbf{i}}_i$, $i = 1$ to 3.

Sometimes it is useful to consider a restricted range of components. We follow the standard practice of using a Greek letter for the subscript, i.e. p_ν with $\nu = 1$ to 2 are two components of the vector $\mathbf{p}$ with components p_i, $i = 1$ to 3. Thus $p_\nu p_\nu = p_1^2 + p_2^2$. The two-dimensional vector formed from these components is denoted by p, a sub-space vector in sans serif font. Thus $\mathsf{p}^{\mathrm{T}}\mathsf{p} = p_\nu p_\nu$. More generally we use the sans serif font to indicate variables in which the dimensionality is restricted, in some sense, e.g. we use S and R to indicate the source and receiver, as in $\mathbf{x}_{\mathsf{S}}$ and $\mathbf{x}_{\mathsf{R}}$, as they are normally restricted to lie on a plane or line.

In general, a matrix of any dimension is denoted by a bold symbol, e.g. $\mathbf{A}$. The dimension should be obvious from the context and is described as $m \times n$, where the matrix has m rows and n columns. An element of the matrix is A_{ij}, or occasionally $(\mathbf{A})_{ij}$ where $i = 1$ to m and $j = 1$ to n. We frequently form larger matrices from vectors or smaller matrices. Thus, for instance, the 3×3 identity matrix can be formed from the unit cartesian vectors, i.e. $\mathbf{I} = (\,\hat{\mathbf{i}}\ \ \hat{\mathbf{j}}\ \ \hat{\mathbf{k}}\,)$. Similarly we write

$$\mathbf{A} = \begin{pmatrix} \mathbf{A}_{11} & \mathbf{A}_{11} \\ \mathbf{A}_{21} & \mathbf{A}_{22} \end{pmatrix}, \tag{0.1.3}$$

where $\mathbf{A}$ is $2n \times 2n$, and $\mathbf{A}_{ij}$ are $n \times n$ sub-matrices. Conversely, we sometimes need to extract sub-matrices. The notation we use is

$$\boxed{\mathbf{A}^{(i_1 \ldots i_\mu) \times (j_1 \ldots j_\nu)}}, \tag{0.1.4}$$

for a $\mu \times \nu$ matrix formed from the intersections of the i_1-th, i_2-th to i_μ-th rows and j_1-th, j_2-th to j_ν-th columns of $\mathbf{A}$. If all the rows or columns are included, we abbreviate this as (.). Thus $\mathbf{A}^{(.)\times(j)}$ is the j-th column of the matrix $\mathbf{A}$, $\mathbf{A}^{(.)\times(.)}$ is the same as $\mathbf{A}$, and $\mathbf{A}^{(i)\times(j)}$ is the element A_{ij}.

0.1.4 Identity matrices

We make frequent use of the identity matrix which we denote by $\mathbf{I}$ whatever the size of the square matrix, i.e. the elements are $(\mathbf{I})_{ij} = \delta_{ij}$, where δ_{ij} is the Kronecker delta. Frequently we need matrices which change the order of rows and columns and possibly the signs. For this purpose, it is useful to define the matrices

$$\mathbf{I}_1 = \begin{pmatrix} \mathbf{0} & \mathbf{I} \\ -\mathbf{I} & \mathbf{0} \end{pmatrix}, \quad \mathbf{I}_2 = \begin{pmatrix} \mathbf{0} & \mathbf{I} \\ \mathbf{I} & \mathbf{0} \end{pmatrix} \quad \text{and} \quad \mathbf{I}_3 = \begin{pmatrix} -\mathbf{I} & \mathbf{0} \\ \mathbf{0} & \mathbf{I} \end{pmatrix}, \tag{0.1.5}$$

where we have used the notation (0.1.3).

0.1.5 Gradient operator

The gradient operator ∇ can be considered as a vector (we break the above rule and do not use a bold symbol here). Thus

$$\boxed{\nabla = \begin{pmatrix} \partial/\partial x_1 \\ \partial/\partial x_2 \\ \partial/\partial x_3 \end{pmatrix}}. \tag{0.1.6}$$

Treating ∇ as a 3×1 vector, $\nabla\phi$, the gradient, is also a $(3 \times 1) \times (1 \times 1) = (3 \times 1)$ vector. The divergence $\nabla \cdot \mathbf{u}$ can be rewritten $\nabla^{\mathrm{T}}\mathbf{u}$ which is $(1 \times 3) \times (3 \times 1) = (1 \times 1)$, i.e. a scalar. Similarly $\nabla\mathbf{u}^{\mathrm{T}}$ is a 3×3 matrix with elements $\partial u_i/\partial x_j$. Thus ∇ can be treated as a vector satisfying the normal rules of vector-matrix algebra. However, algebra using this notation is limited to scalars, vectors and second-order tensors. Thus we can write

$$\nabla\,(\mathbf{a}^{\mathrm{T}}\mathbf{b}) = (\nabla\mathbf{a}^{\mathrm{T}})\,\mathbf{b} + (\nabla\mathbf{b}^{\mathrm{T}})\,\mathbf{a}\,, \tag{0.1.7}$$

for the 3×1 gradient of a scalar product. The 3×3 second derivatives of a scalar ϕ can be written $\nabla (\nabla\phi)^{\mathrm{T}}$ but we cannot expand it for a scalar product, for

$$\frac{\partial^2}{\partial x_i \partial x_j} (\mathbf{a}^{\mathrm{T}}\mathbf{b}) = a_k \frac{\partial^2 b_k}{\partial x_i \partial x_j} + b_k \frac{\partial^2 a_k}{\partial x_i \partial x_j} + \frac{\partial a_k}{\partial x_i}\frac{\partial b_k}{\partial x_j} + \frac{\partial a_k}{\partial x_j}\frac{\partial b_k}{\partial x_i}, \qquad (0.1.8)$$

and the right-hand side involves third-order tensors which are contracted with a vector. When such expressions arise, e.g. in Chapter 10, we must use the full subscript notation rather than the compact vector-matrix notation. We note that the invariant Gibbs notation (Dahlen and Tromp, 1998, §A.3) is elegant for vectors and second-order tensors, but also becomes unwieldy for higher-order tensors, when subscript notation is preferred (Dahlen and Tromp, 1998, pp. 821–822).

Partial derivatives are normally written as, for instance, $\partial\phi/\partial t$. We avoid the notation $\phi_{,t}$ to prevent confusion with other subscripts. In general we try to avoid subscripts by using vector-matrix notation.

0.1.6 The vertical coordinate

Where possible we try to avoid specific coordinates by using vector notation, but frequently it is necessary to use coordinates, usually cartesian. In the Earth, we invariably define a vertical axis, even though gravity is neglected throughout this book, as interfaces, including the surface of the Earth, are approximately horizontal. We follow the usual practice of using the z coordinate for the vertical direction, and the horizontal plane is defined by the x and y coordinates. Much confusion is caused by the choice of the direction of positive z, and we can do nothing to avoid this as both choices are common in the literature. All we can do is to be consistent (justifying including these comments here). Throughout this book, z is measured *positive in the upwards direction*, i.e. depths are in the negative z direction. The choice is arbitrary and we only justify this by noting that when the sphericity of the Earth is important, it is convenient that radius is measured in the same direction as the vertical, so the signs of gradients are the same. Drawing z vertically upwards in diagrams, and x positive to the right, means that in a right-handed system of axes, y is measured *into* the page.

Further confusion arises from the convention used for numbering interfaces and layers in a horizontally layered model. Again, all we can do is to be consistent throughout this book. We define the ℓ-th interface as being at $z = z_\ell$, and number the interfaces from the top downwards. The ℓ-th layer is between the ℓ-th and $(\ell + 1)$-th interfaces, i.e. $z_\ell \geq z \geq z_{\ell+1}$ with the ℓ-th interface at the top of the ℓ-th layer. Normally, the first and n-th layers are half-spaces and the source lies at

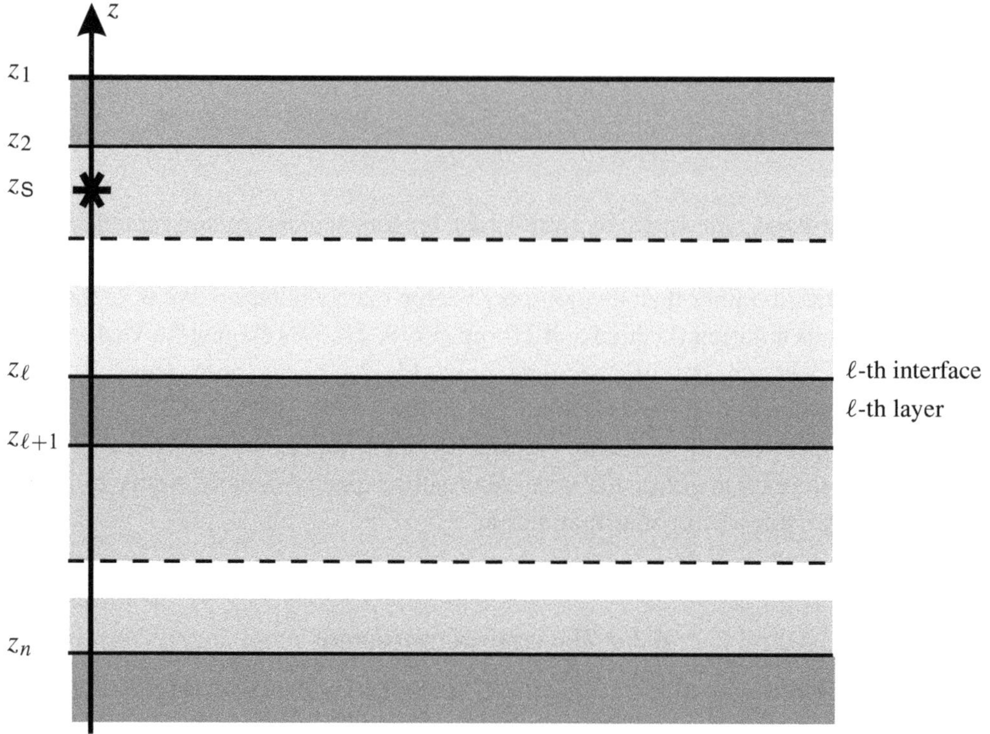

Fig. 0.1. The interface/layer arrangement for a flat, layered model illustrating the free surface at z_1, source at z_S, the ℓ-th interface at z_ℓ, the ℓ-th layer between z_ℓ and $z_{\ell+1}$ and the final layer or half-space below the interface z_n.

$z = z_S$. When needed, the free surface of the Earth is at $z = z_1$. This is illustrated in Figure 0.1.

0.1.7 Acronyms

For brevity, several acronyms are used throughout the text. Most of these are widely used but for completeness we list them here:

ART asymptotic ray theory is the mathematical theory used to describe seismic rays (see Chapter 5). The solution is normally expressed as an asymptotic series in inverse powers of frequency.

AVO amplitude versus offset describes the behaviour of the amplitude of a reflected signal versus the offset from source to receiver. Normally factors of geometrical spreading and attenuation are removed so the amplitude changes are described by the reflection coefficient (see Chapter 6).

FFT fast Fourier transform is the efficient numerical algorithm used to evaluate the discrete Fourier transform, which is usually used as an approximation for a Fourier integral (see Chapter 3).

GRA geometrical ray approximation is the approximation usually used to describe seismic rays (see Chapter 5). Usually it is the zeroth-order term in the ART ansatz, i.e. the amplitude is independent of frequency and the phase depends linearly on frequency.

KMAH the Keller (1958), Maslov (1965, 1972), Arnol'd (1973) and Hörmander (1971) index counts the caustics along a ray (see Chapter 5). In isotropic, elastic media it increments by one at each line caustic and two at each point caustic, but in anisotropic media and for some other types of waves, it may decrement.

NMO normal moveout describes the approximate behaviour of a reflected arrival near zero offset (see Section 2.5.1). The travel time is approximately a parabolic function of offset.

QI quasi-isotropic ray theory is a variant of ART used to describe the coupling of quasi-shear waves that occurs in heterogeneous, anisotropic media when the shear wave velocities are similar (see Section 10.2).

SOFAR a minimum of the acoustic velocity at about 1.5 km depth in the deep ocean (above the velocity increases because of the temperature increase, and below it increases because of the increased hydrostatic pressure) forms the deep ocean sound (or SOFAR) channel. Energy can be trapped in the SOFAR channel, with rays turning above and below the minimum, and sound propagates to far distances (see Section 2.3).

TI a transversely isotropic medium is an anisotropic medium with an axis of symmetry, i.e. the elastic properties only varying as a function of angle from the symmetry axis and are axially symmetric (see Section 4.4.4). It is also known as hexagonal or polar anisotropy.

WKB the Wentzel (1926), Kramers (1926), Brillouin (1926) asymptotic solution is an approximate solution of differential equations. It is widely used for wave equations at high frequencies when the phase varies as the integral of the local wavenumber and the amplitude varies to conserve energy flux. The WKB solution is very useful but, however many terms are taken in the asymptotic solution, it does not describe reflected signals from a smooth but rapid change in properties – the so-called WKB paradox (see Section 7.2.5).

WKBJ a variant of the WKB acronym used by geophysicists to honour Jeffreys' (1924) contribution (another variant is JWKB). The WKBJ asymptotic expansion (Section 7.2.5), WKBJ iterative solution (Section 7.2.6) and WKBJ seismogram (Section 8.4.1 – so-called as it only depends on the WKBJ approximation) are important solutions for studying seismic waves in stratified media.

0.2 Symbols

As the symbols and notation used in seismology have never been successfully standardized, we tabulate in Table 0.1 many of the symbols used in this book, their units, description and reference to an equation where they are first used. The list is not exhaustive and all variants are not included, e.g. forms with alternative subscripts and arguments are not listed. Symbols that are only used locally are not included.

Table 0.1. *Symbols, units, description and first equations used in the text (continued on the following pages).*

Symbols	Units	Description	Equation
$\mathbf{x}, \mathbf{r}$	[L]	position vector	(0.1.1)
$\mathsf{q}, \mathbf{Q}, q_\nu$	[L]	wavefront coordinates	(5.1.21)
$\hat{\mathbf{i}}, \hat{\mathbf{j}}, \hat{\mathbf{k}}$	[0]	unit coordinate vectors	(2.5.26)
$\hat{\mathbf{n}}$	[0]	unit surface normal	(4.1.4)
R	[L]	radial length	(2.2.2)
d	[L]	layer thickness	(8.1.3)
$\mathbf{k}, k$	[L^{-1}]	wavenumber	(2.1.2)
λ	[L]	wavelength	
$s, \mathcal{L}$	[L]	ray length	(5.1.12)
ℓ	[0]	dimension	(5.4.28)
$\mathcal{R}^{(\ell)}$	[L]	effective ray length	(5.4.25)
t	[T]	time	(2.1.1)
ω	[T^{-1}]	circular frequency	(2.1.2)
ν	[T^{-1}]	frequency	
c, α, β	[LT^{-1}]	wave (phase) velocity	(2.1.1)
$\mathbf{V}, V$	[LT^{-1}]	ray (group) velocity	(5.1.14)
$\mathbf{p}, p, q$	[L^{-1}T]	slowness	(2.1.3)
p	[L^{-1}T]	sub-space slowness	(3.2.13)
θ	[0]	phase-normal angle	(2.1.4)
ϕ	[0]	phase-group angle	(5.3.32)
$X, \mathbf{X}, \mathsf{X}$	[L]	range function	(2.3.4)
T	[T]	travel-time function	(2.2.8)
τ	[T]	intercept-time function	(2.3.15)
$\bar{T}, \bar{t}$	[T]	reduced travel time	(2.5.48)
$\widetilde{T}$	[T]	generalized time function	(8.1.2)

Table 0.1. *continued. Note the units of the Green functions are for the three-dimensional case. An extra unit of* [L] *exists in two dimensions. The units of the field variables, e.g.* **u**, *are in the temporal and spatial domain. An extra unit of* [T] *exists in the spectral domain.*

Symbols	Units	Description	Equation
$\mathbf{t}$	$[\mathrm{ML^{-1}T^{-2}}]$	traction	(4.1.1)
σ	$[\mathrm{ML^{-1}T^{-2}}]$	stress	(4.1.5)
P	$[\mathrm{ML^{-1}T^{-2}}]$	pressure	(4.4.1)
$\mathbf{u}$	$[\mathrm{L}]$	particle displacement	(4.2.1)
$\mathbf{v}$	$[\mathrm{LT^{-1}}]$	particle velocity	(4.1.19)
$\mathbf{e}$	$[0]$	strain	(4.2.2)
θ	$[0]$	dilatation	(4.2.8)
ρ	$[\mathrm{ML^{-3}}]$	density	(4.0.1)
κ	$[\mathrm{ML^{-1}T^{-2}}]$	bulk modulus	(4.4.3)
k	$[\mathrm{M^{-1}LT^{2}}]$	compressibility	(4.4.4)
c_{ijkl}, C_{ij}, $\mathbf{c}_{ij}$	$[\mathrm{ML^{-1}T^{-2}}]$	elastic stiffnesses	(4.4.5)
a_{ijkl}, A_{ij}	$[\mathrm{L^{2}T^{-2}}]$	squared-velocity parameters	(5.7.19)
s_{ijkl}, S_{ij}, $\mathbf{s}_{ij}$	$[\mathrm{M^{-1}LT^{2}}]$	elastic compliances	(4.4.40)
λ, μ	$[\mathrm{ML^{-1}T^{-2}}]$	Lamé elastic parameters	(4.4.49)
$\mathbf{f}$	$[\mathrm{ML^{-2}T^{-2}}]$	body force per unit volume	(4.1.7)
$\mathbf{f}_{\mathrm{S}}$	$[\mathrm{MLT^{-1}}]$	point impulse	(4.5.76)
$\mathbf{M}$, $\mathbf{M}_{\mathrm{S}}$	$[\mathrm{ML^{2}T^{-2}}]$	moment tensor	(4.6.5)
$\underline{\mathbf{u}}$	$[\mathrm{M^{-1}T}]$	Green particle displacement	(4.5.17)
$\underline{\mathbf{v}}$	$[\mathrm{M^{-1}}]$	Green particle velocity	(4.5.20)
$\underline{P}$	$[\mathrm{L^{-2}T^{-1}}]$	Green pressure	(4.5.20)

Table 0.1. *continued. Note ℓ is the dimensionality of the solution (2 or 3). $\mathcal{P}^{(\ell)}(t)$ has an extra unit* [T] *compare with $\mathcal{P}^{(\ell)}(\omega)$. The Green amplitude coefficients $\underline{\mathbf{v}}^{(0)}$, $\underline{P}^{(0)}$ and $\underline{\mathbf{t}}_j^{(0)}$ are for three-dimensions. They have extra units of* $[\mathrm{LT}^{-1/2}]$ *in two dimensions.*

Symbols	Units	Description	Equation
$\mathbf{y}$	–	phase space vector	(5.1.28)
$\mathbf{v}^{(m)}$, $v^{(0)}$	–	velocity amplitude coefficients	(5.1.1)
$P^{(m)}$	–	pressure amplitude coefficients	(5.1.1)
$\mathbf{t}_j^{(m)}$	–	traction amplitude coefficients	(5.3.1)
$\hat{\mathbf{g}}$	[0]	normalized polarization	(5.2.4)
$\mathbf{g}$	$[\mathrm{M}^{-1/2}\mathrm{LT}^{1/2}]$	energy flux normalized polarization	(5.4.33)
$\boldsymbol{\mathcal{G}}$	$[\mathrm{M}^{-1}\mathrm{L}^2\mathrm{T}]$	polarization dyadic	(8.0.8)
$H(\mathbf{x}, \mathbf{p})$	[0]	Hamiltonian	(5.1.18)
$Z, \mathbf{Z}_i$	$[\mathrm{ML}^{-2}\mathrm{T}^{-1}]$	impedance	(5.2.8)
$\mathbf{N}$	$[\mathrm{MT}^{-3}]$	energy flux vector	(5.2.11)
σ	[0]	KMAH index	(5.2.70)
$\boldsymbol{\mathcal{L}}, \boldsymbol{\mathcal{M}}, \boldsymbol{\mathcal{N}}$	–	anisotropic ART operators	(5.3.5)
$\mathbf{D}$	–	dynamic differential system	(5.2.19)
J	–	ray tube cross-section	(5.2.12)
D	[0]	Jacobian volume mapping	(5.2.14)
$\mathbf{P}$, P	–	dynamic propagator matrix	(5.2.23)
$\mathbf{J}$	–	dynamic fundamental matrix	(5.2.29)
$\mathbf{M}$, M	$[\mathrm{L}^{-2}\mathrm{T}]$	wavefront curvature matrix	(5.2.46)
$\mathcal{S}^{(\ell)}$	$[\mathrm{L}^{2(\ell-1)}\mathrm{T}^{1-\ell}]$	ray spreading function	(5.2.67)
$\mathcal{T}^{(\ell)}$	$[\mathrm{L}^{1-\ell}\mathrm{T}^{(\ell-1)/2}]$	ray scalar amplitude	(5.4.34)
$\mathcal{P}^{(\ell)}(\omega)$	$[\mathrm{L}^{1-\ell}\mathrm{T}^{(\ell-1)/2}]$	ray propagation term	(5.4.36)
Φ, $\boldsymbol{\Phi}$	[0]	ray phase term	(5.4.28)
$\underline{\mathbf{v}}^{(0)}$	$[\mathrm{M}^{-1}\mathrm{T}^2]$	Green particle amplitude coefficient	(5.4.28)
$\underline{P}^{(0)}$, $\underline{\mathbf{t}}_j^{(0)}$	$[\mathrm{L}^{-2}\mathrm{T}]$	Green stress amplitude coefficient	(5.4.28)
$f^{(\ell)}(\omega)$	$[\mathrm{T}^{(1-\ell)/2}]$	source spectral term	(5.2.64)
$\mathcal{M}_\mathsf{S}$	–	source excitation functions	(8.0.12)
$\boldsymbol{\mathcal{H}}_\mathsf{R}$	–	receiver conversion coefficients	(8.0.12)

Table 0.1. *concluded. The error terms, $\underline{\boldsymbol{\mathcal{E}}}^{\mathrm{N}}$ and $\underline{\boldsymbol{\mathcal{E}}}_j^{\mathrm{H}}$, are for three dimensions. They have an extra unit* [L] *in two dimensions.*

Symbols	Units	Description	Equation
$\hat{\mathbf{l}}$, $\hat{\mathbf{m}}$, $\hat{\mathbf{n}}$	[0]	interface basis vectors	(6.0.1)
$\mathbf{w}$, $\mathbf{W}$	–	velocity-traction vectors	(6.1.1)
$\mathcal{T}_{ij}$, $\boldsymbol{\mathcal{T}}$, $\mathbf{R}$, $\mathbf{T}$	[0]	reflection/transmission coefficients	(6.3.4)
$\mathbf{h}$	$[\mathrm{M}^{-1/2}\mathrm{L}\mathrm{T}^{1/2}]$	interface polarization conversion	(6.6.3)
Ω	$[\mathrm{L}^{-2}\mathrm{T}^{2}]$	$q_\beta^2 - p^2$	(6.3.54)
Δ	–	coefficient denominator	(6.3.61)
Y	$[\mathrm{M}^{-1}\mathrm{L}^{2}\mathrm{T}]$	admittance	(9.1.28)
$\mathbf{A}$	–	differential systems	(6.3.1)
$\mathbf{Q}$	[0]	interface ray discontinuity	(6.3.22)
$\mathbf{C}$	[0]	coupling matrix	(6.7.6)
$[\gamma]$	[0]	perturbation coefficients	(6.7.22)
γ	$[\mathrm{L}^{-1}]$	differential coefficients	(7.2.96)
$\mathbf{r}$	[0]	component vector	(7.2.7)
$\boldsymbol{\mathcal{F}}$	–	transformed source matrix	(7.1.7)
$\mathbf{F}$	–	plane-wave fundamental matrix	(7.2.2)
$\mathbf{P}$	–	plane-wave propagator matrix	(7.2.4)
$\mathbf{L}$	–	Langer matrix	(7.2.132)
$\mathbf{B}$	$[\mathrm{L}^{-1}\mathrm{T}]$	Langer differential matrix	(7.2.134)
$\boldsymbol{\mathcal{A}}$	–	Langer propagator	(7.2.144)
$\mathbf{X}$	–	phase propagator	(7.2.202)
ℓ	[L]	head-wave length	(9.1.41)
γ	$[\mathrm{L}\mathrm{T}^{-1}]$	Rayleigh wave velocity	(9.1.63)
B_{JK}	$[\mathrm{L}^{2}\mathrm{T}^{-2}]$	ray perturbation terms	(10.2.8)
$\underline{\boldsymbol{\mathcal{E}}}^{\mathrm{N}}$	$[\mathrm{L}^{-3}]$	error in equation of motion	(10.3.32)
$\underline{\boldsymbol{\mathcal{E}}}_j^{\mathrm{H}}$	$[\mathrm{M}^{-1}\mathrm{L}^{-1}\mathrm{T}^{2}]$	error in constitutive equation	(10.3.33)
$\underline{\boldsymbol{\mathcal{D}}}^{(\ell)}$	$[\mathrm{M}^{-1}\mathrm{L}^{4-2\ell}\mathrm{T}^{\ell}]$	scattering dyadic	(10.3.53)
Γ^{E}	$[\mathrm{L}^{-1}]$	scalar Born error kernel	(10.3.53)
Γ^{B}	$[\mathrm{L}^{-1}\mathrm{T}]$	scalar Born perturbation kernel	(10.3.71)
Γ^{K}	[0]	scalar Kirchhoff kernel	(10.4.11)

0.3 Special functions

Special functions, both standard and non-standard, are very useful in the mathematics of wave propagation. Many special functions – trigonometrical, hyperbolic, Bessel, Hankel, Airy, etc. – are widely used and the definitions and notation are (almost) standardized. We follow the definitions given in Abramowitz and Stegun (1965). Other functions are not as widely used and some are unique to this book. Table 0.2 lists these and where they are first used or defined.

Table 0.2. *Special functions, description and equations where used first.*

Function	Description	Equation
$[\ldots]$	saltus function	(2.4.1)
$\text{unit}(\ldots) = [\ldots]$	unit function	Section 0.1.3
$[\ldots, \ldots]$	commutator	(6.7.12)
$\{\ldots\}$	second-order minors	(7.2.205)
$\text{Re}(\ldots)$	real part	(3.1.10)
$\text{Im}(\ldots)$	imaginary part	(3.1.13)
$\text{sgn}(\ldots)$	sign (scalar)	(2.1.9)
	normalized (vector)	(0.1.2)
	signature (matrix)	(10.1.44)
$\delta(t)$	Dirac delta function	(3.1.19)
$\Delta(t)$	analytic delta function	(3.1.21)
$\delta^{(m)}(t)$	m-th integral of delta function	(5.1.2)
$H(t)$	Heaviside step function	(4.5.71)
$B(t)$	boxcar function	(8.4.13)
$\lambda(t)$	lambda function $H(t)\,t^{-1/2}$	(4.5.85)
$\Lambda(t)$	analytic lambda function	(5.2.82)
$\mu(t)$	mu function $H(t)\,t^{1/2}$	(9.2.30)
$Aj(x)$, $Bj(x)$	generalized Airy functions	(7.2.144)
$J(t)$	two-dimensional Green function	(8.1.37)
$C(t, x)$	Airy caustic function	(9.2.46)
$Fi(t)$	Fresnel time function	(9.2.60)
$F(t, y)$	Fresnel shadow function	(9.2.61)
$Fr(\omega)$	Fresnel spectral function	(9.3.60)
$Tun(\omega, \psi)$	tunnelling spectral function	(9.3.41)
$Tun(t, \psi)$	tunnelling time function	(9.3.45)
$Sh^{(m)}(t)$	deep shadow funtion	(9.3.101)

0.4 Canonical signals

The main objective of this book is to outline the mathematics and results necessary to describe the canonical signals generated by an impulsive source in an elastic, heterogeneous medium. Table 0.3 lists the first-motion approximations for

Table 0.3. *Signal types with their first-motion approximation and approximate spectra for an impulsive point source.*

Signal	First-Motion	Spectrum
Direct ray (homogeneous) (4.5.75)	$\delta(t - R/c)/R$	$\mathrm{e}^{\mathrm{i}\omega R/c}/R$
Direct ray (stratified) (5.7.17) (9.1.5)	$\delta(t - T)/(\mathrm{d}X/\mathrm{d}p)^{1/2}$	$\mathrm{e}^{\mathrm{i}\omega T}/(\mathrm{d}X/\mathrm{d}p)^{1/2}$
Direct ray (inhomogeneous) (5.4.27)	$\delta(t - T)/\mathcal{R}$	$\mathrm{e}^{\mathrm{i}\omega T}/\mathcal{R}$
Partial reflection (interface) (9.1.5)	$\mathcal{T}\delta(t - T)/(\mathrm{d}X/\mathrm{d}p)^{1/2}$	$\mathcal{T}\mathrm{e}^{\mathrm{i}\omega T}/(\mathrm{d}X/\mathrm{d}p)^{1/2}$
Partial reflection (stratified) (9.1.21)	$\gamma(p, \zeta)\|_{t=T, x=X}$	
Total reflection (interface) (9.1.55)	$\mathrm{Re}(\mathcal{T}\Delta(t - T))/(\mathrm{d}X/\mathrm{d}p)^{1/2}$	$\mathcal{T}\mathrm{e}^{\mathrm{i}\omega T}/(\mathrm{d}X/\mathrm{d}p)^{1/2}$
Turning ray – forward branch (9.2.5)	$\delta(t - T)/(-\mathrm{d}X/\mathrm{d}p)^{1/2}$	$\mathrm{e}^{\mathrm{i}\omega T}/(-\mathrm{d}X/\mathrm{d}p)^{1/2}$
Turning ray – reversed branch (9.2.9)	$\bar{\delta}(t - T)/(\mathrm{d}X/\mathrm{d}p)^{1/2}$	$\mathrm{i}^{-\mathrm{sgn}(\omega)}\mathrm{e}^{\mathrm{i}\omega T}/(\mathrm{d}X/\mathrm{d}p)^{1/2}$
General ray (5.4.38) (6.8.3)	$\mathrm{Re}\left(\mathrm{i}^{-\sigma}\mathcal{T}\Delta(t - T)\right)/\mathcal{R}$	$\mathrm{i}^{-\mathrm{sgn}(\omega)\sigma}\mathcal{T}\mathrm{e}^{\mathrm{i}\omega T}/\mathcal{R}$
Head wave (9.1.52)	$H(t - T_n)/\ell_n^{3/2}x^{1/2}$	$-\mathrm{e}^{\mathrm{i}\omega T_n}/\mathrm{i}\omega\ell_n^{3/2}x^{1/2}$
Interface wave (9.1.75) (9.1.89)	$\mathrm{Re}\left(x^{-1/2}\Lambda(t - T_{\mathrm{pole}})\frac{\partial p}{\partial \mathcal{T}^{-1}}\right)$	
Tunnelling wave (9.1.136)	$\mathrm{Re}\left(\underline{\mathcal{G}}\Delta(t - \widetilde{T})\right)$	
Airy caustic (9.2.46) (9.3.53)	$\frac{\mathrm{d}}{\mathrm{d}t}C\left(t - \widetilde{T}_A, 2\ell_a/(9X''_A)^{1/3}\right)$	$\omega^{1/6}\mathrm{e}^{\mathrm{i}\omega\widetilde{T}_A - \mathrm{i}\pi/4} \times Ai\left(-(3\omega\Delta\widetilde{T}_A/2)^{2/3}\right)$
Fresnel shadow (9.2.61) (9.3.60)	$\frac{\mathrm{d}}{\mathrm{d}t}F\left((t - T)/(\widetilde{T}_1 - T), \pm 1\right)$	$Fr\left((2\omega\Delta T_1)^{1/2}/\pi\right)$
Deep shadow (9.3.97) (9.3.101)	$Sh^{(3)}\left(3(t - \widetilde{T}_1)/(V_1\ell_1)^3\right)$	$\omega^{-2/3}\mathrm{e}^{-\omega^{1/3}V_1\ell_1\mathrm{e}^{-\mathrm{i}\pi/6} - 2\mathrm{i}\pi/3}$

the various signals described in this text. It summarizes their salient functional features, and cross-references the equations in the text where fuller results can be found. The time domain and spectral results are given for the particle displacement assuming an impulsive (delta function), point force in three dimensions. More complete expressions without the first-motion approximation are given in the text which remain valid more generally. The expressions in the table indicate the type of signals expected. For brevity, only the most important factors from the results are included, i.e. the terms that vary most rapidly. The complete expressions can be found in the text in the equations given.

1
Introduction

Numerical simulations of the propagation of elastic waves in realistic Earth models can now be calculated routinely and used as an aid to survey design, interpretation and inversion of data. The theory of elastodynamics is complicated enough, and models depend on enough multiple parameters, that computers are almost essential to evaluate final results numerically. Nevertheless a wide variety of methods have been developed ranging from exact analytical results (in homogeneous media and in homogeneous layered media, e.g. the Cagniard method), through approximations (asymptotic or iterative, e.g. ray theory and the WKBJ method), transform methods in stratified media (propagator matrix methods, e.g. the reflectivity method), to purely numerical methods (e.g. finite-difference, finite-element or spectral-element methods), in one, two and three-dimensional models. Recent extensions of approximate methods, e.g. the Maslov method, quasi-isotropic ray theory, and Born scattering theory and the Kirchhoff surface integral method applied to anisotropic, complex media have extended the range of application and/or validity of the basic methods.

Although the purely numerical methods can now be used routinely in modelling and interpretation, the analytic, asymptotic and approximate methods are still useful. There are three main reasons why the simpler, approximate but less expensive methods are useful and worth studying (and developing further). First, complete numerical calculations in realistic Earth models are as complicated to interpret as real data. Interpretation normally requires different parts of the signal to be identified and used in interpretation. Signals that are easy to interpret are usually well modelled with approximate, inexpensive theories, e.g. geometrical ray theory. Our intuitive understanding of wave propagation normally corresponds to these theories. As no complete, robust, non-linear inverse theory has been developed, we must use simple modelling theories to interpret real data and understand (and check) numerical calculations. Secondly, the analytic and approximate modelling methods allow the properties of different parts of the signals to be analysed independently. Again for survey design, interpretation and inversion, this reduction of the properties and sensitivities of different signals to different parameters of the model is invaluable. Finally, although numerical solutions are possible in realistic Earth models, practical limitations still exist. Calculations in two dimensions are now inexpensive enough that they can be used routinely and complete surveys simulated, but in three dimensions this is only possible with compromises. Although computer speeds and memory have and continue to increase dramatically, this limitation in three dimensions is unlikely to disappear soon. To simulate a three-dimensional survey, the number of sources and receivers normally increases quadratically with the dimension of the survey (apart from the fact that acquisition systems are improving rapidly and the density of independent sources and receivers is also increasing). More importantly, there is a severe frequency limitation on numerical calculations. The expense of numerical methods rises as the fourth power of frequency (three from the spatial dimensions as the number of nodes in the model is related to the shortest wavelengths required, and one through the time steps or bandwidth required to model the highest frequency). Currently and for the foreseeable future, this places a severe limitation on the numerical modelling of high-frequency waves in realistic, three-dimensional Earth models. Analytic, asymptotic and approximate methods, in which the cost is independent or not so highly dependent on frequency, are and will remain useful. This book develops these methods (and does not discuss the purely numerical methods).

Although the analytic, asymptotic and approximate methods have limited ranges of validity, recent extensions of these methods have been very successful in increasing their usefulness. This book discusses four extensions of asymptotic ray theory which are inexpensive to compute in realistic, three-dimension Earth models: Maslov asymptotic ray theory extends ray theory to caustic regions; quasi-isotropic ray theory extends ray theory to the near degeneracies that exist in weakly anisotropic media; Born scattering theory that models signals scattered by small perturbations in the model and importantly allows signals due to errors in the ray solution to be included; and the Kirchhoff surface integral method which allows signals and diffractions from non-planar surfaces to be modelled at least approximately. Although these methods are widely used, limitations exist in the theories and further developments are needed. The future will probably see the development of hybrid methods that combine these and other extensions of ray theory with one another, and with numerical methods.

The foundations of elastic wave propagation were available by the beginning of the 20th century. Hooke's law had been extended to elasticity. Cauchy had developed the theory of stress and strain, each depending on six independent components, and Green had shown that 21 independent elastic parameters were necessary in general anisotropic media. In isotropic media this number reduces to two (the Lamé, 1852, parameters) and the existence of *P* and *S* waves was known. Love (1944, reprinted from 1892) gave an excellent review of the development of elasticity theory. Rayleigh (1885) had explained the existence of the waves now named after him, that propagate along the surface of an elastic half-space. Finally Lamb (1904), in arguably the first paper of theoretical seismology, was able to explain the *excitation* and *propagation* of *P* and *S* rays, head waves and Rayleigh waves from a point source on a homogeneous half-space. The paper contained the first theoretical seismograms.

Developments after Lamb's classic paper were initially slow. Stoneley (1924) established the existence of interface waves, now bearing his name, on interfaces between elastic half-spaces. Only with Cagniard (1939) was a new theoretical method developed which was a significant improvement on Lamb's method, although it was not until de Hoop (1960) that this became widely known and used. Bremmer (e.g. 1939; van der Pol and Bremmer, 1937*a*, *b*, etc.) in papers concerning radio waves developed methods that would become useful in seismology. Pekeris (1948) studied the excitation and propagation of guided waves in a fluid layer, calculating theoretical seismograms (although the existence of the equivalent Love waves had been known before). Lapwood (1949) studied the asymptotics of Lamb's problem in much greater detail, and Pekeris (1955*a*, *b*) developed another analytic method equivalent to Cagniard's. After a slow start in the first half of the 20th century, rapid developments in the second half depended on computers and improvements in acquisition systems to justify numerical simulations. This book describes these developments.

Chapter 2: *Basic wave propagation* introduces our subject by reviewing the basics of wave propagation. In particular, the properties of plane and spherical waves at interfaces are described. Ray results in stratified media are outlined in order to describe the morphology of travel-time curves. The various singularities, discontinuities and degeneracies of these ray results are emphasized, as it is these regions that are of particular interest throughout the rest of the book.

This introductory chapter is followed by two review chapters: Chapter 3: *Transforms* and Chapter 4: *Review of continuum mechanics and elastic waves*. The first of these reviews the various transforms – Fourier, Hilbert, Fourier–Bessel, Legendre, Radon, etc. – used throughout the rest of the book. This material can be found in many textbooks and is included here for completeness and to establish our notation and conventions. Chapter 4 reviews continuum mechanics – stress, strain, elastic parameters, etc. – and the generation of plane and spherical elastic waves in homogeneous media – *P* and *S* waves, point force and stress glut sources, radiation patterns, etc. – together with some fundamental equations and theorems of elasticity – the Navier wave equation, Betti's theorem, reciprocity, etc.

The main body of the book begins with Chapter 5: *Asymptotic ray theory*. This develops asymptotic ray theory in three-dimensional acoustic and anisotropic elastic media, and then specializes these results to isotropic elastic media and one and two-dimensional models. The theory for kinematic ray tracing (time, position and ray direction), dynamic ray tracing (geometrical spreading and paraxial rays) and polarization results is described. These are combined with the results from the previous chapter, to give the ray theory Green functions.

Chapter 6: *Rays at an interface* extends these results to models that include interfaces, discontinuities in material properties. The additions required for kinematic ray tracing (Snell's law), dynamic ray tracing and polarizations (reflection/transmission coefficients) are developed for acoustic, isotropic and anisotropic media, for free surfaces, for fluid media and for differential coefficients. Finally these are combined with the results of the previous chapter to give the full ray theory Green functions for models with interfaces.

The results of these two chapters on ray theory break down at the singularities of ray theory, i.e. caustics, shadows, critical points, etc. The next three chapters develop transform methods for studying signals at these singularities but restricted to stratified media. The first chapter, Chapter 7: *Differential systems for stratified media*, reduces the equation of motion and the constitutive equations to one-dimensional, ordinary differential equations for acoustic, isotropic and anisotropic media. Care is taken to preserve the notation used in the previous chapters on ray theory to emphasize the similarities and reuse results. The chapter then develops various solutions of these equations, in homogeneous and inhomogeneous layers, using the propagator and ray expansion formalisms. The WKBJ and Langer asymptotic methods, and the WKBJ iterative solutions, are included. The second chapter of this group, Chapter 8: *Inverse transforms for stratified media*, then describes the inverse transform methods that can be used with the solutions from the previous chapter. These include the Cagniard method in two and three dimensions, the WKBJ seismogram method and the numerical spectral method. These methods are then used in the final chapter of the group, Chapter 9: *Canonical signals*, which describes approximations to various signals that occur in many simple problems. These range from direct and turning rays, through partial and total reflections, head waves and interface waves, to caustics and shadows. These results link back to the introductory Chapter 2 where the various singularities, discontinuities and degeneracies of ray results had been emphasized. Particular emphasis is placed on describing the signals using simple, 'standard' special functions.

The final chapter, Chapter 10: *Generalizations of ray theory*, describes recent extensions of ray theory which increase the range of application and validity and include some of the advantages of the transform methods. The methods are Maslov asymptotic ray theory which extends ray theory to caustic regions; quasi-isotropic ray theory which extends ray theory to the near degeneracies that exist in weakly anisotropic media; Born scattering theory which models signals scattered by small perturbations in the model and importantly allows signals due to errors in the ray solution to be included; and the Kirchhoff surface integral method which allows signals and diffractions from non-planar surfaces to be modelled at least approximately.

2
Basic wave propagation

This introductory chapter introduces the reader to the concepts of seismic waves – *plane waves*, *point sources*, *rays* and *travel times* – without mathematical detail or analysis. Many different types of rays and seismic signals are illustrated, in order to set the scene for the rest of this book. The objective of the book is to provide the mathematical tools to model and understand the signals described in this chapter.

In this chapter, we introduce the basic concepts of wave propagation in a simple, stratified medium. None of the sophisticated mathematics needed to solve for the complete wave response to an impulsive point source in an elastic medium is introduced nor used. Ideas such as Snell's law, reflection and transmission, wavefronts and rays, travel-time curves and related properties, are introduced. The concepts are all straightforward and should be intuitively obvious. Nevertheless many simple questions are left unanswered, e.g. how are the amplitudes of waves found, and what happens when wavefronts are non-planar or singular. These questions will be answered in the rest of this book. This chapter sets the scene and motivates the more detailed investigations that follow.

2.1 Plane waves

2.1.1 Plane waves in a homogeneous medium

In this section, we introduce the notation and nomenclature that we are going to use to describe waves. We assume that the reader has a basic knowledge of waves and oscillations, and so understands a simple wave equation with the notation of complex variables used for a solution.

The simplest form of wave equation describing waves in three dimensions is the Helmholtz equation

$$\nabla^2 \phi = \frac{1}{c^2} \frac{\partial^2 \phi}{\partial t^2}. \tag{2.1.1}$$

Table 2.1. *Symbols, names and units used to describe plane waves.*

Symbol	Name	Units
ϕ	field variable	
A	(complex) amplitude	
t	time	[T]
$\mathbf{x}$	position vector	[L]
ω	circular frequency	[T^{-1}]
ν	$= \omega/2\pi$, frequency (Hz)	[T^{-1}]
$\mathbf{k}$	wave vector	[L^{-1}]
k	$= \lvert\mathbf{k}\rvert$, wavenumber	[L^{-1}]
c	$= \omega/k$, wave (phase) velocity	[LT^{-1}]
T	$2\pi/\omega = 1/\nu$, period	[T]
λ	$2\pi/k = c/\nu$, wavelength	[L]
$\mathbf{p}$	$= \mathbf{k}/\omega$, slowness vector	[L^{-1}T]
$\lvert\mathbf{p}\rvert$	$= 1/c$, slowness	[L^{-1}T]
$\mathbf{k}\cdot\mathbf{x} - \omega t$	phase	[0]

When the velocity, c, is independent of position, the plane-wave solution of this equation is

$$\phi = A\mathrm{e}^{\mathrm{i}(\mathbf{k}\cdot\mathbf{x}-\omega t)} \tag{2.1.2}$$

$$= A\mathrm{e}^{\mathrm{i}\omega(\mathbf{p}\cdot\mathbf{x}-t)}. \tag{2.1.3}$$

In Table 2.1 we give the definitions, etc. of the variables used in or related to this equation (see Figure 2.1).

Note the signs used in the exponent of the oscillating, travelling wave given by equations (2.1.2) or (2.1.3). The same signs will be used later in Fourier transforms (Chapter 3). Various sign conventions are used in the literature. We prefer this one, a positive sign on the spatial term and a negative sign on the temporal term, as it leads to the simple identification that waves travelling in the positive direction have positive components of slowness. For most purposes, it is convenient to use the slowness vector, **p**, rather than the wave vector, **k**, to indicate the wave direction as it is independent of frequency. Surfaces on which $\mathbf{k}\cdot\mathbf{x}$ is constant, as illustrated in Figure 2.1, are known as *wavefronts*.

2.1.2 Plane waves at an interface

If a plane wave is incident on an interface, we expect reflected and transmitted waves to be generated. Some continuity condition must apply to the field variable,

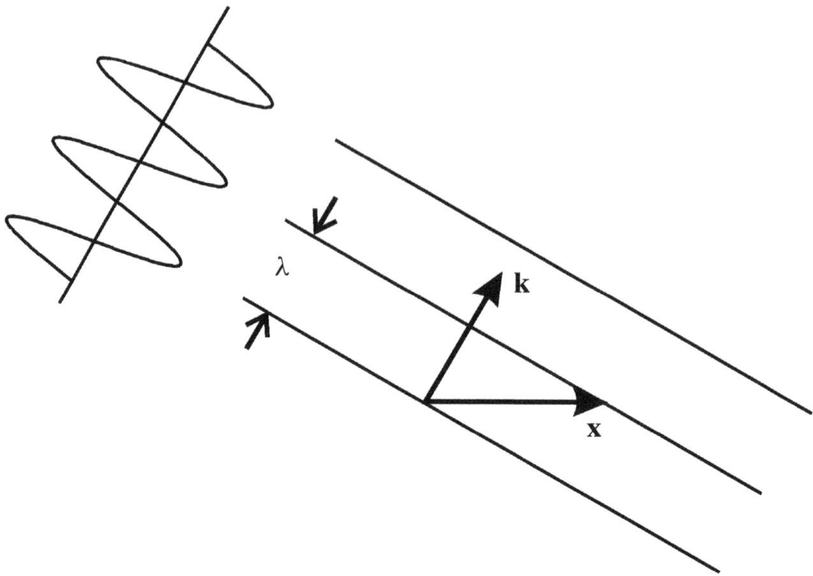

Fig. 2.1. A plane wave in a homogeneous medium. At a fixed time, t, the lines represent wavefronts where $\mathbf{k} \cdot \mathbf{x}$ is constant. Thus if $\mathbf{k} \cdot \mathbf{x} = 2n\pi$, the wavefronts are separated by the wavelength, λ.

ϕ, its derivative or combinations thereof. Even without specifying the details of the boundary condition, it is clear that, for a linear wave equation, all waves must have the same frequency ω, and the same wavelength *along* the interface. Without these conditions, the waves could not match at all times and positions. Let us consider a plane interface at $z = 0$, separating homogeneous half-spaces with velocities c_1 for $z > 0$, and c_2 for $z < 0$ (Figure 2.2). For simplicity we rotate the coordinate system so the incident slowness vector, $\mathbf{p}_{\text{inc}}$, lies in the x–z plane, i.e. $p_y = 0$. Then the wavelength along the interface is $2\pi/k_x = 2\pi/\omega p_x$, so the slowness component p_x must be the same for all waves.

2.1.2.1 *Snell's law*

As $|\mathbf{p}| = 1/c$, the incident wave's slowness vector is

$$\mathbf{p}_{\text{inc}} = \begin{pmatrix} p_x \\ 0 \\ p_z \end{pmatrix} = \begin{pmatrix} p_x \\ 0 \\ -\left(c_1^{-2} - p_x^2\right)^{1/2} \end{pmatrix} = \frac{1}{c_1} \begin{pmatrix} \sin\theta_{\text{inc}} \\ 0 \\ -\cos\theta_{\text{inc}} \end{pmatrix}. \quad (2.1.4)$$

The square root in p_z is taken positive, so the minus sign is included as the incident wave is propagating in the negative z direction (Figure 2.2). The angle θ_{inc} is the

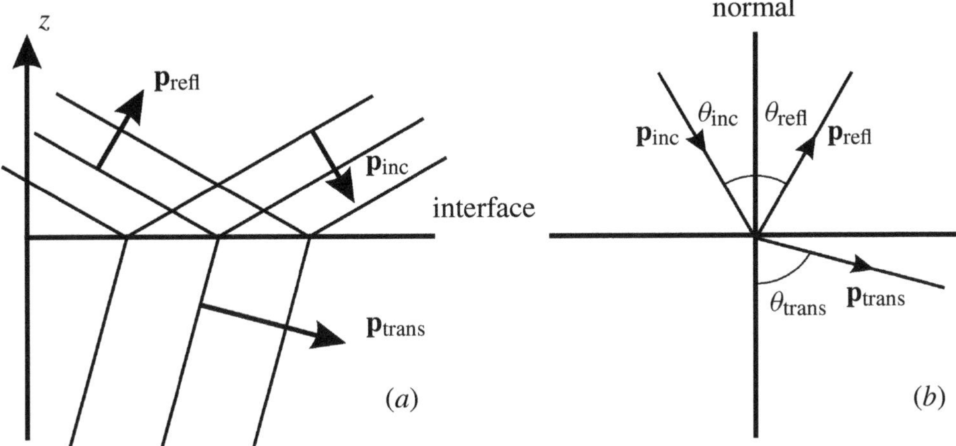

Fig. 2.2. Plane waves incident, reflected and transmitted at a plane interface between homogeneous half-spaces. The situation when $c_2 > c_1$ is illustrated. Part (*a*) shows the wavefronts and (*b*) the slowness vectors.

angle between the slowness vector and the normal to the interface and is taken so $0 \leq \theta_{\text{inc}} \leq \pi/2$. The reflected and transmitted waves must have slowness vectors

$$\mathbf{p}_{\text{refl}} = \begin{pmatrix} p_x \\ 0 \\ +\left(c_1^{-2} - p_x^2\right)^{1/2} \end{pmatrix} = \frac{1}{c_1} \begin{pmatrix} \sin\theta_{\text{refl}} \\ 0 \\ +\cos\theta_{\text{refl}} \end{pmatrix} \tag{2.1.5}$$

$$\mathbf{p}_{\text{trans}} = \begin{pmatrix} p_x \\ 0 \\ -\left(c_2^{-2} - p_x^2\right)^{1/2} \end{pmatrix} = \frac{1}{c_2} \begin{pmatrix} \sin\theta_{\text{trans}} \\ 0 \\ -\cos\theta_{\text{trans}} \end{pmatrix}, \tag{2.1.6}$$

where the angles θ_{refl} and θ_{trans} are similarly defined.

Equality of the x slowness component for these vectors immediately leads to the reflection law

$$\boxed{\theta_{\text{inc}} = \theta_{\text{refl}},} \tag{2.1.7}$$

and the refraction or *Snell's law*

$$\boxed{\frac{\sin\theta_{\text{inc}}}{c_1} = \frac{\sin\theta_{\text{trans}}}{c_2}.} \tag{2.1.8}$$

The situation illustrated in Figure 2.2 is when $c_2 > c_1$ so $\theta_{\text{trans}} > \theta_{\text{inc}}$ and the slowness vector is refracted *away* from the normal. The opposite occurs when $c_2 < c_1$.

2.1.2.2 The critical angle and total reflection

If $c_2 > c_1$ and $\theta_{\text{inc}} = \theta_{\text{crit}} = \sin^{-1}(c_1/c_2)$, then p_z for the transmitted ray is zero. This is known as the critical angle. If $\theta_{\text{inc}} > \theta_{\text{crit}}$, the z slowness component for the transmitted ray becomes imaginary,

$$(\mathbf{p}_{\text{trans}})_z = -\left(\frac{1}{c_2^2} - p_x^2\right)^{1/2} = -\left(\frac{1}{c_2^2} - \frac{\sin^2\theta_{\text{inc}}}{c_1^2}\right)^{1/2}$$
$$= -\,\mathrm{i}\,\mathrm{sgn}(\omega)\left(\frac{\sin^2\theta_{\text{inc}}}{c_1^2} - \frac{1}{c_2^2}\right)^{1/2}. \tag{2.1.9}$$

Note that the sign of the imaginary root is taken positive imaginary, i.e. $\mathrm{Im}(\omega p_z) < 0$, so the transmitted wave decays in the negative z direction (as it must do physically). Thus

$$\mathrm{e}^{\mathrm{i}\omega\mathbf{p}_{\text{trans}}\cdot\mathbf{x}} = \mathrm{e}^{\mathrm{i}\omega p_x x}\mathrm{e}^{\mathrm{i}\omega p_z z} = \mathrm{e}^{\mathrm{i}\omega p_x x}\mathrm{e}^{|\omega p_z|z} \to 0, \tag{2.1.10}$$

as $z \to -\infty$.

In the wavefront diagram (Figure 2.3), the wavefronts for the transmitted wave are perpendicular to the interface (from the x dependence in expression (2.1.10)),

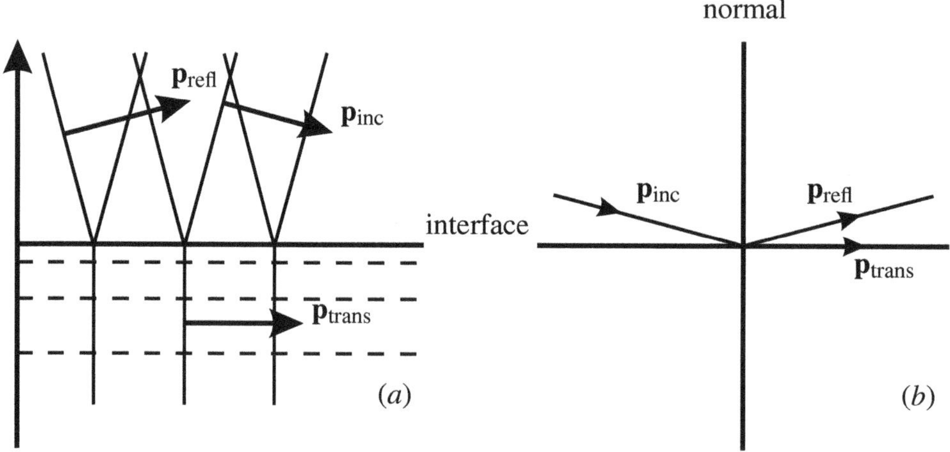

Fig. 2.3. Plane waves incident and totally reflected at a plane interface between homogeneous half-spaces. The transmitted wave is evanescent. Part (*a*) shows the wavefronts and (*b*) the slowness vectors. The dashed lines are constant amplitude lines for the evanescent transmitted waves.

and lines of constant amplitude are parallel to the interface (from the z dependence in expression (2.1.10)). Such a wave is known as an *evanescent* wave (we avoid the term inhomogeneous wave as the word inhomogeneous is overloaded in the subject of wave propagation, being used to describe different features in waves, media and differential equations). This is in contrast to a *travelling* wave, as considered so far, where the constant amplitude and phase surfaces coincide. In the evanescent wave, the propagation is parallel to the interface and the amplitude decays exponentially away from the interface. No energy is transmitted away from the interface in the second medium, so we describe the reflected wave as being a *total reflection*. This contrasts with the situation when $\theta_{\text{inc}} < \theta_{\text{crit}}$ or $c_2 < c_1$, when the transmission propagates energy away from the interface and we describe the reflected wave as a *partial reflection*. Note that because of the signs used in the exponent of expression (2.1.3), we have been able to identify the *positive* real or imaginary root for p_z with propagation in the *positive* z direction (travelling or evanescent), and *negative* with *negative* propagation. Sometimes in the literature, different signs have been taken, making the identification confusingly mixed. We should also mention that Figure 2.3 has been drawn assuming no phase shift between the incident and reflected or transmitted waves. In fact, when total reflection occurs, there is normally a phase shift (e.g. Section 6.3.1, equation (6.3.11)), and the wavefronts should be shifted accordingly.

2.1.3 Time-domain solutions

As we are only considering non-dispersive waves (the velocity, c, is independent of frequency, ω), it is straightforward to write the solution in the time domain. All frequencies can be combined to describe the propagation of an *impulsive* signal. In a homogeneous medium, a solution is

$$\phi = f\Big(t - \mathbf{p}\cdot(\mathbf{x} - \mathbf{x}_0)\Big), \tag{2.1.11}$$

which satisfies the wave equation (2.1.1) for any function f. For

$$\nabla\phi = -\mathbf{p}f' \quad \text{and} \quad \nabla^2\phi = |\mathbf{p}|^2 f'' \tag{2.1.12}$$

$$\frac{\partial\phi}{\partial t} = f' \qquad \text{and} \quad \frac{\partial^2\phi}{\partial t^2} = f'', \tag{2.1.13}$$

so provided $|\mathbf{p}| = 1/c$, equation (2.1.1) is satisfied. (The prime indicates ordinary differentiation of the function with respect to its argument. The arbitrary constant $\mathbf{p}\cdot\mathbf{x}_0$ in the argument of the function $f(t)$ is introduced as it is convenient to measure phase with respect to some origin, $\mathbf{x}_0$.)

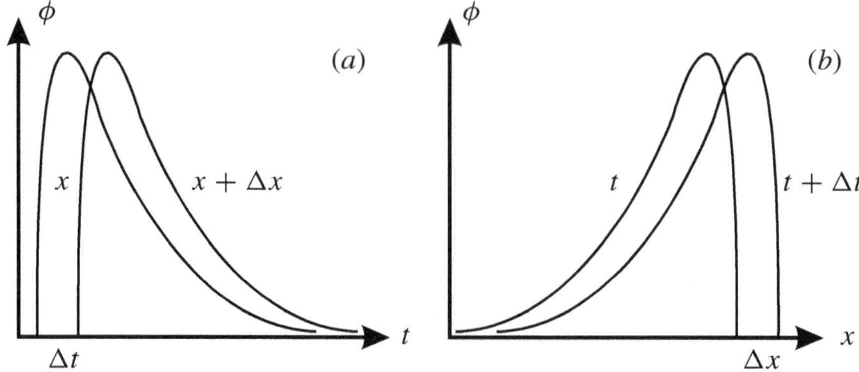

Fig. 2.4. An arbitrary propagating pulse in: (*a*) the time domain; and (*b*) the spatial domain. The pulse is propagating in the x direction.

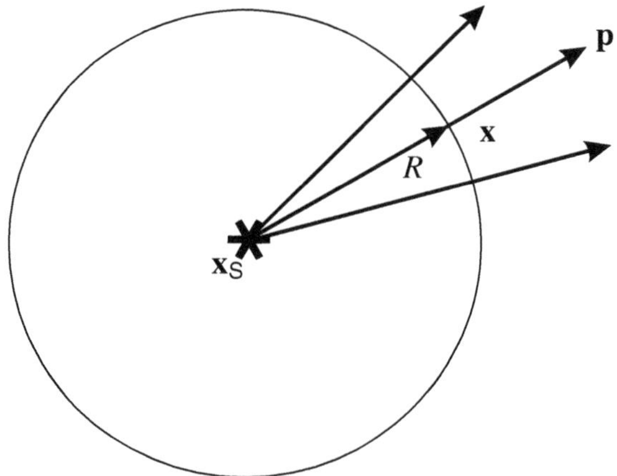

Fig. 2.5. A spherical wavefront and three rays in a homogeneous medium.

In Figure 2.4, we illustrate a propagating impulse in the temporal and spatial domains. Obviously the shape of the pulse is reversed between the two domains.

2.2 A point source

So far we have considered plane waves, necessarily of infinite extent, and therefore not physically realizable. Let us now consider the wavefronts from a point source which might represent a physical source, e.g. an explosion. We only consider the geometry of the wavefronts and postpone a more detailed analysis.

Obviously in a homogeneous medium, the wavefronts from a point source are spherical (Figure 2.5). Denoting the position vector of the point source as $\mathbf{x}_S$, we define the radial vector, etc.

$$\mathbf{r} = \mathbf{x} - \mathbf{x}_S \tag{2.2.1}$$

$$R = |\mathbf{x} - \mathbf{x}_S| = |\mathbf{r}| \tag{2.2.2}$$

$$\hat{\mathbf{r}} = \mathbf{r}/R = \operatorname{sgn}(\mathbf{r}). \tag{2.2.3}$$

Then the slowness vector must be

$$\mathbf{p} = \frac{\hat{\mathbf{r}}}{c}, \tag{2.2.4}$$

and the solution (2.1.11) reduces to

$$\phi = f\left(t - \frac{\mathbf{r}\cdot\mathbf{r}}{cR}\right) \Big/ R = f\left(t - \frac{R}{c}\right) \Big/ R. \tag{2.2.5}$$

In Figure 2.5, we illustrate a spherical wavefront in a homogeneous medium. The wavefront is defined by

$$t = T(\mathbf{x}) = \text{constant}, \tag{2.2.6}$$

where $T(\mathbf{x})$ is the *travel-time* function. Thus the solution is

$$\phi = f\big(t - T(\mathbf{x})\big) \big/ R, \tag{2.2.7}$$

where for a point source in a homogeneous medium, the travel-time function is

$$T(\mathbf{x}) = |\mathbf{x} - \mathbf{x}_S|/c = R/c. \tag{2.2.8}$$

We will describe later, in much greater detail, the definition and methods of solving for the travel-time function. For the moment we note that

$$\boxed{\mathbf{p} = \nabla T,} \tag{2.2.9}$$

which follows as the slowness vector is perpendicular to wavefronts defined by equation (2.2.6), and the variable (2.2.9) has the magnitude of the slowness (time divided by distance). By analogy with potential fields (cf. electric potential and electric field, gravity potential and acceleration due to gravity), given the travel-time function, $T(\mathbf{x})$, we can construct continuous lines perpendicular to wavefronts in the slowness direction, called *ray paths* (e.g. Figure 2.5). As the wavefronts are propagating in the direction of the rays, we can consider the energy as propagating along rays. Again later we will describe in much greater detail the properties of rays in more complicated media (inhomogeneous and/or anisotropic media – Chapter 5). For now we introduce them as a useful, intuitive concept. Given the

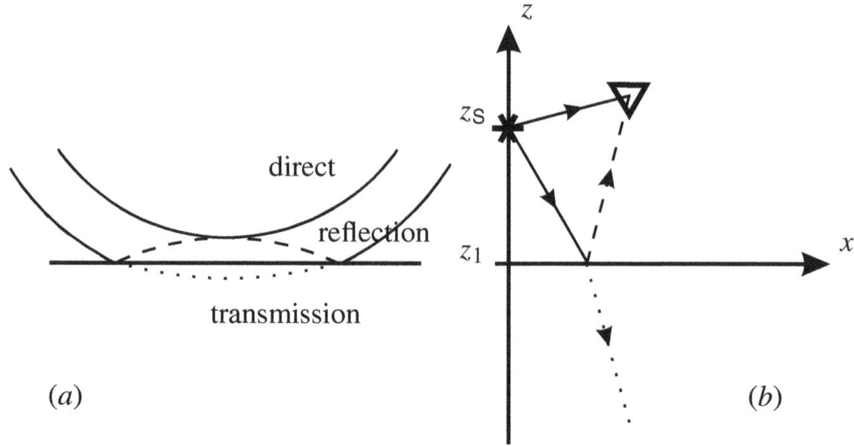

Fig. 2.6. Wavefronts at a plane interface with a velocity decrease. Part (*a*) shows the wavefronts at two times, and part (*b*) the ray paths. The direct wave is the solid line, the reflection is the dashed line, and the transmission is the dotted line.

definition (2.2.9) we can compute travel times as

$$\boxed{T(\mathbf{x}) = \int_{\mathbf{x}_0}^{\mathbf{x}} \mathbf{p} \cdot d\mathbf{x},} \tag{2.2.10}$$

where usually the integration is performed along a ray, although this is not necessary.

2.2.1 A point source with an interface

Again we consider two homogeneous half-spaces with a point source in the first medium. The cases of a velocity increase and decrease are significantly different.

In Figure 2.6, wavefronts from a point source in a model with a velocity decrease are illustrated. The first wavefront is at a time before it reaches the interface. Only the direct, spherical wavefront (the solid line) exists. At a later time, the wavefront has interacted with the interface. A reflected wavefront (the dashed line), also spherical but centred on the image point of the source, exists. The transmitted wavefront (the dotted line) spreads out slower and is not a simple spherical shape. Where the three wavefronts intersect the interface, the wavefronts satisfy Snell's law (as in Figure 2.2 except for a velocity decrease).

In Figure 2.7, wavefronts from a point source in a model with a velocity increase are illustrated. The direct and reflected wavefronts are similar to before

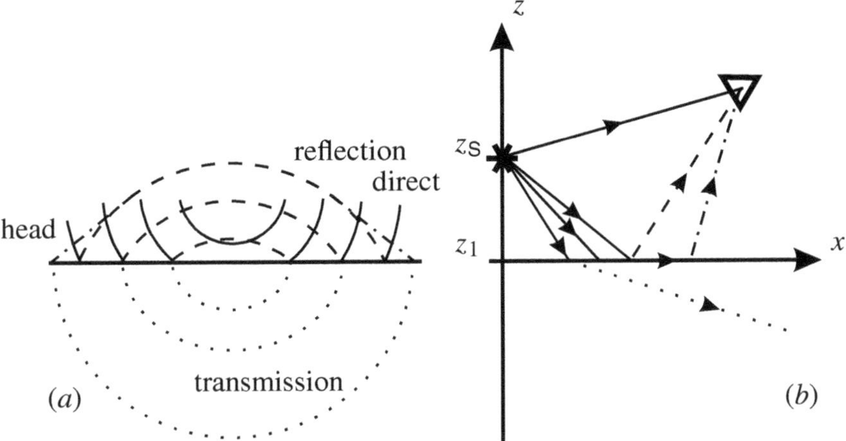

Fig. 2.7. Wavefronts at a plane interface with a velocity increase. As Figure 2.6, with the extra head wave indicated by a dashed-dotted line. The partial and total reflection are shown dashed. Wavefronts at four times are shown: before the direct wave reaches the interface; and before, at and after the critical angle is reached.

(Figure 2.6). Because the velocity increases, the transmitted wavefront spreads out more rapidly and is ahead of the corresponding sphere. Initially when the direct wavefront is incident on the interface, the intersection point moves along the interface with infinite velocity. As the incident angle increases, the velocity of this intersection point drops until at grazing angle, i.e. at infinite range, it has reduced to c_1. Therefore, at some point it will match the velocity in the second medium $c_2 > c_1$. At this point, the critical angle, the transmitted wavefront in the second medium will be perpendicular to the interface. The wavefront in the second medium will now continue to propagate sideways with a velocity c_2, while the incident wavefront will propagate slower. Up to the critical angle, the three wavefronts intersect at the interface satisfying Snell's law as in Figure 2.2. After the critical angle, the transmitted wavefront breaks away as it is propagating faster. Wavefronts don't just stop discontinuously, so the transmitted wavefront continues to be connected with the reflected wavefront, by the so-called *head wave* (the dashed-dotted line), illustrated in Figure 2.7. In the plane illustrated, the head wave is straight. As the diagram is a cross-section of an axially symmetric wavefront, the complete head-wave wavefront is part of a cone (it is sometimes called the *conical wave*). The head wave joins the end of the transmitted wavefront with the critical point on the reflected wavefront. Later we will investigate in detail the generation and properties of the head wave (Chapter 9). The critical point divides the reflected wavefront (the dashed line) into two parts: near normal reflection, the wave is a partial reflection and at wide angles, it is a total reflection.

2.3 Travel-time function in layered media

In the previous section we have seen that rays can be traced in the direction of the slowness vector **p**, orthogonal to the wavefronts defined by $t = T(\mathbf{x})$. We use this concept to calculate the travel-time function, $T(\mathbf{x})$, in a layered medium. The purpose of this section is to outline how this function is calculated, and to describe the morphology of rays, the travel-time and related functions.

In a layered medium, the slowness parallel to the interfaces is conserved. Let us write the slowness vector as

$$\mathbf{p} = \begin{pmatrix} p \\ 0 \\ \pm q \end{pmatrix}, \tag{2.3.1}$$

to avoid subscripts on the components. We have defined the axes so z is perpendicular to the layers and the slowness is in the x–z plane (so $p_y = 0$). The slowness component p is conserved for the ray. In Figure 2.8, we have illustrated the ray (slowness vector) propagating through a layer. We shall refer to the x direction as horizontal and the z direction as vertical, positive upwards, the directions we will always set up axes in a flat, layered Earth, e.g. Figure 0.1.

Let us first consider a model of homogeneous, plane layers. If the velocity in a layer with thickness Δz_i is c_i, in this layer we have

$$p = \frac{1}{c_i} \sin\theta_i \tag{2.3.2}$$

$$q_i = \left(c_i^{-2} - p^2\right)^{1/2} = \frac{1}{c_i} \cos\theta_i, \tag{2.3.3}$$

where θ_i is the angle between the ray and the z axis (Figure 2.8). From the geometry of the ray segment in the layer, we can easily calculate how far it goes

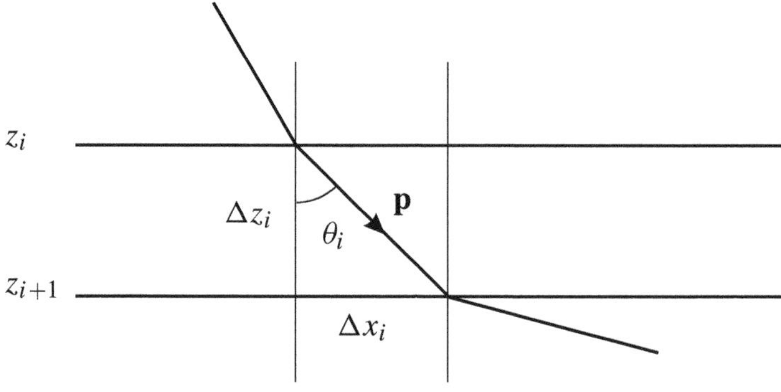

Fig. 2.8. A ray (slowness vector) propagating through the i-th layer.

horizontally (the *range*) and the *travel time*

$$X = \sum_i \Delta x_i = \sum_i \tan\theta_i \, \Delta z_i = \sum_i \frac{p\,\Delta z_i}{q_i} \tag{2.3.4}$$

$$T = \sum_i \Delta T_i = \sum_i \frac{\Delta z_i}{c_i \cos\theta_i} = \sum_i \frac{\Delta z_i}{c_i^2 q_i}, \tag{2.3.5}$$

where the summation is over all layers along the ray. In general the ray may be reflected or transmitted at any interface. All layers traversed are included (with Δz_i positive for either propagation direction) and for rays that reflect, a layer may be included multiple times. Later, we will need the derivative of the range function

$$\frac{\partial X}{\partial p} = \sum_i \frac{\partial(\Delta x_i)}{\partial p} = \sum_i \frac{c_i \Delta z_i}{\cos^3\theta_i} = \sum_i \frac{\Delta z_i}{c_i^2 q_i^3} = \sum_i \frac{\Delta X_i}{p}\left(1 + \frac{p}{q_i}\frac{\Delta x_i}{\Delta z_i}\right). \tag{2.3.6}$$

It is straightforward to extend these results to a continuous stratified velocity function, i.e. $c(z)$. Letting $\Delta z \to \mathrm{d}z$ in equations (2.3.4) and (2.3.5), the range and travel time are

$$\boxed{X(p) = \oint \frac{p\,\mathrm{d}z}{q}} \tag{2.3.7}$$

$$\boxed{T(p) = \oint \frac{\mathrm{d}z}{c^2 q},} \tag{2.3.8}$$

where the slowness components are

$$\boxed{p = \frac{\sin\theta(z)}{c(z)}} \tag{2.3.9}$$

$$\boxed{q = \frac{\cos\theta(z)}{c(z)} = q(p,z) = \left(c^{-2}(z) - p^2\right)^{1/2}.} \tag{2.3.10}$$

The notation $\oint$ is used as a shorthand to indicate integration over all segments of the ray, arranged so as to give positive contributions. Thus for the ray illustrated in Figure 2.9, the complete result is

$$X(p) = \oint \frac{p\,\mathrm{d}z}{q} = \int_{z_\mathrm{R}}^{z_1} + \int_{z_2}^{z_1} + \int_{z_2}^{z_\mathrm{S}} \frac{p\,\mathrm{d}z}{q}. \tag{2.3.11}$$

Typically we write the receiver coordinate as z_R and write the range and travel time as $X(p, z_\mathrm{R})$ and $T(p, z_\mathrm{R})$. Notice that we have not obtained $T(\mathbf{x})$. This would require the elimination of the parameter p. Only in simple circumstances is this possible. Normally we have to be satisfied with the parameterized

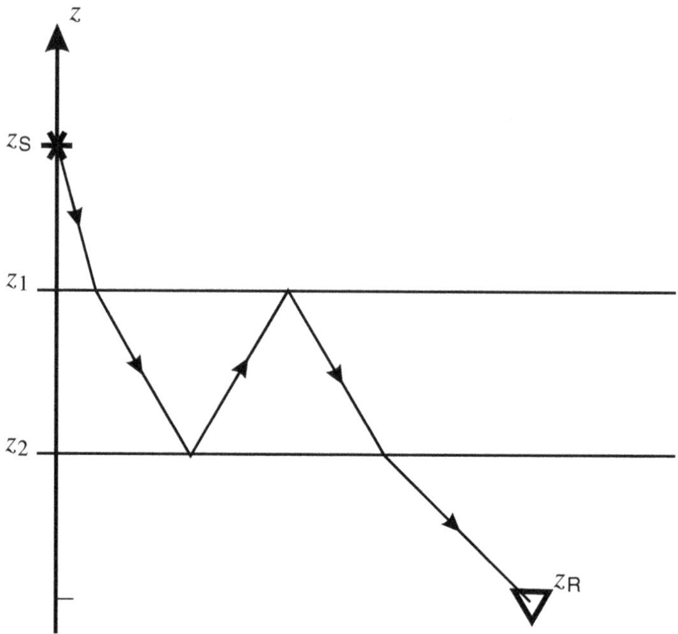

Fig. 2.9. A ray with a reverberation in a layer.

functions – the conserved horizontal slowness (normally the layers are horizontal), p, is commonly called the *ray parameter*. The functions $X(p, z_R)$ and $T(p, z_R)$ are commonly called the *ray integrals*, and $T(X)$, the *travel-time curve*.

In order to describe the possible morphologies of the travel-time and related functions, it is useful to know the derivatives of the ray integrals. Provided the end-points of the integrals are *fixed* (we discuss below in Section 2.3.1 the case when the end-point is a function of the ray parameter p), we can easily differentiate the integrands to obtain

$$\boxed{\frac{\mathrm{d}X}{\mathrm{d}p} = \oint \frac{\mathrm{d}z}{c^2 q^3}} \tag{2.3.12}$$

$$\frac{\mathrm{d}T}{\mathrm{d}p} = \oint \frac{p\,\mathrm{d}z}{c^2 q^3} = p\,\frac{\mathrm{d}X}{\mathrm{d}p}. \tag{2.3.13}$$

From this final result, we obtain

$$\boxed{p = \frac{\mathrm{d}T}{\mathrm{d}X},} \tag{2.3.14}$$

a result that is more generally true (cf. $\mathbf{p} = \nabla T$ (2.2.9)). It can also be proved geometrically. Consider two neighbouring rays with parameters p and $p + \mathrm{d}p$

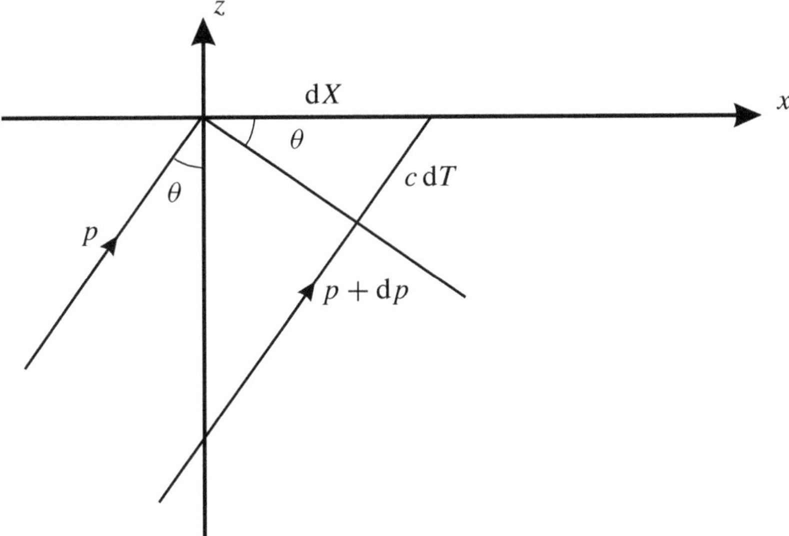

Fig. 2.10. Two rays with parameters p and $p + \mathrm{d}p$, extra range $\mathrm{d}X$ and extra ray length $c\,\mathrm{d}T$.

(Figure 2.10). The extra length of ray is $c\,\mathrm{d}T$ and the extra range $\mathrm{d}X$. From the geometry of the ray and wavefront, we have $c\,\mathrm{d}T/\mathrm{d}X = \sin\theta$ which, with expression (2.3.9), gives result (2.3.14).

A useful function is

$$\boxed{\tau(p, z) = T(p, z) - p\,X(p, z) = \oint q\,\mathrm{d}z.} \tag{2.3.15}$$

Clearly as p is the gradient of the travel-time $T(X)$ function (2.3.14), τ is the intercept of the tangent to the travel-time curve with the time axis (Figure 2.11). The function (2.3.15) is known as the *tau-p curve*, or the *intercept time* (or sometimes the *delay time* although this is open to confusion). Differentiating either expression in the definition (2.3.15), it is straightforward to prove that

$$\boxed{\frac{\mathrm{d}\tau}{\mathrm{d}p} = -X.} \tag{2.3.16}$$

Rearranging the definition so

$$T(p, z) = \tau(p, z) + pX(p, z), \tag{2.3.17}$$

it is clear that the tangent to the tau-p curve intercepts the τ axis at T (Figure 2.12).

The relationship between two functions such as the travel time, $T(X)$, and intercept time, $\tau(p)$, illustrated in Figures 2.11 and 2.12, is known as a *Legendre*,

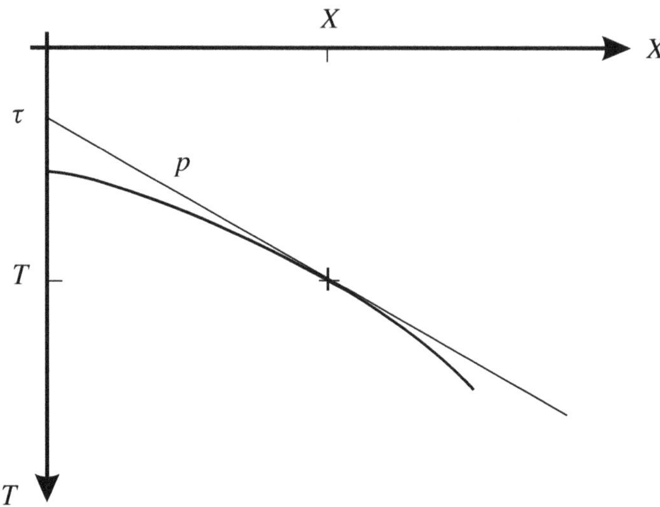

Fig. 2.11. A simple travel-time curve, $T(X)$, and its tangent with slope p and intercept time τ.

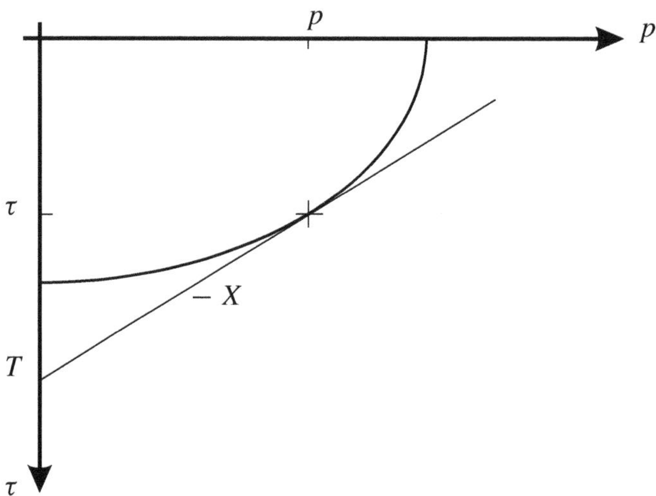

Fig. 2.12. A simple intercept-time function, $\tau(p)$, and its tangent with slope $-X$ and intercept T.

tangent or *contact* transform, discussed more generally in Section 3.4.1. Such transformations arise between thermodynamic energy functions, and between the Lagrangian and Hamitonian in mechanics. We can think of both functions as being generated by the range, $X(p)$, or slowness, $p(X)$, functions (Figure 2.13). From expressions (2.3.14) and (2.3.16), the two functions can be generated by integrating the p v. X with respect to the two variables.

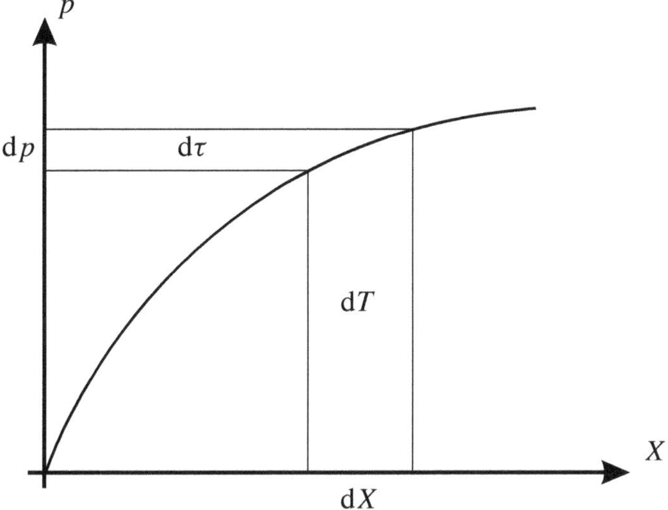

Fig. 2.13. The generating function, $p(X)$, illustrating increments of travel time, $\mathrm{d}T$, and intercept time, $\mathrm{d}\tau$.

2.3.1 The turning point

With the velocity a function of depth, $c = c(z)$, there may be a depth at which $p\,c(z) = 1$ so $q(p, z) = (c^{-2}(z) - p^2)^{1/2} = 0$. This is illustrated in Figure 2.14. Suppose that $u(z) = 1/c(z)$ is the slowness function and we define the inverse slowness function, $z(u)$, i.e. the depth at which the slowness is as specified. Then the depth where $q(p, z) = 0$ is $z = z(p)$, i.e. $c\left(z(p)\right) = p^{-1}$. From equation (2.3.9), $\theta(z) = \pi/2$ at this depth. This depth is called the *turning point*, and a *turning ray* is illustrated in Figure 2.15 (turning rays are sometimes called *diving rays*, but this terminology does not appear to be very descriptive as many non-turning rays dive, and after the crucial turning point, the turning ray is not diving). For simplicity we only consider and illustrate situations in which the ray is turning from above, but it is perfectly straightforward to include rays turning from below if the sign of the gradient is changed (this rarely happens in the solid Earth but is common in the ocean SOFAR channel). The ray integrals for $X(p)$ and $T(p)$, expressions (2.3.7) and (2.3.8), are still valid, with the lower limit of the turning segments given by $z(p)$. The integrals have an inverse square-root singularity at this lower limit which is integrable, i.e.

$$q(p, z) \simeq 2^{1/2} p^{3/2} (-c')^{1/2} (z - z(p))^{1/2}, \qquad (2.3.18)$$

for $z > z(p)$ ($c' = \mathrm{d}c/\mathrm{d}z < 0$ at the turning point), and so

$$\int_{z(p)}^{z^*} \frac{\mathrm{d}z}{q(p, z)} \simeq 2^{1/2} (z^* - z(p))^{1/2} / p^{3/2} (-c')^{1/2}, \qquad (2.3.19)$$

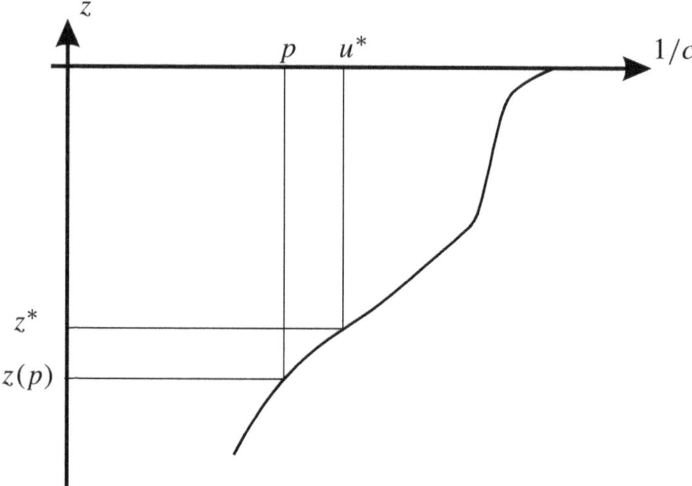

Fig. 2.14. The slowness $1/c$ plotted as a function of depth indicating a turning point where $p = 1/c$. We have also marked a fixed depth z^* above the turning point.

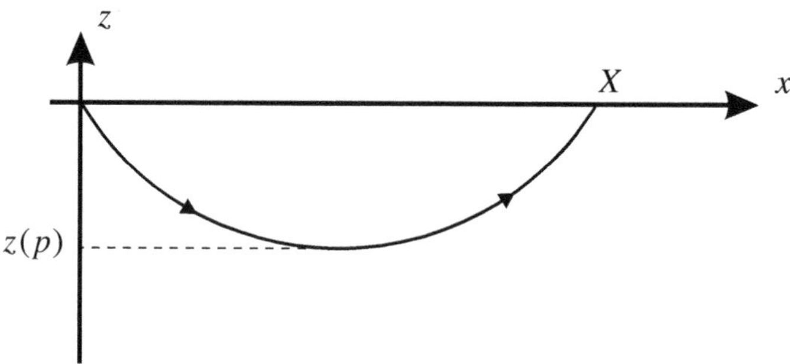

Fig. 2.15. A turning ray with a turning point at depth $z(p)$.

to lowest order. If $c'(z) = 0$ at the turning point, then this argument breaks down. We return to this point below (Section 2.4.9).

Evaluating $\mathrm{d}X/\mathrm{d}p$ for a turning ray is tricky, as the integrand is singular at its lower limit which is a function of p. We have to integrate by parts to remove the singularity and then differentiate. We separate off a small part of the integral just above the turning point from $z(p)$ to z^*. Denoting the slowness as $u = 1/c$, we define

$$g(z) = \frac{u(z)}{u'(z)} = -\frac{c(z)}{c'(z)}. \tag{2.3.20}$$

At the turning point $g\left(z(p)\right) > 0$, and z^* is arbitrary except that $z^* > z(p)$ and $g(z) > 0$ throughout the range $z^* > z > z(p)$ (Figure 2.14). Thus the range integral (2.3.7) is rewritten

$$X(p, z) = 2n \int_{z(p)}^{z^*} \frac{p}{q}\,\mathrm{d}z + \oint_{z^*} \frac{p}{q}\,\mathrm{d}z \tag{2.3.21}$$

$$= 2n \int_{p}^{u^*} \frac{gp}{uq}\,\mathrm{d}u + \oint_{z^*} \frac{p}{q}\,\mathrm{d}z \tag{2.3.22}$$

$$= 2ng \sec^{-1}\left(\frac{u}{p}\right)\Bigg|_{p}^{u^*} - 2n \int_{p}^{u^*} \frac{\mathrm{d}g}{\mathrm{d}u} \sec^{-1}\left(\frac{u}{p}\right)\mathrm{d}u + \oint_{z^*} \frac{p}{q}\,\mathrm{d}z \tag{2.3.23}$$

$$= 2ng^* \sec^{-1}\left(\frac{u^*}{p}\right) - 2n \int_{p}^{u^*} \frac{\mathrm{d}g}{\mathrm{d}u} \sec^{-1}\left(\frac{u}{p}\right)\mathrm{d}u + \oint_{z^*} \frac{p}{q}\,\mathrm{d}z, \tag{2.3.24}$$

where in the first line (2.3.21), n is the number of segments turning at $z(p)$ (normally $n = 1$), $\oint_{z^*}$ is shorthand notation for $\oint$ minus the range $z^* > z > z(p)$, in the second line (2.3.22), we have changed the variable to u and $u^* = u(z^*)$, in the third line (2.3.23), we have integrated by parts using the standard integral (A.0.5) for the inverse secant function (Abramowitz and Stegun, 1964, §4.4.56), and in the final line (2.3.24), $g^* = g(z^*)$. Now the integral is not singular at the turning point and we can differentiate easily. In fact as the integrand is zero, differentiating the variable lower limit makes no contribution. Thus

$$\frac{\mathrm{d}X}{\mathrm{d}p} = -\frac{2ng^*}{q^*} + 2n \int_{p}^{u^*} \frac{\mathrm{d}g}{\mathrm{d}u} \frac{\mathrm{d}u}{q} + \oint_{z^*} \frac{u^2}{q^3}\mathrm{d}z \tag{2.3.25}$$

or

$$\boxed{\frac{\mathrm{d}X}{\mathrm{d}p} = -\frac{2ng^*}{q^*} + 2n \int_{z(p)}^{z^*} \frac{g'}{q}\,\mathrm{d}z + \oint_{z^*} \frac{\mathrm{d}z}{c^2 q^3},} \tag{2.3.26}$$

where $q^* = q(p, z^*)$.

The interesting feature of this result (2.3.26) is that the first term is negative and large if z^* is close to the turning point. The second term has an integrable singularity and is not large. The third term is positive and will be large when z^* is close to the turning point. In many circumstances, the first term will dominate and the derivative will be negative, $\mathrm{d}X/\mathrm{d}p < 0$. This is the normal behaviour for a turning ray in contrast to a reflection (or any ray with segments ending at fixed depths) for which $\mathrm{d}X/\mathrm{d}p > 0$. These behaviours are illustrated in Figure 2.16. However, if the gradient at the turning point is large, g^* is small and the third term

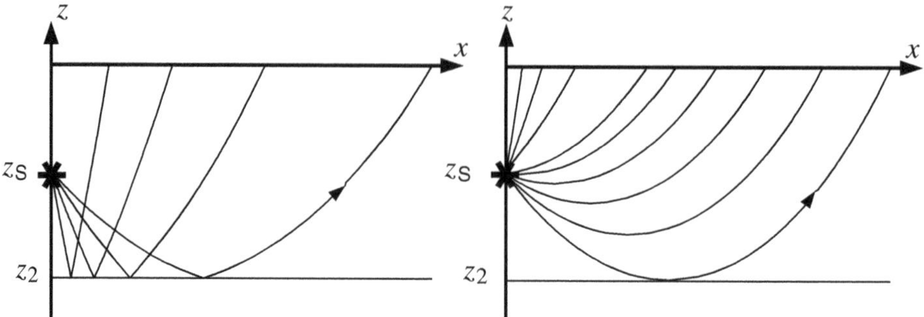

Fig. 2.16. Illustration that $\mathrm{d}X/\mathrm{d}p$ is positive for reflections but negative for turning rays (normally).

may dominate. Then $\mathrm{d}X/\mathrm{d}p > 0$ is possible for a turning ray. We will illustrate this in the next section (Section 2.4.4), where we discuss the morphology of rays and travel-time results.

2.3.2 The Earth flattening transformation

For teleseismic studies in the whole Earth, it is useful to consider the velocity as a function of radius, i.e. $c(r)$, and the range measured as an angle from the source, Δ (Figure 2.17). The results in a spherical geometry can be obtained by an exact conformal mapping from an equivalent cartesian model. This is known as the *Earth flattening transformation*. At some reference radius, r_0, e.g. the receiver radius, the horizontal coordinates are equal, i.e.

$$x = r_0 \Delta . \tag{2.3.27}$$

At other radii, the horizontal element in the spherical model, $r\,\mathrm{d}\Delta$, must be stretched to $\mathrm{d}x = r_0 \mathrm{d}\Delta$, i.e. in a ratio r_0/r. To keep the mapping conformal, so the ray angle with the vertical is the same, $\theta(z) = \theta(r)$ (Figure 2.17), we must also stretch the vertical coordinate

$$\mathrm{d}z = \frac{r_0}{r}\,\mathrm{d}r \tag{2.3.28}$$

(as an aside, we might mention that we prefer to measure z positive upwards, rather than downwards as the depth, so $\mathrm{d}r$, $\mathrm{d}z$ and derivatives have the same sign). Solving this gives

$$\frac{r}{r_0} = \exp\left(\frac{z}{r_0}\right), \tag{2.3.29}$$

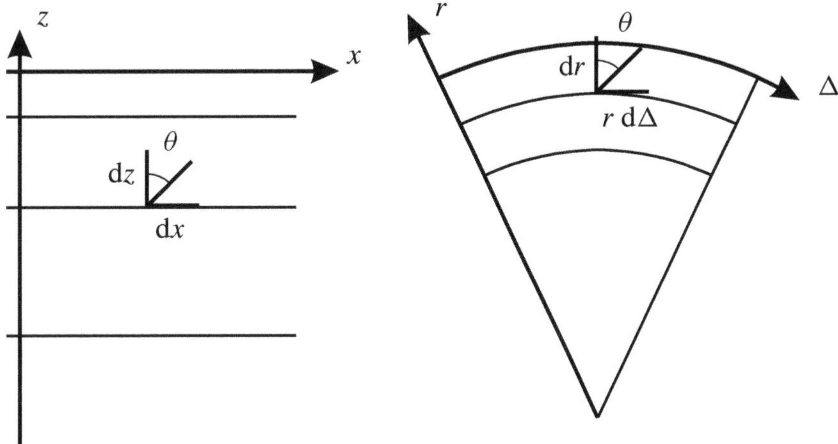

Fig. 2.17. The Earth flattening transformation between a spherical and cartesian model.

arranged so $z = 0$ corresponds to the reference radius $r = r_0$. If the length element is stretched, then the velocity must be increased to compensate, i.e.

$$c(r) = \frac{r}{r_0} c(z). \tag{2.3.30}$$

With these transformations, the results in a cartesian model, with $c(z)$, and a spherical model, with $c(r)$, are exactly equivalent and we have a conformal mapping. We find that

$$p_\Delta = \frac{r \sin\theta(r)}{r_0 c(r)} \tag{2.3.31}$$

is the conserved horizontal slowness and the angular range and travel-time integrals are

$$\Delta(p_\Delta) = \oint \frac{\tan\theta(r)}{r} \mathrm{d}r \tag{2.3.32}$$

$$T(p_\Delta) = \oint \frac{\sec\theta(r)}{c(r)} \mathrm{d}r. \tag{2.3.33}$$

The quantity $r_0 p_\Delta$ is the angular slowness and is commonly called *Bullen's ray parameter* (Bullen, 1963, §7.2.2). The integrals (2.3.32) and (2.3.33) are easily converted into the notation and forms given by Bullen.

Although we will not be greatly concerned with spherical models, they are useful to illustrate the possible forms of travel-time curves (Section 2.4). Straight rays in a homogeneous sphere or spherical shell, map into curved rays in a heterogeneous cartesian model (the velocity increases exponentially with depth, using the velocity mapping (2.3.30) with the depth mapping (2.3.29)).

2.4 Types of ray and travel-time results

In this section we illustrate the forms of rays and travel-time functions for various generic structures. In each case the diagram contains the model, $c(z)$, the ray paths, the travel-time curve, $T(X)$, the inverse range function, $p(X)$, and the intercept function, $\tau(p)$. Although the results have been calculated for a particular model, it is the relationships of the various curves, their gradients and curvatures, etc. rather than the specific values that are important. The forms of the curves can normally be deduced easily from the analytic forms, equations (2.3.7), (2.3.8), (2.3.12), (2.3.15) and (2.3.26), sometimes for simple models, e.g. homogeneous or constant gradient, or just by considering the limiting behaviour. The first five sections, Sections 2.4.1 (Direct and reflected rays), 2.4.2 (Turning rays), 2.4.3 (Refractions), 2.4.4 (Triplication) and 2.4.5 (Low-velocity shadow), describe basic ray types. The sixth section, Section 2.4.6 (Velocity gradient discontinuity) describing a velocity *gradient* discontinuity, is included as numerical models often have this feature. The remaining four sections, Sections 2.4.7 (High-velocity layer), 2.4.8 (Hidden layer), 2.4.9 (Velocity maximum) and 2.4.10 (Focusing layers), have been included to illustrate some unexpected or pathological but valid results.

For each ray type, we mention a specific model for which analytic results can be generated as an illustration. We leave it as an exercise to the reader to investigate these further (see Exercise 2.4).

2.4.1 Direct and reflected rays

Direct rays, reflections, or any rays where segments end at fixed depths, have the same basic form. This is illustrated in Figure 2.18, which includes the direct ray, a reflection and a head wave (as we assume $c_2 > c_1$). In this example, we have taken the source and receiver at the same depth in a homogeneous layer. Notice that the direct ray and reflection are asymptotic at large ranges. The reflection has a partial and total reflection part. In the intercept function, the head wave and direct waves are just points. Notice that $\mathrm{d}X/\mathrm{d}p \to +\infty$ as $p \to c_1$ for the reflection.

Fig. 2.18. The direct (solid lines) and reflected (long-dashed lines) rays in a homogeneous layer. A head wave (short-dashed lines) exists as $c_2 > c_1$ – some ray paths corresponding to the head waves are included. The partial reflections are shown with a thin dashed line and the total reflections with a thick dashed line. Although this figure, and Figures 2.19–2.28, are accurately calculated, they are intended as thumbnail sketches of the forms of travel-time functions, and so numerical values have not been included on the axes. Each figure consists of five sub-figures: the velocity–depth function, $c(z)$; the ray paths; the slowness-range function, $p(X)$; the travel-time curve, $T(X)$; and the intercept function, $\tau(p)$. The depth, range, slowness and times axes are to a common scale in the sub-figures.

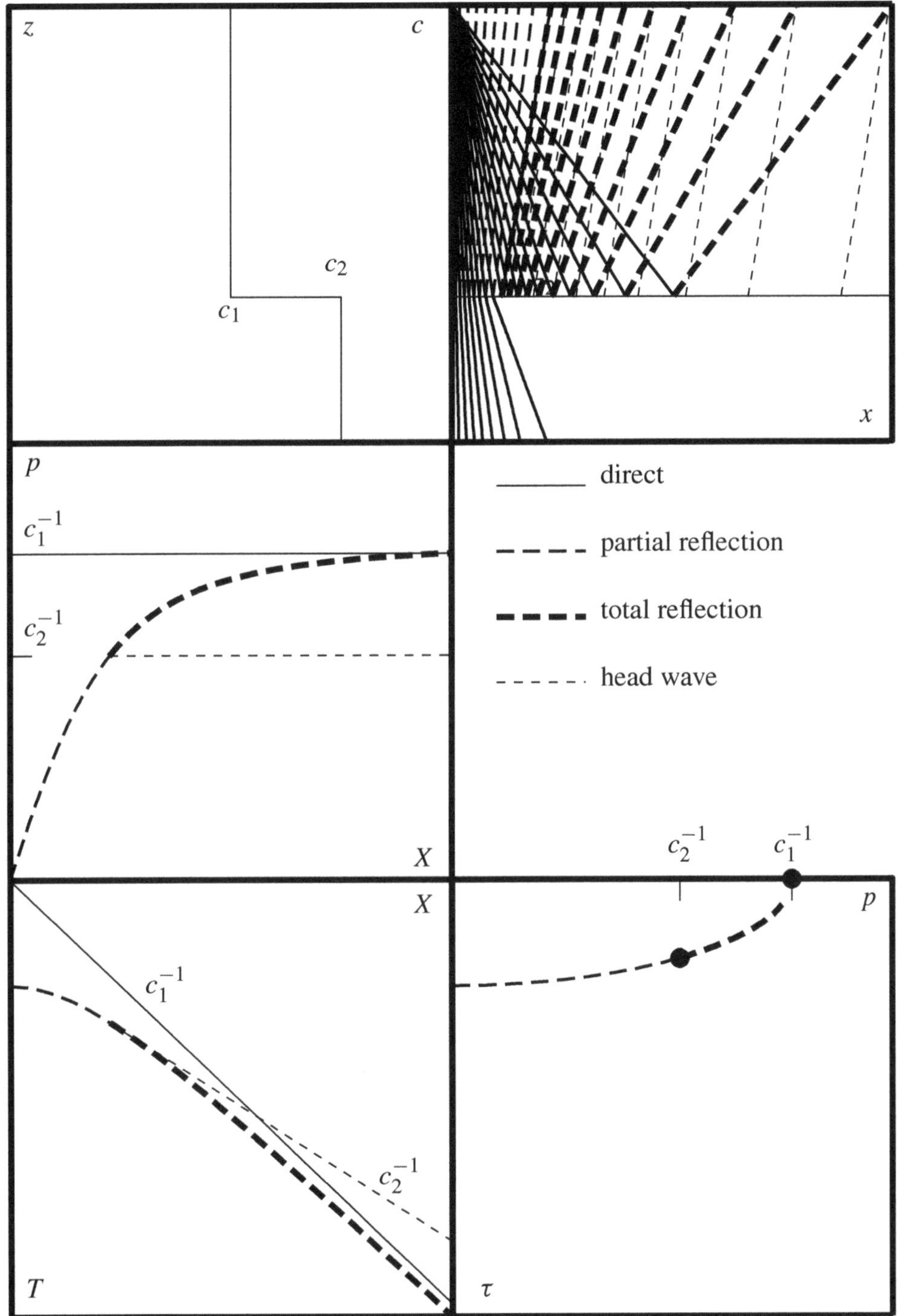
z
c
c_2
c_1
x
p
c_1^{-1}
c_2^{-1}
X
direct
partial reflection
total reflection
head wave
c_2^{-1}
c_1^{-1}
X
p
c_1^{-1}
c_2^{-1}
T
τ

If the source and receiver are not at the same depth, then the direct ray has the same form as the reflection. This is illustrated in Figure 2.19, where we have assumed $c_2 < c_1$ so no head wave exists. If the source and receiver depths tend together, the curves in Figure 2.19 for the direct ray tend to the straight lines in Figure 2.18.

Analytic examples of these rays are easily generated for a homogeneous layered model. The direct ray in a homogeneous medium is discussed in Section 4.5, e.g. approximation (4.5.75). Using the Cagniard transform method, it is described in Section 8.1 and 8.2, e.g. equations (8.1.36) and (8.2.70). This method is ideal for describing reflections as well. The ray-theory approximation (Chapter 5) can also be used to describe direct and reflected waves, provided singularities such as head waves are avoided. The WKBJ and Maslov seismograms methods (Sections 8.4.1 and 10.1) can also be used for direct and reflected waves. They have the advantage of remaining valid at some singularities, but the disadvantage that end-point errors cause acausal artifacts.

2.4.2 Turning rays

If we now introduce a gradient into the upper layer, the ray paths are curved and turning rays exist (Figure 2.20). The reflection now meets the turning ray at the interface *grazing ray* ($p = 1/c_1$), and forms a *shadow edge* (point S in Figure 2.20) – this replaces the asymptote at infinity in Figure 2.19. Beyond the shadow edge, we would expect some diffracted signals (non-geometrical, low-frequency signals) as wavefronts never just stop (as spatial discontinuities do not solve the wave-equation). The direct ray and turning ray meet at a point, H, where the turning point is at the source depth ($p = 1/c_S$). Head waves still exist as we have assumed $c_2 > c_1$ and the second medium is homogeneous. Notice that when the direct and turning rays join with $p \to 1/c_S$ from below, and the ray leaves the source horizontally (point H in Figure 2.20), we have $|\mathrm{d}X/\mathrm{d}p| \to \infty$. At the shadow, S, $\mathrm{d}X/\mathrm{d}p \to +\infty$ as $p \to 1/c_1$ from below for the reflection. The turning ray just terminates, as $p \to 1/c_1$ from above, at the shadow.

Analytic examples of these rays are easily generated by the conformal mapping (Section 2.3.2) of a homogeneous spherical shell, e.g. *P* and *PcP* in the whole Earth. They can also be generated with a linear gradient (Section 2.5.2), which has been used in Figure 2.20.

The ray theory approximation (Chapter 5) can be used to describe turning waves, provided singularities such as caustics are avoided. The WKBJ and Maslov seismogram methods (Sections 8.4.1 and 10.1) can also be used for turning waves. They have the advantage of remaining valid at caustics. Turning rays can be described using spectral methods (Section 8.5), which are necessary to describe fully

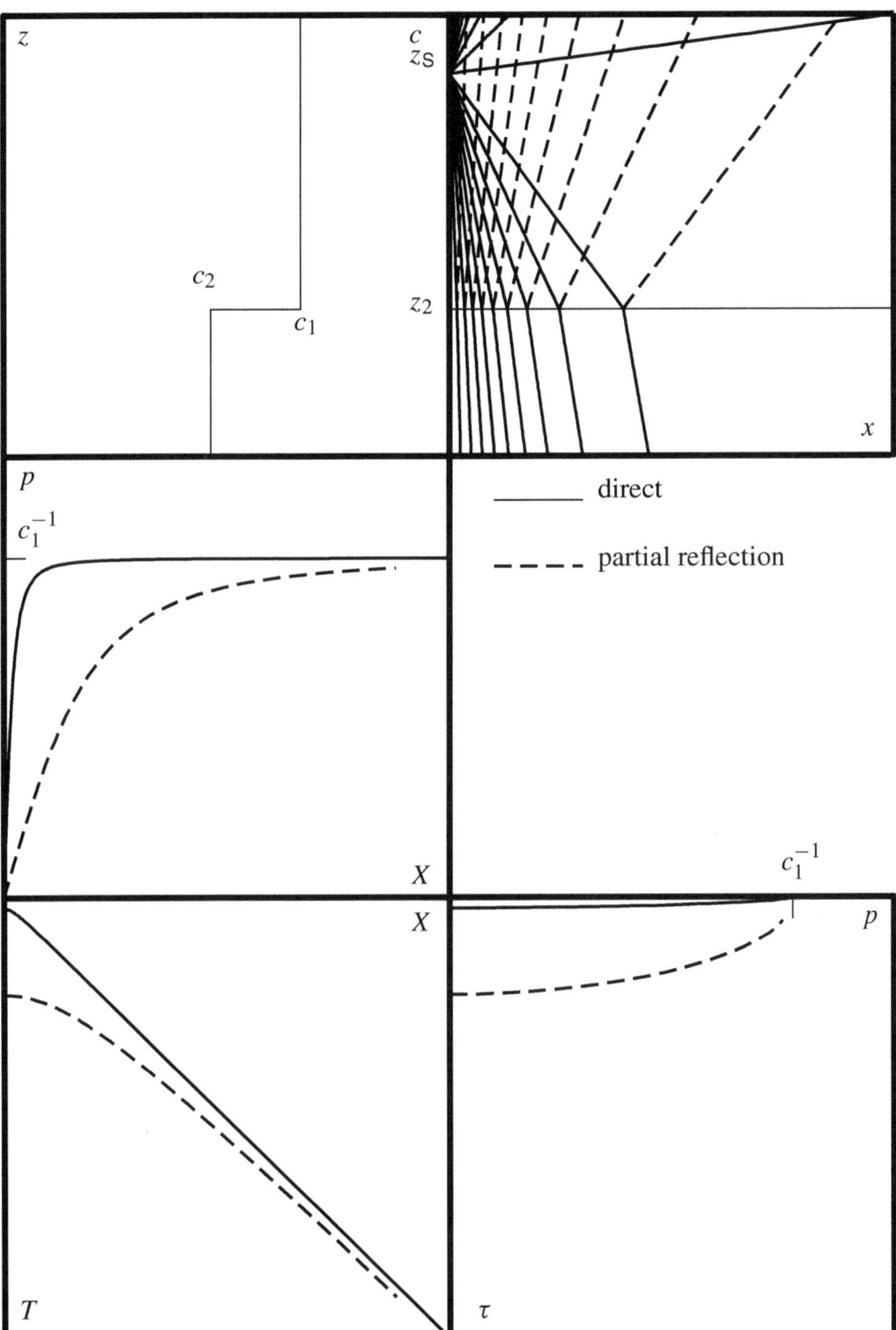

Fig. 2.19. As Figure 2.18 except the source and receiver are at different depths and no head wave exists as $c_2 < c_1$.

the behaviour at and near the shadow $p = 1/c_1$. The WKBJ method partially describes the behaviour at the shadow edge (the Fresnel shadow, Section 8.4.1 and equation (9.2.58)), and an approximate spectral method (Section 8.5) can describe the behaviour in the deep shadow, e.g. equation (9.3.101). Numerical spectral methods are normally necessary for solutions valid at any range including near the shadow edge.

2.4.3 Refractions

If we now introduce a gradient into the second layer, turning rays exist for the signal transmitted through the interface (Figure 2.21). These are often called *refractions*. Refractions with multiple bounces exist below the interface. In acoustics, this is sometimes known as the *whispering gallery mode*. In the limit as the gradient in the second medium decreases, the family of all refractions asymptotically approaches the head wave in Figure 2.20. It is a non-trivial problem to describe the interference of the family of refractions, and to describe the zero-gradient, head-wave limit. The mathematics which describes a head wave and a whispering gallery mode are quite distinct. This is a relevant problem as observed head waves are probably usually refractions, as gradients normally exist due to rock compaction at depth, etc. Notice that $\mathrm{d}X/\mathrm{d}p \to -\infty$ as $p \to 1/c_2$ for the refractions.

Analytic examples of these rays are easily generated by the conformal mapping (Section 2.3.2) of a homogeneous spherical shell surrounding a homogeneous sphere. Examples in the whole Earth are *SKS*, *SKKS*, ... They can also be generated using linear gradients (Section 2.5.2) as in Figure 2.21.

The ray-theory approximation (Chapter 5) can be used to describe refracted waves, provided singularities such as caustics are avoided. The WKBJ and Maslov seismogram methods (Sections 8.4.1 and 10.1) can also be used for refracted waves. They have the advantage of remaining valid at caustics. Refractions can be described using spectral methods (Section 8.5), which are necessary to describe the whispering gallery mode and the interference of the multiple refractions near the end-point, $p = 1/c_2$.

2.4.4 Triplication

Suppose the sharp interface in Figure 2.21 is replace by a smooth velocity function with a high velocity gradient. The partial reflection and multiple refractions disappear as there is no interface. The rest of the curve must become smooth (Figure 2.22). When the gradient is large enough, the curve must have a *triplication*, corresponding to the total reflection part of Figures 2.20 and 2.21. This is a

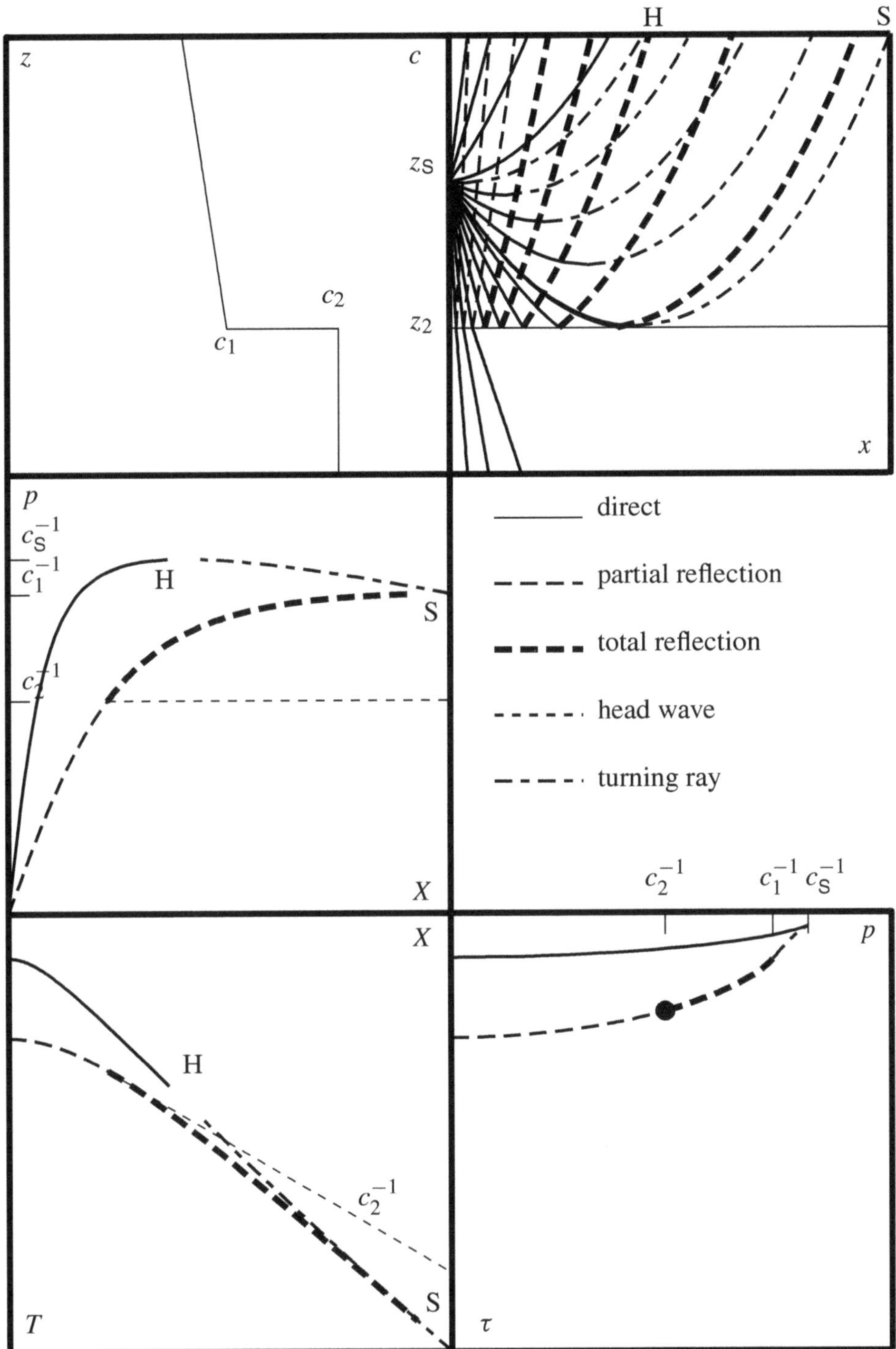

Fig. 2.20. Direct (solid lines), turning (long-short dashed lines) and reflections (long-dashed lines) in a layer with a gradient. As Figures 2.18 and 2.19. The shadow S and ray with a turning point at the source, H, are indicated. Ray paths corresponding to the head waves are not included.

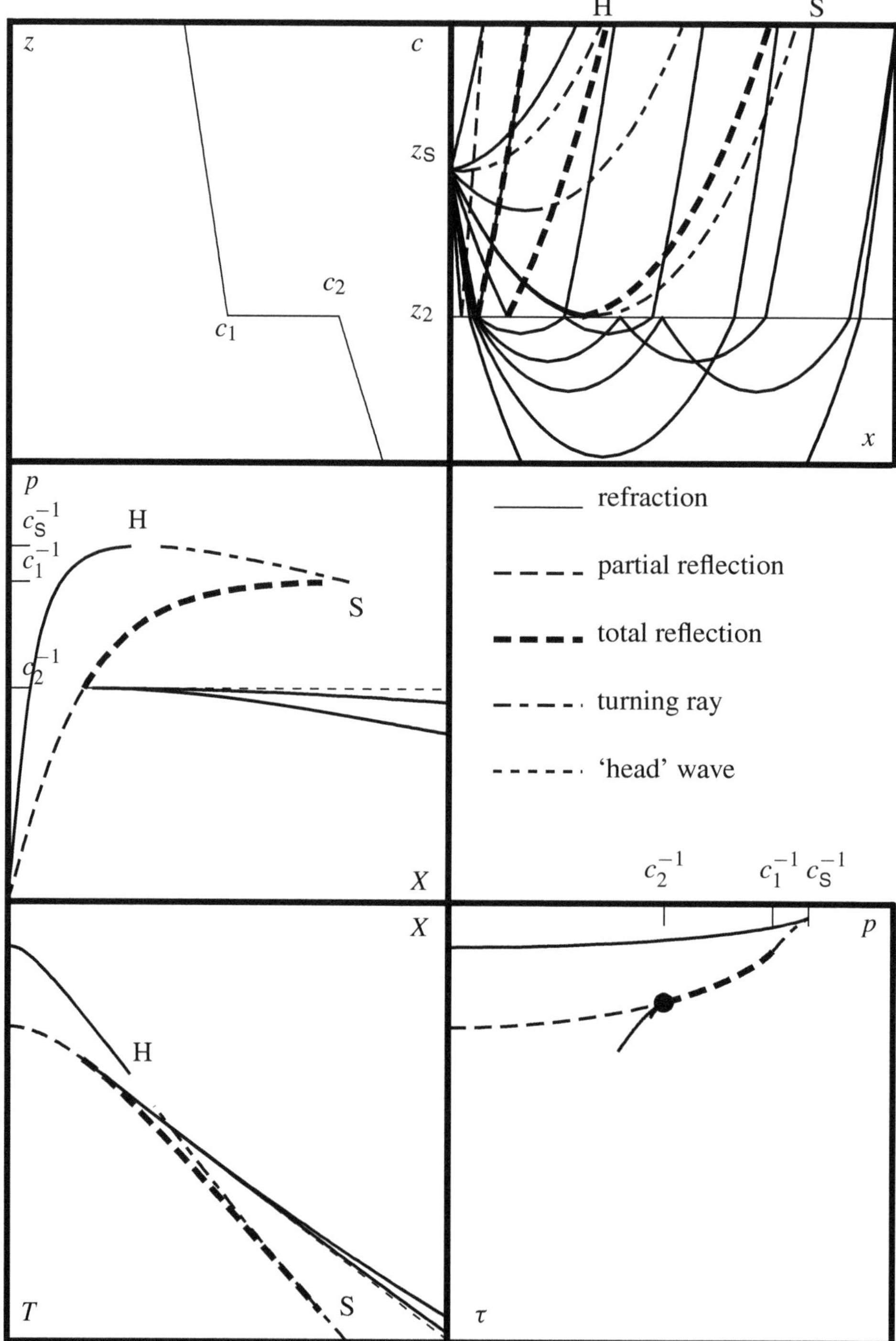

Fig. 2.21. As Figure 2.20, except the gradient in the second medium causes refractions (solid lines). The refraction with an infinite number of bounces has the same kinematic properties as the head wave in Figure 2.20. Ray paths corresponding to the head waves are not included.

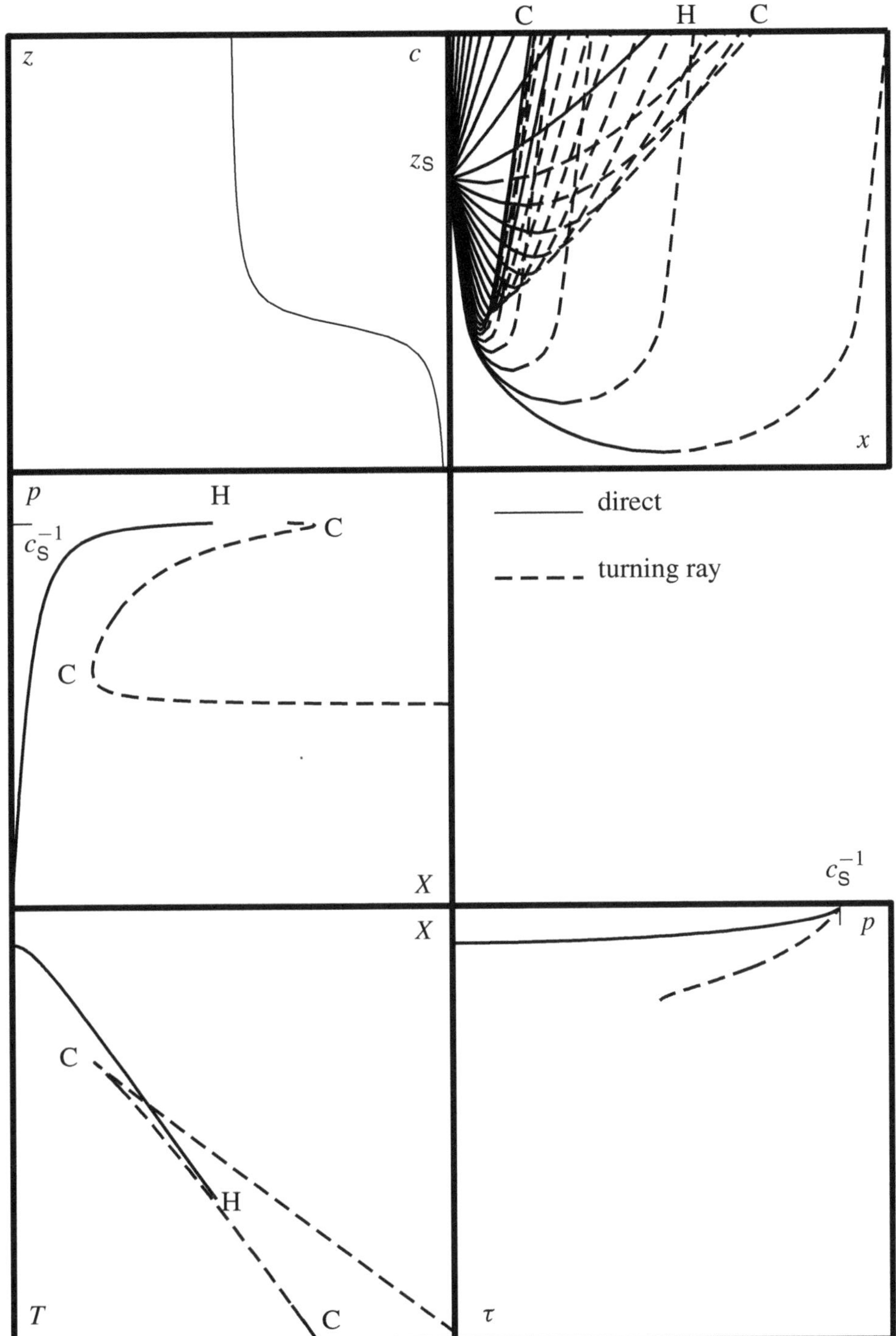

Fig. 2.22. A triplication caused by a high velocity gradient. Only turning rays (dashed lines) exist. The forward or normal branch has $\mathrm{d}X/\mathrm{d}p < 0$ whereas the backward or reversed branch has $\mathrm{d}X/\mathrm{d}p > 0$. The caustic points, C, are indicated.

situation where $\mathrm{d}X/\mathrm{d}p > 0$ (2.3.26) for a turning ray. The points where $\mathrm{d}X/\mathrm{d}p = 0$ are known as *caustics* (points C in Figure 2.22). (Caustic means burning – as in caustic soda – and the term was used in optics because caustics caused by a lens were used for burning. The same term is used in seismics although they don't cause burning, of course!) We also say that the rays are focused, i.e. changing the direction by a small amount at the source only causes a second-order change at the receiver. Only turning rays exist in this model. We call the branches with $\mathrm{d}X/\mathrm{d}p < 0$, the *forward* or *normal branches*, and the branch between the caustics with $\mathrm{d}X/\mathrm{d}p > 0$, the *backward* or *reversed branch*. Again, beyond the caustic points, where the forward and backward branches meet, where no rays exist, we would expect some diffracted signals, and near vertical incidence we would expect some low-frequency, partial reflections (as, for long-wavelength signals, the velocity gradient looks like an interface).

Analytic models that generate a triplication are not so simple to generate. The so-called Epstein layer is an analytic function that can be used to model a high velocity gradient resulting in a triplication (Hron and Chapman, 1974*a*, and references therein). The velocity function used to generate Figure 2.22 is an inverse tangent function.

The WKBJ and Maslov seismogram methods (Sections 8.4.1 and 10.1) are ideal to describe signals at and near a triplication as they remain valid at and near the caustics, e.g. equation (9.2.46). Spectral methods can also be used, e.g. equations (9.3.53) and (9.3.54). The low-frequency, partial reflections can be modelled by the WKBJ iterative solution (Section 9.1.2), and these signals are given by expression (9.1.21).

2.4.5 Low-velocity shadow

If we modify the model with velocity gradients and an interface (Figure 2.21) so that the velocity decreases at the interface $c_2 < c_1$, but then increases with depth above the cap value, c_1 (the case when it doesn't increase again is trivial – we just lose the refractions and the total reflection), then the results are significantly different. The grazing ray again produces a shadow edge (point S_1 in Figure 2.23). The travel time and range for the transmission jump due to the extra segment through the low-velocity zone (LVZ) and form another shadow S_2. At the shadow edge $\mathrm{d}X/\mathrm{d}p \to \infty$ and normally the transmission also forms a caustic (point C). For if the LVZ is small, the gap will be small and although $\mathrm{d}X/\mathrm{d}p \to \infty$ for $p \to 1/c_1$ from below, as p decreases ($p \to 0$) we expect $\mathrm{d}X/\mathrm{d}p < 0$, as existed for the turning ray without the LVZ. Therefore there must be a point where $\mathrm{d}X/\mathrm{d}p = 0$. The gap is best characterized by the saltus of the intercept time, $[\tau]$, which is easily measured. This is given by

$$[\tau(1/c_1)] = \tau(1/c_1 + 0) - \tau(1/c_1 - 0) = -2\int_{z(1/c_1)}^{z_2} \left(\frac{1}{c^2(z)} - \frac{1}{c_1^2}\right)^{1/2} \mathrm{d}z. \tag{2.4.1}$$

The saltus of the range or travel time are difficult to measure as, at finite frequencies, the shadow edges are blurred by diffracted signals.

Analytic examples of these rays are easily generated by the conformal mapping (Section 2.3.2) of a homogeneous spherical shell surrounding a homogeneous sphere. They can also be generated using linear gradients (Section 2.5.2) as in Figure 2.23. Examples in the whole Earth are *PKP*, *PKKP*, ...

The low-velocity shadow behaves like an interface shadow with the addition of a caustic. The same methods can be used as described previously. The ray-theory approximation (Chapter 5) can be used to describe turning waves, reflection and refraction, provided singularities such as caustics and shadows are avoided. The WKBJ and Maslov methods (Sections 8.4.1 and 10.1) can also be used. They have the advantage of remaining valid at caustics. Turning rays can be described using spectral methods (Section 8.5), which are necessary to describe fully the behaviour at and near the shadows $p = 1/c_1$. The WKBJ method partially describes the behaviour at the shadow edge (the Fresnel shadow, Section 8.4.1 and equation (9.2.58)), and an approximate spectral method (Section 8.5) can describe the behaviour in the deep shadow, e.g. equation (9.3.101). Numerical spectral methods are normally necessary for solutions valid at any range including near the shadow edge.

2.4.6 Velocity gradient discontinuity

It is useful to consider the results for a second-order discontinuity, i.e. the velocity is continuous but the velocity gradient is discontinuous. Numerical models often use low-order methods of interpolation and have gradient discontinuities (see Section 2.5.2). Two cases are illustrated in Figure 2.24: a gradient increase and decrease.

If $[c'(z_2)] > 0$, corresponding to a gradient increase with depth (z is upwards so c' is negative), then $\mathrm{d}X/\mathrm{d}p \to +\infty$ as $p \to 1/c_2$ from below. Normally a caustic will form, with $\mathrm{d}X/\mathrm{d}p = 0$ and a small triplication, as $\mathrm{d}X/\mathrm{d}p < 0$ for the normal turning rays. If $[c'(z_2)] < 0$, corresponding to a gradient decrease with depth, then $\mathrm{d}X/\mathrm{d}p \to -\infty$ as $p \to 1/c_2$ from below. No triplication exists as $\mathrm{d}X/\mathrm{d}p < 0$ for the normal turning rays. The singular behaviour near $p = 1/c_2$ will cause numerical problems with two-point ray-tracing and amplitude calculations, even though the gradient discontinuity is normally only a numerical artifact.

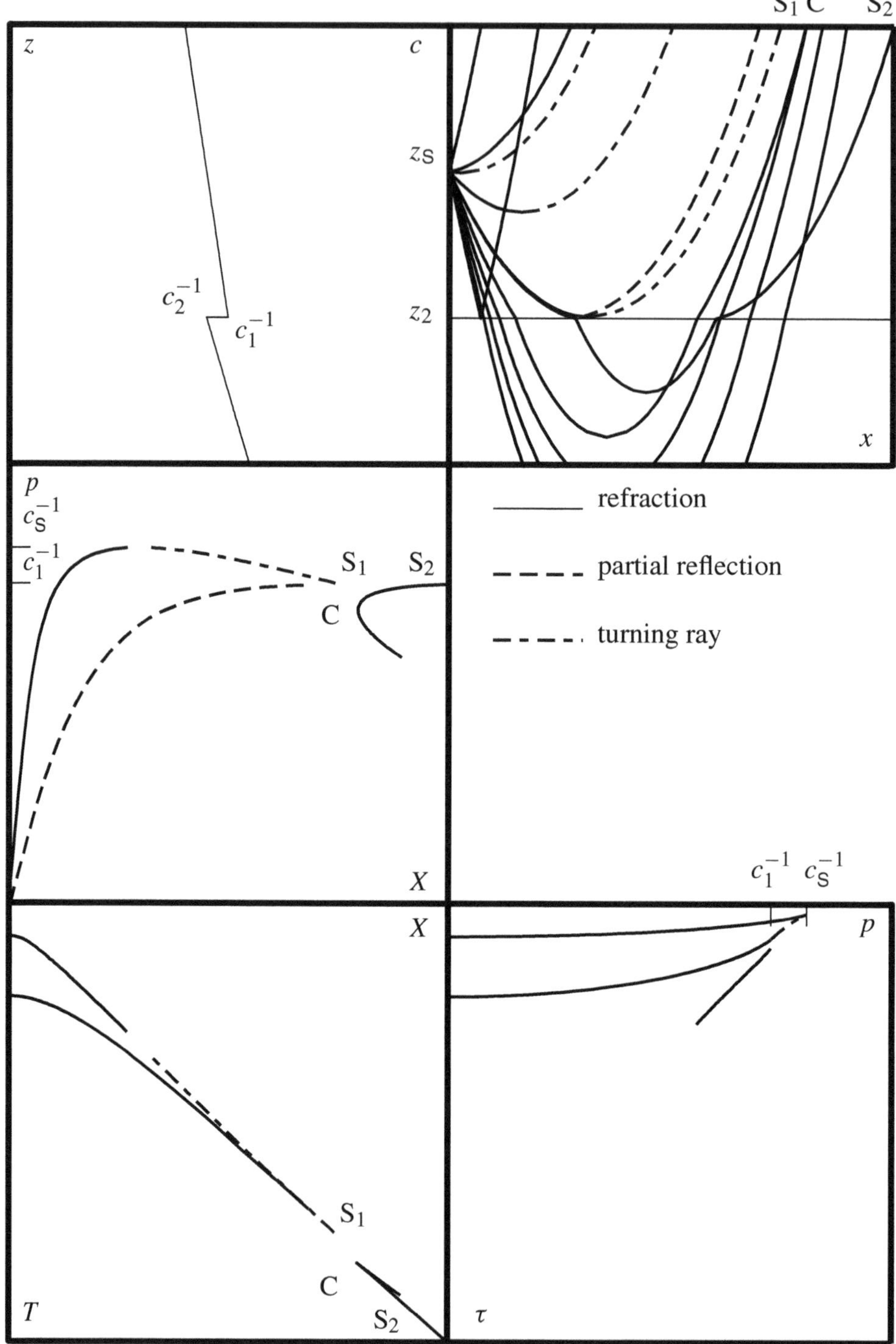

Fig. 2.23. As Figure 2.21, except $c_2 < c_1$. Direct and refracted rays are solid lines, reflected are dashed, and turning rays are dashed-dotted.

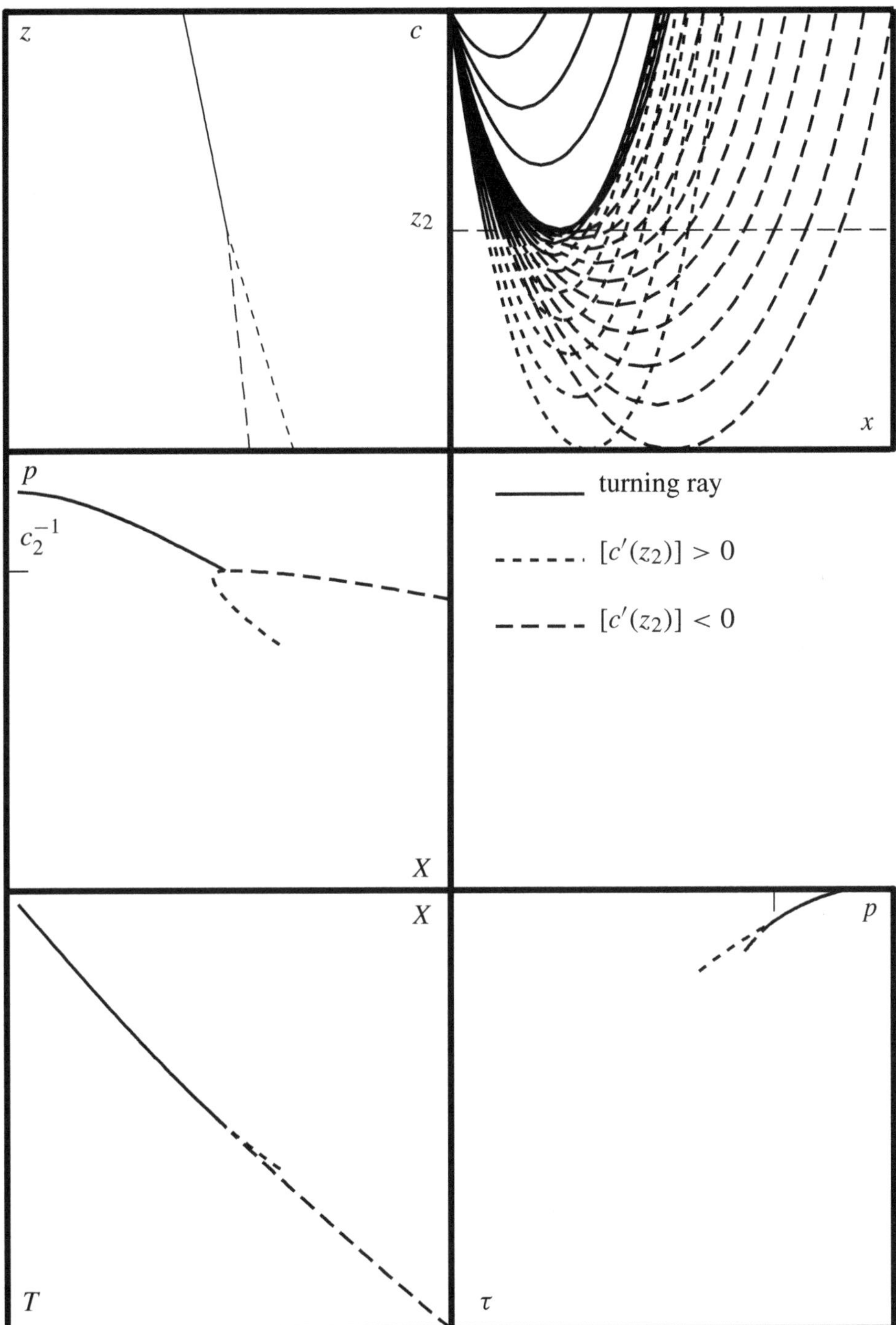

Fig. 2.24. A velocity gradient discontinuity. The results for $[c'(z_2)] > 0$ are shown with long-dashed lines, and for $[c'(z_2)] < 0$ with short-dashed lines.

Analytic results for a second-order velocity discontinuity can be generated using two linear velocity layers (Section 2.5.2), as in Figure 2.24.

In the transform domain, reflections from a gradient discontinuity can be described by matching the WKBJ solution, e.g. equation (7.2.118), or using the WKBJ iterative solution (7.2.125). The inverse transforms can be evaluated using the same techniques as for a reflection from an interface.

2.4.7 High-velocity layer

We now consider models with an embedded, homogeneous layer. The case of a low-velocity layer, e.g. $c_3 > c_1 > c_2$ is not particularly complicated although it has some interesting features. The upper interface causes no head wave and the direct ray and reflection are asymptotic to the slowness $p = 1/c_1$. The reflection from the lower interface is also asymptotic to $p = 1/c_1$, with a head wave at $p = 1/c_3$. Notice that no feature of the travel-time curves is characterized by the slowness $p = 1/c_2$ in the low-velocity layer. The low-velocity layer just causes a delay in the reflection from its lower interface. Of great interest is the ambiguity this causes in the inverse problem.

A high-velocity layer is more interesting. For simplicity let us consider a layer embedded in a homogeneous whole space ($c_2 > c_1 = c_3$) (Figure 2.25). We also consider the source and receiver on opposite sides of the layer. This is a situation that commonly occurs in crosswell seismics, for instance. The only rays that exist are the transmission and reverberations in the high-velocity layer. These rays are asymptotic to the slowness $p = 1/c_2$ and at large offsets have long segments almost parallel to the interfaces in the layer. But if the layer is very thin, it must be almost invisible to finite-frequency waves. We expect a signal very similar to the direct ray in a velocity c_1, i.e. asymptotic to slowness $p = 1/c_1 > 1/c_2$. This signal is shown with dashed lines in Figure 2.25. It must tunnel through the high-velocity layer, as the wavefield is evanescent there (Section 2.1.2.2). The transmission and the reverberations must combine and cancel as the layer gets thinner. In the limit as the layer becomes infinitesimally thin, the rays must disappear and the tunnelling signal must dominate and become a ray. This interesting wave propagation problem is obviously important as most sedimentary sequences have many thin, high-velocity layers as indicated by well logs.

The reflection from a thin high-velocity layer is similarly interesting. The reflection will be totally reflected when the layer is thick and the receiver at a large enough range. But as the layer thickness decreases, energy must tunnel through the thin layer and be transmitted. The reflection must be reduced. Reverberations of the tunnelling wave must cancel with the reflection. In the limit of an infinitesimally thin layer, the tunnelling wave behaves as the direct ray and the reflection

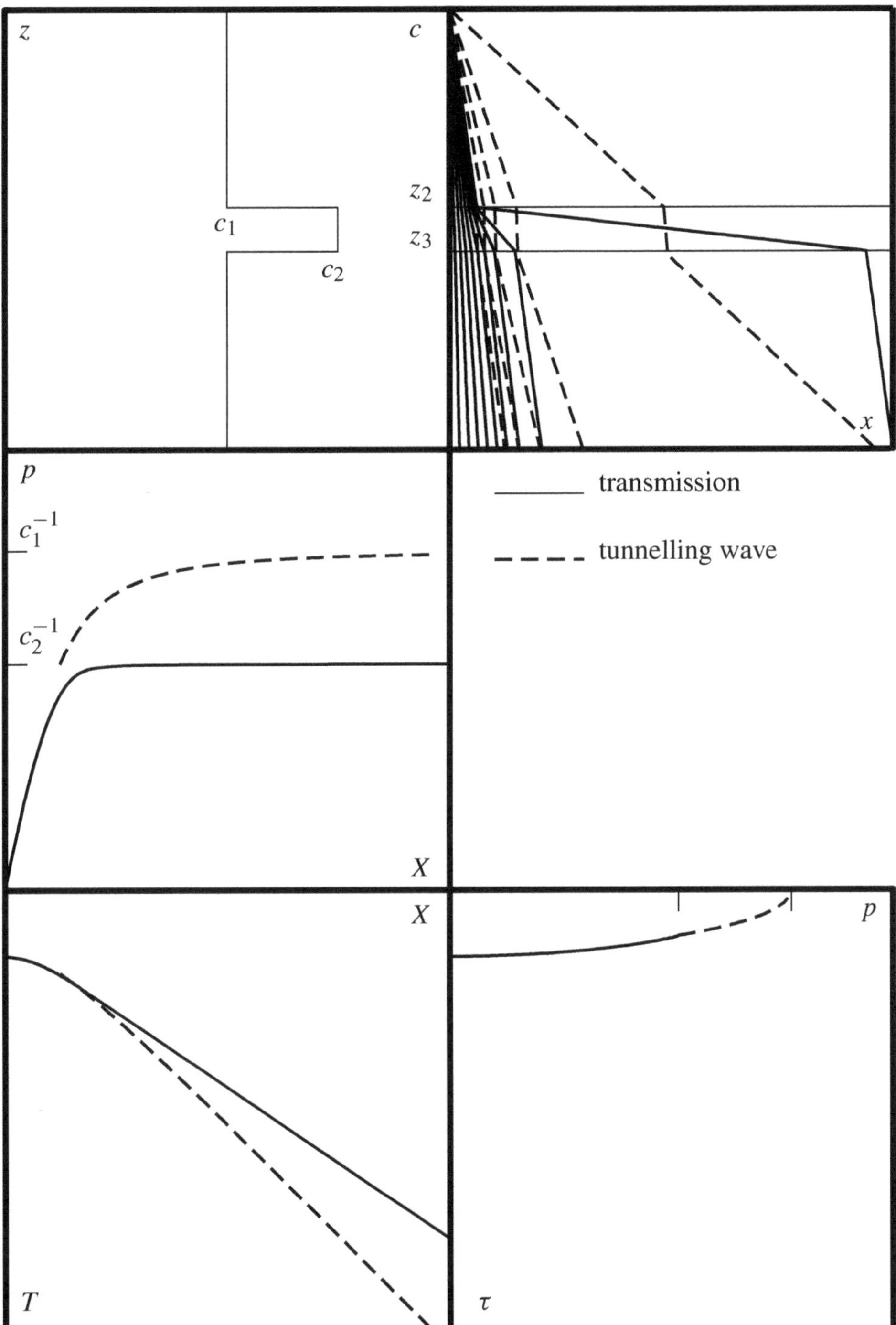

Fig. 2.25. A thin, high-velocity layer. The transmissions are shown with solid lines. Reverberations in the layer will exist, but are not shown. The tunnelling signal that becomes the direct ray in the limit of a zero thickness layer, is a dashed line.

must disappear. This signal has been called a *frustrated total reflection* (Towne, 1967, Section 17–13).

Analytic results are easily generated with models of homogeneous layers.

The Cagniard method (Sections 8.1 and 8.2) together with the ray expansion (Section 7.2.4), e.g. equation (7.2.70) is ideal to describe the rays in a high-velocity layer. Problems associated with the ray expansion in thin layers are also discussed in Section 7.2.4.5. The tunnelling signal is easily analysed (9.1.136).

2.4.8 Hidden layer

Next we consider an intermediate layer, e.g. $c_3 > c_2 > c_1$. The travel-time curves, etc. are easily constructed. The reflection from each interface is asymptotic to the preceding head wave (Figure 2.26). The interesting problem here is the first arrivals. Normally, these are made up of linear segments from the direct wave and head waves, i.e. with slownesses $p = 1/c_1$, $1/c_2$ and $1/c_3$, with intercepts

$$\tau(1/c_1) = 0 \tag{2.4.2}$$

$$\tau(1/c_2) = 2d_1\left(\frac{1}{c_1^2} - \frac{1}{c_2^2}\right)^{1/2} \tag{2.4.3}$$

$$\tau(1/c_3) = 2d_1\left(\frac{1}{c_1^2} - \frac{1}{c_3^2}\right)^{1/2} + 2d_2\left(\frac{1}{c_2^2} - \frac{1}{c_3^2}\right)^{1/2} \tag{2.4.4}$$

($d_1 = z_S - z_2, d_2 = z_2 - z_3$). If the three linear segments are the first arrivals, it is straightforward to interpret the data. However it is simple to show that if

$$d_2 < d_1 \frac{\left(\frac{1}{c_1} - \frac{1}{c_3}\right)\left(\frac{1}{c_1^2} - \frac{1}{c_2^2}\right)^{1/2} - \left(\frac{1}{c_1} - \frac{1}{c_2}\right)\left(\frac{1}{c_1^2} - \frac{1}{c_3^2}\right)^{1/2}}{\left(\frac{1}{c_1} - \frac{1}{c_2}\right)\left(\frac{1}{c_2^2} - \frac{1}{c_3^2}\right)^{1/2}}, \tag{2.4.5}$$

then the head wave from the first layer with slowness $p = 1/c_2$ is never a first arrival. As the head wave from the intermediate interface is not a first arrival, its existence is easily missed in real data. The data are interpreted without the intermediate layer. This is known as the *hidden layer problem*, e.g. Green (1962).

The Cagniard method (Sections 8.1 and 8.2) together with the ray expansion (Section 7.2.4), e.g. equation (7.2.70), is ideal to describe the rays in the hidden layer problem. Although the head wave from one layer is hidden, it will affect the waveforms.

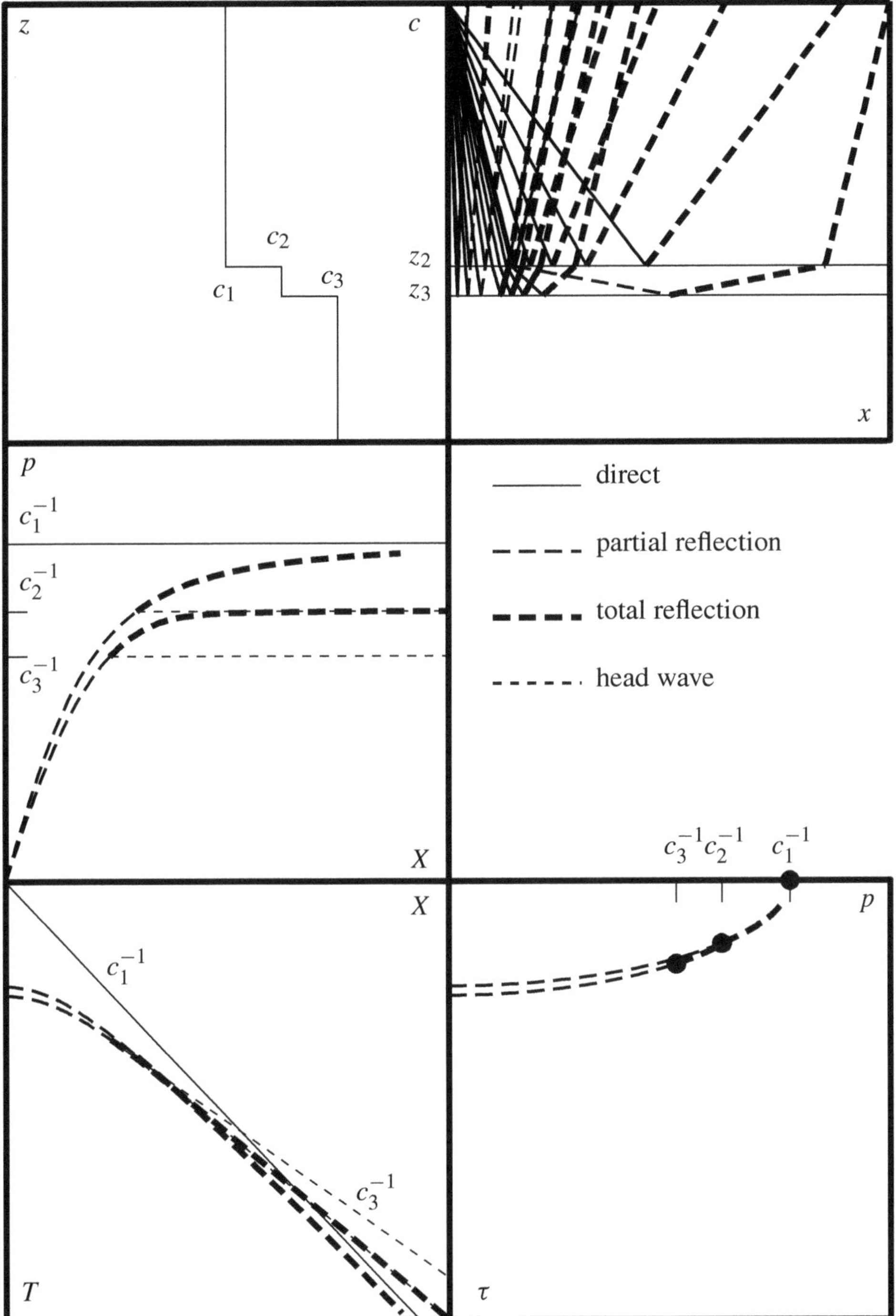

Fig. 2.26. A thin intermediate layer which does not cause a first arrival. The direct rays are shown with solid lines, the reflections with long-dashed lines, and the head waves with short-dashed lines.

2.4.9 Velocity maximum

Above we considered a LVZ (Section 2.4.5) below a high-velocity lid, and a high-velocity layer (Section 2.4.7). Suppose that the high-velocity lid is caused by a smooth velocity function where $c'(z_0) = 0$. The results of Section 2.4.5 and Figure 2.23 still apply except no shadow or reflections exist. The rays tend to infinity as $p \to 1/c_0 = u_0$ (Figure 2.27).

To confirm and model the divergent behaviour for rays turning near the velocity maximum, let us consider a model

$$u^2(z) = u_0^2 + a^2 z^2. \tag{2.4.6}$$

(This can either be considered as a specific model or a leading term in a more general structure. For convenience, we have measured the z coordinate from the minimum in the squared slowness.) Then

$$X(p) = p \int^{z^*} \frac{dz}{(u_0^2 + a^2 z^2 - p^2)^{1/2}} \tag{2.4.7}$$

$$= \frac{p}{a} \cosh^{-1} \frac{az^*}{(p^2 - u_0^2)^{1/2}} \tag{2.4.8}$$

$$\to -\frac{p}{2a} \ln(p^2 - u_0^2), \tag{2.4.9}$$

as $p \to u_0$ from above (using Abramowitz and Stegun, 1964, §4.6.38 and §4.6.21 – see equation (A.0.1)). Similarly

$$X(p) \to -\frac{p}{a} \ln(u_0^2 - p^2), \tag{2.4.10}$$

as $p \to u_0$ from below (using Abramowitz and Stegun, 1964, §4.6.37 and §4.6.20). These ranges diverge logarithmically for the limiting rays. The intercept saltus (2.4.1) is still valid, but obviously the saltus of the range or travel time cannot be calculated.

Analytic models that generate a velocity maximum are possible. The so-called Epstein layer is an analytic function that can be used (Hron and Chapman, 1974*b*, and references therein). The model used to generate Figure 2.27 is a Gaussian function added to a linear gradient. We have not analysed this situation explicitly in this text, but numerical spectral methods can be used to find the waveforms.

2.4.10 Focusing layers

Finally, we mention that although we use the term travel-time curves, special models will create travel times that are points. If the range function $X(p)$ is a series of steps (Figure 2.28), the travel-time curve is points, joined by lines for horizontally

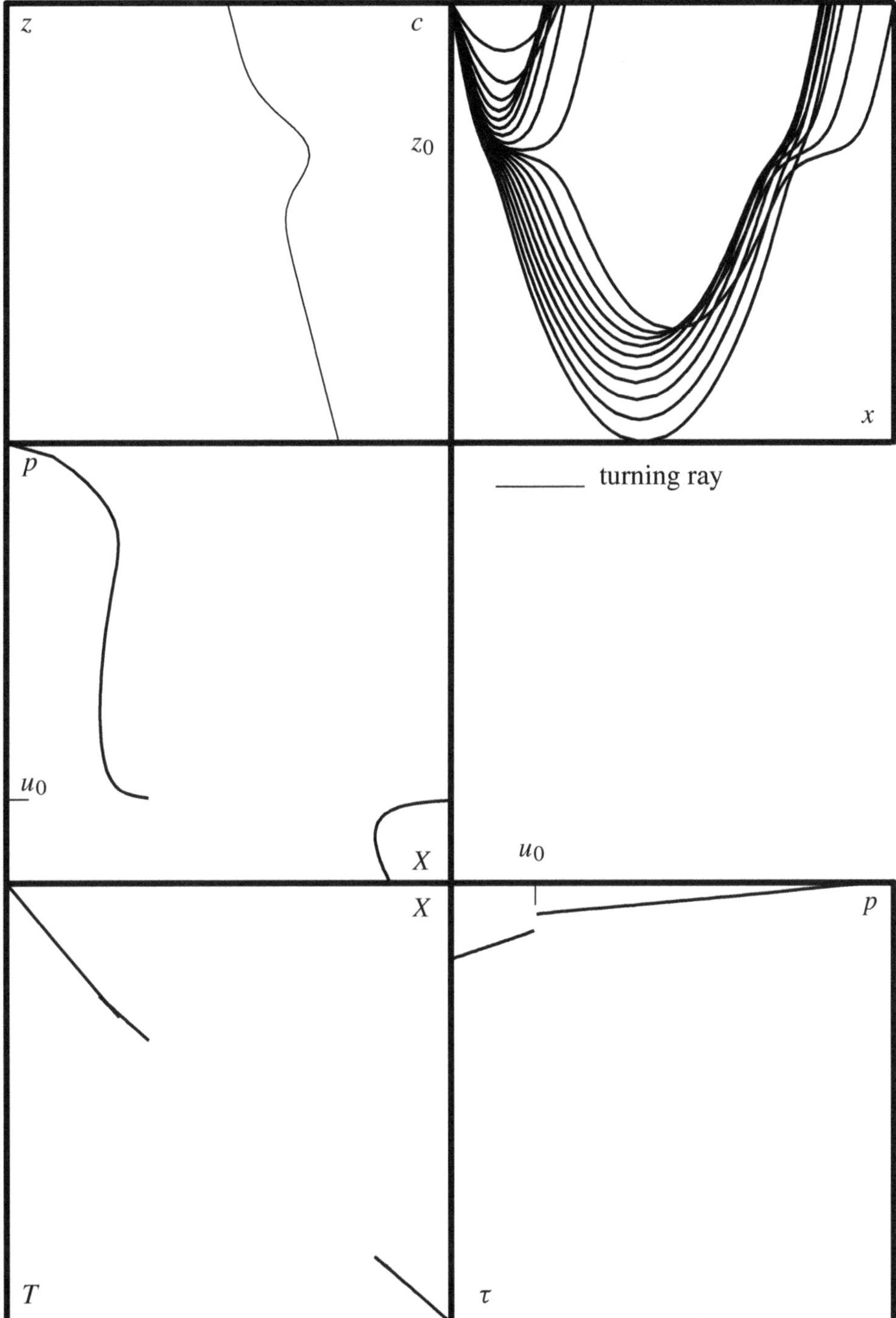

Fig. 2.27. The gap due to a LVZ with a smooth velocity maximum lid. Rays with large range, tending to infinity in the limiting situation, have not been included in the figure.

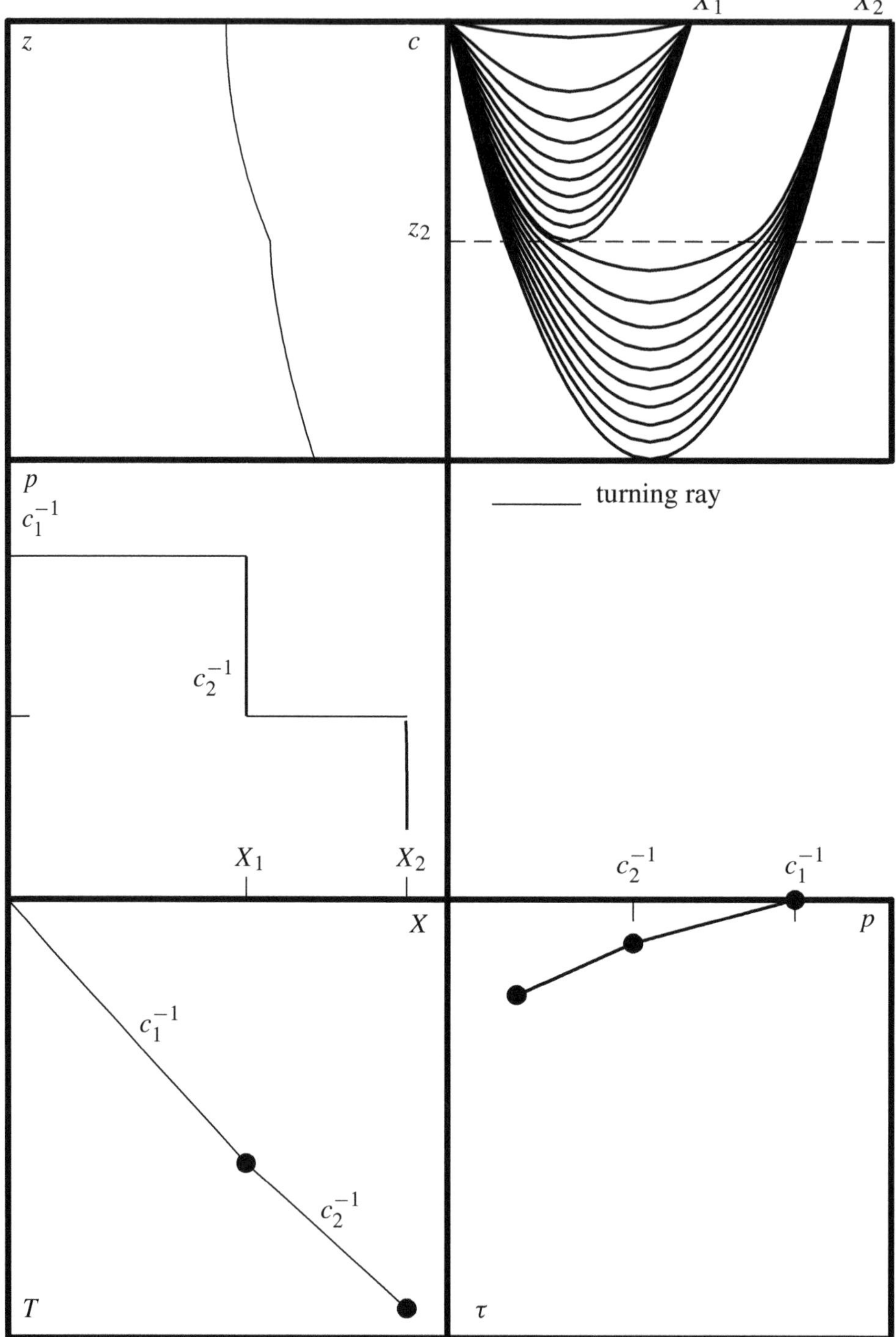

Fig. 2.28. A model producing a series of perfect foci.

travelling ray segments. The intercept time is similarly points and lines, with the roles reversed. The velocity–depth function that produces this result is

$$c(z) = c_1 \cosh\left(-\frac{\pi z}{X_1}\right), \tag{2.4.11}$$

for a single layer and focus. For multiple layers and foci, we cannot solve explicitly for $c(z)$, but can find its inverse function $z(c)$

$$z(c) = -\frac{1}{\pi}\sum_{n=1}^{N} \Delta X_n \cosh^{-1}\left(\frac{c}{c_i}\right), \tag{2.4.12}$$

where $\Delta X_n = X_n - X_{n-1}$ (with $X_0 = 0$), and $1/c_n$ is the slowness between X_{n-1} and X_n. This model is most easily generated considering the travel-time inverse problem with the step function $X(p)$ (the Herglotz–Wiechert–Bateman method – Aki and Richards, 1980, Section 12.1; 2002, Section 9.4.1).

We have not analysed this situation explicitly in this text, but numerical spectral methods can be used to find the waveforms.

2.5 Calculation of travel-time functions

In this section we discuss some methods for calculating or approximating travel-time functions. In the first section (Section 2.5.1) we discuss useful approximate methods for near-vertical reflections. In the second section (Section 2.5.2) we discuss a few methods for parameterizing models that are useful for calculating travel-time results.

2.5.1 Normal moveout

The travel-time integrals (2.3.7) and (2.3.8) are parameterized by the horizontal slowness, p. For vertical reflections, the parameter is zero so it is useful to consider expansions of the integrals for small p. We define the vertical travel time

$$T_0 = \oint \frac{1}{c}\,\mathrm{d}z. \tag{2.5.1}$$

Then the range is

$$\begin{aligned} X(p) &= \oint \frac{p}{q}\,\mathrm{d}z \\ &= \oint cp(1 - c^2p^2)^{-1/2}\mathrm{d}z \\ &= \oint \left(cp + \frac{1}{2}c^3p^3 + O(p^5)\right)\mathrm{d}z \\ &= p\,T_0\,\bar{c}_2^2 + \frac{1}{2}p^3\,T_0\,\bar{c}_4^4 + O(p^5), \end{aligned} \tag{2.5.2}$$

where we have defined

$$\bar{c}_k = \left(\frac{1}{T_0}\oint c^{k-1}\mathrm{d}z\right)^{1/k} = \left(\frac{1}{T_0}\oint c^{k}\mathrm{d}t\right)^{1/k}. \tag{2.5.3}$$

Thus $\bar{c}_2$ is the *root-mean-square velocity* (RMS velocity) although note that it is averaged with respect to time not depth. Similarly, the travel time can be expanded for small p

$$T(p) = T_0 + \frac{1}{2}p^2\, T_0\, \bar{c}_2^2 + \frac{3}{8}p^4\, T_0\, \bar{c}_4^4 + O(p^5). \tag{2.5.4}$$

First we can eliminate p between equations (2.5.2) and (2.5.4) only retaining the lowest term. Then

$$T(X) = T_0 + \frac{X^2}{2\,T_0\,\bar{c}_2^2}, \tag{2.5.5}$$

and the reflection travel-time curve is approximately parabolic. This equation describes the *normal moveout* (NMO) of the reflected arrivals. Reflections at different (small) ranges can be lined up by applying a time shift

$$\Delta T = -\frac{X^2}{2T_0\bar{c}_2^2}, \tag{2.5.6}$$

to the data. Having lined up the arrivals, the data at different ranges can be combined in order to cancel noise and emphasize the reflected signals, e.g. for data $\phi(t,x)$ we compute

$$\sum_x \phi\left(t + \frac{X^2}{2\,T_0\,\bar{c}_2^2}, x\right). \tag{2.5.7}$$

This process is called *NMO stacking*. It is often performed early in the processing chain, before a velocity model is available. The RMS velocity is chosen to optimize the stacking rather than being calculated from a model. In fact as expression (2.5.5) is only an approximation, the optimum velocity will not be exactly the RMS velocity. It is therefore called the *stacking velocity* to distinguish it from the true RMS velocity. Note the *NMO correction* depends on the vertical travel time, T_0, and so the time shift varies along the seismogram. This very simple operation distorts the frequency content and waveform of the data in a complicated manner. The RMS velocity was defined in the time domain (2.5.3), as the correction is normally determined and applied when only the time-domain data are available. A physicist would probably call it the mean velocity with respect to depth.

Equation (2.5.5) is an approximation even in a homogeneous layer. For a homogeneous layer we have exact, simple results, (2.3.4) and (2.3.5), and can eliminate

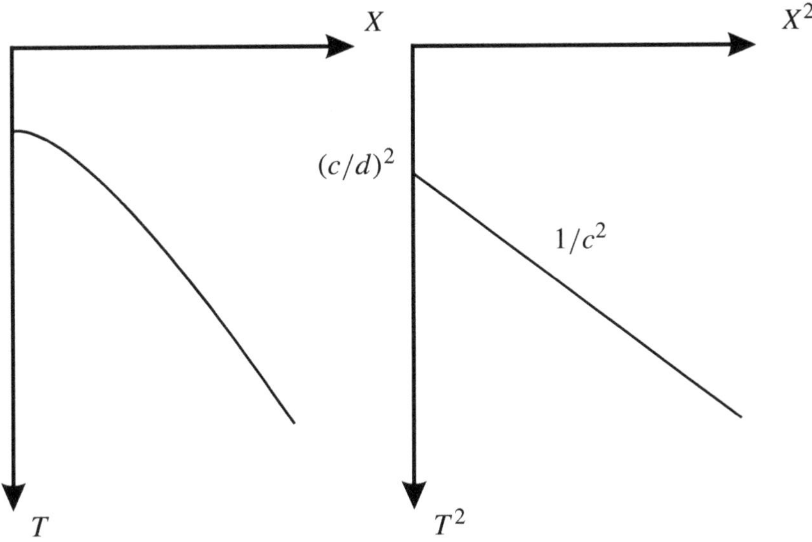

Fig. 2.29. A $X^2 - T^2$ plot.

the ray parameter to obtain

$$c^2T^2 - X^2 = d^2, \tag{2.5.8}$$

where d is the total vertical length of the ray. Thus the travel-time curve is a hyperbola. The $X^2 - T^2$ *method* exploits this by plotting travel-time data as T^2 against X^2 when a straight line of slope c^{-2} and intercept $(c/d)^2$ is obtained (Figure 2.29). This is a simple interpretation method for reflection data, but it is only exact for a homogeneous layer.

Using higher-order terms in the Taylor expansions (2.5.2) and (2.5.4), we can produce a higher-order NMO correction. Substituting the first-order estimate

$$p = \frac{X}{T_0\bar{c}_2^2}, \tag{2.5.9}$$

in the cubic term in expression (2.5.2) we obtain a better estimate

$$p = \frac{X}{T_0\,\bar{c}_2^2} - \frac{\bar{c}_4^4\,X^3}{2\,T_0^3\,\bar{c}_2^8} + O(X^5). \tag{2.5.10}$$

Substituting in (2.5.4), we obtain

$$T = T_0 + \frac{X^2}{2\,T_0\,\bar{c}_2^2} - \frac{\bar{c}_4^4\,X^4}{8\,T_0^3\,\bar{c}_2^8} + O(X^6). \tag{2.5.11}$$

Notice this correction is always negative. For the $X^2 - T^2$ method, we obtain

$$T^2 = T_0^2 + \frac{X^2}{\bar{c}_2^2} - \frac{(\bar{c}_4^4 - \bar{c}_2^4)\,X^4}{4\,T_0^2\,\bar{c}_2^8}. \tag{2.5.12}$$

and again the final term is bound to be negative for heterogeneous media. These expressions can easily be used to estimate geometrical spreading ($\mathrm{d}X/\mathrm{d}p$) to, for instance, correct AVO data (Ursin, 1990). However, the use of higher-order terms is limited as lateral variations and anisotropy will become important with increasing range.

2.5.2 Numerical methods

The travel-time integrals (2.3.7) and (2.3.8) could be evaluated by numerical means for any velocity function, $c(z)$, but more commonly they are evaluated analytically for special forms of the velocity function that can be used in restricted depth ranges. Thus, for instance, the velocity can be linearly interpolated between specified points. The analytic interpolation methods can also be used in two-dimensional or three-dimensional models, when a simple analytic function is used in a small area or volume element. In these cases, the velocity varies in one dimension but the gradient need not be vertical nor aligned in neighbouring elements. Analytic results are available for a reasonable number of functions. In this section we discuss two that are in wide usage: linear interpolation of the velocity and the squared slowness. We also discuss a more general polynomial method. We have, of course, already given the results, (2.3.4) and (2.3.5), for homogeneous layers (constant interpolation).

2.5.2.1 Linear velocity interpolation

Linear velocity interpolation is very widely used as it is a simple numerical interpolation method, and because the ray paths are circular arcs. While we could solve the range integral (2.3.7) for a linear velocity function, it is easier to ask the inverse question – what velocity function gives a circular ray?

Consider the circular ray arc illustrated in Figure 2.30. The ray parameter (2.3.9) is conserved. But from the geometry of the circular arc

$$\sin\theta(z) = -\frac{z}{R},$$

if z is measured from the centre of the arc, so

$$c(z) = -\frac{z}{Rp}. \tag{2.5.13}$$

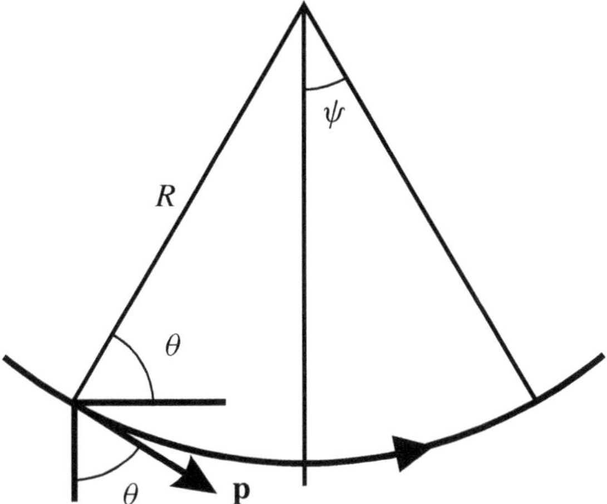

Fig. 2.30. A circular ray arc.

Thus the velocity function is linear. So if the velocity is linear, we can compute at the initial point

$$p = \frac{\sin\theta_0}{c_0}, \tag{2.5.14}$$

and then compute the radius

$$R = -\frac{1}{c'p} = -\frac{c_0}{c'\sin\theta_0}. \tag{2.5.15}$$

We can then compute the centre of the ray arc

$$\mathbf{x}_c = \mathbf{x}_0 + \begin{pmatrix} R\cos\theta_0 \\ 0 \\ R\sin\theta_0 \end{pmatrix}, \tag{2.5.16}$$

where $\mathbf{x}_0$ is the initial point on the ray. Thus the ray path satisfies

$$(\mathbf{x} - \mathbf{x}_c)^2 = R^2, \tag{2.5.17}$$

so

$$x = x_0 + R\cos\theta_0 \pm \sqrt{R^2 - (z - z_0 - R\sin\theta_0)^2}. \tag{2.5.18}$$

As p (2.5.14) and R (2.5.15) are known, this completely describes the ray path.

To compute the travel time, it is convenient to define the angle ψ between the z axis and the radius vector measured in the direction of propagation (Figure 2.29). When the ray is propagating in the positive z direction, ψ is positive, and $\theta = \pi/2 - \psi$. Conversely, when it is propagating in the negative z direction, ψ is negative and $\theta = \pi/2 + \psi$. Then

$$dT = \frac{ds}{c} = \frac{R\,d\psi}{c} = \frac{\sec\psi\,d\psi}{c'}, \tag{2.5.19}$$

and

$$\Delta T = \int dT = \frac{1}{|c'|}\int \sec\psi\,d\psi$$
$$= \frac{1}{|c'|}\tanh^{-1}(\sin\psi)\Big| \tag{2.5.20}$$
$$= \frac{1}{|c'|}\ln\left(\tan\frac{\theta}{2}\right)\Big| \tag{2.5.21}$$

(using Abramowitz and Stegun, 1965, §4.3.117 and §4.6.22 – equations (A.0.3)). Similar results have been given by Gebrande(1976) and Telford, Geldart, Sheriff and Keys (1976, p. 273). A convenient alternative expression for (2.5.20) is obtained from

$$\chi = \tanh^{-1}(\sin\psi) = \ln\left(\frac{1+\sin\psi}{\cos\psi}\right), \tag{2.5.22}$$

so the definite integral is

$$\Delta T = \frac{1}{|c'|}\ln\frac{(1+\sin\psi_1)\cos\psi_0}{(1+\sin\psi_0)\cos\psi_1}$$
$$= \frac{1}{|c'|}\ln\left(\frac{1+y}{1-y}\right) = \frac{1}{|c'|}\operatorname{log1p}\left(\frac{2y}{1-y}\right), \tag{2.5.23}$$

after considerable manipulation, where

$$y = \frac{\sin(\psi_1 - \psi_0)}{\cos\psi_1 + \cos\psi_0}. \tag{2.5.24}$$

The final expression (2.5.23) uses the function $\operatorname{log1p}(\epsilon) = \ln(1+\epsilon)$, which exists in most C libraries. Often ϵ is small and numerical accuracy is lost if $(1+\epsilon)$ and $\ln(1+\epsilon)$ are computed directly. Rather we compute ϵ, and use a power series expansion if ϵ is small, i.e. $\ln(1+\epsilon) \simeq \epsilon$. In the limit $c' \to 0$, $R \to \infty$, $\psi \to 0$, $\Delta\psi = \psi_1 - \psi_0 \to 0$ and $y \to \Delta\psi/2$, this procedure with result (2.5.23) yields

$$\Delta T \simeq \frac{1}{|c'|}\ln(1+\Delta\psi) \simeq \frac{\Delta\psi}{|c'|} \to \Delta T, \tag{2.5.25}$$

using expression (2.5.19).

In two or three dimensions it is useful to express the results in vector notation. We define a set of basis vectors

$$\hat{\mathbf{k}} = \operatorname{sgn}(\nabla c) \tag{2.5.26}$$

$$\hat{\mathbf{j}} = \operatorname{sgn}(\hat{\mathbf{k}} \times \mathbf{p}_0) \tag{2.5.27}$$

$$\hat{\mathbf{i}} = \hat{\mathbf{j}} \times \hat{\mathbf{k}}, \tag{2.5.28}$$

where $\hat{\mathbf{k}}$ is in a local vertical direction aligned with the velocity gradient, $\hat{\mathbf{j}}$ is perpendicular to the ray plane, and $\hat{\mathbf{i}}$ is in a local horizontal direction. The arc radius (2.5.15) is then

$$R = \frac{1}{|\nabla c|(\mathbf{p}_0 \cdot \hat{\mathbf{i}})}, \tag{2.5.29}$$

the centre of the arc (2.5.16) is

$$\mathbf{x}_c = \mathbf{x}_0 + R(\hat{\mathbf{j}} \times \hat{\mathbf{p}}_0), \tag{2.5.30}$$

and the equation of the ray arc (2.5.18)

$$\mathbf{x} = \mathbf{x}_c + R(\sin\psi\,\hat{\mathbf{i}} + \cos\psi\,\hat{\mathbf{k}}). \tag{2.5.31}$$

The path is parameterized in terms of the angle ψ (Figure 2.30), where

$$\tan\psi = -\left(\frac{\mathbf{p}\cdot\hat{\mathbf{k}}}{\mathbf{p}\cdot\hat{\mathbf{i}}}\right). \tag{2.5.32}$$

The increments in the slowness and travel time are then

$$\Delta\mathbf{p} = \frac{\hat{\mathbf{j}} \times \Delta\mathbf{x}}{R}, \tag{2.5.33}$$

and (2.5.23)

$$\Delta T = \frac{1}{|\nabla c|}\,\mathrm{log1p}\left(\frac{2y}{1-y}\right) \quad \text{with} \quad y = \frac{|\nabla c|\hat{\mathbf{p}}_0 \cdot \Delta\mathbf{x}}{c_0 + c_1}. \tag{2.5.34}$$

Linear velocity interpolation is widely used in layered models, and in model elements in three-dimensional models. It allows efficient, accurate ray tracing. The main disadvantage is the features caused by gradient discontinuities (Section 2.4.6 and Figure 2.24).

2.5.2.2 *Linear squared-slowness interpolation*

Another simple interpolation method is with a linear function for the slowness squared (sometimes called the *sloth* – Muir and Dellinger, 1985. Simultaneously they introduced the term alacrity and the symbol w – two v's – for squared velocity, and the symbol m – an inverted w – for the sloth. Only the latter term has caught on), e.g.

$$u^2(z) = \frac{1}{c^2(z)} = u_0^2 + az. \tag{2.5.35}$$

Then

$$\begin{aligned} X(p) &= \oint p(u_0^2 - p^2 + az)^{1/2}\,\mathrm{d}z \\ &= \frac{2p}{a}(u_0^2 - p^2 + az)^{1/2}\bigg| \end{aligned} \tag{2.5.36}$$

$$\begin{aligned} T(p) &= \oint \frac{u_0^2 + az}{(u_0^2 - p^2 + az)^{1/2}}\,\mathrm{d}z \\ &= \frac{2}{3a}(u_0^2 - p^2 + az)^{3/2} + \frac{2p^2}{a}(u_0^2 - p^2 + az)^{1/2}\bigg| . \end{aligned} \tag{2.5.37}$$

From result (2.5.36) we can see that the ray path is a parabola. The simplicity of these results – the only special function required is the square root for q – makes them very attractive, and they have been widely used (e.g. Virieux, Farra and Madariaga, 1988). They can be rewritten

$$\mathbf{x} = \mathbf{x}_0 + \sigma\mathbf{p}_0 + \frac{\sigma^2}{4}\mathbf{a} \tag{2.5.38}$$

$$\mathbf{p} = \mathbf{p}_0 + \frac{\sigma}{2}\mathbf{a} \tag{2.5.39}$$

$$T = T_0 + \sigma u_0^2 + \frac{\sigma^2}{2}\mathbf{a}\cdot\mathbf{p}_0 + \frac{\sigma^3}{12}\mathbf{a}\cdot\mathbf{a}, \tag{2.5.40}$$

where $\mathbf{a} = \nabla u^2$, i.e. $a = |\mathbf{a}|$, and the parameter σ increases along the ray arc and is

$$\sigma = \pm\frac{2}{a}(u_0^2 - p^2 + az)^{1/2}\bigg| . \tag{2.5.41}$$

The equivalence of these results can be proved algebraically, but is easily established once we have the kinematic ray equations (Section 5.7.2). Again this interpolation can be used in layers or elements in three-dimensional models, but suffers from the same disadvantages of gradient discontinuities.

2.5.2.3 Higher-order slowness interpolation

Many other two-parameter interpolation functions exist for which the ray integrals can be evaluated analytically, e.g. $c(r) = a - br^2$, $c(z) = a\exp(bz)$ or equivalently $c(r) = ar^b$, etc. A three-parameter interpolation is discussed in Exercise 2.3. Ideally we would like a higher-order function so that gradient continuity can be imposed, e.g. spline interpolation, but no analytic results are known. However, a simple alternative can be used. Rather than considering velocity as a function of depth, we consider depth as a function of velocity or rather slowness, i.e. the inverse slowness function is

$$z(u) = \sum_{n=0}^{N} Z_n u^n, \tag{2.5.42}$$

where Z_n are the coefficients of this polynomial representation. The disadvantages of using this function are that it excludes homogeneous media (but for that the results (2.3.4) and (2.3.5) are trivial), and stationary values, $u'(z) = 0$ (for which we could use result (2.4.8)). We must take care that $z'(u) = 0$ does not occur in the range of interest as this would lead to the non-physical situation of two slownesses at the same depth.

The intercept time can be evaluated as

$$\begin{aligned} \tau(p) &= \oint (u^2 - p^2)^{1/2} \mathrm{d}z \\ &= \oint (u^2 - p^2)^{1/2} \frac{\mathrm{d}z}{\mathrm{d}u}\, \mathrm{d}u \\ &= \sum_{n=1}^{N} n Z_n u_n(p), \end{aligned} \tag{2.5.43}$$

where

$$u_n(p) = \oint (u^2 - p^2)^{1/2} u^{n-1} \mathrm{d}u \tag{2.5.44}$$

$$= \frac{n-2}{n+1} p^2 u_{n-2}(p) + \frac{u^{n-2}}{n+1} (u^2 - p^2)^{3/2} \Big| \tag{2.5.45}$$

The recurrence relation (2.5.45) for $n > 2$ is obtained integrating (2.5.44) by parts and rearranging. To start the recurrence we need

$$u_1(p) = \frac{1}{2} \left(u(u^2 - p^2)^{1/2} - p^2 \cosh^{-1} \frac{u}{p} \right) \Big| \tag{2.5.46}$$

$$u_2(p) = \frac{1}{3} (u^2 - p^2)^{3/2} \Big| , \tag{2.5.47}$$

where Abramowitz and Stegun(1965, §4.6.38) has been used (A.0.1).

Thus the intercept function can be written analytically, and evaluated efficiently and accurately for any order of polynomial. The travel time, range and spreading function can be simply derived from this function, using relationships (2.3.16) and (2.3.15), as all the parts can be differentiated analytically. Another interesting feature of this method is that the result (2.5.43) is linear in the coefficients, Z_n. This makes the parameterization ideal for inverse problems and very rapid calculations.

Similar results can be obtained if the depth is considered as a function of velocity, $z(c)$. The algebraic details are left as an exercise.

2.5.3 Example – *PKP, PKiKP* and *PKIKP*

Finally as an example of ray tracing in a realistic one-dimensional model, we consider core phases in a whole Earth model. These rays are traced in the model 1066B due to Gilbert and Dziewonski (1975) – the exact details are not important here (nor the fact that this model was derived from normal mode data) and any other Earth model could have been used, e.g. the Preliminary Reference Earth Model (PREM) (Dziewonski and Anderson, 1991). The model is illustrated in Figure 2.31.

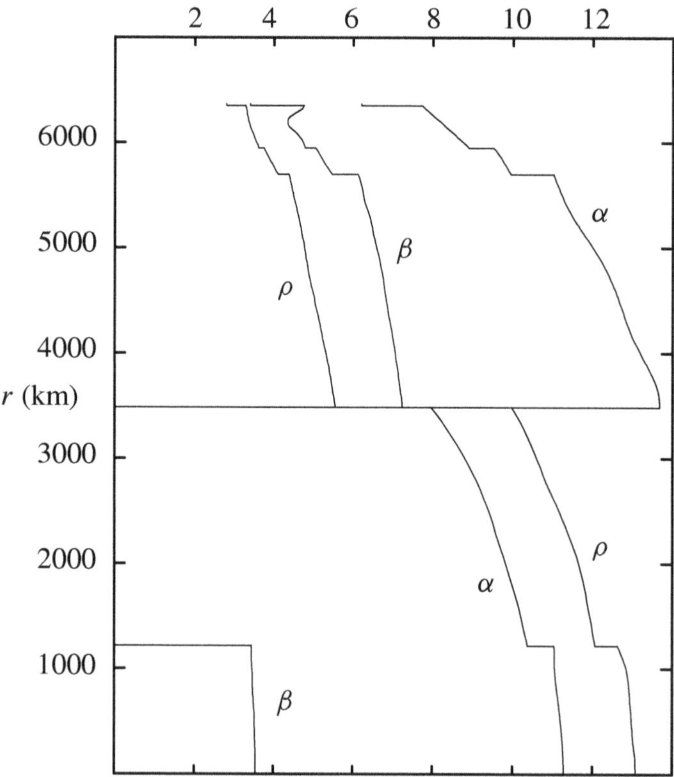

Fig. 2.31. The Earth model 1066B (Gilbert and Dziewonski, 1975). The *P* and *S* wave velocities, α and β, and density, ρ, are plotted as a function of radius. The units are km/s and Mg/m^3, respectively.

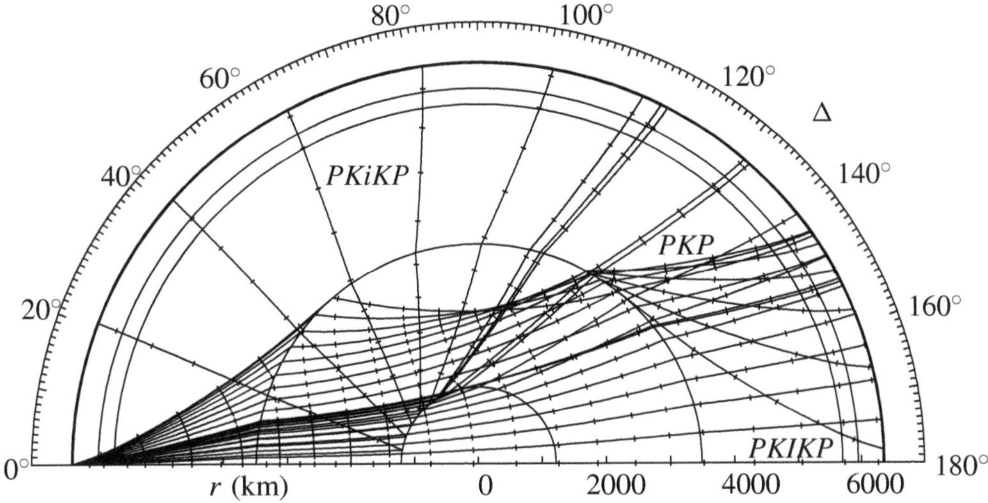

Fig. 2.32. Some ray paths for the core rays *PKP*, *PKiKP* and *PKIKP* in the model 1066B (Figure 2.31). The source is at the surface of the Earth. The tick marks on the rays are on wavefronts 60 s apart.

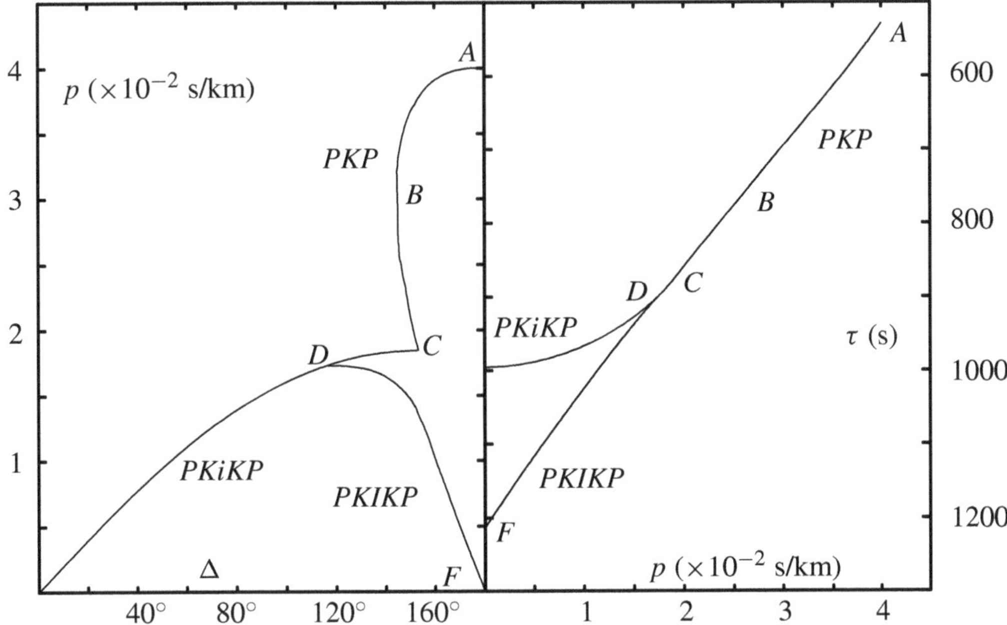

Fig. 2.33. The $p(\Delta)$ and $\tau(p)$ functions for the core rays *PKP*, *PKiKP* and *PKIKP* in the model 1066B (Figure 2.31). The shadow end-points A and C, and the caustic B of *PKP* are marked, together with the critical point D of *PKiKP* and the anti-pode ray F of *PKIKP*.

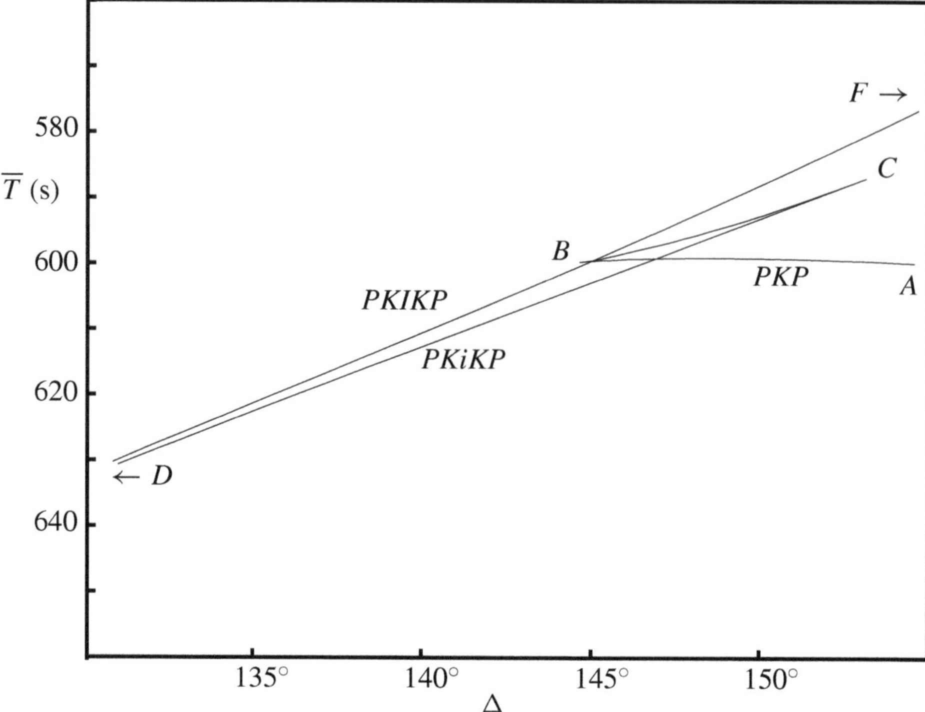

Fig. 2.34. The reduced travel time, $\overline{T}$ (2.5.48) for the core rays *PKP*, *PKiKP* and *PKIKP* in the model 1066B (Figure 2.31). The reducing slowness is $u_0 = 4$ s/°. The same points are marked as in Figure 2.33.

In Figure 2.32, the core rays *PKP*, *PKiKP* and *PKIKP* are illustrated on a cross-section of the Earth. These rays are the transmission through the outer core, the reflection from the inner core, and the transmission through the inner core, respectively. The interfaces in the Earth model 1066B are indicated on the plot, and the tick marks on the rays mark wavefronts 60 s apart. The most striking feature in this figure is the caustic formed by the *PKP* rays at about 144.6°. The ray paths and travel times have been calculated using analytic results for the ray integrals, (2.3.32) and (2.3.33), where the tabulated velocity values from Gilbert and Dziewonski (1975) have been interpolated using the *Mohorovičić velocity* function (see Exercise 2.2).

The functions $p(\Delta)$ and $\tau(p)$ for these rays are illustrated in Figure 2.33. In particular notice the caustic of *PKP* at point B, the shadow between *PKP* and *PKiKP* at point C, and the critical point between *PKiKP* and *PKIKP* at point D.

Finally in Figure 2.34, we plot the travel-time functions for some of these rays (from $\Delta = 130°$ to 150°). In order to separate the interesting features, we have plotted the reduced travel time, where

$$\overline{T} = T - u_0 X. \tag{2.5.48}$$

The reducing slowness, u_0, is 4 s/° in this figure. The same features are labelled as in Figure 2.33. We return to this model and these core rays much later when WKBJ seismograms are illustrated in Figure 8.16.

Exercises

2.1 Confirm that the standard, simple algebraic expression (2.5.20) for the travel time along a circular arc reduces to the numerically robust expression (2.5.23). In turn show that this reduces to the vector expression (2.5.34).

2.2 Examples of other two-parameter velocity functions for which algebraic results are known are $c(r) = a - br^2$ (a circular ray) and $c(z) = a\exp(bz)$. Obtain algebraic expressions for the range and travel time for a ray in these velocity functions.

Show that $c(z) = a\exp(bz)$ is equivalent to $c(r) = ar^b$ via the Earth flattening transformation (this velocity distribution is sometimes known as the *Mohorovičić velocity* function).

Further reading: The constant gradient functions in velocity (Section 2.5.2.1) and slowness-squared (Section 2.5.2.2) can be generalized to constant gradient functions of $c^{-n}(z)$ or $\ln(c(z))$. These have been discussed by Červený (2001, Section 3.7).

2.3 Many two-parameter velocity functions are known with algebraic results (see the previous exercise) which can be used to approximate a general discretized velocity function. However, these suffer from the disadvantage that the function $\mathrm{d}X/\mathrm{d}p$ has singularities due to each gradient discontinuity (see Figure 2.24). A three-parameter velocity function, $c^{-2}(z) = a + bz + cz^2$, a parabolic layer, has been discussed by Červený (2001, Section 3.7) – see also Kravtsov and Orlov (1990), and references therein. This has the advantage that, in principle at least, gradient continuity can be obtained. Obtain the algebraic results for this function.

2.4 *Programming exercise:* Figures 2.18–2.28 were drawn using a relative simple Matlab program. Write a program to compute the travel-time functions in which it is easy to change the velocity function and type of ray.

Hint: The ray can be defined using a structure array (*struct*) which defines the sequence of ray segments – the type, depth limits, etc. The integrals can all be calculated using the function *quad* where the required integrand function is passed as an argument. The velocity function – the model – can be defined by a function whose name is known to the integrand routines.

3
Transforms

This short chapter reviews the transforms – *Fourier*, *Hilbert*, *Fourier–Bessel*, *Legendre* and *Radon* – and related results needed in the rest of this book. As we only consider causal signals, the distinction between the Fourier and Laplace transforms is minor, and we choose to use the Fourier transform as it is easier to implement numerically or when the results are approximate. A more thorough treatment of most of the results in this chapter can be found in many more specialized books, but the important results, and a consistent notation followed later, are reviewed here.

In this chapter we will review the transforms used elsewhere in this book. The main purpose is to introduce notation, etc. and establish sign conventions. The basic theories of the transforms are not given as these are thoroughly covered in many textbooks on mathematical physics, etc., see, for example, Chapter 13 in Riley, Hobson and Bence (2002).

3.1 Temporal Fourier transform

Although we will usually be interested in the impulsive response in the time domain, it is often convenient to work in the frequency domain. For a time function, $f(t)$, we shall use the Fourier transform with respect to time

$$\boxed{f(\omega) = \int_{-\infty}^{\infty} f(t) \mathrm{e}^{\mathrm{i}\omega t} \mathrm{d}t.} \tag{3.1.1}$$

Note we follow the convention common in many mathematical publications of using the same symbol f for the function in the time or frequency domain. The specific function is indicated by the argument, t or ω, and is normally obvious from the context. Note the sign of the exponent and the lack of factors 2π, etc. The angular (circular) frequency, ω (radians/s), is as defined in Section 2.1.1. The function $f(\omega)$ is known as the *complex spectrum*.

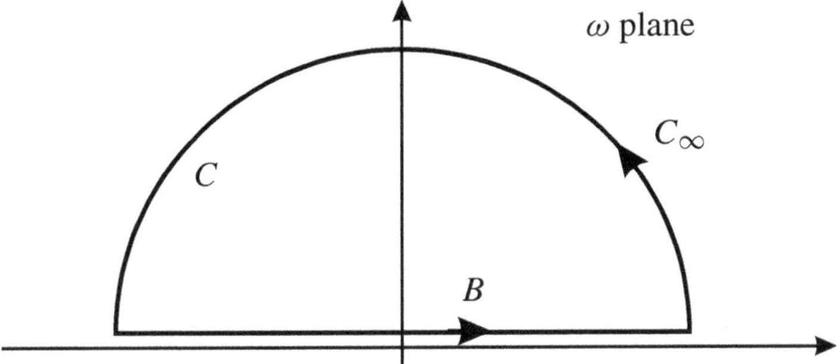

Fig. 3.1. The Bromwich contour B closed at infinity C_∞ to make the closed loop C.

The inverse transform to (3.1.1) is

$$\boxed{f(t) = \frac{1}{2\pi}\int_B f(\omega)\mathrm{e}^{-\mathrm{i}\omega t}\mathrm{d}\omega.} \tag{3.1.2}$$

B is the Bromwich contour and is essentially $\int_{-\infty}^{\infty}$ except the exact position must be chosen to obtain causality.

3.1.1 Causality

The problems we will deal with are all causal, i.e. an impulse source at $t = 0$ will cause a non-zero solution for $t \geq 0$. Thus

$$f(t) = 0 \quad \text{for} \quad t < 0. \tag{3.1.3}$$

The Bromwich contour must be above any singularities in the complex ω plane. The proof that if there are no singularities above B, then the signal is causal (and *vice versa*) follows from closing the contour B at infinity in the half-space where $\mathrm{Im}(\omega) > 0$ (see Figure 3.1).

Then

$$\frac{1}{2\pi}\int_C f(\omega)\mathrm{e}^{-\mathrm{i}\omega t}\mathrm{d}\omega = \sum \text{Residues contained in } C \tag{3.1.4}$$

$$= \frac{1}{2\pi}\int_B f(\omega)\mathrm{e}^{-\mathrm{i}\omega t}\mathrm{d}\omega + \frac{1}{2\pi}\int_{C_\infty} f(\omega)\mathrm{e}^{-\mathrm{i}\omega t}\mathrm{d}\omega, \tag{3.1.5}$$

where the contour C consists of the counter-clockwise circuit of the half-plane with $\mathrm{Im}(\omega) > 0$, i.e. the Bromwich contour B along the real axis and an infinite

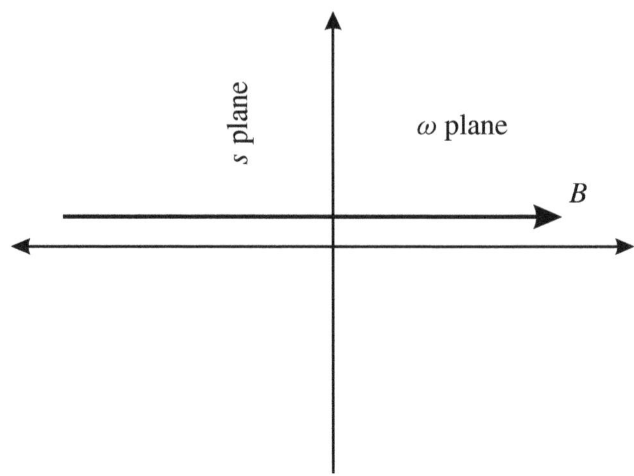

Fig. 3.2. The complex ω and s planes, and the Bromwich contour, B.

loop C_∞ closing the contour (see Figure 3.1). But

$$\frac{1}{2\pi}\int_{C_\infty} f(\omega)\mathrm{e}^{-\mathrm{i}\omega t}\mathrm{d}\omega = 0, \tag{3.1.6}$$

for $\mathrm{Im}(\omega) > 0$ and $t < 0$ (due to the exponential decay). Thus for $t < 0$

$$f(t) = \sum \text{Residues contained in } C. \tag{3.1.7}$$

Hence the result.

In general the Fourier transform is defined for non-causal functions and a unique inverse transform requires a careful definition of the path (see, for instance, van der Pol and Bremmer, 1964). We shall not encounter this complication as the contour must always be above all singularities.

Because we are only considering causal signals, the Fourier transform and the Laplace transform are essentially equivalent with $s = -\mathrm{i}\,\omega$ (see Figure 3.2). The forward Laplace transform

$$f(s) = \int_0^\infty f(t)\mathrm{e}^{-st}\mathrm{d}t, \tag{3.1.8}$$

is equivalent to equation (3.1.1) as causality allows the lower limit to be $t = 0$. Apart from the change of variable $s \leftrightarrow \omega$, the main difference is that the forward Laplace transform is often only considered for s real, i.e. the positive, imaginary ω axis. Because the spectrum is analytic in the upper half ω plane, it can be analytically continued from any line in the half-plane to the Bromwich contour. Thus knowing $f(s)$ for positive, real s, and knowing $f(t)$ is causal, implies that we know $f(s)$ on the Bromwich contour, as needed for the inverse transform. This procedure is valid when we have the exact, analytic function, $f(s)$ or $f(\omega)$, as then the analytic continuation (from the imaginary ω axis to the Bromwich contour) is automatic. But for approximate analytic or numerical results, we have to be very careful. It is well known that analytic continuation is unstable. Analytic approximations often only have limited regions of validity which may not be the complete upper-half ω plane. So neither approximate nor numerical results are easy to analytically continue. Their knowledge on the imaginary ω axis may not be good enough to continue to the Bromwich contour for the inverse transform. The best illustration of this is a simple example. Consider the function $\cos(a\omega)$, with a positive and real. On the positive, imaginary axis, $\exp(-\mathrm{i}a\omega)/2$ is a good approximation (and conversely $\exp(\mathrm{i}a\omega)/2$ is a good approximation on the negative imaginary axis). The real axis is called an *anti-Stokes line* and neither approximation is good (but a linear combination, the sum, is valid, of course). Using the valid approximation from the positive, imaginary axis would obviously give the wrong inverse Fourier transform. This same *Stokes phenomenon* holds, but in a more complicated fashion, for many of the special functions, e.g. Bessel and Airy functions, used in wave theory.

Numerically, analytic continuation is unstable as small errors may grow exponentially. As $\mathrm{Im}(\omega)$ increases, the spectrum obviously gets smoother and yet contains the same information. The analytic continuation from the smooth region to the rough is unstable. Therefore, except for exact analytic results, it is very dangerous to assume that a knowledge of the transformed results on the positive, real s or positive, imaginary ω axis is adequate for the inverse transform. Errors based on inaccurate continuation have appeared in the literature.

As in general we need to know the spectrum on the Bromwich contour, on or near the real ω axis, we prefer to use the Fourier transform to the Laplace transform and just note that for causal signals they are equivalent.

3.1.2 Analytic time series

As we will normally only be considering real signals, it is straightforward to show that

$$\boxed{f^*(\omega) = f(-\omega),} \tag{3.1.9}$$

if ω is real, where the superscript asterisk denotes the complex conjugate. As we can always exploit this symmetry, we will sometimes restrict our discussion to positive frequencies which avoids tricky questions concerning roots such as $\omega^{1/n}$.

This symmetry means we can rewrite (3.1.2)

$$f(t) = \frac{1}{\pi} \operatorname{Re} \int_0^\infty f(\omega) e^{-i\omega t} d\omega, \tag{3.1.10}$$

where for simplicity we have assumed the Bromwich contour is along the real axis. This result suggests we define a complex time series

$$F(t) = \frac{1}{\pi} \int_0^\infty f(\omega) e^{-i\omega t} d\omega \tag{3.1.11}$$

$$\text{or} \quad \boxed{F(t) = f(t) + i\bar{f}(t),} \tag{3.1.12}$$

where

$$\bar{f}(t) = \frac{1}{\pi} \operatorname{Im} \int_0^\infty f(\omega) e^{-i\omega t} d\omega. \tag{3.1.13}$$

This complex times series, $F(t)$, is called the *analytic time series*. In general we will use this notation (capitalizing the real time series). Comparing (3.1.13) with (3.1.10) we must have

$$\boxed{\bar{f}(\omega) = -i\,\mathrm{sgn}(\omega) f(\omega),} \tag{3.1.14}$$

where $\mathrm{sgn}(\omega) = 1$ if $\omega > 0$, 0 if $\omega = 0$, and -1 if $\omega < 0$. This is necessary as $\bar{f}(t)$ is real and its spectrum must satisfy (3.1.9). Note that if $f(t)$ is causal, $\bar{f}(t)$ is non-causal as $\mathrm{sgn}(\omega)$ makes its spectrum non-analytic in the upper ω plane. The spectrum of the analytic time series (3.1.12) is

$$F(\omega) = 2H(\omega) f(\omega), \tag{3.1.15}$$

where $H(\omega)$ is the Heaviside step function.

3.1.3 Convolution

We will make frequent use of the *convolution theorem*, i.e. if we have the product of two spectra

$$f(\omega) = g(\omega) h(\omega), \tag{3.1.16}$$

then the time series are related by

$$f(t) = \int_{-\infty}^{\infty} g(t')\, h(t - t')\, dt' \tag{3.1.17}$$

$$= g(t) * h(t). \tag{3.1.18}$$

This is called a *convolution* and the notation (3.1.18) is used as a shorthand for the integral (3.1.17). Unless explicitly stated otherwise, we will assume the asterisk denotes a convolution in the time domain.

3.1.4 Dirac delta function

We will make frequent use of the Dirac delta function, $\delta(t)$, defined such that

$$\int_{-\infty}^{\infty} g(t')\delta(t-t')\mathrm{d}t' = g(t). \tag{3.1.19}$$

Thus the spectrum of a delta function is

$$\delta(\omega) = \int_{-\infty}^{\infty} \delta(t)\,\mathrm{e}^{\mathrm{i}\omega t}\mathrm{d}t = 1, \tag{3.1.20}$$

and the analytic delta function is

$$\Delta(t) = \delta(t) - \frac{\mathrm{i}}{\pi t}. \tag{3.1.21}$$

In Appendix B.1, we generalize the definition of the Dirac delta function to include complex arguments.

3.1.5 Hilbert transform

The time series $\bar{f}(t)$ (3.1.12) is known as the *Hilbert transform* of $f(t)$, or the *allied function*. The Hilbert transform of the Dirac delta function (3.1.21) is

$$\bar{\delta}(t) = -\frac{1}{\pi t}. \tag{3.1.22}$$

Using the convolution theorem (Section 3.1.3), the Hilbert transform of $f(t)$ can be written

$$\bar{f}(t) = -\frac{1}{\pi}\mathcal{P}\int_{-\infty}^{\infty} \frac{f(t')}{t-t'}\,\mathrm{d}t', \tag{3.1.23}$$

where $\mathcal{P}$ represents the Cauchy principal value of the integral. Note from the spectrum that

$$\bar{\bar{f}}(\omega) = -f(\omega), \tag{3.1.24}$$

and hence

$$f(t) = \frac{1}{\pi}\mathcal{P}\int_{-\infty}^{\infty} \frac{\bar{f}(t')}{t-t'}\,\mathrm{d}t'. \tag{3.1.25}$$

Equations (3.1.23) and (3.1.25) are known as the *Hilbert transform pair*.

Table 3.1. *Operations in the time and frequency domains*

Operation	Time domain	Frequency domain
Fourier transform	$f(t) = \frac{1}{2\pi}\int_B f(\omega)e^{-i\omega t}\,d\omega$	$f(\omega) = \int_{-\infty}^{\infty} f(t)e^{i\omega t}\,dt$
convolution	$f(t) = \int_{-\infty}^{\infty} g(t')h(t-t')\,dt'$	$f(\omega) = g(\omega)h(\omega)$
Hilbert transform	$\bar{f}(t) = -\frac{1}{\pi}\mathcal{P}\int_{-\infty}^{\infty}\frac{f(t')}{t-t'}\,dt'$	$\bar{f}(\omega) = -i\,\mathrm{sgn}(\omega)f(\omega)$
analytic time series	$F(t) = f(t) + i\bar{f}(t)$	$F(\omega) = 2H(\omega)f(\omega)$
shift rule	$f(t+a)$	$f(\omega)e^{-i\omega a}$
scaling rule	$f(t/a)/\lvert a\rvert$	$f(a\omega)$ (a real)
differentiation	$\partial_t^n f(t)$	$(-i\omega)^n f(\omega)$ $(n > 0)$
integration	$\int^t \ldots \int^{t_2} f(t_1)dt_1 \ldots dt_n$	$(-i\omega)^{-n} f(\omega)$ $(n > 0)$

3.1.6 Derivative

The crucial property of the Fourier transform is that a derivative in the time domain can be replaced by a multiplication in the frequency domain. Thus the Fourier transform of the time derivative is

$$\begin{aligned}\left(\frac{\partial f}{\partial t}\right)(\omega) = \int_{-\infty}^{\infty}\frac{\partial f}{\partial t}e^{i\omega t}dt &= f(t)e^{i\omega t}\Big|_{-\infty}^{\infty} - \int_{-\infty}^{\infty} i\omega f(t)e^{i\omega t}dt \\ &= -i\omega f(\omega),\end{aligned} \tag{3.1.26}$$

provided $f(t)e^{i\omega t} \to 0$ as $t \to \infty$ ($f(-\infty) = 0$, from causality). Normally this limit at infinity follows obviously from physical considerations, although sometimes it requires the introduction of a small amount of attenuation. In practice, we will omit making a detailed discussion of this limit.

The operations of convolution, Hilbert transform and derivative are associative and commutative. This follows immediately as in the frequency domain they are just multiplications. Thus

$$\overline{a * b} = \bar{a} * b = a * \bar{b} = b * \bar{a} = \ldots . \tag{3.1.27}$$

Sometimes the operations on particular functions are only defined in the sense of generalized functions, e.g. $\delta'(t)$, the derivative of a delta function. We will adopt a somewhat cavalier attitude when manipulating combined operations and functions, as the validity of the complete expressions can normally be argued on physical grounds

In Table 3.1, we have summarized the basic operations in the time and frequency domains. We will introduce Fourier transforms of particular functions as we need them – some are summarized in Appendix B.

3.2 Spatial Fourier transform

In order to reduce a partial differential equation to an ordinary differential equation, we use the spatial Fourier transform

$$f(k) = \int_{-\infty}^{\infty} f(x)\mathrm{e}^{-\mathrm{i}kx}\mathrm{d}x, \tag{3.2.1}$$

with the inverse transform (cf. equation (2.1.2))

$$f(x) = \frac{1}{2\pi}\int_{-\infty}^{\infty} f(k)\mathrm{e}^{\mathrm{i}kx}\mathrm{d}k. \tag{3.2.2}$$

It is frequently convenient to change the transform variable to p where (cf. equation (2.1.3) and Table 2.1)

$$k = \omega p. \tag{3.2.3}$$

As with the temporal transform, we denote the transform function by its argument – $f(x)$ and $f(k)$ are, of course, different functions. The wavenumber in the x direction is k, and p is the slowness in this direction (Table 2.1). Using the slowness, we write the transform pair as

$$\boxed{f(p) = \int_{-\infty}^{\infty} f(x)\mathrm{e}^{-\mathrm{i}\omega px}\mathrm{d}x} \tag{3.2.4}$$

$$f(x) = \frac{\omega}{2\pi}\int_{C} f(p)\mathrm{e}^{\mathrm{i}\omega px}\mathrm{d}p, \tag{3.2.5}$$

where C is along the line where $\arg(p) = -\arg(\omega)$ (Figure 3.3) as the integral (3.2.2) is for $\arg(k) = 0$. Again we just denote the transform function by the argument ($f(p)$ and $f(k)$ are again different functions).

Combining with the temporal transform, equation (3.1.2), we have

$$f(t, x) = \frac{1}{4\pi^2}\int_{B}\int_{C} \omega f(\omega, p)\mathrm{e}^{\mathrm{i}\omega(px-t)}\mathrm{d}p\,\mathrm{d}\omega. \tag{3.2.6}$$

We can identify positive, real p with waves propagating in the positive x direction.

An important feature of the complex k or p planes are branch points from square roots such as

$$q = (c^{-2} - p^2)^{1/2}, \tag{3.2.7}$$

where c is a velocity. The branch points are at $p = \pm c^{-1}$. The square root appears in phase terms such as

$$\mathrm{e}^{\mathrm{i}\omega qz}. \tag{3.2.8}$$

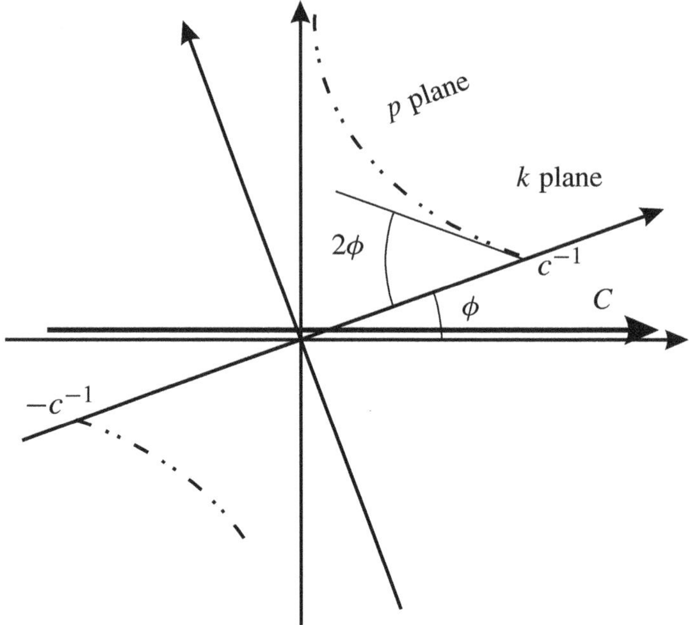

Fig. 3.3. The complex p and k planes, and the contour, C, with $\phi = \arg(\omega)$. Branch cuts defined by $\mathrm{Im}(\omega q) = 0$ are illustrated with the dashed-dotted lines.

We must choose the correct square root in (3.2.7) so that (3.2.8) represents the correct physical solution. When q is real, a positive value for q corresponds to a wave travelling in the positive z direction. When q is complex, the important property is that the wave travelling in the positive z direction decays in the direction of propagation. Thus in (3.2.8) we must have

$$\mathrm{Im}(\omega q) \geq 0. \tag{3.2.9}$$

Therefore the branch cuts from the branch points are defined by $\mathrm{Im}(\omega q) = 0$ and the contour must begin and end on the Riemann sheet defined by (3.2.9).

When ω is real, the branch cuts with $\mathrm{Im}(q) = 0$ are for $-c^{-1} < p < c^{-1}$ on the real axis, and the imaginary p axis (from (3.2.7)). As the contour C is on the real axis, the question arises as to how the contour is arranged with respect to the branch cuts and points. When ω is not real, the situation is illustrated in Figure 3.3 with $\phi = \arg(\omega)$. The branch cuts leave the branch points $p = \pm c^{-1}$ at $\pm 2\phi$ to the real p axis, and are asymptotic to the imaginary k axis. The branch points do not lie on the contour C, and the contour passes between the branch points without encountering the cuts. For $\phi \neq 0$ or π the situation is straightforward. Taking the limit $\phi \to 0$ (ω real and positive), the branch cuts become 'L' shaped on the real and imaginary axes, and the contour C passes between them (Figure 3.4*a*).

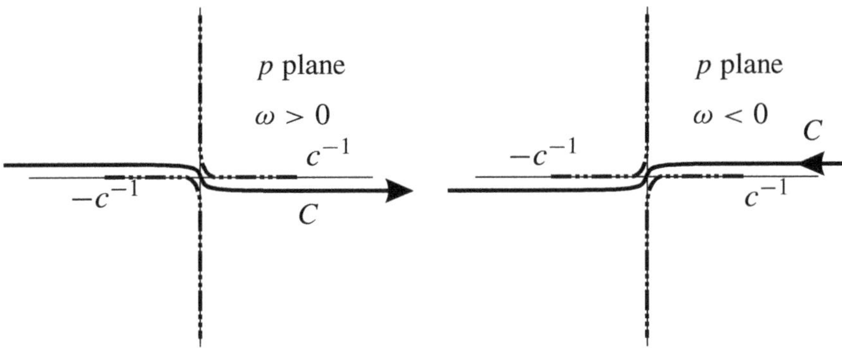

Fig. 3.4. The complex p plane and the contour, C, when: (*a*) ω is real and positive ($\phi = 0$); (*b*) ω is real and negative ($\phi = \pi$), i.e. special cases of Figure 3.3. The branch cuts defined by $\mathrm{Im}(\omega q) = 0$ are illustrated with dashed-dotted lines.

In the limit $\phi \to \pi$ (ω real and negative), the branch cuts are mirrored in the real p axis, and the contour C runs along the real axis in the reversed direction, ∞ to $-\infty$ (Figure 3.4*b*). Thus for ω real, as is needed for the inverse Fourier transform, we can rewrite the inverse slowness transform (3.2.6) as

$$\boxed{f(x) = \frac{|\omega|}{2\pi} \int_{-\infty}^{\infty} f(p)\, \mathrm{e}^{\mathrm{i}\omega p x} \mathrm{d}p.} \tag{3.2.10}$$

When $\omega > 0$, the contour runs infinitesimally below the positive p axis, and above the negative axis. For $\omega < 0$, the situation is reversed (Figure 3.4). This discussion of the exact position of the contour relative to the axis and branch cuts is important as other singularities may lie on the real p axis.

When $\phi = \pi/2$, i.e. ω imaginary or the Laplace variable (3.1.8), s, real, the contour C is along the imaginary p axis, and the branch cuts are on the real p axis for $|p| > c^{-1}$. This situation is sometimes considered when exact analytic results are known (so real s is adequate).

In three-dimensional problems, it is often convenient to apply a two-dimensional Fourier transform. Typically, this may be applied to the horizontal coordinates, e.g. x_1 and x_2. Thus generalizing (3.2.4) and (3.2.5), we have

$$f(p_1, p_2) = \iint_{-\infty}^{\infty} f(x_1, x_2)\, \mathrm{e}^{-\mathrm{i}\omega(p_1 x_1 + p_2 x_2)} \mathrm{d}x_1 \mathrm{d}x_2 \tag{3.2.11}$$

$$f(x_1, x_2) = \frac{\omega^2}{4\pi^2} \iint_{-\infty}^{\infty} f(p_1, p_2)\, \mathrm{e}^{\mathrm{i}\omega(p_1 x_1 + p_2 x_2)} \mathrm{d}p_1 \mathrm{d}p_2. \tag{3.2.12}$$

It is frequently convenient to indicate components of a sub-space by x_ν, where the Greek subscript indicates the restricted range $\nu = 1$ to 2. A vector in such a sub-space is written in sans serif font, e.g. $\mathsf{x} = (x_1, x_2)$. Thus we can write

$$p_1 x_1 + p_2 x_2 = p_\nu x_\nu = \mathsf{p} \cdot \mathsf{x}, \tag{3.2.13}$$

and equations (3.2.11) and (3.2.12)

$$f(\mathsf{p}) = \iint_{-\infty}^{\infty} f(\mathsf{x})\, \mathrm{e}^{-\mathrm{i}\omega \mathsf{p}\cdot\mathsf{x}} \mathrm{d}\mathsf{x} \tag{3.2.14}$$

$$f(\mathsf{x}) = \frac{\omega^2}{4\pi^2} \iint_{-\infty}^{\infty} f(\mathsf{p})\, \mathrm{e}^{\mathrm{i}\omega \mathsf{p}\cdot\mathsf{x}} \mathrm{d}\mathsf{p}. \tag{3.2.15}$$

3.3 Fourier–Bessel transform

In problems with axial symmetry, or low-order azimuthal variation, an alternative to a two-dimensional Fourier transform is to expand in a Fourier series and to use a Fourier–Bessel transform.

3.3.1 Fourier series

If the spatial coordinate is written as polar components, e.g. $\mathsf{x} = (r, \chi)$, then we can expand the azimuthal variation in a Fourier series, i.e.

$$f(r, \ell) = \frac{1}{2\pi} \int_0^{2\pi} f(\mathsf{x})\, \mathrm{e}^{-\mathrm{i}\ell\chi} \mathrm{d}\chi. \tag{3.3.1}$$

Sometimes as ℓ can only take discrete values, it is written as a subscript of the transformed function rather than an argument. The inverse transform is the Fourier series

$$f(\mathsf{x}) = \sum_{\ell=-\infty}^{\infty} f(r, \ell)\, \mathrm{e}^{\mathrm{i}\ell\chi}. \tag{3.3.2}$$

3.3.2 Fourier–Bessel transform

Having expanded in the Fourier series coefficients, $f(r, \ell)$, we can then transform the radial coordinate using the Fourier–Bessel transform, i.e.

$$f(k, \ell) = \int_0^{\infty} f(r, \ell)\, r\, J_\ell(kr)\, \mathrm{d}r, \tag{3.3.3}$$

although we normally make the substitution $k = \omega p$, as in (3.2.3), so

$$f(p, \ell) = \int_0^{\infty} f(r, \ell)\, r\, J_\ell(\omega p r)\, \mathrm{d}r. \tag{3.3.4}$$

The inverse transform is then

$$f(r,\ell) = \omega^2 \int_0^\infty f(p,\ell)\, p J_\ell(\omega p r)\, \mathrm{d}p. \tag{3.3.5}$$

In these equations, $J_\ell(z)$ is the Bessel function of order ℓ.

It is sometimes convenient to rewrite this inverse transform using Hankel functions, to make the result more analogous to the Fourier transform. We note that as $J_\ell(-z) = \mathrm{e}^{\mathrm{i}\ell\pi} J_\ell(z)$ (Abramowitz and Stegun, 1965, §9.1.35)

$$f(-p,\ell) = \mathrm{e}^{\mathrm{i}\ell\pi} f(p,\ell). \tag{3.3.6}$$

Expanding using $2J_\ell(z) = H_\ell^{(1)}(z) + H_\ell^{(2)}(z)$ (Abramowitz and Stegun, 1965, §9.1.3 and §9.1.4), and noting $H_\ell^{(2)}(-z) = -\mathrm{e}^{\mathrm{i}\ell\pi} H_\ell^{(1)}(z)$ (Abramowitz and Stegun, 1965, §9.1.39), we can rewrite (3.3.6) as

$$f(r,\ell) = \frac{\omega|\omega|}{2} \int_{-\infty}^\infty f(p,\ell)\, p H_\ell^{(1)}(\omega p r)\, \mathrm{d}p. \tag{3.3.7}$$

We have used the Hankel function of the first kind in this integral as asymptotically it is (Abramowitz and Stegun, 1965, §9.2.3)

$$H_\ell^{(1)}(z) \simeq \left(\frac{2}{\pi z}\right)^{1/2} \mathrm{e}^{\mathrm{i}(z-\ell\pi/2-\pi/4)}, \tag{3.3.8}$$

and the integral is like the inverse Fourier transform (3.2.10). Although the Hankel function is singular at the origin $p = 0$, it can be shown that the integral is finite due to the symmetries of the integral. The p contour passes infinitesimally above the branch point at the origin and the branch cut on the negative real p axis.

3.4 Tau-p transform

3.4.1 Legendre transform

There are two dual possibilities to describe a function in $\mathbf{x}$ space. For generality, we divide the $\mathbf{x}$ space into two sub-spaces, x and y. We can either use a function $f(\mathsf{x},\mathsf{y})$ describing a surface in the x–f space, or we may regard the surface as the envelope of its tangent planes. Certain problems are more simply described by the dual description. The transformation between the two descriptions is known as the *Legendre transform* (Courant and Hilbert, 1966, Chapter 1, §6). The transform can be applied to an arbitrary number of variables and we use the notation $f(\mathsf{x},\mathsf{y})$ to indicate the domain in which the transform will be applied, x, and the domain left untransformed, y.

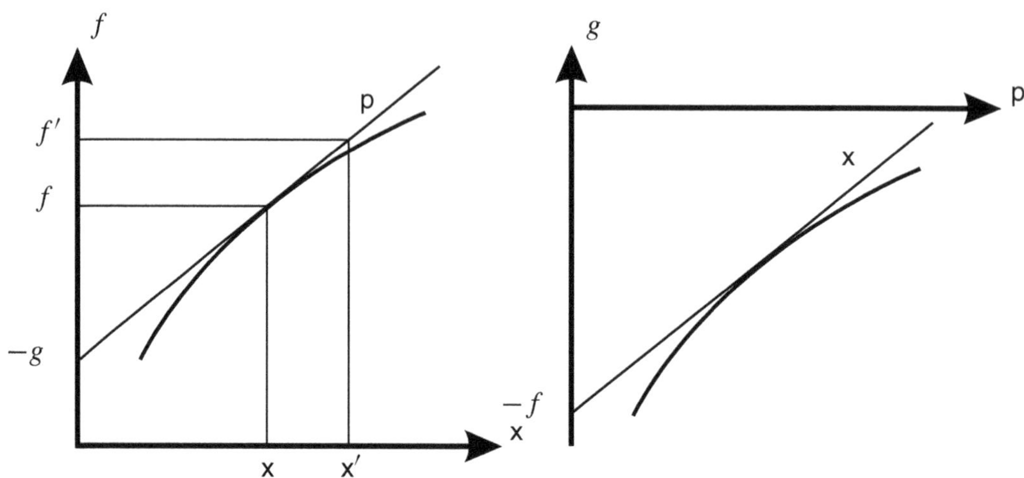

Fig. 3.5. The function $f(\mathsf{x}, \mathsf{y})$ in one dimension, with the slope p and intercept $-g$, and the dual function $g(\mathsf{p}, \mathsf{y})$ with slope x and intercept $-f$.

Let us denote the gradient of the tangent plane in x space by

$$\mathsf{p} = \nabla_{\mathsf{x}} f, \tag{3.4.1}$$

i.e. the vector with components $\partial f/\partial x_i$. The equation of the tangent plane can be written

$$f' - f - (\mathsf{x}' - \mathsf{x}) \cdot \mathsf{p} = 0, \tag{3.4.2}$$

where f' and x' are the variables, and the plane passes through f and x (see Figure 3.5). We call p and

$$g = \mathsf{x} \cdot \mathsf{p} - f, \tag{3.4.3}$$

the coordinates of the tangent plane. The function $g(\mathsf{p}, \mathsf{y})$ can be determined from $f(\mathsf{x}, \mathsf{y})$ and is called the Legendre transform (with respect to x) and describes the tangent planes, i.e. $-g$ is the intercept of the plane with the f axis at $\mathsf{x} = 0$ and p is its slope. Clearly from (3.4.2)

$$f(\mathsf{x}, \mathsf{y}) + g(\mathsf{p}, \mathsf{y}) = \mathsf{p} \cdot \mathsf{x}, \tag{3.4.4}$$

and the transform is symmetric. The gradient of the surface $g(\mathsf{p}, \mathsf{y})$ in the p domain is

$$\mathsf{x} = \nabla_{\mathsf{p}} g, \tag{3.4.5}$$

and the intercept of the tangent plane to the $g(\mathsf{p}, \mathsf{y})$ surface with the g axis at $\mathsf{p} = 0$ is $-f$ (Figure 3.5).

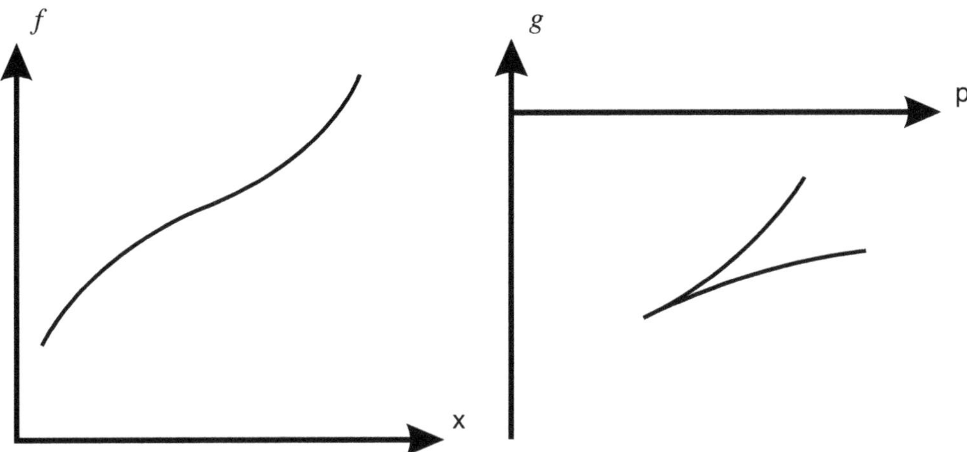

Fig. 3.6. The functions $f(\mathsf{x}, \mathsf{y})$ and $g(\mathsf{p}, \mathsf{y})$ when f has an inflexion point, and g is multi-valued.

The Legendre transformation is always feasible if the equation (3.4.1) can be solved for x. This is possible provided the Jacobian

$$\nabla(\nabla f)^{\mathrm{T}} = \begin{vmatrix} \frac{\partial^2 f}{\partial x_1^2} & \frac{\partial^2 f}{\partial x_1 x_2} & \cdots & \frac{\partial^2 f}{\partial x_1 x_n} \\ \frac{\partial^2 f}{\partial x_1 x_2} & \frac{\partial^2 f}{\partial x_2^2} & \cdots & \frac{\partial^2 f}{\partial x_2 x_n} \\ \cdots & \cdots & \cdots & \cdots \\ \frac{\partial^2 f}{\partial x_1 x_n} & \frac{\partial^2 f}{\partial x_2 x_n} & \cdots & \frac{\partial^2 f}{\partial x_n^2} \end{vmatrix} \neq 0. \tag{3.4.6}$$

If this determinant is zero, then tangents are tangent along lines or inflexion points not just at points. This is illustrated in Figure 3.6, where in one dimension, the function $f(\mathsf{x}, \mathsf{y})$ has an inflexion point and the function $g(\mathsf{p}, \mathsf{y})$ is multi-valued. The Jacobians for the dual functions are related by

$$\begin{pmatrix} \frac{\partial^2 f}{\partial x_1^2} & \frac{\partial^2 f}{\partial x_1 x_2} & \cdots & \frac{\partial^2 f}{\partial x_1 x_n} \\ \frac{\partial^2 f}{\partial x_1 x_2} & \frac{\partial^2 f}{\partial x_2^2} & \cdots & \frac{\partial^2 f}{\partial x_2 x_n} \\ \cdots & \cdots & \cdots & \cdots \\ \frac{\partial^2 f}{\partial x_1 x_n} & \frac{\partial^2 f}{\partial x_2 x_n} & \cdots & \frac{\partial^2 f}{\partial x_n^2} \end{pmatrix} \begin{pmatrix} \frac{\partial^2 g}{\partial p_1^2} & \frac{\partial^2 g}{\partial p_1 p_2} & \cdots & \frac{\partial^2 g}{\partial p_1 p_n} \\ \frac{\partial^2 g}{\partial p_1 p_2} & \frac{\partial^2 g}{\partial p_2^2} & \cdots & \frac{\partial^2 g}{\partial p_2 p_n} \\ \cdots & \cdots & \cdots & \cdots \\ \frac{\partial^2 g}{\partial p_1 p_n} & \frac{\partial^2 g}{\partial p_2 p_n} & \cdots & \frac{\partial^2 g}{\partial p_n^2} \end{pmatrix} = \mathbf{I}. \tag{3.4.7}$$

Legendre transforms arise in several circumstances, e.g. in thermodynamic energy functions, in Hamiltonian and Lagrangian functions, in travel-time functions, etc. They have already been encountered in Section 2.3 connecting the travel and intercept times ($f \leftrightarrow T$ and $g \leftrightarrow -\tau$).

3.4.2 Radon transform

Radon transforms arise in several circumstances in seismology. In general, a Radon transform of a function in several dimensions is the function integrated along lines or surfaces in the multi-dimensional space. The reduced dimensionality of the space is replaced by transform variables defining the lines or surfaces. Radon transforms arise in tomography, where model properties are integrated along ray paths, to give, for instance, the travel times or absorption. In earthquake studies with extended fault sources, the waveform at any time, can be represented as an integral of elementary sources on the fault along a line defined with the appropriate travel time. In other problems, the Radon transform can be used as an alternative to two-dimensional Fourier transforms, either between the space of two horizontal coordinates and slownesses, or between the affine space of time and slowness. The latter is of interest here.

With one spatial dimension, the combined temporal (3.1.1) and spatial (3.2.4) Fourier transforms are

$$f(\omega, p) = \iint_{-\infty}^{\infty} f(t, x)\, \mathrm{e}^{\mathrm{i}\omega(t-px)} \mathrm{d}t\, \mathrm{d}x. \tag{3.4.8}$$

Taking the inverse Fourier transform (3.1.2) of this

$$f(\tau, p) = \frac{1}{2\pi} \int_B f(\omega, p)\, \mathrm{e}^{-\mathrm{i}\omega\tau} \mathrm{d}\omega \tag{3.4.9}$$

$$= \frac{1}{2\pi} \iint_{-\infty}^{\infty} \int_B f(t, x)\, \mathrm{e}^{\mathrm{i}\omega(t-px-\tau)} \mathrm{d}\omega\, \mathrm{d}t\, \mathrm{d}x \tag{3.4.10}$$

$$= \iint_{-\infty}^{\infty} f(t, x)\, \delta(t - px - \tau)\, \mathrm{d}t\, \mathrm{d}x, \tag{3.4.11}$$

where the order of integration has been reversed in (3.4.10) and the inverse frequency transform of the exponential gives the delta function in (3.4.11). Evaluating the delta function, we obtain

$$f(\tau, p) = \int_{-\infty}^{\infty} f(t = \tau + px, x)\, \mathrm{d}x. \tag{3.4.12}$$

The function $f(t, x)$ is integrated along lines $t = \tau + px$ to obtain the transformed function $f(\tau, p)$, i.e. a Radon transform.

To obtain the inverse Radon transform, we combine the inverse transforms (3.1.2) and (3.2.10)

$$f(t,x) = \frac{1}{4\pi^2} \int_B |\omega| \int_{-\infty}^{\infty} f(\omega, p)\, e^{i\omega(px-t)} dp\, d\omega. \tag{3.4.13}$$

The factor $|\omega|$ is equivalent to minus the derivative of the Hilbert transform, i.e.

$$|\omega| = -(-i\omega)(-i\,\mathrm{sgn}(\omega)), \tag{3.4.14}$$

so changing the order of integration and inverting the frequency transform, we obtain

$$f(t,x) = -\frac{1}{2\pi}\frac{d}{dt} \int_{-\infty}^{\infty} \bar{f}(\tau = t - px, p)\, dp. \tag{3.4.15}$$

Again, apart from the time series operations of differentiation and Hilbert transformation, this is a Radon transform – the function $f(\tau, p)$ is integrated along lines defined by $\tau = t - px$.

The pair (3.4.12) and (3.4.15) transform between the (t, x) domain and the (τ, p) domain, i.e. between space and slowness without introducing frequency. It can be regarded as decomposing the time series $f(t, x)$ into plane waves without also decomposing into frequency components. The advantage of the Radon transform pair is that the transformed results are still real, and that $f(\tau, p)$ has a simple form, particularly in theoretical problems. In processing, equation (3.4.12) is referred to as a *slant* or *velocity stacking*.

The symmetry of the transform pair can be enhanced by slightly changing the definition of the transformed function. Defining

$$g(\omega, p) = \left(\frac{|\omega|}{2\pi}\right)^{1/2} e^{i\pi/4} f(\omega, p), \tag{3.4.16}$$

and factoring the $|\omega|$ in equation (3.4.13)

$$|\omega| = \left(|\omega|^{1/2} e^{-i\pi/4}\right)\left(|\omega|^{1/2} e^{i\pi/4}\right), \tag{3.4.17}$$

we have

$$g(\tau, p) = -\frac{1}{2^{1/2}\pi}\frac{d}{dt}\,\bar{\lambda}(\tau) * \int_{-\infty}^{\infty} f(\tau + px, x)\, dx \tag{3.4.18}$$

$$f(t,x) = \frac{1}{2^{1/2}\pi}\frac{d}{dt}\,\lambda(t) * \int_{-\infty}^{\infty} g(t - px, p)\, dp, \tag{3.4.19}$$

where the time function $\lambda(t) = H(t)\,t^{-1/2}$ is discussed in Appendix B.2. If the original function $f(t, x)$ contains delta functions on curves, e.g. $f(t, x) = A(x)\delta(t - T(x))$, then it is straightforward to show that the transformed functions

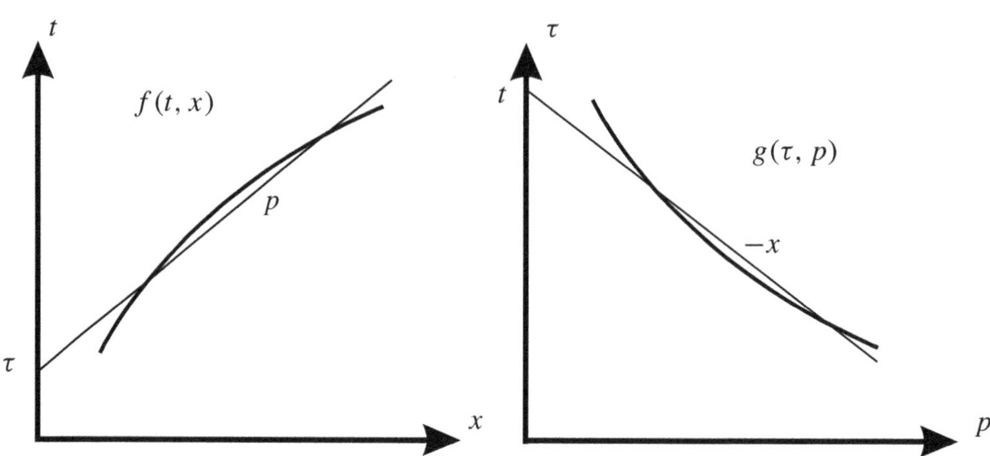

Fig. 3.7. The functions $f(t, x)$ and $g(\tau, p)$ and the line integrals of the Radon transforms (3.4.18) and (3.4.19). The transformation of a line singularity is illustrated.

$g(\tau, p)$ also contains delta functions or their Hilbert transform (depending on the sign of the curvature of $T(x)$) (Exercise 3.3). The operator $(\mathrm{d}/\mathrm{d}t)\lambda(t)*$ is important in this symmetry. The relationship between $f(t, x)$ and $g(\tau, p)$ using relationships (3.4.18) and (3.4.19) is illustrated in Figure 3.7. The close connection with the Legendre transform of the previous section (Section 3.4.1) is evident.

Note that (3.4.18) and (3.4.19) are an exact transform pair. A similar result is also obtained as an asymptotic approximation to the Fourier–Bessel transforms (3.3.4) and (3.3.5). Rewriting (3.3.4) and (3.3.5)

$$g(\omega, p, \ell) = |\omega| \int_0^\infty f(\omega, r, \ell)\, r\, J_\ell(\omega p r)\, \mathrm{d}r \tag{3.4.20}$$

$$f(\omega, r, \ell) = |\omega| \int_0^\infty g(\omega, p, \ell)\, p\, J_\ell(\omega p r)\, \mathrm{d}p, \tag{3.4.21}$$

we can take the inverse Fourier transforms using (B.4.1), to obtain

$$g(t, p, \ell) = -\frac{\mathrm{d}}{\mathrm{d}t} \int_0^\infty \bar{f}(t, r, \ell) * \frac{2(-\mathrm{i})^\ell T_\ell(t/pr)}{(p^2 r^2 - t^2)^{1/2}}\, r\, \mathrm{d}r \tag{3.4.22}$$

$$f(t, r, \ell) = -\frac{\mathrm{d}}{\mathrm{d}t} \int_0^\infty \bar{g}(t, p, \ell) * \frac{2(-\mathrm{i})^\ell T_\ell(t/pr)}{(p^2 r^2 - t^2)^{1/2}}\, p\, \mathrm{d}p, \tag{3.4.23}$$

where $T_\ell(x)$ are the Chebyshev polynomials (B.4.2). Within the integral we have a temporal convolution with the inverse Fourier transform of the Bessel function (B.4.1). Although this is more complicated than a Radon transform, it is closely related. The inverse Fourier transforms can be approximated by the singularities

$$\frac{2(-\mathrm{i})^{\ell} T_{\ell}(t/pr)}{(p^2r^2 - t^2)^{1/2}} \simeq \left(\frac{2}{pr}\right)^{1/2} \left((+\mathrm{i})^{\ell} \lambda(t + pr) + (-\mathrm{i})^{\ell} \bar{\lambda}(t - pr)\right). \quad (3.4.24)$$

Substituting in (3.4.22) and (3.4.23), we obtain

$$g(t, p, \ell) \simeq \frac{\mathrm{d}}{\mathrm{d}t} \lambda(t) * \int_0^{\infty} \left((-\mathrm{i})^{\ell} f(t - pr, r, \ell) - (+\mathrm{i})^{\ell} \bar{f}(t + pr, r, \ell)\right) \left(\frac{2r}{p}\right)^{1/2} \mathrm{d}r \quad (3.4.25)$$

$$f(t, r, \ell) \simeq \frac{\mathrm{d}}{\mathrm{d}t} \lambda(t) * \int_0^{\infty} \left((-\mathrm{i})^{\ell} g(t - pr, p, \ell) - (+\mathrm{i})^{\ell} \bar{g}(t + pr, p, \ell)\right) \left(\frac{2p}{r}\right)^{1/2} \mathrm{d}p. \quad (3.4.26)$$

Normally only one of the line integrals contributes significantly as the features of the function $f(t, r, \ell)$ constructively combine, and we obtain

$$g(t, p, \ell) \simeq -\frac{\mathrm{d}}{\mathrm{d}t} \bar{\lambda}(t) * \int_0^{\infty} (+\mathrm{i})^{\ell} f(t + pr, r, \ell) \left(\frac{2r}{p}\right)^{1/2} \mathrm{d}r \quad (3.4.27)$$

$$f(t, r, \ell) \simeq \frac{\mathrm{d}}{\mathrm{d}t} \lambda(t) * \int_0^{\infty} (-\mathrm{i})^{\ell} g(t - pr, p, \ell) \left(\frac{2p}{r}\right)^{1/2} \mathrm{d}p, \quad (3.4.28)$$

which are similar to the pair (3.4.18) and (3.4.19).

Exercises

3.1 The results in Table 3.1 can be found in many textbooks, but confirm the proofs from first principles. What assumptions are necessary for each result?

3.2 Several useful Fourier transforms are given in Appendix B. Confirm these results.

3.3 Evaluate the Radon transform as defined in equation (3.4.18) for the function $f(t, x) = A(x)\,\delta(t - T(x))$. Approximate this about the singularities, and show that the original function is recovered when substituted in the inverse transform (3.4.19).

4

Review of continuum mechanics and elastic waves

In this chapter we review the results of continuum mechanics for infinitesimal deformations. We introduce the concepts of *traction*, *stress*, *strain* and *stress glut* to describe the forces in, and deformation of, a fluid or solid. The physics is then described by the *boundary conditions*, the *constitutive relations* and the *equations of motion*. The *Navier wave equation* is solved for a point source in a homogeneous, isotropic medium in order to obtain the corresponding *Green function*. The development is limited to infinitesimal deformations and the complications of pre-stress and finite deformations are ignored.

In order to discuss elastic waves in seismology (with wavelengths from metres to infinity) we need a mechanism to describe average properties of a material without reference to the detailed atomic or even crystalline structure. This is called *continuum mechanics*.

We are used to the concept of density, i.e. mass per unit volume, and rarely question it. But what if we have an inhomogeneous medium and density varies with position? What is the density at a point? We are tempted to write

$$\rho(\mathbf{x}) = \lim_{\Delta V \to 0} \frac{\Delta m}{\Delta V} \tag{4.0.1}$$

where Δm is the mass of material contained in the volume ΔV which contains the point $\mathbf{x}$ (Figure 4.1).

But with this definition, a plot of $\Delta m/\Delta V$ against ΔV might look like Figure 4.2.

The rough behaviour at small volumes would occur due to crystalline and atomic (and ultimately sub-atomic) structure. At large volumes the slow variation is due to changing material properties. Continuum mechanics assumes that there is a significant range of intermediate volumes where the ratio $\Delta m/\Delta V$ is approximately constant and we can treat the medium as a homogeneous continuum. Thus the mathematical limit (4.0.1) is replaced by the continuum limit where ΔV

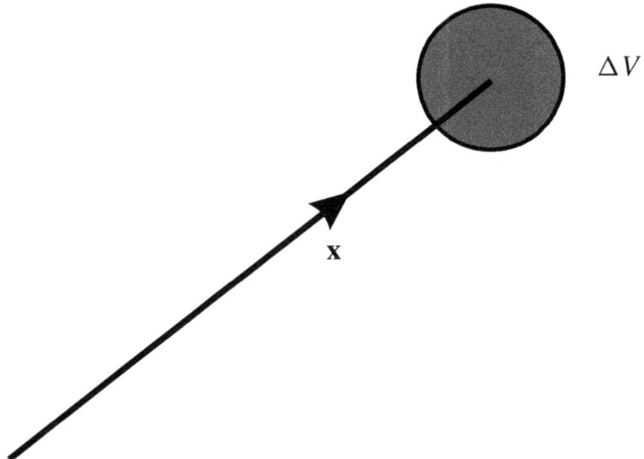

Fig. 4.1. A small volume ΔV at $\mathbf{x}$ containing mass Δm.

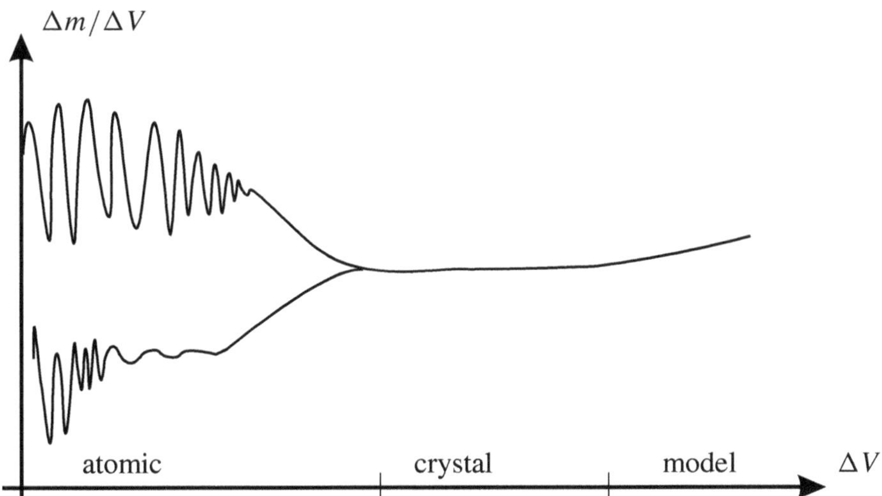

Fig. 4.2. Behaviour of $\Delta m/\Delta V$ against ΔV. The volume scale covers many orders of magnitude and is only diagrammatic. Two continuum regions are indicated. At smaller volumes where two constituents have different properties, two separate curves are obtained depending on the exact location of the volume. At larger volumes an effective medium theory applies.

is small compared with model size, wavelength and heterogeneities in the continuum model (e.g. $\Delta V \ll (\rho/|\nabla\rho|)^3$ except we allow discontinuities in properties), but large compared with the crystalline or atomic structure. We do not discuss here the important subject of *effective media theories*, where two or more scales exist. At a small scale, super-atomic but sub-wavelength, e.g. crystal size, the medium is relatively homogeneous and the continuum limit is well defined. But at a larger scale, but still sub-wavelength, the medium is heterogeneous, maybe made up of several homogeneous constituents, e.g. polycrystalline or homogeneous thin layers. Nevertheless a continuum limit should exist at these scales which can be derived from the smaller-scale properties. This is illustrated in Figure 4.2 for a medium with two constituents. The transition from a small to a larger scale is the subject of effective medium theories. A difficulty is that the transition may require a more complex continuum (e.g. the transition from isotropic to anisotropic media) and is usually frequency/wavelength and application dependent.

We assume the continuum limit applies, and treat properties and waves as through they are defined at points.

In this chapter we only consider the simplest theories. We only consider cartesian vector and tensor components. The distinction between covariant and contravariant, tensor and physical components is not important here as we do not consider general curvilinear coordinates. The deformation is assumed to be small so the distinction between Eulerian and Lagrangian coordinates is not significant, and we restrict ourselves to infinitesimal strain. Finite strain and pre-stressed theories are not considered. More rigorous and complete theories are developed in many textbooks (e.g. Fung, 1965; Dahlen and Tromp, 1998) but are not needed here.

4.1 Infinitesimal stress tensor and traction

The stress tensor generalizes the concept of a force acting on a particle to a continuum. Consider a surface in a volume of material. Suppose we cut the material on this surface, breaking completely the atomic bonds. Then a certain force is necessary to maintain equilibrium and keep the material in its original position or state. Suppose the force is $\Delta\mathbf{f}$ on a surface area ΔS. In the continuum sense, this force must be proportional to the area and we define the *traction* as the continuum limit of the force per unit area, i.e.

$$\mathbf{t} = \lim_{\Delta S \to 0} \frac{\Delta\mathbf{f}}{\Delta S}. \tag{4.1.1}$$

Notice that the concept of cutting the material is only a hypothetical thought experiment. In reality, internal forces cannot be destroyed without significantly altering the material.

The forces may be such that the material tends to pull apart or hypothetically to overlap, i.e. a material in tension or compression. Traction will be equal and opposite on the two sides of the cut as otherwise we would have a finite force acting on a zero mass (the cut) leading to infinite accelerations. The traction must also be a function of the orientation of the cut. For instance in a bar hanging under

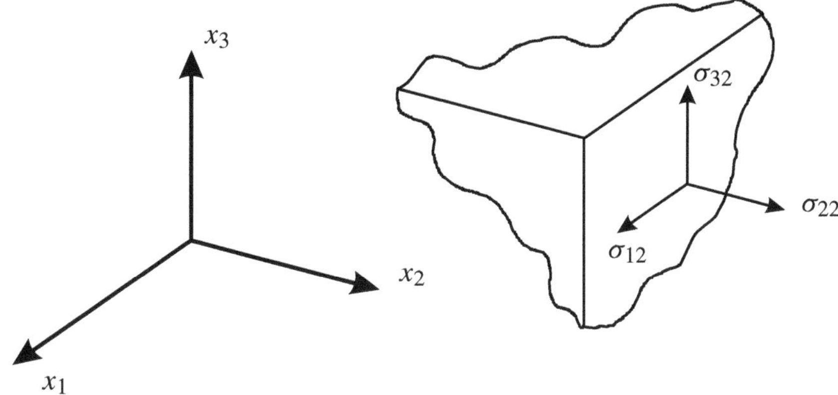

Fig. 4.3. Traction components on a cartesian surface – the components of $\mathbf{t}_2$ are illustrated.

its own weight, the traction on horizontal cuts will be significant whereas on a vertical cut it will be small or zero. Obviously it will vary with position (on the hanging bar it will be greater at the top than the bottom). Thus in general we must have $\mathbf{t}(\mathbf{x}, \hat{\mathbf{n}})$, where $\hat{\mathbf{n}}$ is a unit normal to the surface ΔS (which can be taken plane). We take $\hat{\mathbf{n}}$ pointing out of the medium on the surface of the cut.

Can the dependence on direction $\hat{\mathbf{n}}$ be simply represented? Suppose we do three experiments with surfaces aligned with cartesian axes at the same point and determine the traction vectors (Figure 4.3). Thus we measure $\mathbf{t}_i$ for $i = 1$ to 3 for $\hat{\mathbf{n}} = \hat{\mathbf{i}}_i$, the unit cartesian vectors. Let us denote the components of $\mathbf{t}_i$ by σ_{ji}, i.e.

$$\mathbf{t}_i = \begin{pmatrix} \sigma_{1i} \\ \sigma_{2i} \\ \sigma_{3i} \end{pmatrix}. \tag{4.1.2}$$

We have measured nine force components but are they sufficient to allow us to define the traction on any surface?

Suppose we consider the equilibrium of a small tetrahedral element of material (Figure 4.4) where the sloping face is defined by an arbitrary vector $\hat{\mathbf{n}}$. The cartesian faces have areas ΔS_i and the sloping face ΔS, say. These are related by $\Delta S_i = \hat{n}_i \Delta S$ (proved by projection).

Suppose that the traction does not depend on position. The equilibrium of the tetrahedral element requires the forces on the four faces to balance. Thus for the j-th component, we must have

$$t_j \Delta S - \sigma_{ji} \Delta S_i = 0, \tag{4.1.3}$$

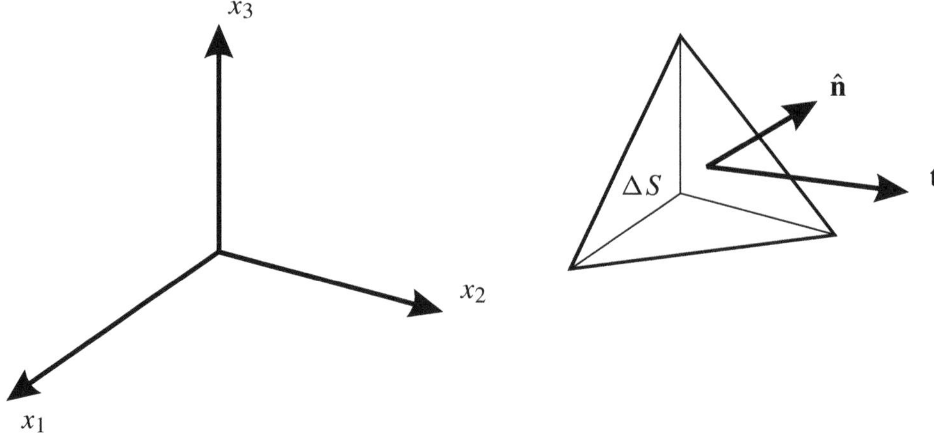

Fig. 4.4. A tetrahedral volume element with sloping face ΔS.

where the forces are given by the traction components (4.1.2) times the areas of the faces. We are using the Einstein summation convention (summation over the repeated index i, with $i = 1$ to 3). The minus sign occurs because the cartesian surfaces in Figure 4.4 are reversed compared with Figure 4.3. Substituting for the areas of the faces ($\Delta S_i = \hat{n}_i \Delta S$) and cancelling the ΔS, we obtain

$$t_j = \sigma_{ji}\hat{n}_i. \tag{4.1.4}$$

Thus the traction on any surface can be derived from the components on the cartesian surfaces (Figure 4.3). More generally we can write (4.1.4) as

$$\boxed{\mathbf{t}(\mathbf{x}, \hat{\mathbf{n}}) = \boldsymbol{\sigma}(\mathbf{x})\hat{\mathbf{n}},} \tag{4.1.5}$$

which is sometimes known as the *Cauchy formula*. As $\boldsymbol{\sigma}$ connects two vectors, it must be a second-order tensor. We call it the *stress tensor*. For our purposes, we can consider it as a 3×3 matrix

$$\boldsymbol{\sigma} = \begin{pmatrix} \sigma_{11} & \sigma_{12} & \sigma_{13} \\ \sigma_{21} & \sigma_{22} & \sigma_{23} \\ \sigma_{31} & \sigma_{32} & \sigma_{33} \end{pmatrix}. \tag{4.1.6}$$

4.1.1 Static equilibrium

The stress may vary with position in a body due to internal forces acting on the medium, e.g. gravity. The equation for static equilibrium of any volume V with bounding surface S is

$$\int_V \mathbf{f} \, \mathrm{d}V + \int_S \mathbf{t} \, \mathrm{d}S = \mathbf{0}. \tag{4.1.7}$$

The internal forces are the forces per unit volume, $\mathbf{f}$. Equation (4.1.7) can be written as

$$\int_V \mathbf{f} \, \mathrm{d}V + \int_S \boldsymbol{\sigma} \hat{\mathbf{n}} \, \mathrm{d}S = \mathbf{0}, \tag{4.1.8}$$

where $\hat{\mathbf{n}}$ is the *outward* normal to the surface S. Applying the divergence theorem,

$$\int_V \mathbf{f} \, \mathrm{d}V + \int_V \nabla \cdot \boldsymbol{\sigma} \, \mathrm{d}V = \mathbf{0}. \tag{4.1.9}$$

As this must apply for any volume however small, we must have

$$\mathbf{f} + \nabla \cdot \boldsymbol{\sigma} = \mathbf{0}, \tag{4.1.10}$$

or in component form

$$f_i + \frac{\partial \sigma_{ij}}{\partial x_j} = 0, \tag{4.1.11}$$

at all points. This is the equation of *static equilibrium.*

A more elementary proof is to consider a rectangular volume element $\mathrm{d}x_1\mathrm{d}x_2\mathrm{d}x_3$. Tractions on one face, e.g. $\mathrm{d}x_2\mathrm{d}x_3$, are slightly out of balance with tractions on the opposite face because of gradients of stress, i.e. $-\sigma_{i1}\mathrm{d}x_2\mathrm{d}x_3$ on one face compared with $\left(\sigma_{i1} + (\partial\sigma_{i1}/\partial x_1)\,\mathrm{d}x_1\right)\mathrm{d}x_2\mathrm{d}x_3$ on the opposite face. Combining all faces, and cancelling the volume $\mathrm{d}x_1\mathrm{d}x_2\mathrm{d}x_3$, we obtain equation (4.1.11).

The definition of stress used here is more properly called the *Cauchy stress tensor* and corresponds to an Eulerian treatment of the coordinates. More properly, we should have defined the first Piola–Kirchhoff stress tensor with a Lagrangian treatment of coordinates. There is a third, the second Piola–Kirchhoff stress tensor, which is used in the constitutive relation (Section 4.4). The connection between the three depends on the deformation tensor (Dahlen and Tromp, 1998, Section 2.5), but for an infinitesimal deformation the distinction is not important.

4.1.2 Symmetry

Static equilibrium also requires there to be no net torque, i.e.

$$\int_V (\mathbf{x} \times \mathbf{f}) \, \mathrm{d}V + \int_S (\mathbf{x} \times \mathbf{t}) \, \mathrm{d}S = \mathbf{0}, \tag{4.1.12}$$

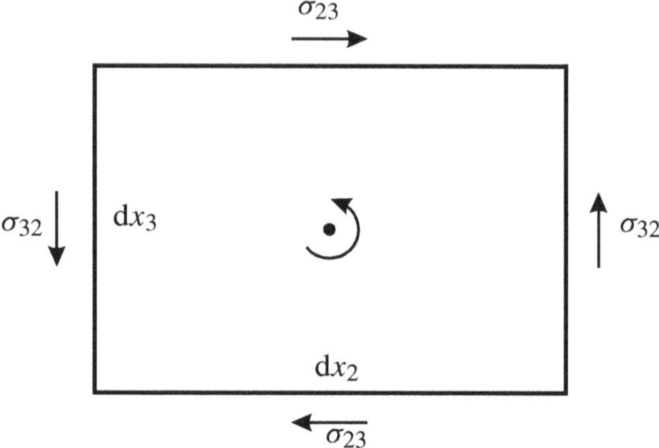

Fig. 4.5. Cross-section of a rectangular volume element perpendicular to the x_1 axis.

where $\mathbf{x}$ is the position vector. Consider a rectangular volume element $\mathrm{d}x_1\mathrm{d}x_2\mathrm{d}x_3$ and moments about an axis parallel to the x_1 axis through the centre of the x_2–x_3 face (Figure 4.5).

To lowest order, the significant traction components causing moments about the axis are t_2 on the faces perpendicular to the x_3 axis, and t_3 on the faces perpendicular to the x_2 axis. The net moment is

$$(\sigma_{32}\mathrm{d}x_1\mathrm{d}x_3)\mathrm{d}x_2 - (\sigma_{23}\mathrm{d}x_1\mathrm{d}x_2)\mathrm{d}x_3. \tag{4.1.13}$$

Moments due to variations of the stress, body forces or other components of stress are lower order, $O(\mathrm{d}x^4)$, so as $\mathrm{d}V = \mathrm{d}x_1\mathrm{d}x_2\mathrm{d}x_3 \rightarrow 0$ only the above term is significant. Thus for no net moment, we must have

$$\sigma_{32} - \sigma_{23} = 0, \tag{4.1.14}$$

or more generally

$$\boxed{\boldsymbol{\sigma} = \boldsymbol{\sigma}^{\mathrm{T}}.} \tag{4.1.15}$$

The stress tensor is symmetric and only has six independent components.

It is tempting and not unknown to refer to the diagonal components of the stress tensor, i.e. σ_{11}, etc., as the *normal stresses*, and the off-diagonal components, i.e. σ_{23}, etc., as the *shear stresses*. It is not unknown in older publications to use different symbols for the two, e.g. σ_x, etc. for the normal stress and τ_{yz} for the shear stress. Such terminology only makes sense if connected with a surface, e.g. the normal stresses with respect to cartesian surfaces. In general, suppose that in one coordinate system, the stress tensor is

$$\boldsymbol{\sigma} = \begin{pmatrix} 1 & 0 & 0 \\ 0 & -1 & 0 \\ 0 & 0 & 0 \end{pmatrix}, \tag{4.1.16}$$

i.e. tension in the x_1 direction and compression in the x_2 direction. Rotating the coordinate system by $\pi/4$ about the x_3 axis, the stress tensor becomes

$$\boldsymbol{\sigma} = \begin{pmatrix} 0 & -1 & 0 \\ -1 & 0 & 0 \\ 0 & 0 & 0 \end{pmatrix}, \tag{4.1.17}$$

i.e. shear stress in the x_1–x_2 plane. Thus normal stresses in one coordinate system are shear stresses in another, while the tensor $\boldsymbol{\sigma}$ physically remains the same. In general, one should therefore be very careful using the terminology normal and shear stresses.

Although the tensor components depend on the coordinate system, a useful invariant of the stress tensor is

$$P = -\frac{1}{3}\operatorname{tr}(\boldsymbol{\sigma}). \tag{4.1.18}$$

The factor $-1/3$ is introduced as this corresponds to the *hydrostatic pressure* (note pressure and stress are measured in opposite directions).

4.1.3 Equation of motion

If equation (4.1.7) describes the condition for static equilibrium, then when the medium is not in equilibrium, Newton's second law of motion must give

$$\int_V \mathbf{f}\,\mathrm{d}V + \int_S \mathbf{t}\,\mathrm{d}S = \frac{\mathrm{D}}{\mathrm{D}t}\int_V \rho\,\mathbf{v}\,\mathrm{d}V, \tag{4.1.19}$$

where $\mathbf{v}$ is the velocity of the medium (particle velocity). Proceeding as before, we obtain (cf. equation (4.1.10))

$$\boxed{\mathbf{f} + \nabla\cdot\boldsymbol{\sigma} = \rho\frac{\partial\mathbf{v}}{\partial t}.} \tag{4.1.20}$$

We ignore important details about the nature of the time derivatives in equations (4.1.19) and (4.1.20). For a more rigorous derivation it is important to consider the distinction between Lagrangian and Eulerian derivatives (fixed in space or following the material particles). If we assume that the medium only undergoes an infinitesimal deformation, then the distinction is not important. The time derivative in equation (4.1.19) is properly the *material derivative* (following the volume),

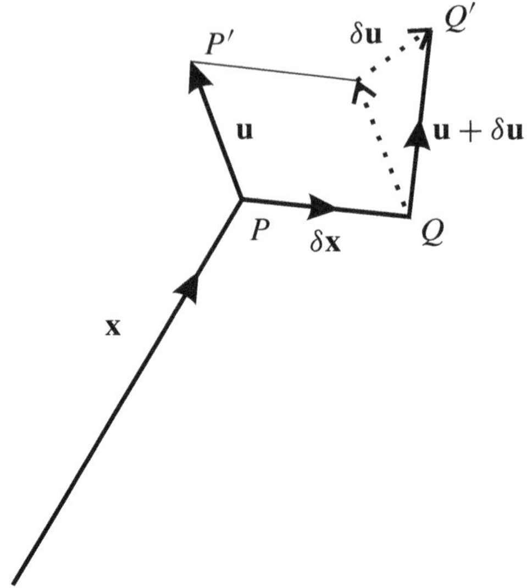

Fig. 4.6. The infinitesimal deformation of neighbouring points PQ to $P'Q'$.

but again for infinitesimal deformations we can use the partial derivative (as in equation (4.1.20)).

Finally we note that in equations (4.1.20), etc., the units are unit($\boldsymbol{\sigma}$) = unit(P) = unit($\mathbf{t}$) = $[\mathrm{ML}^{-1}\mathrm{T}^{-2}]$, unit($\mathbf{f}$) = $[\mathrm{ML}^{-2}\mathrm{T}^{-2}]$, unit($\rho$) = $[\mathrm{ML}^{-3}]$, and the units in equation (4.1.20) balance.

4.2 Infinitesimal strain tensor

The deformation of a medium is described by considering the particle motion of two neighbouring points, P and Q, separated in the undeformed medium by $\delta\mathbf{x}$ (see Figure 4.6). The point P is deformed to P' by $\mathbf{u}$. The point Q moves to Q' and the motion is $\mathbf{u}+\delta\mathbf{u}$. The motion $\mathbf{u}$ is a uniform translation and is described by the rigid body motion.

The deformation is described by $\delta\mathbf{u}$. We write the first-order deformation as

$$\delta u_i = \frac{\partial u_i}{\partial x_j}\,\delta x_j \tag{4.2.1}$$

(with the summation convention), and separate the gradient into symmetric and anti-symmetric parts, i.e.

$$\delta u_i = e_{ij}\,\delta x_j - \xi_{ij}\,\delta x_j, \tag{4.2.2}$$

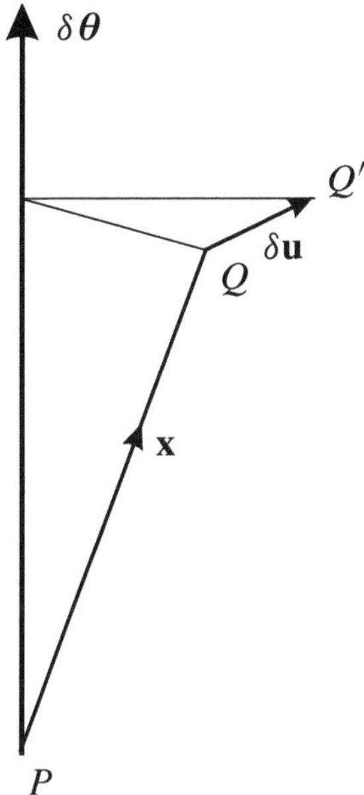

Fig. 4.7. The rotation of the point Q to Q' by the vector $\delta\boldsymbol{\theta}$ through the point P.

where

$$\boxed{e_{ij} = \frac{1}{2}\left(\frac{\partial u_j}{\partial x_i} + \frac{\partial u_i}{\partial x_j}\right) = e_{ji}} \tag{4.2.3}$$

$$\xi_{ij} = \frac{1}{2}\left(\frac{\partial u_j}{\partial x_i} - \frac{\partial u_i}{\partial x_j}\right) = -\xi_{ji}. \tag{4.2.4}$$

The three independent, non-zero components of ξ_{ij} can be considered as a vector

$$\delta\boldsymbol{\theta} = \begin{pmatrix} \xi_{23} \\ \xi_{31} \\ \xi_{12} \end{pmatrix}, \tag{4.2.5}$$

and if $e_{ij} = 0$, the deformation (4.2.2) can be written as

$$\delta\mathbf{u} = \delta\boldsymbol{\theta} \times \delta\mathbf{x}. \tag{4.2.6}$$

This is an infinitesimal rotation of angle $|\delta\boldsymbol{\theta}|$ about the axis $\delta\boldsymbol{\theta}$ (see Figure 4.7).

The rotation of Q about P given by ξ_{ij} does not deform the medium. It is described by the rigid body rotation of the medium. The total motion of the medium is described by rigid body translation and rotation, and deformation. So by default, the components e_{ij}, given by definition (4.2.3), must represent the deformation of the medium. Thus the terms e_{ij} are the cartesian components of the *infinitesimal*

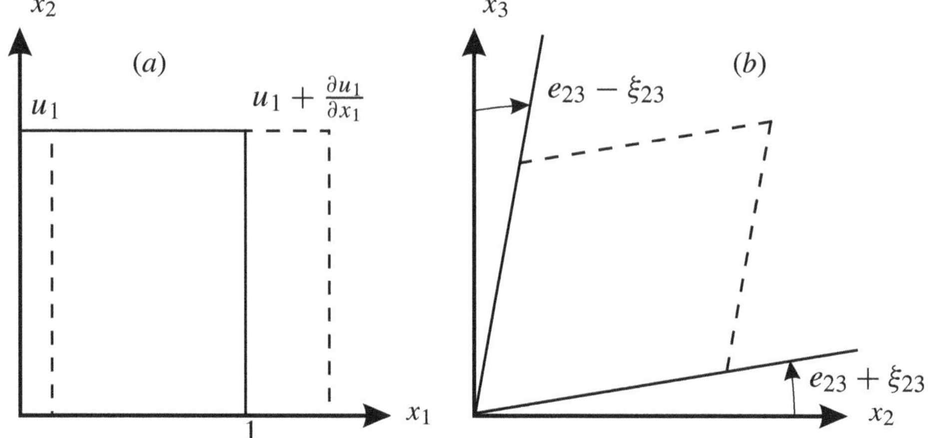

Fig. 4.8. The deformation of a unit square due to strain components (*a*) e_{11}, and (*b*) e_{23}.

strain tensor, **e**. It connects two vectors, $\delta\mathbf{x}$ and $\delta\mathbf{u}$, and so must be a second-order tensor. By definition, it is symmetric

$$\mathbf{e} = \mathbf{e}^{\mathrm{T}}, \tag{4.2.7}$$

and so has six independent components.

If only component e_{11} is non-zero, the medium is stretched in the x_1 direction (Figure 4.8*a*). If components e_{23} and ξ_{23} are non-zero, then the medium is rotated about the x_1 axis by the angle ξ_{23} and sheared in the x_2–x_3 plane (Figure 4.8*b*). Again the terms normal and shear strain are used, for Figures 4.8*a* and *b*, but they only make sense in a particular coordinate system. Normal strains of opposite signs along the x_1 and x_2 axes become shear strains, if the axes are rotated by $\pi/4$ (cf. stress (4.1.16) and (4.1.17)).

A useful invariant of the strain tensor is

$$\boxed{\theta = \mathrm{tr}(\mathbf{e}) = e_{ii} = \frac{\partial u_1}{\partial x_1} + \frac{\partial u_2}{\partial x_2} + \frac{\partial u_3}{\partial x_3} = \nabla \cdot \mathbf{u}.} \tag{4.2.8}$$

It is known as the *dilatation* and is the fractional change in volume.

Finally we note that $\mathrm{unit}(\mathbf{e}) = \mathrm{unit}(\theta) = 0$.

4.3 Boundary conditions

At an interface between two different media, we must consider boundary conditions as some components of the stress and/or particle velocity may be discontinuous.

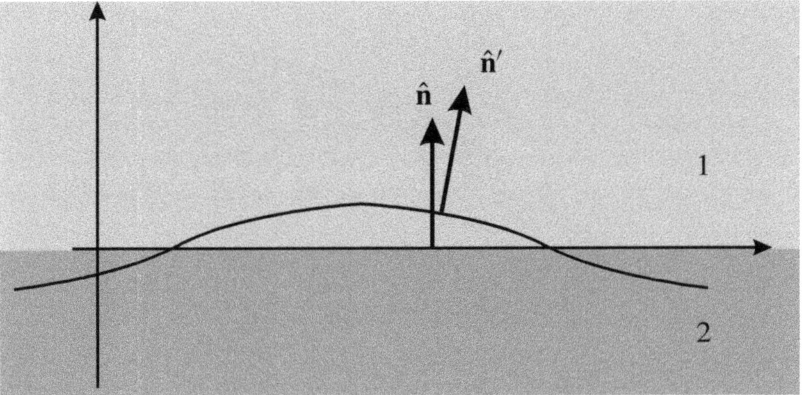

Fig. 4.9. An interface between media 1 and 2 with normal $\hat{\mathbf{n}}$ and the perturbed normal $\hat{\mathbf{n}}'$.

Consider two media in contact. In general they can slide, pull apart, etc. However, under confining pressures, as exist in the Earth, this does not normally occur. The exception may be a fault, particularly if fluid is present. Initially we consider a *welded interface*, where the different materials have no relative motion, i.e. they are welded together.

4.3.1 A welded interface

Labelling the two media 1 and 2, as in Figure 4.9, and indicating a property in each medium at the interface by a corresponding subscript, the welded-interface boundary condition is

$$\boxed{\mathbf{v}_1 = \mathbf{v}_2.} \tag{4.3.1}$$

In this section, a subscript on the traction also indicates the medium, not the normal to the surface. These *kinematic continuity* conditions apply on the deformed interface. As we are assuming only infinitesimal displacements, we can apply these conditions at the undeformed interface.

There must also be *dynamic boundary conditions* concerning the forces. Consider the pill box of material intersecting the interface (Figure 4.10). The forces on the pill box cause it to accelerate

$$\frac{\mathrm{D}}{\mathrm{D}t}\left(\int_{V_1+V_2} \rho\,\mathbf{v}\,\mathrm{d}V\right) = \int_{V_1+V_2} \mathbf{f}\,\mathrm{d}V + \int_{S_1+S_2+S_s} \mathbf{t}\,\mathrm{d}S, \tag{4.3.2}$$

where V_1 and V_2 are the volumes in media 1 and 2 respectively, and S_1 and S_2 are the surfaces parallel to the interface, and S_s are the sides of the pill box. If we let

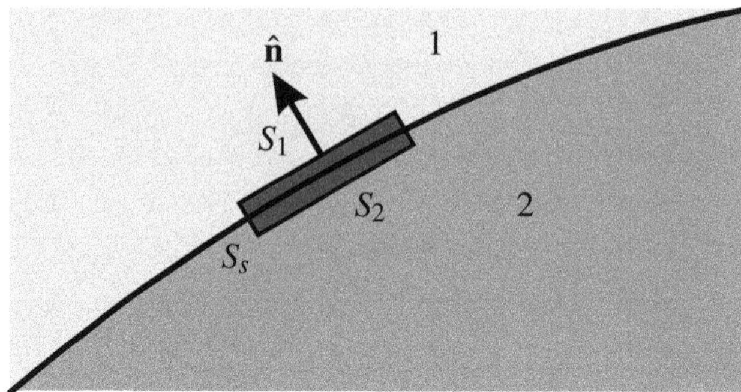

Fig. 4.10. A pill box intersecting the interface indicating the surfaces S_1, S_2 and S_s and the volumes V_1 and V_2.

the thickness of the box go to zero, $S_s \to 0$, $V_1 \to 0$ and $V_2 \to 0$ while the areas S_1 and S_2 are finite and equal. Thus

$$\int_{S_1+S_2} \mathbf{t}\, \mathrm{d}S = 0, \tag{4.3.3}$$

so $\mathbf{t}_1 + \mathbf{t}_2 = \mathbf{0}$ as S_1 and S_2 are equal (assume the area is small enough that the stress can be treated as constant). Remember that in this section, the subscript on the traction indicates the medium, not the normal to the surface. Therefore

$$\boxed{\boldsymbol{\sigma}_1 \hat{\mathbf{n}} = \boldsymbol{\sigma}_2 \hat{\mathbf{n}},} \tag{4.3.4}$$

as $\hat{\mathbf{n}} = -\hat{\mathbf{n}}_1 = \hat{\mathbf{n}}_2$. Thus if the interface is normal to the x_3 axis, the components σ_{31}, σ_{23} and σ_{33} of the stress tensor are continuous. In total, at a welded interface, we have six boundary conditions, continuity of particle velocity and traction on the interface surface. Note that other components of stress, i.e. normal stress components on surfaces perpendicular to the interface, and shear stress parallel to the interface on these planes (e.g. σ_{11}, σ_{22} and σ_{12} if the surface is normal to the x_3 axis), do not satisfy any boundary condition and may be discontinuous.

4.3.2 A fluid–solid or fluid–fluid interface

With a perfect fluid or a lubricated interface, sliding can occur. Defining a unit normal to the interface, $\hat{\mathbf{n}}$, pointing from media 2 to 1, then only the normal component of velocity is continuous, i.e.

$$\mathbf{v}_1 \cdot \hat{\mathbf{n}} = \mathbf{v}_2 \cdot \hat{\mathbf{n}}. \tag{4.3.5}$$

The tangential components of the velocity may be discontinuous.

The argument used above for the dynamic conditions still applies but certain components of the traction must be zero. In a fluid, the only stress is due to the hydrostatic pressure. At a fluid–fluid interface, the hydrostatic pressure is continuous. At a fluid–solid interface, the tangential components of the traction on the interface surface are zero in the fluid, and therefore must be zero in the solid, i.e. if the interface is normal to the x_3 axis, σ_{33} in the solid is $-P$ in the fluid, and $\sigma_{23} = \sigma_{31} = 0$ in the solid at the interface.

4.3.3 A free interface

At a free interface, i.e. the boundary between a solid or fluid and a vacuum, the boundary condition is that the traction component is zero. Taking medium 1 as the vacuum, the condition is that

$$\boxed{\mathbf{t}_2 = \boldsymbol{\sigma}_2\,\hat{\mathbf{n}} = \mathbf{0}.} \tag{4.3.6}$$

The particle velocity is unconstrained and is determined from the incident and reflected waves. The amplitudes of the reflected waves can be determined from the condition (4.3.6).

4.4 Constitutive relations

In the previous two sections, we have introduced the stress and strain tensors. These describe the forces in a continuum and the deformation. We now introduce an empirical, physical law relating these variables. As we have only considered infinitesimal strain, we consider the lowest-order, linear law. First we consider the simplest case of an acoustic medium or fluid (Section 4.4.1), then a general anisotropic, elastic medium (Section 4.4.2) which we then specialize to isotropic (Section 4.4.3) and transversely isotropic (Section 4.4.4) media.

4.4.1 Acoustic medium

Acoustic waves propagate in a fluid. A perfect fluid is defined as a medium in which shear stresses are always zero. Whatever shear strain occurs, there is no shear stress. Since the shear stresses depend on the coordinate system (cf. equations (4.1.16) and (4.1.17)), it is necessary that the stress tensor be isotropic, i.e.

$$\sigma_{11} = \sigma_{22} = \sigma_{33} = -P, \tag{4.4.1}$$

where P is the hydrostatic pressure defined in (4.1.18). Thus the stress tensor is

$$\boldsymbol{\sigma} = -P\,\mathbf{I}, \tag{4.4.2}$$

where $\mathbf{I}$ is the identity tensor. We assume that the fluid is isotropic so the linear constitutive law must be

$$\boxed{P = -\kappa\,\theta,} \tag{4.4.3}$$

where θ is the dilatation (4.2.8). The parameter κ is known as the *bulk modulus*. The *constitutive relation* (4.4.3) can also be written in the reverse order

$$\nabla \cdot \mathbf{u} = -k\,P, \tag{4.4.4}$$

where $k = 1/\kappa$, and is known as the *compressibility*. Remember that equations (4.4.3) and (4.4.4) are only approximate, empirical laws, normally valid for small pressures and deformations ($\theta \ll 1$).

4.4.2 Elastic, anisotropic medium

The simplest law connecting the stress and strain is a linear relationship, i.e.

$$\boxed{\sigma_{ij} = c_{ijkl}\,e_{kl}} \tag{4.4.5}$$

(summation over repeated indices). The expression (4.4.5) contains nine equations for the stress components and each equation contains nine strain components. This is a generalization of Hooke's law – that the extension of a wire or spring is proportional to the applied load – and is called a *constitutive relation* or *law*. Experimental evidence shows that for small deformations ($e_{ij} \ll 1$) acting for short times, it is a reasonable physical law. But it is important to realize that it is only an empirical relationship and for large deformations or after long times, is unlikely to be accurate. In general, non-linear terms or terms depending on (strain) rates may be important. However, as a first approximation and for our purposes, the constitutive relation (4.4.5) is an excellent empirical law.

Because the stress and strain are second-order tensors, the parameters c_{ijkl} form a fourth-order tensor. We shall refer to this as the elastic parameter tensor or the *stiffness* tensor (we avoid the expression 'elastic *constants*' as seismology is concerned with *variations* in the parameters with position!). In general, a fourth-order tensor would have $3^4 = 81$ independent parameters, but this number is much reduced for the stiffness tensor. The stress tensor is symmetric (4.1.15) so we must have

$$c_{ijkl} = c_{jikl}. \tag{4.4.6}$$

The strain tensor is also symmetric (4.2.7). While this does not require the parameters to be symmetric for $k \leftrightarrow l$, any asymmetry would be undetectable. Thus if

$c_{ijkl} \neq c_{ijlk}$ we can replace the parameters by

$$c'_{ijkl} = (c_{ijkl} + c_{ijlk})/2, \tag{4.4.7}$$

without altering the constitutive relation (4.4.5). Then the modified parameters are symmetric

$$c'_{ijkl} = c'_{ijlk}. \tag{4.4.8}$$

We assume this modification has been made and drop the prime with no loss in generality. Substituting the definition of the strain tensor (4.2.3) in equation (4.4.5), the constitutive relation can be rewritten

$$\sigma_{ij} = c_{ijkl} \frac{\partial u_k}{\partial x_l}. \tag{4.4.9}$$

Note that this reduction does not require $\partial u_k/\partial x_l = \partial u_l/\partial x_k$, but follows as any asymmetric part of the partial derivative, i.e. rotation ξ_{kl}, does not contribute to the stress.

Thus of the 81 parameters, many are equal. Only six arrangements of the indices i and j need be considered:

$$\{ij\} = \{11\}, \{22\}, \{33\}, \{23\}, \{31\}, \{12\}, \tag{4.4.10}$$

and similarly for the pair $\{kl\}$. Thus the stiffness tensor will only have $6 \times 6 = 36$ independent parameters. In order to display these parameters it is convenient to map the index pairs into a single index, i.e. $\{ij\} \to m$ where

$$\{11\} \to 1, \{22\} \to 2, \{33\} \to 3, \{23\} \to 4, \{31\} \to 5, \{12\} \to 6. \tag{4.4.11}$$

This prescription for reducing and ordering the indices is sometimes named after Voigt (1910) (notice the cyclic order of the first and last three pairs). We then define new parameters C_{mn} where

$$C_{mn} = c_{ijkl} \tag{4.4.12}$$

where $\{ij\} \to m$ and $\{kl\} \to n$ according to the *Voigt notation*.

The constitutive relation can be rewritten as a vector-matrix equation

$$\begin{pmatrix} \sigma_{11} \\ \sigma_{22} \\ \sigma_{33} \\ \sigma_{23} \\ \sigma_{31} \\ \sigma_{12} \end{pmatrix} = \begin{pmatrix} C_{11} & C_{12} & C_{13} & C_{14} & C_{15} & C_{16} \\ C_{21} & C_{22} & C_{23} & C_{24} & C_{25} & C_{26} \\ C_{31} & C_{32} & C_{33} & C_{34} & C_{35} & C_{36} \\ C_{41} & C_{42} & C_{43} & C_{44} & C_{45} & C_{46} \\ C_{51} & C_{52} & C_{53} & C_{54} & C_{55} & C_{56} \\ C_{61} & C_{62} & C_{63} & C_{64} & C_{65} & C_{66} \end{pmatrix} \begin{pmatrix} e_{11} \\ e_{22} \\ e_{33} \\ 2e_{23} \\ 2e_{31} \\ 2e_{12} \end{pmatrix} = \mathbf{C} \begin{pmatrix} e_{11} \\ e_{22} \\ e_{33} \\ 2e_{23} \\ 2e_{31} \\ 2e_{12} \end{pmatrix}. \tag{4.4.13}$$

The matrix **C** is the 6×6 matrix with elements C_{mn}. Obviously this matrix contains the necessary 36 independent parameters. Note that there is no suggestion that the six components of stress (or strain) form a physical vector – this equation (4.4.13) is just a useful notation exploiting the symmetries of the stress and strain tensors. In particular, the 6×6 matrix **C** is a useful and widely used method to display the elastic parameters, see, e.g. the textbooks Nye (1957), Musgrave (1970) and Auld (1973). In the 'strain vector' in equation (4.4.13), the factors of 2 are necessary in the last three components as e_{kl} and e_{lk} both contribute (equally) in equation (4.4.5) when $k \neq l$ (but see Exercise 4.12).

However, a small rearrangement of (4.4.13) allows the equation to be treated as a valid vector-matrix equation in a six-dimensional space. This is useful for more than just display, and allows studies of the eigenstresses and eigenstrains (Thomson, 1856, 1877; Cowin, Mehrabadi and Sadegh, 1991). Equation (4.4.13) is rewritten (Fedorov, 1968)

$$\begin{pmatrix} \sigma_{11} \\ \sigma_{22} \\ \sigma_{33} \\ \sqrt{2}\sigma_{23} \\ \sqrt{2}\sigma_{31} \\ \sqrt{2}\sigma_{12} \end{pmatrix} = \begin{pmatrix} C_{11} & C_{12} & C_{13} & \sqrt{2}C_{14} & \sqrt{2}C_{15} & \sqrt{2}C_{16} \\ C_{21} & C_{22} & C_{23} & \sqrt{2}C_{24} & \sqrt{2}C_{25} & \sqrt{2}C_{26} \\ C_{31} & C_{32} & C_{33} & \sqrt{2}C_{34} & \sqrt{2}C_{35} & \sqrt{2}C_{36} \\ \sqrt{2}C_{41} & \sqrt{2}C_{42} & \sqrt{2}C_{43} & 2C_{44} & 2C_{45} & 2C_{46} \\ \sqrt{2}C_{51} & \sqrt{2}C_{52} & \sqrt{2}C_{53} & 2C_{54} & 2C_{55} & 2C_{56} \\ \sqrt{2}C_{61} & \sqrt{2}C_{62} & \sqrt{2}C_{63} & 2C_{64} & 2C_{65} & 2C_{66} \end{pmatrix} \begin{pmatrix} e_{11} \\ e_{22} \\ e_{33} \\ \sqrt{2}e_{23} \\ \sqrt{2}e_{31} \\ \sqrt{2}e_{12} \end{pmatrix}, \tag{4.4.14}$$

which we write compactly as

$$\tilde{\boldsymbol{\sigma}} = \widetilde{\mathbf{C}}\,\tilde{\mathbf{e}}. \tag{4.4.15}$$

This notation is convenient to obtain compact expressions for the rotation of the second-order stress and strain tensors, and the fourth-order elastic parameter tensor. These transformations are described in some detail in Auld (1973, Chapter 3) using the notation in equation (4.4.13). We follow the same procedure but use the somewhat neater equation (4.4.14).

Components of a second-order tensor are transformed according to the rule

$$\sigma'_{i'j'} = b_{i'i} b_{j'j} \sigma_{ij}, \tag{4.4.16}$$

where $b_{i'i}$ are elements of the 3×3 transformation matrix, and a summation over repeated indices is implied in equation (4.4.16). Equation (4.4.16) can be rewritten as

$$\tilde{\sigma}' = \widetilde{\mathbf{B}}\tilde{\sigma}, \tag{4.4.17}$$

where the matrix $\widetilde{\mathbf{B}}$ is

$$\widetilde{\mathbf{B}} = \begin{pmatrix} b_{11}^2 & b_{12}^2 & b_{13}^2 & \sqrt{2}b_{12}b_{13} & \sqrt{2}b_{13}b_{11} & \sqrt{2}b_{11}b_{12} \\ b_{21}^2 & b_{22}^2 & b_{23}^2 & \sqrt{2}b_{22}b_{23} & \sqrt{2}b_{23}b_{21} & \sqrt{2}b_{21}b_{22} \\ b_{31}^2 & b_{32}^2 & b_{33}^2 & \sqrt{2}b_{32}b_{33} & \sqrt{2}b_{33}b_{31} & \sqrt{2}b_{31}b_{32} \\ \sqrt{2}b_{21}b_{31} & \sqrt{2}b_{22}b_{32} & \sqrt{2}b_{23}b_{33} & b_{22}b_{33}+b_{23}b_{32} & b_{21}b_{33}+b_{23}b_{31} & b_{21}b_{32}+b_{22}b_{31} \\ \sqrt{2}b_{31}b_{11} & \sqrt{2}b_{32}b_{12} & \sqrt{2}b_{33}b_{13} & b_{32}b_{13}+b_{33}b_{12} & b_{31}b_{13}+b_{33}b_{11} & b_{31}b_{12}+b_{32}b_{11} \\ \sqrt{2}b_{11}b_{21} & \sqrt{2}b_{12}b_{22} & \sqrt{2}b_{13}b_{23} & b_{12}b_{23}+b_{13}b_{22} & b_{11}b_{23}+b_{13}b_{21} & b_{11}b_{22}+b_{12}b_{21} \end{pmatrix}. \tag{4.4.18}$$

The strain tensor can be similarly transformed into a vector-matrix equation

$$\tilde{\mathbf{e}}' = \widetilde{\mathbf{B}}\,\tilde{\mathbf{e}}. \tag{4.4.19}$$

Combining transformations (4.4.17) and (4.4.19) in equation (4.4.15), we find

$$\tilde{\sigma}' = \widetilde{\mathbf{C}}'\,\tilde{\mathbf{e}}', \tag{4.4.20}$$

where the transformed elastic parameters can be obtained from

$$\widetilde{\mathbf{C}}' = \widetilde{\mathbf{B}}\,\widetilde{\mathbf{C}}\,\widetilde{\mathbf{B}}^{-1} = \widetilde{\mathbf{B}}\,\widetilde{\mathbf{C}}\,\widetilde{\mathbf{B}}^{\mathrm{T}}. \tag{4.4.21}$$

The simplification $\widetilde{\mathbf{B}}^{-1} = \widetilde{\mathbf{B}}^{\mathrm{T}}$ follows as the transformation matrix is orthogonal, i.e. $\mathbf{b}^{-1} = \mathbf{b}^{\mathrm{T}}$. Equation (4.4.21) must be equivalent to the tensor transformation rule

$$c_{i'j'k'l'} = b_{i'i}b_{j'j}b_{k'k}b_{l'l}c_{ijkl}, \tag{4.4.22}$$

but computationally is much more compact and efficient. Equation (4.4.21) is known as the *Bond transformation* after Bond (1943).

For computational purposes, worthy of consideration as each equation (4.4.22) contains 324 multiplications, we note that the matrix $\widetilde{\mathbf{B}}$ (4.4.18) can be rewritten

$$\widetilde{\mathbf{B}} = \mathbf{E}\,\widetilde{\mathbf{B}}'\,\mathbf{E}, \tag{4.4.23}$$

where the elements of the matrix $\widetilde{\mathbf{B}}'$ are

$$\widetilde{B}'_{ij} = b_{lm}b_{kn} + b_{ln}b_{km}, \tag{4.4.24}$$

with the Voigt mapping (4.4.11) $i \leftrightarrow \{lk\}$ and $j \leftrightarrow \{mn\}$, and

$$\mathbf{E} = \begin{pmatrix} \mathbf{I}/\sqrt{2} & \mathbf{0} \\ \mathbf{0} & \mathbf{I} \end{pmatrix}. \tag{4.4.25}$$

The matrix $\mathbf{C}$ (4.4.13) can be related to the matrix $\widetilde{\mathbf{C}}$ by

$$\widetilde{\mathbf{C}} = 2\mathbf{E}\,\mathbf{C}\,\mathbf{E}, \tag{4.4.26}$$

and substituting in equation (4.4.21), we obtain

$$\mathbf{C}' = \widetilde{\mathbf{B}}''\,\mathbf{C}\,\widetilde{\mathbf{B}}''^{\mathrm{T}}, \tag{4.4.27}$$

where

$$\widetilde{\mathbf{B}}'' = \widetilde{\mathbf{B}}'\,\mathbf{E}^2. \tag{4.4.28}$$

With definition (4.4.24), this is compact and efficient to compute.

The final symmetry of the stiffness tensor depends on the existence of a unique, internal strain energy function, W. Consider the work done by the stresses σ_{ij} in increasing the strains infinitesimally by de_{ij}. Each component does work and the infinitesimal increase in the energy function per unit volume is

$$W = \sigma_{ij}\,\mathrm{d}e_{ij}. \tag{4.4.29}$$

The total strain energy per unit volume (defining the zero at zero strain, $\mathbf{e} = \mathbf{0}$) is

$$W = \int_{\mathbf{0}}^{\mathbf{e}} \sigma_{ij}\,\mathrm{d}e_{ij}. \tag{4.4.30}$$

The stresses are, of course, functions of the strains and if the constitutive equation (4.4.5) applies, we have

$$W = \int_{\mathbf{0}}^{\mathbf{e}} c_{ijkl}\,e_{kl}\,\mathrm{d}e_{ij} \tag{4.4.31}$$

$$= \frac{1}{2} c_{ijkl}\,e_{kl}\,e_{ij}, \tag{4.4.32}$$

provided the result is independent of the route to the final strain $\mathbf{e}$. Thus interchanging the dummy index pairs $\{ij\}$ and $\{kl\}$, we must have

$$c_{ijkl} = c_{klij}. \tag{4.4.33}$$

This symmetry implies that the matrix $\mathbf{C}$ (4.4.13) is symmetric, and that there are 21 independent parameters. Thus the most general stiffness tensor for an elastic medium contains 21 independent parameters.

Although the 21 independent parameters are conveniently displayed as the upper triangle of a 6×6 matrix $\mathbf{C}$ (see, for instance, Nye, 1950; Musgrave, 1970; Auld, 1973), the vector-matrix equation (4.4.13) is not particularly convenient for investigating the elastic waves (largely because the 'vectors' have no direct physical significance). To simplify the equation of motion (4.1.20) and constitutive relations (4.4.9), it is convenient to introduce another notation (close to that of Woodhouse, 1974). We decompose the stress tensor into cartesian traction vectors, $\mathbf{t}_j = \boldsymbol{\sigma}\hat{\mathbf{i}}_j$, where (cf. equation (4.1.2))

$$(\mathbf{t}_j)_i = \sigma_{ij}, \tag{4.4.34}$$

so that $\mathbf{t}_j$ is the traction on a surface perpendicular to the j-th axis. Similarly, we decompose the strain tensor into cartesian vectors, $\mathbf{e}_k$, where

$$(\mathbf{e}_k)_l = e_{lk} = \frac{1}{2}\left(\frac{\partial u_l}{\partial x_k} + \frac{\partial u_k}{\partial x_l}\right). \tag{4.4.35}$$

The constitutive relation (4.4.9) can then be rewritten

$$\boxed{\mathbf{t}_j = \mathbf{c}_{jk}\mathbf{e}_k = \mathbf{c}_{jk}\frac{\partial \mathbf{u}}{\partial x_k},} \tag{4.4.36}$$

where the $\mathbf{c}_{jk}$ are 3×3 matrices such that

$$(\mathbf{c}_{jk})_{il} = c_{ijlk}. \tag{4.4.37}$$

Obviously from the symmetries of the stiffness tensor we obtain

$$\mathbf{c}_{kj} = \mathbf{c}_{jk}^{\mathrm{T}}. \tag{4.4.38}$$

It is useful to write the elements of the matrices $\mathbf{c}_{jk}$ in terms of the parameters C_{mn} (4.4.12). We have

$$\mathbf{c}_{11} = \begin{pmatrix} C_{11} & C_{16} & C_{15} \\ C_{16} & C_{66} & C_{56} \\ C_{15} & C_{56} & C_{55} \end{pmatrix} \quad \mathbf{c}_{22} = \begin{pmatrix} C_{66} & C_{26} & C_{46} \\ C_{26} & C_{22} & C_{24} \\ C_{46} & C_{24} & C_{44} \end{pmatrix}$$
$$\mathbf{c}_{33} = \begin{pmatrix} C_{55} & C_{45} & C_{35} \\ C_{45} & C_{44} & C_{34} \\ C_{35} & C_{34} & C_{33} \end{pmatrix} \quad \mathbf{c}_{23} = \begin{pmatrix} C_{56} & C_{46} & C_{36} \\ C_{25} & C_{24} & C_{23} \\ C_{45} & C_{44} & C_{34} \end{pmatrix}$$
$$\mathbf{c}_{31} = \begin{pmatrix} C_{15} & C_{56} & C_{55} \\ C_{14} & C_{46} & C_{45} \\ C_{13} & C_{36} & C_{35} \end{pmatrix} \quad \mathbf{c}_{12} = \begin{pmatrix} C_{16} & C_{12} & C_{14} \\ C_{66} & C_{26} & C_{46} \\ C_{56} & C_{25} & C_{45} \end{pmatrix}. \tag{4.4.39}$$

Occasionally it is useful to pose the constitutive relationship in the opposite direction and use the compliance parameters, s_{ijkl}, i.e.

$$e_{ij} = s_{ijkl}\sigma_{kl} \tag{4.4.40}$$

(we follow the standard terminology while noting the unfortunate choice of letter – s_{ijkl} for the *c*ompliances and c_{ijkl} for the *s*tiffnesses!). The compliance tensor has all the same symmetries as the stiffness tensor, and therefore has 21 independent components.

We introduce a vector-matrix notation like equation (4.4.36) using the compliance parameters, i.e.

$$\mathbf{e}_j = \mathbf{s}_{jk}\mathbf{t}_k, \tag{4.4.41}$$

where $(\mathbf{s}_{jk})_{il} = s_{ijlk}$. Note that on the left-hand side we must have the strain vector not $\partial\mathbf{u}/\partial x_j$ as the constitutive equation only depends on the symmetric part of the deformation, and the asymmetric part, i.e. ξ_{ij}, is not determined by the traction.

The compliance parameters can be obtained from the stiffness parameters by inverting the ninth-order linear system (4.4.5). However, numerically it is easier to invert the 6×6 Voigt system $\mathbf{C}$ (4.4.13), or rather (4.4.14) as then the 6×6 matrix of compliances has exactly the same form as the stiffness matrix (with similar factors of $\sqrt{2}$ and 2) and is obtained directly by matrix inversion. While this matrix inversion is most efficient for numerical work, the fundamental inter-relationship comes from

$$\sigma_{ij} = c_{ijpq}e_{pq} = c_{ijpq}s_{pqrs}\sigma_{rs}. \tag{4.4.42}$$

Various expressions for $c_{ijpq}s_{pqrs}$ would make the right-hand side reduce to the left, e.g. $\delta_{ir}\delta_{js}$. However, the expression must be symmetric for $i \leftrightarrow j$ and $r \leftrightarrow s$, and the correct result is

$$c_{ijpq}s_{pqrs} = \frac{1}{2}(\delta_{ir}\delta_{js} + \delta_{is}\delta_{jr}). \tag{4.4.43}$$

From this expression we can obtain

$$\mathbf{c}_{jk}\mathbf{s}_{kj} = \mathbf{s}_{kj}\mathbf{c}_{jk} = 2\mathbf{I}. \tag{4.4.44}$$

4.4.3 Elastic, isotropic medium

In the previous section we have discussed the constitutive relation for a general, anisotropic elastic solid which has 21 independent parameters. For many purposes it is desirable to simplify this model by introducing various symmetries into the continuum. The simplest elastic solid is isotropic, i.e. the properties are the same in all directions.

In an isotropic solid, many elastic parameters must be equal or zero. Some of these equalities are obvious, e.g. $c_{1111} = c_{2222} = c_{3333}$ as all axes must be equivalent, but others require consideration of parameters in directions between the axes. The full analysis is somewhat tedious but has appeared in several textbooks, e.g. Musgrave (1970). For brevity we quote a general mathematical result for fourth-order isotropic tensors (Exercise 4.8). According to Jeffreys (1931, p. 66), any isotropic fourth-order tensor can be written as

$$c_{ijkl} = \lambda\, \delta_{ij}\delta_{kl} + \mu\, \delta_{ik}\delta_{jl} + \nu\, \delta_{il}\delta_{jk}, \tag{4.4.45}$$

where λ, μ and ν are three independent parameters. Interchanging k and l we have

$$c_{ijlk} = \lambda\, \delta_{ij}\delta_{kl} + \mu\, \delta_{il}\delta_{jk} + \nu\, \delta_{ik}\delta_{jl}. \tag{4.4.46}$$

For the elastic tensor, these components c_{ijkl} and c_{ijlk} must be equal even if $k \neq l$. Subtracting (4.4.45) and (4.4.46), we obtain

$$(\mu - \nu)(\delta_{ik}\delta_{jl} - \delta_{il}\delta_{jk}) = 0 \tag{4.4.47}$$

which must be true for any i, j, k and l. Choosing $i = k = 1$ and $j = l = 2$, expression (4.4.47) reduces to

$$\mu = \nu. \tag{4.4.48}$$

Thus

$$\boxed{c_{ijkl} = \lambda\delta_{ij}\delta_{kl} + \mu(\delta_{ik}\delta_{jl} + \delta_{il}\delta_{jk}).} \tag{4.4.49}$$

It is easily established that this tensor satisfies the required symmetries of the elastic stiffness tensor and that no further reduction is necessary. Thus the stiffness tensor for the most general isotropic media has two independent parameters, λ and μ, and can be written as in equation (4.4.49). The two parameters are called the *Lamé elastic parameters* (Lamé, 1852) and the notation λ and μ is now almost universal (although $G = \mu$ is used in some older books and engineering).

The simple form of the stiffness tensor (4.4.49) allows the constitutive relation (4.4.5) to be written in a compact manner. It is straightforward to show that relation (4.4.5) reduces to

$$\boxed{\sigma_{ij} = \lambda\,\theta\,\delta_{ij} + 2\,\mu\,e_{ij},} \tag{4.4.50}$$

where θ is the dilatation (4.2.8), or

$$\boldsymbol{\sigma} = \lambda\,\theta\,\mathbf{I} + 2\,\mu\,\mathbf{e}. \tag{4.4.51}$$

An elastic solid in which $\lambda = \mu$, i.e. a one-parameter model, is known as a *Poisson solid*. Many elastic solids approximate a Poisson solid, but we emphasize that it is not required in a general, isotropic solid and is often introduced as an algebraic convenience rather than a physical fact.

The isotropic compliance tensor, s_{ijkl}, must be of the same form (4.4.49) (with different parameters, of course). In terms of the Lamé parameters it is

$$s_{ijkl} = -\frac{\lambda}{2\mu(3\lambda + 2\mu)}\delta_{ij}\delta_{kl} + \frac{1}{4\mu}(\delta_{ik}\delta_{jl} + \delta_{il}\delta_{jk}). \tag{4.4.52}$$

The validity of this expression is most easily established by substituting equations (4.4.49) and (4.4.52) in result (4.4.43) and confirming the identity.

Using the Voigt notation, the stiffness matrix $\mathbf{C}$ (4.4.13) for an isotropic medium is

$$\mathbf{C} = \begin{pmatrix} \lambda + 2\mu & \lambda & \lambda & 0 & 0 & 0 \\ \lambda & \lambda + 2\mu & \lambda & 0 & 0 & 0 \\ \lambda & \lambda & \lambda + 2\mu & 0 & 0 & 0 \\ 0 & 0 & 0 & \mu & 0 & 0 \\ 0 & 0 & 0 & 0 & \mu & 0 \\ 0 & 0 & 0 & 0 & 0 & \mu \end{pmatrix}, \tag{4.4.53}$$

and the compliance matrix $\mathbf{S}$ is

$$\mathbf{S} = \begin{pmatrix} \frac{\lambda+\mu}{\mu(3\lambda+2\mu)} & -\frac{\lambda}{2\mu(3\lambda+2\mu)} & -\frac{\lambda}{2\mu(3\lambda+2\mu)} & 0 & 0 & 0 \\ -\frac{\lambda}{2\mu(3\lambda+2\mu)} & \frac{\lambda+\mu}{\mu(3\lambda+2\mu)} & -\frac{\lambda}{2\mu(3\lambda+2\mu)} & 0 & 0 & 0 \\ -\frac{\lambda}{2\mu(3\lambda+2\mu)} & -\frac{\lambda}{2\mu(3\lambda+2\mu)} & \frac{\lambda+\mu}{\mu(3\lambda+2\mu)} & 0 & 0 & 0 \\ 0 & 0 & 0 & \frac{1}{4\mu} & 0 & 0 \\ 0 & 0 & 0 & 0 & \frac{1}{4\mu} & 0 \\ 0 & 0 & 0 & 0 & 0 & \frac{1}{4\mu} \end{pmatrix}. \tag{4.4.54}$$

It is trivial to confirm that the modified forms of these matrices are inverse.

Finally, we will need the 3×3 matrices $\mathbf{c}_{jk}$ and $\mathbf{s}_{jk}$. These are

$$\mathbf{c}_{11} = \begin{pmatrix} \lambda + 2\mu & 0 & 0 \\ 0 & \mu & 0 \\ 0 & 0 & \mu \end{pmatrix}, \tag{4.4.55}$$

and

$$\mathbf{c}_{23} = \begin{pmatrix} 0 & 0 & 0 \\ 0 & 0 & \lambda \\ 0 & \mu & 0 \end{pmatrix}, \tag{4.4.56}$$

with obvious cyclic transpositions for the other matrices. Similarly

$$\mathbf{s}_{11} = \begin{pmatrix} \frac{\lambda+\mu}{\mu(3\lambda+2\mu)} & 0 & 0 \\ 0 & \frac{1}{4\mu} & 0 \\ 0 & 0 & \frac{1}{4\mu} \end{pmatrix}, \tag{4.4.57}$$

and

$$\mathbf{s}_{23} = \begin{pmatrix} 0 & 0 & 0 \\ 0 & 0 & -\frac{\lambda}{2\mu(3\lambda+2\mu)} \\ 0 & \frac{1}{4\mu} & 0 \end{pmatrix}. \tag{4.4.58}$$

Again it is trivial to confirm the result (4.4.44).

4.4.4 Transversely isotropic medium

A simple form of anisotropy, which is very useful, is a transversely isotropic (TI) medium. The medium has axial symmetry and is isotropic in a plane perpendicular to the axis. This form of anisotropy is also called *polar anisotropy* as the parameters only vary with the polar angle from the symmetry axis, or *hexagonal anisotropy* as it exists in media with hexagonal symmetry (Musgrave, 1970).

Choosing x_3 as the axis of symmetry (if the axis is vertical, the medium is commonly called TIV), the elastic constants are (using the matrix **C**, (4.4.13))

$$\mathbf{C} = \begin{pmatrix} C_{11} & C_{12} & C_{13} & 0 & 0 & 0 \\ C_{12} & C_{11} & C_{13} & 0 & 0 & 0 \\ C_{13} & C_{13} & C_{33} & 0 & 0 & 0 \\ 0 & 0 & 0 & C_{44} & 0 & 0 \\ 0 & 0 & 0 & 0 & C_{44} & 0 \\ 0 & 0 & 0 & 0 & 0 & C_{66} \end{pmatrix} \tag{4.4.59}$$

$$= \begin{pmatrix} \lambda_\perp + 2\mu_\perp & \lambda_\perp & \nu & 0 & 0 & 0 \\ \lambda_\perp & \lambda_\perp + 2\mu_\perp & \nu & 0 & 0 & 0 \\ \nu & \nu & \lambda_\parallel + 2\mu_\parallel & 0 & 0 & 0 \\ 0 & 0 & 0 & \mu_\parallel & 0 & 0 \\ 0 & 0 & 0 & 0 & \mu_\parallel & 0 \\ 0 & 0 & 0 & 0 & 0 & \mu_\perp \end{pmatrix}, \tag{4.4.60}$$

as an alternative notation, with the constraint

$$C_{12} = C_{11} - 2C_{66}. \tag{4.4.61}$$

There are five independent parameters $\lambda_\perp$, $\lambda_\|$, $\mu_\perp$, $\mu_\|$ and ν, or C_{11}, C_{33}, C_{44}, C_{66} and C_{13}. The matrices $\mathbf{c}_{jk}$ are

$$\mathbf{c}_{11} = \begin{pmatrix} C_{11} & 0 & 0 \\ 0 & C_{66} & 0 \\ 0 & 0 & C_{44} \end{pmatrix} \quad \mathbf{c}_{22} = \begin{pmatrix} C_{66} & 0 & 0 \\ 0 & C_{11} & 0 \\ 0 & 0 & C_{44} \end{pmatrix}$$
$$\mathbf{c}_{33} = \begin{pmatrix} C_{44} & 0 & 0 \\ 0 & C_{44} & 0 \\ 0 & 0 & C_{33} \end{pmatrix} \quad \mathbf{c}_{23} = \begin{pmatrix} 0 & 0 & 0 \\ 0 & 0 & C_{13} \\ 0 & C_{44} & 0 \end{pmatrix}$$
$$\mathbf{c}_{31} = \begin{pmatrix} 0 & 0 & C_{44} \\ 0 & 0 & 0 \\ C_{13} & 0 & 0 \end{pmatrix} \quad \mathbf{c}_{12} = \begin{pmatrix} 0 & C_{12} & 0 \\ C_{66} & 0 & 0 \\ 0 & 0 & 0 \end{pmatrix}. \tag{4.4.62}$$

Exercises 4.3, 4.4 and 4.5 investigate results for TI media.

Finally we note that the units of elastic parameters are the same as stress, i.e.

$$\text{unit}(\kappa) = \text{unit}(c_{ijkl}) = \text{unit}(\lambda) = \text{unit}(\mu) = [\text{ML}^{-1}\text{T}^{-2}], \tag{4.4.63}$$

as strain has no units. The units of the compliance parameters are

$$\text{unit}(k) = \text{unit}(s_{ijkl}) = [\text{M}^{-1}\text{LT}^{2}]. \tag{4.4.64}$$

4.5 Navier wave equation and Green functions

In the previous sections we have developed the equations of motion (4.1.20) and the constitutive relations for anisotropic (4.4.36), isotropic (4.4.51) and transversely isotropic (4.4.60) media, and for acoustic media (4.4.3).

4.5.1 Acoustic waves

First we investigate acoustic waves. Substituting for the stress tensor (4.4.2) in the equations of motion (4.1.20) we obtain

$$\boxed{\frac{\partial \mathbf{v}}{\partial t} = -\frac{1}{\rho}\nabla P + \frac{1}{\rho}\mathbf{f}.} \tag{4.5.1}$$

The constitutive relations (4.4.3) can be rewritten

$$\boxed{\frac{\partial P}{\partial t} = -\kappa \nabla \cdot \mathbf{v}.} \tag{4.5.2}$$

We have written these equations to emphasize the symmetry – time derivatives on the left-hand side and spatial derivatives times model parameters on the right-hand side. The symmetry can be further enhanced by introducing a *volume injection* source term in the second equation (4.5.2). This would be appropriate for an air-gun source. However, an air-gun can be modelled as a combination of derivatives of force sources, so we do not include this generalization.

4.5.1.1 Betti's theorem

Before considering specific solutions of the equations of motion and the constitutive relation, it is useful to investigate the relationship between various solutions. For the moment, let us rewrite the equations of motion (4.5.1) and constitutive relation (4.5.2) in the less symmetric, but more usual form

$$\rho\,\ddot{\mathbf{u}} = -\nabla P + \mathbf{f} \tag{4.5.3}$$

$$P = -\kappa\,\nabla\cdot\mathbf{u}, \tag{4.5.4}$$

where $\ddot{\mathbf{u}}$ is the particle acceleration. Let us consider another solution in the same model due to a force $\mathbf{f}'$, where we denote the displacement by $\mathbf{u}'$ and the pressure by P', i.e.

$$\rho\,\ddot{\mathbf{u}}' = -\nabla P' + \mathbf{f}' \tag{4.5.5}$$

$$P' = -\kappa\,\nabla\cdot\mathbf{u}'. \tag{4.5.6}$$

We consider these solutions over a volume V surrounded by a surface S (and denote a surface element with outward normal as $\mathrm{d}\mathbf{S} = \hat{n}\,\mathrm{d}S$). We combine the two solutions in a volume and surface integral

$$\begin{aligned}
&\int_V (\mathbf{f} - \rho\,\ddot{\mathbf{u}})\cdot\mathbf{u}'\,\mathrm{d}V - \int_S P\,\mathbf{u}'\cdot\mathrm{d}\mathbf{S} \\
&\quad = \int_V \nabla P\cdot\mathbf{u}'\,\mathrm{d}V - \int_V \nabla\cdot(P\,\mathbf{u}')\,\mathrm{d}V \\
&\quad = -\int_V P\,\nabla\cdot\mathbf{u}'\,\mathrm{d}V \\
&\quad = \int_V (\nabla\cdot\mathbf{u})\,\kappa\,\left(\nabla\cdot\mathbf{u}'\right)\,\mathrm{d}V,
\end{aligned} \tag{4.5.7}$$

where we first substitute equation (4.5.3) and use the divergence theorem to convert the surface integral to a volume integral, and then expand the divergence and substituted equation (4.5.4) to obtain result (4.5.7). The final expression is symmetrical in the two solutions, so we obtain

$$\int_V (\mathbf{f} - \rho\,\ddot{\mathbf{u}})\cdot\mathbf{u}'\,\mathrm{d}V - \int_S P\,\mathbf{u}'\cdot\mathrm{d}\mathbf{S} = \int_V (\mathbf{f}' - \rho\,\ddot{\mathbf{u}}')\cdot\mathbf{u}\,\mathrm{d}V - \int_S P'\,\mathbf{u}\cdot\mathrm{d}\mathbf{S}. \tag{4.5.8}$$

This result is an example of *Betti's theorem*.

It is significant that this result does not involve the initial conditions of the two solutions, nor the times at which they are evaluated. The arguments of the functions $\mathbf{f}$, $\mathbf{u}$ and P might be $(t, \mathbf{x})$, and the functions $\mathbf{f}'$, $\mathbf{u}'$ and P', $(t', \mathbf{x})$. Let us substitute $t = \tau$ and $t' = t - \tau$ and consider the time integral of the terms including just the time derivatives, i.e.

$$\int_0^\tau \left\{\ddot{\mathbf{u}}(\tau) \cdot \mathbf{u}'(t-\tau) - \ddot{\mathbf{u}}'(t-\tau) \cdot \mathbf{u}(\tau)\right\} \mathrm{d}\tau$$

$$= \int_0^t \frac{\partial}{\partial \tau} \left\{\dot{\mathbf{u}}(\tau) \cdot \mathbf{u}'(t-\tau) + \mathbf{u}(\tau) \cdot \dot{\mathbf{u}}'(t-\tau)\right\} \mathrm{d}\tau \tag{4.5.9}$$

$$= \dot{\mathbf{u}}(t) \cdot \mathbf{u}'(0) + \mathbf{u}(t) \cdot \dot{\mathbf{u}}'(0) - \dot{\mathbf{u}}(0) \cdot \mathbf{u}'(t) - \mathbf{u}(0) \cdot \dot{\mathbf{u}}'(t). \tag{4.5.10}$$

If before some time t_{S}, $\mathbf{u}$ and $\mathbf{u}'$ are zero everywhere in V (and necessarily the derivatives will be zero too), then the convolution integral

$$\int_{-\infty}^{\infty} \left\{\ddot{\mathbf{u}}(\tau) \cdot \mathbf{u}'(t-\tau) - \ddot{\mathbf{u}}'(t-\tau) \cdot \mathbf{u}(\tau)\right\} \mathrm{d}\tau = 0, \tag{4.5.11}$$

is zero. The remaining terms in expression (4.5.8) then give

$$\int_{-\infty}^{\infty} \int_V \left\{\mathbf{u}(\tau, \mathbf{x}) \cdot \mathbf{f}'(t-\tau, \mathbf{x}) - \mathbf{u}'(t-\tau, \mathbf{x}) \cdot \mathbf{f}(\tau, \mathbf{x})\right\} \mathrm{d}V \, \mathrm{d}\tau =$$

$$\int_{-\infty}^{\infty} \int_S \left\{P'(t-\tau, \mathbf{x})\, \mathbf{u}(\tau, \mathbf{x}) - P(\tau, \mathbf{x})\, \mathbf{u}'(t-\tau, \mathbf{x})\right\} \cdot \mathrm{d}\mathbf{S} \, \mathrm{d}\tau, \tag{4.5.12}$$

or

$$\int_V \left\{\mathbf{u}^{\mathrm{T}}(t, \mathbf{x}) * \mathbf{f}'(t, \mathbf{x}) - \mathbf{u}'^{\mathrm{T}}(t, \mathbf{x}) * \mathbf{f}(t, \mathbf{x})\right\} \mathrm{d}V =$$

$$\int_S \left\{P'(t, \mathbf{x}) * \mathbf{u}^{\mathrm{T}}(t, \mathbf{x}) - P(t, \mathbf{x}) * \mathbf{u}'^{\mathrm{T}}(t, \mathbf{x})\right\} \mathrm{d}\mathbf{S}, \tag{4.5.13}$$

using the convolution symbol (3.1.18). This is Betti's theorem for causal solutions.

4.5.1.2 Reciprocity and the Green function

We now consider two unit point forces. For simplicity we take these in coordinate directions, i.e.

$$\mathbf{f}(t, \mathbf{x}) = \hat{\mathbf{i}}_m \, \delta(t - t_{\mathrm{S}})\, \delta(\mathbf{x} - \mathbf{x}_{\mathrm{S}}) \tag{4.5.14}$$

$$\mathbf{f}'(t, \mathbf{x}) = \hat{\mathbf{i}}_n \, \delta(t - t'_{\mathrm{S}})\, \delta(\mathbf{x} - \mathbf{x}'_{\mathrm{S}}). \tag{4.5.15}$$

It is then trivial to evaluate the volume integrals in expression (4.5.12). Additionally, we assume that the boundary conditions on the surface S are homogeneous. This means that either the displacement or the pressure is zero on the surface. Then the surface integral is zero. This corresponds to *rigid* or *free* boundary conditions. Alternatively, the volume integral can be taken over the whole space, and as the surface S tends to infinity, the surface integral decays to zero (at least if a small amount of attenuation is included). With these conditions, expression (4.5.12) reduces to

$$u_n(t - t'_S, \mathbf{x}'_S) = u'_m(t - t_S, \mathbf{x}_S). \tag{4.5.16}$$

We refer to the solution for a unit, component, point force, i.e. as (4.5.14) and (4.5.15), as the *Green function*. It is convenient to introduce the notation $u_{mn}(t, \mathbf{x}; t_S, \mathbf{x}_S)$ for the m-th component of the Green function due to a source at time t_S and position $\mathbf{x}_S$ with unit force in the n-th direction, and to write the complete set of solutions as $\underline{\mathbf{u}}(t, \mathbf{x}; t_S, \mathbf{x}_S)$ where $\underline{\mathbf{u}}$ can be considered as a 3×3 matrix (each column is the vector displacement for a force component). Then equation (4.5.16) is equivalent to

$$\underline{\mathbf{u}}(t - t'_S, \mathbf{x}'_S; t_S, \mathbf{x}_S) = \underline{\mathbf{u}}^{\mathrm{T}}(t - t_S, \mathbf{x}_S; t'_S, \mathbf{x}'_S). \tag{4.5.17}$$

This symmetry is known as the (spatial) *reciprocity* of the solution. The importance of spatial reciprocity was discussed in seismology by Knopoff and Gangi (1959) although the fundamental theorems are much older, of course. It implies that the solution is the same if the source and receiver are interchanged, i.e. the direction of wave propagation is reversed. The reciprocal theorem is a very powerful and useful result but care must be taken applying it. As well as interchanging the positions of the source and receiver, the components must be interchanged, i.e. in one direction we have the force in the m-th direction and the n-th component of the displacement, and in the reciprocal experiment the force is in the n-th direction and the m-th component of displacement is used. In fact, the component is fixed at the same position in space, whether it be the force or displacement. This is illustrated in Figure 4.11.

Provided the model and boundary conditions are independent of time, the solution is independent of absolute time, and we have

$$\underline{\mathbf{u}}(t, \mathbf{x}'_S; t_S, \mathbf{x}_S) = \underline{\mathbf{u}}(t - t_S, \mathbf{x}'_S; 0, \mathbf{x}_S) = \underline{\mathbf{u}}(-t_S, \mathbf{x}'_S; -t, \mathbf{x}_S). \tag{4.5.18}$$

This result is known as temporal reciprocity. As it is essentially trivial compared with the spatial reciprocity result, we simplify the notation by taking the source at time zero. Thus result (4.5.17) is rewritten ($t_S = t'_S = 0$)

$$\boxed{\underline{\mathbf{u}}(t, \mathbf{x}_R; \mathbf{x}_S) = \underline{\mathbf{u}}^{\mathrm{T}}(t, \mathbf{x}_S; \mathbf{x}_R),} \tag{4.5.19}$$

where we have changed super/subscripts.

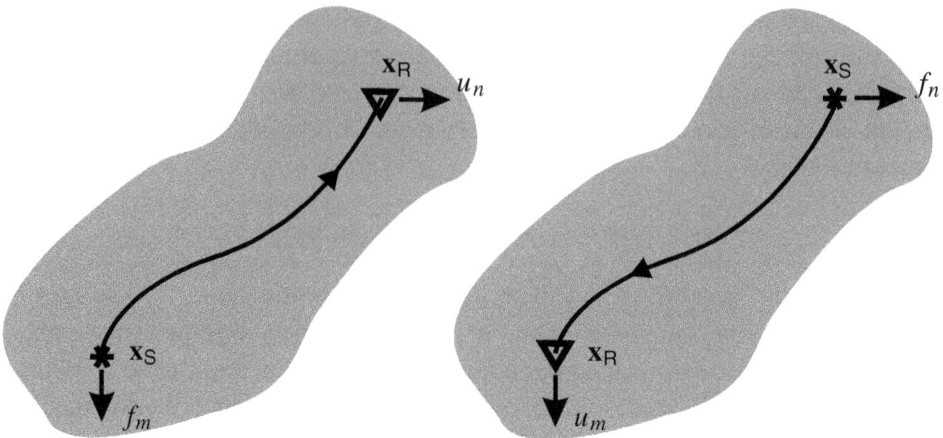

Fig. 4.11. The displacement component n due to a force in the m-th direction, and the reciprocal experiment with a source in the n-th direction and the m-th displacement component.

The Green functions are solutions of the equations of motion (4.5.1) and the constitutive relation (4.5.2) that can be written as

$$\boxed{\frac{\partial \underline{\mathbf{v}}}{\partial t} = -\frac{1}{\rho}\,\nabla \underline{P} + \frac{1}{\rho}\,\mathbf{I}\,\delta(\mathbf{x}-\mathbf{x}_S)\,\delta(t)} \tag{4.5.20}$$

$$\boxed{\frac{\partial \underline{P}}{\partial t} = -\kappa\,\nabla\cdot\underline{\mathbf{v}},} \tag{4.5.21}$$

where the Green functions are indicated by the underscore. We understand the arguments $(t, \mathbf{x}; \mathbf{x}_S)$ of the velocity and pressure Green functions. The three columns of $\mathbf{I}$ represent the three component forces, and the Green functions have an extra dimension over the physical variables. Thus $\underline{\mathbf{v}}$ has dimensions 3×3 and $\underline{P}$ is 1×3. We prefer this notation to the more usual G_{ij} as it avoids too many subscripts, using too many letters for Green functions of different variables, and keeps the physical variable obvious.

We can eliminate either $\underline{P}$ or $\underline{\mathbf{v}}$ between equations (4.5.20) and (4.5.21) to obtain a second-order wave equation. Eliminating $\underline{\mathbf{v}}$ we obtain

$$\frac{\partial^2 \underline{P}}{\partial t^2} = \kappa\,\nabla\cdot\left(\frac{1}{\rho}\,\nabla\underline{P}\right) - \kappa\,\nabla\cdot\left(\frac{1}{\rho}\,\mathbf{I}\,\delta(\mathbf{x}-\mathbf{x}_S)\right)\,\delta(t) \tag{4.5.22}$$

$$= c^2\nabla^2\underline{P} - c^2\nabla^{\mathsf{T}}\delta(\mathbf{x}-\mathbf{x}_S)\,\delta(t), \tag{4.5.23}$$

where in equation (4.5.23) we assume the density is constant (∇ is treated as a 3×1 vector). Without the source, this is the Helmholtz equation (2.1.1).

Alternatively, eliminating $\underline{P}$ we have

$$\rho \frac{\partial^2 \underline{\mathbf{u}}}{\partial t^2} = \nabla \left(\kappa \, \nabla \cdot \underline{\mathbf{u}} \right) + \mathbf{I} \, \delta(\mathbf{x} - \mathbf{x}_\mathrm{S}) \, \delta(t), \tag{4.5.24}$$

where $\underline{\mathbf{u}}$ is the particle displacement Green function. We prefer this form of the wave equation as the source is simpler and it is analogous to the Navier wave equation for elastic waves (4.5.47). These equations, (4.5.22), (4.5.23) and (4.5.24), are so easily rewritten in the frequency domain that we have not included them explicitly.

4.5.1.3 Representation theorem

We now consider the case where the Green function is known, and we require the solution for a more general source. In expression (4.5.12) we take $\mathbf{f}'$ as the source for the Green functions $\underline{\mathbf{u}}'$, i.e.

$$\mathbf{f}'(t, \mathbf{x}) = \hat{\mathbf{i}}_n \, \delta(\mathbf{x} - \mathbf{x}'_\mathrm{S}) \, \delta(t). \tag{4.5.25}$$

Then expression (4.5.12) reduces to

$$\begin{aligned} u_n(t, \mathbf{x}'_\mathrm{S}) = & \int_{-\infty}^{\infty} \int_V \left(\underline{\mathbf{u}}'\right)_{mn} (t - \tau, \mathbf{x}; \mathbf{x}'_\mathrm{S}) \, f_m(\tau, \mathbf{x}) \, \mathrm{d}V \, \mathrm{d}\tau \\ & + \int_{-\infty}^{\infty} \int_S \Big\{ \left(\underline{P}'\right)_n (t - \tau, \mathbf{x}; \mathbf{x}'_\mathrm{S}) \, u_j(\tau, \mathbf{x}) \\ & - P(\tau, \mathbf{x}) \left(\underline{\mathbf{u}}'\right)_{jn} (t - \tau, \mathbf{x}; \mathbf{x}'_\mathrm{S}) \Big\} \, \mathrm{d}S_j \, \mathrm{d}\tau, \end{aligned} \tag{4.5.26}$$

which can be rewritten as

$$\boxed{\begin{aligned} \mathbf{u}(t, \mathbf{x}_\mathrm{R}) = & \int_V \underline{\mathbf{u}}^\mathrm{T}(t, \mathbf{x}; \mathbf{x}_\mathrm{R}) * \mathbf{f}(t, \mathbf{x}) \, \mathrm{d}V \\ & + \int_S \Big\{ \underline{P}^\mathrm{T}(t, \mathbf{x}; \mathbf{x}_\mathrm{R}) \, \hat{\mathbf{i}}_j^\mathrm{T} * \mathbf{u}(t, \mathbf{x}) - \underline{\mathbf{u}}^\mathrm{T}(t, \mathbf{x}; \mathbf{x}_\mathrm{R}) * \hat{\mathbf{i}}_j \, P(t, \mathbf{x}) \Big\} \, \mathrm{d}S_j \end{aligned}} \tag{4.5.27}$$

(reversing the roles of $\mathbf{x}_\mathrm{S}$ and $\mathbf{x}_\mathrm{R}$), where the time integral is a convolution (as in definition (3.1.17) with notation (3.1.18)).

This is a *representation theorem*. It represents the solution at a point in the volume V as being due to the force sources $\mathbf{f}$ throughout the volume plus contributions due to the displacement $\mathbf{u}$ and pressure P on the surface S. Provided the Green function is reciprocal, which depends on the boundary conditions on the surface, these terms can be converted into a more satisfactory form where the Green function propagates from the integration point $\mathbf{x}$ to the receiver $\mathbf{x}_\mathrm{R}$ (rather than the reverse in expression (4.5.27)). The simplest result occurs when the volume V is the whole space and the surface integral can be neglected as it tends to infinity. Then expression (4.5.27) reduces to

$$\boxed{\mathbf{u}(t, \mathbf{x}_\mathrm{R}) = \int_V \underline{\mathbf{u}}(t, \mathbf{x}_\mathrm{R}; \mathbf{x}) * \mathbf{f}(t, \mathbf{x})\, \mathrm{d}V.} \tag{4.5.28}$$

The integral is over the support of the force **f**. Apart from the time convolution, a normal matrix-vector product occurs between the displacement Green function and the force. This intuitive result can also be deduced from the *linearity* of the equations of motion (4.5.3) and constitutive relations (4.5.4), which together with the time invariance of the model and boundary conditions, allow the *superposition principle* to be used to combine elementary point-source solutions (the Green functions) to model any distributed source **f**.

More generally, if the Green functions satisfy reciprocity, the representation theorem (4.5.27) can be rewritten

$$\begin{aligned}\mathbf{u}(t, \mathbf{x}_\mathrm{R}) = &\int_V \underline{\mathbf{u}}(t, \mathbf{x}_\mathrm{R}; \mathbf{x}) * \mathbf{f}(t, \mathbf{x})\, \mathrm{d}V \\ &- \int_S \left\{\kappa(\mathbf{x})\big(\underline{\mathbf{u}}\nabla\big)\,(t, \mathbf{x}_\mathrm{R}; \mathbf{x})\hat{\mathbf{i}}_j^\mathrm{T} * \mathbf{u}(t, \mathbf{x}) + \underline{\mathbf{u}}(t, \mathbf{x}_\mathrm{R}; \mathbf{x}) * \hat{\mathbf{i}}_j P(t, \mathbf{x})\right\} \mathrm{d}S_j,\end{aligned} \tag{4.5.29}$$

where the shorthand notation $\big(\underline{\mathbf{u}}\nabla\big)$ is more properly $\big(\nabla^\mathrm{T}\underline{\mathbf{u}}^\mathrm{T}\big)^\mathrm{T}$, and the gradient operator ∇ is with respect to **x**. If the Green functions satisfy reciprocity because of homogeneous boundary conditions on the surface S, then one term in the surface integral in (4.5.29) will be zero, but this is not essential as the surface S need not correspond to that for the Green function, e.g. the Green function can be for the whole space.

It is worth commenting on the units of Green functions. There are two alternatives. If we want the Green functions to have the same units as the solutions, then it is necessary to introduce a unit constant in the identity matrix in equation (4.5.20) with units $[\mathrm{MLT}^{-1}]$. The inverse of this constant then appears in equation (4.5.28). Alternatively, we can assume that the identity matrix in equation (4.5.20) has no units, when the Green functions have different units from the solutions. This is the approach we will use. For the body forces we have unit(**f**) $= [\mathrm{ML}^{-2}\mathrm{T}^{-2}]$, and from equation (4.5.28) the units of the Green function displacement are

$$\mathrm{unit}(\underline{\mathbf{u}}) = [\mathrm{M}^{-1}\mathrm{T}]. \tag{4.5.30}$$

Similarly the Green function pressure has units

$$\mathrm{unit}(\underline{P}) = [\mathrm{L}^{-2}\mathrm{T}^{-1}]. \tag{4.5.31}$$

Remembering that the positional Dirac delta in equation (4.5.20) has units [L^{-3}] and the convolution operator in equation (4.5.28) units [T], then the units of all equations balance.

For many purposes it is convenient to work in the frequency domain. Taking the Fourier transforms of equations (4.5.20) and (4.5.21), we have

$$-\mathrm{i}\omega\,\underline{\mathbf{v}} = -\frac{1}{\rho}\,\nabla\underline{P} + \frac{1}{\rho}\,\mathbf{I}\,\delta(\mathbf{x}-\mathbf{x}_\mathrm{S}) \tag{4.5.32}$$

$$-\mathrm{i}\omega\,\underline{P} = -\kappa\,\nabla\cdot\underline{\mathbf{v}}, \tag{4.5.33}$$

where the argument $(\omega, \mathbf{x}; \mathbf{x}_\mathrm{S})$ is understood. Equation (4.5.28) then becomes

$$\boxed{\mathbf{u}(\omega,\mathbf{x}_\mathrm{R}) = \int_V \underline{\mathbf{v}}(\omega,\mathbf{x}_\mathrm{R};\mathbf{x})\,\mathbf{f}(\omega,\mathbf{x})\,\mathrm{d}V.} \tag{4.5.34}$$

Remember that the Fourier transform of any variable introduces an extra unit [T] in the transformed variable. The matrix-vector product in equation (4.5.34) introduces no extra units whereas the convolution in (4.5.28) introduced an extra unit [T].

4.5.2 Anisotropic elastic waves

In elastic media, the equations of motion (4.1.20) and the constitutive relations (4.4.36) can be rewritten

$$\boxed{\frac{\partial\mathbf{v}}{\partial t} = \frac{1}{\rho}\,\frac{\partial\mathbf{t}_j}{\partial x_j} + \frac{1}{\rho}\,\mathbf{f}} \tag{4.5.35}$$

$$\boxed{\frac{\partial\mathbf{t}_j}{\partial t} = \mathbf{c}_{jk}\,\frac{\partial\mathbf{v}}{\partial x_k}.} \tag{4.5.36}$$

We proceed as with the acoustic equations, to develop Betti's theorem, reciprocity and the Green functions, and representation theorems.

4.5.2.1 Betti's theorem

As with the acoustic equations, we rewrite the equations of motion (4.5.35) and the constitutive relations (4.5.36) as

$$\rho\,\ddot{\mathbf{u}} = \frac{\partial\mathbf{t}_j}{\partial x_j} + \mathbf{f} \tag{4.5.37}$$

$$\mathbf{t}_j = \mathbf{c}_{jk}\,\frac{\partial\mathbf{u}}{\partial x_k}, \tag{4.5.38}$$

and similar equations for a second solution $\mathbf{u}'$ and $\mathbf{t}'_j$ due to a force $\mathbf{f}'$. A similar volume and surface integral to (4.5.7) reduces to a symmetrical form, i.e.

$$\int_V (\mathbf{f} - \rho\,\ddot{\mathbf{u}}) \cdot \mathbf{u}'\,\mathrm{d}V + \int_S \mathbf{t}_j \cdot \mathbf{u}'\,\mathrm{d}S_j = \int_V \frac{\partial \mathbf{u}^{\mathrm{T}}}{\partial x_k} \mathbf{c}_{kj} \frac{\partial \mathbf{u}'}{\partial x_j}\,\mathrm{d}V, \qquad (4.5.39)$$

so we obtain Betti's theorem

$$\int_V (\mathbf{f} - \rho\,\ddot{\mathbf{u}}) \cdot \mathbf{u}'\,\mathrm{d}V + \int_S \mathbf{t}_j \cdot \mathbf{u}'\,\mathrm{d}S_j = \int_V \left(\mathbf{f}' - \rho\,\ddot{\mathbf{u}}'\right) \cdot \mathbf{u}\,\mathrm{d}V + \int_S \mathbf{t}'_j \cdot \mathbf{u}\,\mathrm{d}S_j. \qquad (4.5.40)$$

Result (4.5.11) still applies, so this can be rewritten (cf. result (4.5.12))

$$\int_{-\infty}^{\infty} \int_V \left\{\mathbf{u}(\tau, \mathbf{x}) \cdot \mathbf{f}'(t - \tau, \mathbf{x}) - \mathbf{u}'(t - \tau, \mathbf{x}) \cdot \mathbf{f}(\tau, \mathbf{x})\right\}\,\mathrm{d}V\,\mathrm{d}\tau =$$

$$\int_{-\infty}^{\infty} \int_S \left\{\mathbf{u}'(t - \tau, \mathbf{x}) \cdot \mathbf{t}_j(\tau, \mathbf{x}) - \mathbf{u}(\tau, \mathbf{x}) \cdot \mathbf{t}'_j(t - \tau, \mathbf{x})\right\}\,\mathrm{d}S_j\,\mathrm{d}\tau, \qquad (4.5.41)$$

when the sources and solutions are causal, i.e. quiescent before a certain time.

4.5.2.2 *Reciprocity and Green functions*

The reciprocity results for acoustic waves are identical for elastic waves provided homogeneous boundary conditions apply on the surface S. Thus result (4.5.19) still applies

$$\boxed{\underline{\mathbf{u}}(t, \mathbf{x}_{\mathrm{R}}; \mathbf{x}_{\mathrm{S}}) = \underline{\mathbf{u}}^{\mathrm{T}}(t, \mathbf{x}_{\mathrm{S}}; \mathbf{x}_{\mathrm{R}}),} \qquad (4.5.42)$$

where the Green functions are solutions of the equations

$$\boxed{\frac{\partial \underline{\mathbf{v}}}{\partial t} = \frac{1}{\rho} \frac{\partial \underline{\mathbf{t}}_j}{\partial x_j} + \frac{1}{\rho} \mathbf{I}\,\delta(\mathbf{x} - \mathbf{x}_{\mathrm{S}})\,\delta(t)} \qquad (4.5.43)$$

$$\boxed{\frac{\partial \underline{\mathbf{t}}_j}{\partial t} = \mathbf{c}_{jk} \frac{\partial \underline{\mathbf{v}}}{\partial x_k}.} \qquad (4.5.44)$$

In the frequency domain, these equations are

$$\boxed{-\mathrm{i}\,\omega\,\underline{\mathbf{v}} = \frac{1}{\rho} \frac{\partial \underline{\mathbf{t}}_j}{\partial x_j} + \frac{1}{\rho} \mathbf{I}\,\delta(\mathbf{x} - \mathbf{x}_{\mathrm{S}})} \qquad (4.5.45)$$

$$\boxed{-\mathrm{i}\,\omega\,\underline{\mathbf{t}}_j = \mathbf{c}_{jk} \frac{\partial \underline{\mathbf{v}}}{\partial x_k},} \qquad (4.5.46)$$

or eliminating the tractions between (4.5.43) and (4.5.44) we obtain

$$\rho \frac{\partial^2 \underline{\mathbf{u}}}{\partial t^2} = \frac{\partial}{\partial x_j}\left(\mathbf{c}_{jk}\frac{\partial \underline{\mathbf{u}}}{\partial x_k}\right) + \mathbf{I}\,\delta(\mathbf{x}-\mathbf{x}_{\mathrm{S}})\,\delta(t). \tag{4.5.47}$$

This is known as the *Navier wave equation.*

4.5.2.3 Representation Theorem

Using the Green function $\underline{\mathbf{u}}'$ with a source (4.5.25), the representation theorem for elastic waves is (cf. equation (4.5.26))

$$\begin{aligned} u_n(t, \mathbf{x}'_{\mathrm{S}}) = & \int_{-\infty}^{\infty}\int_V \left(\underline{\mathbf{u}}'\right)_{mn}(t-\tau, \mathbf{x}; \mathbf{x}'_{\mathrm{S}})\, f_m(\tau, \mathbf{x})\,\mathrm{d}V\,\mathrm{d}\tau \\ & + \int_{-\infty}^{\infty}\int_S \Big\{ \left(\mathbf{t}_j\right)_m(\tau, \mathbf{x})\, \left(\underline{\mathbf{u}}'\right)_{mn}(t-\tau, \mathbf{x}; \mathbf{x}'_{\mathrm{S}}) \\ & - \left(\underline{\mathbf{t}}'_j\right)_{mn}(t-\tau, \mathbf{x}; \mathbf{x}'_{\mathrm{S}})\, u_m(\tau, \mathbf{x})\Big\}\,\mathrm{d}S_j\,\mathrm{d}\tau, \end{aligned} \tag{4.5.48}$$

which again can be rewritten as

$$\boxed{\begin{aligned} \mathbf{u}(t, \mathbf{x}_{\mathrm{R}}) = & \int_V \underline{\mathbf{u}}^{\mathrm{T}}(t, \mathbf{x}; \mathbf{x}_{\mathrm{R}}) * \mathbf{f}(t, \mathbf{x})\,\mathrm{d}V \\ & + \int_S \Big\{ \underline{\mathbf{u}}^{\mathrm{T}}(t, \mathbf{x}_{\mathrm{R}}; \mathbf{x}) * \mathbf{t}_j(t, \mathbf{x}) - \underline{\mathbf{t}}_j^{\mathrm{T}}(t, \mathbf{x}; \mathbf{x}_{\mathrm{R}}) * \mathbf{u}(t, \mathbf{x})\Big\}\,\mathrm{d}S_j. \end{aligned}} \tag{4.5.49}$$

Again, in the appropriate circumstances, this reduces to the superposition result (4.5.28)

$$\boxed{\mathbf{u}(t, \mathbf{x}_{\mathrm{R}}) = \int_V \underline{\mathbf{u}}(t, \mathbf{x}_{\mathrm{R}}; \mathbf{x}) * \mathbf{f}(t, \mathbf{x})\,\mathrm{d}V,} \tag{4.5.50}$$

or more generally

$$\begin{aligned} \mathbf{u}(t, \mathbf{x}_{\mathrm{R}}) = & \int_V \underline{\mathbf{u}}(t, \mathbf{x}_{\mathrm{R}}; \mathbf{x}) * \mathbf{f}(t, \mathbf{x})\,\mathrm{d}V \\ & + \int_S \left\{ \underline{\mathbf{u}}(t, \mathbf{x}_{\mathrm{R}}; \mathbf{x}) * \mathbf{t}_j(t, \mathbf{x}) - \frac{\partial \underline{\mathbf{u}}(t, \mathbf{x}_{\mathrm{R}}; \mathbf{x})}{\partial x_k}\,\mathbf{c}_{jk}(\mathbf{x}) * \mathbf{u}(t, \mathbf{x})\right\}\,\mathrm{d}S_j. \end{aligned} \tag{4.5.51}$$

4.5.3 Isotropic elastic waves

In homogeneous, isotropic media, substituting (4.4.55) and (4.4.56), it is straightforward to show that equation (4.5.47) can be written

$$\rho \frac{\partial^2 \underline{\mathbf{u}}}{\partial t^2} = (\lambda + 2\mu)\nabla(\nabla \cdot \underline{\mathbf{u}}) - \mu\nabla \times (\nabla \times \underline{\mathbf{u}}) + \mathbf{I}\,\delta(\mathbf{x} - \mathbf{x}_S)\,\delta(t). \qquad (4.5.52)$$

This is proved by simple expansion of the gradient operators in cartesian coordinates, but as the equation is a vector equation, it is valid independent of the coordinate system. We have not bothered to write equation (4.5.52) in a form valid in inhomogeneous media (i.e. retaining derivatives of λ and μ), as in inhomogeneous media it is normally simpler to work with the first-order coupled equations of motion (4.5.35) and constitution relation (4.5.36). Equation (4.5.52) is sometimes rewritten using the identity

$$\nabla^2 \underline{\mathbf{u}} = \nabla(\nabla \cdot \underline{\mathbf{u}}) - \nabla \times (\nabla \times \underline{\mathbf{u}}), \qquad (4.5.53)$$

but this is only true in cartesian coordinates.

4.5.4 Plane isotropic, elastic waves

Let us confirm that the Navier wave equation for isotropic, homogeneous elastic media supports plane waves as discussed in Chapter 2.

As the medium is homogeneous and isotropic, we can take the direction of propagation as x_1 without loss in generality. We look for a plane-wave solution without body forces in the form (cf. equation (2.1.11))

$$\mathbf{v}(t, x_1) = \mathbf{a}\, f\left(t - \frac{x_1}{c}\right), \qquad (4.5.54)$$

where $\mathbf{a}$ is a constant vector and $f(t)$ is an arbitrary pulse shape (cf. Figure 2.4). Differentiating we have

$$\frac{\partial v_i}{\partial t} = a_i f'\left(t - \frac{x_1}{c}\right) \qquad (4.5.55)$$

$$\frac{\partial v_i}{\partial x_1} = -\frac{a_i}{c} f'\left(t - \frac{x_1}{c}\right), \qquad (4.5.56)$$

where $f'(t)$ indicates differentiation with respect to the argument. The important feature is that derivatives with respect to t and x_1 both contain the same function $f'(t)$ (and derivatives with respect to x_2 and x_3 are zero). Then the wave equation

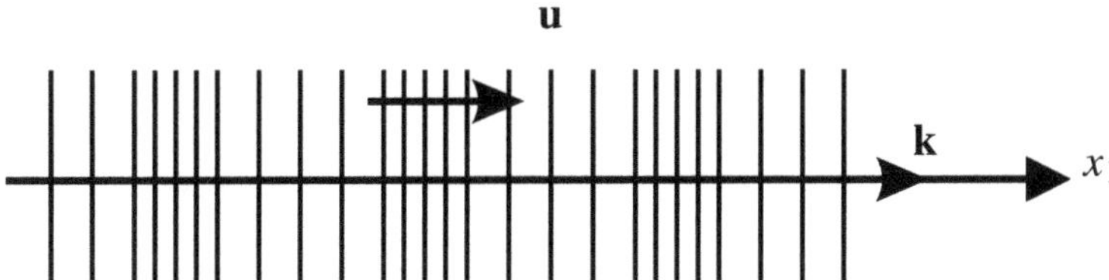

Fig. 4.12. The disturbance of a uniform grid as a P wave propagates in the x_1 direction.

becomes

$$\rho\, a_1 f'' = (\lambda + 2\mu)\frac{a_1}{c^2} f'' \quad (4.5.57)$$

$$\rho\, a_2 f'' = \mu \frac{a_2}{c^2} f'' \quad (4.5.58)$$

$$\rho\, a_3 f'' = \mu \frac{a_3}{c^2} f''. \quad (4.5.59)$$

The equations for the a_i's are independent so there are three possible solutions ($f'' = 0$ is not an interesting solution and it can be cancelled).

If component $a_1 \neq 0$ in equation (4.5.57), then

$$\boxed{c = \pm\sqrt{\frac{\lambda + 2\mu}{\rho}} = \pm\alpha,} \quad (4.5.60)$$

say. Then the other components must be zero, $a_2 = a_3 = 0$, if the equations (4.5.58) and (4.5.59) are to be consistent. Thus this is a *longitudinal* wave with velocity α. It is called a P (primary, push–pull, ...) wave and is illustrated in Figure 4.12.

If component $a_2 \neq 0$ in (4.5.58), then

$$\boxed{c = \pm\sqrt{\frac{\mu}{\rho}} = \pm\beta,} \quad (4.5.61)$$

say. Then component $a_1 = 0$ in equation (4.5.57) to be consistent, and component a_3 is arbitrary in equation (4.5.59). Thus this is a *transverse* wave with velocity β and two independent transverse components. It is called a S (secondary, sideways, shear, ...) wave and is illustrated in Figure 4.13.

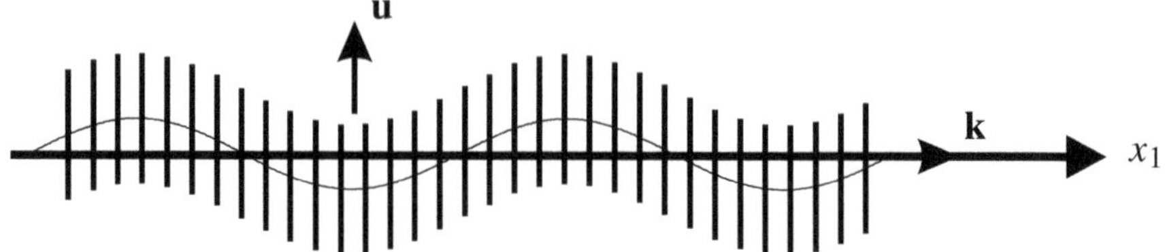

Fig. 4.13. The disturbance of a uniform grid as a S wave propagates in the x_1 direction.

4.5.5 Isotropic, homogeneous Green function

In Chapter 2 we also discussed spherical waves from a point source. Let us establish how such waves are excited by finding the Green function solution of equation (4.5.52).

In the frequency domain, equation (4.5.52) becomes

$$(\lambda + 2\mu)\nabla(\nabla \cdot \underline{\mathbf{u}}) - \mu\nabla \times (\nabla \times \underline{\mathbf{u}}) + \rho\,\omega^2 \underline{\mathbf{u}} = -\mathbf{I}\,\delta(\mathbf{x} - \mathbf{x}_S). \quad (4.5.62)$$

In order to find solutions of this equation, the important identity is

$$\left(\nabla^2 + k^2\right)\left(\frac{e^{ikR}}{R}\right) = -4\pi\,\delta(\mathbf{x} - \mathbf{x}_S), \quad (4.5.63)$$

i.e. the spherical wave due to a point source, where R is the length of the radius vector (2.2.2). Letting $k = 0$, we can write the spatial delta function in terms of the Laplacian of R^{-1}, which in turn can be expanded using identity (4.5.53), i.e.

$$\mathbf{I}\,\delta(\mathbf{x} - \mathbf{x}_S) = -\frac{1}{4\pi}\left\{\nabla\left(\nabla \cdot \frac{\mathbf{I}}{R}\right) - \nabla \times \left(\nabla \times \frac{\mathbf{I}}{R}\right)\right\}. \quad (4.5.64)$$

If the solution of equation (4.5.62) can be written

$$\underline{\mathbf{u}} = \nabla(\nabla \cdot (\mathbf{I}\,\phi)) - \nabla \times (\nabla \times (\mathbf{I}\,\psi)), \quad (4.5.65)$$

then this satisfies equation (4.5.62) provided ϕ and ψ satisfy

$$\alpha^2\nabla^2\phi + \omega^2\phi = \frac{1}{4\pi\rho R} \quad (4.5.66)$$

$$\beta^2\nabla^2\psi + \omega^2\psi = \frac{1}{4\pi\rho R}. \quad (4.5.67)$$

Solutions of these equations are (using identity (4.5.63) with $k = 0, \omega/\alpha$ and ω/β)

$$\phi = \frac{1}{4\pi\rho\,\omega^2}\left(\frac{1 - e^{i\omega R/\alpha}}{R}\right) \quad (4.5.68)$$

$$\psi = \frac{1}{4\pi\rho\,\omega^2}\left(\frac{1 - e^{i\omega R/\beta}}{R}\right). \quad (4.5.69)$$

Substituting in expression (4.5.65), after some manipulation the spectral Green solution is

$$\underline{\mathbf{u}}(\omega, \mathbf{x}_\mathrm{R}; \mathbf{x}_\mathrm{S}) = \frac{1}{4\pi\rho\,\omega^2}\left\{\nabla\times\left(\nabla\times\left(\mathbf{I}\frac{\mathrm{e}^{\mathrm{i}\omega R/\beta}}{R}\right)\right) - \nabla\left(\nabla\cdot\left(\mathbf{I}\frac{\mathrm{e}^{\mathrm{i}\omega R/\alpha}}{R}\right)\right)\right\}. \tag{4.5.70}$$

In the time domain this can be written

$$\begin{aligned}\underline{\mathbf{u}}(t, \mathbf{x}_\mathrm{R}; \mathbf{x}_\mathrm{S}) = {} & \frac{\delta(t - R/\alpha)}{4\pi\rho\,\alpha^2 R}\hat{\mathbf{r}}\,\hat{\mathbf{r}}^\mathrm{T} + \frac{\delta(t - R/\beta)}{4\pi\rho\,\beta^2 R}(\mathbf{I} - \hat{\mathbf{r}}\,\hat{\mathbf{r}}^\mathrm{T}) \\ & + \frac{t}{4\pi\rho R^3}\{H(t - R/\alpha) - H(t - R/\beta)\}\,(3\hat{\mathbf{r}}\,\hat{\mathbf{r}}^\mathrm{T} - \mathbf{I}),\end{aligned} \tag{4.5.71}$$

where $\hat{\mathbf{r}}$ is the unit direction vector (2.2.3). This solution for a point force source, now called the *dyadic Green function*, was first obtained by Stokes (1851). More complete and detailed proofs can be found in many textbooks, e.g. Hudson (1980, Section 2.4), Aki and Richards (1980, 2002, Chapter 4), Ben-Menahem and Singh (1984, Chapter 4), Pujol (2003, Chapter 9), etc. It is readily confirmed that the units of equation (4.5.71) agree with result (4.5.30), remembering that unit$(\delta(t)) = [\mathrm{T}^{-1}]$.

The first terms in solution (4.5.71) decay as R^{-1}, i.e. the expected *spherical spreading*. The discontinuities of the final term in (4.5.71) decay more rapidly as R^{-2} and so in the far-field we can use just

$$\underline{\mathbf{u}}(t, \mathbf{x}_\mathrm{R}; \mathbf{x}_\mathrm{S}) \simeq \frac{\delta(t - R/\alpha)}{4\pi\rho\,\alpha^2 R}\hat{\mathbf{r}}\,\hat{\mathbf{r}}^\mathrm{T} + \frac{\delta(t - R/\beta)}{4\pi\rho\,\beta^2 R}(\mathbf{I} - \hat{\mathbf{r}}\,\hat{\mathbf{r}}^\mathrm{T}), \tag{4.5.72}$$

the *far-field approximation*. The final term in solution (4.5.71), the *near-field term*, decays more rapidly and is lower in frequency, being the time integral of the discontinuity in the far-field term. It also has different polarization and excitation than the near-field term. It is normally ignored but, of course, in some circumstances may be important.

The factors $\hat{\mathbf{r}}\,\hat{\mathbf{r}}^\mathrm{T}$ and $\mathbf{I} - \hat{\mathbf{r}}\,\hat{\mathbf{r}}^\mathrm{T}$ in approximation (4.5.72) describe the *radiation patterns* for P and S waves in the far-field term. These are illustrated in Figures 4.14 and 4.15. If we consider the unit, basis vectors $\hat{\mathbf{r}}$, $\hat{\boldsymbol{\theta}}$ and $\hat{\boldsymbol{\phi}}$ in spherical coordinates, then

$$\mathbf{I} = \hat{\mathbf{r}}\,\hat{\mathbf{r}}^\mathrm{T} + \hat{\boldsymbol{\theta}}\,\hat{\boldsymbol{\theta}}^\mathrm{T} + \hat{\boldsymbol{\phi}}\,\hat{\boldsymbol{\phi}}^\mathrm{T} \tag{4.5.73}$$

(this identity is obvious as applying the right-hand side to any vector resolves the vector into its spherical components, or it can be proved by expanding the basis vectors in their cartesian components). Thus the radiation pattern for S waves can

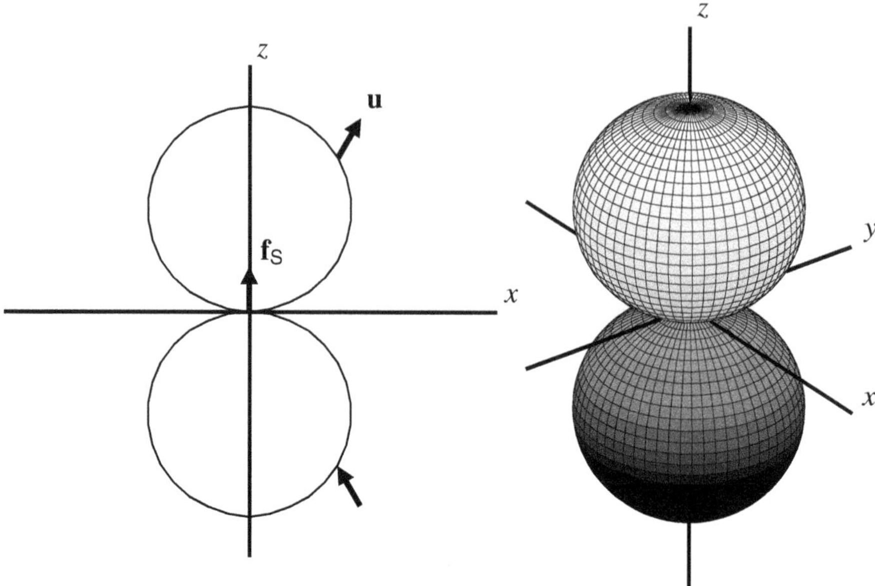

Fig. 4.14. The radiation patterns for P waves due to a point force, f_z. The plots show the relative amplitudes in different directions. On the left is a cross-section of the pattern in a plane containing the force, showing the force and polarization directions. On the right is a view of the three-dimensional radiation pattern. The arrows at the centre of the left plot represent the force, and the arrows on the perimeter, the polarization directions.

be rewritten

$$\mathbf{I} - \hat{\mathbf{r}}\,\hat{\mathbf{r}}^{\mathrm{T}} = \hat{\boldsymbol{\theta}}\,\hat{\boldsymbol{\theta}}^{\mathrm{T}} + \hat{\boldsymbol{\phi}}\,\hat{\boldsymbol{\phi}}^{\mathrm{T}}, \tag{4.5.74}$$

where $\hat{\boldsymbol{\theta}}$ and $\hat{\boldsymbol{\phi}}$ are in the transverse directions.

4.5.5.1 *Point force, Green dyadic*

Using expression (4.5.72) with identity (4.5.74) for the far-field radiation, we obtain the important dyadic result that any wave in it can be written

$$\boxed{\underline{\mathbf{u}}(t, \mathbf{x}_R; \mathbf{x}_S) \simeq \frac{\delta(t - R/c)}{4\pi\rho\, c^2 R}\, \hat{\mathbf{g}}\,\hat{\mathbf{g}}^{\mathrm{T}},} \tag{4.5.75}$$

where $c = \alpha$ or β is the P or S velocity, and $\hat{\mathbf{g}}$ is the corresponding wave polarization (the sign of the polarization doesn't matter provided it is defined in a consistent manner at the source and receiver).

For a point force

$$\mathbf{f}(t, \mathbf{x}) = \mathbf{f}_S\, f(t)\, \delta(\mathbf{x} - \mathbf{x}_S), \tag{4.5.76}$$

the solution (4.5.28) for each wave becomes

$$\mathbf{u}(t, \mathbf{x}_\mathrm{R}) \simeq \frac{f(t - R/c)}{4\pi\rho\, c^2 R}\, \hat{\mathbf{g}}\, \hat{\mathbf{g}}^\mathrm{T}\, \mathbf{f}_\mathrm{S}, \tag{4.5.77}$$

where $R = |\mathbf{x} - \mathbf{x}_\mathrm{S}|$. The factor $\hat{\mathbf{g}}^\mathrm{T}\, \mathbf{f}_\mathrm{S}$ is known as the *source radiation pattern* as it describes the directional behaviour of the displacement magnitude $|\mathbf{u}|$.

For a P wave, the polarization is longitudinal and

$$\hat{\mathbf{g}}_P = \hat{\mathbf{r}} = \begin{pmatrix} \sin\theta\, \cos\phi \\ \sin\theta\, \sin\phi \\ \cos\theta \end{pmatrix}, \tag{4.5.78}$$

where θ and ϕ are the polar co-latitude and longitude, i.e. the angles from the z and x axes. The radiation pattern is given by

$$\hat{\mathbf{g}}^\mathrm{T}\, \mathbf{f}_\mathrm{S} = \big(f_x\, \cos\phi + f_y\, \sin\phi\big) \sin\theta + f_z\, \cos\theta. \tag{4.5.79}$$

The amplitude varies as the cosine of the angle from the force, i.e. due to the factor $\hat{\mathbf{g}}^\mathrm{T}\, \mathbf{f}_\mathrm{S}$, with a maximum on the axis of the force, positive in the force direction and negative in the opposite direction. This is illustrated in Figure 4.14.

For an S wave, the polarization is transverse. The amplitude varies as the sine of the angle from the force, i.e. due to the factor $\hat{\mathbf{g}}^\mathrm{T}\, \mathbf{f}_\mathrm{S}$, with zero on the axis of the force, and maximum in the force direction on the equator. This is illustrated in Figure 4.15. The polarization of an S wave is in the plane perpendicular to the

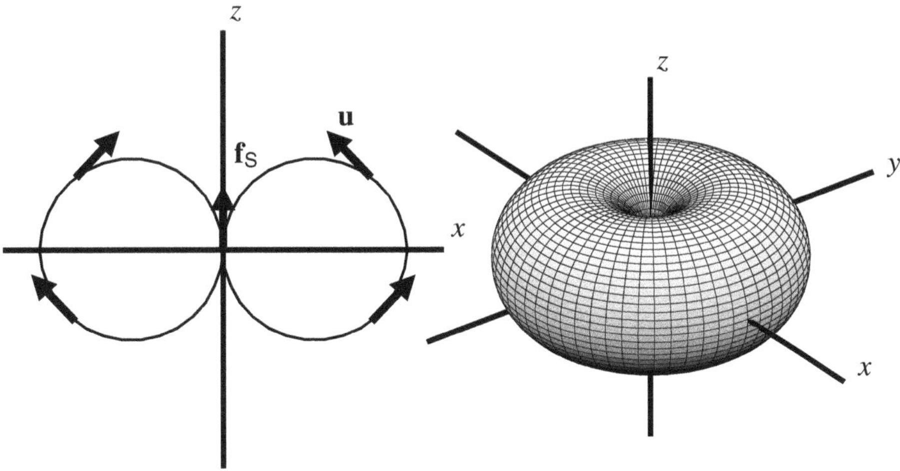

Fig. 4.15. The radiation patterns for S waves due to a point force, f_z, as Figure 4.14.

propagation, $\hat{\mathbf{r}}$. It is convenient to decompose this into a horizontal component, e.g.

$$\hat{\mathbf{g}}_H = \hat{\boldsymbol{\phi}} = \begin{pmatrix} -\sin\phi \\ \cos\phi \\ 0 \end{pmatrix}, \tag{4.5.80}$$

if the z axis is vertical, and

$$\hat{\mathbf{g}}_V = \hat{\boldsymbol{\theta}} = \begin{pmatrix} \cos\theta\,\cos\phi \\ \cos\theta\,\sin\phi \\ -\sin\theta \end{pmatrix}, \tag{4.5.81}$$

orthogonal in the vertical plane. The nomenclature *SV* and *SH* is standard if the x–y plane is horizontal. Note that $\hat{\mathbf{g}}_P$, $\hat{\mathbf{g}}_V$ and $\hat{\mathbf{g}}_H$ are orthonormal and $\hat{\mathbf{g}}_P = \hat{\mathbf{r}}$, i.e. the propagation direction.

The source radiation pattern for *SH* waves is

$$\hat{\mathbf{g}}_H^{\mathrm{T}}\,\mathbf{f}_{\mathrm{S}} = f_y\,\cos\phi - f_x\,\sin\phi, \tag{4.5.82}$$

and for *SV* waves

$$\hat{\mathbf{g}}_V^{\mathrm{T}}\,\mathbf{f}_{\mathrm{S}} = \left(f_x\,\cos\phi + f_y\,\sin\phi\right)\cos\theta - f_z\,\sin\theta. \tag{4.5.83}$$

If the force is in the z direction ($f_x = f_y = 0$), then only the *SV* wave is excited (as expected from the axial symmetry).

Thus we have established that in a homogeneous, isotropic medium, two types of plane waves exist with velocities α and β ($\alpha > \beta$). In Chapter 5 we discuss how these waves propagate in inhomogeneous media.

We have not introduced the *seismic potentials* (except implicitly in equation (4.5.65)), e.g. $\mathbf{u} = \nabla\phi + \nabla \times \boldsymbol{\psi}$, as we will not find them useful. Only in the simplest problems can they be used to represent elastic waves exactly. In contrast to electromagnetic theory, where the potentials have physical significance and are used directly in boundary conditions, etc., in elastodynamics they have no direct physical significance and introduce an extra derivative into the boundary conditions.

4.5.5.2 Line force, Green dyadic

If, in three dimensions, we consider a uniform distribution of point sources along a line, then we generate the equivalent two-dimensional problem. We can either consider the problem as a point source in two dimensions or a line source in three dimensions, where the solution is translationally invariant in the direction of the line. In a uniform medium, the wavefronts will be cylindrical rather than spherical. We may construct the cylindrical solutions by superposition of point sources along the line (see Exercise 4.7).

We consider cylindrical polar coordinates, (r, ϕ, z), where the sources are distributed along the z axis. We use unit, basis vectors $\hat{\mathbf{r}}$, $\hat{\phi}$ and $\hat{\mathbf{z}}$ (remember that here $\hat{\mathbf{r}}$ and $\hat{\phi}$ lie in the plane perpendicular to the z axis). The algebra to integrate point sources along the axis is straightforward but is omitted for brevity (see, for instance, Hudson, 1980, Section 2.5). The two-dimensional, dyadic Green function corresponding to the three-dimensional result (4.5.71) is

$$
\begin{aligned}
\underline{\mathbf{u}}(t, \mathbf{x}_{\mathrm{R}}; \mathbf{x}_{\mathrm{S}}) = {} & \frac{H(t - r/\alpha)}{2\pi\rho\,\alpha^2 \left(t^2 - r^2/\alpha^2\right)^{1/2}} \hat{\mathbf{r}}\hat{\mathbf{r}}^{\mathrm{T}} \\
& + \frac{H(t - r/\beta)}{2\pi\rho\,\beta^2 \left(t^2 - r^2/\beta^2\right)^{1/2}} (\hat{\phi}\hat{\phi}^{\mathrm{T}} + \hat{\mathbf{z}}\hat{\mathbf{z}}^{\mathrm{T}}) \\
& + \frac{1}{2\pi\rho r^2} \left\{ H(t - r/\alpha) \left(t^2 - r^2/\alpha^2\right)^{1/2} \right. \\
& \left. - H(t - r/\beta) \left(t^2 - r^2/\beta^2\right)^{1/2} \right\} (\hat{\mathbf{r}}\hat{\mathbf{r}}^{\mathrm{T}} - \hat{\phi}\hat{\phi}^{\mathrm{T}})
\end{aligned}
\tag{4.5.84}
$$

(see equations (2.34) and (2.36) in Hudson, 1980). The cylindrical radius from the line source is r.

Again the first terms in expression (4.5.84) contain the far-field, radiation terms. They are no longer delta function impulses, but contain a tail $\left(t^2 - r^2/c^2\right)^{-1/2}$, where c is either the P or S velocity, α or β. Simulating a line source in three dimensions with a distribution of point sources along the line, it is clear that this tail is due to sources out of the symmetry plane of propagation (see Figure 4.16). This tail can also be obtained from the inverse Fourier transform of the Hankel function (B.4.13) which represent the cylindrical waves in the frequency domain.

The singularity at the geometrical arrival, $t = r/c$, can be approximated. The three far-field terms in (4.5.84) can all be written as (cf. result (4.5.75))

$$
\boxed{\underline{\mathbf{u}}(t, \mathbf{x}_{\mathrm{R}}; \mathbf{x}_{\mathrm{S}}) \simeq \frac{\lambda(t - r/c)}{2^{3/2}\pi\rho\, c^{3/2} r^{1/2}} \hat{\mathbf{g}}\hat{\mathbf{g}}^{\mathrm{T}},}
\tag{4.5.85}
$$

where the singularity, $\lambda(t)$, is defined in (B.2.1). Compared with the three-dimensional Green function (4.5.75), the two-dimensional function (4.5.85) only spreads cylindrically and the decay is as $r^{-1/2}$. It is straightforward to confirm that the units of this expression are unit($\underline{\mathbf{u}}$) = [M^{-1}LT]. (For physical reality, we consider a line source in three dimensions rather than a true two-dimensional problem. The density ρ is still the three-dimensional density, and expression (4.5.85) must be convolved with the force per unit volume. Alternatively, we might consider the true two-dimensional problem and replace the density by the mass per *unit area*

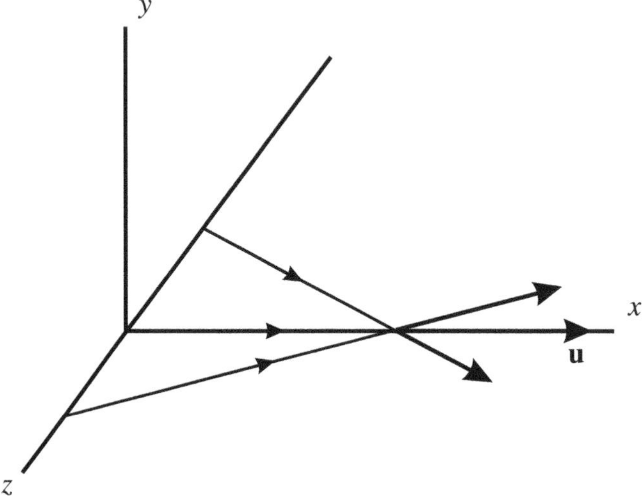

Fig. 4.16. Point sources along a line, the z axis, contributing to the signal in two-dimensional wave propagation.

and the body force by the force per *unit area*. Then unit($\underline{\mathbf{u}}$) = $[\mathrm{M}^{-1}\mathrm{T}]$ as in three dimensions.) The discontinuities of near-field terms in expression (4.5.84) decay more rapidly as $r^{-3/2}$ and are lower frequency (the integral of $\lambda(t - r/c)$). These are contained in the final terms in expression (4.5.84) and in the difference between expressions (4.5.85) and the true tail.

4.6 Stress glut source

In the previous section, we have investigated the solution due to a point force source (4.5.52). Extended force sources can be generated by spatial and/or temporal convolution. While a force is the simplest source, ideal from a mathematical point of view, it is often not physically realistic. The Earth is a closed system, and isolated forces cannot exist. In a closed system there must be an equal and opposite reaction. Earthquakes, explosions, fractures, etc. cannot be represented by a force. Only some surface sources, e.g. vibrators (the reaction force acts against the mass of the vibrator truck, which need not be considered part of the Earth), can be represented as force sources, at least approximately.

In a closed system, no unbalanced forces are possible so the equation of motion (4.5.35) must reduce to

$$\frac{\partial \mathbf{v}}{\partial t} = \frac{1}{\rho}\frac{\partial \mathbf{t}_j}{\partial x_j}. \tag{4.6.1}$$

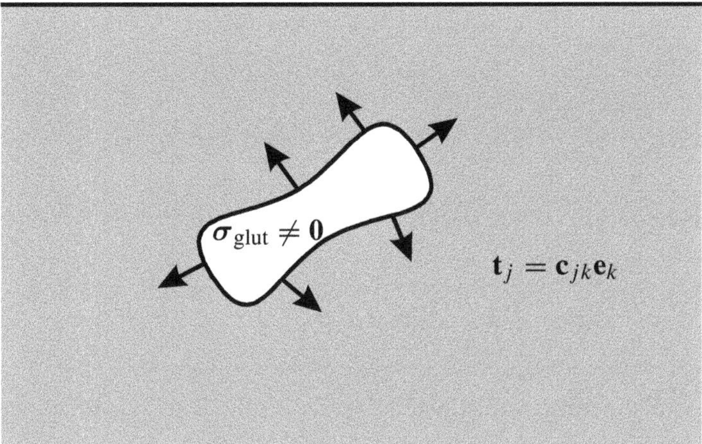

Fig. 4.17. The model with a source region where the stress glut is non-zero, surrounded by a region with zero stress glut where the model stress, from the constitutive law, is the true stress.

The only solution of this and the constitutive law (4.5.36) is identically zero! What causes seismic waves? Equation (4.6.1) is unlikely to be wrong – it is just a generalization of Newton's second law. The constitutive law (4.5.36) is, however, only an empirical relationship. Backus and Mulcahy (1976*a*, *b*) introduced an important new concept for seismic sources. The source, e.g. earthquake, explosion, etc., is a localized, transient failure of the constitutive law. In general the constitutive relation (4.5.36) can be taken to be valid, but 'in' the source it fails. We call the difference between the model stress that satisfies the constitutive law and the true stress the *stress glut*. Thus

$$\boxed{\boldsymbol{\sigma}_{\rm glut} = \boldsymbol{\sigma}_{\rm model} - \boldsymbol{\sigma}_{\rm true}.} \tag{4.6.2}$$

The model stress, $\boldsymbol{\sigma}_{\rm model}$, satisfies the constitutive law (4.5.36). The true stress, $\boldsymbol{\sigma}_{\rm true}$, satisfies the equation of motion (4.6.1). The stress glut, $\boldsymbol{\sigma}_{\rm glut}$, is non-zero in an explosion or in the fault zone of an earthquake (Figure 4.17). Substituting definition (4.6.2) in equation (4.6.1), we obtain

$$\frac{\partial \mathbf{v}}{\partial t} = \frac{1}{\rho}\frac{\partial \mathbf{t}_{{\rm true}\,j}}{\partial x_j} \tag{4.6.3}$$

$$= \frac{1}{\rho}\frac{\partial \mathbf{t}_{{\rm model}\,j}}{\partial x_j} - \frac{1}{\rho}\nabla \cdot \boldsymbol{\sigma}_{\rm glut}, \tag{4.6.4}$$

i.e. $\boxed{\mathbf{f} = -\nabla \cdot \boldsymbol{\sigma}_{\text{glut}}}$ is the *equivalent body-force density* for the stress glut (in addition there may be $\mathbf{t} = \hat{\mathbf{n}} \cdot \boldsymbol{\sigma}_{\text{glut}}$ *equivalent surface-force density* sources, but we do not consider this explicitly).

For a small source, in terms of the wavelength and propagation distance, it is convenient to consider the volume average of the stress glut change in the source, i.e. the volume and time integral of the rate of change of stress glut over the support of the source

$$\boxed{\mathbf{M} = \iint \frac{\partial\, \boldsymbol{\sigma}_{\text{glut}}}{\partial t}\, \mathrm{d}V\, \mathrm{d}t = \int \left[\boldsymbol{\sigma}_{\text{glut}}\right] \mathrm{d}V,} \tag{4.6.5}$$

where $\left[\,\boldsymbol{\sigma}_{\text{glut}}\right]$ is the temporal saltus of the stress glut. This is known as the *moment tensor*. The name arises as it has the units of force times distance, i.e. unit($\mathbf{M}$) $=$ [$\mathrm{ML^2T^{-2}}$]. Assuming the source can be represented as an infinitesimal point, we have the glut rate

$$\frac{\partial\, \boldsymbol{\sigma}_{\text{glut}}}{\partial t} = \mathbf{M}_{\mathrm{S}}\, \delta(\mathbf{x} - \mathbf{x}_{\mathrm{S}})\, \delta(t - t_{\mathrm{S}}), \tag{4.6.6}$$

and the body force equivalent is

$$\boxed{\mathbf{f} = -\mathbf{M}_{\mathrm{S}} \nabla \delta(\mathbf{x} - \mathbf{x}_{\mathrm{S}})\, H(t - t_{\mathrm{S}}).} \tag{4.6.7}$$

The volume integral (4.5.50) is then evaluated using the standard result

$$\int \nabla \delta(\mathbf{x} - \mathbf{x}_{\mathrm{S}}) f(\mathbf{x})\, \mathrm{d}\mathbf{x} = -\nabla f(\mathbf{x}_{\mathrm{S}}). \tag{4.6.8}$$

Often the stress glut is restricted to a small volume surrounding a surface, a *fault*. In an idealized model, the stress glut is restricted to the surface. Consider an element of an idealized fault defined by an area ΔA, a unit normal $\hat{\mathbf{n}}$, a thickness Δn ($\to 0$) and a displacement discontinuity $[\mathbf{u}]$ across the fault (Figure 4.18). On a slip fault we must have $\hat{\mathbf{n}} \cdot [\mathbf{u}] = 0$ but in general we need not make this restriction, as, for instance, in an explosive or hydro-fracturing situation the sides of a fault may be forced apart. The displacement discontinuity does not satisfy the constitutive law and a stress glut will exist within the fault volume $\Delta n\, \Delta A$. Some components of the model stress (from equation (4.4.36))

$$\mathbf{t}_{\text{model}\, j} = \mathbf{c}_{jk} \frac{\partial \mathbf{u}}{\partial x_k}, \tag{4.6.9}$$

will be large (and singular as $\Delta n \to 0$) and physically meaningless.

Integrating across the fault, we have the slip discontinuity

$$[\mathbf{u}] = \int \frac{\partial \mathbf{u}}{\partial x_k} \hat{n}_k\, \mathrm{d}n, \tag{4.6.10}$$

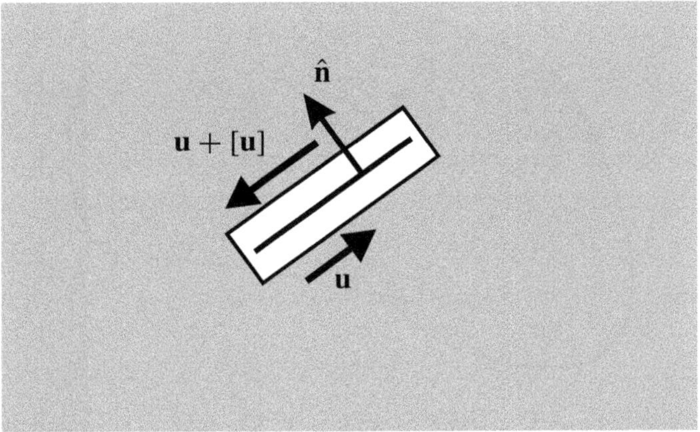

Fig. 4.18. A small element of fault, in which the stress glut is non-zero due to the displacement discontinuity [**u**] across it.

so integrating the stress glut we find

$$\int \left(\sigma_{\text{glut}}\right)_{ij} \mathrm{d}n = \int (\,\boldsymbol{\sigma}_{\text{model}} - \boldsymbol{\sigma}_{\text{true}})_{ij}\, \mathrm{d}n \to c_{ijkl}\, \hat{n}_k\, [u_l]\,, \qquad (4.6.11)$$

as $\Delta n \to 0$, because $\boldsymbol{\sigma}_{\text{true}}$ remains finite. This is the *stress-glut density on the fault surface*. Backus and Mulcahy (1976*a*) develop these results more rigorously using distribution theory.

If the medium is isotropic, and the displacement $[\mathbf{u}] = [u]\ \hat{\boldsymbol{\nu}}$, then the moment tensor is

$$\mathbf{M}_{\mathrm{S}} = M_{\mathrm{S}}(\,\hat{\boldsymbol{\nu}}\,\hat{\mathbf{n}}^{\mathrm{T}} + \hat{\mathbf{n}}\,\,\hat{\boldsymbol{\nu}}^{\mathrm{T}}), \qquad (4.6.12)$$

where M_{S} is the *scalar seismic moment*

$$M_{\mathrm{S}} = \int \mu\ [u]\ \mathrm{d}A. \qquad (4.6.13)$$

4.6.1 The slip-discontinuity fault

The concept of the stress glut reveals that the seismic source must be due to a local failure of the constitutive laws or external forces acting on the Earth. An alternative approach is to exclude the regions where the stress glut is non-zero from the model, and consider the boundary conditions on the surface surrounding the region where the constitutive laws fail. Then throughout the model, the constitutive relations hold and the seismic source is represented by the boundary conditions. This model for a fault surface was developed by Burridge and Knopoff (1964) and preceded the concept of the stress glut.

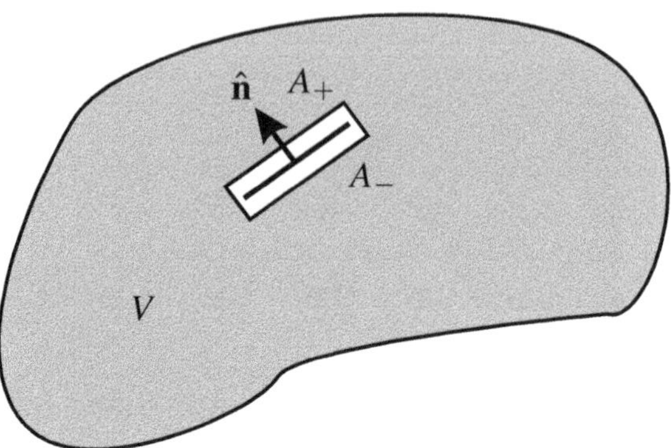

Fig. 4.19. The volume V containing a fault A. The fault is surrounded by a surface to exclude the volume where the stress glut is non-zero, so the surface of V includes the surface A_+ on the positive side of A, and A_- on the negative side (with respect to the unit normal, $\hat{\mathbf{n}}$).

The representation theorem (4.5.49) is used for the solution where the volume V is the whole Earth excluding the region where the stress glut is non-zero. For a real fault, this will be a small region surrounding the fault zone where the laws of elasticity fail. For an idealized fault, an infinitesimal thin volume including the fault surface, A, is excluded. Thus the surface of the volume V consists of the surface of the Earth, plus surfaces on both sides of the fault (Figure 4.19).

For simplicity, we assume that the body forces are zero, so the body-force integral can be ignored. On the surface of the Earth, we assume that both the solution $\mathbf{u}$ and the Green functions $\underline{\mathbf{u}}$ and $\underline{\mathbf{t}}_j$ satisfy homogeneous boundary conditions so the surface integral can be ignored. Thus the representation theorem (4.5.49) reduces to

$$\mathbf{u}(t, \mathbf{x}_{\mathrm{R}}) = \int_{A_+ + A_-} \left\{ \underline{\mathbf{u}}^{\mathrm{T}}(t, \mathbf{x}; \mathbf{x}_{\mathrm{R}}) * \mathbf{t}_j(t, \mathbf{x}) - \underline{\mathbf{t}}_j^{\mathrm{T}}(t, \mathbf{x}; \mathbf{x}_{\mathrm{R}}) * \mathbf{u}(t, \mathbf{x}) \right\} \mathrm{d}S_j \qquad (4.6.14)$$

$$= \int_A \left[\underline{\mathbf{t}}_j^{\mathrm{T}}(t, \mathbf{x}; \mathbf{x}_{\mathrm{R}}) * \mathbf{u}(t, \mathbf{x}) - \underline{\mathbf{u}}^{\mathrm{T}}(t, \mathbf{x}; \mathbf{x}_{\mathrm{R}}) * \mathbf{t}_j(t, \mathbf{x}) \right] \hat{n}_j \,\mathrm{d}S, \qquad (4.6.15)$$

where the square brackets indicate the saltus across the surface, i.e. the difference in values on A_+ and A_- (the minus sign is needed as the normal $\hat{\mathbf{n}}$ is inwards on the positive side, A_+, whereas the representation theorem (4.5.49) was developed with the surface normal pointing outwards). For simplicity, we take the Green function to be continuous across the surface A. We define a slip on the fault surface $[\mathbf{u}]$ but the tractions must be continuous. No boundary conditions on A are needed to determine the Green function. Waves that are diffracted by the fault surface are contained in the determination of the slip, $[\mathbf{u}]$ (although this detail is not normally included in the specification of the slip function). Thus integral (4.6.15) reduces to

$$\mathbf{u}(t, \mathbf{x}_\mathrm{R}) = \int_A \underline{\mathbf{t}}_j^{\mathrm{T}}(t, \mathbf{x}; \mathbf{x}_\mathrm{R}) * [\mathbf{u}(t, \mathbf{x})]\, \hat{n}_j \,\mathrm{d}S, \tag{4.6.16}$$

which is equivalent to the above result (4.6.7) with definition (4.6.5) and result (4.6.11).

4.6.2 A moment-tensor source

The body-force equivalent of a point, stress-glut source is given by equation (4.6.7). Combining the dyadic Green function (4.5.75) for a ray in a homogeneous medium with this source, in the convolution integral (4.5.28), we obtain

$$\mathbf{u}(t, \mathbf{x}_\mathrm{R}; \mathbf{x}_\mathrm{S}) \simeq \frac{\delta(t - R/c)}{4\pi\rho\, c^3 R}\, \hat{\mathbf{g}}\, \hat{\mathbf{g}}_\mathrm{S}^{\mathrm{T}}\, \mathbf{M}_\mathrm{S}\, \hat{\mathbf{p}}_\mathrm{S}. \tag{4.6.17}$$

This result is most easily obtained first in the frequency domain. A factor $-\mathrm{i}\omega\,\mathbf{p} = -\mathrm{i}\omega\,\hat{\mathbf{p}}/c$ arises from differentiating $\exp(\mathrm{i}kR) = \exp(\mathrm{i}\omega\, R/c)$ with respect to the origin $\mathbf{x}_\mathrm{S}$. The unit vector, $\hat{\mathbf{p}}$, is in the propagation direction, i.e. $\hat{\mathbf{p}} = \hat{\mathbf{r}}$, although we use the slowness direction, $\hat{\mathbf{p}}$, to emphasize the origin of the term from $\mathbf{p} = \nabla T$. The factor $-\mathrm{i}\omega$ cancels with the integration in expression (4.6.7). Other derivative terms are ignored in the spirit of the far-field approximation as they are $O(1/\omega)$ and decay faster.

Again the factor $\hat{\mathbf{g}}_\mathrm{S}^{\mathrm{T}}\, \mathbf{M}_\mathrm{S}\, \hat{\mathbf{p}}_\mathrm{S}$ is called the source radiation pattern. Using polarizations (4.5.78), (4.5.80) and (4.5.81), we can obtain the radiation pattern for *P* waves

$$\begin{aligned} \hat{\mathbf{g}}_P^{\mathrm{T}}\, \mathbf{M}_\mathrm{S}\, \hat{\mathbf{p}}_\mathrm{S} = {} & \left(M_{xx} \cos^2\phi + M_{yy} \sin^2\phi + M_{xy} \sin 2\phi \right) \sin^2\theta \\ & + M_{zz} \cos^2\theta + \left(M_{zx} \cos\phi + M_{yz} \sin\phi \right) \sin 2\theta, \end{aligned} \tag{4.6.18}$$

for *SV* waves

$$\begin{aligned} \hat{\mathbf{g}}_V^{\mathrm{T}}\, \mathbf{M}_\mathrm{S}\, \hat{\mathbf{p}}_\mathrm{S} = {} & \frac{1}{2} \left(M_{xx} \cos^2\phi + M_{yy} \sin^2\phi - M_{zz} + M_{xy} \sin 2\phi \right) \sin 2\theta \\ & + \left(M_{zx} \cos\phi + M_{yz} \sin\phi \right) \cos 2\theta, \end{aligned} \tag{4.6.19}$$

and for *SH* waves

$$\begin{aligned} \hat{\mathbf{g}}_H^{\mathrm{T}}\, \mathbf{M}_\mathrm{S}\, \hat{\mathbf{p}}_\mathrm{S} = {} & \left(\frac{1}{2} \left(M_{yy} - M_{xx} \right) \sin 2\phi + M_{xy} \cos 2\phi \right) \sin\theta \\ & + \left(M_{yz} \cos\phi - M_{zx} \sin\phi \right) \cos\theta. \end{aligned} \tag{4.6.20}$$

These expressions allow us to consider various simple stress-glut sources: an explosion, a dipole and a double couple.

4.6.2.1 An explosion

A point explosion can be represented by an isotropic moment tensor

$$\mathbf{M}_{\mathrm{S}} = M_{\mathrm{S}}\,\mathbf{I}, \tag{4.6.21}$$

where M_{S} is the product of the volume times the pressure change in the source. If we require a pressure impulse, then an extra time derivative must be introduced into the solution. Notice that as the volume of the explosion cavity is reduced to zero, the pressure must be increased indefinitely to keep the product finite.

With radiation pattern (4.6.18), the P ray is

$$\mathbf{u}(t, \mathbf{x}_{\mathrm{R}}; \mathbf{x}_{\mathrm{S}}) \simeq \frac{\delta(t - R/c)}{4\pi\rho\, c^3 R}\,\hat{\mathbf{g}}\, M_{\mathrm{S}}. \tag{4.6.22}$$

For an S ray, the polarization is transverse to the radial vector and the result is zero, confirmed by expressions (4.6.19) and (4.6.20).

4.6.2.2 A dipole

Suppose on a fault surface, the two surfaces move apart. To be specific, we choose the x–y plane as the fault plane, and z as the axis of symmetry. Thus in expression (4.6.12), we take $\hat{\mathbf{n}} = \hat{\nu} = \hat{\mathbf{k}}$. Thus

$$\mathbf{M}_{\mathrm{S}} = 2M_{\mathrm{S}} \begin{pmatrix} 0 & 0 & 0 \\ 0 & 0 & 0 \\ 0 & 0 & 1 \end{pmatrix}. \tag{4.6.23}$$

For a P ray, expression (4.6.18) is

$$\hat{\mathbf{g}}_P^{\mathrm{T}}\, \mathbf{M}_{\mathrm{S}}\, \hat{\mathbf{p}}_{\mathrm{S}} = 2M_{\mathrm{S}} \cos^2\theta. \tag{4.6.24}$$

This is the radiation pattern for P rays from a *dipole source* (Figure 4.20). Notice that it is more directive than the radiation from a force (Figure 4.14), and has the same sign in both directions.

For an S ray, the radiation pattern for the SV component is

$$\hat{\mathbf{g}}_V^{\mathrm{T}}\, \mathbf{M}_{\mathrm{S}}\, \hat{\mathbf{p}} = -M_{\mathrm{S}} \sin 2\theta. \tag{4.6.25}$$

The radiation pattern is shown in Figure 4.21. The pattern is axially symmetric, like

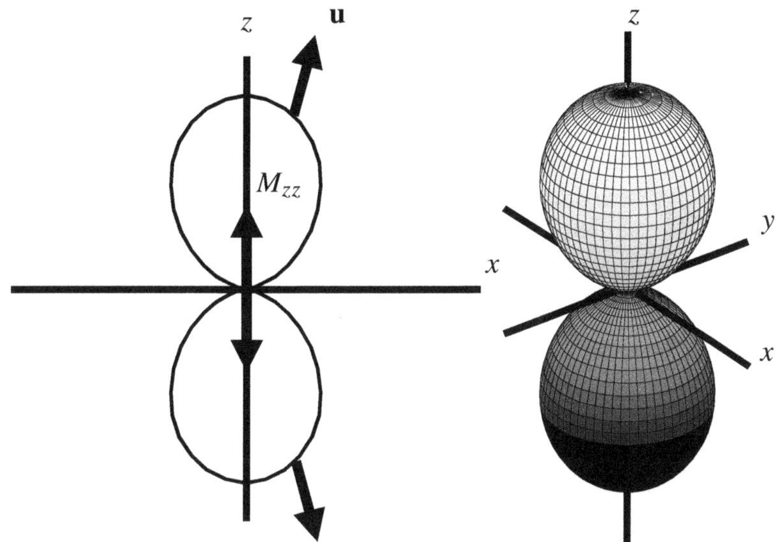

Fig. 4.20. The radiation patterns for *P* waves due to a dipole, M_{zz}. As Figure 4.14.

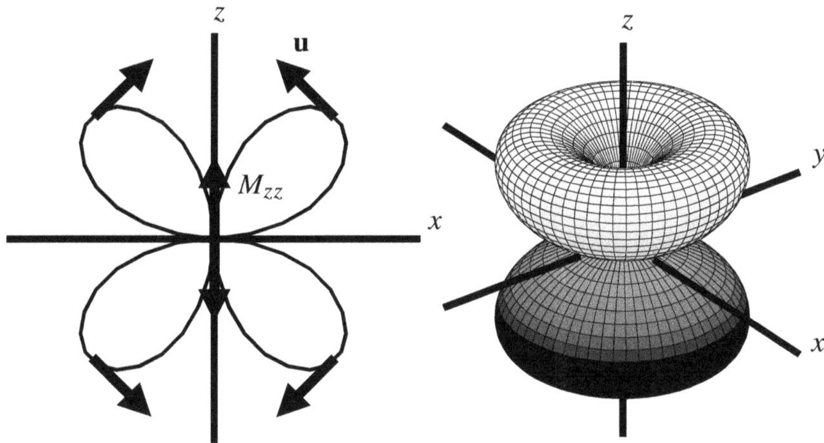

Fig. 4.21. The radiation patterns for *SV* waves due to a dipole, M_{zz}. As Figure 4.14.

two doughnuts, as expected. The radiation pattern for the *SH* polarization (4.6.20) is zero as expected from the axial symmetry.

4.6.2.3 A double couple

The other type of source is due to the off-diagonal elements of the moment tensor. Consider a fault in the x–z plane, with slip in the x direction, i.e. $\hat{\boldsymbol{\nu}} = \hat{\mathbf{j}}$ and $\hat{\mathbf{n}} = \hat{\mathbf{i}}$.

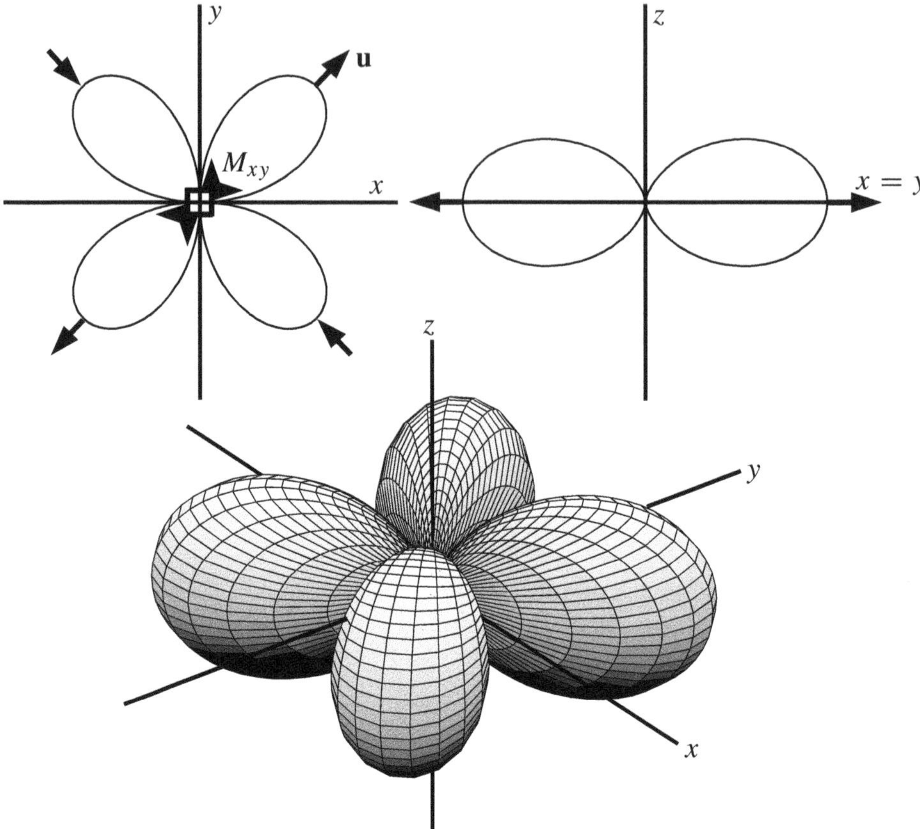

Fig. 4.22. The radiation patterns for P waves due to a double couple, M_{xy}. The plots show the relative amplitudes in different directions: the upper-left plot is in the cross-section in the x–y plane; the upper-right plot is in the plane $x = y$; and the lower plot is a view of the three-dimensional radiation pattern. Otherwise as Figure 4.14.

The moment tensor is

$$\mathbf{M}_\mathrm{S} = M_\mathrm{S} \begin{pmatrix} 0 & 1 & 0 \\ 1 & 0 & 0 \\ 0 & 0 & 0 \end{pmatrix}. \tag{4.6.26}$$

Notice that the moment tensor, and hence the radiation patterns, are identical if the fault normal and slip are interchanged. Hence this is called a *double couple.*

For the P ray, the directivity term (4.6.18) reduces to

$$\hat{\mathbf{g}}_P^\mathrm{T}\, \mathbf{M}_\mathrm{S}\, \hat{\mathbf{p}}_\mathrm{S} = M_\mathrm{S}\, \sin 2\phi\, \sin^2\theta. \tag{4.6.27}$$

This is illustrated in Figure 4.22. In the x–y plane, the radiation pattern has the form of a four-leafed clover, with alternating sign. In a vertical plane, it has lobes in the x–y plane.

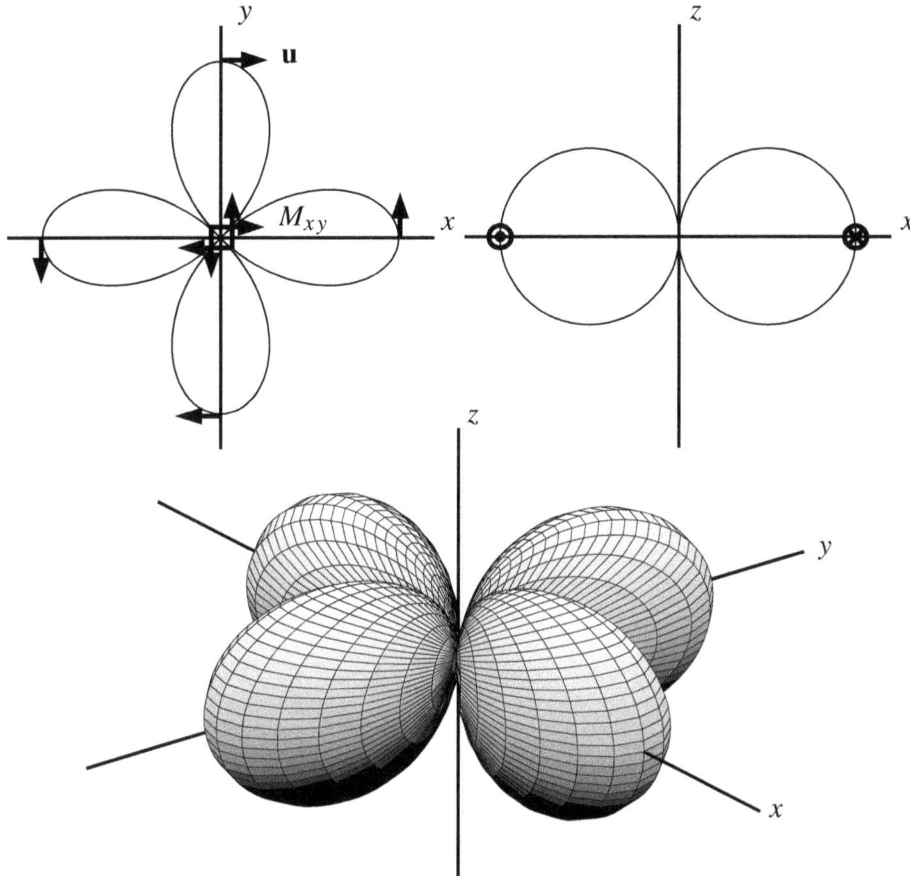

Fig. 4.23. The radiation patterns for *SH* waves due to a double couple, M_{xy}. As Figure 4.22.

The radiation patterns for the two *S* ray polarizations differ. For the *SH* polarization parallel to the x–y plane, the result (4.6.20) is

$$\hat{\mathbf{g}}_H^{\mathrm{T}} \, \mathbf{M}_{\mathrm{S}} \, \hat{\mathbf{p}}_{\mathrm{S}} = M_{\mathrm{S}} \, \cos 2\phi \, \sin \theta. \tag{4.6.28}$$

Again this has the form of a four-leafed clover (Figure 4.23). For the *SV* polarization in the z and ray plane, the directivity function (4.6.19) is

$$\hat{\mathbf{g}}_V^{\mathrm{T}} \, \mathbf{M}_{\mathrm{S}} \, \hat{\mathbf{p}}_{\mathrm{S}} = \frac{1}{2} M_{\mathrm{S}} \, \sin 2\phi \, \sin 2\theta. \tag{4.6.29}$$

The radiation pattern has eight lobes, one in each quadrant (Figure 4.24).

It is straightforward to rotate and combine these results to find the radiation from any moment-tensor source.

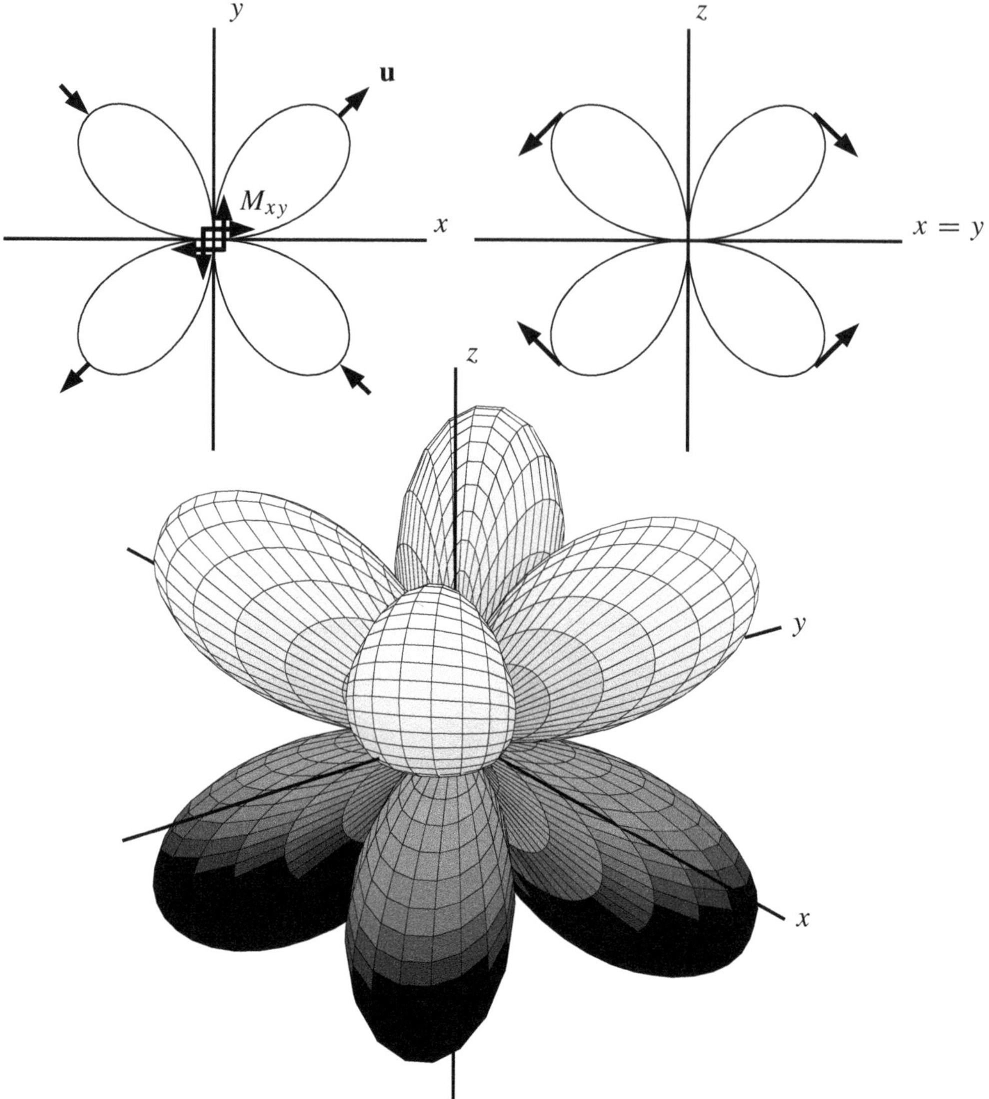

Fig. 4.24. The radiation patterns for *SV* waves due to a double couple, M_{xy}. As Figure 4.22.

Exercises

4.1 Various other elastic parameters are used to describe isotropic, elastic media apart from the Lamé parameters, λ and μ:

a) If a solid is compressed by a hydrostatic pressure, then the bulk modulus, κ, is defined as minus the ratio of the hydrostatic pressure to the dilatation, e.g. $\kappa = -P/\theta$ where $\sigma_{xx} = \sigma_{yy} = \sigma_{zz} = -P$. Show that

$$\kappa = \lambda + \frac{2}{3}\mu.$$

b) If a solid is stretched with just one non-zero normal stress component, e.g. stretching a wire where the sides are unconstrained, then Young's modulus, E, is defined as the ratio of the stress to the same strain component, e.g. $E = \sigma_{xx}/e_{xx}$ with $\sigma_{yy} = \sigma_{zz} = 0$. Show that

$$E = \frac{\mu(3\lambda + 2\mu)}{\lambda + \mu}.$$

c) With the same experiment as in part **b)**, Poisson's ratio, ν, is defined as minus the ratio of the transverse strain and the longitudinal strain, e.g. $\nu = -e_{yy}/e_{xx}$. Show that

$$\nu = \frac{\lambda}{2(\lambda + \mu)} = \frac{\alpha^2 - 2\beta^2}{2(\alpha^2 - \beta^2)}.$$

4.2 Show that if a plane wave

$$\mathbf{u} = \mathbf{g}\, e^{\mathrm{i}\omega\hat{\mathbf{p}}\cdot\mathbf{x}/c - \mathrm{i}\omega t},$$

where $\hat{\mathbf{p}}$ is the normalized phase direction and c the phase velocity, propagates in a general, homogeneous anisotropic medium described by the matrices $\mathbf{c}_{jk}$ (4.4.39), then the polarization, $\mathbf{g}$, and phase velocity, c, must satisfy the eigen-equation

$$\left(\hat{p}_j \hat{p}_k \mathbf{c}_{jk}/\rho\right)\mathbf{g} = c^2\mathbf{g}.$$

This is commonly called the *Christoffel equation* (see Section 5.3.2). As the matrix $\hat{\mathbf{\Gamma}} = \hat{p}_j \hat{p}_k \mathbf{c}_{jk}/\rho$ is symmetric the eigenvalues are real and the non-degenerate eigenvectors are orthogonal.

Show that for an isotropic medium, the solutions reduce to the longitudinal P and the degenerate transverse S waves, with phase velocities given by equations (4.5.60) and (4.5.61).

4.3 Using the results of the previous question, show that for a transversely isotropic medium with an x_3 axis of symmetry (TIV), the eigenvalues (squared phase velocities) are given by

$$\rho c^2 = C_{66}\sin^2\phi + C_{44}\cos^2\phi$$

or

$$\rho c^2 = \frac{1}{2}\Big(C_{44} + C_{33}\cos^2\phi + C_{11}\sin^2\phi \pm \sqrt{\left[(C_{33} - C_{44})\cos^2\phi - (C_{11} - C_{44})\sin^2\phi\right]^2 + 4a^2\cos^2\phi\,\sin^2\phi}\Big),$$

where $\hat{\mathbf{p}} = (\sin\phi, 0, \cos\phi)$ is the phase direction (so the polar angle from the symmetry axis is ϕ), and

$$a = C_{13} + C_{44}.$$

Show that the corresponding polarizations are

$$\hat{\mathbf{g}} = \begin{pmatrix} 0 \\ 1 \\ 0 \end{pmatrix},$$

which is commonly called the *qSH* wave, and

$$\hat{\mathbf{g}} = \operatorname{sgn}\begin{pmatrix} 2a\hat{p}_1\hat{p}_3 \\ 0 \\ (C_{33} - C_{44})\hat{p}_3^2 - (C_{11} - C_{44})\hat{p}_1^2 \pm \\ \left[\left((C_{33} - C_{44})\hat{p}_3^2 - (C_{11} - C_{44})\hat{p}_1^2\right)^2 + 4\hat{p}_1^2\hat{p}_3^2a^2\right]^{1/2} \end{pmatrix}.$$

The wave with the upper sign in the phase velocity and polarization is commonly called the *qP* wave, and the lower sign *qSV*.

4.4 Verify that if the elastic parameters of two TIV media are equal, except for the substitution

$$C_{13} \to -C_{13} - 2C_{44},$$

then the phase velocities are identical, but the polarizations differ. Physically, the anomalous negative value for C_{13} is extremely unlikely with very unusual polarizations (verify by numerical example), but it is possible (Helbig and Schoenberg, 1987).

4.5 A TIV material is conveniently described by two axial velocities

$$\alpha_0 = \sqrt{C_{33}/\rho}$$
$$\beta_0 = \sqrt{C_{44}/\rho},$$

and three dimensionless parameters

$$\delta = \frac{(C_{13} + C_{44})^2 - (C_{33} - C_{44})^2}{2C_{33}(C_{33} - C_{44})}$$
$$\epsilon = \frac{C_{11} - C_{33}}{2C_{33}}$$
$$\gamma = \frac{C_{66} - C_{44}}{2C_{44}}.$$

These useful parameters were introduced by Thomsen (1986). Verify that the elastic parameters, C_{ij}, can be determined from these, except for an arbitrary sign in the equation for C_{13} (see previous exercise). The anomalous negative sign is normally ignored. In a general TI medium, two additional parameters are needed to specify the direction of the symmetry axis, e.g. the spherical polar angles, for a total of seven parameters.

Show that if the dimensionless parameters are small (compared with unity), the phase velocities (see the previous exercises) can be approximated by

$$
\begin{aligned}
c_{qSH} &= \beta_0(1 + \gamma \sin^2 \phi) \\
c_{qSV} &= \beta_0 \left(1 + \frac{\alpha_0^2}{\beta_0^2}(\epsilon - \delta) \sin^2 \phi \cos^2 \phi \right) \\
c_{qP} &= \alpha_0(1 + \delta \sin^2 \phi \cos^2 \phi + \epsilon \sin^4 \phi).
\end{aligned}
$$

To first order, the polarizations are in the phase direction and transverse, i.e. as in an isotropic medium. See Thomsen (1986) for more details.

4.6 *Further reading:* Interpreting general anisotropic elastic parameters is difficult. If all 21 parameters are non-zero, is the medium in fact one with a high-order of symmetry, e.g. TI, but with tilted axes or planes of symmetry? In other words, would a simple rotation reduce the number of non-zero parameters significantly? This question has been addressed by several authors who develop decompositions that do not depend on the coordinate system. A useful review is by Baerheim (1993), who compares his results with those by earlier authors (Backus, 1970 and Cowin, 1989, 1993). The general results are too complicated to be included here and the paper by Baerheim (1993) is suggested for further reading.

One particularly useful result is for the mean-squared velocities, averaged over all directions. This can be used to define an isotropic medium that approximates an anisotropic medium. Defining the dilatational modulus tensor

$$D_{ij} = c_{ijkk} \,,$$

and the Voigt tensor

$$V_{ij} = c_{ikjk},$$

the mean squared P and S velocities are

$$
\begin{aligned}
\overline{\alpha^2} &= \Big(\mathrm{tr}(\mathbf{D}) + 2\,\mathrm{tr}(\mathbf{V})\Big) \Big/ \, 15\rho \\
\overline{\beta^2} &= \Big(3\,\mathrm{tr}(\mathbf{V}) - \mathrm{tr}(\mathbf{D})\Big) \Big/ \, 30\rho.
\end{aligned}
$$

Show that in isotropic media, these expressions reduce to

$$\overline{\alpha^2} = \alpha^2$$
$$\overline{\beta^2} = \beta^2 \; ;$$

in transversely isotropic media to

$$\overline{\alpha^2} = \frac{8C_{11} + 3C_{33} + 8C_{44} + 4C_{31}}{15\rho}$$
$$\overline{\beta^2} = \frac{C_{11} + C_{33} + 6C_{44} + 5C_{66} - 2C_{13}}{15\rho} \; ;$$

and in general anisotropic media to

$$\overline{\alpha^2} = \frac{3(C_{11} + C_{22} + C_{33}) + 4(C_{44} + C_{55} + C_{66}) + 2(C_{23} + C_{31} + C_{12})}{15\rho}$$
$$\overline{\beta^2} = \frac{(C_{11} + C_{22} + C_{33}) + 3(C_{44} + C_{55} + C_{66}) - (C_{23} + C_{31} + C_{12})}{15\rho}.$$

In a weak transversely isotropic medium (see previous exercise), show that the results are

$$\overline{\alpha^2} = \alpha_0^2 \left(1 + \frac{16}{15}\epsilon + \frac{4}{15}\delta\right)$$
$$\overline{\beta^2} = \beta_0^2 \left(1 + \frac{2}{3}\gamma + \frac{2}{15}(\epsilon - \delta)\frac{\alpha_0^2}{\beta_0^2}\right).$$

Show that these two results agree with averaging the approximate squared velocities given in the previous exercise over a sphere, and averaging the two shear wave velocities.

4.7 Confirm that the two-dimensional result (4.5.84) or (4.5.85) can be derived from the equivalent three-dimensional results by dividing the line source into infinitesimal point elements and integrating along the line (see Figure 4.16).

Further reading: See, for instance, Hudson (1980, Section 2.5).

4.8 Prove the general form of an isotropic fourth-order tensor (4.4.45) (Jeffreys, 1931).

4.9 The dipole, explosion and double-couple results have only been given for the far-field radiation patterns. Investigate the near-field terms by differentiating the exact point source results (see, for instance, Aki and Richards (1980, 2002) or Pujol (2003) for details). Are the signals exactly zero on the node planes?

4.10 *Further reading:* The homogeneous, far-field radiation pattern has been given here for an isotropic medium (4.5.75). Lighthill (1960), Buchwald

(1959) and Burridge (1967) have investigated the equivalent result in an anisotropic medium. The result, while more complicated to derive, is remarkably similar, i.e. in the frequency domain

$$\underline{\mathbf{u}}(\omega, \mathbf{x}_\mathrm{R}; \mathbf{x}_\mathrm{S}) \simeq \frac{\mathrm{e}^{\mathrm{i}\omega\mathbf{p}\cdot(\mathbf{x}_\mathrm{R}-\mathbf{x}_\mathrm{S})}}{4\pi\rho\,|\mathbf{V}|K^{1/2}(\mathbf{p})R}\,\hat{\mathbf{g}}\,\hat{\mathbf{g}}^\mathrm{T},$$

where $K(\mathbf{p})$ is the Gaussian curvature of the slowness surface at $\mathbf{p}$, and $\mathbf{V}$ is the ray (group) velocity. See, for instance, Burridge (1967, Section 6.7). This result is useful in anisotropic ray theory (see Section 5.4.2 and Kendall, Guest and Thomson, 1992).

4.11 *Further reading:* We have only developed the stress and strain tensors, and the elastic constitutive equations and equations of motion in cartesian coordinates. For some problems, the equations in cylindrical or spherical coordinates are useful. These results can be found in several textbooks, for instance, the classic book by Love (1944) or the more modern treatments by Fung (1965) or Takeuchi and Saito (1972).

4.12 Muir (personal communication, 2003) has pointed out that although the Voigt notation is compact (Section 4.4.2), many of the awkward factors of 2 or $\sqrt{2}$, e.g. equation (4.4.14), can be avoided if we retain all 9 components in the stress and strain vectors. Writing a stress vector as

$$\sigma = \begin{pmatrix} \mathbf{t}_1 \\ \mathbf{t}_2 \\ \mathbf{t}_3 \end{pmatrix},$$

and similarly for a strain vector, the constitutive equation can be written

$$\sigma = \mathbf{C}\mathbf{e} = \begin{pmatrix} \mathbf{c}_{11} & \mathbf{c}_{12} & \mathbf{c}_{13} \\ \mathbf{c}_{21} & \mathbf{c}_{22} & \mathbf{c}_{23} \\ \mathbf{c}_{31} & \mathbf{c}_{32} & \mathbf{c}_{33} \end{pmatrix} \mathbf{e},$$

where the 3×3 sub-matrices of the 9×9 matrix $\mathbf{C}$ are defined in equations (4.4.36)–(4.4.39).

The symmetry of the stress and strain tensors are imposed by the equality of rows and elements in the matrix $\mathbf{C}$. Although this makes the matrix C singular, show by numerical example or otherwise, that the compliance matrix $\mathbf{S}$ can be obtained from the *generalized inverse* of the stiffness matrix $\mathbf{C}$.

It should also be commented that Muir (personal communication, 2003) suggested an alternative order for the components of the stress and strain vectors that emphasizes symmetries in the matrix $\mathbf{C}$ for symmetric media, i.e. isotropic, transversely isotropic, etc.

5
Asymptotic ray theory

Ray theory is the cornerstone of high-frequency, body-wave seismology. Without it, seismic signals in realistic, complex media would be extremely difficult to describe and interpret. The mathematical technique of asymptotic ray theory is developed in this chapter. Although the details appear complicated, e.g. the higher-order terms in dynamic ray theory, it must be remembered that usually only the lowest-order terms are needed and used. These are relatively easy to understand and compute. Ray theory in a continuous medium consists of three parts which are developed in this chapter for acoustic, isotropic and anisotropic elastic media: *kinematic ray theory* that describes the geometry and times of rays and wavefronts; *dynamic ray theory* that describes the geometrical spreading of rays and the displacement magnitude; and *polarization theory* describing the displacement direction. The chapter concludes with a pretty example of ray tracing in an anisotropic medium with a linear gradient, which illustrates that even in this simple example, complicated non-intuitive results occur.

In this chapter, we develop asymptotic ray theory (ART) for acoustic, anisotropic and isotropic media. We use the equations of motion, and the constitutive relations already discussed in Chapter 4, without the source terms, and match the ray solutions to the point-source Green function given in Section 4.5.5. We discuss anisotropic, elastic waves before specializing to isotropic media, as the development is actually more straightforward. The degenerate shear waves in isotropic media require special treatment. In this chapter we assume that the media properties are continuous, and in Chapter 6 we generalize this with the ray expansion to media with discontinuities – interfaces.

5.1 Acoustic kinematic ray theory

The equations of motion (4.5.1) and constitutive relations (4.5.2) for an acoustic medium were given in Section 4.5.1. In this section we develop the kinematic ray equations for acoustic waves.

5.1.1 Acoustic ray ansatz

From the discussion of waves (and rays) in Chapter 2, we have many intuitive ideas about how the solutions of equations (4.5.1) and (4.5.2) might behave at high frequencies. By high frequencies we mean that the wavelength is small compared with the propagation distances and the heterogeneities. Under these circumstances, we might expect the waves to behave locally as plane waves. Although the wavelengths are assumed to be small compared with the heterogeneities, we do allow for discontinuities – interfaces – in the model by applying boundary conditions (Section 4.3), although we delay this until the next chapter (Chapter 6). The solution procedure is to guess a probable form of the solution, known mathematically as an *ansatz*, and then find the conditions for it to satisfy the equations.

Thus the ray ansätze are

$$\boxed{\begin{pmatrix} \mathbf{v} \\ -P \end{pmatrix}(\omega, \mathbf{x}_\mathrm{R}) = f(\omega) \sum_n \mathrm{e}^{\mathrm{i}\omega T(\mathbf{x}_\mathrm{R}, \mathcal{L}_n)} \sum_{m=0}^{\infty} \frac{1}{(-\mathrm{i}\omega)^m} \begin{pmatrix} \mathbf{v}^{(m)} \\ -P^{(m)} \end{pmatrix}(\mathbf{x}_\mathrm{R}, \mathcal{L}_n).} \tag{5.1.1}$$

This expression contains lots of features (notation) which express our physical intuition about the solution. First for compactness, we have written the ansätze for the velocity and minus the pressure as a 4×1 vector as from the symmetry of equations (4.5.1) and (4.5.2), we would expect the velocity and pressure to have similar solutions (we use *minus* the pressure purely because it has the opposite sign from stress used in elastic solutions). Note that for compactness we have written the vectors as functions of $(\omega, \mathbf{x}_\mathrm{R})$ and $(\mathbf{x}_\mathrm{R}, \mathcal{L}_n)$ on the left and right-hand sides, respectively, indicating that each component is a function of the same arguments. The approach used here including velocity and pressure ansätze in equation (5.1.1) differs slightly from that commonly taken where the ansatz for displacement, for instance, is substituted in the Navier wave equation (4.5.24) and the ansatz for pressure is never considered explicitly.

We have written the ansätze in the frequency domain. The function $f(\omega)$ is an arbitrary spectrum depending on the source. The notation $\mathcal{L}_n$ and the arbitrary summation over n is used to indicate the 'ray paths'. Various paths may exist to the point $\mathbf{x}_\mathrm{R}$ and these are enumerated by the index n. The path in general is a function of the source, receiver and other parameters along the ray, e.g. the reflection/transmission history. These properties are summarized in the term $\mathcal{L}_n$ which contains enough information to describe the *ray's history* or *ray descriptor*. At this stage we need not be specific about how this information is parameterized. As soon as the model has interfaces, and even without, multiple rays are possible (e.g. Section 2.4.4). At interfaces, multiple rays are necessary to satisfy the boundary

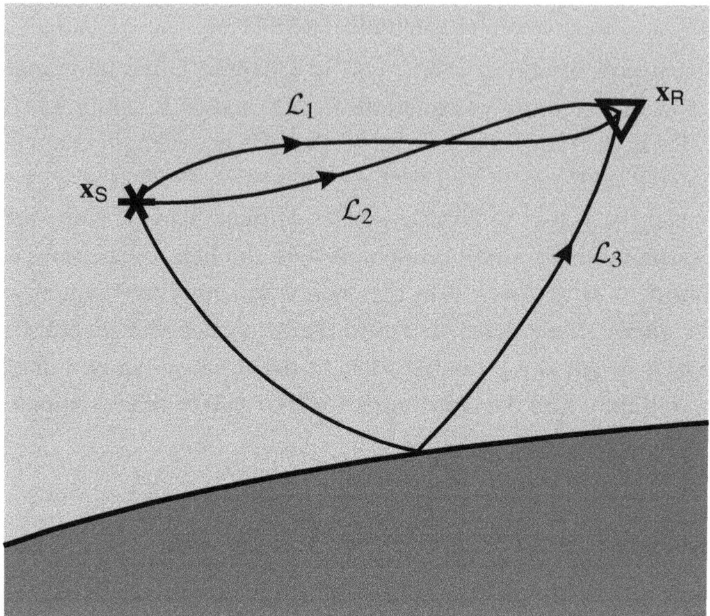

Fig. 5.1. Three ray paths, $\mathcal{L}_1$, $\mathcal{L}_2$ and $\mathcal{L}_3$, between a source $\mathbf{x}_S$ and receiver $\mathbf{x}_R$.

conditions. The fact that the solution can be represented as the sum of different rays is known as the *ray expansion*. A situation with three rays is illustrated in Figure 5.1.

From studies in homogeneous media, it is well known that acoustic and elastic waves propagate (approximately) without dispersion with a velocity independent of frequency. At interfaces between homogeneous media, plane waves satisfy Snell's law (Section 2.1.2.1). In inhomogeneous media, we expect similar behaviour at high frequencies when the wavelength is short compared with the scale of heterogeneities in the medium. The solution is therefore written as a series in *amplitude coefficients*, $\mathbf{v}^{(m)}$ and $P^{(m)}$, which are independent of frequency, and a phase factor linearly dependent on the frequency ω and the *travel time*, T, which again is independent of frequency. At high frequencies, the solution is therefore non-dispersive, but the series in inverse powers of $-\mathrm{i}\omega$ allows for pulse distortion at low frequencies.

Although we have expressed one function, e.g. $\mathbf{v}(\omega, \mathbf{x}_R)$, in terms of more than one function, $T(\mathbf{x}_R, \mathcal{L}_n)$ and $\mathbf{v}^{(m)}(\mathbf{x}_R, \mathcal{L}_n)$, we expect to be able to determine the functions uniquely through the frequency dependence. Without the frequency behaviour, this process would be non-unique, e.g. if the definition were $\mathbf{v}(\mathbf{x}_R) = \mathbf{v}^{(0)}(\mathbf{x}_R) \exp(\mathrm{i}T(\mathbf{x}_R))$, the functions $\mathbf{v}^{(0)}(\mathbf{x}_R)$ and $T(\mathbf{x}_R)$ would not have unique solutions.

Thus the frequency, or time dependence, only enters through the linear phase term and inverse powers in the expansion. We allow the amplitude coefficients to be complex but require the travel time to be real. Strictly, if the amplitude coefficients are complex, they must depend on the sign of the frequency. For generality we have included an arbitrary function of frequency, $f(\omega)$, which satisfies relationship (3.1.9). This function will be closely related to the source spectrum. We will specify it exactly later for the ray Green functions (equations (5.2.64) and (5.2.78)). To obtain a real time series for the solution, we must also have $\mathbf{v}^{(m)*}(\omega) = \mathbf{v}^{(m)}(-\omega)$. Since this is the only frequency dependence, we omit it from the amplitude coefficients and where necessary understand the value for positive frequencies.

It is straightforward to write the ansatz of asymptotic ray theory in the time domain. Inverting the Fourier transform, we obtain

$$\boxed{\begin{aligned}&\begin{pmatrix}\mathbf{v}\\ -P\end{pmatrix}(t,\mathbf{x}_\mathrm{R}) =\\ &\mathrm{Re}\left(F(t) * \sum_n \sum_{m=0}^{\infty} \begin{pmatrix}\mathbf{v}^{(m)}\\ -P^{(m)}\end{pmatrix}(\mathbf{x}_\mathrm{R},\mathcal{L}_n)\,\delta^{(m)}\left(t - T(\mathbf{x}_\mathrm{R},\mathcal{L}_n)\right)\right),\end{aligned}} \tag{5.1.2}$$

where $F(t)$ is the analytic time series corresponding to $f(t)$ (Section 3.1.2), and $\delta^{(m)}(t)$ is the m-th integral of the Dirac delta function, i.e. $\delta^{(0)}(t) = \delta(t)$ the Dirac delta function, $\delta^{(1)}(t) = H(t)$ the Heaviside function, and $\delta^{(m+1)}(t) = t^m H(t)/m!$ for $m > 1$.

In the time domain, it is obvious that the series (5.1.2) is not convergent (for arbitrary $F(t)$, unless all terms for $m > 1$ are zero). If the series (5.1.2) is terminated at $m = M$, then at large times the M-th term will dominate and the series will diverge as $t^{M-1}H(t)/(M-1)!$. This is physically impossible (to conserve energy the velocity must decay or least be constant, as $t \to \infty$). Therefore the ray ansätze, (5.1.1) or (5.1.2), cannot converge to the exact solution, except in the special case when the series terminates. Although ansatz (5.1.1) is a good guess for the solution of the acoustic wave equations, our intuition also suggests that it cannot be complete. In Section 2.4.7 we have discussed signals that tunnel through a high-velocity layer. We expect these signals to decay exponentially in the frequency domain, a feature that cannot be modelled with the series (5.1.1). Although we have not discussed them yet, some diffracted signals depend on fractional powers of frequency, e.g. $\omega^{-1/2}$, a property that again cannot be modelled by the series (5.1.1). We will not pursue the difficult mathematical question as to whether the ansätze is actually an asymptotic series.

Although the ray ansätze, (5.1.1) and (5.1.2), cannot be exact, they will prove to be extremely useful. It would be no exaggeration to claim that the use of seismology as a remote sensing tool depends on the ray ansätze being a very useful approximation with very wide application. Generally only the first term in the series is considered and we have

$$\boxed{\begin{pmatrix} \mathbf{v} \\ -P \end{pmatrix}(t, \mathbf{x}_R) \simeq \sum_n \mathrm{Re}\left(\begin{pmatrix} \mathbf{v}^{(0)} \\ -P^{(0)} \end{pmatrix}(\mathbf{x}_R, \mathcal{L}_n)\, F\left(t - T(\mathbf{x}_R, \mathcal{L}_n)\right)\right),} \tag{5.1.3}$$

an approximation which is normally called the *geometrical ray approximation* (GRA – we avoid the expression geometrical ray theory as the acronym GRT has been overused, also referring to generalized ray theory (Helmberger, 1968) and the generalized Radon transform (Burridge, de Hoop, Miller and Spencer, 1998 and references therein)).

5.1.2 The acoustic ray series

Substituting the ansätze (5.1.1) in the equations (4.5.1) and (4.5.2) (without the body forces in equation (4.5.1)), and setting the coefficient of each power of ω zero (as they must be if the equations are true for arbitrary frequency), we obtain for $m \geq 0$ (defining $\mathbf{v}^{(-1)} = \mathbf{0}$ and $P^{(-1)} = 0$)

$$-\nabla P^{(m-1)} = \rho\, \mathbf{v}^{(m)} - \mathbf{p} P^{(m)} \tag{5.1.4}$$

$$\kappa \nabla \cdot \mathbf{v}^{(m-1)} = \kappa\, \mathbf{p} \cdot \mathbf{v}^{(m)} - P^{(m)}, \tag{5.1.5}$$

where (cf. equation (2.2.9))

$$\mathbf{p} = \nabla T, \tag{5.1.6}$$

is the *slowness vector*, and we have omitted the argument $(\mathbf{x}, \mathcal{L}_n)$. Eliminating $\mathbf{v}^{(m)}$ or $P^{(m)}$ between (5.1.4) and (5.1.5), we obtain

$$(\kappa \mathbf{p}^2 - \rho) P^{(m)} = \kappa \left(\mathbf{p} \cdot \nabla P^{(m-1)} + \rho \nabla \cdot \mathbf{v}^{(m-1)}\right) \tag{5.1.7}$$

$$(\kappa \mathbf{p}\,\mathbf{p}^{\mathrm{T}} - \rho\, \mathbf{I})\mathbf{v}^{(m)} = \nabla P^{(m-1)} + \kappa \left(\nabla \cdot \mathbf{v}^{(m-1)}\right) \mathbf{p}. \tag{5.1.8}$$

These equations can be solved for the travel time and amplitude coefficients.

5.1.3 The acoustic eikonal equation ($m = 0$)

For $m = 0$, equation (5.1.7) reduces to

$$\left(\alpha^2 \mathbf{p}^2 - 1\right) P^{(0)} = 0, \tag{5.1.9}$$

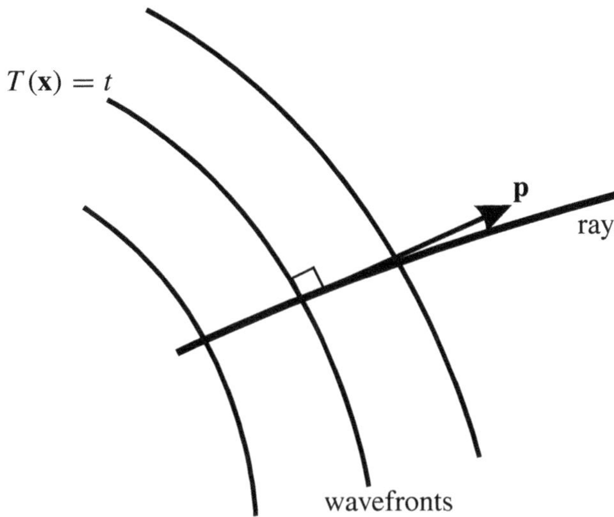

Fig. 5.2. A ray path in the direction of the slowness vector, **p**, with orthogonal wavefronts, $T(\mathbf{x}) = t$.

where

$$\alpha = \sqrt{\frac{\kappa}{\rho}}, \tag{5.1.10}$$

or equivalently

$$\boxed{(\nabla T)^2 = \frac{1}{\alpha^2},} \tag{5.1.11}$$

as we can assume without loss in generality that $P^{(0)} \neq 0$ – we can always redefine $f(\omega)$ in ansatz (5.1.1) so $m = 0$ is the first non-zero term. This is known as the *eikonal equation*. Surfaces where $T(\mathbf{x}) = t$ are called *wavefronts*. The slowness vector (5.1.6) is perpendicular to the wavefronts (Figure 5.2).

Suppose we consider a ray which is the orthogonal trajectory to the wavefronts and measure arc length, s, along this ray. Then the eikonal equation is equivalent to

$$\frac{\mathrm{d}T}{\mathrm{d}s} = \pm\frac{1}{\alpha}. \tag{5.1.12}$$

Provided we choose to measure s in the direction of increasing T, we can take the positive sign and

$$T(s) = T(s_0) + \int_{s_0}^{s} \frac{\mathrm{d}s}{\alpha}, \tag{5.1.13}$$

where the integration is along the ray path. Now we need equations to find the ray paths.

5.1.4 The acoustic kinematic ray equations

For acoustic waves it is relatively simple to reduce the partial differential eikonal equation (5.1.11), with definition (5.1.6), to a system of ordinary differential equations which can be solved for the ray paths, i.e. trajectories orthogonal to the wavefronts.

The ray path is orthogonal to the wavefront so $\mathrm{d}\mathbf{x}/\mathrm{d}T$ must be in the direction of the slowness vector $\mathbf{p} = \nabla T$. Using equation (5.1.12) with $|\mathrm{d}\mathbf{x}| = \mathrm{d}s$ we must have

$$\boxed{\frac{\mathrm{d}\mathbf{x}}{\mathrm{d}T} = \alpha^2 \mathbf{p} = \mathbf{V},} \tag{5.1.14}$$

where $\mathbf{V}$ is the *ray velocity* ($|\mathbf{V}| = \alpha$). For the rate of change of the slowness direction we have

$$\boxed{\frac{\mathrm{d}\mathbf{p}}{\mathrm{d}T} = \alpha \frac{\mathrm{d}}{\mathrm{d}s} (\nabla T) = \alpha \nabla \left(\frac{\mathrm{d}T}{\mathrm{d}s} \right) = \alpha \nabla \left(\frac{1}{\alpha} \right) = -\frac{\nabla \alpha}{\alpha},} \tag{5.1.15}$$

using definition (5.1.6) and result (5.1.12). The crucial step in this derivation is the commutation of the operators $\mathrm{d}/\mathrm{d}s$ and ∇. Equations (5.1.14) and (5.1.15) are known as the *kinematic ray equations*. They are a set of coupled, ordinary differential equations, of sixth order, for the position, $\mathbf{x}$, and slowness, $\mathbf{p}$, vectors. They allow the ray paths to be found since given initial conditions, position $\mathbf{x}_\mathrm{S}$, and direction $\hat{\mathbf{p}}_\mathrm{S} = \alpha(\mathbf{x}_\mathrm{S})\mathbf{p}_\mathrm{S}$, the ordinary differential equations can be solved to *trace* the ray. The solution, as a result of *ray shooting*, would be written as $\mathbf{x}_\mathrm{R}(T, \mathbf{x}_\mathrm{S}, \mathbf{p}_\mathrm{S})$ and $\mathbf{p}_\mathrm{R}(T, \mathbf{x}_\mathrm{S}, \mathbf{p}_\mathrm{S})$, i.e. for a given initial position and direction, we find the ray position and slowness as a function of time along the ray. Although the kinematic ray equations form a sixth-order differential system, not all are independent. There are two constraints:

$$\boxed{\alpha |\mathbf{p}| = 1} \tag{5.1.16}$$

$$\boxed{\mathbf{p} \cdot \mathbf{V} = 1.} \tag{5.1.17}$$

The first follows from the eikonal equation, (5.1.6) and (5.1.11), and the second from equation (5.1.14), although it is a general result, valid for any wave propagation (see Figure 5.7 below).

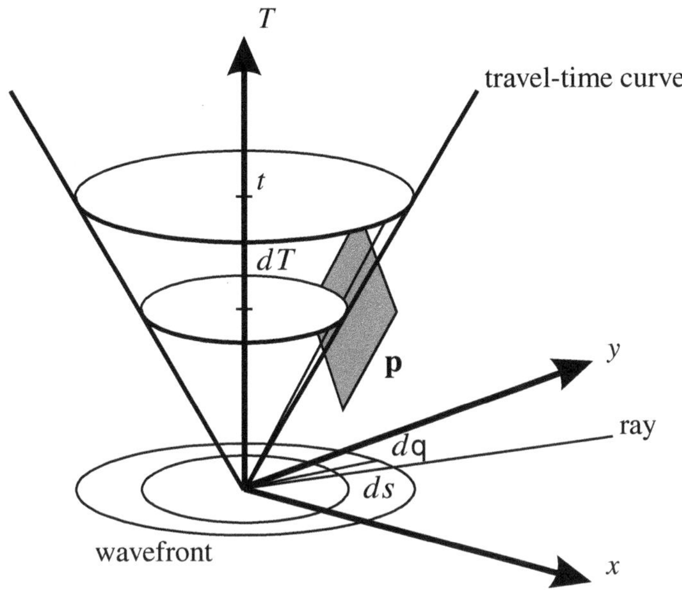

Fig. 5.3. A ray cone in two-dimensional space. The projection of the $T = t$ curve onto the $\mathbf{x}$ plane is the wavefront. The distances $\mathrm{d}T$, $\mathrm{d}q$ and $\mathrm{d}s$ are indicated.

Although this simple derivation is adequate for acoustic waves, it is useful to mention a more comprehensive mathematical method which can be used in anisotropic media. It emphasizes the close relationship to methods used in other fields. Consider the Hamiltonian

$$\boxed{H(\mathbf{x}, \mathbf{p}) = \frac{1}{2}\,\alpha^2(\mathbf{x})\,\mathbf{p}^2.} \tag{5.1.18}$$

The eikonal equation (5.1.9) is equivalent to $H = 1/2$. With definition (5.1.6), this partial differential equation

$$H(\mathbf{x}, \nabla T) = \frac{1}{2}\,\alpha^2(\mathbf{x})\,\frac{\partial T}{\partial x_i}\,\frac{\partial T}{\partial x_i} = \frac{1}{2}, \tag{5.1.19}$$

can be solved by the method of characteristics (Courant and Hilbert, 1966, Vol. 2, Chapters 2 and 6). It is equivalent to the *Hamilton–Jacobi equations*. The solution $T(\mathbf{x})$ is known as an integral surface in the $\mathbf{x}$–T space. The integral surface through a point $\mathbf{x}_S$ is known as the *ray cone*. If $\mathbf{x}$ is restricted to two dimensions, this is easily visualized (Figure 5.3).

For a homogeneous medium, the surface is a simple cone. A cross-section parallel to the $\mathbf{x}$ plane is a wavefront ($T(\mathbf{x}) = t$, a circle in a homogeneous medium), and a cross-section on a plane containing the T axis is a travel-time

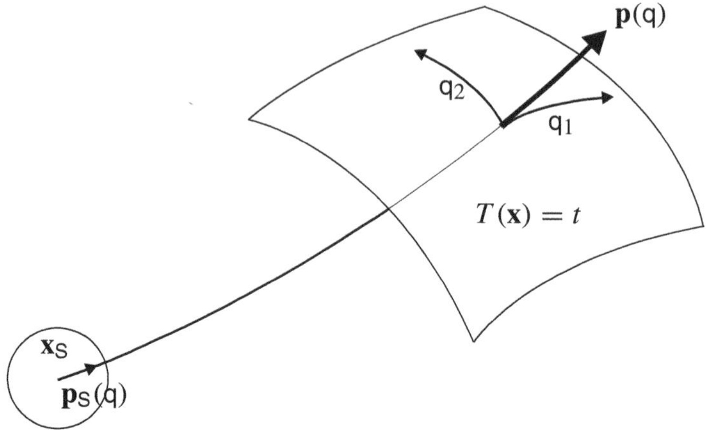

Fig. 5.4. A wavefront at time T indicating ray parameters, q.

curve (Chapter 2). A tangent plane to the ray cone has gradient $\mathbf{p}$. From equation (5.1.6), $\mathrm{d}T = \mathbf{p} \cdot \mathrm{d}\mathbf{x}$. The ray cone is generated by a two-parameter family of characteristic curves, or rays, through the vertex of the cone. Thus we can consider the rays through the vertex as having tangent planes $\mathbf{p}(\mathsf{q})$, where q is a vector of the parameters. In two dimensions it would be a single parameter, e.g. the ray angle, and in three dimensions q is a 2–vector and the parameters might be the polar angles. In general q can be considered as curvilinear coordinates on the wavefront. We call its components the *ray parameters*. This is illustrated in Figure 5.4.

If s is the distance along the ray from the vertex

$$\frac{\mathrm{d}T}{\mathrm{d}s} = \mathbf{p}(\mathsf{q}) \cdot \frac{\mathrm{d}\mathbf{x}}{\mathrm{d}s}, \tag{5.1.20}$$

from definition (5.1.6). Differentiating this with respect to the parameters, q_ν, we have

$$0 = \frac{\partial p_i}{\partial \mathsf{q}_\nu} \frac{\mathrm{d}x_i}{\mathrm{d}s} \tag{5.1.21}$$

($\nu = 1$ or 2 with summation over i) as parameters, q_ν, do not vary along the ray, only in the wavefront. Differentiating the Hamiltonian (5.1.18) with respect to the parameters we have

$$\frac{\partial H}{\partial p_i} \frac{\partial p_i}{\partial q_\nu} = 0. \tag{5.1.22}$$

Comparing equations (5.1.21) and (5.1.22) with equation (5.1.20), the relation

$$\mathrm{d}x_i : \mathrm{d}T = \frac{\partial H}{\partial p_i} : p_j \frac{\partial H}{\partial p_j} \tag{5.1.23}$$

holds for the generators of the cone. But

$$p_j \frac{\partial H}{\partial p_j} = \alpha^2 p_j p_j = 1, \tag{5.1.24}$$

so

$$\boxed{\frac{\mathrm{d}x_i}{\mathrm{d}T} = \frac{\partial H}{\partial p_i}.} \tag{5.1.25}$$

Now differentiating the Hamiltonian with respect to components of $\mathbf{x}$,

$$\frac{\partial H}{\partial p_i}\frac{\partial p_i}{\partial x_j} + \frac{\partial H}{\partial x_j} = 0, \tag{5.1.26}$$

which with equation (5.1.25) reduces to

$$\boxed{\frac{\mathrm{d}p_i}{\mathrm{d}T} = -\frac{\partial H}{\partial x_i}.} \tag{5.1.27}$$

Equations (5.1.25) and (5.1.27) are equivalent to the results (5.1.14) and (5.1.15). Thus the *kinematic ray equations* are equivalent to the *Hamilton equations*, widely used in mechanics.

Many other forms for the ray Hamiltonian are possible and have appeared in the literature, e.g. the definition (5.1.18) can be replaced by $H(\mathbf{x}, \mathbf{p}) = (\mathbf{p}^2 - \alpha^{-2})/2 = 0$, resulting in independent variables other than T. However, expression (5.1.18) seems the most natural, particularly when we generalize to anisotropy. Červený (2002) has also shown that only if the Hamiltonian is of second degree in slowness, as definition (5.1.18) clearly is, will the Lagrangian and Hamiltonian be related by the Legendre transform (Section 3.4.1). For other choices, the Legendre transform breaks down as the Jacobian is zero.

The six-dimensional space of position and slowness, $\mathbf{x} \times \mathbf{p}$, is called the *phase space*. For notational brevity, it is convenient to define a six-dimensional phase space vector

$$\boxed{\mathbf{y} = \begin{pmatrix} \mathbf{x} \\ \mathbf{p} \end{pmatrix}.} \tag{5.1.28}$$

The kinematic ray equations (5.1.25) and (5.1.27) can then be written compactly as

$$\boxed{\frac{\mathrm{d}\mathbf{y}}{\mathrm{d}T} = \mathbf{I}_1 \nabla_{\mathbf{y}} H,} \tag{5.1.29}$$

where the matrix $\mathbf{I}_1$ is defined in (0.1.5), and $\nabla_{\mathbf{y}}$ is the obvious extension of the gradient operator (0.1.6) to the phase space.

Many standard results from Hamiltonian mechanics are also useful in ray theory. The Lagrangian of the system is defined by the Legendre transform (Section 3.4.1)

$$\boxed{L(\mathbf{x}, \dot{\mathbf{x}}) = \mathbf{p} \cdot \dot{\mathbf{x}} - H(\mathbf{x}, \mathbf{p}) = \frac{\dot{\mathbf{x}}^2}{2\alpha^2(\mathbf{x})},} \tag{5.1.30}$$

where we have used the shorthand $\dot{\mathbf{x}} = \mathrm{d}\mathbf{x}/\mathrm{d}T = \mathbf{V}$. We have

$$\dot{x}_i \frac{\partial L}{\partial \dot{x}_i} = p_i \frac{\partial H}{\partial p_i} = 1 \tag{5.1.31}$$

$$\frac{\partial L}{\partial x_i} = -\frac{\partial H}{\partial x_i} = \dot{p}_i = -\frac{1}{\alpha}\frac{\partial \alpha}{\partial x_i}. \tag{5.1.32}$$

The Lagrangian satisfies the Euler–Lagrange equations

$$\boxed{\frac{\mathrm{d}}{\mathrm{d}T}\left(\frac{\partial L}{\partial \dot{x}_i}\right) - \frac{\partial L}{\partial x_i} = 0,} \tag{5.1.33}$$

and so its integral along the ray path is stationary (*Hamilton's principle*). But on the ray, the Lagrangian is constant ($L = 1/2$) and its integral is the travel time (*action*)

$$T(\mathbf{x}) = 2\int_{T_0}^{T} L(\mathbf{x}, \dot{\mathbf{x}})\,\mathrm{d}T = \int_{T_0}^{T} \mathrm{d}T \tag{5.1.34}$$

$$= \int_{T_0}^{T} (L + H)\,\mathrm{d}T = \mathrm{Ext}\int_{\mathbf{x}_0}^{\mathbf{x}} \mathbf{p} \cdot \mathrm{d}\mathbf{x}. \tag{5.1.35}$$

In ray theory, Hamilton's principle is known as *Fermat's principle*. The last result is equivalent to integrating the constraint (5.1.17) with definition (5.1.6) on the ray path.

5.1.4.1 Example of kinematic ray tracing

As an example of ray tracing in a simple, three-dimensional model we consider a model used in Brandsberg-Dahl, de Hoop and Ursin (2003). This is an idealized model of a gas cloud in the Valhall Field located in the Norwegian part of the North Sea. In a background model of a linear vertical velocity gradient, the velocity is reduced in the 'gas-cloud' with a negative perturbation in the form of a spherically symmetric Gaussian function. The model is defined as

$$\alpha(\mathbf{x}) = 1.6 - 0.45z - 0.8\exp(-r^2/r_0^2) \tag{5.1.36}$$

$$\beta(\mathbf{x}) = 0.6 - 0.55z + 0.1\exp(-r^2/r_0^2) \tag{5.1.37}$$

$$\rho(\mathbf{x}) = 2.0 - 0.30z - 0.2\exp(-r^2/r_0^2) \tag{5.1.38}$$

(for our purposes, only the P velocity is needed and the model can be treated as acoustic) where we have changed the direction of the z axis to correspond to our convention of measuring it positive upwards and the units are km/s and Mg/m^3, with

$$r = |\mathbf{x} - \mathbf{x}_0| \tag{5.1.39}$$

$$r_0 = 0.3 \tag{5.1.40}$$

$$\mathbf{x}_0 = (4.6,\ 0,\ -0.6), \tag{5.1.41}$$

where the units are km (see Table 1 in Brandsberg-Dahl, de Hoop and Ursin, 2003). In Figure 5.5, rays from a typical image point at $\mathbf{x} = (\,4.68\,,\ 0,\ -1.5\,)$ km are illustrated (equivalent to Figure 5 in Brandsberg-Dahl, de Hoop and Ursin, 2003). For clarity the rays have been traced in the symmetry plane, so in fact the ray paths are two dimensional although the model is three dimensional. Notice the focusing due to the low-velocity cloud and the formation of caustics. We will return to this model in Chapter 10 where we will use the Maslov method to model the signals through the caustics.

5.2 Acoustic dynamic ray theory

Having developed ordinary differential equations for the position, $\mathbf{x}$, and slowness, $\mathbf{p}$, of a ray, the kinematic ray equations, and implicitly for the travel time, T, the independent variable, we now need differential equations for the amplitude coefficients. First we consider the solution for the zeroth-order amplitude coefficients, $\mathbf{v}^{(0)}$ and $P^{(0)}$, and later we return to solving iteratively for higher-order terms (Section 5.2.4).

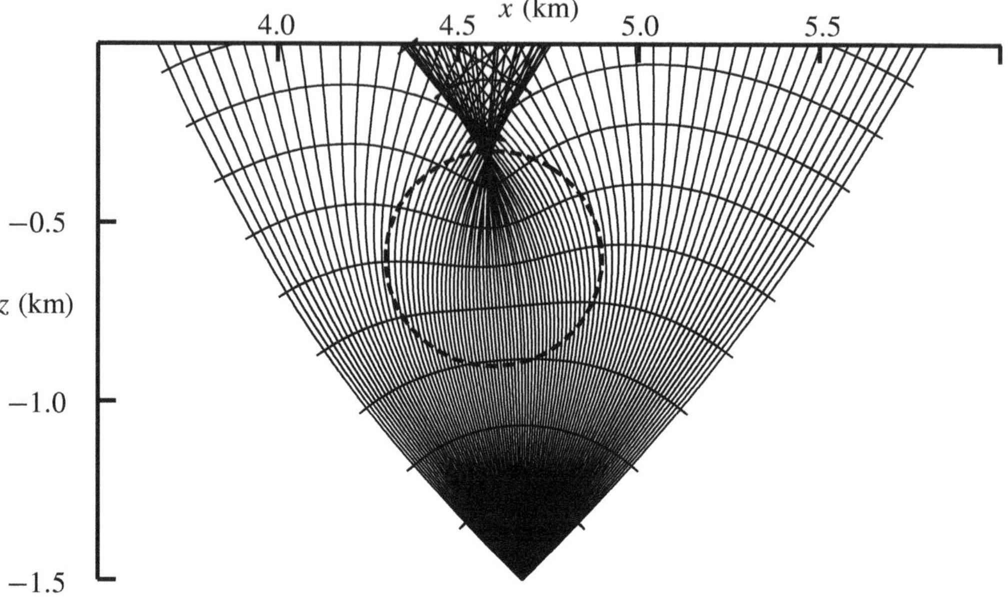

Fig. 5.5. A diagram of rays traced in the $y = 0$ km plane from a source at $\mathbf{x}_S = (4.68,\ 0, -1.5)$ km in the Valhall gas-cloud model (equations (5.1.36)–(5.1.41)). Rays are traced at angles $-45°$ to $45°$ from the vertical in $1°$ intervals. The wavefronts are marked at 0.1 s intervals. The gas cloud is indicated by the dashed circle centred on $\mathbf{x}_0$ (5.1.41) with radius r_0 (5.1.40).

5.2.1 The acoustic transport equation

First we find the *polarization* of the zeroth-order term, i.e. the direction of $\mathbf{v}^{(0)}$. For $m = 0$, equations (5.1.4) and (5.1.5) reduce to

$$\rho\, \mathbf{v}^{(0)} = P^{(0)}\mathbf{p} \tag{5.2.1}$$

$$P^{(0)} = \kappa\, \mathbf{p} \cdot \mathbf{v}^{(0)}. \tag{5.2.2}$$

Equation (5.1.8) with $m = 0$ reduces to

$$\left(\alpha^2 \mathbf{p}\mathbf{p}^{\mathrm{T}} - \mathbf{I}\right) \mathbf{v}^{(0)} = \mathbf{0}, \tag{5.2.3}$$

which is equivalent to the eikonal equation (5.1.9) except that it is an eigenvector rather than scalar equation. By inspection of equation (5.2.3), or from equation (5.2.1), the solution for $\mathbf{v}^{(0)}$ must be parallel to the slowness vector $\mathbf{p}$, but the amplitude is undetermined. The geometrical ray approximation for an acoustic wave is longitudinal, as expected, i.e. the unit polarization, $\hat{\mathbf{g}}$, and the ray velocity, $\mathbf{V}$, are parallel. Let us define the normalized or unit polarization vector

$$\boxed{\hat{\mathbf{g}} = \alpha \mathbf{p},} \tag{5.2.4}$$

and the scalar amplitude function $v^{(0)}$ such that

$$\boxed{\mathbf{v}^{(0)}(\mathbf{x}, \mathcal{L}_n) = v^{(0)}(\mathbf{x}, \mathcal{L}_n)\, \hat{\mathbf{g}}(\mathbf{x}, \mathcal{L}_n)} \tag{5.2.5}$$

(note that all terms depend on the location and the ray path).

In order to find $v^{(0)}$, we consider equation (5.1.8) with $m = 1$. Pre-multiplying by $\mathbf{v}^{(0)^{\mathrm{T}}}$, the left-hand side contains the transpose of equation (5.2.3) and is zero. The right-hand side can be simplified using equation (5.2.2) and is simply

$$\nabla \cdot \left(P^{(0)} \mathbf{v}^{(0)} \right) = 0. \tag{5.2.6}$$

Substituting equations (5.2.2), (5.2.4) and (5.2.5), this reduces to

$$\nabla \cdot \left(Z\, v^{(0)^2} \hat{\mathbf{g}} \right) = 0, \tag{5.2.7}$$

where

$$\boxed{Z = \kappa/\alpha = \rho\, \alpha,} \tag{5.2.8}$$

is the *scalar impedance* (unit(Z) = [$\mathrm{ML^{-2}T^{-1}}$]). The impedance connects the pressure and velocity amplitude coefficients, i.e.

$$P^{(0)} = Z\, \hat{\mathbf{p}} \cdot \mathbf{v}^{(0)} = Z \left| \mathbf{v}^{(0)} \right| . \tag{5.2.9}$$

The vector $\mathbf{N} = P^{(0)} \mathbf{v}^{(0)} = Z v^{(0)^2} \hat{\mathbf{g}}$ is the energy flux vector, analogous to the Poynting vector in electromagnetism, and equation (5.2.6) states that $\nabla \cdot \mathbf{N} = 0$, i.e. energy is conserved as the net flux in and out of any volume is zero.

Substituting $\mathbf{V} = \alpha \hat{\mathbf{g}}$ in equation (5.2.7) with definition (5.2.8), it can be rewritten

$$\frac{\mathrm{d}}{\mathrm{d}T} \ln \left(\rho v^{(0)^2} \right) = -\nabla \cdot \mathbf{V}, \tag{5.2.10}$$

using $\mathbf{V} \cdot \nabla = \mathrm{d}/\mathrm{d}T$ along the ray. This ordinary differential equation (5.2.10), or variants thereof, is known as the *transport equation*.

We introduce the concept of a *ray tube* defined as the volume swept out by the wavefront between neighbouring rays with slightly different initial conditions (Figure 5.6). Suppose that the position on the wavefront is parameterized by two variables q_1 and q_2, which we called the ray parameters. We define the cross-sectional area of a ray tube defined by perturbations dq_1 and dq_2 as $J \mathrm{dq}_1 \mathrm{dq}_2$. $J(T, \mathrm{q}_1, \mathrm{q}_2)$ is the cross-sectional area function. Considering the volume formed by the ray tube, we must have

$$\int_{\text{raytube}} \nabla \cdot \mathbf{N}\, \mathrm{d}V = 0 = \int \mathbf{N} \cdot \mathrm{d}\mathbf{S}. \tag{5.2.11}$$

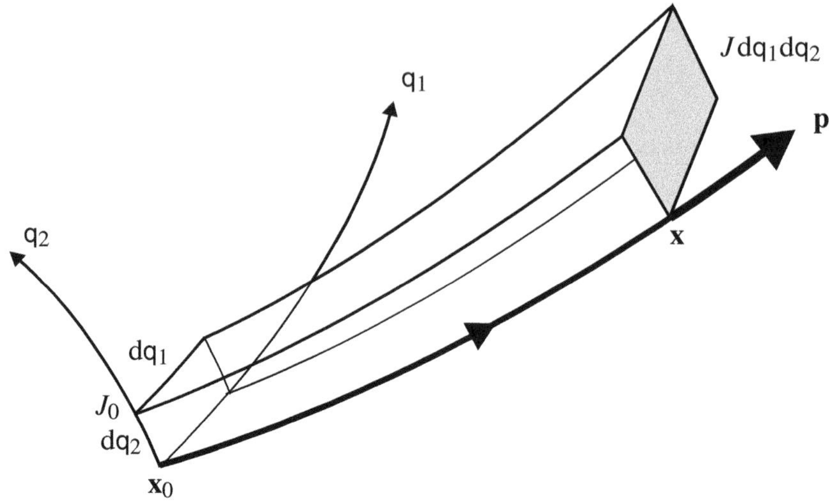

Fig. 5.6. A ray tube formed by perturbation dq_1 and dq_2 in the ray parameters. The cross-section of the 'rectangular' ray tube is $J\mathsf{dq}_1\mathsf{dq}_2$.

As the vector **N** is parallel to the rays, the surface integral is only non-zero on the wavefront elements at the ends of the tube. Thus result (5.2.7) is equivalent to

$$-\left(Zv^{(0)^2}J\right)_0 \mathsf{dq}_1\mathsf{dq}_2 + \left(Zv^{(0)^2}J\right)\mathsf{dq}_1\mathsf{dq}_2 = 0, \qquad (5.2.12)$$

for these two elements, where the subscript $_0$ indicates the initial point where **N** is inwards – hence the minus sign. Thus

$$v^{(0)} = \frac{\text{constant}}{\sqrt{ZJ}}. \qquad (5.2.13)$$

We postpone establishing a value for the constant, which is found by matching with a known solution, e.g. the point-source results (Section 4.5.5), until we have investigated the paraxial ray equations used for finding the tube cross-sectional area function, $J(T, \mathsf{q}_1, \mathsf{q}_2)$.

An alternative, non-geometrical proof of the result (5.2.13) is based on Smirnov's lemma (Appendix C.2), and can be used for more complicated transport equations. The kinematic ray equation (5.1.14) is of the form (C.2.1) and we can apply Smirnov's lemma (C.2.8). Thus

$$\frac{\mathrm{d}}{\mathrm{d}T}\ln D = \nabla\cdot\mathbf{V}, \qquad (5.2.14)$$

where D is the Jacobian mapping a volume element from an initial point $\mathbf{x}_0$ to $\mathbf{x}$. To define a volume element at $\mathbf{x}_0$, we perturb the initial point on the wavefront, and in the ray direction perpendicular to the wavefront. In the ray direction, the perturbation is $\alpha\,\mathrm{d}T$. Using these perturbations to define a volume element, the Jacobian is

$$D = \frac{\alpha J}{\alpha_0 J_0}, \tag{5.2.15}$$

so equation (5.2.14) is

$$\frac{\mathrm{d}}{\mathrm{d}T}\ln\left(\frac{\alpha J}{\alpha_0 J_0}\right) = \nabla\cdot\mathbf{V}, \tag{5.2.16}$$

and combining with equation (5.2.10), we obtain

$$\boxed{\frac{\mathrm{d}}{\mathrm{d}T}\ln\left(Z v^{(0)^2} J\right) = 0.} \tag{5.2.17}$$

The solution of this equation is, of course, the same result (5.2.13). In these equations, (5.2.15) and (5.2.16), the initial point, $\mathbf{x}_0$, is usually taken on a small (infinitesimal) sphere (wavefront) about the source, $\mathbf{x}_S$ (to avoid the singularity at the source), where the amplitude is known from the point source solutions (Sections 4.5.5 and 4.6.2).

5.2.2 Paraxial ray equations

The cross-sectional area function of a ray tube, $J(T, \mathrm{q}_1, \mathrm{q}_2)$, can be calculated numerically by tracing neighbouring rays with perturbed parameters, $\mathrm{q}_\nu + \delta\mathrm{q}_\nu$. In general, however, it is more satisfactory to develop the paraxial ray equations which directly determine the cross-section without numerical perturbation and differencing. Both the ray position and slowness are perturbed, i.e.

$$\mathrm{d}\mathbf{y} = \begin{pmatrix}\mathrm{d}\mathbf{x}\\ \mathrm{d}\mathbf{p}\end{pmatrix} = \begin{pmatrix}\partial\mathbf{x}/\partial\mathrm{q}_i\\ \partial\mathbf{p}/\partial\mathrm{q}_i\end{pmatrix}\mathrm{d}\mathrm{q}_i, \tag{5.2.18}$$

using the phase-space vector (5.1.28). We have used the notation $\mathrm{d}\mathbf{y}$ for an infinitesimal perturbation of the position-slowness vector, $\mathbf{y}$ (5.1.28). These perturbations or partial derivatives satisfy the *paraxial ray* or *dynamic ray equations*. Using the compact notation (5.1.29) for the kinematic ray equations, these are

$$\boxed{\frac{\mathrm{d}\,\mathrm{d}\mathbf{y}}{\mathrm{d}T} = \mathbf{I}_1\,\nabla_\mathbf{y}\left(\nabla_\mathbf{y} H\right)^{\mathrm{T}}\mathrm{d}\mathbf{y}} = \mathbf{D}\,\mathrm{d}\mathbf{y}, \text{ say.} \tag{5.2.19}$$

As the dynamic ray equations are linear in the perturbation, the variable of the infinitesimal perturbation, $\mathrm{d}\mathbf{y}$, can be replaced by differentials with respect to a ray parameter, $\partial\mathbf{y}/\partial\mathrm{q}_i$ or, assuming first-order perturbation theory, by a finite perturbation, $\delta\mathbf{y}$. Dividing the 6×6 matrix $\nabla_{\mathbf{y}}\left(\nabla_{\mathbf{y}}H\right)^{\mathrm{T}}$ into its 3×3 sub-matrices

$$\boxed{\mathbf{D}=\mathbf{I}_1\nabla_{\mathbf{y}}\left(\nabla_{\mathbf{y}}H\right)^{\mathrm{T}}=\begin{pmatrix}\mathbf{T}^{\mathrm{T}} & \mathbf{R}\\ -\mathbf{S} & -\mathbf{T}\end{pmatrix},}\quad \text{say,} \tag{5.2.20}$$

where the elements are $T_{ij}=\partial^2 H/\partial x_i\partial p_j$, $R_{ij}=\partial^2 H/\partial p_i\partial p_j$ and $S_{ij}=\partial^2 H/\partial x_i\partial x_j$. The matrices $\mathbf{R}$ and $\mathbf{S}$ are symmetric. For the acoustic rays, the matrix $\mathbf{D}$ reduces to

$$\mathbf{D}=\begin{pmatrix}\mathbf{0} & \alpha^2\mathbf{I}\\ \mathbf{C} & \mathbf{0}\end{pmatrix} \tag{5.2.21}$$

where

$$C_{ij}=\left(\frac{1}{\alpha}\frac{\partial\alpha}{\partial x_i}\right)\left(\frac{1}{\alpha}\frac{\partial\alpha}{\partial x_j}\right)-\frac{1}{\alpha}\frac{\partial^2\alpha}{\partial x_i\partial x_j}. \tag{5.2.22}$$

Equation (5.2.19) is easily solved numerically, and for some special cases analytically. Systems of linear, ordinary differential equations, such as (5.2.19), arise in many applications. General results and terminology are discussed in Appendix C.1. In particular, the concept of the propagator matrix solution is introduced. In general the solutions at two positions on a ray can be related by the propagator of the matrix $\mathbf{D}$, which we denote by $\mathbf{P}$. In Appendix C.1, we discuss the use of the propagator to solve a system of ordinary differential equations. With the independent variable T, we would normally denote the propagator from T_0 to T by $\mathbf{P}(T, T_0)$ (Section C.1). However, it is important to remember the dependence of the solution on the ray path and possible multi-pathing, although we do not include it in our notation. We simply write

$$\boxed{\mathrm{d}\mathbf{y}=\mathbf{P}(T, T_0)\,\mathrm{d}\mathbf{y}_0.} \tag{5.2.23}$$

It is convenient to divide the propagator into 3×3 sub-matrices containing the different partial derivatives (cf. equation (5.2.18)). Thus we write the full dynamic propagator as

$$\boxed{\mathbf{P}=\begin{pmatrix}\mathbf{P}_{xx} & \mathbf{P}_{xp}\\ \mathbf{P}_{px} & \mathbf{P}_{pp}\end{pmatrix},} \tag{5.2.24}$$

where, for instance, $\mathbf{P}_{xp}$ contains derivatives of the position, $\mathbf{x}$, with respect to slowness, $\mathbf{p}_0$. The notation is intended to remind us that the sub-matrix, $\mathbf{P}_{xp}$, of the propagator, $\mathbf{P}$, contains the partial derivatives, $\partial\mathbf{x}/\partial\mathbf{p}_0$, etc. This sub-matrix describes the spreading of rays from a point source and later we will see how the cross-sectional area function, J, can be derived from it. We call it the *spreading matrix.*

We should comment that the sub-matrices of the dynamic propagator (5.2.24) are commonly written

$$\mathbf{P} = \begin{pmatrix} \mathbf{Q}_1 & \mathbf{Q}_2 \\ \mathbf{P}_1 & \mathbf{P}_2 \end{pmatrix}$$

e.g. Červený (2001, p. 279). We prefer the notation in equation (5.2.24) as an *aide mémoire* of the derivatives in the sub-matrices.

In practice, we never compute the full 6×6 propagator $\mathbf{P}$ as only two or four solutions are needed (for perturbations of the source position or ray direction). Instead we directly find part of a fundamental matrix, say $\mathbf{J}$. As initial conditions for $\mathbf{dy}$ at a source point, $\mathbf{x}_S$, we consider perturbations in the location satisfying constraint (5.1.16), and perturbations in the slowness keeping the location fixed. The former requires a simultaneous perturbation in the slowness in order to compensate for the perturbation in velocity due to the position change. As the dynamic ray equations (5.2.19) are linear, we can use a finite perturbation for the differential, $\mathbf{dy}$, in the numerical solution. Suppose $\mathbf{q}_\nu$ are any two vectors normal to the ray slowness at $\mathbf{x}_S$, i.e. $\mathbf{q}_\nu \cdot \mathbf{p}_S = 0$ (we use Greek indices when the range is less than 3). Typically such vectors can be defined as

$$\mathbf{q}_\nu = \frac{\partial \mathbf{p}_S}{\partial q_\nu}. \tag{5.2.25}$$

For a point source in an isotropic medium, such as the acoustic medium under consideration, these vectors lie in the spherical slowness surface orthogonal to the ray direction. Two such vectors can be constructed as

$$\mathbf{q}_2 = \mathrm{sgn} \begin{pmatrix} -p_2 \\ p_1 \\ 0 \end{pmatrix} \qquad \text{and} \qquad \mathbf{q}_1 = \mathrm{sgn}\,(\mathbf{q}_2 \times \mathbf{p}_0)\,, \tag{5.2.26}$$

corresponding to perturbations in the polar angles θ and ϕ (measured with respect to the z axis), on a unit sphere around the source, $\mathbf{x}_S$. These vectors have been normalized and are orthogonal. This is a convenient choice but not essential. Defining a 3×2 matrix of these wavefront unit vectors

$$\mathbf{Q}_S = \begin{pmatrix} \mathbf{q}_1 & \mathbf{q}_2 \end{pmatrix}, \tag{5.2.27}$$

suitable initial conditions for the required solutions of (5.2.20) are then

$$\mathbf{J}_S = \begin{pmatrix} \mathbf{Q}_S & \mathbf{0} \\ -\mathbf{p}_S\,(\nabla\alpha/\alpha)_S^{\mathrm{T}}\,\mathbf{Q}_S & \mathbf{Q}_S \end{pmatrix}, \tag{5.2.28}$$

a 6×4 matrix. The first two columns correspond to a perturbation of the source position, and the last two correspond to a perturbation of the source ray direction. In the first two columns, the slowness is also perturbed so the Hamiltonian constraint (5.1.18) is still satisfied. The solutions that are found are

$$\mathbf{J} = \begin{pmatrix} \mathbf{J}_{xx} & \mathbf{J}_{xp} \\ \mathbf{J}_{px} & \mathbf{J}_{pp} \end{pmatrix} = \mathbf{P}\,\mathbf{J}_\mathsf{S}, \tag{5.2.29}$$

which is a partial dynamic fundamental matrix. The equations are linear, so any initial perturbation is allowed, e.g.

$$\mathsf{dy} = \mathbf{J}\,\mathsf{dy}_\mathsf{S} = (\mathbf{P}\,\mathbf{J}_\mathsf{S})\,\mathsf{dy}_\mathsf{S}, \tag{5.2.30}$$

where dy_S is a 4×1 vector, specifying the source position and slowness perturbation in wavefront coordinates. The matrix $\mathbf{J} = \mathbf{P}\mathbf{J}_\mathsf{S}$ is 6×4, and its sub-matrices, e.g. $\mathbf{J}_{xp}$, are 3×2. For the geometrical ray results of a point source, we will find we only need the third and fourth columns of equations (5.2.28) and (5.2.29). As the initial conditions satisfy the constraints (5.1.16) and (5.1.17), and the independent variable is time, T, the perturbations remain in the wavefront and continue to satisfy the constraints. It is sometimes convenient, at least formally, to express the wavefront perturbation in terms of a wavefront basis, $\mathbf{Q}$, a 3×2 matrix of unit vectors in the wavefront. Then the wavefront perturbations are

$$\mathsf{dy} = \mathbf{Q}^\mathsf{T}\mathbf{J}\,\mathsf{dy}_\mathsf{S} = \mathbf{Q}^\mathsf{T}\left(\mathbf{P}\,\hat{\mathbf{J}}_\mathsf{S}\right)\,\mathsf{dy}_\mathsf{S} = \mathsf{P}\,\mathsf{dy}_\mathsf{S},\ \text{say}, \tag{5.2.31}$$

where P is the 4×4 wavefront propagator.

As at most we only require the four solutions of the dynamic equations (5.2.20), the equations can be reduced to a fourth-order system using wavefront coordinates (which enforce the constraints (5.1.16) and (5.1.17) on the solutions). However, reduced forms of the equations are algebraically more complicated and in general it is easier to use the full system at least for fundamental theoretical manipulations. Formally, we use the full sixth-order propagator, $\mathbf{P}$, which maps a volume in the six-dimensional phase space from T_S to T, but numerically we only compute the partial fundamental matrix $\mathbf{J} = \mathbf{P}\mathbf{J}_\mathsf{S}$. The two extra solutions are easily investigated without solving the dynamic equations (5.2.20) explicitly. First if the initial point is perturbed in the ray direction, i.e. $\mathrm{d}\mathbf{x}_\mathsf{S} = \mathrm{d}T_\mathsf{S}\mathbf{V}_\mathsf{S}$, then the final point is also perturbed in the ray direction by the same time shift, i.e. $\mathrm{d}\mathbf{x} = \mathrm{d}T_\mathsf{S}(\alpha/\alpha_\mathsf{S})\mathbf{V}$ (with a slowness perturbation from the velocity gradient). The other solution is similar but more subtle, corresponding to a perturbation in the slowness magnitude. We outline an argument due to Burridge (private communication). If the slowness is perturbed in the magnitude, i.e. $\mathrm{d}\mathbf{p}_\mathsf{S} = \epsilon\,\mathbf{p}_\mathsf{S}$, then if the travel time $T - T_\mathsf{S}$ is increased in the same ratio $1 + \epsilon$ and the slowness by the same ratio everywhere, then the Hamiltonian system and the ray paths remain the same. The perturbation at time T is therefore $\mathrm{d}\mathbf{x} = \epsilon(T - T_\mathsf{S})\mathbf{V}$. This argument follows simply as the Hamiltonian is second degree in slowness, further justifying our choice of Hamiltonian (5.1.18).

We emphasize again that while formally we use the full dynamic propagator matrix, $\mathbf{P}$ (5.2.24), numerically we only calculate a partial dynamic fundamental matrix, $\mathbf{J}$ (5.2.29). It would be straightforward to add two solutions to this to form a complete fundamental matrix, and to compute the propagator (C.1.4), but this is not necessary.

5.2.2.1 Symplectic symmetries

As already noted the sub-matrices of the differential system (5.2.19) contain the symmetries $\mathbf{D}_{11} = -\mathbf{D}_{22}^{\mathrm{T}}$, $\mathbf{D}_{12} = \mathbf{D}_{12}^{\mathrm{T}}$ and $\mathbf{D}_{21} = \mathbf{D}_{21}^{\mathrm{T}}$ (5.2.20). These apply to any Hamiltonian system and result in important inter-relationships within the propagator matrix $\mathbf{P}$. In this section we investigate these symmetries. The results apply to the general system (5.2.20) without the acoustic simplification (5.2.21). In particular, they remain valid for the general anisotropic system considered in Section 5.4.

We define a *symplectic transform* of a matrix $\mathbf{A}$ as

$$\boxed{\mathbf{A}^{\dagger} = \mathbf{I}_1^{\mathrm{T}} \mathbf{A}^{\mathrm{T}} \mathbf{I}_1,} \tag{5.2.32}$$

where $\mathbf{I}_1$ is defined in (0.1.5). Note that $\mathbf{I}_1^{\mathrm{T}} = -\mathbf{I}_1 = \mathbf{I}_1^{-1}$. Sub-dividing the matrix $\mathbf{A}$ into sub-matrices, we have

$$\mathbf{A}^{\dagger} = \begin{pmatrix} \mathbf{A}_{22}^{\mathrm{T}} & -\mathbf{A}_{12}^{\mathrm{T}} \\ -\mathbf{A}_{21}^{\mathrm{T}} & \mathbf{A}_{11}^{\mathrm{T}} \end{pmatrix} \quad \text{where} \quad \mathbf{A} = \begin{pmatrix} \mathbf{A}_{11} & \mathbf{A}_{12} \\ \mathbf{A}_{21} & \mathbf{A}_{22} \end{pmatrix}. \tag{5.2.33}$$

It is obvious that the matrix $\mathbf{D}$ defined in (5.2.20) satisfies $\mathbf{D} = -\mathbf{D}^{\dagger}$, i.e. it is anti-symplectic.

Now let us consider the propagator $\mathbf{P}$ and the combination $\mathbf{P}^{\mathrm{T}}\mathbf{I}_1\mathbf{P}$. We have

$$\begin{aligned} \frac{\mathrm{d}}{\mathrm{d}T}\left(\mathbf{P}^{\mathrm{T}}\mathbf{I}_1\mathbf{P}\right) &= \frac{\mathrm{d}\mathbf{P}^{\mathrm{T}}}{\mathrm{d}T}\mathbf{I}_1\mathbf{P} + \mathbf{P}^{\mathrm{T}}\mathbf{I}_1\frac{\mathrm{d}\mathbf{P}}{\mathrm{d}T} \\ &= \mathbf{P}^{\mathrm{T}}\left(\mathbf{D}^{\mathrm{T}}\mathbf{I}_1 + \mathbf{I}_1\mathbf{D}\right)\mathbf{P} \\ &= \mathbf{0}, \end{aligned} \tag{5.2.34}$$

as $\mathbf{D} = -\mathbf{D}^{\dagger}$. Thus

$$\mathbf{P}^{\mathrm{T}}(T, T_0)\mathbf{I}_1\mathbf{P}(T, T_0) = \text{ constant } = \mathbf{I}_1, \tag{5.2.35}$$

as $\mathbf{P}(T_0, T_0) = \mathbf{I}$. Thus we obtain

$$\boxed{\mathbf{P}^{-1}(T, T_0) = \mathbf{P}^{\dagger}(T, T_0) = \mathbf{P}(T_0, T).} \tag{5.2.36}$$

The inverse of the 6×6 propagator matrix can be obtained by a symplectic transform. This result is of considerable practical and theoretical importance as it means that rays can be traced in either direction without the expense of inverting a 6×6 matrix. It is often more convenient to trace rays in the reversed direction, e.g. in the Kirchhoff integral (Section 10.4), to trace rays from the receiver to the reflector.

It is important to note that the propagator, $\mathbf{P}(T_0, T)$, is not identical to the propagator for the reversed ray from $\mathbf{x}_0$ to $\mathbf{x}$. It is the reversed propagator but still for the same system of equations (5.2.19) with independent variable measured in the original direction along the ray from $\mathbf{x}$ to $\mathbf{x}_0$. The system of equations (5.2.19) depend on ray properties other than the travel time (as already mentioned, the simple notation $\mathbf{P}(T, T_0)$ obscures the fact that the propagator depends on the ray path, etc.). Thus $\mathbf{p} = \nabla T$ is the same for the forward or reversed propagator. In the propagator for the reversed ray, the slowness vector must change sign. Thus results that are of odd order in slowness, e.g. $\mathbf{P}_{xp}$ and $\mathbf{P}_{px}$, will change sign for the reversed ray, compared with the results given here for the reversed propagator.

Using results (5.2.36) and (5.2.33), we have that

$$\mathbf{P}^{-1}(T, T_0) = \begin{pmatrix} \mathbf{P}^{\mathrm{T}}_{pp} & -\mathbf{P}^{\mathrm{T}}_{xp} \\ -\mathbf{P}^{\mathrm{T}}_{px} & \mathbf{P}^{\mathrm{T}}_{xx} \end{pmatrix}, \tag{5.2.37}$$

where the sub-matrices are understood to have the same argument (T, T_0). Thus we have the reciprocal relationships

$$\mathbf{P}_{xx}(T_0, T) = \mathbf{P}^{\mathrm{T}}_{pp}(T, T_0) \tag{5.2.38}$$

$$\mathbf{P}_{pp}(T_0, T) = \mathbf{P}^{\mathrm{T}}_{xx}(T, T_0) \tag{5.2.39}$$

$$\mathbf{P}_{xp}(T_0, T) = -\mathbf{P}^{\mathrm{T}}_{xp}(T, T_0) \tag{5.2.40}$$

$$\mathbf{P}_{px}(T_0, T) = -\mathbf{P}^{\mathrm{T}}_{pp}(T, T_0). \tag{5.2.41}$$

The third relationship (5.2.40) with the spreading matrix is of great importance since it is the crucial result to establish reciprocity for the geometrical ray approximation (5.2.71). Combining the inverse propagator (5.2.37) with the propagator (5.2.24), we obtain the inter-relationships between the partial derivatives

$$\mathbf{P}^{\mathrm{T}}_{pp}\mathbf{P}_{xx} - \mathbf{P}^{\mathrm{T}}_{xp}\mathbf{P}_{px} = \mathbf{I} \tag{5.2.42}$$

$$\mathbf{P}^{\mathrm{T}}_{pp}\mathbf{P}_{xp} = \mathbf{P}^{\mathrm{T}}_{xp}\mathbf{P}_{pp} \tag{5.2.43}$$

$$\mathbf{P}^{\mathrm{T}}_{px}\mathbf{P}_{xx} = \mathbf{P}^{\mathrm{T}}_{xx}\mathbf{P}_{px}, \tag{5.2.44}$$

where a fourth relationship is equivalent to (the transpose of) the first, and the same argument is understood for all the matrices. These equations are equivalent to the fundamental Poisson and Lagrangian brackets of classical mechanics. The last two expressions (5.2.43) and (5.2.44) simply state that the matrices $\mathbf{P}^{\mathrm{T}}_{pp}\mathbf{P}_{xp}$ and $\mathbf{P}^{\mathrm{T}}_{px}\mathbf{P}_{xx}$ are symmetric.

5.2.2.2 Wavefront curvature

If we consider neighbouring rays from a point source at $\mathbf{x}_S$ related by perturbation $\delta\mathbf{p}_S$ with $\delta\mathbf{x}_S = \mathbf{0}$, position and slowness perturbations on a wavefront are given by

$$\delta\mathbf{x} \simeq \mathbf{P}_{xp}\delta\mathbf{p}_\mathrm{S} \tag{5.2.45}$$

$$\delta\mathbf{p} \simeq \mathbf{P}_{pp}\delta\mathbf{p}_\mathrm{S} = \mathbf{M}\,\delta\mathbf{x}, \tag{5.2.46}$$

where we have defined the matrix

$$\boxed{\mathbf{M} = \mathbf{P}_{pp}\mathbf{P}_{xp}^{-1}.} \tag{5.2.47}$$

The matrix

$$\mathbf{K} = c\mathbf{M}, \tag{5.2.48}$$

is known as the *curvature matrix* of the wavefront. Its elements have units of inverse length, i.e. unit($\mathbf{K}$) = [L^{-1}]. As $\mathbf{p}$ is normal to the wavefront (5.1.6) where the travel time is constant, we can integrate (5.1.17) with (5.1.14) along any path to obtain

$$T(\mathbf{x}, \mathbf{x}_\mathrm{S}) = \int_{\mathbf{x}_\mathrm{S}}^{\mathbf{x}} \mathbf{p}\cdot \mathrm{d}\mathbf{x}. \tag{5.2.49}$$

In particular, we can integrate in a plane tangent to the wavefront to obtain the time on a neighbouring ray

$$T(\mathbf{x}+\delta\mathbf{x}, \mathbf{x}_\mathrm{S}) = T(\mathbf{x}, \mathbf{x}_\mathrm{S}) + \int_{\mathbf{x}}^{\mathbf{x}+\delta\mathbf{x}} (\mathbf{p}+\delta\mathbf{p})\cdot \mathrm{d}\mathbf{x}, \tag{5.2.50}$$

where $\delta\mathbf{p}$ is related to $\delta\mathbf{x}$ by (5.2.46). Thus we obtain

$$T(\mathbf{x}+\delta\mathbf{x}, \mathbf{x}_\mathrm{S}) \simeq T(\mathbf{x}, \mathbf{x}_\mathrm{S}) + \frac{1}{2}\delta\mathbf{x}^T\mathbf{M}\delta\mathbf{x}. \tag{5.2.51}$$

As we are generally only concerned with perturbations in the plane tangent to the wavefront, this equation can be rewritten in terms of the 2×2 wavefront differentials (5.2.31), e.g.

$$T(\mathbf{x}+\delta\mathbf{x}, \mathbf{x}_\mathrm{S}) \simeq T(\mathbf{x}, \mathbf{x}_\mathrm{S}) + \frac{1}{2}\delta\mathsf{q}^T\mathsf{M}\delta\mathsf{q}, \tag{5.2.52}$$

where $\mathsf{M} = \mathsf{P}_{pp}\mathsf{P}_{xp}^{-1}$ is a 2×2 matrix. The travel-time perturbation in expressions (5.2.51) and (5.2.52) is due to the difference between the curved wavefront and a plane. It finds many uses.

5.2.2.3 *String rule for the spreading matrix*

Finally we can obtain a useful string rule for the geometrical spreading matrix $\mathbf{P}_{xp}$. The string rule for the propagator is (C.1.5), i.e.

$$\mathbf{P}(T_1, T_0) = \mathbf{P}(T_1, T)\mathbf{P}(T, T_0), \tag{5.2.53}$$

where we have assumed that the intermediate point $\mathbf{x}$ lies on the ray $(\mathbf{x}_1, \mathbf{x}_0)$. Expanding the string rule (5.2.53) with the inverse of $\mathbf{P}(T_1, T)$, we obtain

$$\mathbf{P}_{xp}(T_1, T_0) = \mathbf{P}^{\mathrm{T}}_{pp}(T, T_1)\mathbf{P}_{xp}(T, T_0) - \mathbf{P}^{\mathrm{T}}_{xp}(T, T_1)\mathbf{P}_{pp}(T, T_0). \quad (5.2.54)$$

Using the symmetry (5.2.43), we can rewrite this as

$$\mathbf{P}_{xp}(T_1, T_0) = \mathbf{P}^{\mathrm{T}}_{xp}(T, T_1)\left\{\mathbf{M}(T, T_1) - \mathbf{M}(T, T_0)\right\}\mathbf{P}_{xp}(T, T_0), \quad (5.2.55)$$

where we have used definition (5.2.47). Expression (5.2.55) together with the reciprocity relation (5.2.40) connects the matrix $\mathbf{P}_{xp}$ for two segments with the whole. As we are generally only concerned with perturbations in the wavefront, these equations can be rewritten in terms of the 2×2 wavefront differentials (5.2.31), e.g.

$$\mathsf{P}_{xp}(T_1, T_0) = \mathsf{P}^{\mathrm{T}}_{xp}(T, T_1)\left\{\mathsf{M}(T, T_1) - \mathsf{M}(T, T_0)\right\}\mathsf{P}_{xp}(T, T_0). \quad (5.2.56)$$

As mentioned above, care is needed in interpreting terms from the reversed propagator. The terms in expressions (5.2.55) and (5.2.56) are all for the propagator from the forward ray ($T_1 > T_0$). With $T_0 < T < T_1$, terms that are odd in slowness, e.g. both P_{xp} and M, change sign in the propagator of the reversed ray. Let us indicate these by replacing the arguments by the positions, i.e. $\mathsf{P}_{xp}(\mathbf{x}, \mathbf{x}_1)$ is from the propagator of the ray traced from $\mathbf{x}_1$ to $\mathbf{x}$ and is $-\mathsf{P}_{xp}(T, T_1)$. Thus expression (5.2.56) becomes

$$\boxed{\mathsf{P}_{xp}(\mathbf{x}_1, \mathbf{x}_0) = \mathsf{P}^{\mathrm{T}}_{xp}(\mathbf{x}, \mathbf{x}_1)\left\{\mathsf{M}(\mathbf{x}, \mathbf{x}_1) + \mathsf{M}(\mathbf{x}, \mathbf{x}_0)\right\}\mathsf{P}_{xp}(\mathbf{x}, \mathbf{x}_0).} \quad (5.2.57)$$

5.2.2.4 Liouville's theorem

Because of the symmetry (5.2.20), the paraxial ray equations always satisfy

$$\mathrm{tr}(\mathbf{D}) = 0. \quad (5.2.58)$$

Using the Jacobi identity (C.1.16), this implies that

$$\boxed{|\mathbf{P}(T, T_0)| = 1.} \quad (5.2.59)$$

In other words, the volume mapping in the six-dimensional phase space, $\mathbf{y}$, conserves volume. This is a well-known result of Hamiltonian systems and is known as *Liouville's theorem*. It is an important result when we come to discuss caustics (Section 10.1).

5.2.3 Geometrical Green dyadic

For a point source, we only need to perturb the ray source direction, not its position. Therefore we only need the third and fourth columns of initial conditions (5.2.28) and solutions (5.2.29). The solutions required are

$$\mathbf{J}_{xp} = \left(\partial \mathbf{x}/\partial \mathrm{q}_1 \; \partial \mathbf{x}/\partial \mathrm{q}_2 \right) = \mathbf{P}_{xp} \left(\partial \mathbf{p}_\mathrm{S}/\partial \mathrm{q}_1 \; \partial \mathbf{p}_\mathrm{S}/\partial \mathrm{q}_2 \right), \tag{5.2.60}$$

i.e. perturbations of the ray source direction, $\mathbf{p}_\mathrm{S}$, propagated to give wavefront perturbations at the receiver. At a point source, the initial conditions (5.2.25) perturb the ray direction. As already mentioned, this solution can be found solving the full sixth-order system (5.2.20) or a reduced fourth-order system. The equation can also be reduced to a non-linear equation for J, but again it is probably easier to work with system (5.2.20) which is relatively simple algebraically and computationally. Anyway results other than just J are needed for Maslov asymptotic ray theory (Section 10.1).

The cross-sectional area of the ray tube is given by

$$J \, \mathrm{dq}_1 \mathrm{dq}_2 = \left| \frac{\partial \mathbf{x}}{\partial \mathrm{q}_1} \times \frac{\partial \mathbf{x}}{\partial \mathrm{q}_2} \right| \mathrm{dq}_1 \mathrm{dq}_2, \tag{5.2.61}$$

so using solutions (5.2.60), we can calculate the necessary function J.

To find the constant in result (5.2.13), we need to consider the wave solution for a point source. Near a point source, we can use the solution in a homogeneous medium (4.5.72), which reduces for acoustic waves to

$$\mathbf{u}(t, \mathbf{x}_\mathrm{R}; \mathbf{x}_\mathrm{S}) = \frac{\delta(t - R/\alpha_\mathrm{S})}{4\pi \rho_\mathrm{S} \, \alpha_\mathrm{S}^2 R} \, \hat{\mathbf{g}}_\mathrm{R} \, \hat{\mathbf{g}}_\mathrm{S}^\mathrm{T} \mathbf{f}_\mathrm{S}. \tag{5.2.62}$$

We only need the far-field approximation as we only take the geometrical approximation in ray theory. Thus

$$v_\mathrm{S}^{(0)}(\mathbf{x}_\mathrm{R}; \mathbf{x}_\mathrm{S}) = \frac{\hat{\mathbf{g}}_\mathrm{S}^\mathrm{T} \mathbf{f}_\mathrm{S}}{2\rho_\mathrm{S} \, \alpha_\mathrm{S}^2 R}, \tag{5.2.63}$$

with

$$f(\omega) = -\frac{\mathrm{i}\omega}{2\pi} = f^{(3)}(\omega), \text{ say,} \tag{5.2.64}$$

in ansatz (5.1.1), where the subscript on $v_\mathrm{S}^{(0)}$ indicates validity near the point source. The superscript on the function $f^{(3)}(\omega)$ indicates the three-dimensional wave propagation from a point source. Near a point source, the approximate solution of the differential equation (5.2.19) gives $\mathbf{P}_{xp} \simeq \mathbf{R}_\mathrm{S}(T - T_\mathrm{S}) = \alpha_\mathrm{S} R \, \mathbf{I}$. Combining this with expressions (5.2.60), (5.2.61) and (5.2.63) in result (5.2.13), we obtain

$$v^{(0)}(\mathbf{x}_\mathrm{R};\mathbf{x}_\mathrm{S}) = v_\mathrm{S}^{(0)}(\mathbf{x}_\mathrm{R};\mathbf{x}_0)\sqrt{\frac{Z_0 J_0}{Z_\mathrm{R} J_\mathrm{R}}} \tag{5.2.65}$$

$$= \frac{1}{2}\left(\left|\frac{\partial \mathbf{p}_\mathrm{S}}{\partial \mathsf{q}_1} \times \frac{\partial \mathbf{p}_\mathrm{S}}{\partial \mathsf{q}_2}\right| \Big/ Z_\mathrm{R} Z_\mathrm{S} J_\mathrm{R}\right)^{1/2} \hat{\mathbf{g}}_\mathrm{S}^\mathrm{T}\mathbf{f}_\mathrm{S} \tag{5.2.66}$$

(in expression (5.2.65), we have matched the solutions at $\mathbf{x}_0$, as at $\mathbf{x}_\mathrm{S}$, $v_\mathrm{S}^{(0)}$ would be infinite and J_S zero). At the receiver, we use (5.2.61) to calculate J, completing the solution.

It is convenient to define

$$\boxed{\mathcal{S}^{(3)}(\mathbf{x}_\mathrm{R},\mathbf{x}_\mathrm{S}) = J \Big/ \left|\frac{\partial \mathbf{p}_\mathrm{S}}{\partial \mathsf{q}_1} \times \frac{\partial \mathbf{p}_\mathrm{S}}{\partial \mathsf{q}_2}\right| = \left|\frac{\partial \mathbf{x}_\mathrm{R}}{\partial \mathsf{q}_1} \times \frac{\partial \mathbf{x}_\mathrm{R}}{\partial \mathsf{q}_2}\right| \Big/ \left|\frac{\partial \mathbf{p}_\mathrm{S}}{\partial \mathsf{q}_1} \times \frac{\partial \mathbf{p}_\mathrm{S}}{\partial \mathsf{q}_2}\right|} \tag{5.2.67}$$

(compared with Kendall, Guest and Thomson, 1992, we omit the velocity from the definition of $\mathcal{S}^{(3)}$ – we use the symbol $\mathcal{S}$, as this is known as the *spreading function*, and the suffix to indicate the dimension). Using the wavefront differentials (5.2.31), this can be written

$$\mathcal{S}^{(3)}(\mathbf{x}_\mathrm{R},\mathbf{x}_\mathrm{S}) = \left\|\mathsf{P}_{xp}(\mathbf{x}_\mathrm{R},\mathbf{x}_\mathrm{S})\right\| . \tag{5.2.68}$$

We have already indicated how to compute this expression using solutions (5.2.60). We postpone until the section on anisotropic ray tracing (Section 5.4), reducing result (5.2.67) to an expression in terms of the propagator $\mathbf{P}_{xp}$ and independent of the ray parameters q, as that more general expression for $\mathcal{S}^{(3)}$ will apply to both the acoustic, isotropic and anisotropic systems.

Combining these results (5.2.5), (5.2.66) and (5.2.67) in ansatz (5.1.3), the dyadic Green function is

$$\underline{\mathbf{u}}(t,\mathbf{x}_\mathrm{R};\mathbf{x}_\mathrm{S}) = \frac{\delta(t - T(\mathbf{x}_\mathrm{R},\mathcal{L}_n))\,\hat{\mathbf{g}}_\mathrm{R}\,\hat{\mathbf{g}}_\mathrm{S}^\mathrm{T}}{4\pi(\rho(\mathbf{x}_\mathrm{R})\rho(\mathbf{x}_\mathrm{S})\alpha(\mathbf{x}_\mathrm{R})\alpha(\mathbf{x}_\mathrm{S})\,\mathcal{S}^{(3)})^{1/2}}. \tag{5.2.69}$$

For brevity, we have not indicated explicitly the dependence of the polarization, $\hat{\mathbf{g}}$, and spreading function, $\mathcal{S}^{(3)}$, on the ray path, e.g. more completely we have $\hat{\mathbf{g}}_\mathrm{S} = \hat{\mathbf{g}}(\mathbf{x}_\mathrm{S},\mathcal{L}_n)$ where $\mathcal{L}_n$ defines the ray direction and type at position $\mathbf{x}_\mathrm{S}$. To check the units, note unit$(\mathcal{S}^{(3)}) = [\mathrm{L}^4\mathrm{T}^{-2}]$ giving unit$(\underline{\mathbf{u}}) = [\mathrm{M}^{-1}\mathrm{T}]$ consistent with (4.5.30). In general, the term J, (5.2.67), need not be positive. The Green function in the frequency domain is generalized to

$$\underline{\mathbf{u}}(\omega,\mathbf{x}_\mathrm{R};\mathbf{x}_\mathrm{S}) = \frac{\mathrm{e}^{\mathrm{i}\,\omega T(\mathbf{x}_\mathrm{R},\mathcal{L}_n) - \mathrm{i}\,\pi\,\mathrm{sgn}(\omega)\sigma(\mathbf{x}_\mathrm{R},\mathcal{L}_n)/2}\,\hat{\mathbf{g}}_\mathrm{R}\,\hat{\mathbf{g}}_\mathrm{S}^\mathrm{T}}{4\pi(\rho(\mathbf{x}_\mathrm{R})\rho(\mathbf{x}_\mathrm{S})\alpha(\mathbf{x}_\mathrm{R})\alpha(\mathbf{x}_\mathrm{S})\,\mathcal{S}^{(3)})^{1/2}}, \tag{5.2.70}$$

where $\sigma(\mathbf{x}_S, \mathcal{L}_n)$ is the KMAH index. The spreading function, $\mathcal{S}^{(3)}$ (5.2.68), has been defined to be positive. In the time domain we obtain

$$\boxed{\underline{\mathbf{u}}(t, \mathbf{x}_R; \mathbf{x}_S) = \frac{\mathrm{Re}\left(\Delta(t - T(\mathbf{x}_R, \mathcal{L}_n))\mathrm{e}^{-\mathrm{i}\pi\,\sigma(\mathbf{x}_R, \mathcal{L}_n)/2}\right)\hat{\mathbf{g}}_R\,\hat{\mathbf{g}}_S^{\mathrm{T}}}{4\pi(\rho(\mathbf{x}_R)\rho(\mathbf{x}_S)\alpha(\mathbf{x}_R)\alpha(\mathbf{x}_S)\,\mathcal{S}^{(3)})^{1/2}},} \tag{5.2.71}$$

the geometrical ray approximation for the Green function. The time series is the σ-th Hilbert transform of a delta function. Later we will justify this generalization (Section 10.1). The KMAH index counts the caustics along the ray (incrementing by one at each first-order zero – a line caustic – of $\mathbf{P}_{xp}$, by two at a second-order zero – a point caustic). It is named after contributions by Keller (1958), Maslov (1965, 1972), Arnol'd (1973) and Hörmander (1971). At caustics, the amplitude coefficient is singular and ray theory breaks down. Something more than ray theory is needed to connect the ray results through a caustic and obtain the KMAH index (see Section 10.1 on Maslov asymptotic ray theory). As J can change sign, we should replace J by $|J|$ throughout this section, e.g. results (5.2.13), (5.2.16) and (5.2.17), or $J^{-1/2}$ by $|J|^{-1/2}\exp(-\mathrm{i}\pi\,\sigma/2)$ where appropriate, e.g. result (5.2.66).

5.2.3.1 *Effective ray length*

It is instructive to introduce an alternative variable to the spreading function, $\mathcal{S}^{(3)}$ (5.2.67). The spreading function has been used as it can be derived directly from the dynamic propagator, (5.2.68), but it has the disadvantage that it depends on the magnitude of the slowness at the source, $\mathbf{p}_S$, and is not purely geometrical. We replace it by

$$\mathcal{R}^{(3)}(\mathbf{x}_R, \mathbf{x}_S) = \mathcal{S}^{(3)\,1/2}(\mathbf{x}_R, \mathbf{x}_S)\Big/\alpha(\mathbf{x}_S). \tag{5.2.72}$$

This variable has unit $\left(\mathcal{R}^{(3)}\right) = [\mathrm{L}]$ and we call it the *effective ray length*. In a homogeneous medium

$$\mathcal{R}^{(3)}(\mathbf{x}_R, \mathbf{x}_S) = |\mathbf{x}_R - \mathbf{x}_S| = R, \tag{5.2.73}$$

the actual ray length. With this definition (5.2.72), the dyadic Green function (5.2.71) becomes

$$\boxed{\underline{\mathbf{u}}(t, \mathbf{x}_R; \mathbf{x}_S) = \frac{\mathrm{Re}\left(\Delta(t - T(\mathbf{x}_R, \mathcal{L}_n))\mathrm{e}^{-\mathrm{i}\pi\,\sigma(\mathbf{x}_R, \mathcal{L}_n)/2}\right)\hat{\mathbf{g}}_R\,\hat{\mathbf{g}}_S^{\mathrm{T}}}{4\pi(\rho(\mathbf{x}_R)\rho(\mathbf{x}_S)\alpha(\mathbf{x}_R)\alpha^3(\mathbf{x}_S))^{1/2}\,\mathcal{R}^{(3)}}.} \tag{5.2.74}$$

5.2.3.2 *Two-dimensional acoustic ray theory*

As some more complicated methods, e.g. transform methods and finite-difference methods, are more easily applied in two dimensions, we summarize the results for the dyadic Green function in two dimensions. Physically, this is equivalent to a line source in three dimensions (Section 4.5.5.2). We emphasize that this is different from three-dimensional ray propagation in a two-dimensional model considered in Section 5.7, where the rays spread in the third dimension. The latter is sometimes referred to as 2.5D propagation. Here the problem is strictly two dimensional and there is no spreading in the third dimension.

Most of the results in Sections 5.1 and 5.2 remain valid in two dimensions provided the third dimension is ignored. In other words, if we replace the position and slowness vectors, $\mathbf{x}$ and $\mathbf{p}$, by two-dimensional vectors, the results still apply. Only when we come to discuss the dynamic results, ray spreading and the Green function are some modifications necessary.

The velocity amplitude coefficient of the ray series is still given by expression (5.2.13) in two dimensions, but the cross-section function, J, is now for spreading in one dimension. Thus equation (5.2.61) is simplified to

$$J\,\mathrm{dq}_1 = \left|\frac{\partial \mathbf{x}}{\partial \mathrm{q}_1}\right| \mathrm{dq}_1, \tag{5.2.75}$$

where only one ray parameter, q_1, is necessary to specify the ray.

In order to fully determine the amplitude coefficient, $v^{(0)}$, the solution is connected to the Green function for a line source (4.5.85). Thus result (5.2.62) is replaced by

$$\mathbf{u}(t, \mathbf{x}_\mathrm{R}; \mathbf{x}_\mathrm{S}) = \frac{\lambda(t - r/\alpha_\mathrm{S})}{2^{3/2}\pi \rho_\mathrm{S}\, \alpha_\mathrm{S}^{3/2} r^{1/2}}\, \hat{\mathbf{g}}_\mathrm{R}\, \hat{\mathbf{g}}_\mathrm{S}^\mathrm{T} \mathbf{f}_\mathrm{S}. \tag{5.2.76}$$

and near the source, we must have

$$v_\mathrm{S}^{(0)}(\mathbf{x}_\mathrm{R}; \mathbf{x}_\mathrm{S}) = \frac{\hat{\mathbf{g}}_\mathrm{S}^\mathrm{T} \mathbf{f}_\mathrm{S}}{2\rho_\mathrm{S}\, \alpha_\mathrm{S}^{3/2} r^{1/2}}, \tag{5.2.77}$$

replacing result (5.2.63), with

$$f(\omega) = -\frac{\mathrm{i}\omega\, \lambda(\omega)}{2^{1/2}\pi} = f^{(2)}(\omega), \text{ say,} \tag{5.2.78}$$

in ansatz (5.1.1). The superscript on the function $f^{(2)}(\omega)$ indicates the two-dimensional wave propagation from a line source. Equation (5.2.66) remains almost identical

$$v^{(0)}(\mathbf{x}_\mathrm{R}; \mathbf{x}_\mathrm{S}) = \frac{1}{2}\left(\left|\frac{\partial \mathbf{p}_\mathrm{S}}{\partial \mathrm{q}_1}\right| \Big/ Z_\mathrm{R} Z_\mathrm{S} J_\mathrm{R}\right)^{1/2} \hat{\mathbf{g}}_\mathrm{S}^\mathrm{T} \mathbf{f}_\mathrm{S}, \tag{5.2.79}$$

except J is defined by equation (5.2.75). We define a two-dimensional version of the function $\mathcal{S}$ to replace definition (5.2.67)

$$\mathcal{S}^{(2)}(\mathbf{x}_{\mathrm{R}}, \mathbf{x}_{\mathrm{S}}) = |J_{\mathrm{R}}| \bigg/ \left|\frac{\partial \mathbf{p}_{\mathrm{S}}}{\partial \mathrm{q}_1}\right| = \left|\frac{\partial \mathbf{x}}{\partial \mathrm{q}_1}\right| \bigg/ \left|\frac{\partial \mathbf{p}_{\mathrm{S}}}{\partial \mathrm{q}_1}\right| . \tag{5.2.80}$$

Finally, with the definition

$$\mathcal{R}^{(2)}(\mathbf{x}_{\mathrm{R}}, \mathbf{x}_{\mathrm{S}}) = \mathcal{S}^{(2)}(\mathbf{x}_{\mathrm{R}}, \mathbf{x}_{\mathrm{S}}) \Big/ \alpha(\mathbf{x}_{\mathrm{S}}), \tag{5.2.81}$$

replacing definition (5.2.72) for the effective length, the dyadic, two-dimensional Green function is

$$\boxed{\underline{\mathbf{u}}(t, \mathbf{x}_{\mathrm{R}}; \mathbf{x}_{\mathrm{S}}) = \frac{\mathrm{Re}\left(\Lambda(t - T(\mathbf{x}_{\mathrm{R}}, \mathcal{L}_n))\mathrm{e}^{-\mathrm{i}\pi\,\sigma(\mathbf{x}_{\mathrm{R}}, \mathcal{L}_n)/2}\right) \hat{\mathbf{g}}_{\mathrm{R}}\, \hat{\mathbf{g}}_{\mathrm{S}}^{\mathrm{T}}}{2^{3/2}\pi(\rho(\mathbf{x}_{\mathrm{R}})\rho(\mathbf{x}_{\mathrm{S}})\alpha(\mathbf{x}_{\mathrm{R}})\alpha(\mathbf{x}_{\mathrm{S}})\, \mathcal{S}^{(2)})^{1/2}}} \tag{5.2.82}$$

or

$$\boxed{\underline{\mathbf{u}}(t, \mathbf{x}_{\mathrm{R}}; \mathbf{x}_{\mathrm{S}}) = \frac{\mathrm{Re}\left(\Lambda(t - T(\mathbf{x}_{\mathrm{R}}, \mathcal{L}_n))\mathrm{e}^{-\mathrm{i}\pi\,\sigma(\mathbf{x}_{\mathrm{R}}, \mathcal{L}_n)/2}\right) \hat{\mathbf{g}}_{\mathrm{R}}\, \hat{\mathbf{g}}_{\mathrm{S}}^{\mathrm{T}}}{2^{3/2}\pi(\rho(\mathbf{x}_{\mathrm{R}})\rho(\mathbf{x}_{\mathrm{S}})\alpha(\mathbf{x}_{\mathrm{R}})\alpha^2(\mathbf{x}_{\mathrm{S}}))^{1/2}\, \mathcal{R}^{(2)\,1/2}}.} \tag{5.2.83}$$

We postpone until the section on anisotropic ray tracing (Section 5.4), discussing the reciprocity of results (5.2.70), (5.2.71), (5.2.74) and (5.2.82), i.e. result (4.5.42), and writing them in a more compact form for a general source (5.4.35).

5.2.4 Higher-order terms

In order to find higher-order amplitude coefficients, we consider equation (5.1.8). The equations are solved iteratively, so we assume $\mathbf{v}^{(n)}$ and $P^{(n)}$ are known for $n < m$. We write the amplitude coefficient $\mathbf{v}^{(m)}$ as

$$\mathbf{v}^{(m)} = v_i^{(m)} \hat{\mathbf{g}}_i, \tag{5.2.84}$$

where $\hat{\mathbf{g}}_1 = \hat{\mathbf{g}} = \alpha \mathbf{p}$, the polarization, and $\hat{\mathbf{g}}_2$ and $\hat{\mathbf{g}}_3$ lie in the wavefront completing an orthogonal basis (the orientation in the wavefront is not important). Substituting expression (5.2.84) in equation (5.1.8), we obtain

$$\rho\left(v_2^{(m)} \hat{\mathbf{g}}_2 + v_3^{(m)} \hat{\mathbf{g}}_3\right) = -\nabla P^{(m-1)} - \kappa\left(\nabla \cdot \mathbf{v}^{(m-1)}\right) \mathbf{p}. \tag{5.2.85}$$

Pre-multiplying by $\hat{\mathbf{g}}_N^{\mathrm{T}}$ ($N = 2$ or 3), we have

$$v_N^{(m)} = -\frac{1}{\rho} \hat{\mathbf{g}}_N^{\mathrm{T}} \nabla P^{(m-1)}, \tag{5.2.86}$$

which can be calculated, at least in principle, as $P^{(m-1)}$ is assumed known. These are known as the *additional components*, and we write them as

$$\mathbf{v}_A^{(m)} = v_N^{(m)}\hat{\mathbf{g}}_N = v_2^{(m)}\hat{\mathbf{g}}_2 + v_3^{(m)}\hat{\mathbf{g}}_3. \tag{5.2.87}$$

The remaining term, called the *principal components*, is found by pre-multiplying equation (5.2.85) by $\hat{\mathbf{g}}_1^{\mathrm{T}}$ and substituting m for $m-1$. Thus

$$\hat{\mathbf{g}}_1^{\mathrm{T}}\left(\nabla P^{(m)} + \kappa\left(\nabla\cdot\mathbf{v}^{(m)}\right)\mathbf{p}\right) = 0. \tag{5.2.88}$$

Using equation (5.1.5) to eliminate $P^{(m)}$,

$$\hat{\mathbf{g}}_1^{\mathrm{T}}\left(\nabla\left(\kappa\mathbf{p}^{\mathrm{T}}\mathbf{v}^{(m)} - \kappa\nabla\cdot\mathbf{v}^{(m-1)}\right) + \kappa\left(\nabla\cdot\mathbf{v}^{(m)}\right)\mathbf{p}\right) = 0, \tag{5.2.89}$$

and separating into known and unknown parts

$$\begin{aligned}&\hat{\mathbf{g}}_1^{\mathrm{T}}\left(\nabla\left(\kappa\mathbf{p}^{\mathrm{T}}v_1^{(m)}\hat{\mathbf{g}}_1\right) + \kappa\left(\nabla\cdot\left(v_1^{(m)}\hat{\mathbf{g}}_1\right)\right)\mathbf{p}\right)\\ &\quad = -\hat{\mathbf{g}}_1^{\mathrm{T}}\left(\nabla\left(\kappa\mathbf{p}^{\mathrm{T}}\mathbf{v}_A^{(m)} - \kappa\nabla\cdot\mathbf{v}^{(m-1)}\right) + \kappa\left(\nabla\cdot\mathbf{v}_A^{(m)}\right)\mathbf{p}\right) = \xi^{(m)},\end{aligned} \tag{5.2.90}$$

say. The right-hand side is known and is denoted by the function $\xi^{(m)}$, which reduces to

$$\xi^{(m)} = \hat{\mathbf{g}}_1^{\mathrm{T}}\nabla\left(\kappa\nabla\cdot\mathbf{v}^{(m-1)}\right) - Z\,\nabla\cdot\mathbf{v}_A^{(m)}. \tag{5.2.91}$$

Simplifying the left-hand side as before (Section 5.2.1), we obtain

$$\nabla\cdot\left(Zv_1^{(m)2}\hat{\mathbf{g}}_1\right) = v_1^{(m)}\xi^{(m)} \tag{5.2.92}$$

(cf. equation (5.2.7)), and analogous to differential equation (5.2.10)

$$\frac{\mathrm{d}}{\mathrm{d}T}\ln\left(\rho v_1^{(m)2}\right) = \frac{\xi^{(m)}}{\rho v_1^{(m)}} - \nabla\cdot\mathbf{V}. \tag{5.2.93}$$

Combining with result (5.2.16), the equation becomes

$$\frac{\mathrm{d}}{\mathrm{d}T}\left(v_1^{(m)}\sqrt{\rho\alpha|J|}\right) = \frac{1}{2}\sqrt{\frac{\alpha|J|}{\rho}}\,\xi^{(m)}, \tag{5.2.94}$$

and the solution of this equation is

$$v_1^{(m)}(T) = \frac{1}{\sqrt{Z|J|}}\left(\sqrt{Z_0|J_0|}\,v_1^{(m)}(T_0) + \frac{1}{2}\int_{T_0}^{T}\sqrt{\frac{\alpha|J|}{\rho}}\xi^{(m)}\mathrm{d}T\right), \tag{5.2.95}$$

where the integral is along the ray path.

The main cause of breakdown of the geometrical ray approximation is caustics ($J = 0$) and discontinuities in the wavefield, e.g. interfaces, critical points and shadows. In these cases, either the geometrical approximation or higher-order terms are singular. Higher-order terms in the asymptotic ray series are of little use and the ray theory must be generalized. The analysis in this section for the higher-order terms is mainly needed to indicate how they might be found, establishing that the ray ansatz is appropriate, and when higher-order terms might be large, invalidating the geometrical approximation. Note that heterogeneity in the wavefield or model will make the higher-order terms large (derivatives in expressions (5.2.86) and (5.2.91)). The main circumstances in which the higher-order terms are useful is when the first term, $v^{(0)}$, is zero or small. This occurs for head waves and on the nodes of the source radiation pattern. Note that the additional components are transverse even though the medium is acoustic.

The discontinuities in the wavefield that would occur at an interface for a single ray can be investigated by considering all the rays generated – reflected and transmitted. In the next chapter, we generalize the results of asymptotic ray theory by solving the interface boundary conditions for the generated rays and exploiting the ray expansion.

5.3 Anisotropic kinematic ray theory

In this section, we generalize the results of asymptotic ray theory to anisotropic media. The procedure is very similar to that for acoustic waves except for the necessary complication that we must deal with vector equations. We use the equations of motion (4.5.35) and constitutive relations (4.5.36) from Section 4.5.2. The operators $\boldsymbol{\mathcal{L}}$, $\boldsymbol{\mathcal{M}}$ and $\boldsymbol{\mathcal{N}}$ are close to those in Červený (1972), etc.

5.3.1 The ray series

The ray ansätze for the velocity and tractions are

$$\boxed{\begin{pmatrix} \mathbf{v} \\ \mathbf{t}_1 \\ \mathbf{t}_2 \\ \mathbf{t}_3 \end{pmatrix}(\omega, \mathbf{x}_\mathrm{R}) = f(\omega) \sum_n \mathrm{e}^{\mathrm{i}\omega T(\mathbf{x}_\mathrm{R}, \mathcal{L}_n)} \sum_{m=0}^{\infty} \frac{1}{(-\mathrm{i}\omega)^m} \begin{pmatrix} \mathbf{v}^{(m)} \\ \mathbf{t}_1^{(m)} \\ \mathbf{t}_2^{(m)} \\ \mathbf{t}_3^{(m)} \end{pmatrix}(\mathbf{x}_\mathrm{R}, \mathcal{L}_n).} \quad (5.3.1)$$

This is identical in form to that used for acoustic ray theory (5.1.1) with the extra amplitude coefficients for traction rather than just pressure.

Substituting ansätze (5.3.1) in equations (4.5.35) and (4.5.36), and setting the coefficient of each power of ω zero, we obtain for $m \geq 0$ (defining

$\mathbf{v}^{(-1)} = \mathbf{t}_j^{(-1)} = \mathbf{0}$)

$$\frac{1}{\rho}\frac{\partial \mathbf{t}_j^{(m-1)}}{\partial x_j} = \mathbf{v}^{(m)} + \frac{1}{\rho} p_j \mathbf{t}_j^{(m)} \tag{5.3.2}$$

$$\mathbf{c}_{jk}\frac{\partial \mathbf{v}^{(m-1)}}{\partial x_k} = \mathbf{t}_j^{(m)} + p_k \mathbf{c}_{jk} \mathbf{v}^{(m)}. \tag{5.3.3}$$

Eliminating $\mathbf{t}_j^{(m)}$ between equations (5.3.2) and (5.3.3), we obtain

$$(p_j p_k \mathbf{c}_{jk} - \rho\, \mathbf{I})\, \mathbf{v}^{(m)} = p_j \mathbf{c}_{jk} \frac{\partial \mathbf{v}^{(m-1)}}{\partial x_k} - \frac{\partial \mathbf{t}_j^{(m-1)}}{\partial x_j}, \tag{5.3.4}$$

which can be solved for the travel time and amplitude coefficients. This equation can be rewritten using shorthand for the anisotropic ART operators

$$\boldsymbol{\mathcal{N}}\left(\mathbf{v}^{(m)}\right) - \overline{\boldsymbol{\mathcal{M}}}\left(\mathbf{v}^{(m-1)}, \mathbf{t}_j^{(m-1)}\right) = \mathbf{0}, \tag{5.3.5}$$

where

$$\boldsymbol{\mathcal{N}}\left(\mathbf{v}^{(m)}\right) = (p_j p_k \mathbf{c}_{jk} - \rho \mathbf{I}) \mathbf{v}^{(m)} \tag{5.3.6}$$

$$\overline{\boldsymbol{\mathcal{M}}}\left(\mathbf{v}^{(m)}, \mathbf{t}_j^{(m)}\right) = p_j \mathbf{c}_{jk} \frac{\partial \mathbf{v}^{(m)}}{\partial x_k} - \frac{\partial \mathbf{t}_j^{(m)}}{\partial x_j}. \tag{5.3.7}$$

Substituting for $\mathbf{t}_j^{(m)}$ using equation (5.3.3), we write

$$\overline{\boldsymbol{\mathcal{M}}}\left(\mathbf{v}^{(m)}, \mathbf{t}_j^{(m)}\right) = \boldsymbol{\mathcal{M}}\left(\mathbf{v}^{(m)}\right) - \boldsymbol{\mathcal{L}}\left(\mathbf{v}^{(m-1)}\right), \tag{5.3.8}$$

where

$$\boldsymbol{\mathcal{M}}\left(\mathbf{v}^{(m)}\right) = p_j \mathbf{c}_{jk} \frac{\partial \mathbf{v}^{(m)}}{\partial x_k} + \frac{\partial}{\partial x_j}\left(p_j \mathbf{c}_{jk} \mathbf{v}^{(m)}\right) \tag{5.3.9}$$

$$\boldsymbol{\mathcal{L}}\left(\mathbf{v}^{(m)}\right) = \frac{\partial}{\partial x_j}\left(\mathbf{c}_{jk} \frac{\partial \mathbf{v}^{(m)}}{\partial x_k}\right). \tag{5.3.10}$$

Note $\overline{\boldsymbol{\mathcal{M}}}\left(\mathbf{v}^{(0)}, \mathbf{t}_j^{(0)}\right) = \boldsymbol{\mathcal{M}}\left(\mathbf{v}^{(0)}\right)$.

5.3.2 The eikonal equation $(m = 0)$

For $m = 0$, equation (5.3.5) reduces to the eigenvector (*Christoffel*) equation

$$\frac{1}{\rho}\boldsymbol{\mathcal{N}}\left(\mathbf{v}^{(0)}\right) = (p_j p_k \mathbf{c}_{jk}/\rho - \mathbf{I}) \mathbf{v}^{(0)} = \mathbf{0} \tag{5.3.11}$$

cf. Exercise 4.2. The 3×3 real, symmetric matrix

$$\boldsymbol{\Gamma} = p_j p_k \mathbf{a}_{jk}, \tag{5.3.12}$$

where we have defined the density-normalized matrices

$$\mathbf{a}_{jk} = \mathbf{c}_{jk}/\rho \tag{5.3.13}$$

(unit($\mathbf{a}_{jk}$) = [$\mathrm{L}^2\mathrm{T}^{-2}$]), is known as the *Christoffel matrix*. It has real eigenvalues, G_I say, i.e.

$$\left(G_I \mathbf{I} - \boldsymbol{\Gamma}\right) \hat{\mathbf{g}}_I = \mathbf{0}, \tag{5.3.14}$$

where $I = 1, 2$ or 3. We denote the orthonormal eigenvectors as $\hat{\mathbf{g}}_I(\mathbf{x}, \mathbf{p})$, which are defined at all $(\mathbf{x}, \mathbf{p})$. The required eigenvalue is unity, $G_I = 1$, which constrains the permitted values of slowness, $\mathbf{p}$. Alternatively, for a given slowness direction, $\hat{\mathbf{p}}$, we can define

$$\hat{\boldsymbol{\Gamma}} = \hat{p}_j \hat{p}_k \mathbf{a}_{jk}, \tag{5.3.15}$$

and the eigen-equation (5.3.14) can be rewritten

$$\left(c_I^2 \mathbf{I} - \hat{\boldsymbol{\Gamma}}\right) \hat{\mathbf{g}}_I = \mathbf{0}, \tag{5.3.16}$$

and the three eigenvalues, c_I^2, define the permitted slownesses, $\mathbf{p} = \hat{\mathbf{p}}/c_I$. The slowness must satisfy

$$\boxed{|\,\boldsymbol{\Gamma} - \mathbf{I}| = 0,} \tag{5.3.17}$$

the *eikonal* equation, defining the *slowness surface*. We can define a Hamiltonian

$$\boxed{H_I(\mathbf{x}, \mathbf{p}) = \frac{1}{2} p_j p_k \hat{\mathbf{g}}_I^{\mathrm{T}} \mathbf{a}_{jk} \hat{\mathbf{g}}_I = \hat{\mathbf{g}}_I^{\mathrm{T}} \boldsymbol{\Gamma} \hat{\mathbf{g}}_I} \tag{5.3.18}$$

(with no summation over I). Other choices for the Hamiltonian are possible and have appeared in the literature but the definition used here is natural and straightforward. It also is second degree in slowness, allowing the Lagrangian to be defined by Legendre transform (Section 3.4.1 – Červený, 2002), and the slowness-perturbation solution of the paraxial ray equations to be discussed simply (Section 5.2.2). Equation (5.3.17) is equivalent to the constraint

$$\boxed{H_I(\mathbf{x}, \mathbf{p}) = \frac{1}{2},} \tag{5.3.19}$$

just as in the acoustic system (Section 5.1). Substituting the slowness definition (5.1.6) in the Hamiltonian (5.3.19), it is equivalent to the Hamilton–Jacobi equations. Thus the position and slowness must satisfy the Hamilton equations (5.1.29)

(the *kinematic ray equations*, cf. Section 5.1)

$$\boxed{\frac{\mathrm{d}x_i}{\mathrm{d}T} = \frac{\partial H}{\partial p_i} = a_{ijkl} p_k \hat{g}_j \hat{g}_l = p_k \hat{\mathbf{g}}_I^{\mathrm{T}} \mathbf{a}_{ik} \hat{\mathbf{g}}_I = \hat{\mathbf{g}}_I^{\mathrm{T}} \mathbf{Z}_i \hat{\mathbf{g}}_I / \rho = V_i} \quad (5.3.20)$$

$$\boxed{\frac{\mathrm{d}p_i}{\mathrm{d}T} = -\frac{\partial H}{\partial x_i} = -\frac{1}{2}\frac{\partial a_{jklm}}{\partial x_i} p_j p_m \hat{g}_k \hat{g}_l,} \quad (5.3.21)$$

where $a_{ijkl} = c_{ijkl}/\rho$ (note that derivatives of $\hat{\mathbf{g}}_I(\mathbf{x}, \mathbf{p})$ do not occur in these expressions as for a normalized vector, the derivative is necessarily orthogonal to the unit vector – see Exercise 5.1). In equation (5.3.20), we have defined a *tensor impedance*, $\mathbf{Z}_i$, where

$$\mathbf{Z}_i = p_k \mathbf{c}_{ik}, \quad (5.3.22)$$

and the *ray* or *group velocity*

$$\mathbf{V} = \frac{\mathrm{d}\mathbf{x}}{\mathrm{d}T}. \quad (5.3.23)$$

The tensor impedance (5.3.22) is analogous to the acoustic impedance where the zeroth-order amplitude coefficients are related by the equation $P^{(0)} = Z\left|\mathbf{v}^{(0)}\right|$ (5.2.9) and $\mathrm{unit}(\mathbf{Z}_i) = [\mathrm{ML}^{-2}\mathrm{T}^{-1}]$ again. For elastic waves we have

$$\mathbf{t}_i^{(0)} = -\mathbf{Z}_i \mathbf{v}^{(0)}, \quad (5.3.24)$$

connecting the traction and velocity amplitude coefficients (from equation (5.3.3) with $m = 0$). Again the ray equations are sixth-order ordinary differential equations, which are straightforward to solve by the ray-shooting method. Compared with the acoustic ray equations, (5.1.14) and (5.1.15), they are algebraically more complicated and in general require the numerical eigen-solution for the polarization, $\hat{\mathbf{g}}_I$. In general, the polarization, $\hat{\mathbf{g}}_I$, the slowness vector, $\mathbf{p}$, and the ray velocity, $\mathbf{V}$, are not in the same direction. Figure 5.7 illustrates a ray path found by solving the kinematic ray equations, a wavefront defined by constant travel time, T, and the slowness, group and polarization vectors, $\mathbf{p}$, $\mathbf{V}$ and $\hat{\mathbf{g}}$, respectively. It is important to remember that if rays of different types coincide, e.g. straight rays in a homogeneous medium, then the polarizations will not, in general, be orthogonal as even though the ray velocities, $\mathbf{V}_i$, are in the same direction, the slowness directions, $\hat{\mathbf{p}}_i$, will differ.

5.3.3 *The slowness surface and wavefront*

The anisotropic ray equations (5.3.20) and (5.3.21) are subject to the constraint (5.3.19) and relationship (5.1.17). At a fixed point, $\mathbf{x}$, or in a homogeneous

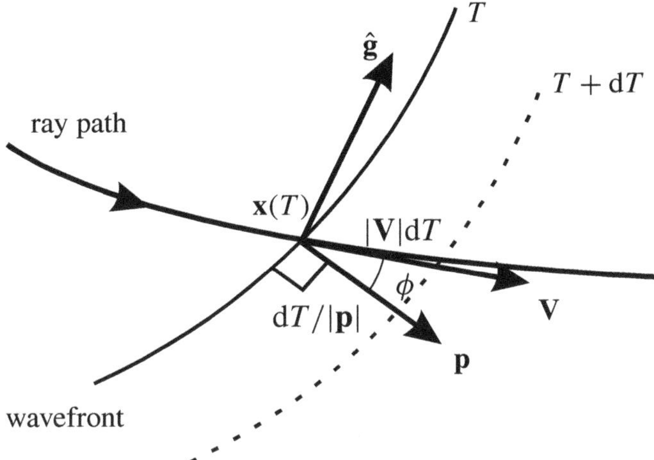

Fig. 5.7. A ray path and wavefront, and slowness, group velocity and polarization vectors (for clarity the polarization is for a quasi-shear ray). The distance between wavefronts dT apart is $|\mathbf{V}|dT$ in the ray direction $\mathbf{V}$, and $dT/|\mathbf{p}|$ in the slowness direction $\mathbf{p}$ – hence result (5.1.17).

medium, the Hamiltonian constraint (5.3.19) defines the *slowness surfaces* (5.3.17). For a given slowness direction, $\hat{\mathbf{p}} = \text{sgn}(\mathbf{p})$, equation (5.3.17) can be rewritten

$$\boxed{|\hat{\mathbf{\Gamma}} - c^2\mathbf{I}| = 0,} \tag{5.3.25}$$

where c is the phase velocity, $c = |\mathbf{p}|^{-1}$. This equation reduces to a cubic in c^2, so in general there are three solutions and three slowness surfaces. For a given slowness direction, $\hat{\mathbf{p}}$, there will be phase velocities, $\pm c$. Alternatively, as equation (5.3.25) is quadratic in $\hat{\mathbf{p}}$, slowness solutions are the same for $\pm\hat{\mathbf{p}}$ and the slowness surface has point symmetry. The slowness solutions are illustrated in Figure 5.8.

Ordering the solutions $c_1^2 \le c_2^2 \le c_3^2$, we label these qS_2 and qS_1 and qP, by analogy with the waves that exist in an isotropic medium (Section 4.5.4), i.e. qS_ν are the two quasi-shear rays, and qP is the quasi-P ray. Unfortunately the convention is to call the slower quasi-shear wave qS_2 which, as we order the velocities, has velocity c_1 (and the faster quasi-shear wave is qS_1 and has velocity c_2). The slowness surface always has point symmetry, i.e. $\mathbf{p}_I = \pm\hat{\mathbf{p}}/c_I$ are solutions. In isotropic waves we can anticipate that the solutions degenerate and the two shear waves have the same velocity, cf. equation (4.5.61) and $c_1 = c_2 = \beta$ ($c_3 = \alpha$, equation (4.5.60)). In considering the eigen-equation

$$\hat{\mathbf{\Gamma}}\hat{\mathbf{g}}_I = c_I^2\hat{\mathbf{g}}_I \tag{5.3.26}$$

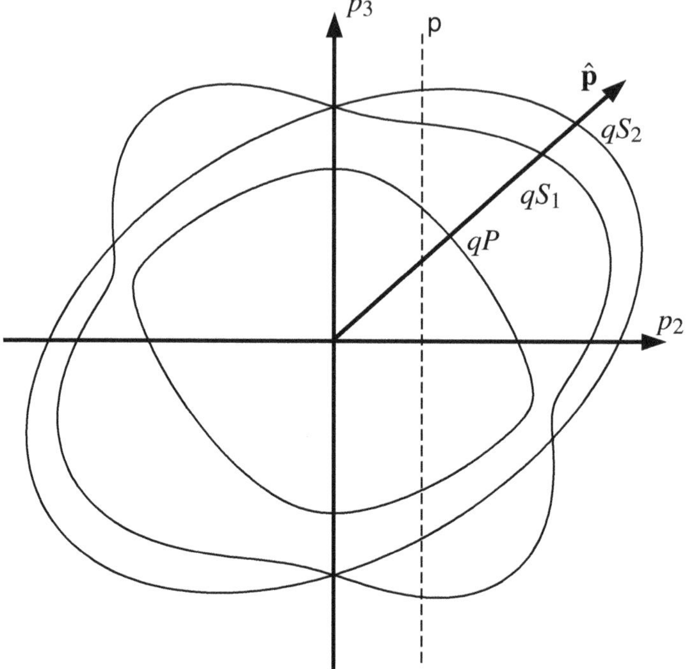

Fig. 5.8. The $p_2 - p_3$ cross-section of the three slowness surfaces qP, qS_1 and qS_2 for α-quartz. For a given direction, $\hat{\mathbf{p}}$, there are three solutions with point symmetry. For a fixed p (p_2), there can at most be six solutions (with no symmetries). The figure is based on the elastic constants from Bechmann (1958) as used by Shearer and Chapman (1988, p. 579). See also Figure 10.2.1(i) in Musgrave (1970).

(no summation over I), the three eigenvectors are orthogonal as the matrix is symmetric. At a point on a ray, only one, $\hat{\mathbf{g}}_I$, is the actual polarization. We reiterate that the other orthogonal eigenvectors normally do not correspond to the other rays, even if the paths coincide, as the slowness directions $\hat{\mathbf{p}}$ must be identical.

In any slowness direction, $\hat{\mathbf{p}}$, there can only be three positive, and the symmetric negative, solutions. The slowness surfaces cannot be folded, although, of course, they can touch or intersect or be degenerate. The quasi-shear surfaces can be concave or convex but the fastest surface must be convex. For if we fix two components of the slowness, $\mathbf{p}$, in equation (5.3.17), i.e. $\mathsf{p} = (p_1, p_2)$, and solve equation (5.3.17) for the third component p_3, this is equivalent to finding the intersection of a line with the three slowness surfaces. Equation (5.3.17) reduces to a sextic and so at most it has six real solutions. If a line intersects the innermost slowness surface it will have intersected the outer two slowness surfaces at, at least, four points, and so can only intersect the innermost surface at, at most, two more points. Therefore, the innermost surface, qP, must be convex. Note that the line may not intersect the innermost surface(s) so it is permitted to intersect the quasi-shear surfaces in more than two points, i.e. the quasi-shear surfaces may be concave. This is illustrated in Figure 5.8. Note that the solutions for p_3 need not be symmetric, e.g. both solutions on one surface may have the same sign.

From equation (5.3.20), we have

$$V_i = \frac{\partial H}{\partial p_i}, \tag{5.3.27}$$

i.e. the normal to the slowness surface defined by (5.3.19) is in the ray or group velocity direction. The equation

$$T(\mathbf{x}, \mathbf{p}) = t, \tag{5.3.28}$$

defines a wavefront, and for a point source in a homogeneous medium the solution is

$$\mathbf{x} - \mathbf{x}_S = t\,\mathbf{V}. \tag{5.3.29}$$

Defining the Lagrangian by the Legendre transform (Section 3.4.1 – this is possible as the Hamiltonian (5.3.18) is second degree in slowness, Červený, 2002)

$$L(\mathbf{x}, \dot{\mathbf{x}}) = \mathbf{p} \cdot \dot{\mathbf{x}} - H(\mathbf{x}, \mathbf{p}) = \frac{1}{2}. \tag{5.3.30}$$

Just as $H(\mathbf{x}, \mathbf{p}) = 1/2$ at a point defines the slowness surface, $L(\mathbf{x}, \mathbf{V}) = 1/2$ defines the group velocity surface or, for a point source at $\mathbf{x}$, the wavefront surface at unit time. From the definition of the slowness, equation (5.1.6)

$$p_i = \frac{\partial T}{\partial x_i}, \tag{5.3.31}$$

the normal to the wavefront is the slowness vector, $\mathbf{p}$. This is illustrated in Figure 5.9. The slowness surface defined by $H = 1/2$ (5.3.19) and wavefront $L = 1/2$ (5.3.30) in a homogeneous medium are polar reciprocal. The coordinates on the surfaces are $\mathbf{V}$ and $\mathbf{p}$, with normals $\mathbf{p}$ and $\mathbf{V}$, respectively.

The constraint (5.1.17), $\mathbf{p} \cdot \mathbf{V} = 1$ is no longer trivial as the slowness and ray velocity vectors are no longer parallel. Nevertheless, the constraint follows algebraically from equation (5.3.20) using the Hamiltonian (5.3.18) with constraint (5.3.19). It can also be obtained from the geometry of the wavefront (Figure 5.7). From the definition of slowness, $\mathbf{p} = \nabla T$ (5.1.6), the slowness is normal to the wavefront and so the separation of two wavefronts with a time difference of $\mathrm{d}T$ is $c\,\mathrm{d}T$. If the ray velocity, $\mathbf{V}$, is at an angle ϕ to the slowness, then the length of the ray arc between the wavefronts is $\mathrm{d}s = c\,\mathrm{d}T/\cos\phi$. The travel time along the ray must be $\mathrm{d}T = \mathrm{d}s/V$, where $V = |\mathbf{V}|$, the ray speed. If these results are consistent,

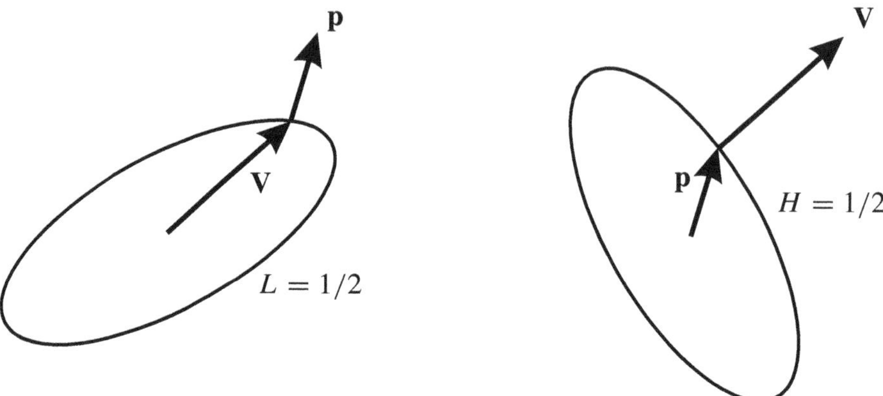

Fig. 5.9. The wavefront ($L = 1/2$) and slowness surface ($H = 1/2$). The coordinates on the surfaces are $\mathbf{V}$ and $\mathbf{p}$, with normals $\mathbf{p}$ and $\mathbf{V}$, respectively.

we must have

$$V \cos\phi = c, \tag{5.3.32}$$

which is equivalent to constraint (5.1.17).

5.4 Anisotropic dynamic ray theory

In this section we derive the equations needed to calculate the amplitude coefficients. The method and many of the results are very similar to the acoustic results (Section 5.2), so we only describe differences.

5.4.1 The transport equation ($m \geq 1$)

First we consider the general transport equation and then specialize to find the zeroth-order amplitude coefficients. The equations are solved iteratively so we assume we have solved for the amplitude coefficients to order $m - 1$, and are solving for order m.

Let us write $\mathbf{v}^{(m)}$ in terms of the orthogonal eigenvectors

$$\mathbf{v}^{(m)} = v_i^{(m)} \hat{\mathbf{g}}_i . \tag{5.4.1}$$

Pre-multiplying equation (5.3.5) by $\hat{\mathbf{g}}_I^{\mathrm{T}}$ and substituting expression (5.4.1), we obtain

$$\hat{\mathbf{g}}_I^{\mathrm{T}} \overline{\boldsymbol{\mathcal{M}}} \left(\mathbf{v}^{(m-1)}, \mathbf{t}_j^{(m-1)} \right) = \rho\, v_I^{(m)} G_I , \tag{5.4.2}$$

using the orthogonality of the eigenvectors and defining

$$G_I = 2H_I - 1. \tag{5.4.3}$$

In general G_I will be zero when I corresponds to the ray type, and non-zero for the other indices. In degenerate cases, $G_I = 0$ for two indices. We denote indices for which $G_I = 0$ by $I = E$, and $G_I \neq 0$ by $I = N$ (summation over an uppercase index indicates restriction to one or two terms. For instance, E might be the index 2 and N the set of indices 1 and 3. In a degenerate case, E might be the set 1 and 2, and N just 3).

We can solve equation (5.4.2) for the $v_N^{(m)}$'s, i.e.

$$v_N^{(m)} = \frac{1}{\rho G_N} \hat{\mathbf{g}}_N^{\mathrm{T}} \overline{\boldsymbol{\mathcal{M}}} \left(\mathbf{v}^{(m-1)}, \mathbf{t}_j^{(m-1)} \right) \tag{5.4.4}$$

(no summation over N). These terms are called the *additional components* and we write them as

$$\mathbf{v}_A^{(m)} = v_N^{(m)} \hat{\mathbf{g}}_N \tag{5.4.5}$$

(summation over N). They can be found from the known, lower-order terms (5.4.4). Thus equation (5.4.1) is

$$\mathbf{v}^{(m)} = v_E^{(m)} \hat{\mathbf{g}}_E + \mathbf{v}_A^{(m)}, \tag{5.4.6}$$

and the other term(s), $v_E^{(m)}$, are called the *principal components*.

First let us consider the non-degenerate case, where there is only one E index. Later we show how, with the correct choice of eigenvectors, the same method can be used to find the principal components in the degenerate case (where there are two E indices).

Setting $I = E$ and $m - 1 \rightarrow m$ in expression (5.4.2), we have

$$\hat{\mathbf{g}}_E^{\mathrm{T}} \overline{\boldsymbol{\mathcal{M}}} \left(\mathbf{v}^{(m)}, \mathbf{t}_j^{(m)} \right) = 0.$$

Separating the part that is known using definition (5.3.8) and expression (5.4.6), we obtain

$$\hat{\mathbf{g}}_E^{\mathrm{T}} \boldsymbol{\mathcal{M}} \left(v_E^{(m)} \hat{\mathbf{g}}_E \right) = \hat{\mathbf{g}}_E^{\mathrm{T}} \left(\boldsymbol{\mathcal{L}} \left(\mathbf{v}^{(m-1)} \right) - \boldsymbol{\mathcal{M}} \left(\mathbf{v}_A^{(m)} \right) \right) = \xi^{(m)}, \text{ say.} \tag{5.4.7}$$

The right-hand side, $\xi^{(m)}$, can, at least in principle, be calculated from the known terms, $\mathbf{v}^{(m-1)}$ and $\mathbf{v}_A^{(m)}$. Overall this equation is a scalar and so we can transpose the second term in $\boldsymbol{\mathcal{M}}$ on the left-hand side, and multiply by $v_E^{(m)}$ to reduce it to a simple differential

$$\nabla \cdot \left(\rho v_E^{(m)2} \mathbf{V} \right) = v_E^{(m)} \xi^{(m)}, \tag{5.4.8}$$

using differential equation (5.3.20). This can be rewritten

$$\boxed{\frac{\mathrm{d}}{\mathrm{d}T} \ln \left(\rho v_E^{(m)2} \right) = \frac{\xi^{(m)}}{\rho v_E^{(m)}} - \nabla \cdot \mathbf{V},} \tag{5.4.9}$$

using $\mathbf{V} \cdot \nabla = \mathrm{d}/\mathrm{d}T$. This differential equation (5.4.9), or variants thereof, is known as the *transport equation* and is identical in form to equation (5.2.93).

The solution is found as in the acoustic case (Section 5.2), except care must be taken to distinguish the phase and group velocities. As before we use Smirnov's lemma (Section C.2) to connect volume elements which are defined by a ray tube and time perturbations (dq_1, dq_2 and $\mathrm{d}T$). The Jacobian is obtained from

$$\frac{\mathrm{d}\mathbf{x}}{\mathrm{d}T} \cdot \left(\frac{\partial \mathbf{x}}{\partial \mathrm{q}_1} \times \frac{\partial \mathbf{x}}{\partial \mathrm{q}_2} \right) = V J \cos \phi \tag{5.4.10}$$

$$= cJ, \tag{5.4.11}$$

where in equation (5.4.10), ϕ is the angle between the normal to the wavefront and the ray velocity, $\mathbf{V}$ ($V = |\mathbf{V}|$), and in equation (5.4.11), c is the phase velocity ($c|\mathbf{p}| = 1$ and equation (5.1.17) gives $V \cos \phi = c$ – equation (5.3.32)) (see Figure 5.10). Thus in equation (5.2.14),

$$D = \frac{cJ}{c_0 J_0}, \tag{5.4.12}$$

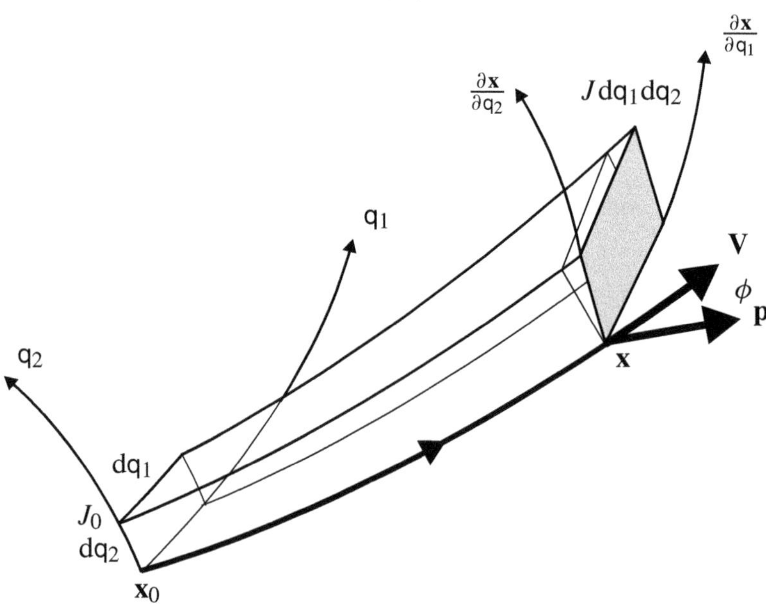

Fig. 5.10. A ray tube formed by perturbations dq_1 and dq_2. The shaded area in the wavefront, $J\mathrm{dq}_1\mathrm{dq}_2$, is normal to the slowness vector, $\mathbf{p}$, whereas the ray is in the direction of the ray velocity, $\mathbf{V}$.

and combining equations (5.4.9) and (5.2.14), we obtain

$$\frac{\mathrm{d}}{\mathrm{d}T}\ln\left(\rho\, c\, v_E^{(m)^2} J\right) = \frac{\xi^{(m)}}{\rho v_E^{(m)}}. \tag{5.4.13}$$

The solution is as result (5.2.95):

$$v_E^{(m)}(T) = \frac{1}{\sqrt{\rho\, c|J|}}\left(\sqrt{\rho\, c_0|J_0|}\, v_E^{(m)}(T_0) + \frac{1}{2}\int_{T_0}^{T}\sqrt{\frac{c|J|}{\rho}}\,\xi^{(m)}\mathrm{d}T\right). \tag{5.4.14}$$

It is now trivial to specialize these results to the zeroth-order coefficients. With $m = 0$, the additional terms (5.4.4) are zero and $\mathbf{v}_A^{(0)} = \mathbf{0}$. Thus the polarization of the leading term in the ray expansion (5.3.1) is parallel to the corresponding eigenvector, $\hat{\mathbf{g}}_E$. For the principal component, $\xi^{(0)} = 0$ and result (5.4.14) reduces to

$$v_E^{(0)} = \frac{\text{constant}}{\sqrt{\rho\, c|J|}}. \tag{5.4.15}$$

The zeroth-order transport equation (5.4.8) can be written

$$\nabla \cdot \mathbf{N} = 0, \tag{5.4.16}$$

where

$$N_j = -\mathbf{v}^{(0)^{\mathrm{T}}}\mathbf{t}_j^{(0)} = \rho\, v^{(0)^2} V_j, \tag{5.4.17}$$

are components of the *energy flux vector*, $\mathbf{N}$ (cf. the Poynting vector in electromagnetic theory) in the ray direction. Note that the amplitude coefficients may be complex due to caustics or total reflections, etc. Equation (5.4.16) does not hold at caustics or reflection/transmission points.

The paraxial ray equations for anisotropic rays are identical in form to those discussed for acoustic rays (Section 5.2.2), although, of course, the elements in the matrix $\mathbf{D}$ (5.2.20) are more complicated. For brevity, we do not repeat all the equations of Section 5.2.2 but will use the results calculated with the appropriate matrix $\mathbf{D}$ for anisotropic media. All the results concerning the dynamic propagator and symmetries still hold.

In anisotropic ray theory, we assume that the Christoffel equation (5.3.14) is non-degenerate, i.e. the velocities of the three ray types differ. In Section 5.6 below, we consider the special isotropic case when the two shear velocities are equal. It is important to realize, however, that general anisotropic media degenerate in certain directions (see Figure 5.8), and in weakly anisotropic media the two quasi-shear ray velocities will be similar. In near-degenerate situations, the factor G_N (5.4.3) will be small and the additional component, (5.4.4) and (5.4.5), large. This will invalidate anisotropic ray theory, as the first-order term may be larger than the leading term. In heterogeneous media, anisotropic ray theory must be used with care as it will break down in degenerate directions, and in weakly anisotropic, heterogeneous media it may break down generally except at very high frequencies (the exact condition will depend on the pulse length compared with the separation of the shear rays). These very important limitations of anisotropic ray theory will be discussed further in Section 10.2 on quasi-isotropic ray theory.

5.4.2 Geometrical Green dyadic

Equation (5.2.60) still applies in anisotropic media and the solution can be found solving (5.2.19) with appropriate initial conditions and the anisotropic Hamiltonian (5.3.18). Kendall, Guest and Thomson (1992) have shown, by comparing with the exact solution for a point force $\mathbf{f}_\mathsf{S}$ in homogeneous media (Buchwald, 1959; Lighthill, 1960; Duff, 1960; Burridge, 1967 – see Exercise 4.10), that (5.4.15) becomes

$$v_E^{(0)}(\mathbf{x}_\mathsf{R}, \mathbf{x}_\mathsf{S}) = \frac{1}{2}\left(\left|\frac{\partial \mathbf{p}_\mathsf{S}}{\partial \mathsf{q}_1} \times \frac{\partial \mathbf{p}_\mathsf{S}}{\partial \mathsf{q}_2}\right| \Big/ \rho_\mathsf{S}\rho_\mathsf{R}\, c_\mathsf{R} V_\mathsf{S} J_\mathsf{R}\right)^{1/2} \hat{\mathbf{g}}_\mathsf{S}^\mathsf{T}\, \mathbf{f}_\mathsf{S}, \qquad (5.4.18)$$

where, compared with result (5.2.66), the distinction between phase and group velocities is important. In expression (5.3.1), $f(\omega) = f^{(3)}(\omega)$ (5.2.64). Numerically, this is all that is needed as solutions (5.2.60) give the required factors for cJ (5.4.11). Although sufficient for the numerical solution, theoretically it is useful to analyse the system further to eliminate the arbitrary parameterization q_ν, and prove reciprocity.

In anisotropic media, we generalize the definition of the spreading function (5.2.67) to

$$\begin{aligned} \mathcal{S}^{(3)}(\mathbf{x}_\mathsf{R}, \mathbf{x}_\mathsf{S}) &= c_\mathsf{R}|J_\mathsf{R}| \Big/ V_\mathsf{R}\left|\frac{\partial \mathbf{p}_\mathsf{S}}{\partial \mathsf{q}_1} \times \frac{\partial \mathbf{p}_\mathsf{S}}{\partial \mathsf{q}_2}\right| \\ &= \left|\widehat{\mathbf{V}}_\mathsf{R} \cdot \left(\frac{\partial \mathbf{x}_\mathsf{R}}{\partial \mathsf{q}_1} \times \frac{\partial \mathbf{x}_\mathsf{R}}{\partial \mathsf{q}_2}\right)\right| \Big/ \left|\frac{\partial \mathbf{p}_\mathsf{S}}{\partial \mathsf{q}_1} \times \frac{\partial \mathbf{p}_\mathsf{S}}{\partial \mathsf{q}_2}\right| \end{aligned} \qquad (5.4.19)$$

(again compared with Kendall, Guest and Thomson, 1992, the ray speed is not included in $\mathcal{S}^{(3)}$). The derivatives $\partial \mathbf{x}_\mathsf{R}/\partial \mathsf{q}_\nu$ lie in the wavefront. Their cross-product is normal to the wavefront, in the direction of the slowness vector, $\mathbf{p}_\mathsf{R}$. We have used the relationship (5.3.32) to write the numerator as a dot-product of this cross-product with the ray direction $\widehat{\mathbf{V}}_\mathsf{R} = \mathrm{sgn}(\mathbf{V}_\mathsf{R})$. Kendall, Guest and Thomson (1992) have shown how to simplify expression (5.4.19). Now

$$\widehat{\mathbf{V}}_\mathsf{R} \cdot \left(\frac{\partial \mathbf{x}_\mathsf{R}}{\partial \mathsf{q}_1} \times \frac{\partial \mathbf{x}_\mathsf{R}}{\partial \mathsf{q}_2}\right) = \epsilon_{ijk} \widehat{V}_i \left(P_{jm}\frac{\partial p_m}{\partial \mathsf{q}_1}\right)\left(P_{kn}\frac{\partial p_n}{\partial \mathsf{q}_2}\right), \qquad (5.4.20)$$

where for simplicity we have written $\widehat{V}_i$, P_{ij} and p_i for components of the ray direction, $\widehat{\mathbf{V}}_\mathsf{R}$, the propagator, $\mathbf{P}_{xp}$, and the source slowness, $\mathbf{p}_\mathsf{S}$, respectively. This can be converted to

$$\widehat{\mathbf{V}}_\mathsf{R}\cdot\left(\frac{\partial \mathbf{x}_\mathsf{R}}{\partial \mathsf{q}_1}\times\frac{\partial \mathbf{x}_\mathsf{R}}{\partial \mathsf{q}_2}\right) = \frac{1}{2}\epsilon_{ijk}\widehat{V}_{\mathsf{R}_i}P_{jm}P_{kn}\left(\frac{\partial p_m}{\partial \mathsf{q}_1}\frac{\partial p_n}{\partial \mathsf{q}_2}-\frac{\partial p_n}{\partial \mathsf{q}_1}\frac{\partial p_m}{\partial \mathsf{q}_2}\right) \quad (5.4.21)$$

$$= \frac{1}{2}\left|\frac{\partial \mathbf{p}_\mathsf{S}}{\partial \mathsf{q}_1}\times\frac{\partial \mathbf{p}_\mathsf{S}}{\partial \mathsf{q}_2}\right|\epsilon_{ijk}\epsilon_{lmn}\widehat{V}_{\mathsf{R}_i}P_{jm}P_{kn}\widehat{V}_{\mathsf{S}_l}, \quad (5.4.22)$$

where we have used the fact that the slowness perturbations at the source, $\partial\mathbf{p}_\mathsf{S}/\partial\mathsf{q}_\nu$, lie in the slowness surface and so their cross-product is in the ray direction $\widehat{\mathbf{V}}_\mathsf{S}$.

The product of elements of the propagator, $\mathbf{P}_{xp}$, in expression (5.4.22) reduce to cofactors of the matrix, and substituting expression (5.4.22) in equation (5.4.19), we obtain

$$\boxed{\mathcal{S}^{(3)}(\mathbf{x}_\mathsf{R},\mathbf{x}_\mathsf{S}) = \left|\widehat{\mathbf{V}}_\mathsf{R}^\mathsf{T}\,\mathbf{P}_{xp}^\ddagger\,\widehat{\mathbf{V}}_\mathsf{S}\right|,} \quad (5.4.23)$$

where the adjoint of a matrix $\mathbf{A}$ is defined as

$$\mathbf{A}^\ddagger = \mathrm{adj}(\mathbf{A}) = |\mathbf{A}|\,\mathbf{A}^{-1}. \quad (5.4.24)$$

It is also useful to define an effective ray length analogous to expression (5.2.72)

$$\boxed{\mathcal{R}^{(3)}(\mathbf{x}_\mathsf{R},\mathbf{x}_\mathsf{S}) = {\mathcal{S}^{(3)}}^{1/2}(\mathbf{x}_\mathsf{R},\mathbf{x}_\mathsf{S})\Big/\,c(\mathbf{x}_\mathsf{S}).} \quad (5.4.25)$$

Combining the results (5.4.18) and (5.4.19), the dyadic Green function is given by (cf. the acoustic result (5.2.71))

$$\boxed{\underline{\mathbf{u}}(t,\mathbf{x}_\mathsf{R};\mathbf{x}_\mathsf{S}) = \frac{\mathrm{Re}\left(\Delta(t-T(\mathbf{x}_\mathsf{R},\mathcal{L}_n))\mathrm{e}^{-\mathrm{i}\pi\,\sigma(\mathbf{x}_\mathsf{R},\mathcal{L}_n)/2}\right)\hat{\mathbf{g}}_\mathsf{R}\,\hat{\mathbf{g}}_\mathsf{S}^\mathsf{T}}{4\pi(\rho(\mathbf{x}_\mathsf{R})\rho(\mathbf{x}_\mathsf{S})V(\mathbf{x}_\mathsf{R})V(\mathbf{x}_\mathsf{S})\mathcal{S}^{(3)})^{1/2}}} \quad (5.4.26)$$

or

$$\boxed{\underline{\mathbf{u}}(t,\mathbf{x}_\mathsf{R};\mathbf{x}_\mathsf{S}) = \frac{\mathrm{Re}\left(\Delta(t-T(\mathbf{x}_\mathsf{R},\mathcal{L}_n))\mathrm{e}^{-\mathrm{i}\pi\,\sigma(\mathbf{x}_\mathsf{R},\mathcal{L}_n)/2}\right)\hat{\mathbf{g}}_\mathsf{R}\,\hat{\mathbf{g}}_\mathsf{S}^\mathsf{T}}{4\pi(\rho(\mathbf{x}_\mathsf{R})\rho(\mathbf{x}_\mathsf{S})V(\mathbf{x}_\mathsf{R})V(\mathbf{x}_\mathsf{S}))^{1/2}c(\mathbf{x}_\mathsf{S})\mathcal{R}^{(3)}},} \quad (5.4.27)$$

where $\mathcal{S}^{(3)}$ is defined by equation (5.4.23), $\mathcal{R}^{(3)}$ by equation (5.4.25) and $\hat{\mathbf{g}} = \hat{\mathbf{g}}_I$ is chosen to be the appropriate eigen-solution in the Christoffel equation (5.3.11). For brevity, we have not indicated explicitly the dependence of the polarizations, $\hat{\mathbf{g}}$, group velocity, $\mathbf{V}$, phase velocity, c, spreading function, $\mathcal{S}^{(3)}$, and effective ray length, $\mathcal{R}^{(3)}$ on the ray path $\mathcal{L}_n$. In anisotropic media, the velocities depend on direction as well as position. Again, the KMAH index counts the caustics along the ray (incrementing by one at each first-order zero – line caustic – of $\mathbf{P}_{xp}$, by two at a second-order zero – point caustic). In anisotropic media, it may also decrease (Klimeš, 1997; Bakker, 1998; Garmany, 2000).

The reciprocity of result (5.4.26), i.e. result (4.5.42), is straightforward to establish using relationships (5.2.40) and (5.4.23). The travel time is clearly the same for the forward or reversed ray. The polarization changes sign, but overall the dyadic Green function remains the same as it contains the polarizations at source and receiver. The polarizations must be defined in a consistent manner, uniformly along the ray path. The reciprocity of the KMAH index is not trivial as the caustics occur at different positions on the forward and reversed rays. Nevertheless, a simple argument establishes the reciprocity. At a caustic, the matrix $\mathbf{P}_{xp}$ is singular. The signature of the matrix is related to the KMAH index but only in modulo arithmetic. In fact this is sufficient to establish the reciprocity of the Green function, as it only depends on the modulo of the KMAH index. However, the KMAH index is reciprocal, without modulo arithmetic. We must count all the caustics along the ray to find the KMAH index, not just the signature of $\mathbf{P}_{xp}$ at the end-point. Consider the simple experiment where we start with the receiver near the source and then move it along the ray to the final position. Initially, the KMAH index must be zero. As we separate the receiver from the source, at each caustic matrix $\mathbf{P}_{xp}$ is singular and the KMAH index increments. Similarly for the reversed ray, the index increments although the caustic occurs at the opposite point. As the ray length increases through a caustic, caustics are created at either end of the ray, at the receiver for the forward ray and at the source for the reciprocal ray. At this length, a pair of paraxial rays is 'focused' at both source and receiver, symmetrically (Figure 5.11*a*). This contrasts with the situation when the ray length is slightly longer,

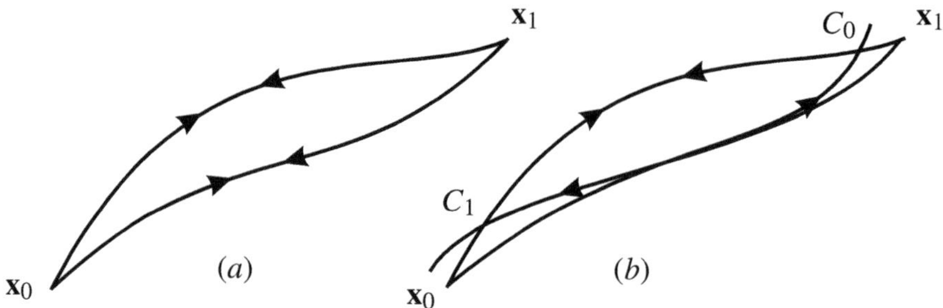

Fig. 5.11. (*a*) Paraxial rays when a caustic forms at the receiver, or at the 'source' on the reciprocal ray; (*b*) paraxial rays when the ray length is slightly longer than in (*a*) so the caustics are in different positions for the ray and its reciprocal. The rays from $\mathbf{x}_0$ form a caustic at C_0, and from $\mathbf{x}_1$ at C_1.

when the caustics are asymmetrical (Figure 5.11*b*). Thus, whatever the length of the ray, (5.2.40) holds as caustics are introduced simultaneously for the extending forward or reversed rays. Thus the KMAH index is reciprocal.

5.4.2.1 General geometrical Green dyadic notation

It is convenient to summarize the results for the geometrical dyadic Green functions for acoustic and elastic media, and in two and three dimensions, in a compact notation. First, in the frequency domain, we write the geometrical ray approximation (the leading term of the asymptotic series ansätze, (5.1.1) and (5.3.1)) for a single ray as

$$\begin{pmatrix} \underline{\mathbf{v}} \\ -\underline{P} \end{pmatrix}(\omega, \mathbf{x}_{\mathsf{R}}, \mathcal{L}_n) = f^{(\ell)}(\omega)\, \Phi(\omega, \mathbf{x}_{\mathsf{R}}, \mathcal{L}_n) \begin{pmatrix} \underline{\mathbf{v}}^{(0)} \\ -\underline{P}^{(0)} \end{pmatrix}(\mathbf{x}_{\mathsf{R}}, \mathcal{L}_n) \qquad (5.4.28)$$

$$\begin{pmatrix} \underline{\mathbf{v}} \\ \underline{\mathbf{t}}_j \end{pmatrix}(\omega, \mathbf{x}_{\mathsf{R}}, \mathcal{L}_n) = f^{(\ell)}(\omega)\, \Phi(\omega, \mathbf{x}_{\mathsf{R}}, \mathcal{L}_n) \begin{pmatrix} \underline{\mathbf{v}}^{(0)} \\ \underline{\mathbf{t}}_j^{(0)} \end{pmatrix}(\mathbf{x}_{\mathsf{R}}, \mathcal{L}_n), \qquad (5.4.29)$$

for the acoustic and anisotropic cases, respectively. The Green spectral functions, $f^{(\ell)}(\omega)$, have been defined in equations (5.2.78) and (5.2.64) for $\ell = 2$ and 3 (in general anisotropic media, it is pointless to consider two-dimensional wave propagation, but in isotropic media or simple anisotropic media such as TIV (Section 4.4.4), it may be useful). The phase part of the propagation has been written as the function

$$\boxed{\Phi(\omega, \mathbf{x}_{\mathsf{R}}, \mathcal{L}_n) = \mathrm{e}^{\mathrm{i}\omega T(\mathbf{x}_{\mathsf{R}}, \mathcal{L}_n)},} \qquad (5.4.30)$$

which depends simply on the travel time and frequency.

Comparing the results (5.2.71) and (5.4.26), the amplitude coefficients in expressions (5.4.28) and (5.4.29) can then be written in a dyadic form

$$\begin{pmatrix} \underline{\mathbf{v}}^{(0)} \\ -\underline{P}^{(0)} \end{pmatrix}(\mathbf{x}_{\mathsf{R}}, \mathcal{L}_n) = \mathcal{T}^{(\ell)}(\mathbf{x}_{\mathsf{R}}, \mathcal{L}_n) \begin{pmatrix} \mathbf{g} \\ -(Z/2)^{1/2} \end{pmatrix}(\mathbf{x}_{\mathsf{R}}, \mathcal{L}_n)\, \mathbf{g}^{\mathsf{T}}(\mathbf{x}_{\mathsf{S}}, \mathcal{L}_n) \qquad (5.4.31)$$

$$\begin{pmatrix} \underline{\mathbf{v}}^{(0)} \\ \underline{\mathbf{t}}_j^{(0)} \end{pmatrix}(\mathbf{x}_{\mathsf{R}}, \mathcal{L}_n) = \mathcal{T}^{(\ell)}(\mathbf{x}_{\mathsf{R}}, \mathcal{L}_n) \begin{pmatrix} \mathbf{g} \\ -\mathbf{Z}_j\, \mathbf{g} \end{pmatrix}(\mathbf{x}_{\mathsf{R}}, \mathcal{L}_n)\, \mathbf{g}^{\mathsf{T}}(\mathbf{x}_{\mathsf{S}}, \mathcal{L}_n), \qquad (5.4.32)$$

where the scalar and tensor impedances, Z and $\mathbf{Z}_j$, are defined in equations (5.2.8) and (5.3.22), respectively. We have used a non-normalized form for the polarization

$$\boxed{\mathbf{g}(\mathbf{x}, \mathcal{L}_n) = (2\rho V)^{-1/2}\hat{\mathbf{g}},} \qquad (5.4.33)$$

where V is the ray velocity. We refer to this as the *energy-flux normalized polarization*. The scalar amplitude part of the propagation is contained in the function

$$\boxed{\mathcal{T}^{(\ell)}(\mathbf{x}_\mathrm{R}, \mathcal{L}_n) = \mathcal{S}^{(\ell)^{-1/2}} \mathrm{e}^{-\mathrm{i}\pi\,\mathrm{sgn}(\omega)\sigma(\mathbf{x}_\mathrm{R}, \mathcal{L}_n)/2},} \tag{5.4.34}$$

which is independent of frequency (except through the sign). It depends on the geometry of the ray path through the spreading functions $\mathcal{S}^{(\ell)}$ which have been defined in equations (5.2.68), (5.2.80) and (5.4.23) (later $\mathcal{T}^{(\ell)}$ is generalized to contain the product of reflection and transmission coefficients along the ray (6.8.2), but is still independent of frequency except through the sign).

We are normally interested in the displacement Green function from results (5.4.28) and (5.4.29), which can be written compactly as

$$\boxed{\underline{\mathbf{u}}(t, \mathbf{x}_\mathrm{R}; \mathbf{x}_\mathrm{S}) = \frac{1}{2\pi} \mathbf{g}(\mathbf{x}_\mathrm{R}, \mathcal{L}_n)\, \mathcal{P}^{(\ell)}(t, \mathbf{x}_\mathrm{R}, \mathcal{L}_n)\, \mathbf{g}^\mathrm{T}(\mathbf{x}_\mathrm{S}, \mathcal{L}_n),} \tag{5.4.35}$$

where in the frequency domain the scalar propagation function is

$$\mathcal{P}^{(\ell)}(\omega, \mathbf{x}_\mathrm{R}, \mathcal{L}_n) = -\frac{2\pi f^{(\ell)}(\omega)}{\mathrm{i}\omega} \mathcal{T}^{(\ell)}(\mathbf{x}_\mathrm{R}, \mathcal{L}_n)\, \Phi(\omega, \mathbf{x}_\mathrm{R}, \mathcal{L}_n). \tag{5.4.36}$$

The corresponding time functions are

$$\mathcal{P}^{(2)}(t, \mathbf{x}_\mathrm{R}, \mathcal{L}_n) = 2^{1/2}\,\mathrm{Re}\left(\mathcal{T}^{(2)}(\mathbf{x}_\mathrm{R}, \mathcal{L}_n)\, \Lambda\left(t - T(\mathbf{x}_\mathrm{R}, \mathcal{L}_n)\right)\right) \tag{5.4.37}$$

$$\mathcal{P}^{(3)}(t, \mathbf{x}_\mathrm{R}, \mathcal{L}_n) = \mathrm{Re}\left(\mathcal{T}^{(3)}(\mathbf{x}_\mathrm{R}, \mathcal{L}_n)\, \Delta\left(t - T(\mathbf{x}_\mathrm{R}, \mathcal{L}_n)\right)\right), \tag{5.4.38}$$

where the analytic delta and lambda functions are defined in equations (B.1.7) and (B.2.5), respectively.

For a point force and moment tensor source (4.6.17), the dyadic (5.4.35) is applied to

$$\mathbf{f}_\mathrm{S} + \mathbf{M}_\mathrm{S}\mathbf{p}_\mathrm{S}. \tag{5.4.39}$$

Thus

$$\mathbf{u}(t, \mathbf{x}_\mathrm{R}; \mathbf{x}_\mathrm{S}) = \frac{1}{2\pi} \mathbf{g}(\mathbf{x}_\mathrm{R}, \mathcal{L}_n)\, \mathcal{P}^{(\ell)}(t, \mathbf{x}_\mathrm{R}; \mathbf{x}_\mathrm{S})\, \mathbf{g}^\mathrm{T}(\mathbf{x}_\mathrm{S}, \mathcal{L}_n)\, (\mathbf{f}_\mathrm{S} + \mathbf{M}_\mathrm{S}\mathbf{p}_\mathrm{S}), \tag{5.4.40}$$

i.e. $\mathbf{g}^\mathrm{T}(\mathbf{f}_\mathrm{S} + \mathbf{M}_\mathrm{S}\mathbf{p}_\mathrm{S})$ is the source directivity function.

5.5 Isotropic kinematic ray theory

In Section 4.4.3 we have discussed the simplifications of the constitutive relation that occur in isotropic, elastic media. We use these results to specialize the anisotropic results of Section 5.3.

As the medium is isotropic, we can choose any propagation direction to determine the Hamiltonian from the anisotropic definition (5.3.18). Let us choose

$$\mathbf{p} = \begin{pmatrix} 0 & 0 & p \end{pmatrix}^{\mathrm{T}}. \tag{5.5.1}$$

Then

$$p_j p_k \mathbf{c}_{jk} = p^2 \mathbf{c}_{33} = \begin{pmatrix} \mu p^2 & 0 & 0 \\ 0 & \mu p^2 & 0 \\ 0 & 0 & (\lambda + 2\mu) p^2 \end{pmatrix}. \tag{5.5.2}$$

One solution of equation (5.3.11) requires $\mathbf{v}^{(0)}$ to be longitudinal, i.e. with direction (5.5.1), $\hat{\mathbf{g}} = \hat{\mathbf{i}}_3$, and $c = V = \alpha$, the P wave velocity (4.5.60). Alternatively, the solution requires $\mathbf{v}^{(0)}$ to be transverse, e.g. $\hat{\mathbf{g}}_\nu = \hat{\mathbf{i}}_\nu$, and $c = V = \beta$, the S wave velocity (4.5.61). Generalizing to any direction, the Hamiltonians are

$$H_1 = H_2 = \frac{1}{2} \beta^2 p_i p_i, \quad H_3 = \frac{1}{2} \alpha^2 p_i p_i \tag{5.5.3}$$

(for definiteness, we always choose $I = 3$ as the P wave).

Using the isotropic Hamiltonian, (5.5.3), the kinematic ray equations (5.3.20) and (5.3.21) simplify to

$$\frac{\mathrm{d}\mathbf{x}}{\mathrm{d}T} = V^2 \mathbf{p} \tag{5.5.4}$$

$$\frac{\mathrm{d}\mathbf{p}}{\mathrm{d}T} = -\frac{\nabla V}{V}, \tag{5.5.5}$$

where $V = \alpha$ or β. These are identical in form to the acoustic kinematic equations, (5.1.14) and (5.1.15).

The eigen-equation (5.3.11) is degenerate for shear waves and does not determine the S polarization uniquely. To find the shear wave polarizations in isotropic media, we must reconsider the equation (5.3.5) with $m = 1$.

For future use let us give some useful relationships involving the polarization vectors and elastic matrices ($\nu = 1$ or 2):

$$p_j \mathbf{c}_{jk} \hat{\mathbf{g}}_3 = (\lambda + \mu) p_k \hat{\mathbf{g}}_3 + \frac{\mu}{c} \hat{\mathbf{i}}_k \tag{5.5.6}$$

$$p_j \mathbf{c}_{kj} \hat{\mathbf{g}}_3 = 2\mu p_k \hat{\mathbf{g}}_3 + \frac{\lambda}{c} \hat{\mathbf{i}}_k \tag{5.5.7}$$

$$p_j \mathbf{c}_{jk} \hat{\mathbf{g}}_\nu = \mu p_k \hat{\mathbf{g}}_\nu + \lambda (\hat{\mathbf{g}}_\nu \cdot \hat{\mathbf{i}}_k) \mathbf{p} \tag{5.5.8}$$

$$p_j \mathbf{c}_{kj} \hat{\mathbf{g}}_\nu = \mu p_k \hat{\mathbf{g}}_\nu + \mu (\hat{\mathbf{g}}_\nu \cdot \hat{\mathbf{i}}_k) \mathbf{p}, \tag{5.5.9}$$

where $\hat{\mathbf{i}}_k$ are the unit cartesian coordinate vectors. These results can easily be derived using equations (4.4.55) and (4.4.56), etc. (Exercise 5.6). Expressions (5.5.7) and (5.5.9) are useful for the traction vectors, $\mathbf{t}_k$ (5.3.24), and are equivalent to the constitutive relationship (4.4.51).

5.6 Isotropic dynamic ray theory

5.6.1 Shear ray polarization

In the equation (5.3.5) with $m = 1$, we expand $\mathbf{v}^{(1)}$ according to equation (5.4.1). For a shear ray, the equation reduces to

$$\overline{\mathcal{M}}(\mathbf{v}^{(0)}, \mathbf{t}_j^{(0)}) = \rho \left(\frac{\beta^2}{\alpha^2} - 1 \right) v_3^{(1)} \hat{\mathbf{g}}_3. \tag{5.6.1}$$

Pre-multiplying by $\hat{\mathbf{g}}_\nu^{\mathrm{T}}$, and considering one shear wave, $\mathbf{v}^{(0)} = v_1^{(0)} \hat{\mathbf{g}}_1$ say, without loss in generality, we obtain

$$\hat{\mathbf{g}}_\nu^{\mathrm{T}} \mathcal{M}(v_1^{(0)} \hat{\mathbf{g}}_1) = 0. \tag{5.6.2}$$

Using results (5.5.8) and (5.5.9) in equation (5.6.2) and expanding, we obtain

$$\delta_{1\nu} \left(2\mu \mathbf{p} \cdot \nabla v_1^{(0)} + v_1^{(0)} \mathbf{p} \cdot \nabla \mu + \mu v_1^{(0)} \nabla \cdot \mathbf{p} \right) + 2\rho v_1^{(0)} \hat{\mathbf{g}}_\nu^{\mathrm{T}} \frac{\mathrm{d}\hat{\mathbf{g}}_1}{\mathrm{d}T} = 0, \tag{5.6.3}$$

where we have used the orthonormality of the eigenvectors, and the derivative $\beta^2 \mathbf{p} \cdot \nabla = \mathrm{d}/\mathrm{d}T$ to simplify the final term.

When $\nu = 1$, this gives

$$2\mu \mathbf{p} \cdot \nabla v_1^{(0)} + v_1^{(0)} \mathbf{p} \cdot \nabla \mu + \mu v_1^{(0)} \nabla \cdot \mathbf{p} = \frac{1}{v_1^{(0)}} \nabla \cdot \left(\rho \beta^2 v_1^{(0)\,2} \mathbf{p} \right) = 0, \tag{5.6.4}$$

which is the isotropic transport equation equivalent to result (5.4.8) with $\mathbf{V} = \beta^2 \mathbf{p}$ and $\xi^{(0)} = 0$.

When $\nu = 2$, equation (5.6.3) is simply

$$2\rho v_1^{(0)} \hat{\mathbf{g}}_2^{\mathrm{T}} \frac{\mathrm{d}\hat{\mathbf{g}}_1}{\mathrm{d}T} = 0. \tag{5.6.5}$$

For $\hat{\mathbf{g}}_1$ and $\hat{\mathbf{g}}_2$ to be the polarizations of the independent shear waves, they must satisfy this differential equation (5.6.5). Thus the change in $\hat{\mathbf{g}}_1$, which is necessarily orthogonal to itself (as the vector is normalized), is also orthogonal to $\hat{\mathbf{g}}_2$. It can only be in the $\hat{\mathbf{g}}_3$ direction, i.e.

$$\frac{\mathrm{d}\hat{\mathbf{g}}_\nu}{\mathrm{d}T} = a \hat{\mathbf{g}}_3 \tag{5.6.6}$$

(generalized for both $\nu = 1$ and 2). Thus the shear polarization only changes in the ray direction and does not twist about it (in electromagnetic theory this is known as Rytov's field-vectors rotation law and is a particular case of the more general phenomenon, Berry's topological phase – see Kravtsov and Orlov, 1990, p. 233; Berry, 1982, 1988). It remains to find a in (5.6.6) which is determined from the geometrical condition that $\hat{\mathbf{g}}_\nu$ remains transverse, i.e.

$$a = V\,\mathbf{p}^{\mathrm{T}}\,\frac{\mathrm{d}\hat{\mathbf{g}}_\nu}{\mathrm{d}T} = -V\,\hat{\mathbf{g}}_\nu^{\mathrm{T}}\,\frac{\mathrm{d}\mathbf{p}}{\mathrm{d}T} = \hat{\mathbf{g}}_\nu^{\mathrm{T}}\,\nabla V, \tag{5.6.7}$$

using equation (5.5.5). Thus

$$\boxed{\frac{\mathrm{d}\hat{\mathbf{g}}_\nu}{\mathrm{d}T} = \left(\hat{\mathbf{g}}_\nu^{\mathrm{T}}\nabla V\right)\hat{\mathbf{g}}_3} \tag{5.6.8}$$

(the factor $a = \hat{\mathbf{g}}_\nu^{\mathrm{T}}\nabla V$ is a velocity gradient component in the wavefront). This result has been given by Popov and Pšenčík (1978) and Petrashen and Kashtan (1984). In the literature it has usually been proved indirectly using the equations for the normal and binormal of the ray path, and relating the shear polarization to the torsion of the ray (Popov and Pšenčík, 1978; Červený and Hron, 1980; Červený, 1985). The above direct proof is due to Pšenčík (personal communication, 1998) and avoids ever having to discuss these other matters.

It is important to remember that all the orthogonal eigenvectors, $\hat{\mathbf{g}}_i$, exist for any ray type. The eigenvector, $\hat{\mathbf{g}}_I$, is the polarization. Thus for a P ray, equation (5.6.8) with $V = \alpha$ is used to define the wavefront basis vectors, $\hat{\mathbf{g}}_1$ and $\hat{\mathbf{g}}_2$. For S rays, equation (5.6.8) with $V = \beta$ is used to define the polarization solutions, $\hat{\mathbf{g}}_1$ and $\hat{\mathbf{g}}_2$ – the eigenvector $\hat{\mathbf{g}}_3$ is tangent to the ray.

Apart from using results (5.5.6)–(5.5.9), (5.6.8), the orthonormality of the polarization vectors and kinematic ray results, it is useful to note

$$\frac{\mathrm{d}\hat{\mathbf{g}}_3}{\mathrm{d}T} = -\left(\hat{\mathbf{g}}_\nu^{\mathrm{T}}\nabla V\right)\hat{\mathbf{g}}_\nu. \tag{5.6.9}$$

The divergence of the polarizations, useful in Born scattering theory (Section 10.3) are

$$\nabla\cdot\hat{\mathbf{g}}_\nu = \frac{1}{V}\left(\hat{\mathbf{g}}_\nu\cdot\nabla V\right), \tag{5.6.10}$$

and

$$\nabla\cdot\hat{\mathbf{g}}_3 = K, \tag{5.6.11}$$

where K is the wavefront curvature (see Exercise 5.5).

5.6.2 Higher-order amplitude coefficients

Having achieved the decoupling of the shear-wave amplitude coefficients by using the shear-wave polarization vectors as the basis, the analysis for anisotropic media for higher-order remains valid for isotropic media (Červený and Hron, 1980): for the P ray, we have two additional terms and one principal component; for S rays, we have one additional term, and two principal components, both determined using result (5.4.14).

In isotropic media, the first-order additional terms (equation (5.4.4) with $m = 1$) can be written in relatively simple forms (Eisner and Pšenčík, 1996). For P waves, we obtain

$$\begin{aligned} v_\nu^{(1)} &= \frac{\alpha^2 \hat{\mathbf{g}}_\nu^{\mathrm{T}}}{\rho(\beta^2 - \alpha^2)} \boldsymbol{\mathcal{M}}(v_3^{(0)} \hat{\mathbf{g}}_3) \\ &= -\hat{\mathbf{g}}_\nu^{\mathrm{T}} \left(\alpha \frac{\partial v_3^{(0)}}{\partial g_\nu} + \frac{v_3^{(0)}}{\alpha^2 - \beta^2} \Big((\alpha^2 - \beta^2) \nabla \alpha - 4\alpha\beta \nabla \beta \right. \\ &\quad \left. + \alpha(\alpha^2 - 2\beta^2) \nabla (\ln \rho) \Big) \right) \end{aligned} \tag{5.6.12}$$

(g_ν is a coordinate in the direction $\hat{\mathbf{g}}_\nu$). Similarly for S waves, the additional term is

$$\begin{aligned} v_\nu^{(1)} &= \frac{\beta^2 \hat{\mathbf{g}}_3^{\mathrm{T}}}{\rho(\alpha^2 - \beta^2)} \boldsymbol{\mathcal{M}}(v_\nu^{(0)} \hat{\mathbf{g}}_\nu) \\ &= \hat{\mathbf{g}}_3^{\mathrm{T}} \left(\beta \frac{\partial v_\nu^{(0)}}{\partial g_\nu} + \frac{v_\nu^{(0)}}{\alpha^2 - \beta^2} \Big((\alpha^2 + 3\beta^2) \nabla \beta + \beta^3 \nabla (\ln \rho) \Big) \right). \end{aligned} \tag{5.6.13}$$

5.7 One and two-dimensional media

By convention we describe a medium where the material properties only vary in one or two-dimensions, e.g. $\rho(\mathbf{x}) = \rho(x_3)$ or $\rho(\mathbf{x}) = \rho(x_1, x_3)$, as 1D or 2D. This terminology is ambiguous as the model is still three dimensional and we consider wave propagation in three dimensions from a point source. This is distinct from the truly two-dimensional propagation considered in Section 5.2.3.2. For a 2D model, wave propagation in three dimensions is sometimes referred to as 2.5D wave propagation, but a similar term does not exist for a 1D model.

Naturally the ray equations simplify if the heterogeneity in the model is restricted to one or two dimensions. The appropriate ray equation, (5.1.15), (5.3.21) or (5.5.5), immediately reduces to the conservation of the slowness component(s)

$$p_\nu = \text{constant}, \tag{5.7.1}$$

in the other dimensions. In an anisotropic medium, although this results in some simplifications of the other equations, the ray paths are still three dimensional. Even in a 1D, anisotropic model, the rays can be non-planar.

Although there is some simplification of the other equations in 2D, only in 1D media are these really significant, circumventing the need to solve ordinary differential equations. Thus if the medium only depends on x_3, then equation (5.1.14), (5.3.20) or (5.5.4) is

$$\frac{\mathrm{d}x_i}{\mathrm{d}T} = V_i(p_1, p_2, x_3), \tag{5.7.2}$$

as p_3 can be found from the constraint (5.1.18), (5.3.19) or (5.5.3). Then for $\nu = 1$ or 2 we have

$$x_\nu(p_1, p_2, x_3) = \oint \frac{V_\nu}{V_3}\,\mathrm{d}x_3 \tag{5.7.3}$$

$$T(p_1, p_2, x_3) = \oint \frac{1}{V_3}\,\mathrm{d}x_3, \tag{5.7.4}$$

where the depth integral is over all segments of the ray arranged so $\mathrm{d}x_3/V_3$ is positive (see the equivalent equations (2.3.7) and (2.3.8) in Section 2.3; in anisotropic media it is possible for V_ν to be negative – Section 5.7.2.2 and Shearer and Chapman, 1988).

In isotropic media, we have axial symmetry and without loss in generality we can take $p_2 = x_2 = 0$. Then integrals (5.7.3) and (5.7.4) reduce to the simple *ray integrals*

$$x_1(p_1, 0, x_3) = \oint \frac{p_1}{p_3}\,\mathrm{d}x_3 = \oint \tan\theta(x_3)\,\mathrm{d}x_3 \tag{5.7.5}$$

$$T(p_1, 0, x_3) = \oint \frac{1}{V^2 p_3}\,\mathrm{d}x_3 = \oint \frac{\sec\theta(x_3)}{V(x_3)}\,\mathrm{d}x_3, \tag{5.7.6}$$

used in layered media. The slowness vector is

$$\mathbf{p} = \frac{1}{V(x_3)}\begin{pmatrix}\sin\theta(x_3)\\ 0\\ \cos\theta(x_3)\end{pmatrix}, \tag{5.7.7}$$

where p_1 is conserved and $\theta(x_3)$ is the angle the ray makes with the vertical.

In 1D and 2D models, it is convenient to use the horizontal slowness components, p_x and p_y, to parameterize the rays (for q_ν). We consider an isotropic medium (so $\mathbf{p}$ is in the ray direction) and propagation in the plane $y = 0$. For the tube cross-section near the source, we have

$$J_\mathrm{S}\,\mathrm{d}p_{x\mathrm{S}}\,\mathrm{d}p_{y\mathrm{S}} = R^2 \sin\theta_\mathrm{S}\,\mathrm{d}\theta_\mathrm{S}\,\mathrm{d}\phi_\mathrm{S}, \tag{5.7.8}$$

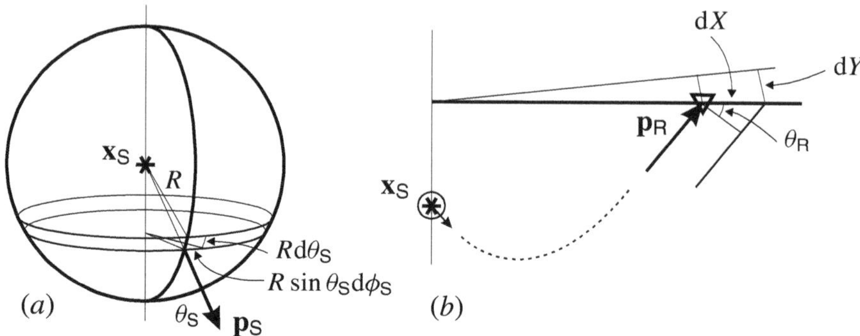

Fig. 5.12. The cross-section of a ray tube: (*a*) on a sphere about the source, as in equation (5.7.8); and (*b*) at the receiver, as in equation (5.7.11).

where θ_S and ϕ_S are polar angles with respect the vertical axis (Figure 5.12). On a unit sphere about the source

$$\alpha_S^2 \, \mathrm{d}p_{xS} \, \mathrm{d}p_{yS} = \tan\theta_S \, \mathrm{d}\theta_S \, \mathrm{d}\phi_S. \tag{5.7.9}$$

This is easily seen as $\alpha_S p_{xS} = \sin\theta_S$ so $\mathrm{d}\theta_S = \alpha_S \sec\theta_S \mathrm{d}p_{xS}$, and $\mathrm{d}\phi_S = \mathrm{d}p_{yS}/p_{xS}$. Thus

$$J_S = \frac{\alpha_S^2 R^2}{\cos\theta_S}. \tag{5.7.10}$$

At the receiver, where X and Y are the horizontal range functions, the ray tube cross-section is given by

$$J \, \mathrm{d}p_{xS}\mathrm{d}p_{yS} = \cos\theta_R \, \mathrm{d}X \, \mathrm{d}Y = \cos\theta_R \left(\frac{\partial X}{\partial p_{xS}}\mathrm{d}p_{xS}\right)\left(\frac{\partial Y}{\partial p_{yS}}\mathrm{d}p_{yS}\right), \tag{5.7.11}$$

where θ_R is the angle between the ray and the vertical (Figure 5.12). Thus the Green dyadic (5.4.27) can be written

$$\boxed{\underline{\mathbf{u}}(t, \mathbf{x}_R; \mathbf{x}_S) = \frac{\mathrm{Re}\left(\Delta(t - T(\mathbf{x}_R, \mathcal{L}_n))\mathrm{e}^{-\mathrm{i}\pi\,\sigma(\mathbf{x}_R,\mathcal{L}_n)/2}\right) \hat{\mathbf{g}}_R \, \hat{\mathbf{g}}_S^{\mathrm{T}}}{4\pi \left(\rho_R \rho_S \alpha_R \alpha_S \cos\theta_R \cos\theta_S \left|\frac{\partial X}{\partial p_{xS}} \frac{\partial Y}{\partial p_{yS}}\right|\right)^{1/2}}.} \tag{5.7.12}$$

In isotropic 1D models, there are more simplifications as the ray path becomes planar and similarly, when rays propagate in the symmetry plane of a 2D model. In both cases only the lower-order kinematic and dynamic ray equations need be solved. In the dynamic equations, the transverse components, e.g. in the y direction, separate and can be solved simply. From the dynamic ray equations (5.2.19), with (5.2.20), (5.2.21) and (5.2.22), the transverse spreading term is (Brokešova, 1992)

$$\frac{\partial Y}{\partial p_{y\mathrm{S}}} = \oint V^2 \mathrm{d}T = \oint V\,\mathrm{d}s. \tag{5.7.13}$$

Sometimes a 2D model is considered as having cylindrical rather than cartesian symmetry (although this is unlikely to be physically realistic as the axis of symmetry must be through the source). In such a model, rays propagate on planes of constant azimuth (constant ϕ), so

$$\mathrm{d}\phi = \frac{\mathrm{d}p_{y\mathrm{S}}}{p_{x\mathrm{S}}} = \frac{\mathrm{d}p_y}{p_x} = \frac{\mathrm{d}Y}{X}, \tag{5.7.14}$$

and

$$\frac{\partial Y}{\partial p_{y\mathrm{S}}} = \frac{X}{p_{x\mathrm{S}}}. \tag{5.7.15}$$

In a 1D model

$$\boxed{\frac{\partial Y}{\partial p_{y\mathrm{S}}} = \oint V\,\mathrm{d}s = \oint \frac{V\,\mathrm{d}z}{\cos\theta} = \frac{X}{p_x},} \tag{5.7.16}$$

using integral (2.3.7), and expressions (5.7.13) and (5.7.15) are equal. Thus in a 1D model, the Green dyadic is

$$\underline{\mathbf{u}}(t, \mathbf{x}_\mathrm{R}; \mathbf{x}_\mathrm{S}) = \frac{\mathrm{Re}\left(\Delta(t - T(\mathbf{x}_\mathrm{R}, \mathcal{L}_n))\mathrm{e}^{-\mathrm{i}\pi\,\sigma(\mathbf{x}_\mathrm{R},\mathcal{L}_n)/2}\right) \hat{\mathbf{g}}_\mathrm{R}\,\hat{\mathbf{g}}_\mathrm{S}^\mathrm{T}}{4\pi\left(\rho_\mathrm{R}\rho_\mathrm{S}\alpha_\mathrm{R}\alpha_\mathrm{S}\cos\theta_\mathrm{R}\cos\theta_\mathrm{S}\left|\frac{X}{p_x}\frac{\partial X}{\partial p_x}\right|\right)^{1/2}}, \tag{5.7.17}$$

where expressions were given for $\mathrm{d}X/\mathrm{d}p_x$ in Section 2.3. The expression (5.7.17) remains valid at $X = 0$ using l'Hopital's rule to replace $X/p_x \rightarrow \mathrm{d}X/\mathrm{d}p_x$ which is necessarily equal to $\mathrm{d}Y/\mathrm{d}p_y$ (as in equation (5.7.12)) in a 1D model.

5.7.1 Transversely isotropic media

Normally for anisotropic media, we have to solve the ray equations numerically. Even in homogeneous media, we have to solve the Christoffel equation (5.3.11) numerically. Fortunately, in transversely isotropic media (Section 4.4.4), the Christoffel equation can be solved easily, giving the results required for homogeneous media and simplifying numerical solutions in inhomogeneous media. The results in this section can be found in many textbooks, e.g. Musgrave (1970). Some of the results in this section were included in Exercise 4.3.

For simplicity, we assume that the axis of symmetry coincides with the x_3 axis (as in Section 4.4.4). If the orientation of the axis is different, the results can be converted by rotation. As the medium is axially symmetric, we assume propagation in the x_1–x_3 plane, i.e. the x_2 components are zero. With these simplifications, the Christoffel equation (5.3.11) reduces to

$$\begin{pmatrix} A_{11}p_1^2 + A_{44}p_3^2 - 1 & 0 & ap_1p_3 \\ 0 & A_{66}p_1^2 + A_{44}p_3^2 - 1 & 0 \\ ap_1p_3 & 0 & A_{44}p_1^2 + A_{33}p_3^2 - 1 \end{pmatrix} \hat{\mathbf{g}} = \mathbf{0}, \tag{5.7.18}$$

where

$$A_{ij} = C_{ij}/\rho, \tag{5.7.19}$$

the squared-velocity parameters, and

$$a = A_{13} + A_{44}. \tag{5.7.20}$$

One solution has polarization in the transverse, horizontal direction, e.g.

$$\hat{\mathbf{g}} = \hat{\mathbf{i}}_2. \tag{5.7.21}$$

We refer to this as the *qSH* ray. The Hamiltonian that defines the slowness surface for the *qSH* ray is

$$H(\mathbf{x}, \mathbf{p}) = \frac{1}{2}\left(A_{66}p_1^2 + A_{44}p_3^2\right), \tag{5.7.22}$$

which is an ellipse. Given a phase direction, $\hat{\mathbf{p}}$, it is trivial to find the phase slowness

$$c^2 = A_{66}\hat{p}_1^2 + A_{44}\hat{p}_3^2, \tag{5.7.23}$$

or given a horizontal component of slowness, as occurs in ray tracing in a one-dimensional model, to find the vertical slowness component

$$p_3 = \pm\left(\frac{1}{A_{44}} - \frac{A_{66}}{A_{44}}p_1^2\right)^{1/2}. \tag{5.7.24}$$

This equation replaces the isotropic equation (2.3.10).

The group velocity for *qSH* rays is easily found from equation (5.3.20), i.e.

$$V_1 = A_{66}p_1 \tag{5.7.25}$$

$$V_3 = A_{44}p_3. \tag{5.7.26}$$

Substituting in integrals (5.7.3) and (5.7.4), we obtain the ray integrals for a one-dimensional medium. For a homogeneous, layered medium, we have

$$\Delta x_1 = \frac{A_{66}}{A_{44}}\frac{p_1}{p_3}\Delta x_3 \tag{5.7.27}$$

$$\Delta T = \frac{1}{A_{44}}\frac{1}{p_3}\Delta x_3, \tag{5.7.28}$$

generalizing the isotropic results, (2.3.4) and (2.3.5). After some algebra, the range derivative can be written

$$\frac{\partial(\Delta x_i)}{\partial p_1} = \frac{A_{66}}{A_{44}^2 p_3^3}\Delta x_3 = \frac{\Delta x_1}{p_1}\left(1 + \frac{p_1}{p_3}\frac{\Delta x_1}{\Delta x_3}\right). \tag{5.7.29}$$

In an isotropic medium, $A_{44} = A_{66} = \beta^2$, the first expression reduces to the isotropic result and the second expression is identical to equation (2.3.6).

The other solutions of the eigen-equation (5.7.18) are more interesting. The polarization is in the plane of propagation and requires

$$A_{11}A_{44}p_1^4 + A_{33}A_{44}p_3^4 + Ap_1^2p_3^2 - (A_{11} + A_{44})p_1^2 - (A_{33} + A_{44})p_3^2 + 1 = 0, \tag{5.7.30}$$

where

$$A = A_{11}A_{33} + A_{44}^2 - a^2. \tag{5.7.31}$$

For a given horizontal slowness, p_1, we can solve this for the vertical slowness

$$p_3^2 = \frac{B \mp \left[B^2 - 4A_{33}A_{44}(A_{11}p_1^2 - 1)(A_{44}p_1^2 - 1)\right]^{1/2}}{2A_{33}A_{44}}, \tag{5.7.32}$$

where

$$B = A_{33} + A_{44} - Ap_1^2, \tag{5.7.33}$$

or for a given slowness direction, $\hat{\mathbf{p}}$, for the phase velocity

$$c^2 = \frac{1}{2}\left(A_{44} + A_{33}\hat{p}_3^2 + A_{11}\hat{p}_1^2 \pm \sqrt{\left((A_{33} - A_{44})\hat{p}_3^2 - (A_{11} - A_{44})\hat{p}_1^2\right)^2 + 4\hat{p}_1^2\hat{p}_3^2a^2}\right). \tag{5.7.34}$$

By definition, the upper sign corresponds to the qP surface, and we refer to the other solution as qSV. The corresponding eigenvectors are

$$\hat{\mathbf{g}} = \operatorname{sgn}\begin{pmatrix} 2a\hat{p}_1\hat{p}_3 \\ 0 \\ (A_{33} - A_{44})\hat{p}_3^2 - (A_{11} - A_{44})\hat{p}_1^2 \\ \pm\left[\left((A_{33} - A_{44})\hat{p}_3^2 - (A_{11} - A_{44})\hat{p}_1^2\right)^2 + 4\hat{p}_1^2\hat{p}_3^2a^2\right]^{1/2} \end{pmatrix}. \tag{5.7.35}$$

Using equation (5.7.30) in definition (5.3.20), we can determine the group velocity. After some algebra, we obtain

$$V_1 = p_1 \left(2A_{11}A_{44}p_1^2 + Ap_3^2 - A_{11} - A_{44}\right) \Big/ D \qquad (5.7.36)$$

$$V_3 = p_3 \left(Ap_1^2 + 2A_{33}A_{44}p_3^2 - A_{33} - A_{44}\right) \Big/ D, \qquad (5.7.37)$$

where

$$D = (A_{11} + A_{44})p_1^2 + (A_{33} + A_{44})p_3^2 - 2. \qquad (5.7.38)$$

These expressions can be used in integrals (5.7.3) and (5.7.4) for the ray integrals in a one-dimensional model. In a homogeneous layer, we obtain

$$\Delta x_1 = \frac{p_1}{p_3} \frac{2A_{11}A_{44}p_1^2 + Ap_3^2 - A_{11} - A_{44}}{Ap_1^2 + 2A_{33}A_{44}p_3^2 - A_{33} - A_{44}} \Delta x_3 \qquad (5.7.39)$$

$$\Delta T = \frac{D}{p_3\left[Ap_1^2 + 2A_{33}A_{55}p_3^2 - A_{33} - A_{44}\right]} \Delta x_3, \qquad (5.7.40)$$

for the range and travel time. Expression (5.7.39) can be differentiated

$$\frac{\partial(\Delta x_i)}{\partial p_1} = \frac{\Delta x_1}{p_1}\left(1 + \frac{p_1}{p_3}\frac{\Delta x_1}{\Delta x_3}\right) + \frac{4}{p_3 \Delta x_3} \frac{A_{11}A_{44}p_1^2\Delta x_3^2 - Ap_1p_3\Delta x_1\Delta x_3 + A_{33}A_{44}p_3^2\Delta x_1^2}{Ap_1^2 + 2A_{33}A_{44}p_3^2 - A_{33} - A_{44}}, \qquad (5.7.41)$$

for the spreading function. These expressions provide all the terms for the Green function in a homogeneous, layered, transversely isotropic medium.

An example of the slowness surfaces and wavefronts for transversely isotropic media is shown in Figure 5.13. Note that on the x_3 symmetry axis, the qP velocity is $\sqrt{A_{33}}$ and the qSV and qSH velocities are equal and equal to $\sqrt{A_{44}}$. On the x_1 axis (and by axial symmetry, in any direction in the x_1–x_2 plane), the qP velocity is $\sqrt{A_{11}}$, the qSH velocity is $\sqrt{A_{66}}$ and the qSV velocity is $\sqrt{A_{44}}$ again. It is beyond the scope of this section to investigate all the possible forms the slowness surfaces and wavefronts can take, but it is worth commenting that even in fine layered or fracture media, the qSV slowness surface may be concave at about $\pi/4$ to the symmetry axis and the wavefront may form a cusp (as in Figure 5.13). The qSV and qSH slowness surfaces typically cross, and the identification with qS_1 and qS_2 is mixed.

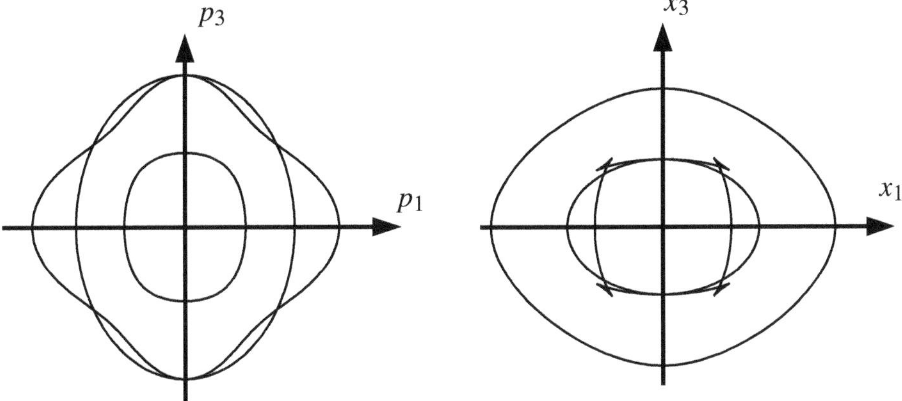

Fig. 5.13. The slowness surfaces and wavefronts for a typical, transversely isotropic medium. The surfaces are for Green Horn shale (Jones and Wang, 1981) although, as the surfaces are typical of many TI media, we have not indicated numerical values.

5.7.2 Constant gradient media

5.7.2.1 Linear squared-slowness interpolation

In Section 2.5.2.2, we considered the ray integrals in a medium with constant gradient in the squared slowness. The results are easily confirmed now we know the kinematic ray equations. Equation (5.1.15) becomes

$$\frac{\mathrm{d}\mathbf{p}}{\mathrm{d}T} = -\frac{\nabla\alpha}{\alpha} = \frac{\nabla u^2}{2u^2} = \frac{\mathbf{a}}{2u^2}, \tag{5.7.42}$$

where $\mathbf{a} = \nabla u^2$. Defining σ by

$$\frac{\mathrm{d}T}{\mathrm{d}\sigma} = u^2, \tag{5.7.43}$$

we have

$$\frac{\mathrm{d}\mathbf{p}}{\mathrm{d}\sigma} = \frac{1}{2}\mathbf{a}, \tag{5.7.44}$$

which integrates to give equation (2.5.39). The other kinematic equation (5.1.14) becomes

$$\frac{\mathrm{d}\mathbf{x}}{\mathrm{d}\sigma} = \mathbf{p}, \tag{5.7.45}$$

which with result (2.5.39) integrates to give result (2.5.38). Finally using

$$u^2 = u_0^2 + \mathbf{a}\cdot(\mathbf{x} - \mathbf{x}_0) = u_0^2 + \sigma\,\mathbf{a}\cdot\mathbf{p}_0 + \frac{\sigma^2}{4}\mathbf{a}\cdot\mathbf{a}, \tag{5.7.46}$$

from equations (2.5.35) and (2.5.38), equation (5.7.43) can be integrated to give result (2.5.40). The variable σ has no particular physical meaning except that it increases monotonically along the ray and so can be used as an independent variable to parameterize position on the ray, i.e. from equation (5.7.43)

$$\sigma = \int \alpha^2 \mathrm{d}T. \tag{5.7.47}$$

5.7.2.2 Linear anisotropic velocity

In general, even for the simplest anisotropic velocity function, the kinematic ray equations must be solved numerically. However, Shearer and Chapman (1988) showed that rays in a linear velocity function have a particularly simple geometry (the results for a linear isotropic velocity function have already been investigated – Section 2.5.2.1 – and the ray path is a circular arc).

Consider an anisotropic medium where the squared velocities are

$$c_{ijkl}(\mathbf{x})/\rho(\mathbf{x}) = a_{ijkl}\, x_3^2, \tag{5.7.48}$$

with a_{ijkl} constant. As the medium only varies in the x_3 direction, p_ν ($\nu = 1$ and 2) are constant – result (5.7.1). Without loss in generality, we can take $p_2 = 0$, so the Hamiltonian (5.3.18) can be written

$$H(\mathbf{x}, \mathbf{p}) = H(x_3, p_1, p_3) = \frac{1}{2}. \tag{5.7.49}$$

As the terms in the Hamiltonian are quadratic in x_3 only, it is obvious that

$$x_3 \frac{\partial H}{\partial x_3} = 1 = x_i \frac{\partial H}{\partial x_i}, \tag{5.7.50}$$

where in the final expression there is summation over i, but two terms are zero. Using equation (5.3.21), this is equivalent to

$$x_i \frac{\mathrm{d}p_i}{\mathrm{d}T} = -1, \tag{5.7.51}$$

and combining with equation (5.1.17), this gives

$$\frac{\mathrm{d}}{\mathrm{d}T}(\mathbf{x} \cdot \mathbf{p}) = 0. \tag{5.7.52}$$

The origin of $\mathbf{x}$ is restricted to the plane $x_3 = 0$ where the velocities would be zero by definition (5.7.48), but we are free to choose the x_1 origin so that for an initial point, $\mathbf{x}$ and $\mathbf{p}$ are perpendicular. Then

$$\mathbf{x} \cdot \mathbf{p} = 0, \tag{5.7.53}$$

everywhere on the ray (Figure 5.14).

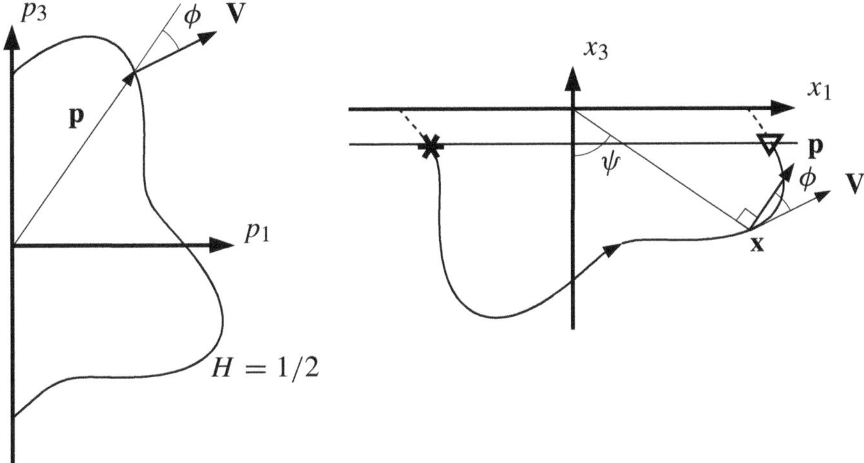

Fig. 5.14. The slowness surface defined by $H(\mathbf{x}, \mathbf{p}) = 1/2$ with $p_2 = 0$, and the ray path projected onto the x_1–x_3 plane. The angle between the slowness vector and ray velocity is ϕ, and the position vector makes an angle ψ with the x_3 axis. Physically, the ray cannot exist at $x_3 = 0$ as that corresponds to the elastic parameters being zero. Possible source and receiver locations are illustrated.

Because all the terms (5.7.48) scale together, the slowness surface defined by the Hamiltonian (5.7.49) scales inversely with x_3. Thus if we take the Hamiltonian at fixed $x_3 = X_3$, we have

$$H\left(X_3, \frac{x_3}{X_3} p_1, \frac{x_3}{X_3} p_3\right) = \frac{1}{2}, \tag{5.7.54}$$

satisfied by p_1 and p_3 at x_3. Using equation (5.7.53) with $p_2 = 0$, this can be rewritten

$$H\left(X_3, \frac{p_1}{X_3} x_3, -\frac{p_1}{X_3} x_1\right) = \frac{1}{2}. \tag{5.7.55}$$

But p_1 is fixed for one ray, so this equation connects x_1 and x_3, i.e. it defines the geometry of the ray path. The same function that defines the slowness surface defines the ray path. The shape of the slowness surface (in the p_1–p_3 plane) must be the same as the shape of the ray, with appropriate scaling (the slowness values for the surface at $x_3 = X_3$ are multiplied by $|X_3/p_1|$). The role of the components is interchanged – p_1 becomes x_3, and p_3 becomes $-x_1$. Thus the surface is rotated through $\pi/2$ (Figure 5.14). A similar result was obtained by Bennett (1968) for the special case of transversely isotropic media. In an isotropic medium, the slowness surface is spherical ((5.1.18) or (5.5.3)), and the ray path is a circular arc (cf. Section 2.5.2.1).

Equation (5.7.55) does not imply that the ray path has $x_2 = 0$, only that the projection of the path onto the x_1–x_3 plane is similar to the slowness curve. The transverse horizontal coordinate, x_2, is undetermined by this equation. It can be found by integrating the appropriate ray equation (5.3.20). Only on symmetry planes is the integrand zero and the ray does not deviate from the plane.

In order to evaluate the travel time, we consider equation (5.1.17). The slowness vector can be written

$$\mathbf{p} = \begin{pmatrix} p_1 \\ 0 \\ -\dfrac{x_1}{x_3} p_1 \end{pmatrix}, \tag{5.7.56}$$

in order to satisfy equation (5.7.53). Thus from equation (5.1.17) we obtain

$$\mathrm{d}T = p_1 \mathrm{d}x_1 + p_3 \mathrm{d}x_3 = \frac{p_1}{|x_3|} |\mathbf{x} \times \mathbf{dx}| = \frac{p_1}{|x_3|} |\mathbf{x}|^2 \mathrm{d}\psi, \tag{5.7.57}$$

where $\mathrm{d}\psi$ is the angle subtended at the origin by the ray segment $\mathbf{dx}$ (Figure 5.14). Thus

$$\mathrm{d}T = p_1 |\mathbf{x}| \sec\psi \, \mathrm{d}\psi, \tag{5.7.58}$$

where ψ is the angle between $\mathbf{x}$ and the x_3 axis (Figure 5.14). The travel time between two points can be obtained from

$$T = p_1 \int_{\chi_1}^{\chi_2} |\mathbf{x}| \, \mathrm{d}\chi, \tag{5.7.59}$$

where χ is defined in equation (2.5.22) (so $\mathrm{d}\chi = \sec\psi \, \mathrm{d}\psi$). In an isotropic medium, $|\mathbf{x}|$ is constant and this equation (5.7.59) simply reduces to

$$T = p_1 |\mathbf{x}| (\chi_2 - \chi_1), \tag{5.7.60}$$

which is exactly equivalent to result (2.5.20). In anisotropic media, $|\mathbf{x}|$ will normally vary slowly and the integral (5.7.59) is easily evaluated.

It is easy to show that expression (5.7.58) can be reduced to distance divided by velocity. With ϕ the angle between the slowness vector and ray velocity (Figure 5.14), we have

$$|\mathbf{dx}| = |\mathbf{x}| \mathrm{d}\psi \, \sec\phi = V |\mathbf{x}| \mathrm{d}\psi / c, \tag{5.7.61}$$

as $V \cos\phi = c$ from equation (5.1.17). But

$$p_1 \sec\psi = |\mathbf{p}| = 1/c, \tag{5.7.62}$$

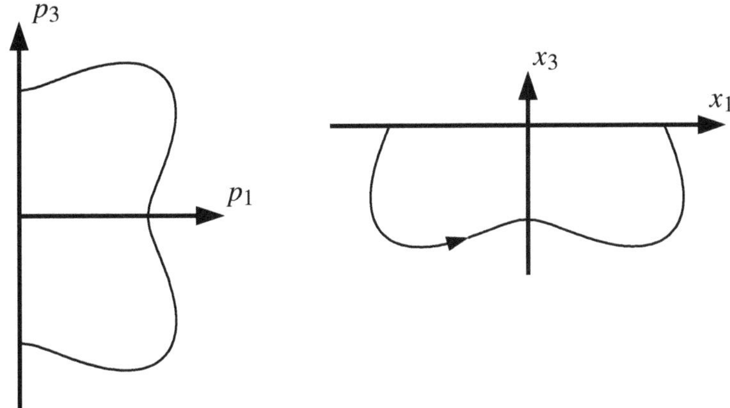

Fig. 5.15. The slowness surface and a ray path for iron with a linear velocity gradient. The elastic parameters can be found in Musgrave (1970) as used by Shearer and Chapman (1988, p. 580). See the caption of Figure 5.14 about possible source and receiver locations.

so substituting results (5.7.61) and (5.7.62) in result (5.7.58)

$$\mathrm{d}T = |\mathrm{d}\mathbf{x}|/V, \tag{5.7.63}$$

as required.

Shearer and Chapman (1988) show some interesting ray paths for anisotropic media with a linear gradient, illustrating and confirming the above theory. In particular, rays with multiple turning points, and propagating backwards in the x_1 direction are possible. In Figure 5.15, we illustrate one of the more interesting ray paths for iron with a linear gradient, showing multiple turning points, and a backward propagating path. This theory and results are cute but not of much practical use. They give pretty results that are educational. The practical difficulties are that the linear velocity model is very restrictive. All velocities scale together, i.e. they are all zero at the same depth ($x_3 = 0$ in the derivation). The form of the anisotropy does not change, so we cannot, for instance, have a gradient between isotropy and anisotropy. In addition, although we find the in-plane geometry easily, the out-plane geometry and travel time must still be obtained numerically.

Exercises

5.1 Confirm that in equation (5.3.20) the terms due to derivatives of the polarizations, $\partial\hat{\mathbf{g}}_I/\partial p_i$, cancel as the polarizations are normalized.

By differentiating the eigen-equation (5.3.14), show that the partial derivative of the polarization is

$$\frac{\partial \hat{\mathbf{g}}_I}{\partial p_k} = \sum_{\nu \neq I} \left(\frac{2 p_j \hat{\mathbf{g}}_\nu^{\mathrm{T}} \mathbf{a}_{jk}^S \hat{\mathbf{g}}_I}{1 - G_\nu} \right) \hat{\mathbf{g}}_\nu,$$

where the summation is for values of ν different from I (we assume non-degeneracy), and $\mathbf{a}_{jk}^S$ is the symmetric part of $\mathbf{a}_{jk}$, i.e. $\mathbf{a}_{jk}^S = (\mathbf{a}_{jk} + \mathbf{a}_{kj})/2$.

Using the above result, obtain an expression for the partial derivatives $\partial V_i / \partial p_j$ required for the elements of matrix $\mathbf{R}$ in equation (5.2.20). Confirm that it reduces to the isotropic result, $\mathbf{R} = \alpha^2 \mathbf{I}$ (5.2.21).

5.2 Show that in isotropic media

$$\nabla^2 T = \frac{1}{cJ} \frac{\mathrm{d}}{\mathrm{d}T} \left(\frac{J}{c} \right),$$

where T is the travel time, J is the ray tube cross-section and c is the velocity.

5.3 Show that in isotropic media, the transport equation, e.g. equation (5.2.10) can be written

$$\frac{\mathrm{d}v^{(0)}}{\mathrm{d}T} + \frac{1}{2} \left(c^2 \nabla^2 T + \frac{\mathrm{d}}{\mathrm{d}T} \ln(\rho c^2) \right) v^{(0)} = 0$$

(which can be found in classic developments, e.g. Červený and Hron, 1980).

5.4 Show that in isotropic media, the matrix $\mathbf{M}$ (defined in equation (5.2.47)), satisfies a Ricatti differential equation

$$\frac{\mathrm{d}\mathbf{M}}{\mathrm{d}T} + c^2 \mathbf{M}^2 = \mathbf{C},$$

where the matrix $\mathbf{C}$ is defined in equation (5.2.22).

5.5 If a wavefront has principal radii of curvature r_1 and r_2, show that

$$\nabla \cdot \hat{\mathbf{p}} = \frac{1}{r_1} + \frac{1}{r_2} = K,$$

say, in isotropic media as $\hat{\mathbf{p}}$ is normal to the wavefront (this result is the equivalent of equation (5.6.11) – also prove the result (5.6.10)). $K = \mathrm{tr}(\mathbf{K})$ is the curvature of the wavefront, where matrix $\mathbf{K}$ is defined in equation (5.2.48). Show that this is consistent with the differential equation

$$\frac{\mathrm{d}J}{\mathrm{d}s} = JK,$$

where J is the ray tube cross-section and s the ray length.

5.6 Confirm the isotropic polarization results at the end of Section 5.5 (equations (5.5.6) to (5.5.9)).

5.7 Confirm that the expressions in Section 5.7.1 for a transversely isotropic medium reduce to those for an isotropic medium.

In an isotropic medium, if p_1 is real, then p_3 is either real or imaginary. Demonstrate that this is not necessarily so in transversely isotropic media, and that p_3 may be complex.

5.8 If a ray-tracing program is available for two- or-three dimensional models, set up a model with random velocity variations. Trace rays through this model and confirm numerically the dynamic reciprocity results, e.g. result (5.2.36), and the KMAH index (it is assumed that the ray-tracing program is good enough to satisfy kinematic reciprocity!).

5.9 *Programming exercise:* Write a computer program to compute the slowness surfaces and wavefronts for an anisotropic, homogeneous medium, e.g. Figure 5.9, together with the polarization vectors, e.g. given the slowness direction $\hat{\mathbf{p}}$, compute the vector slowness, $\mathbf{p}$, the ray velocity, $\mathbf{V}$, and the polarization, $\hat{\mathbf{g}}$. Try the program for realistic values of the anisotropic parameters (from, for instance, Musgrave, 1970).

Hint: Although given the slowness direction, $\hat{\mathbf{p}}$, the solution for the slowness reduces to a cubic polynomial, it is better to find the slowness from the eigen-equation (5.3.11). The 3×3 Christoffel matrix is symmetric and so three, real eigenvalues are guaranteed, whereas rounding errors may make the solutions of a cubic polynomial complex (especially near the degenerate, isotropic case).

5.10 *Further reading:* The dependence of the Hamiltonian, $H_I(\mathbf{x}, \mathbf{p})$ (5.3.18), on the slowness, $\mathbf{p}$, occurs explicitly in the Christoffel matrix, $\mathbf{\Gamma} = p_j p_k \mathbf{a}_{jk}$, and implicitly in the polarization, $\hat{\mathbf{g}}_I$, through the solution of the eigen-equation (5.3.26). Alternatively, using standard matrix methods, it can be written explicitly in terms of the slowness.

The matrix

$$\mathbf{G}(G) = G\,\mathbf{I} - \mathbf{\Gamma},$$

in equation (5.3.14) is known as the characteristic matrix of matrix $\mathbf{\Gamma}$. Its determinant,

$$\Gamma(G) = |\mathbf{G}(G)| = |G\,\mathbf{I} - \mathbf{\Gamma}|,$$

is the characteristic function or polynomial in G, and

$$\Gamma(G) = 0,$$

is the characteristic equation, cf. equation (5.3.17). The roots, G_I, of the characteristic equation are the eigenvalues of the matrix $\mathbf{\Gamma}$.

Denoting the adjoint (5.4.24) of the characteristic matrix, $\mathbf{G}$,

$$\mathbf{G}^{\ddagger}(G) = \text{adj}\left(\mathbf{G}(G)\right),$$

we have

$$\mathbf{G}(G)\mathbf{G}^{\ddagger}(G) = \Gamma(G)\,\mathbf{I},$$

and

$$\mathbf{G}(G_I)\mathbf{G}^{\ddagger}(G_I) = 0,$$

for an eigenvalue. Provided G_I is not a degenerate root, $\mathbf{G}(G_I)$ is simply degenerate, and as $\mathbf{G}^{\ddagger}(G_I)$ is not null, it is of unit rank. It can be expanded in terms of the left and right eigenvectors of matrix $\mathbf{\Gamma}$ (this is a specialization of the well-known singular value decomposition (SVD, e.g. Golub and Loan, 1996; Riley, Hobson and Bence, 2002, Chapter 8), where any matrix can be expanded as an outer product of its left and right singular vectors). As the matrix $\mathbf{\Gamma}$ is symmetric, the left and right singular vectors are the eigenvectors and

$$\mathbf{G}^{\ddagger}(G_I) = \text{tr}\left(\mathbf{G}^{\ddagger}(G_I)\right)\hat{\mathbf{g}}_I\hat{\mathbf{g}}_I^{\mathsf{T}}.$$

Červený (1972) has used this result in the kinematic ray equations (5.3.20) and (5.3.21). The Hamiltonian (5.3.18) becomes

$$H_I(\mathbf{x}, \mathbf{p}) = \frac{1}{2} p_j p_k a_{ijlk} G_{il} / G_{mm},$$

where G_{il} are elements of the matrix $\mathbf{G}^{\ddagger}(1)$ (Červený, 1972, used the symbol D_{il}). The elements of $\mathbf{G}^{\ddagger}(G)$ are simply

$$G^{\ddagger}_{11}(G) = (\Gamma_{22} - G)(\Gamma_{33} - G) - \Gamma_{23}^2$$

$$G^{\ddagger}_{23}(G) = \Gamma_{12}\Gamma_{31} - \Gamma_{23}(\Gamma_{11} - G),$$

etc. with cyclic permutations, or generally

$$G^{\ddagger}_{jk}(G) = \frac{1}{2}\epsilon_{jil}\epsilon_{krs}(\Gamma_{ir} - G\delta_{ir})(\Gamma_{ls} - G\delta_{ls}),$$

where ϵ_{jil} is the Levi-Civitta symbol.

This result expresses the Hamiltonian, $H_I(\mathbf{x}, \mathbf{p})$, explicitly in terms of the slowness without the polarizations. However, this is not necessary in order to obtain the differentials with respect to slowness (see Exercise 5.1), and the polarizations will probably be known from the solution of the Christoffel equation (see Exercise 5.9), e.g. for equations (5.3.20) and (5.3.21), anyway.

5.11 *Further reading:* In anisotropic media, the KMAH index, σ, may decrease instead of increase at a caustic. Klimeš (1997), Bakker (1998) and Garmany (2000) give details of when this occurs (for a more general description, see Lewis, 1965).

6
Rays at an interface

At a discontinuity in material properties – an *interface* – multiple rays or waves are generated. For an individual ray, the ray properties – direction, amplitudes and polarizations – are discontinuous. This chapter describes the discontinuities in these results, i.e. *Snell's law* (direction), the *dynamic ray discontinuity* (geometrical spreading), and *reflection/transmission coefficients* (amplitude and polarizations). The reflection/transmission coefficients are developed in a general manner for acoustic, elastic and fluid–solid interfaces which, with the correct normalizations, emphasizes relationships between different coefficients. At an interface, several rays combine and the response is not given by a simple ray polarization. The necessary *receiver conversion coefficients*, particularly at a free surface, are derived. A procedure for linearly perturbing the coefficients is developed. These *perturbation* or *differential coefficients* are particularly useful for obtaining approximate coefficients from a weak-contrast interface. In the final section, the concept of an individual ray is generalized to a *ray table* characterized by a *ray signature*. This is used to generalize the ray Green dyadic to include interfaces.

The results in the previous chapter, Chapter 5, describe rays in media without discontinuities. If the medium contains interfaces, i.e. discontinuities in the material properties, density or elastic parameters, then the ray theory solution breaks down due to discontinuities in the solution or its derivatives. It is necessary to impose the boundary conditions on the solution at the interface before continuing the ray solution, and this is investigated in this chapter.

Let us define the interface by $S(\mathbf{x}) = 0$ separating two different media. For simplicity, let us label these 1 and 2, where the wave is incident from medium 1. A normal to the interface is defined by $\hat{\mathbf{n}} = \pm\,\mathrm{sgn}(\nabla S)$, where by convention the sign is taken so that the vector $\hat{\mathbf{n}}$ points into medium 1. It is necessary to complete an orthogonal basis at the point where the ray intersects the interface by defining two vectors, $\hat{\mathbf{l}}$ and $\hat{\mathbf{m}}$, in the tangent plane to the interface. The orientation of these is arbitrary but it is convenient to choose one, $\hat{\mathbf{l}}$, in the plane of the ray

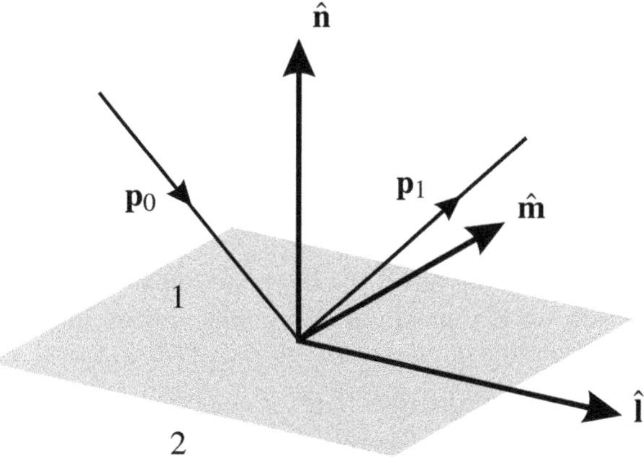

Fig. 6.1. A ray incident on an interface from medium 1. The unit normal to the interface is $\hat{\mathbf{n}}$ pointing into medium 1. The unit vector $\hat{\mathbf{m}}$ is normal to the ray plane, and $\hat{\mathbf{l}}$ completes the basis. The vectors $\hat{\mathbf{l}}$ and $\hat{\mathbf{m}}$ lie in the plane of the interface.

slowness and the interface normal, and one, $\hat{\mathbf{m}}$, normal to this ray plane, i.e. (see Figure 6.1)

$$\hat{\mathbf{m}} = \operatorname{sgn}(\hat{\mathbf{n}} \times \hat{\mathbf{p}}) \quad \text{and} \quad \hat{\mathbf{l}} = \hat{\mathbf{m}} \times \hat{\mathbf{n}}. \tag{6.0.1}$$

In isotropic, elastic media, these will separate the *SV* and *SH* components of the shear wave. In anisotropic media, we might as well use the same basis as using the slowness vector in definition (6.0.1) the ray slownesses are all restricted to the $\hat{\mathbf{l}}-\hat{\mathbf{n}}$ plane (see Section 6.2.1). If definition (6.0.1) degenerates, i.e. when the ray slowness is normal to the interface, any choice of vectors in the interface will do. Defining an orthonormal transformation matrix $\mathbf{L} = (\ \hat{\mathbf{l}}\ \ \hat{\mathbf{m}}\ \ \hat{\mathbf{n}}\ )$, we can transform any vector/tensor components to the interface basis, e.g. for the anisotropic parameters,

$$c'_{i'j'k'l'} = L_{i'i} L_{j'j} L_{k'k} L_{l'l} c_{ijkl}, \tag{6.0.2}$$

where the prime indicates values in this interface basis and L_{mn} are elements of the matrix $\mathbf{L}$. For the acoustic and isotropic elastic systems, it is unnecessary, of course, to transform the elastic parameters. For the rest of this chapter, we assume the elastic parameters have been transformed into the interface basis (6.0.2). We emphasize again that in anisotropic media, we use the slowness vector, $\mathbf{p}$, to define the interface basis, $\mathbf{L}$, not the ray velocity vectors, $\mathbf{V}$.

For a model of plane layers, if the interfaces are defined so $\hat{\mathbf{n}} = \pm\hat{\mathbf{k}}$, the vertical unit vector (the sign depends whether the incident ray is travelling in the negative or positive z direction), then if the ray slowness direction $\hat{\mathbf{p}}$ is restricted to the x–z plane, with a positive x component, then $\hat{\mathbf{l}}$ and $\hat{\mathbf{m}}$ are $\hat{\mathbf{\imath}}$ and $\pm\,\hat{\mathbf{\jmath}}$, respectively. The transformation matrix is the identity matrix, i.e. $\mathbf{L} = \mathbf{I}$ or $\mathbf{L} = (\,\hat{\mathbf{\imath}}\ \ -\hat{\mathbf{\jmath}}\ \ -\hat{\mathbf{k}}\,)$.

6.1 Boundary conditions

The interface boundary conditions have been discussed in general terms in Section 4.3. In this section we discuss their application in ray theory.

In general it is impossible to satisfy the boundary conditions with one incident and one generated ray leaving the interface, as the boundary conditions require the continuity of components of displacement and traction. This boundary condition requires multiple generated rays. We denote any property of these rays by two subscripts, e.g. $\mathbf{p}_{ij}$ are the slowness vectors with $i = 1$ or 2 indicating the medium, and j indicating the ray type, e.g. just $j = 1$ in acoustic media, but $j = 1, 2$ or 3 in elastic media (see Figure 6.2). If there are multiple ray types, the ordering is not crucial but it is convenient to use $j = 1$ to indicate the slower qS wave, 2 the faster qS wave, and 3 the qP wave (i.e. in order of increasing velocity). The nomenclature used for the qP and qS rays depends on the application and anisotropy. In TIV media, it is convenient to use qSV and qSH. In shear wave splitting studies, the nomenclature qS_1 and qS_2 is often used so $j = 1$ corresponds to qS_2, the slower shear wave, $j = 2$ to qS_1, the faster shear wave, and $j = 3$ to the qP wave.

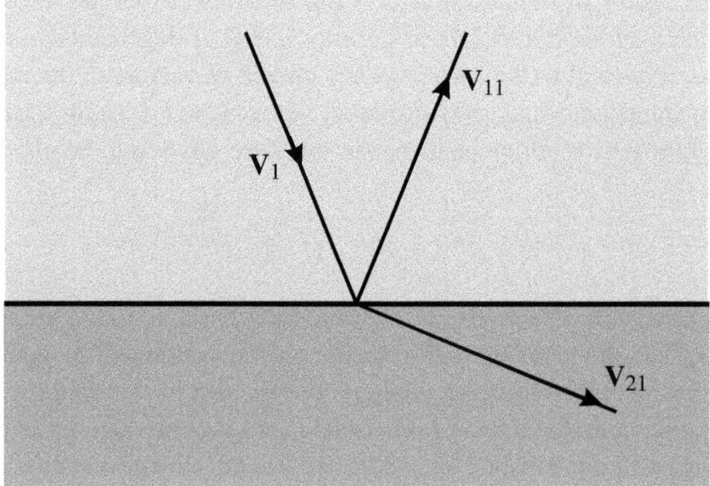

Fig. 6.2. A ray incident on an interface from medium 1 with velocity $\mathbf{V}_1$, generating a reflected ray with velocity $\mathbf{V}_{11}$ and a transmitted ray $\mathbf{V}_{21}$.

We indicate the incident wave with a single subscript, i.e. $\mathbf{p}_j$. All rays will intersect the interface at the same point, $\mathbf{x} = \mathbf{x}_j = \mathbf{x}_{ij}$. In general, the existence of generated rays is necessary to satisfy the boundary conditions. In practice, having solved for the starting conditions of the generated rays, we follow each generated ray separately. A complete ray consists of a sequence of *ray segments* between interfaces, and the *ray history* or *signature* (or *code*) describes the sequence of choices defining the generated ray type at each interface. The complete response consists of a sum of rays with all possible histories – this is enumerated by the index n in (5.1.1). An individual ray will not satisfy the continuity conditions at interfaces, but the complete response will.

6.1.1 Acoustic boundary conditions

The boundary conditions at a fluid–fluid interface were described in Section 4.3.2. The normal component of the particle vector, $v_n = \hat{\mathbf{n}} \cdot \mathbf{v}$, and the pressure, P, should be continuous. In ray theory these boundary conditions must apply to each order of amplitude coefficient, $v_n^{(m)}$ and $P^{(m)}$ in the ray series (5.1.1). We define a vector of these field variables

$$\mathbf{w} = \begin{pmatrix} v_n \\ -P \end{pmatrix} \tag{6.1.1}$$

(the minus sign is included in the pressure to make it analogous to stress).

6.1.2 Elastic boundary conditions

Assuming the interface is welded, the boundary condition at the interface is that the particle vector, $\mathbf{v}$, and the normal traction, $\mathbf{t}'_n = \hat{n}_j \mathbf{t}_j$, should be continuous. As stated above, we assume that the variables are in the interface basis, e.g. (6.0.2), and for brevity omit the prime. For later use, we define a 6-vector of these continuous field variables

$$\boxed{\mathbf{w} = \begin{pmatrix} \mathbf{v} \\ \mathbf{t}_n \end{pmatrix}.} \tag{6.1.2}$$

6.2 Continuity of the ray equations

6.2.1 Snell's law

The continuity conditions at an interface require that the travel time, T, is the same for all rays. As at any point on the interface the travel time is the same for all rays, the gradient of the travel time, $\mathbf{p} = \nabla T$, in the plane of the interface

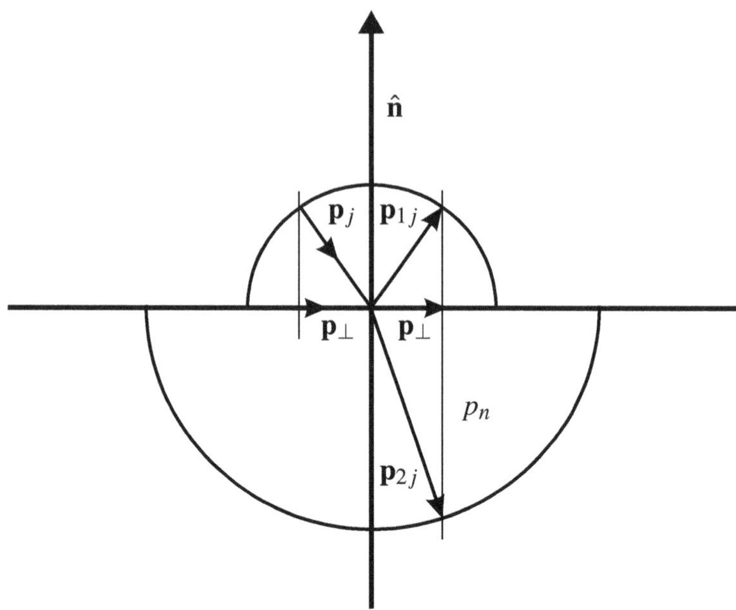

Fig. 6.3. A ray incident on an interface with slowness $\mathbf{p}_j$ and generated rays with slowness $\mathbf{p}_{ij}$. The component in the interface $\mathbf{p}_\perp$ is the same for all rays, but the normal components $p_n\hat{\mathbf{n}}$ differ as they satisfy the Hamiltonian constraint (6.2.3). The circles are the slowness surfaces defined for the j-th ray type in the i-th medium, $H_{ij}(\mathbf{x}, \mathbf{p}) = 1/2$ (6.2.3).

must also be the same for all rays. Continuity of the travel time is trivial, just imposing an initial condition on the new ray segment. Continuity of the gradient of the travel time (5.1.6) in the plane of the interface leads to the slowness vectors for the generated rays and the initial conditions for the kinematic ray equations. In Section 2.1.2, we derived Snell's law for plane waves. The results for rays will be identical. We derive them using vector notation as the geometry of the interface is arbitrary.

The slowness in the plane of the interface is given by

$$\mathbf{p}_\perp = \mathbf{p} - (\mathbf{p} \cdot \hat{\mathbf{n}})\hat{\mathbf{n}} = (\mathbf{p} \cdot \hat{\mathbf{l}})\hat{\mathbf{l}}. \tag{6.2.1}$$

This must be the same for all the rays at the interface and can be calculated for the incident ray. Thus the generated rays will have slownesses (see Figure 6.3)

$$\boxed{\mathbf{p}_{ij} = p_n\hat{\mathbf{n}} + \mathbf{p}_\perp = p_n\hat{\mathbf{n}} + (\mathbf{p}_j \cdot \hat{\mathbf{l}})\hat{\mathbf{l}},} \tag{6.2.2}$$

where p_n must be found from the constraint, $H = 1/2$, on the Hamiltonian (5.1.18), (5.3.18) or (5.5.3) for the generated ray i.e.

$$H_{ij}(\mathbf{x}, \mathbf{p}_{ij}) = 1/2, \tag{6.2.3}$$

is solved for p_n (the subscript on the Hamiltonian indicates the medium and ray type). The vectors $\mathbf{p}_j$, $\mathbf{p}_{ij}$, $\hat{\mathbf{n}}$ and $\hat{\mathbf{l}}$ must be coplanar. We must choose the solutions corresponding to the waves propagating away from the interface. Thus

$$\hat{\mathbf{n}} \cdot \mathbf{V}_{1j} > 0 \quad \text{and} \quad \hat{\mathbf{n}} \cdot \mathbf{V}_{2j} < 0, \tag{6.2.4}$$

which follow from the convention used to define the sign of $\hat{\mathbf{n}}$, where $\mathbf{V}_{ij}$ are the ray velocities.

The relationship between the slowness vectors, $\mathbf{p}_j$ and $\mathbf{p}_{ij}$, is known as *Snell's law*. This and the continuity of $\mathbf{x}$ and T give the required conditions for kinematic ray tracing through an interface.

The above results have been written in vector notation for generality, and apply to acoustic, isotropic and anisotropic elastic media. In anisotropic media, we give below a sixth-order eigen-equation for $\mathbf{w}$ (6.3.14) which can be solved for the eigenvalue p_n. From the $\mathbf{p}_{ij}$'s, we derive the ray velocities $\mathbf{V}_{ij}$ using definition (5.3.20). In acoustic and isotropic elastic media the results are simpler, of course.

Snell's law can be written in terms of the incident, reflected and transmitted ray angles (Section 2.1.2)

$$\boxed{|\mathbf{p}_\perp| = \frac{\sin\theta_{\text{inc}}}{c_1} = \frac{\sin\theta_{\text{refl}}}{c_1} = \frac{\sin\theta_{\text{trans}}}{c_2}.} \tag{6.2.5}$$

For acoustic waves $c = \alpha$; for isotropic elastic waves c may be α or β. The slowness components normal to the interface, p_n, are

$$\hat{\mathbf{n}} \cdot \mathbf{p}_{1j} = \frac{\cos\theta_{\text{refl}}}{c_1} \tag{6.2.6}$$

$$\hat{\mathbf{n}} \cdot \mathbf{p}_{2j} = -\frac{\cos\theta_{\text{trans}}}{c_2}. \tag{6.2.7}$$

It is convenient to denote the normal slownesses, $\pm p_n$, in isotropic elastic media by

$$q_\alpha = \left(1/\alpha^2 - \mathbf{p}_\perp^2\right)^{1/2} \tag{6.2.8}$$

$$q_\beta = \left(1/\beta^2 - \mathbf{p}_\perp^2\right)^{1/2}. \tag{6.2.9}$$

In anisotropic media, it is important to remember that θ is the angle that the phase slowness, $\mathbf{p}$, makes with the normal to the interface, $\hat{\mathbf{n}}$, and c is the phase velocity, but in choosing the correct solutions using condition (6.2.4), $\mathbf{V}$ is the ray group velocity (5.3.20).

In some circumstances, a solution for the normal slowness, p_n, may be complex, i.e. the transmitted wave is evanescent when the incident wave is beyond critical. While these waves must be included to solve for the amplitude continuity condition, the corresponding rays are not normally included in the ray solution.

6.2.2 Dynamic ray discontinuity

In order to continue the solution of the dynamic ray equations (5.2.19) through an interface, we must connect the derivatives of $\mathbf{x}$, $\mathbf{p}_j$ and $\mathbf{p}_{ij}$. Results have been given by Červený, Langer and Pšenčík (1974) (for isotropic media) and Gajewski and Pšenčík (1990) (for anisotropic media). These have been modified by Farra and Le Bégat (1995) to maintain the symplectic symmetries (5.2.36) and in this section we summarize their results.

In order to study the derivatives, it is crucial to distinguish the values in the wavefront (as given by the dynamic ray equations (5.2.19) with constant travel time, T), and those on the interface ($S = 0$) as illustrated in Figure 6.4. These are related by

$$\left(\frac{\partial \mathbf{y}}{\partial \mathrm{q}_\nu}\right)_S = \left(\frac{\partial \mathbf{y}}{\partial \mathrm{q}_\nu}\right)_T + \left(\frac{\partial T}{\partial \mathrm{q}_\nu}\right)_S \dot{\mathbf{y}}, \tag{6.2.10}$$

where the vector $\mathbf{y}$ is defined in equation (5.1.28) and for brevity we have written $\mathrm{d}\mathbf{y}/\mathrm{d}T = \dot{\mathbf{y}}$, given by the Hamiltonian equations (5.1.29) (Greek subscripts, e.g. ν, are restricted to 1 and 2). The position derivative $(\partial \mathbf{x}/\partial \mathrm{q}_\nu)_S$ must be in

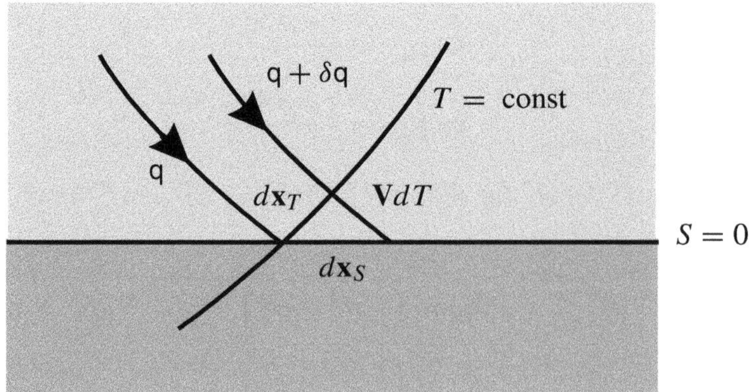

Fig. 6.4. A ray and a paraxial ray at an interface, showing the connection between wavefront, $\mathrm{d}\mathbf{x}_T$, and interface perturbations, $\mathrm{d}\mathbf{x}_S$, i.e. equation (6.2.10).

the interface, so pre-multiplying by $\hat{\mathbf{n}}^{\mathrm{T}}$, we can solve for the travel-time interface derivative

$$\left(\frac{\partial T}{\partial \mathrm{q}_\nu}\right)_S = -\frac{\hat{\mathbf{n}}^{\mathrm{T}}}{\dot{\mathbf{x}}_j^{\mathrm{T}}\hat{\mathbf{n}}}\left(\frac{\partial \mathbf{x}}{\partial \mathrm{q}_\nu}\right)_T . \tag{6.2.11}$$

Substituting in equation (6.2.10) we can convert the wavefront perturbation to the interface

$$\left(\mathrm{d}\mathbf{y}_j\right)_S = \begin{pmatrix} \mathbf{\Pi}_1 & \mathbf{0} \\ \mathbf{\Pi}_2 & \mathbf{I} \end{pmatrix} \left(\mathrm{d}\mathbf{y}_j\right)_T , \tag{6.2.12}$$

where

$$\mathbf{\Pi}_1 = \mathbf{I} - \frac{\dot{\mathbf{x}}_j\hat{\mathbf{n}}^{\mathrm{T}}}{\dot{\mathbf{x}}_j^{\mathrm{T}}\hat{\mathbf{n}}} \quad \text{and} \quad \mathbf{\Pi}_2 = -\frac{\dot{\mathbf{p}}_j\hat{\mathbf{n}}^{\mathrm{T}}}{\dot{\mathbf{x}}_j^{\mathrm{T}}\hat{\mathbf{n}}}. \tag{6.2.13}$$

The position of the incident ray on the interface and the generated ray are identical, i.e.

$$\left(\mathbf{x}_j\right)_S = \left(\mathbf{x}_{ij}\right)_S , \tag{6.2.14}$$

so perturbations are also equal. A useful expression can be obtained using equation (6.2.10) in the expression for the ray-tube cross-section (result (5.4.11) with $\dot{\mathbf{x}} = \mathbf{V}$ (5.3.23))

$$\frac{c_j J_j}{\mathbf{V}_j^{\mathrm{T}}\hat{\mathbf{n}}} = \frac{c_{ij} J_{ij}}{\mathbf{V}_{ij}^{\mathrm{T}}\hat{\mathbf{n}}} \tag{6.2.15}$$

(no summation over i and j). However, the connection between the slowness vectors is more complicated than result (6.2.14). The difference between the generated and incident slownesses is normal to the interface equation (6.2.2), i.e.

$$\mathbf{p}_{ij} - \mathbf{p}_j = \left(\hat{\mathbf{n}}^{\mathrm{T}}(\mathbf{p}_{ij} - \mathbf{p}_j)\right)\hat{\mathbf{n}}. \tag{6.2.16}$$

This can be differentiated with respect to q_ν. To evaluate the differentials of the right-hand side, we use the differentials of the Hamiltonians

$$\frac{\partial H}{\partial \mathrm{q}_\nu} = -\left(\dot{\mathbf{p}}^{\mathrm{T}}\frac{\partial \mathbf{x}}{\partial \mathrm{q}_\nu}\right)_S + \left(\dot{\mathbf{x}}^{\mathrm{T}}\frac{\partial \mathbf{p}}{\partial \mathrm{q}_\nu}\right)_S = 0, \tag{6.2.17}$$

for either the incident, H_j, or generated ray Hamiltonians, H_{ij}, with the kinematic ray equations (5.3.20) and (5.3.21). Pre-multiplying the differential of equation (6.2.16) by $\dot{\mathbf{x}}_{ij}^{\mathrm{T}}$ and using equation (6.2.17) for H_{ij}, we can solve for $\left(\partial\left(\hat{\mathbf{n}}^{\mathrm{T}}(\mathbf{p}_{ij} - \mathbf{p}_j)\right)/\partial \mathrm{q}_\nu\right)_S$. With

$$\left(\frac{\partial \hat{\mathbf{n}}}{\partial \mathrm{q}_\nu}\right)_S = \nabla^{\mathrm{T}}\hat{\mathbf{n}}\left(\frac{\partial \mathbf{x}}{\partial \mathrm{q}_\nu}\right)_S , \tag{6.2.18}$$

the differentials of equations (6.2.16) and (6.2.14) can be written

$$\left(\mathrm{d}\mathbf{y}_{ij}\right)_S = \begin{pmatrix} \mathbf{I} & \mathbf{0} \\ \boldsymbol{\Pi}_3 & \boldsymbol{\Pi}_4 \end{pmatrix} \left(\mathrm{d}\mathbf{y}_j\right)_S, \tag{6.2.19}$$

where

$$\boldsymbol{\Pi}_3 = \frac{\hat{\mathbf{n}}}{\dot{\mathbf{x}}_{ij}^{\mathrm{T}}\hat{\mathbf{n}}}(\dot{\mathbf{p}}_{ij} - \dot{\mathbf{p}}_j)^{\mathrm{T}} + \hat{\mathbf{n}}^{\mathrm{T}}(\mathbf{p}_{ij} - \mathbf{p}_j)\left(\mathbf{I} - \frac{\hat{\mathbf{n}}\dot{\mathbf{x}}_{ij}^{\mathrm{T}}}{\dot{\mathbf{x}}_{ij}^{\mathrm{T}}\hat{\mathbf{n}}}\right)\nabla^{\mathrm{T}}\hat{\mathbf{n}} \tag{6.2.20}$$

$$\boldsymbol{\Pi}_4 = \mathbf{I} - \frac{\hat{\mathbf{n}}(\dot{\mathbf{x}}_{ij} - \dot{\mathbf{x}}_j)^{\mathrm{T}}}{\dot{\mathbf{x}}_{ij}^{\mathrm{T}}\hat{\mathbf{n}}}. \tag{6.2.21}$$

Note that these matrices contain factors which combine to give equation (6.2.17) for Hamiltonian H_j, i.e. are zero in the differential, but are included in the propagator (6.2.19) so it satisfies the symplectic symmetry (5.2.36) (Farra and Le Bégat, 1995). Finally, we use equations (6.2.10) and (6.2.11), to convert the differential (6.2.19) onto the wavefront, i.e.

$$\left(\mathrm{d}\mathbf{y}_{ij}\right)_T = \left(\mathrm{d}\mathbf{y}_{ij}\right)_S + \begin{pmatrix} \boldsymbol{\Pi}_5 & \mathbf{0} \\ \boldsymbol{\Pi}_6 & \mathbf{0} \end{pmatrix} \left(\mathrm{d}\mathbf{y}_j\right)_T, \tag{6.2.22}$$

where

$$\boldsymbol{\Pi}_5 = \frac{\dot{\mathbf{x}}_{ij}\hat{\mathbf{n}}^{\mathrm{T}}}{\dot{\mathbf{x}}_j^{\mathrm{T}}\hat{\mathbf{n}}} \quad \text{and} \quad \boldsymbol{\Pi}_6 = \frac{\dot{\mathbf{p}}_{ij}\hat{\mathbf{n}}^{\mathrm{T}}}{\dot{\mathbf{x}}_j^{\mathrm{T}}\hat{\mathbf{n}}}. \tag{6.2.23}$$

Overall

$$\boxed{\left(\mathrm{d}\mathbf{y}_{ij}\right)_T = \boldsymbol{\Pi}\left(\mathrm{d}\mathbf{y}_j\right)_T,} \tag{6.2.24}$$

where

$$\boxed{\begin{aligned} \boldsymbol{\Pi} &= \begin{pmatrix} \mathbf{I} & \mathbf{0} \\ \boldsymbol{\Pi}_3 & \boldsymbol{\Pi}_4 \end{pmatrix}\begin{pmatrix} \boldsymbol{\Pi}_1 & \mathbf{0} \\ \boldsymbol{\Pi}_2 & \mathbf{I} \end{pmatrix} + \begin{pmatrix} \boldsymbol{\Pi}_5 & \mathbf{0} \\ \boldsymbol{\Pi}_6 & \mathbf{0} \end{pmatrix} \\ &= \begin{pmatrix} \boldsymbol{\Pi}_1 + \boldsymbol{\Pi}_5 & \mathbf{0} \\ \boldsymbol{\Pi}_3\boldsymbol{\Pi}_1 + \boldsymbol{\Pi}_4\boldsymbol{\Pi}_2 + \boldsymbol{\Pi}_6 & \boldsymbol{\Pi}_4 \end{pmatrix}, \end{aligned}} \tag{6.2.25}$$

propagates the perturbation from the incident to the generated ray. It satisfies the symplectic symmetry (5.2.36) (Farra and Le Bégat, 1995).

The results in this section apply to rays in acoustic, elastic and anisotropic media as we have been careful to use the slowness vector, **p**, and the velocity vector, **V**, as appropriate.

6.3 Reflection/transmission coefficients

In order to find the reflection/transmission coefficients connecting the incident and generated rays, we use the continuity of the vector $\mathbf{w}$ (equations (6.1.1) or (6.1.2)). First we consider the acoustic case as the algebra is so simple, and we generalize to the anisotropic elastic case using a similar notation. We then specialize these results to isotropic elastic media.

6.3.1 Acoustic coefficients

At an interface, the two-dimensional vector $\mathbf{w}$ (6.1.1) is continuous. We can use this condition to calculate the magnitudes of the generated rays, a reflection and a transmission, relative to the incident ray, i.e. the *reflection/transmission coefficients*. In the ray series (5.1.1), continuity must apply to terms of each order – here we consider only the zeroth-order amplitude coefficients. Eliminating the discontinuous components of particle velocity between equations (5.2.1) and (5.2.2), we obtain

$$\mathbf{A}\mathbf{w}^{(0)} = p_n \mathbf{w}^{(0)}, \tag{6.3.1}$$

where

$$\mathbf{A} = \begin{pmatrix} 0 & -q_\alpha^2/\rho \\ -\rho & 0 \end{pmatrix}, \tag{6.3.2}$$

and q_α is given by equation (6.2.8), or in terms of the ray angle (Section 2.1.2) by equations (6.2.6) or (6.2.7). Let us write the eigen-solutions as

$$\boxed{\mathbf{A}\mathbf{W} = \mathbf{W}\mathbf{p}_n,} \tag{6.3.3}$$

where the columns of $\mathbf{W}$ are the eigenvectors $\mathbf{w}^{(0)}$, and $\mathbf{p}_n$ is the diagonal matrix of eigenvalues. We order the eigenvectors so the first corresponds to the wave propagating in the positive $\hat{\mathbf{n}}$ direction, i.e. $\hat{\mathbf{n}} \cdot \mathbf{V} > 0$, and the last is travelling in the opposite direction, i.e. $\hat{\mathbf{n}} \cdot \mathbf{V} < 0$ (cf. condition (6.2.4)).

Allowing the source to be in either medium, the two reflection/transmission experiments can be described by a matrix equation

$$\mathbf{W}_1 \begin{pmatrix} \mathcal{T}_{11} & \mathcal{T}_{12} \\ 1 & 0 \end{pmatrix} = \mathbf{w}(\mathbf{x}_j) = \mathbf{W}_2 \begin{pmatrix} 0 & 1 \\ \mathcal{T}_{21} & \mathcal{T}_{22} \end{pmatrix}, \tag{6.3.4}$$

where $\mathbf{W}_1$ and $\mathbf{W}_2$ are the eigenvectors in the two media at the interface point, $\mathbf{x}_j$, $\boldsymbol{\mathcal{T}}$ is a 2×2 matrix of reflection/transmission coefficients, $\mathcal{T}_{ij}$, where the incident wave's medium is indicated by j, and the generated wave's medium is indicated by i. Both sides of this equation are equal to $\mathbf{w}$ on the interface. It should

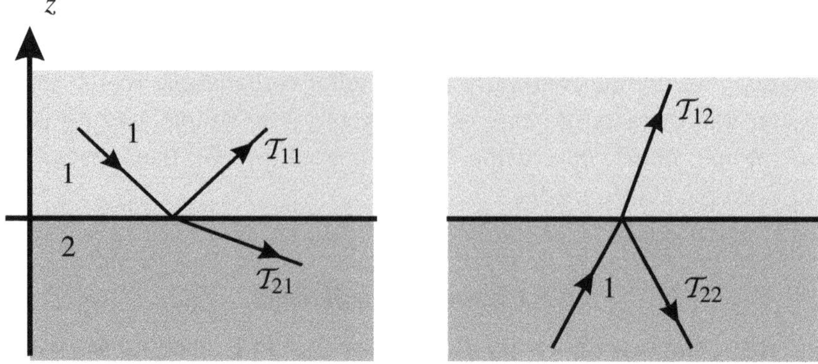

Fig. 6.5. Reflection/transmission experiments for acoustic rays with unit rays incident from medium 1 or medium 2.

be noted that while the ray is travelling from j to i, the subscripts are in the reversed order. While it would be convenient to read $\mathcal{T}_{ij}$ as the coefficient for $i \to j$, particularly later when we generalize the subscript notation to indicate more than two media, the price we have to pay for using matrix-vector algebra is that the subscripts are interpreted as $j \to i$. In fact, we shall see that with the correct normalization the matrix $\boldsymbol{\mathcal{T}}$ is symmetric, so the distinction becomes unimportant. Figure 6.5 illustrates the reflection/transmission experiments represented by equation (6.3.4).

The eigenvectors of the matrix $\mathbf{A}$ are

$$\mathbf{w}^{(0)} = \frac{1}{\sqrt{2\rho q_\alpha}} \begin{pmatrix} \pm q_\alpha \\ -\rho \end{pmatrix} = \frac{1}{\sqrt{\pm 2\rho V_n}} \begin{pmatrix} \pm \alpha q_\alpha \\ -\alpha\rho \end{pmatrix}, \tag{6.3.5}$$

where V_n is the ray velocity normal to the interface (the reason for the normalization will become obvious later). For later use it is useful to define a polarization with the same normalization (cf. definition (5.4.33))

$$\boxed{\mathbf{g} = \frac{1}{\sqrt{\pm 2\rho V_n}} \begin{pmatrix} \alpha p \\ \pm \alpha q_\alpha \end{pmatrix} = \frac{1}{\sqrt{\pm 2\rho V_n}} \hat{\mathbf{g}}.} \tag{6.3.6}$$

Solving the simultaneous equations (6.3.4), we obtain

$$\boxed{\mathcal{T}_{11} = -\mathcal{T}_{22} = \frac{\rho_2 q_{\alpha 1} - \rho_1 q_{\alpha 2}}{\rho_2 q_{\alpha 1} + \rho_1 q_{\alpha 2}}} \tag{6.3.7}$$

$$\boxed{\mathcal{T}_{12} = \mathcal{T}_{21} = \frac{2\sqrt{\rho_1 \rho_2 q_{\alpha 1} q_{\alpha 2}}}{\rho_2 q_{\alpha 1} + \rho_1 q_{\alpha 2}},} \tag{6.3.8}$$

where the subscripts on ρ and q_α indicate the medium. Note

$$\mathcal{T}_{11}^2 + \mathcal{T}_{21}^2 = 1 \tag{6.3.9}$$

$$\mathcal{T}_{22}^2 + \mathcal{T}_{12}^2 = 1. \tag{6.3.10}$$

If the slownesses, $q_{\alpha i}$, are real, the coefficient magnitudes are clearly less than unity. If the incident angle is beyond critical, the transmitted slowness will be imaginary, e.g. $q_{\alpha 2}$. The reflection coefficient becomes complex with unit magnitude, e.g.

$$\mathcal{T}_{11} = \frac{\rho_2 q_{\alpha 1} - \mathrm{i}\,\mathrm{sgn}(\omega)\rho_1 |q_{\alpha 2}|}{\rho_2 q_{\alpha 1} + \mathrm{i}\,\mathrm{sgn}(\omega)\rho_1 |q_{\alpha 2}|} = \mathrm{e}^{-2\mathrm{i}\,\mathrm{sgn}(\omega)\phi}, \tag{6.3.11}$$

where

$$\tan\phi = \frac{\rho_1 |q_{\alpha 2}|}{\rho_2 q_{\alpha 1}}, \tag{6.3.12}$$

and we have been careful to include the necessary dependence on the sign of frequency in (6.3.11). The transmission coefficients will be

$$\mathcal{T}_{12} = \mathcal{T}_{21} = (2\,\mathrm{i}\,\mathrm{sgn}(\omega)\sin 2\phi)^{1/2}\,\mathrm{e}^{-\mathrm{i}\,\mathrm{sgn}(\omega)\phi}. \tag{6.3.13}$$

Finally the reciprocity of the reflection/transmission coefficients is obvious as $\mathcal{T}_{12} = \mathcal{T}_{21}$. The matrix $\boldsymbol{\mathcal{T}}$ is symmetric (the normalization in definition (6.3.5) is introduced to obtain this symmetry). We should emphasize that the expression for the transmission coefficients (6.3.8) depends on the normalization and signs of the eigenvectors (6.3.5). With different normalizations, e.g. normalizing the particle velocity, v_n, to unity, or changing the signs, the coefficients will differ and the matrix $\boldsymbol{\mathcal{T}}$ will not be symmetric. Coefficients must be used *together with the appropriate eigenvectors* and are useless in isolation – coefficients are sometimes quoted without making it clear how the eigenvectors are defined.

6.3.2 Anisotropic coefficients

For elastic waves at an interface, the six-dimensional vector $\mathbf{w}$ (6.1.2) is continuous. We can use this condition to calculate the magnitudes of the generated rays, three reflected rays and three transmitted rays, relative to the incident ray. The method is identical to the acoustic case with the complication that three ray types can exist. The technique we follow here was described by Fryer and Frazer (1984). We assume that we have transformed into the interface basis with the x_3 axis normal to the interface, i.e. $p_n = p_3$. Again, we only consider the zeroth-order amplitude coefficients. Eliminating the discontinuous $\mathbf{t}_\nu$'s ($\nu = 1$ or 2) from equations

(5.3.2) and (5.3.3), we obtain

$$\mathbf{A}\mathbf{w}^{(0)} = p_n \mathbf{w}^{(0)}, \tag{6.3.14}$$

where

$$\mathbf{A}_{22} = \mathbf{A}_{11}^{\mathrm{T}} = -p_\eta \mathbf{c}_{\eta 3} \mathbf{c}_{33}^{-1} \tag{6.3.15}$$

$$\mathbf{A}_{12} = -\mathbf{c}_{33}^{-1} \tag{6.3.16}$$

$$\mathbf{A}_{21} = p_\eta p_\nu \mathbf{c}_{\eta\nu} - \rho \mathbf{I} - p_\eta p_\nu \mathbf{c}_{\eta 3} \mathbf{c}_{33}^{-1} \mathbf{c}_{3\nu}. \tag{6.3.17}$$

This eigen-equation (6.3.14) can be solved for six eigenvalues p_n, which is equivalent to solving equation (6.2.3). The eigen-solutions can be written as equation (6.3.3), where now the matrices are 6×6. The six columns of matrix $\mathbf{W}$ are the eigenvectors $\mathbf{w}^{(0)}$, and matrix $\mathbf{p}_n$ is the diagonal matrix of the six eigenvalues. We order the eigenvectors so the first three correspond to waves propagating in the positive $\hat{\mathbf{n}}$ direction, i.e. $\hat{\mathbf{n}} \cdot \mathbf{V} > 0$, and the last three are travelling in the opposite direction, i.e. $\hat{\mathbf{n}} \cdot \mathbf{V} < 0$ (cf. condition (6.2.4)). Within the triplets, we order them with increasing velocity as suggested above.

The various reflection/transmission experiments for six different source rays can be described by a matrix equation

$$\mathbf{W}_1 \begin{pmatrix} \mathcal{T}_{11} & \mathcal{T}_{12} \\ \mathbf{I} & \mathbf{0} \end{pmatrix} = \mathbf{w}(\mathbf{x}_j) = \mathbf{W}_2 \begin{pmatrix} \mathbf{0} & \mathbf{I} \\ \mathcal{T}_{21} & \mathcal{T}_{22} \end{pmatrix}, \tag{6.3.18}$$

where matrices $\mathbf{W}_1$ and $\mathbf{W}_2$ are the eigenvector matrices in the two media at the interface point, $\mathbf{x}_j$. This equation has exactly the same form as equation (6.3.4) except $\mathcal{T}$ is a 6×6 matrix of reflection/transmission coefficients, and the 3×3 sub-matrices in (6.3.18), $\mathcal{T}_{ij}$, are the coefficients where the incident wave's medium is indicated by j, and the generated wave's medium is indicated by i (the unit matrices $\mathbf{I}$ represent the incident waves). Within the 3×3 sub-matrices, the elements $(\mathcal{T}_{ij})_{kl}$ are the individual coefficients. The incident ray type is indicated by l, and the generated ray type by k (with k and l in the range 1 to 3). Elements of the complete 6×6 matrix $\mathcal{T}$ are $\mathcal{T}_{mn}$ where $m = k + 3(i-1)$ and $n = l + 3(j-1)$. As discussed above, in order to use the matrix notation (6.3.18), it is necessary that the incident ray is the second subscript and the generated ray, the first. Figure 6.6 illustrates the reflection/transmission experiments represented by equation (6.3.18). Being higher order, it is sensible to solve equation (6.3.18) in matrix notation. If we expand equation (6.3.18) and rearrange, it can be rewritten

$$\begin{pmatrix} \mathbf{W}_1^{(.)\times(123)} & -\mathbf{W}_2^{(.)\times(456)} \end{pmatrix} \mathcal{T} = \begin{pmatrix} -\mathbf{W}_1^{(.)\times(456)} & \mathbf{W}_2^{(.)\times(123)} \end{pmatrix}, \tag{6.3.19}$$

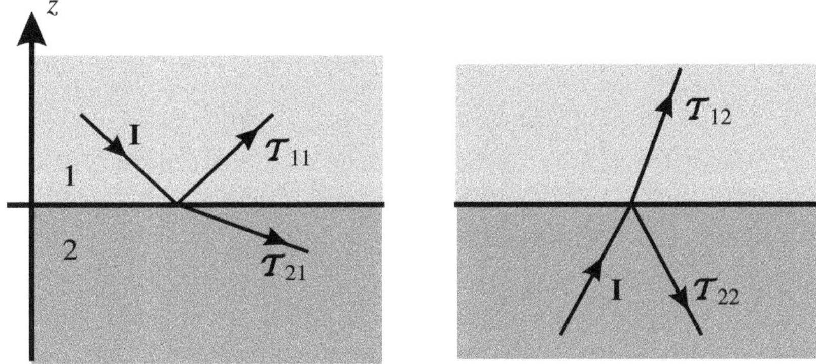

Fig. 6.6. Reflection/transmission experiments for elastic waves. As Figure 6.5 except that each symbol represents three ray types. The diagram is only symbolic as the three rays propagate in different directions not on one ray path as shown.

using the notation (0.1.4). This can be solved as

$$\mathcal{T} = \left(\mathbf{W}_1^{(.)\times(123)} \quad -\mathbf{W}_2^{(.)\times(456)}\right)^{-1}\left(-\mathbf{W}_1^{(.)\times(456)} \quad \mathbf{W}_2^{(.)\times(123)}\right). \tag{6.3.20}$$

Although, in principle this solves the problem, in practice we prefer to follow a different approach which only requires the inversion of a 3×3 matrix, rather than the 6×6 matrix in equation (6.3.20).

Equation (6.3.18) can be rearranged as

$$\begin{pmatrix} \mathcal{T}_{11} & \mathcal{T}_{12} \\ \mathbf{I} & \mathbf{0} \end{pmatrix}\begin{pmatrix} \mathbf{0} & \mathbf{I} \\ \mathcal{T}_{21} & \mathcal{T}_{22} \end{pmatrix}^{-1} = \begin{pmatrix} \mathcal{T}_{12} - \mathcal{T}_{11}\mathcal{T}_{21}^{-1}\mathcal{T}_{22} & \mathcal{T}_{11}\mathcal{T}_{21}^{-1} \\ -\mathcal{T}_{21}^{-1}\mathcal{T}_{22} & \mathcal{T}_{21}^{-1} \end{pmatrix} = \mathbf{Q}, \tag{6.3.21}$$

where

$$\boxed{\mathbf{Q} = \mathbf{W}_1^{-1}\mathbf{W}_2.} \tag{6.3.22}$$

Equation (6.3.21) can be solved for the coefficient matrix

$$\boxed{\mathcal{T} = \begin{pmatrix} \mathcal{T}_{11} & \mathcal{T}_{12} \\ \mathcal{T}_{21} & \mathcal{T}_{22} \end{pmatrix} = \begin{pmatrix} \mathbf{Q}_{12}\mathbf{Q}_{22}^{-1} & \mathbf{Q}_{11} - \mathbf{Q}_{12}\mathbf{Q}_{22}^{-1}\mathbf{Q}_{21} \\ \mathbf{Q}_{22}^{-1} & -\mathbf{Q}_{22}^{-1}\mathbf{Q}_{21} \end{pmatrix}.} \tag{6.3.23}$$

The matrix $\mathbf{Q}$ is defined by equation (6.3.22) in terms of the known eigenvectors in the two media, so this allows us to solve for all the reflection/transmission coefficients. Computing $\mathbf{Q}$ still seems to involve inverting a 6×6 matrix $\mathbf{W}_1$, but fortunately this can be avoided (see below – equation (6.3.28) or (6.3.32)). Otherwise, equation (6.3.23) only requires the inversion of a 3×3 matrix (and for isotropic media, 2×2), a less daunting task. We now find the inverse of the matrix $\mathbf{W}_1$ without explicitly inverting the matrix.

6.3.2.1 The inverse eigenvector matrix

We note that the matrix $\mathbf{I}_2\mathbf{A}$ is symmetric, where matrix $\mathbf{A}$ is defined in equations (6.3.15), (6.3.16) and (6.3.17), and taking the transpose of the product of matrix $\mathbf{I}_2$ (0.1.5) and equation (6.3.3), we can reduce it to

$$(\mathbf{W}^{\mathrm{T}}\mathbf{I}_2)\mathbf{A} = \mathbf{p}_3(\mathbf{W}^{\mathrm{T}}\mathbf{I}_2).$$

Post-multiplying this by the matrix $\mathbf{W}$, pre-multiplying equation (6.3.3) by matrix $\mathbf{W}^{\mathrm{T}}\mathbf{I}_2$, and subtracting, we find that

$$\mathbf{W}^{\mathrm{T}}\mathbf{I}_2\mathbf{W} = \mathbf{K}, \tag{6.3.24}$$

say, must be diagonal. The eigenvector columns of matrix $\mathbf{W}$ are

$$\mathbf{w}^{(0)} = w_E \begin{pmatrix} \hat{\mathbf{g}}_E \\ -p_k \mathbf{c}_{3k} \hat{\mathbf{g}}_E \end{pmatrix}, \tag{6.3.25}$$

where w_E is an arbitrary normalization (no summation over E). Again, for future use, it is useful to define the polarization (cf. equation (6.3.6))

$$\mathbf{g} = w_E \hat{\mathbf{g}}_E. \tag{6.3.26}$$

From result (6.3.24), we find that the diagonal elements of matrix $\mathbf{K}$ are

$$K_E = -(2\rho V_3 w^2)_E, \tag{6.3.27}$$

where equation (5.3.20) defines the component of the ray velocity, V_3. Thus the required inverse matrix in definition (6.3.22) can be computed simple as

$$\mathbf{W}^{-1} = \mathbf{K}^{-1}\mathbf{W}^{\mathrm{T}}\mathbf{I}_2. \tag{6.3.28}$$

If $w_E = 1/\sqrt{\pm 2\rho V_3}$,

$$\mathbf{w}^{(0)} = \frac{1}{\sqrt{\pm 2\rho V_3}} \begin{pmatrix} \hat{\mathbf{g}}_E \\ -p_k \mathbf{c}_{3k} \hat{\mathbf{g}}_E \end{pmatrix}, \tag{6.3.29}$$

with the positive sign for $E = 1$ to 3, and the negative sign for $E = 4$ to 6, then $\mathbf{K} = \mathbf{K}^{-1} = \mathbf{I}_3$. For numerical purposes, especially for evanescent waves, it is simpler to take $w_E = 1$, when $\mathbf{K}$ is defined by equation (6.3.27). It is important to remember that the normalization w_E affects the numerical values of the reflection/transmission coefficients, but not, of course, the resultant amplitude of the field variables. The reflection/transmission coefficients are with respect to the basis vectors, (6.3.25) $\mathbf{w}^{(0)}$, and changes in one are compensated by changes in the other so the appropriate products remain independent of the normalization factor, w_E.

For future use it is convenient to introduce a notation for the sub-matrices in the eigenmatrix $\mathbf{W}$. Thus we write

$$\mathbf{W} = \begin{pmatrix} \acute{\mathbf{W}} & \grave{\mathbf{W}} \end{pmatrix} = \begin{pmatrix} \mathbf{W}_{11} & \mathbf{W}_{12} \\ \mathbf{W}_{21} & \mathbf{W}_{22} \end{pmatrix}, \tag{6.3.30}$$

where the 6×3 sub-matrices $\acute{\mathbf{W}}$ and $\grave{\mathbf{W}}$ contain the up and down-going eigenvectors, respectively, and $\mathbf{W}_{ij}$ are 3×3 sub-matrices. We define a symplectic transform of the velocity-traction vector, $\mathbf{w}$ (6.1.2)

$$\boxed{\mathbf{w}^{\ddagger} = -\mathbf{w}^{\mathrm{T}}\mathbf{I}_2 = -\begin{pmatrix} \mathbf{t}_n^{\mathrm{T}} & \mathbf{v}^{\mathrm{T}} \end{pmatrix}} \tag{6.3.31}$$

(this is a different symplectic transform from that used for the dynamic ray system, (5.2.32)). Then with the normalization $w_E = 1/\sqrt{\pm 2\rho V_3}$, the inverse matrix (6.3.24) is

$$\boxed{\mathbf{W}^{-1} = \mathbf{I}_3\mathbf{W}^{\mathrm{T}}\mathbf{I}_2 = \begin{pmatrix} -\mathbf{W}_{21}^{\mathrm{T}} & -\mathbf{W}_{11}^{\mathrm{T}} \\ \mathbf{W}_{22}^{\mathrm{T}} & \mathbf{W}_{12}^{\mathrm{T}} \end{pmatrix} = \begin{pmatrix} \acute{\mathbf{W}}^{\ddagger} \\ -\grave{\mathbf{W}}^{\ddagger} \end{pmatrix}.} \tag{6.3.32}$$

For eigenvectors $\mathbf{w}_i$ of the matrix $\mathbf{A}$, we have the orthonormal relationship

$$\boxed{\mathbf{w}_i^{\ddagger}\,\mathbf{w}_j = \pm\delta_{ij},} \tag{6.3.33}$$

where the sign depends on the propagation direction, i.e. positive for $i = 1$ to 3, and negative for $i = 4$ to 6, with our ordering convention. For propagating rays, this normalization is connected with the energy flux in the $\hat{\mathbf{n}}$ direction. For evanescent rays, this connection breaks down, but the normalization is still useful. An orthonormality relationship like result (6.3.33) was appreciated by Herrera (1964) and Alsop (1968), but without the connection to the symplectic symmetry of the differential system. Biot (1957) discussed energy flux results. It is important to remember that the orthonormality (6.3.33) applies to rays with the same slowness, $\mathbf{p}_{\perp}$, parallel to the surface used to define the traction, $\mathbf{t}_n$. For different ray types, these will be propagating in different directions. It does *not* apply to different rays propagating in the same direction or with the same total slowness.

Thus the inverse matrix $\mathbf{W}^{-1}$ in equation (6.3.22) is known without inverting any matrix, so

$$\boxed{\mathbf{Q} = \mathbf{I}_3\,\mathbf{W}_1^{\mathrm{T}}\,\mathbf{I}_2\,\mathbf{W}_2,} \tag{6.3.34}$$

and the coefficients (6.3.23) can be calculated by inverting only one 3×3 matrix.

6.3.2.2 The reciprocity of coefficients

Finally we need to prove the reciprocity of the reflection/transmission coefficients $\mathcal{T}$ (6.3.23). In the reciprocal rays, the slownesses in the plane of the interface are reversed and the matrix $\mathbf{A}$ becomes

$$\mathbf{A}' = \mathbf{A}(-p_\nu). \tag{6.3.35}$$

The eigen-solution becomes

$$\mathbf{A}'\mathbf{W}' = -\mathbf{W}'\mathbf{p}_n, \tag{6.3.36}$$

where the change of sign occurs as the slowness surface has point symmetry. The revised eigenvectors, $\mathbf{W}'$, are related by

$$\mathbf{W}' = -\mathbf{I}_3\mathbf{W}, \tag{6.3.37}$$

i.e. the traction components change sign. Importantly the propagation directions of the columns of $\mathbf{W}'$ are reversed, so equation (6.3.18) becomes

$$\mathbf{W}'_1\begin{pmatrix}\mathbf{I} & \mathbf{0}\\ \mathcal{T}'_{11} & \mathcal{T}'_{12}\end{pmatrix} = \mathbf{W}'_2\begin{pmatrix}\mathcal{T}'_{21} & \mathcal{T}'_{22}\\ \mathbf{0} & \mathbf{I}\end{pmatrix}. \tag{6.3.38}$$

Taking the transpose of this equation (6.3.38) and multiplying by $\mathbf{I}_1$ times equation (6.3.18), we obtain

$$\begin{pmatrix}\mathbf{I} & \mathcal{T}'^{\mathrm{T}}_{11}\\ \mathbf{0} & \mathcal{T}'^{\mathrm{T}}_{12}\end{pmatrix}\mathbf{W}'^{\mathrm{T}}_1\mathbf{I}_1\mathbf{W}_1\begin{pmatrix}\mathcal{T}_{11} & \mathcal{T}_{12}\\ \mathbf{I} & \mathbf{0}\end{pmatrix} = \begin{pmatrix}\mathcal{T}'^{\mathrm{T}}_{21} & \mathbf{0}\\ \mathcal{T}'^{\mathrm{T}}_{22} & \mathbf{I}\end{pmatrix}\mathbf{W}'^{\mathrm{T}}_2\mathbf{I}_1\mathbf{W}_2\begin{pmatrix}\mathbf{0} & \mathbf{I}\\ \mathcal{T}_{21} & \mathcal{T}_{22}\end{pmatrix}. \tag{6.3.39}$$

Using definition (6.3.37), it is straightforward to simplify this as

$$\mathbf{W}'^{\mathrm{T}}\mathbf{I}_1\mathbf{W} = -\mathbf{W}^{\mathrm{T}}\mathbf{I}_3^{\mathrm{T}}\mathbf{I}_1\mathbf{W} = \mathbf{W}^{\mathrm{T}}\mathbf{I}_2\mathbf{W} = \mathbf{K} = \begin{pmatrix}\acute{\mathbf{K}} & \mathbf{0}\\ \mathbf{0} & \grave{\mathbf{K}}\end{pmatrix}, \tag{6.3.40}$$

where we have expanded the matrix (6.3.24) into diagonal 3×3 sub-matrices for the positive and negative travelling waves. Then expanding equation (6.3.39), it is seen to be equivalent to

$$\mathcal{T}' = \begin{pmatrix}\grave{\mathbf{K}}_1^{-1} & \mathbf{0}\\ \mathbf{0} & -\acute{\mathbf{K}}_2^{-1}\end{pmatrix}\mathcal{T}^{\mathrm{T}}\begin{pmatrix}-\acute{\mathbf{K}}_1 & \mathbf{0}\\ \mathbf{0} & \grave{\mathbf{K}}_2\end{pmatrix}, \tag{6.3.41}$$

which is the reciprocity result for reflection/transmission coefficients. If the eigenvectors are normalized so $\mathbf{K} = \mathbf{I}_3$, then this simplifies to $\mathcal{T}' = \mathcal{T}^{\mathrm{T}}$, or

$$\boxed{\mathcal{T}(-p_1, -p_2) = \mathcal{T}^{\mathrm{T}}(p_1, p_2).} \tag{6.3.42}$$

This equation describes the fact that if the source and receiver are interchanged, requiring the swapping of subscript indices on $\mathcal{T}_{ij}$ and the reversal of the slowness components parallel to the interface, then the reflection/transmission coefficients are equal (with suitable normalization of the eigenvectors), i.e. they satisfy reciprocity. This result is far from trivial and does not correspond to just reversing time. Apart from the reversal of the source and receiver rays, the other generated rays are completely different in the reciprocal experiments. The reciprocal result (6.3.42) does depend on the polarizations being defined in a consistent manner. Changes in sign of the polarization, permitted by the eigen-equation (6.3.14), will result in changes in the sign of coefficients and reciprocity will only be satisfied for the combination of coefficient times polarization. Equation (6.3.42) will contain sign mismatches.

6.3.2.3 Energy flux conservation

The orthonormality relation (6.3.33), which resulted in the simple reciprocity relationship (6.3.42), is connected with the energy flux of the eigenvectors when the waves are propagating, not evanescent. It leads to another relationship between the coefficients.

Expanding the continuity equation (6.3.18), the first row is

$$\acute{\mathbf{W}}_1 \boldsymbol{\mathcal{T}}_{11} + \grave{\mathbf{W}}_1 = \mathbf{w} = \grave{\mathbf{W}}_2 \boldsymbol{\mathcal{T}}_{21}. \tag{6.3.43}$$

Multiplying both sides by the transform (6.3.31), we obtain

$$-\left(\boldsymbol{\mathcal{T}}^{\mathrm{T}} \acute{\mathbf{W}}_1^{\mathrm{T}} + \grave{\mathbf{W}}^{\mathrm{T}}\right) \mathbf{I}_2 \left(\acute{\mathbf{W}}_1 \boldsymbol{\mathcal{T}}_{11} + \grave{\mathbf{W}}_1\right) = \mathbf{w}^{\ddagger}\, \mathbf{w} = -\boldsymbol{\mathcal{T}}_{21}^{\mathrm{T}} \grave{\mathbf{W}}_2^{\mathrm{T}} \mathbf{I}_2 \grave{\mathbf{W}}_2 \boldsymbol{\mathcal{T}}_{21}. \tag{6.3.44}$$

Expanding, and using the orthonormality (6.3.33), we have

$$\boxed{\boldsymbol{\mathcal{T}}_{11}^{\mathrm{T}} \boldsymbol{\mathcal{T}}_{11} + \boldsymbol{\mathcal{T}}_{21}^{\mathrm{T}} \boldsymbol{\mathcal{T}}_{21} = \mathbf{I}.} \tag{6.3.45}$$

The second row of equation (6.3.18) leads to the similar result

$$\boxed{\boldsymbol{\mathcal{T}}_{22}^{\mathrm{T}} \boldsymbol{\mathcal{T}}_{22} + \boldsymbol{\mathcal{T}}_{12}^{\mathrm{T}} \boldsymbol{\mathcal{T}}_{12} = \mathbf{I}.} \tag{6.3.46}$$

Scalar examples of results (6.3.45) and (6.3.46) have already been noted, results (6.3.9) and (6.3.10). The simple result and proof break down at a fluid–solid interface – equation (6.3.43) no longer applies as the tangential displacement is discontinuous and tangential traction is zero in $\mathbf{w}$ (see Section 6.5).

Physically, these results, (6.3.45) and (6.3.46), express the conservation of energy flux across the interface when the waves are propagating, and the coefficients are real. The results remain true when any waves are evanescent, and coefficients complex, but no longer express conservation of energy flux.

6.3.3 Isotropic coefficients

In isotropic media, we follow exactly the same procedure as in anisotropic media (Section 6.3.2) but significant simplifications are possible, and explicit expressions can be obtained for the coefficients. These were first obtained by Knott (1899) and later by Zoeppritz (1919) and are known by both names although usually the latter. Their results differ only in that Knott's coefficients are normalized with respect to potential amplitudes and Zoeppritz's with respect to displacement. Here we follow the same normalization and method used for the anisotropic coefficients.

Using the isotropic, stiffness matrices (4.4.55) and (4.4.56) in equations (6.3.15)–(6.3.17), we obtain the matrix

$$\mathbf{A} = \begin{pmatrix} 0 & 0 & -p & -1/\mu & 0 & 0 \\ 0 & 0 & 0 & 0 & -1/\mu & 0 \\ -p\lambda/(\lambda+2\mu) & 0 & 0 & 0 & 0 & -1/(\lambda+2\mu) \\ \eta p^2 - \rho & 0 & 0 & 0 & 0 & -p\lambda/(\lambda+2\mu) \\ 0 & \mu p^2 - \rho & 0 & 0 & 0 & 0 \\ 0 & 0 & -\rho & -p & 0 & 0 \end{pmatrix}, \tag{6.3.47}$$

where

$$\eta = \frac{4\mu(\lambda+\mu)}{\lambda+2\mu}. \tag{6.3.48}$$

Note that the components one, three, four and six form a separate system from components two and five: physically this is the separation of $P-SV$ rays from SH rays. The slowness component in the interface is p, i.e.

$$\mathbf{p} = p\,\hat{\mathbf{l}} + p_n\hat{\mathbf{n}}, \tag{6.3.49}$$

as $\hat{\mathbf{l}}$ and $\hat{\mathbf{n}}$ define the ray plane (note from the definition (6.0.1), p is positive).

The eigen-equation (6.3.3) can be solved explicitly. The diagonal matrix of eigenvalues is

$$\mathbf{p}_3 = \begin{pmatrix} q_\beta & 0 & 0 & 0 & 0 & 0 \\ 0 & q_\beta & 0 & 0 & 0 & 0 \\ 0 & 0 & q_\alpha & 0 & 0 & 0 \\ 0 & 0 & 0 & -q_\beta & 0 & 0 \\ 0 & 0 & 0 & 0 & -q_\beta & 0 \\ 0 & 0 & 0 & 0 & 0 & -q_\alpha \end{pmatrix}. \tag{6.3.50}$$

The non-zero components of the eigenvector matrix **W** are

$$\boxed{\begin{aligned} W_{11} &= W_{14} = w_1 q_\beta \\ W_{31} &= -W_{34} = -w_1 p \\ W_{41} &= -W_{44} = -w_1 \mu\, \Omega \\ W_{61} &= W_{64} = w_1 2\mu\, p\, q_\beta \end{aligned}} \tag{6.3.51}$$

$$\boxed{\begin{aligned} W_{22} &= W_{25} = w_2 \\ W_{52} &= -W_{55} = -w_2 \mu\, q_\beta \end{aligned}} \tag{6.3.52}$$

$$\boxed{\begin{aligned} W_{13} &= W_{16} = w_3 p \\ W_{33} &= -W_{36} = w_3 q_\alpha \\ W_{43} &= -W_{46} = -w_3 2\mu\, p\, q_\alpha \\ W_{63} &= W_{66} = -w_3 \mu\, \Omega, \end{aligned}} \tag{6.3.53}$$

where

$$\Omega = q_\beta^2 - p^2 = (\rho - 2\mu\, p^2)/\mu. \tag{6.3.54}$$

With the normalizations factors

$$w_1 = 1/(2\rho\, q_\beta)^{1/2} \tag{6.3.55}$$

$$w_2 = 1/(2\mu\, q_\beta)^{1/2} \tag{6.3.56}$$

$$w_3 = 1/(2\rho\, q_\alpha)^{1/2}, \tag{6.3.57}$$

expression (6.3.24) holds (note $w_i = c w_E$, where w_E was used in definition (6.3.25) and $c = \alpha$ or β, as appropriate).

As before, the coefficients can be found directly from the matrices **W** (6.3.20). These separate into a 4×4 system for $P - SV$ rays and a 2×2 system for SH rays. Thus using the notation (0.1.4), we have

$$\begin{aligned} \mathcal{T}^{(1346)\times(1346)} = &\left(\mathbf{W}_1^{(1346)\times(13)} \quad -\mathbf{W}_2^{(1346)\times(46)}\right)^{-1} \\ &\left(-\mathbf{W}_1^{(1346)\times(46)} \quad \mathbf{W}_2^{(1346)\times(13)}\right), \end{aligned} \tag{6.3.58}$$

for the $P - SV$ system, and

$$\mathcal{T}^{(25)\times(25)} = \left(\mathbf{W}_1^{(25)\times(2)} \quad -\mathbf{W}_2^{(25)\times(5)}\right)^{-1} \left(-\mathbf{W}_1^{(25)\times(5)} \quad \mathbf{W}_2^{(25)\times(2)}\right), \tag{6.3.59}$$

for the SH system. These can also be rewritten as equation (6.3.23) where **Q** can be calculated with result (6.3.34).

Being lower order, it is straightforward, if tedious, to obtain expressions for the reflection/transmission coefficients, requiring at most the inversion of a 2×2 matrix. These coefficients have been published by many authors. We use the notation from Chapman, Chu Jen-Yi and Lyness (1988) (which, in common with many publications, contained a typographical error!). The non-zero coefficients are

$$
\boxed{\begin{aligned}
\mathcal{T}_{33} &= (A_{\alpha-}A_{\beta+} + C_{1-}C_{2+} - D)/\Delta^{PV} \\
\mathcal{T}_{66} &= (-A_{\alpha-}A_{\beta+} + C_{1+}C_{2-} - D)/\Delta^{PV} \\
\mathcal{T}_{11} &= (-A_{\alpha+}A_{\beta-} - C_{1-}C_{2+} + D)/\Delta^{PV} \\
\mathcal{T}_{44} &= (A_{\alpha+}A_{\beta-} - C_{1+}C_{2-} + D)/\Delta^{PV} \\
\mathcal{T}_{36} &= \mathcal{T}_{63} = F_{\alpha 1}F_{\alpha 2}(q_{\beta 1}E_2 + q_{\beta 2}E_1)/\Delta^{PV} \\
\mathcal{T}_{14} &= \mathcal{T}_{41} = F_{\beta 1}F_{\beta 2}(q_{\alpha 1}E_2 + q_{\alpha 2}E_1)/\Delta^{PV} \\
\mathcal{T}_{13} &= \mathcal{T}_{31} = -pF_{\alpha 1}F_{\beta 1}\left(2q_{\alpha 2}q_{\beta 2}E_1B_2 + E_2(E_2 - \rho_1)\right)/\rho_1\Delta^{PV} \\
\mathcal{T}_{46} &= \mathcal{T}_{64} = -pF_{\alpha 2}F_{\beta 2}\left(2q_{\alpha 1}q_{\beta 1}E_2B_1 + E_1(E_1 - \rho_2)\right)/\rho_2\Delta^{PV} \\
\mathcal{T}_{34} &= \mathcal{T}_{43} = -pF_{\alpha 1}F_{\beta 2}(2B_2q_{\beta 1}q_{\alpha 2} + E_1 - \rho_2)/\Delta^{PV} \\
\mathcal{T}_{16} &= \mathcal{T}_{61} = -pF_{\alpha 2}F_{\beta 1}(2B_1q_{\beta 2}q_{\alpha 1} + E_2 - \rho_1)/\Delta^{PV} \\
\mathcal{T}_{22} &= G_{\beta-}/\Delta^H \\
\mathcal{T}_{55} &= -G_{\beta-}/\Delta^H \\
\mathcal{T}_{25} &= \mathcal{T}_{52} = H_{\beta 1}H_{\beta 2}/\Delta^H,
\end{aligned}} \tag{6.3.60}
$$

where

$$
\boxed{\begin{aligned}
\Delta^{PV} &= A_{\alpha+}A_{\beta+} - C_{1+}C_{2+} + D \\
\Delta^{H} &= G_{\beta+},
\end{aligned}} \tag{6.3.61}
$$

and

$$
\boxed{\begin{aligned}
A_{\substack{\alpha \\ \beta}\substack{+ \\ -}} &= \rho_2 q_{\substack{\alpha \\ \beta}1} \pm \rho_1 q_{\substack{\alpha \\ \beta}2} \\
B_{\substack{1 \\ 2}} &= \mu_{\substack{1 \\ 2}} - \mu_{\substack{2 \\ 1}} \\
C_{\substack{1 \\ 2}\substack{+ \\ -}} &= 2p\left(B_{\substack{1 \\ 2}}(\pm q_{\alpha \substack{1 \\ 2}}q_{\beta \substack{1 \\ 2}} + p^2) - \rho_{\substack{1 \\ 2}}\right) \\
D &= p^2(\rho_1 + \rho_2)^2 \\
E_{\substack{1 \\ 2}} &= \rho_{\substack{1 \\ 2}} - 2p^2 B_{\substack{1 \\ 2}} \\
F_{\substack{\alpha \\ \beta}\substack{1 \\ 2}} &= \left(2\rho_{\substack{1 \\ 2}} q_{\substack{\alpha \\ \beta}\substack{1 \\ 2}}\right)^{1/2} \\
G_{\beta \substack{+ \\ -}} &= \mu_1 q_{\beta 1} \pm \mu_2 q_{\beta 2} \\
H_{\beta \substack{1 \\ 2}} &= \left(2\mu_{\substack{1 \\ 2}} q_{\beta \substack{1 \\ 2}}\right)^{1/2}.
\end{aligned}} \tag{6.3.62}
$$

The denominator Δ^{PV} (6.3.61) of the $P-SV$ coefficients is known as the Stoneley radix, Δ^{Stoneley}, say. Stoneley (1924) interface waves exist when $\Delta^{PV} = \Delta^{\text{Stoneley}} = 0$ on a solid–solid interface (interface waves will be discussed in Section 9.1.5).

In an isotropic medium, the reflection/transmission coefficients do not depend on the signs of p_ν as we have reflection symmetry (in fact they only depend on the magnitude $p = (p_1^2 + p_2^2)^{1/2}$ as we have rotational symmetry). The reciprocal relationship (6.3.42) therefore simplifies to

$$\mathcal{T}(p) = \mathcal{T}^{\mathrm{T}}(p). \tag{6.3.63}$$

The explicit results (6.3.60) satisfy this symmetry.

An alternative method for calculating the isotropic reflection coefficients, that is computationally very simple, is discussed in Section 7.2.8.

6.3.4 Transversely isotropic coefficients

Although the reflection/transmission coefficients in transversely isotropic media (Section 4.4.4) are algebraically complicated, some of the intermediate results are simple enough, assuming the axis of symmetry is normal to the interface. It is convenient and worthwhile to include the simple results here. A straightforward algorithm for calculating the coefficients in transversely isotropic media is given in Section 7.2.8.

The matrix **A**, (6.3.14)–(6.3.17), reduces to

$$\mathbf{A} = \left(\begin{array}{ccc} 0 & 0 & -p \\ 0 & 0 & 0 \\ -pC_{13}/C_{33}) & 0 & 0 \\ p^2\left(C_{11} - C_{13}^2/C_{33}\right) - \rho & 0 & 0 \\ 0 & p^2 C_{66} - \rho & 0 \\ 0 & 0 & -\rho \end{array} \right. \cdots$$

$$\cdots \left. \begin{array}{ccc} -1/C_{44} & 0 & 0 \\ 0 & -1/C_{44} & 0 \\ 0 & 0 & -1/C_{33} \\ 0 & 0 & -pC_{13}/C_{33} \\ 0 & 0 & 0 \\ -p & 0 & 0 \end{array} \right), \tag{6.3.64}$$

i.e. similar to the isotropic matrix (6.3.47) with the same elements zero. Thus the system separates into the fourth-order $qP - qSV$ system (first, third, fourth and sixth rows and columns), and the second-order qSH system (second and fifth rows and columns). The results in Section 5.7.1 provide the eigenvalues and eigenvectors, without considering the higher-order system. Thus the *qSH* eigenvalues, $\pm q_{qSH}$, say, are given by equation (5.7.24) (with $p_1 = p$), and the *qP* eigenvalues, $\pm q_{qP}$, and *qSV* eigenvalues, $\pm q_{qSV}$, by equation (5.7.32) (with the minus sign for *qP* and the plus sign for *qSV*). The normalized, polarization vectors are given by result (5.7.21) for the *qSH* wave, and equation (5.7.35) for the *qP* and *qSV* waves. The traction parts of the eigenvectors, $\mathbf{w}_n$, can then be calculated from

$$\mathbf{t}_3 = -\begin{pmatrix} 0 \\ p_3 C_{44} \\ 0 \end{pmatrix}, \tag{6.3.65}$$

for the *qSH* eigenvectors, and

$$\mathbf{t}_3 = -\begin{pmatrix} \left(p_1 \hat{g}_3 + p_3 \hat{g}_1\right) C_{44} \\ 0 \\ p_1 \hat{g}_1 C_{13} + p_3 \hat{g}_3 C_{33} \end{pmatrix}, \tag{6.3.66}$$

for the *qP* and *qSV* eigenvectors. The normalization required to obtain orthonormal eigenvectors (6.3.33) can be achieved using result (5.7.37). It is then straightforward to calculate the coefficients using result (6.3.23), where the *qSH* and *qP* – *qSV* systems can be considered separately so only a 2×2 matrix needs to be inverted.

6.3.5 Examples

It is difficult if not impossible to summarize the numerical behaviour of the various reflection/transmission coefficients as they depend on many model parameters as well as the angle of incidence or interface slowness. Even in acoustic media, an interface is characterized by two ratios (although the media require four parameters, velocities and densities, the reflection/transmission coefficients are dimensionless and only depend on ratios, e.g. α_2/α_1 and ρ_2/ρ_1 – they cannot depend on the units of velocity or density). It is convenient to use the impedance ratio, $Z_2/Z_1 = \rho_2\alpha_2/\rho_1\alpha_1$, which defines the coefficients at normal incidence, and the velocity ratio α_2/α_1, which defines whether and where a critical angle exists (rather than express the coefficients as a function of interface slowness, $p_\perp$, which requires another parameter, velocity, to define the units of slowness, we would use angle or the dimensionless slowness $\alpha_1 p_\perp$). In isotropic media we would require four parameters, e.g. in addition to the *P* ray impedance and velocity ratios, we could use Poisson's ratio in the two media. In the most general anisotropic media, we would require 42 parameters, and the coefficients would depend on the direction as well as the magnitude of the interface slowness, $\mathbf{p}_\perp$! Even in TI media with the axes normal to the interface, we would require 10 parameters to describe the media at the interface (although the coefficients would be independent of azimuth).

As there are so many model parameters, it is difficult to characterize all the possible behaviours of the reflection/transmission coefficients. Nevertheless, using the above formulae it is straightforward to calculate the coefficients for any required model. Here we just indicate some simple numerical results. As an elastic model, we consider one normalized so $\beta_1 = 1$ and $\beta_2 = 1.1$ (so $\beta_2/\beta_1 = 1.1$). In the first medium, we have a Poisson's ratio of $\nu_1 = 1/4$ (so $\alpha_1 = \sqrt{3}\beta_1 = \sqrt{3}$), and in the second medium $\nu_2 = 1/3$ (so $\alpha_2 = 2\beta_2 = 1.2$). The density ratio is $\rho_2/\rho_1 \simeq 1.072$.

First we consider an acoustic model, setting $\beta_1 = \beta_2 = 0$, as the coefficients are relatively straightforward. In Figure 6.7, the reflection coefficient $\mathcal{T}_{11}$ (6.3.7) is illustrated for this interface. The velocity ratio is $\alpha_2/\alpha_1 \simeq 1.270 > 1$ so a critical point exists at $\theta_1 \simeq 51.9^\circ$. In acoustic media, the reflection coefficient is always $\mathcal{T}_{11} = +1$ at a critical point, and as the impedance ratio is $Z_2/Z_1 \simeq 1.361 > 1$ the reflection coefficient is also positive at $\theta_1 = 0$. Between normal incidence and the critical angle the coefficient increases (near normal incidence it behaves quadratically as the coefficient is an even function of $p_\perp$). Beyond the critical angle, the coefficient is complex with unit magnitude (6.3.11). At grazing angle, $\theta_1 = 90^\circ$, the coefficient becomes $\mathcal{T}_{11} = -1$ (which always applies at grazing angle whether a critical angle exists of not). There basically are four different forms of the acoustic reflection coefficient: the sign of the coefficient at normal incidence depends on whether Z_2/Z_1 is greater or less than unity; and the coefficient is always $\mathcal{T}_{11} = +1$ at a critical point if $\alpha_2/\alpha_1 > 1$, or $\mathcal{T}_{11} = -1$ at grazing if $\alpha_2/\alpha_1 < 1$. These conditions are independent so there are four choices, and in two the real coefficient changes sign between normal incidence and the critical or grazing ray.

The reflection coefficients for the isotropic, elastic media are illustrated in Figure 6.8. For clarity, these are plotted in three panels. First the coefficients for an incident *P* ray, $\mathcal{T}_{33}$ and $\mathcal{T}_{13}$ (6.3.60), plotted against the incident angle, $\theta_1 = \sin^{-1}(\alpha_1 p_\perp)$. The acoustic coefficient from Figure 6.7 has been plotted with a dashed line for comparison – the *P* reflection coefficients are similar in the elastic and acoustic cases. Next for an incident *SV* ray, $\mathcal{T}_{31}$ and $\mathcal{T}_{11}$ (6.3.60), are plotted against the incident angle, $\theta_1 = \sin^{-1}(\beta_1 p_\perp)$. Note that although $\mathcal{T}_{13}(p_\perp) = \mathcal{T}_{31}(p_\perp)$ (6.3.42), as functions of the incident angle they are not equal. The behaviour of these coefficients is much more complicated as critical angles exist at $\theta_1 = \sin^{-1}(\beta_1/\alpha_2) \simeq 27.0^\circ$, $\sin^{-1}(\beta_1/\alpha_1) \simeq 35.3^\circ$ and $\sin^{-1}(\beta_1/\beta_2) \simeq 65.4^\circ$. Finally for an incident *SH* ray, the coefficient $\mathcal{T}_{22}$ is plotted, which is simple with a critical angle at $\theta_1 = \sin^{-1}(\beta_1/\beta_2) \simeq 65.4^\circ$ (in fact the acoustic and *SH* coefficients are identical with the substitution $\rho \leftrightarrow 1/\mu$).

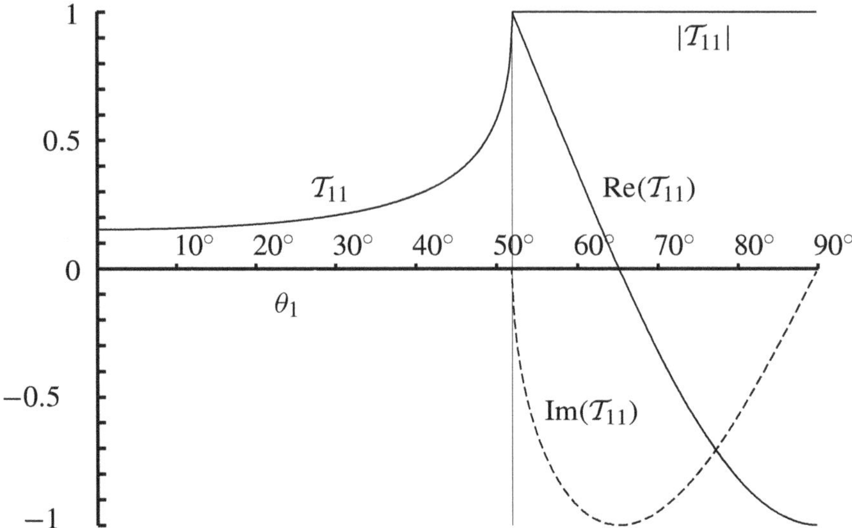

Fig. 6.7. The acoustic reflection coefficient, $\mathcal{T}_{11}$ (6.3.7). The velocity ratio is $\alpha_2/\alpha_1 \simeq 1.270$ and the impedance ratio, $Z_2/Z_1 \simeq 1.361$. Beyond the critical angle at $\theta_1 \simeq 51.9°$, the real and imaginary parts and magnitude of $\mathcal{T}_{11}$ are plotted.

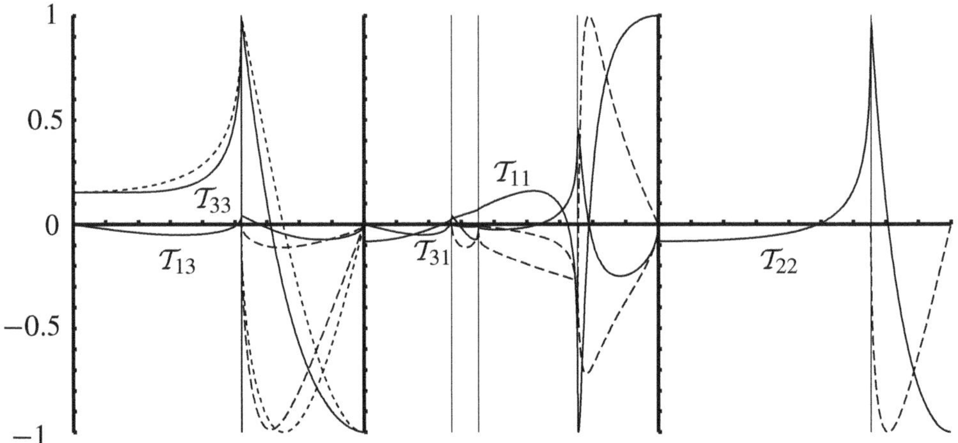

Fig. 6.8. The isotropic reflection coefficients (6.3.60). The velocity ratio is $\beta_2/\beta_1 \simeq 1.1$ with density ratio $\rho_2/\rho_1 \simeq 1.072$ and Poisson's ratios of $\nu_1 = 1/4$ and $\nu_2 = 1/3$. The left panel contains $\mathcal{T}_{33}$ and $\mathcal{T}_{13}$ against the incident *P* ray angle (with the acoustic $\mathcal{T}_{11}$ coefficient denoted by a dashed line); the central panel contains $\mathcal{T}_{31}$ and $\mathcal{T}_{11}$ against the incident *SV* ray angle; the right panel contains $\mathcal{T}_{22}$ against the incident *SH* ray angle. The critical angles $\sin^{-1}(\alpha_1/\alpha_2) \simeq 51.9°$, $\sin^{-1}(\beta_1/\alpha_2) \simeq 27.0°$, $\sin^{-1}(\beta_1/\alpha_1) \simeq 35.3°$ and $\sin^{-1}(\beta_1/\beta_2) \simeq 65.4°$ are indicated. In all cases, the incident angles run from 0° to 90°.

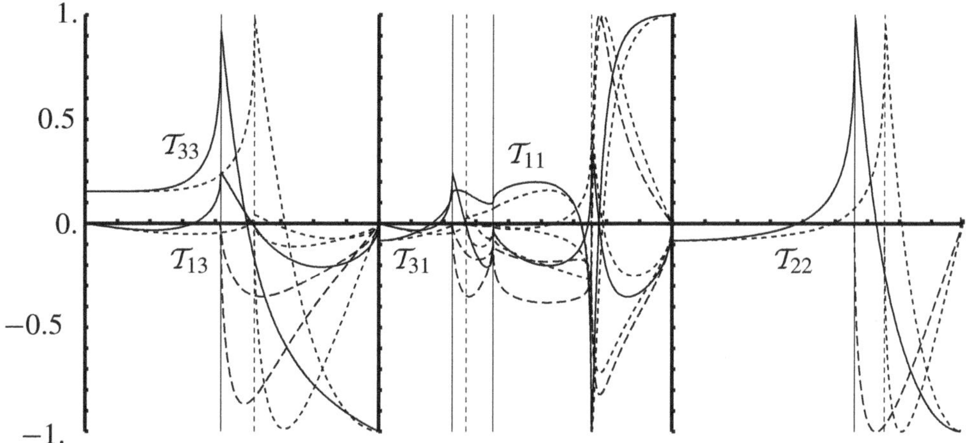

Fig. 6.9. The anisotropic reflection coefficients. The model parameters are as Figure 6.8 except that the second medium is TIV with $\epsilon_2 = 0.2$, $\delta = 0$ and $\gamma_2 = 0.1$. The critical angles are now approximately 41.7°, 22.6°, 35.3° and 56.1°. The isotropic results from Figure 6.8 are plotted with dashed lines.

Finally, the second medium is modified to be anisotropic. For simplicity, the first medium is kept isotropic as above. The second medium is TIV (Section 4.4.4) and the coefficients are calculated using the general expression (6.3.23). The TIV medium can be described by the Thomsen (1986) dimensionless parameters (Exercise 4.5) – the isotropic velocities α_2 and β_2 apply on the vertical symmetry axis. The dimensionless parameters are taken as $\epsilon_2 = 0.2$, $\delta = 0$ and $\gamma_2 = 0.1$. The same coefficients as in the isotropic case (Figure 6.8) are plotted in Figure 6.9. Although the indices of the *qSV* and *qSH* rays change – near normal incidence the *qSV* ray is faster (at normal incidence they have the same velocity, β_2), whereas near the horizontal direction the *qSH* ray is faster – for clarity and simplicity we have used the same indices at all angles (1 for *qSV* and 2 for *qSH*). The isotropic coefficients have also been included in Figure 6.9 for comparison. The major changes are caused by the changes in the critical angles as the horizontal velocities in the second medium are increased to approximately 2.603 for *qP* and 1.205 for *qSH*.

We have presented coefficients for only one acoustic, one isotropic and one anisotropic elastic interface. Although Figures 6.8 and 6.9 contain too much information, the details are not important. The media parameters are of no particular significance and for any specific interface it is straightforward to calculate the coefficients explicitly. It is difficult to characterize the behaviour of all coefficients as they depend on multiple parameters. Nevertheless, for some restricted situations it is possible to characterize the behaviour of the coefficients. For instance, for reservoir rocks, shales and sandstones, the P ray reflection coefficients at small angles can be divided into three classes (Rutherford and Williams, 1989 – Young and LoPiccolo, 2003, have extended this to define a comprehensive classification). This is useful for AVO studies. The three classes are: Class 1, high-impedance sands so $Z_2/Z_1 > 1$ and the reflection coefficient at normal incidence is positive; Class 2, near-zero impedance contrast sands so $Z_2/Z_1 \simeq 1$ and the normal incidence coefficient is approximately zero; and Class 3, low-impedance sands so $Z_2/Z_1 < 1$ so the normal incidence coefficient is negative. The behaviour of the coefficient for small angles is largely controlled by the contrast in Poisson's ratio. For gas sands, the Poisson's ratio is decreased and the coefficients decrease with angle. For Class 1 sands, this leads to a sign change in the coefficient and a characteristic AVO response. This has been described and explained by Koetoed (1955), Bortfeld (1961) and Shuey (1985) in terms of the rigidity factor or contrast in Poisson's ratio. The theory depends on approximating the coefficient for a small contrast and at small angles. We return to the techniques for doing this below (Section 6.7).

6.4 Free surface reflection coefficients

The interface between the ocean or ground and the atmosphere is normally treated as a free surface as the density of the atmosphere is so small compared with water or the solid Earth. Only when we specifically want to study the coupling of elastic waves with atmospheric waves is it necessary to consider the atmosphere. In most seismic experiments either the source or receivers, or both, are close to this free surface and reflections from the free surface are important. In this section, we discuss the special results for reflections from a free surface. As discussed in Section 4.3.3, the boundary condition is that the traction on the free surface is zero, equation (4.3.6). For definiteness, we take the unit normal to the surface as pointing into the vacuum and with the convention used above, the medium is medium 2.

6.4.1 Acoustic coefficients

At a free interface the pressure must be zero and equation (6.3.4) becomes

$$\mathbf{w}(x_j) = \begin{pmatrix} v_n \\ 0 \end{pmatrix} = \mathbf{W}_2 \begin{pmatrix} 1 \\ \mathcal{T}_{22} \end{pmatrix}. \tag{6.4.1}$$

This equation is trivial to solve for the reflection coefficient

$$\boxed{\mathcal{T}_{22} = -1,} \tag{6.4.2}$$

i.e. the acoustic ray is perfectly reflected. Because of the sign convention used to define the eigenvectors (6.3.5), the coefficient $\mathcal{T}_{22}$ is minus unity.

6.4.2 Anisotropic coefficients

At a free–solid interface, equation (6.3.18) is revised as in equation (6.4.1) to

$$\mathbf{w}(x_j) = \begin{pmatrix} \mathbf{v} \\ \mathbf{0} \end{pmatrix} = \mathbf{W}_2 \begin{pmatrix} \mathbf{I} \\ \boldsymbol{\mathcal{T}}_{22} \end{pmatrix}. \tag{6.4.3}$$

Decomposing eigenmatrix $\mathbf{W}_2$ into its 3×3 sub-matrices, we can solve for the reflection coefficients

$$\boldsymbol{\mathcal{T}}_{22} = -\mathbf{W}_2^{(456)\times(456)^{-1}} \mathbf{W}_2^{(456)\times(123)}. \tag{6.4.4}$$

6.4.3 Isotropic coefficients

Expression (6.4.4) is valid for elastic waves in isotropic media. As the eigenvectors are known explicitly, results (6.3.51) through (6.3.57), and as the system separates into fourth-order and second-order systems, it is easily solved. The non-zero coefficients are

$$\boxed{\begin{aligned} &\mathcal{T}_{66} = -\mathcal{T}_{44} = (4p^2 q_{\alpha 2} q_{\beta 2} - \Omega_2^2)/\Delta^{PV} \\ &\mathcal{T}_{46} = \mathcal{T}_{64} = 4p\Omega_2 (q_{\alpha 2} q_{\beta 2})^{1/2}/\Delta^{PV} \\ &\mathcal{T}_{55} = 1, \end{aligned}} \tag{6.4.5}$$

where

$$\boxed{\Delta^{PV} = 4p^2 q_{\alpha 2} q_{\beta 2} + \Omega_2^2.} \tag{6.4.6}$$

The denominator, Δ^{PV} (6.4.6), is known as the Rayleigh radix, Δ^{Rayleigh}, say. Rayleigh (1885) interface waves exist when $\Delta^{PV} = \Delta^{\text{Rayleigh}} = 0$ on a free surface (interface waves will be discussed in Section 9.1.5). *SH* rays are perfectly reflected and because of the sign convention used to define the eigenvectors (6.3.52), the coefficient $\mathcal{T}_{55}$ is unity.

6.5 Fluid–solid reflection/transmission coefficients

At a fluid–solid interface, the tangential components of particle velocity are discontinuous, and the tangential components of the traction must be zero (Section 4.3.2). In the fluid, only an acoustic wave is possible so there are only four generated rays (not six). The continuity of the normal component of particle velocity and traction, and the zero value for the two tangential components of traction, provide the four boundary conditions. Compared with the solid–solid interface, the system of equations is lower order (fourth not sixth) but the mixture of boundary conditions make the notation more complicated. We set up the system of equations for an anisotropic solid, and then specialize to isotropy. Mallick and Frazer (1991) extended Fryer and Frazer (1984) to cover this case. For definiteness, we take medium 1 to be the fluid.

6.5.1 Anisotropic coefficients

Equation (6.3.18) can still be used to express the velocity/traction at the interface for the reflection/transmission experiments (Figure 6.6), i.e.

$$\mathbf{w}_1 = \mathbf{W}_1 \begin{pmatrix} \mathcal{T}_{11} & \mathcal{T}_{12} \\ \mathbf{I} & \mathbf{0} \end{pmatrix} \tag{6.5.1}$$

$$\mathbf{w}_2 = \mathbf{W}_2 \begin{pmatrix} \mathbf{0} & \mathbf{I} \\ \mathcal{T}_{21} & \mathcal{T}_{22} \end{pmatrix}, \tag{6.5.2}$$

but $\mathbf{w}_1 \neq \mathbf{w}_2$. The first and second components are discontinuous and the fourth and fifth must be zero. In equation (6.5.1) only the third and sixth columns of the full elastic system are well defined in a fluid, but coefficients of the degenerate columns can be taken zero. Using the notation in definition (0.1.4), the equation for the continuity of the normal particle velocity and traction is

$$\mathbf{W}_1 \begin{pmatrix} \mathcal{T}^{(3)\times(3)} & \mathcal{T}^{(3)\times(456)} \\ 1 & \mathbf{0} \end{pmatrix} = \mathbf{W}_2^{(36)\times(.)} \begin{pmatrix} \mathbf{0} & \mathbf{I} \\ \mathcal{T}^{(456)\times(3)} & \mathcal{T}^{(456)\times(456)} \end{pmatrix}. \tag{6.5.3}$$

The four columns of these matrices describe the four experiments possible – an acoustic wave incident from the first medium, and three types of elastic waves from the second medium. On the left-hand side, the matrix $\mathbf{W}_1$ is the 2×2 acoustic, eigenvector matrix (as in equation (6.3.4)), and the coefficient matrix is 2×4 (retaining only the significant coefficients in the fluid). On the right-hand side, the eigenvector matrix is 2×6, and the coefficient matrix is 6×4, as only one source ray is possible in the fluid (the first column). The zero traction components are expressed as

$$\mathbf{0} = \mathbf{W}_2^{(45)\times(.)} \begin{pmatrix} \mathbf{0} & \mathbf{I} \\ \mathcal{T}^{(456)\times(3)} & \mathcal{T}^{(456)\times(456)} \end{pmatrix}, \tag{6.5.4}$$

which is 2×4. Expanding these equations and rearranging, we obtain

$$\mathcal{T}^{(3456)\times(3456)} = \begin{pmatrix} \mathbf{W}_1^{(12)\times(1)} & -\mathbf{W}_2^{(36)\times(456)} \\ \mathbf{0} & -\mathbf{W}_2^{(45)\times(456)} \end{pmatrix}^{-1} \times \begin{pmatrix} -\mathbf{W}_1^{(12)\times(2)} & \mathbf{W}_2^{(36)\times(123)} \\ \mathbf{0} & \mathbf{W}_2^{(45)\times(123)} \end{pmatrix}, \tag{6.5.5}$$

which in principle can be solved for the 4×4 matrix of coefficients. It involves inverting a 4×4 matrix, so an alternative solution is desirable.

The zero-traction equation (6.5.4) is a pair of equations connecting three coefficients (for each incident ray). This can be solved so that once one coefficient is known, the other two can be determined, e.g.

$$\mathcal{T}^{(45)\times(3456)} = \mathbf{W}_2^{(45)\times(45)^{-1}}\left(\left(\mathbf{0}\ -\mathbf{W}_2^{(45)\times(123)}\right) - \mathbf{W}_2^{(45)\times(6)}\mathcal{T}^{(6)\times(3456)}\right), \tag{6.5.6}$$

determines the fourth and fifth rows from the sixth. The composite 2×4 matrix on the right-hand side is made up of a 2×1 zero vector, and a 2×3 matrix. The continuity equation (6.5.3) can then be expanded to separate the fourth and fifth rows of $\mathcal{T}$ so result (6.5.6) can be substituted. The result is

$$\mathcal{T}^{(36)\times(3456)} = \left(\mathbf{W}_1^{(12)\times(1)}\ \ \left(\mathbf{W}_2^{(36)\times(45)}\mathbf{W}_2^{(45)\times(45)^{-1}}\mathbf{W}_2^{(45)\times(6)} - \mathbf{W}_2^{(36)\times(6)}\right)\right)^{-1}$$

$$\left(-\mathbf{W}_1^{(12)\times(2)}\ \ \left(\mathbf{W}_2^{(36)\times(123)} - \mathbf{W}_2^{(36)\times(45)}\mathbf{W}_2^{(45)\times(45)^{-1}}\mathbf{W}_2^{(45)\times(123)}\right)\right), \tag{6.5.7}$$

which only requires the inversion of 2×2 matrices. Note both the matrices on the right-hand side are composite: the first matrix which must be inverted is 2×2, made up of two 2×1 sub-matrices; the second is 2×4 and has sub-matrices 2×1 and 2×3. Having solved equation (6.5.7) for the 2×4 coefficients, equation (6.5.6) gives the other 2×4 coefficients.

6.5.2 Isotropic coefficients

In an isotropic medium, the $P - SV$ and SH parts of the system separate. For SH rays, the fluid–solid interface is equivalent to a free interface, and the result in Section 6.4.3 applies, i.e. $\mathcal{T}_{55} = 1$. The remaining $P - SV$ equations are obtained from the general equations (6.5.6) and (6.5.7) by eliminating the second and fifth rows and columns from the second medium. Using equation (6.3.5) in the first medium, and equations (6.3.51) and (6.3.53) in the second medium, it is straightforward to find the coefficients $\mathcal{T}^{(346)\times(346)}$. The coefficients are

$$\boxed{\begin{aligned}
\mathcal{T}_{33} &= (4p^2 q_{\alpha 2} q_{\beta 2} + \Omega_2^2 - \rho\, q_{\alpha 2}/\beta_2^4 q_{\alpha 1})/\Delta^{PV}\\
\mathcal{T}_{44} &= (-4p^2 q_{\alpha 2} q_{\beta 2} + \Omega_2^2 + \rho\, q_{\alpha 2}/\beta_2^4 q_{\alpha 1})/\Delta^{PV}\\
\mathcal{T}_{66} &= (4p^2 q_{\alpha 2} q_{\beta 2} - \Omega_2^2 + \rho\, q_{\alpha 2}/\beta_2^4 q_{\alpha 1})/\Delta^{PV}\\
\mathcal{T}_{36} &= \mathcal{T}_{63} = F_{\alpha 1} F_{\alpha 2}\Omega_2/q_{\alpha 1}\mu_2\Delta^{PV}\\
\mathcal{T}_{46} &= \mathcal{T}_{64} = 2pF_{\alpha 2}F_{\beta_2}\Omega_2/\rho_2\Delta^{PV}\\
\mathcal{T}_{34} &= \mathcal{T}_{43} = -2pF_{\alpha 1}F_{\beta 2}q_{\alpha 2}/q_{\alpha 1}\mu_2\Delta^{PV},
\end{aligned}} \tag{6.5.8}$$

where

$$\boxed{\Delta^{PV} = 4p^2 q_{\alpha 2} q_{\beta 2} + \Omega_2^2 + \rho\, q_{\alpha 2}/\beta_2^4 q_{\alpha 1},} \tag{6.5.9}$$

and

$$\rho = \rho_1/\rho_2, \tag{6.5.10}$$

and we have used F from definitions (6.3.62). The denominator Δ^{PV} (6.5.9) is known as the Scholte radix, Δ^{Scholte}, as Scholte (1958) or pseudo-Rayleigh interface waves exist on a fluid–solid interface when

$$\Delta^{PV} = \Delta^{\text{Scholte}} = \Delta^{\text{Rayleigh}} + \rho\, q_{\alpha 2}/\beta_2^4 q_{\alpha 1} = 0 \tag{6.5.11}$$

(interface waves will be discussed in Section 9.1.5).

6.6 Interface polarization conversions

Throughout these two chapters on ray theory, we have treated each ray separately, i.e. we have assumed the ray expansion. If a receiver is on an interface, the incident and generated rays all have the same arrival time, and are coincident in the response. It is normally convenient to consider them together (although some processing techniques are specially designed to separate these signals, e.g. up–down separation, exploiting different field components, e.g. velocity and pressure, or differentials of components using multiple receivers). The simple polarizations for the individual waves are combined, and the polarization for the incident ray is replaced by an *interface polarization conversion*. In this section, we discuss the polarization conversions.

6.6.1 Acoustic coefficients

Using equation (6.3.4), we can solve for $\mathbf{w}$ on the interface. For completeness, we extend the vector $\mathbf{w}$ to include the tangential particle velocity, v_l. We consider an incident wave from the first medium. Thus the sum of the incident and reflected waves gives

$$\begin{aligned}\begin{pmatrix} v_l \\ v_n \\ -P \end{pmatrix} &= \frac{1}{\sqrt{2\rho_1 q_{\alpha 1}}}\begin{pmatrix} p \\ -q_{\alpha 1} \\ -\rho_1 \end{pmatrix} + \mathcal{T}_{11}\frac{1}{\sqrt{2\rho_1 q_{\alpha 1}}}\begin{pmatrix} p \\ q_{\alpha 1} \\ -\rho_1 \end{pmatrix} \\ &= \frac{\sqrt{2\rho_1 q_{\alpha 1}}}{\rho_2 q_{\alpha 1} + \rho_1 q_{\alpha 2}}\begin{pmatrix} p\rho_2/\rho_1 \\ -q_{\alpha 2} \\ -\rho_2 \end{pmatrix},\end{aligned} \tag{6.6.1}$$

on the interface in the first medium. In the second medium, the transmitted wave gives

$$\begin{pmatrix} v_l \\ v_n \\ -P \end{pmatrix} = \mathcal{T}_{21} \frac{1}{\sqrt{2\rho_2 q_{\alpha 2}}} \begin{pmatrix} p \\ -q_{\alpha 2} \\ -\rho_2 \end{pmatrix} = \frac{\sqrt{2\rho_1 q_{\alpha 1}}}{\rho_2 q_{\alpha 1} + \rho_1 q_{\alpha 2}} \begin{pmatrix} p \\ -q_{\alpha 2} \\ -\rho_2 \end{pmatrix}, \tag{6.6.2}$$

on the interface. Note the required continuity in the normal particle velocity, v_n, and pressure, P, and discontinuity in the tangential particle velocity, v_l. Of course, the particle velocity is no longer longitudinal. Similar results hold for a wave incident from the second medium.

The incident wave, given by the first vector in expression (6.6.1), has polarization, **g**. At the interface it is converted into the polarization given by the final vector in expression (6.6.1). It is convenient to denote this by a new vector, the interface polarization conversion, **h**. The incident ray with polarization **g** can be converted on the interface to the total polarization by replacing **g** by **h**, i.e.

$$\mathbf{g}_P = \frac{1}{\sqrt{2\rho_1 q_{\alpha 1}}} \begin{pmatrix} p \\ 0 \\ -q_{\alpha 1} \end{pmatrix} \to \mathbf{h}_P = \frac{\sqrt{2\rho_1 q_{\alpha 1}}}{\rho_2 q_{\alpha 1} + \rho_1 q_{\alpha 2}} \begin{pmatrix} p\rho_2/\rho_1 \\ 0 \\ -q_{\alpha 2} \end{pmatrix}, \tag{6.6.3}$$

where the components are in the interface coordinate system, i.e. with respect to the basis vectors $\hat{\mathbf{l}}$, $\hat{\mathbf{m}}$ and $\hat{\mathbf{n}}$. This result applies for a receiver located on the interface in the first medium. At a fluid interface, a different result applies in the second medium, i.e. from expression (6.6.2), because of the discontinuity in the tangential component.

We are often concerned with the response at a free surface. For instance, at a fluid free-surface (Section 6.4.1), the particle velocity is

$$\mathbf{h}_P = \frac{1}{\sqrt{2\rho_2 q_{\alpha 2}}} \begin{pmatrix} 0 \\ 0 \\ 2q_{\alpha 2} \end{pmatrix}, \tag{6.6.4}$$

where the components are in the interface coordinate system, i.e. with respect to the basis vectors $\hat{\mathbf{l}}$, $\hat{\mathbf{m}}$ and $\hat{\mathbf{n}}$.

6.6.2 Anisotropic coefficients

The response at an interface is given by

$$\mathbf{w} = \mathbf{W}_1 \begin{pmatrix} \boldsymbol{\mathcal{T}}_{11} \\ \mathbf{I} \end{pmatrix} = \mathbf{W}_2 \begin{pmatrix} \mathbf{0} \\ \boldsymbol{\mathcal{T}}_{21} \end{pmatrix}, \tag{6.6.5}$$

for a ray incident from medium 1. We can substitute for the coefficients (6.3.23) but not much simplification is possible. One possible expression for the conversion coefficient is

$$\mathbf{h} = \mathbf{W}_2^{(123)\times(456)} \boldsymbol{\mathcal{T}}_{21}, \tag{6.6.6}$$

where the matrix $\mathbf{h}$ is made of the three interface polarization conversions, i.e.

$$\mathbf{h} = \begin{pmatrix} \mathbf{h}_1 & \mathbf{h}_2 & \mathbf{h}_3 \end{pmatrix}, \tag{6.6.7}$$

where polarization $\mathbf{g}_I$ converts to $\mathbf{h}_I$. For an incident wave from medium 2, the conversion coefficients are

$$\mathbf{h} = \mathbf{W}_1^{(123)\times(456)} \boldsymbol{\mathcal{T}}_{12}. \tag{6.6.8}$$

At a free surface, the reflection coefficients are (6.4.4) and the conversion coefficients are

$$\mathbf{h} = \mathbf{W}_2^{(123)\times(123)} - \mathbf{W}_2^{(123)\times(456)} \mathbf{W}_2^{(456)\times(456)^{-1}} \mathbf{W}_2^{(456)\times(123)}. \tag{6.6.9}$$

6.6.3 *Isotropic coefficients*

In an isotropic medium, the anisotropic results simplify but the method is fundamentally the same. The results at a free surface are particularly straightforward. The free-surface conversion polarizations are

$$\mathbf{h}_1 = \mathbf{h}_V = \frac{1}{w_1 \mu \Delta^{\mathrm{Rayleigh}}} \begin{pmatrix} \Omega \\ 0 \\ -2pq_\alpha \end{pmatrix} \tag{6.6.10}$$

$$\mathbf{h}_2 = \mathbf{h}_H = \begin{pmatrix} 0 \\ 2w_2 \\ 0 \end{pmatrix} \tag{6.6.11}$$

$$\mathbf{h}_3 = \mathbf{h}_P = \frac{1}{w_3 \mu \Delta^{\mathrm{Rayleigh}}} \begin{pmatrix} 2pq_\beta \\ 0 \\ \Omega \end{pmatrix}, \tag{6.6.12}$$

where $\Delta^{\mathrm{Rayleigh}}$ was given by (6.4.6) and the terms are evaluated in medium 2. The particle velocity components are given with respect to the interface basis, i.e. $\hat{\mathbf{l}}$, $\hat{\mathbf{m}}$ and $\hat{\mathbf{n}}$, and the three polarizations are for an incident *SV*, *SH* and *P* wave, respectively. It is interesting to note that for an incident *SV* ray, the generated *P* ray may be beyond critical (q_α imaginary) and the free-surface conversion polarization (6.6.10) complex, causing a distortion of the incident pulse.

6.7 Linearized coefficients

Although it is straightforward to calculate exact reflection/transmission coefficients, it is often useful to consider approximations based on perturbation theory. Results for the coefficients can be linearized for small changes in the material properties. Thus the perturbed coefficients, $\boldsymbol{T} + \delta\boldsymbol{T}$, are calculated from a first-order Taylor expansion, i.e.

$$\delta\boldsymbol{T} \simeq \frac{\partial\boldsymbol{T}}{\partial\gamma_i}\delta\gamma_i, \tag{6.7.1}$$

where γ_i are medium parameters (elastic stiffnesses, density, velocity, impedances, etc. – a summation over parameters enumerated by the subscript i is implied in equation (6.7.1)). We refer to coefficients calculated using perturbations (6.7.1) as *linearized coefficients*. For notational simplicity, avoiding an extra subscript, we write the basic theory in this section in terms of the perturbation, $\delta\boldsymbol{T}$, assuming linearity. More rigorously, the equations should be for the partial differentials, $\partial\boldsymbol{T}/\partial\gamma_i$, and wherever there is a perturbation, a partial differential can be substituted. The specific choice of parameters is not important for the basic theory and will depend on the application. Within the linear approximation, differentials with respect to different parameters can be related by the normal rules of partial derivatives, so the choice of parameters is not critical to the theory. Nevertheless, the accuracy of the linear approximation will be altered with different choices of parameters, i.e. the magnitude of higher-order terms is altered. In some situations, the linear approximation may break down for certain parameters, e.g. the behaviour may be like a square root from a zero value, and these should be avoided. How to choose the optimum parameters to linearize the perturbation is an interesting question that is not discussed here.

Linearized coefficients can be used in various circumstances. The reference model used may be:

- a null interface, i.e. the perturbation coefficients are the coefficients for an interface with a small contrast;
- an isotropic interface, when the perturbation might be anisotropic and the perturbation coefficients are the difference between the weak anisotropic coefficients and the isotropic coefficients;
- or any perturbation from a completely general model, for a generalized inversion.

In principle it is straightforward to differentiate or perturb the coefficients, but as they are algebraically complicated it is worthwhile to follow a general procedure.

The isotropic results were first obtained by Scholte (1962) and have been given by Aki and Richards (1980, 2002, Section 5.2.6). Shuey (1985) has investigated these particularly for AVO studies to simplify the parameterization at small angles to a rigidity factor depending on the contrast in Poisson's ratio. Various authors (Bortfeld, 1961; Wright, 1986; Smith and Gidlow, 1987; Thomsen, 1990; Fatti, Vail, Smith, Strauss and Levitt , 1994) have presented alternative expressions sometimes by including constraints such as fixed Poisson's ratio. Recently many authors have extended the isotropic results to some anisotropic media. Zillmer, Gajewski and Kashtan (1997) followed an approach close to that used here for the coefficients at a strong-contrast interface with a weak anisotropic perturbation and Zillmer, Gajewski and Kashtan (1998) simplified this to a weak-contrast interface. Ruger (1997) solved the specific problem of a weak contrast and weak TIV or TIH media for qP rays, and Ruger (1998) extended this to describe the azimuthal variation due to weak orthorhombic media. Vavrycuk and Pšenčík (1998) solved the problem for general, weak anisotropy but just for the qP reflection coefficient, while Pšenčík and Vavrycuk (1998) extended this to the transmission coefficient and Jilek (2002) to $qPqS$ coefficients. Vavrycuk (1999) extended these to transmissions using an approach close to that used here. Finally Shaw and Sen (2004) have used the Born scattering integral method to obtain the linearized coefficients for weak anisotropy. We investigate the Born scattering integral method in Section 10.3 and suggest their method as Exercise 10.3.

For any system, the coefficients are found by solving the vector-matrix equation (6.3.18). However, if this equation is differentiated, the inversion of a 6×6 matrix is needed for the linearized coefficients. Instead, we consider the solution (6.3.23), where the matrix $\mathbf{Q}$ is given by equation (6.3.22) or (6.3.34). To differentiate (6.3.23), we need the differential of an inverse matrix, i.e.

$$(\mathbf{X}+\delta\mathbf{X})^{-1} = (\mathbf{I}+\mathbf{X}^{-1}\delta\mathbf{X})^{-1}\mathbf{X}^{-1} \simeq \mathbf{X}^{-1} - \mathbf{X}^{-1}\delta\mathbf{X}\,\mathbf{X}^{-1}. \qquad (6.7.2)$$

Thus

$$\delta\left(\mathbf{X}^{-1}\right) = -\mathbf{X}^{-1}\delta\mathbf{X}\,\mathbf{X}^{-1}. \qquad (6.7.3)$$

Applying this to equation (6.3.23), we obtain

$$\delta\boldsymbol{T} = \begin{pmatrix} \begin{matrix}\delta\mathbf{Q}_{12}\mathbf{Q}_{22}^{-1} - \\ \mathbf{Q}_{12}\mathbf{Q}_{22}^{-1}\delta\mathbf{Q}_{22}\mathbf{Q}_{22}^{-1}\end{matrix} & \begin{matrix}\delta\mathbf{Q}_{11} - \delta\mathbf{Q}_{12}\mathbf{Q}_{22}^{-1}\mathbf{Q}_{21} - \\ \mathbf{Q}_{12}\mathbf{Q}_{22}^{-1}\delta\mathbf{Q}_{21} + \mathbf{Q}_{12}\mathbf{Q}_{22}^{-1}\delta\mathbf{Q}_{22}\mathbf{Q}_{22}^{-1}\mathbf{Q}_{21}\end{matrix} \\ \\ \mathbf{Q}_{22}^{-1}\delta\mathbf{Q}_{22}\mathbf{Q}_{22}^{-1} & \begin{matrix}\mathbf{Q}_{22}^{-1}\delta\mathbf{Q}_{22}\mathbf{Q}_{22}^{-1}\mathbf{Q}_{21} - \\ \mathbf{Q}_{22}^{-1}\delta\mathbf{Q}_{21}\end{matrix} \end{pmatrix}. \qquad (6.7.4)$$

This equation gives the perturbed coefficients in terms of perturbations to the matrix $\mathbf{Q}$. Next we need to relate this to perturbations in the eigenvectors.

Differentiating equation (6.3.22), we obtain

$$\delta\mathbf{Q} = \mathbf{W}_1^{-1}\delta\mathbf{W}_2 - \mathbf{W}_1^{-1}\delta\mathbf{W}_1\mathbf{W}_1^{-1}\mathbf{W}_2 \quad (6.7.5)$$
$$= \delta\mathbf{C}_1\mathbf{Q} - \mathbf{Q}\,\delta\mathbf{C}_2, \quad (6.7.6)$$

where

$$\boxed{\delta\mathbf{C} = -\mathbf{W}^{-1}\delta\mathbf{W} = -\mathbf{I}_3\mathbf{W}^{\mathrm{T}}\mathbf{I}_2\,\delta\mathbf{W},} \quad (6.7.7)$$

if the appropriate normalization (6.3.29) is used so (6.3.32) applies. Equation (6.7.6) gives the perturbation to the matrix $\mathbf{Q}$, in terms of the perturbation to the eigenvectors, $\mathbf{W}$. Next, we need to relate the perturbation to the eigenvectors, $\delta\mathbf{W}$, or rather the perturbation matrix, $\delta\mathbf{C}$ (6.7.7), to perturbations of the matrix $\mathbf{A}$, defined in equations (6.3.15)–(6.3.17).

The symmetry properties of the matrix $\delta\mathbf{C}$ (6.7.7) are easily established using the orthonormality conditions of $\mathbf{W}$. Differentiating expression

$$\mathbf{W}^{-1}\mathbf{W} = \mathbf{I}_3\mathbf{W}^{\mathrm{T}}\mathbf{I}_2\mathbf{W} = \mathbf{I}, \quad (6.7.8)$$

we have

$$\mathbf{I}_3\,\delta\mathbf{W}^{\mathrm{T}}\mathbf{I}_2\mathbf{W} + \mathbf{I}_3\mathbf{W}^{\mathrm{T}}\mathbf{I}_2\,\delta\mathbf{W} = \mathbf{0}. \quad (6.7.9)$$

Rearranging, we have

$$\mathbf{I}_3\,\delta\mathbf{C} = -\left(\mathbf{W}^{\mathrm{T}}\mathbf{I}_2\,\delta\mathbf{W}\right) = \left(\mathbf{W}^{\mathrm{T}}\mathbf{I}_2\,\delta\mathbf{W}\right)^{\mathrm{T}}, \quad (6.7.10)$$

i.e. an anti-symmetric matrix. Dividing $\delta\mathbf{C}$ into its sub-matrices, this implies that the diagonal blocks are anti-symmetric, i.e. $\delta\mathbf{C}_{11} = -\delta\mathbf{C}_{11}^{\mathrm{T}}$ and $\delta\mathbf{C}_{22} = -\delta\mathbf{C}_{22}^{\mathrm{T}}$, and the off-diagonal blocks are symmetrically related, i.e. $\delta\mathbf{C}_{12} = \delta\mathbf{C}_{21}^{\mathrm{T}}$.

It is not necessary to perturb the eigenvectors explicitly to obtain $\delta\mathbf{W}$, as we can perturb the eigen-equation (6.3.3). Then the matrix $\delta\mathbf{C}$ can be obtained directly in terms of model perturbations rather than the eigenvectors (which in the case of anisotropy, we only know numerically). Perturbing the eigen-equation (6.3.3), we have

$$\delta\mathbf{A}\,\mathbf{W} + \mathbf{A}\,\delta\mathbf{W} = \delta\mathbf{W}\,\mathbf{p}_n + \mathbf{W}\,\delta\mathbf{p}_n. \quad (6.7.11)$$

Multiplying by $\mathbf{W}^{-1}$, rearranging using equation (6.3.3) and the definition of $\delta\mathbf{C}$ (6.7.7), we obtain

$$\mathbf{W}^{-1}\delta\mathbf{A}\,\mathbf{W} - \delta\mathbf{p}_n = [\mathbf{p}_n\,,\ \delta\mathbf{C}]\,, \quad (6.7.12)$$

where we have used the shorthand notation

$$[\mathbf{A}\,,\,\mathbf{B}] = \mathbf{A}\,\mathbf{B} - \mathbf{B}\,\mathbf{A}, \tag{6.7.13}$$

for the commutator. Now the diagonal terms in the commutator (6.7.12) are always zero

$$\mathrm{diag}\,[\mathbf{p}_n, \delta\mathbf{C}] = \mathbf{0}, \tag{6.7.14}$$

so

$$\delta\mathbf{p}_n = \mathrm{diag}\left(\mathbf{W}^{-1}\delta\mathbf{A}\,\mathbf{W}\right) = \mathrm{diag}\left(\mathbf{I}_3\mathbf{W}^{\mathrm{T}}\mathbf{I}_2\,\delta\mathbf{A}\,\mathbf{W}\right), \tag{6.7.15}$$

giving the perturbation to the eigenvalues. From the off-diagonal terms, we have

$$\text{off-diag}\,[\mathbf{p}_n, \delta\mathbf{C}] = \text{off-diag}\left(\mathbf{W}^{-1}\,\delta\mathbf{A}\,\mathbf{W}\right) = \text{off-diag}\left(\mathbf{I}_3\mathbf{W}^{\mathrm{T}}\mathbf{I}_2\,\delta\mathbf{A}\,\mathbf{W}\right), \tag{6.7.16}$$

and the off-diagonal elements ($i \neq j$) of $\delta\mathbf{C}$ are

$$\boxed{\delta C_{ij} = \left(\mathbf{W}^{-1}\,\delta\mathbf{A}\,\mathbf{W}\right)_{ij} \Big/ (p_i - p_j) = (\mathbf{I}_3\mathbf{W}^{\mathrm{T}}\mathbf{I}_2\,\delta\mathbf{A}\,\mathbf{W})_{ij} / (p_i - p_j).} \tag{6.7.17}$$

We have already established that the diagonal elements are zero as the diagonal blocks of $\delta\mathbf{C}$ are anti-symmetric. This establishes the connection between the perturbation of the matrix $\mathbf{A}$ with the required perturbation matrix, $\delta\mathbf{C}$. Finally, we can connect the perturbation of the matrix $\mathbf{A}$ with the elastic parameters.

Using result (6.7.3), for the sub-matrices (6.3.15)–(6.3.17), we obtain

$$\boxed{\delta\mathbf{A}_{12} = -\delta\left(\mathbf{c}_{33}^{-1}\right) = \mathbf{c}_{33}^{-1}\delta\mathbf{c}_{33}\,\mathbf{c}_{33}^{-1}.} \tag{6.7.18}$$

The other sub-matrices reduce to

$$\boxed{\begin{aligned} \delta\mathbf{A}_{22} &= \delta\mathbf{A}_{11}^{\mathrm{T}} = -p_\eta\left(\delta\mathbf{c}_{\eta 3}\,\mathbf{c}_{33}^{-1} - \mathbf{c}_{\eta 3}\,\delta\mathbf{A}_{12}\right) \\ \delta\mathbf{A}_{21} &= p_\eta\,p_\nu\,\delta\mathbf{c}_{\eta\nu} - \delta\rho\,\mathbf{I} \\ &\quad -p_\eta\,p_\nu\left(\delta\mathbf{c}_{\eta 3}\,\mathbf{c}_{33}^{-1}\,\mathbf{c}_{3\nu} - \mathbf{c}_{\eta 3}\,\delta\mathbf{A}_{12}\,\mathbf{c}_{3\nu} + \mathbf{c}_{\eta 3}\,\mathbf{c}_{33}^{-1}\,\delta\mathbf{c}_{3\nu}\right). \end{aligned}} \tag{6.7.19}$$

Overall, we now have a complete description of the perturbation of the reflection/transmission coefficients, in terms of the perturbations to the model parameters, the elastic stiffness matrices, $\delta\mathbf{c}_{ij}$ (4.4.39) and the density, ρ. Equations (6.7.18) and (6.7.19) are substituted into equation (6.7.17). This in turn is substituted into equation (6.7.6) and finally this is substituted into equation (6.7.4). Although the algebra is extensive, it is all linear and straightforward in principle.

The differentials of the reflection/transmission coefficients with respect to the model parameters, the elastic stiffness matrices, $\delta\mathbf{c}_{ij}$ (4.4.39) and the density, ρ, in the two media, can be extracted by considering each parameter in turn, i.e. substituting $\delta\mathbf{c}_{jk} \to \partial\mathbf{c}_{jk}/\partial\gamma_i$ and $\delta\rho \to \partial\rho/\partial\gamma_i$ in equations (6.7.18) and (6.7.19). In general, no particular algebraic simplification is possible, so we do not make these substitutions. The necessary linear algebra is easily evaluated numerically. Although for any particular parameter, many of these terms will be zero, it is important to remember that the same parameter may be repeated in several terms, e.g. C_{45} appears in $\mathbf{c}_{33}$ twice, and in $\mathbf{c}_{23}$, $\mathbf{c}_{32}$, $\mathbf{c}_{31}$, $\mathbf{c}_{13}$, $\mathbf{c}_{12}$ and $\mathbf{c}_{21}$. In anisotropic media with symmetries, the number of parameters is reduced and they appear in various terms, e.g. in the isotropic case, all the matrices $\mathbf{c}_{jk}$, (4.4.55) and (4.4.56), depend on the Lamé parameters, λ and μ. Within the linearized approximation, it is straightforward to convert the perturbations or differentials with respect to the elastic stiffnesses and density, (6.7.18) and (6.7.19), to any other elastic parameter, e.g. velocity, impedance, etc., by introducing one more level of linear substitution. The directional dependence of the perturbation coefficients enters through the factors of p_η in the terms (6.7.18) and (6.7.19), and through the eigenvectors $\mathbf{W}$, etc. in expression (6.7.17).

For a null interface, i.e. in the reference medium, we have no contrast so the eigen-matrices are equal, $\mathbf{W}_1 = \mathbf{W}_2$. The reflection coefficients are zero and the transmission coefficients unity, i.e. $\boldsymbol{\mathcal{T}} = \mathbf{I}_2$. The matrix $\mathbf{Q}$ is the identity matrix so equation (6.7.6) reduce to

$$\delta\mathbf{Q} = \delta\mathbf{C}_1 - \delta\mathbf{C}_2 = [\delta\mathbf{C}] = -\mathbf{W}^{-1}[\mathbf{W}], \tag{6.7.20}$$

where $[\mathbf{W}] = \delta\mathbf{W}_1 - \delta\mathbf{W}_2 = \mathbf{W}_1 - \mathbf{W}_2$, the eigenvector saltus across the interface (in the direction of increasing normal coordinate, n). The perturbation coefficients reduce to

$$\delta\boldsymbol{\mathcal{T}} = \begin{pmatrix} \delta\mathbf{Q}_{12} & \delta\mathbf{Q}_{11} \\ -\delta\mathbf{Q}_{22} & -\delta\mathbf{Q}_{21} \end{pmatrix} = -\mathbf{I}_3\delta\mathbf{Q}\mathbf{I}_2 = \mathbf{W}^{\mathrm{T}}\mathbf{I}_2[\mathbf{W}]\mathbf{I}_2. \tag{6.7.21}$$

This simple result can be used to approximate the reflection/transmission coefficients at any small contrast interface, be it isotropic or anisotropic. Alternative expressions for $\delta\boldsymbol{\mathcal{T}}$ in terms of $[\mathbf{A}]$, or $[\rho]$ and $[\mathbf{c}_{jk}]$ (or Born perturbations ρ^{B} and $\mathbf{c}^{\mathrm{B}}_{jk}$ – see Section 10.3.3.4), are given in Exercise 10.3.

6.7.1 Small-contrast acoustic coefficients

In the previous section, we have developed the linear algebra system for the perturbation coefficients. It is straightforward to numerically compute the linear equations connecting perturbations of the coefficients with perturbations in the model.

In this section and the next, we investigate the small-contrast results (6.7.21) for acoustic and isotropic elastic media, where the algebra is relatively simple and explicit expressions can be obtained.

For acoustic media, the eigenvectors (6.3.5) can be perturbed and it is straightforward to evaluate the expression (6.7.21). The result is

$$\boxed{\boldsymbol{\mathcal{T}} = \mathbf{I}_2 + \delta\boldsymbol{\mathcal{T}} = \begin{pmatrix} [\gamma_A] & 1 \\ 1 & -[\gamma_A] \end{pmatrix},} \tag{6.7.22}$$

where

$$\boxed{[\gamma_A] = \frac{[q_\alpha]}{2q_\alpha} - \frac{[\rho]}{2\rho} = -\frac{[\alpha]}{2\alpha}\sec^2\theta - \frac{[\rho]}{2\rho}.} \tag{6.7.23}$$

In this expression (6.7.23), θ is the ray angle to the normal to the interface. It is, of course, trivial to show that this result can be obtained using result (6.7.21) or by approximation of the exact coefficients (6.3.7) and (6.3.8).

6.7.2 Small-contrast isotropic coefficients

The isotropic coefficients separate into the $P - SV$ and SH systems. Again it is simple to perturb the eigenvectors (6.3.52) and obtain the SH results. They are

$$\boxed{\boldsymbol{\mathcal{T}}^{(25)\times(25)} = \mathbf{I}_2 + \delta\boldsymbol{\mathcal{T}} = \begin{pmatrix} [\gamma_H] & 1 \\ 1 & -[\gamma_H] \end{pmatrix},} \tag{6.7.24}$$

where

$$\boxed{[\gamma_H] = \frac{[q_\beta]}{2q_\beta} + \frac{[\mu]}{2\mu} = -\frac{[\beta]}{2\beta}\sec^2\theta + \frac{[\mu]}{2\mu}.} \tag{6.7.25}$$

The $P - SV$ system is algebraically more complicated, but again it is straightforward to perturb the eigenvectors (6.3.51) and (6.3.53). The result for the coefficients is

$$\boxed{\boldsymbol{\mathcal{T}}^{(1346)\times(1346)} = \begin{pmatrix} -[\gamma_S] & [\gamma_R] & 1 & [\gamma_T] \\ [\gamma_R] & [\gamma_P] & -[\gamma_T] & 1 \\ 1 & -[\gamma_T] & [\gamma_S] & -[\gamma_R] \\ [\gamma_T] & 1 & -[\gamma_R] & -[\gamma_P] \end{pmatrix},} \tag{6.7.26}$$

where

$$\boxed{\begin{aligned}[\gamma_P] &= [q_\alpha]/2q_\alpha - [\rho]/2\rho + 2\beta^2 p^2 [\mu]/\mu \\ [\gamma_S] &= [q_\beta]/2q_\beta - [\rho]/2\rho + 2\beta^2 p^2 [\mu]/\mu \\ [\gamma_R] &= p(q_\alpha q_\beta)^{-1/2} \left(\beta^2(p^2 - q_\alpha q_\beta) [\mu]/\mu - [\rho]/2\rho\right) \\ [\gamma_T] &= p(q_\alpha q_\beta)^{-1/2} \left(\beta^2(p^2 + q_\alpha q_\beta) [\mu]/\mu - [\rho]/2\rho\right).\end{aligned}} \tag{6.7.27}$$

These perturbation coefficients are widely used for AVO (amplitude *versus* offset) studies, and reference is normally made to Aki and Richards (1980, 2002). They in turn reference Chapman (1976), but a more appropriate and much earlier reference is Scholte (1962). It is particularly elegant that only four expressions (6.7.27) are needed: $[\gamma_P]$ for reflected P rays; $[\gamma_S]$ for reflected SV rays; $[\gamma_R]$ for all converted reflections; and $[\gamma_T]$ for all converted transmissions.

Many different forms have been published for the perturbation coefficients, $[\gamma]$, from a small discontinuity, e.g. Aki and Richards (1980, 2002, Section 5.2.6). Although they are all equivalent to first-order, some subject to constraints reducing the number of parameters, they differ numerically. The accuracy of a first-order Taylor expansion depends on the parameters used.

6.8 Geometrical Green dyadic with interfaces

6.8.1 Ray signature and tables

In order to generate the geometrical Green dyadic for a ray in a model with discontinuities, it is necessary to define the ray signature, i.e. a description of which ray type to generate each time a ray intersects an interface (in general in an anisotropic medium, from a choice of six). The *ray signature* is used to characterize a set of rays with similar properties. It is sometimes called the *ray code*.

A ray consists of a sequence of segments between events at interfaces. The *ray event* may be a reflection or a transmission between specified ray types. Each ray segment will have a specific ray type, specifying which eigenvalue should be used in the eikonal equation, i.e. subscript I in equation (5.3.19) ($I = 1$ to 3). In the case of shear waves in an isotropic medium, the polarization solution (5.6.8) is used, with initial conditions defined at the source or at an event. A *ray event* joins two segments and must be compatible with the segment types.

For each ray segment, degenerate situations in which the choice of eigenvalue is ambiguous are a problem but this goes deeper than the signature and the choice of eigen-solution. Degeneracies occur in anisotropic media when two eigenvalues are equal or near equal. This commonly occurs for quasi-shear waves. In isotropic media, the shear waves are always degenerate and an alternative method is used (Section 5.6.1 and equation (5.6.8)). In anisotropic media, degeneracies can occur at isolated points and directions. At degeneracies, ray theory breaks down (see Section 5.4.1 – as the degeneracy is approached, $G_N \to 0$, and the amplitude coefficients diverge – see equation (5.4.4)). The two degenerate rays no longer propagate independently and at (and near) the degeneracy, couple together (see quasi-isotropic ray theory in Section 10.2). Generalizing the concept of events to include 'coupling events', and dividing segments at such events, the signature remains adequate to control the ray tracing. The ray signature must allow the segment types and events to be determined. A ray is traced by specifying its signature, and source point and direction (in anisotropic media it is simplest to specify the initial slowness direction, $\hat{\mathbf{p}}$, as the ray direction, $\widehat{\mathbf{V}}$, can be determined from this). Each segment is traced by solving the kinematic ray equations (5.1.29) from its initial point (the source or an event) until the ray intersects an interface. Assuming the interface is compatible with the event specified in the signature, the event can be solved using Snell's law (Section 6.2.1), and a new segment traced from the event with the new direction and type.

As well as describing how a ray should be traced, the signature can also be used to characterize a set of rays. An individual ray is a mathematical convenience and has no physical significance. It is used to describe the propagation direction between wavefronts. Wavefronts must have a finite area (if they don't, ray theory wouldn't be valid anyway), and a ray exists through every point on the wavefront. In practice a finite number of rays are traced to describe the properties on the wavefront. Normally the rays will fill a cone (solid angle) at the source. Directions, $\hat{\mathbf{p}}$, within this cone result in rays which match the signature; rays outside the cone fail due to a mismatch with the signature. Determining the boundary of this cone may be a difficult practical problem but in principle it must exist. At any point on a wavefront, results must be determined by interpolation between the discrete rays traced. The set of rays defining one wavefront form a *ray table*, all with the same signature. Rays with different signatures must be in different tables as it would be foolish to interpolate between rays with different properties. These concepts are illustrated in Figure 6.10.

In a general, three-dimensional, heterogeneous model with non-planar interfaces, the ray signature must specify the segment types and interfaces. It is sensible to assign to each interface an identifier and an orientation, i.e. a positive and a negative side. It does not matter how the identifiers are generated provided they are unique. It simplifies the signature if each surface separates only two volumes, i.e. a unique volume lies on the positive side, and a unique volume lies on the negative side. If physically a surface separates several volumes, e.g. a normal fault cutting many layers, it is convenient to divide the surface into parts and assign a different identifier to each part. This is illustrated in Figure 6.11. Only in a flat layered model, is it a trivial problem to set up the interface identifiers (Figure 0.1).

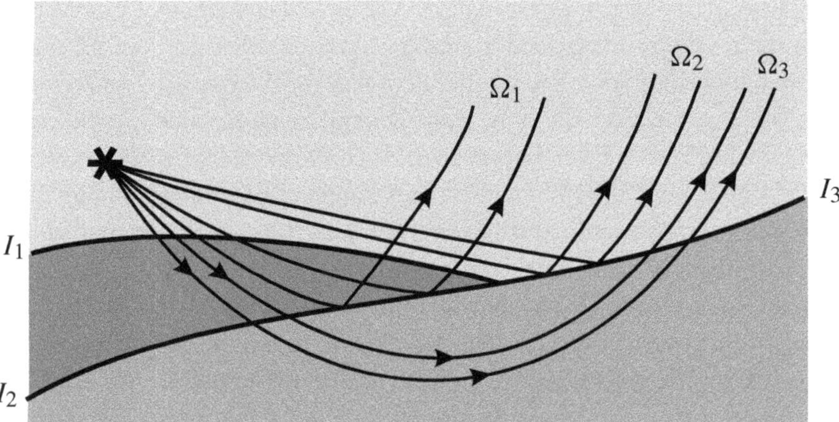

Fig. 6.10. Rays divided into three tables due to different events on the surface. The different ray cones are indicated by Ω_1, Ω_2 and Ω_3. The model has interfaces I_1, I_2 and I_3: rays in cone Ω_1 are reflected from interface I_2 with transmissions through interface I_1; rays in cone Ω_2 are reflected from interface I_3; and rays in cone Ω_3 are transmitted through interfaces I_1, I_2 and I_3.

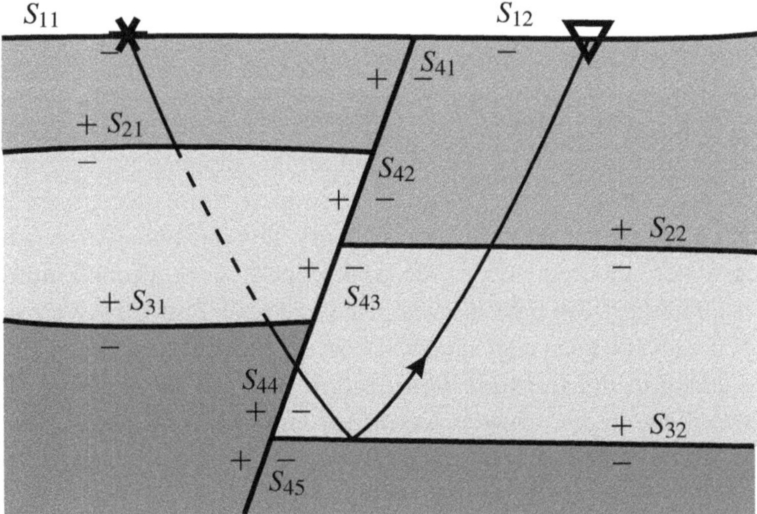

Fig. 6.11. The cross-section of a model with three layers and a fault, where each part of the surfaces is given a unique identifier, $S_\ell = S_{mn}$. A P ray with a converted segment is illustrated. The signature would be $(+S_{21}, \mathcal{T}_{43})$, $(+S_{31}, \mathcal{T}_{61})$, $(+S_{44}, \mathcal{T}_{63})$, $(+S_{32}, \mathcal{T}_{33})$, $(-S_{22}, \mathcal{T}_{36})$ where each event is described by the signed surface identifier, $\pm S_\ell$, and the event code, i.e. the indices ij where $\mathcal{T}_{ij}$ is the appropriate coefficient in the coefficient matrix, $\boldsymbol{\mathcal{T}}$ (6.3.23).

A simple way to define a ray signature is to have a sequence of ray events. Each event will be a (signed) interface identifier and a coefficient type. As the sign of the interface identifier specifies the incident side of the surface, only coefficients from one side are needed, and in the most general, anisotropic case, the coefficient type is one of 18. The coefficient types will specify the ray type of the segments on either side of the event. Obviously there is redundant information in this ray signature, as neighbouring events must have consistent ray types. Alternative methods of specifying the ray signature exist using the volume identifiers and ray types directly, but the sequence of events is convenient. It allows a compressed form of the signature to be used to control ray tracing, in which default events defined as unconverted transmissions can be omitted. The ray signature for a ray is illustrated in Figure 6.11. The ray signature consists of a sequence of signed surface identifiers, e.g. $\pm S_\ell$, and event codes, e.g. the indices ij of the appropriate reflection/transmission coefficient $\mathcal{T}_{ij}$ in the coefficient matrix (6.3.23). Default events can be omitted from the sequence, e.g. if $\mathcal{T}_{ij}$ is an unconverted transmission. In the illustration, the surface index, ℓ, of the surface identifier, $S_\ell = S_{mn}$, is made up of two parts, an interface index, m, and a sub-part, n. As a shorthand, we assume that the signature is contained in the ray descriptor, $\mathcal{L}_n$, used in the ray ansatz (5.1.1) or (5.3.1). Thus typically the ray descriptor may be the source position and direction, plus the signature

$$\mathcal{L}_n = \left\{\mathbf{x}_S, \hat{\mathbf{p}}_S, \{S_\ell, \mathcal{T}_{ij}\}\right\}. \tag{6.8.1}$$

The definition of an event may depend on the application. Normally, interface events occur at surfaces where the material properties are discontinuous. However, if we include higher-order terms in ray theory, this must be generalized to include interfaces where the gradients of material properties are discontinuous (a second-order discontinuity). We have already discussed how if ray theory breaks down due to a degeneracy, or near degeneracy, of the qS eigenvalues, then coupling events must be added to the signature (see Section 10.2 for the coupling theory). With zeroth-order ray theory, caustics would normally be considered as events too, as we cannot interpolate through a caustic, but if we use the generalization of Maslov asymptotic ray theory (see Section 10.1), then caustics can be excluded.

As an example of ray tracing in a three-dimensional model, we consider a simple model that has been widely used to test modelling and migration algorithms, the *French model*. The model is based on an early physical model experiment by French (1974) used to demonstrate two- and three-dimensional migration. We shall use this same model in Chapter 10 to demonstrate the Born and Kirchhoff methods of modelling. An idealization of the physical model is shown in Figure 6.12. This diagram is drawn to scale. Basically the model consists of two interfaces. The first has a plane, horizontal interface at two levels separated by a planar ramp, with two

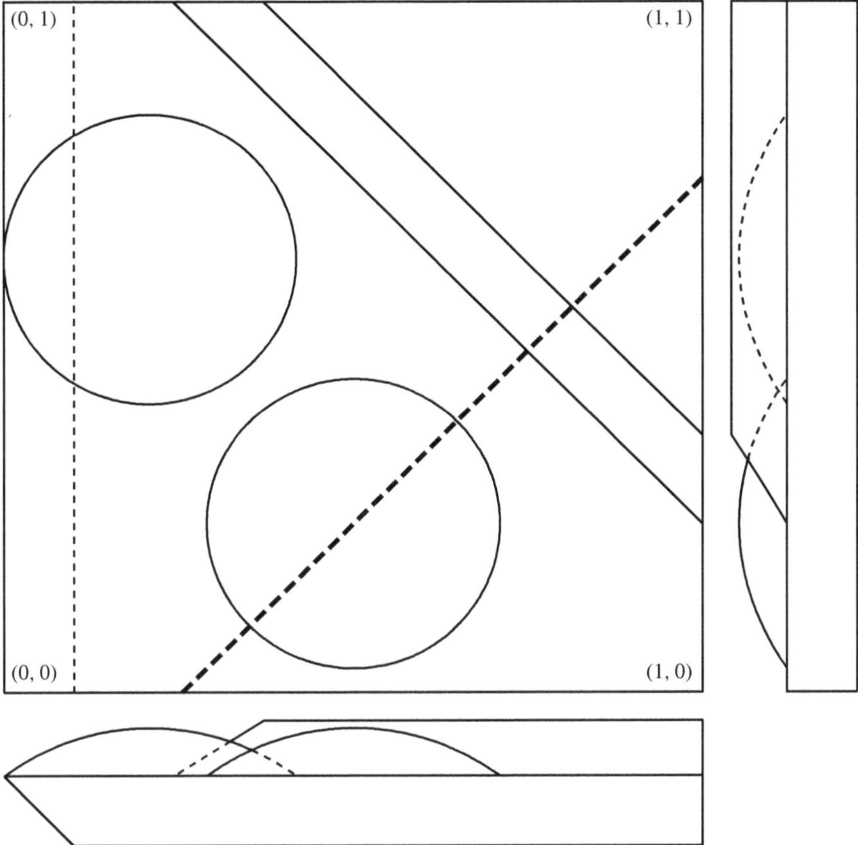

Fig. 6.12. The French model (French, 1974). The diagrams are drawn to scale to idealize the physical model in the original publication. The main diagram is a plan view of the model, and on the right and below are side views. In the calculations, the model is normalized so the plan view is a unit square. The heavy-dashed line indicates the profile used in Figures 6.13, 6.14 and 6.15.

domes that are parts of spheres on the lower level. The second interface is plane and horizontal apart from a ramp at the edge. The numerical model has been scaled so that the horizontal extent is a unit square (the physical model was 9.848 inches square). The physical model measurements are given in French (1974) or have been estimated from diagrams in that paper. In the normalized numerical model, the source and receiver are taken as $z_S = z_R = 0$. The upper interface has horizontal planes at $z \simeq -0.417$ and $z \simeq -0.497$. The horizontal plane of the lower interface is at $z \simeq -0.599$. The velocity in the upper layer is normalized, $\alpha_1 = 1$. The velocity in the intermediate layer is taken as $\alpha_2 = 1.1$ and in the lower half-space as $\alpha_3 = 1.2$.

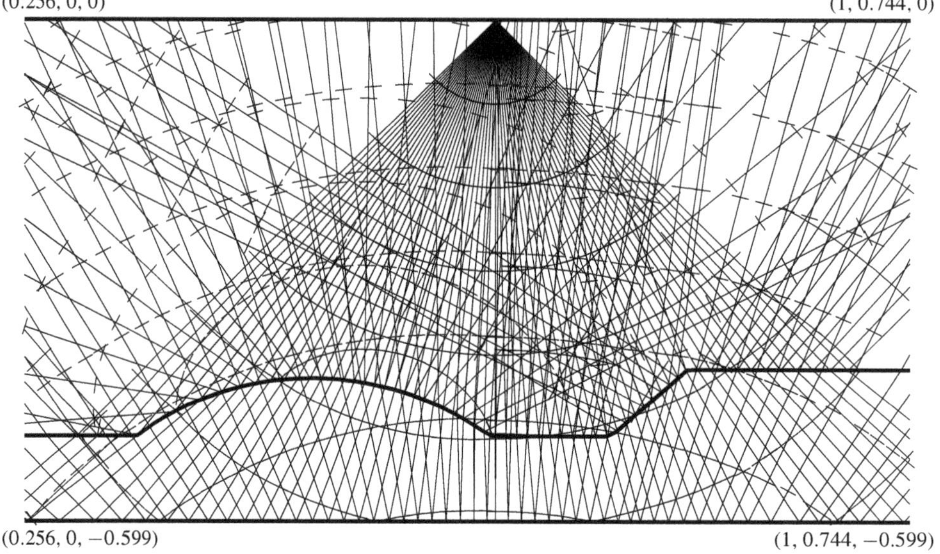

Fig. 6.13. Rays traced in the normalized French model (French, 1974). The source is at $\mathbf{x}_S \simeq (0.653,\ 0.397,\ 0)$ and the rays have been traced in the profile indicated in Figure 6.12. The rays have been traced in a uniform fan with take-off angles from $-45°$ to $+45°$ in $1.5°$ intervals. For clarity we have taken a local symmetry plane of the model, so the ray tracing is two dimensional. The left-hand top corner of the plot is at $\mathbf{x} \simeq (0.256,\ 0,\ 0)$ and the right-hand top corner is at $\mathbf{x} \simeq (1,\ 0.744,\ 0)$. The wavefronts are marked at intervals of 0.1.

In Figure 6.13, rays traced in the three-dimensional French model are illustrated. These have been traced from a source at $\mathbf{x}_S = (0.653,\ 0.397,\ 0)$ on the profile indicate in Figure 6.12. This profile runs over the centre of one of the domes and is perpendicular to the ramp. Locally the model is symmetric about this profile, and rays traced in this plane are essentially two dimensional. For clarity, we only illustrate rays traced in the plane of the profile, although it is straightforward to trace rays in any direction. In Figure 6.13 we have only included primary reflections from the two interfaces.

The second ray-tracing plot in Figure 6.14 is of *zero-offset rays* or *normal rays*. For each ray, the source and receiver are coincident, $\mathbf{x}_S = \mathbf{x}_R$, on the surface $z_R = 0$. At the reflecting interface, the rays are normal to the interface so the reflected ray re-traces the incident ray path. Rays have been traced so the reflecting points are uniformly distributed along the profile.

Finally we plot in Figure 6.15 the one-way travel time for the zero-offset rays in Figure 6.14. As the velocity is normalized ($\alpha_1 = 1$) the vertical one-way travel time maps directly into the depth. Notice that the one-way travel time partly indicates the model profile but features are distorted and displaced. Reflections from

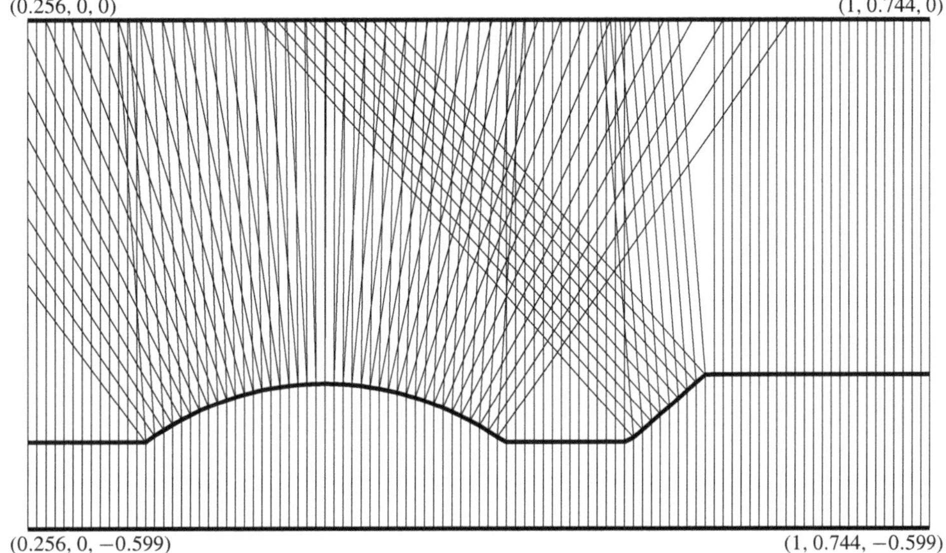

Fig. 6.14. Zero-offset or normal rays traced in the French model (French, 1974) along the profile indicated in Figure 6.12 (as used in Figure 6.13). For clarity, the wavefronts have not been marked.

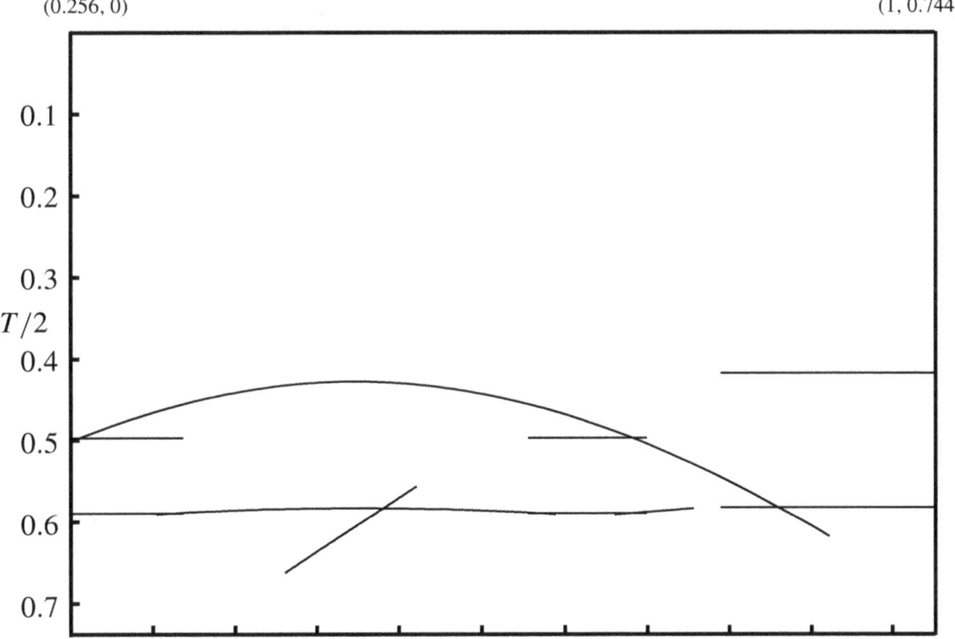

Fig. 6.15. The one-way travel time, $T/2$, for the zero-offset rays illustrated in Figure 6.14 on the profile indicated in Figure 6.12.

the top, plane, horizontal interface are accurately positioned. The centre of the dome is at the correct depth, but the travel-time dome is much wider that the true dome. Non-horizontal surfaces are misplaced. More dramatically, because the ramp is steep, its reflection is completely misplaced. The lower, horizontal interface is flat on this profile, but is misplaced in the travel-time plot. Because the velocity in the layer is higher ($\alpha_2 = 1.1$), the interface is misplaced and is too shallow. In regions below the higher features on the top surface, the pull-up is larger, and the reflector appears non-planar. Migration is the process whereby these reflectors are moved to the correct positions of the reflectors, although this subject is beyond the scope of this book.

6.8.2 Green Dyadic with interfaces

Once the ray signature is specified, with or without default events, together with the source position and direction, then the ray can be traced. In order to determine the Green dyadic, the amplitude coefficient must be modified to contain the product of the appropriate reflection/transmission coefficients for the ray. Thus the geometrical part of the propagation term in the Green dyadic, (5.4.34), is modified

$$\boxed{\mathcal{T}^{(\ell)}(\mathbf{x}_{\mathsf{R}}, \mathcal{L}_n) = \mathcal{S}^{(\ell)^{-1/2}} \mathrm{e}^{-\mathrm{i}\pi\,\mathrm{sgn}(\omega)\sigma(\mathbf{x}_{\mathsf{R}}, \mathcal{L}_n)/2} \left(\prod_k {}_{\ell_k}\mathcal{T}_{i_k j_k} \right),} \tag{6.8.2}$$

where $\left(\prod_k {}_{\ell_k}\mathcal{T}_{i_k j_k}\right)$ is the product of reflection/transmission coefficients along the ray. In the time domain, we obtain

$$\boxed{\underline{\mathbf{u}}(t, \mathbf{x}_{\mathsf{R}}; \mathbf{x}_{\mathsf{S}}) = \frac{1}{2\pi} \mathrm{Re}\left(\mathcal{T}^{(3)}(\mathbf{x}_{\mathsf{R}}, \mathcal{L}_n) \Delta\left(t - T(\mathbf{x}_{\mathsf{R}}, \mathcal{L}_n)\right)\right) \mathbf{g}(\mathbf{x}_{\mathsf{R}}, \mathcal{L}_n) \mathbf{g}^{\mathrm{T}}(\mathbf{x}_{\mathsf{S}}, \mathcal{L}_n),} \tag{6.8.3}$$

the complete geometrical ray approximation for the Green function. The types of coefficients, ${}_{\ell_k}\mathcal{T}_{i_k j_k}$, depend on the ray signature, but the values depend on the ray direction at the interfaces. While the physical concept of the ray signature is straightforward and easy to describe, it is difficult to develop a comprehensive notation. For simplicity, we assume that the signature is defined by the ray descriptor, $\mathcal{L}_n$. The sequence of events is enumerated by the index k in equation (6.8.2). The pre-subscript on the coefficients, ℓ_k, are the interface identifiers in the events. The subscripts $i_k j_k$ are the coefficient identifier. The values of i_k and j_k will determine the coefficient type, whether the event is a reflection or transmission, and therefore whether the ray changes volume. The ray type of the preceding ray segment must match j_k, and of the next segment, i_k. These ray types must be compatible between neighbouring events, and determine the eigenvalue used to define the Hamiltonian and hence to find the ray path and travel time, T. Finally, the ray signature may also contain caustic events specifying on which segments σ may change. Note that when a coefficient is complex, e.g. result (6.3.11), i.e. total reflection occurs, the geometrical pulse is a combination of the source pulse and its Hilbert transform (6.8.3).

Note that the coefficients are calculated with respect to normalized eigenvectors (e.g. acoustic eigenvectors (6.3.5), anisotropic eigenvectors (6.3.29), or isotropic eigenvectors (6.3.51), (6.3.52) and (6.3.53)), so this factor must be accounted for at events, i.e. an incident ray with unit polarization converts to a generated ray with polarization as

$$\hat{\mathbf{g}}_j \rightarrow \frac{\sqrt{(\rho V_n)_j}}{\sqrt{(\rho V_n)_i}} \mathcal{T}_{ij}\, \hat{\mathbf{g}}_i, \tag{6.8.4}$$

or with the polarization (6.3.26)

$$\mathbf{g}_j \rightarrow \mathcal{T}_{ij}\, \mathbf{g}_i \tag{6.8.5}$$

(with no summation over subscripts). It is very important to remember that reflection/transmission coefficients are defined with respect to a basis solution, e.g. the eigenvectors (6.3.5), (6.3.29), (6.3.51), (6.3.52) or (6.3.53). Other choices of basis are possible, e.g. as used by Knott (1899), normalized with respect to potential, or Zoeppritz (1919), normalized with respect to displacement. The coefficients defined here are with respect to eigenvectors (6.3.5), (6.3.29), (6.3.51), (6.3.52) or (6.3.53), *not* with respect to unit displacement. In the Green dyadic, the amplitude immediately before and after an interface changes due to the coefficient, the density and the discontinuity in $\mathcal{S}^{(\ell)}$. The relevant factors in expressions (6.8.2) and (6.8.3) describing the change in amplitude at an interface are

$$\frac{\hat{\mathbf{g}}_j}{\left(\rho_j V_j \mathcal{S}_j^{(\ell)}\right)^{1/2}} \;\text{(before)} \rightarrow \mathcal{T}_{ij} \frac{\hat{\mathbf{g}}_i}{\left(\rho_i V_i \mathcal{S}_i^{(\ell)}\right)^{1/2}} \;\text{(after)}, \tag{6.8.6}$$

where here the subscripts indicate values just before and just after the interface (incident and generated ray). Using expressions (6.2.15) and (5.4.19), it is straightforward to see that this is exactly equivalent to change (6.8.4) as

$$\frac{\mathcal{S}_i^{(\ell)}}{\mathcal{S}_j^{(\ell)}} = \frac{c_i |J_i| V_j}{c_j |J_j| V_i}. \tag{6.8.7}$$

Using the energy-flux normalized polarization (5.4.33), equation (6.8.6) can be rewritten

$$\mathbf{g}_j \rightarrow \frac{\sqrt{\mathcal{S}_j^{(\ell)}}}{\sqrt{\mathcal{S}_i^{(\ell)}}} \mathcal{T}_{ij}\, \mathbf{g}_i = \frac{\sqrt{(\hat{\mathbf{V}} \cdot \hat{\mathbf{n}})_i}}{\sqrt{(\hat{\mathbf{V}} \cdot \hat{\mathbf{n}})_j}} \mathcal{T}_{ij}\, \mathbf{g}_i . \tag{6.8.8}$$

In equation (6.8.5), the polarizations are normalized with respect to energy flux *normal to the interface* (e.g. definition (6.3.29)), whereas in equation (6.8.8) the normalization is with respect to energy flux *along the ray* (5.4.33). The former is commonly used in the plane-wave, transform domain (Chapters 7–9), and the latter in the ray domain (Chapter 10), reducing possible confusion.

Exercises

6.1 *Programming exercise:* Write a computer program to solve Snell's law for an anisotropic medium. (Finding the slowness vectors is more complicated than Exercise 5.9 in Chapter 5. Given the slowness direction, $\hat{\mathbf{p}}$, and finding the slowness, was equivalent to solving a cubic polynomial or finding the eigenvalues of a symmetric 3×3 matrix, whereas given the interface slowness, $\mathbf{p}_{\perp}$, and finding the normal slowness component, p_n, is equivalent to solving a sextic polynomial or finding the eigenvalues of a non-symmetric 6×6 matrix.)

Hint: The simplest method is probably to use a library routine to solve the eigen-equation (6.3.14).

6.2 *Programming exercise:* Using the results of the previous question (or otherwise), compute the eigenvectors, $\mathbf{w}$, of the matrix, $\mathbf{A}$ (6.3.14). Confirm numerically that with the normalization (6.3.29), the eigenvectors satisfy the orthonormality condition (6.3.33).

6.3 *Programming exercise:* Using the results of the previous question (or otherwise), compute the reflection/transmission coefficients (6.3.23) with (6.3.34) for anisotropic media. Confirm the reciprocity (6.3.42) numerically. Confirm numerically that for isotropic media, the results agree with expressions (6.3.60) with (6.3.61) and (6.3.62).

Comment: The sign of the coefficients depends on the sign of the eigenvectors. A computer program for solving the eigen-equation (6.3.14) may give the eigenvectors with arbitrary signs. If the eigenvector signs are not chosen consistently in the two media, the reciprocity result may have sign inconsistencies. It is vital to remember that *reflection/transmission coefficients only have meaning when given together with the eigenvectors.*

7

Differential systems for stratified media

To obtain results that are better than ray theory and remain valid at singularities, solutions of the full wave equations are needed. In a one-dimensional or stratified medium, there is an exact procedure to obtain these – transformation of the wave equation to reduce the partial differential equation to an ordinary differential equation; solution of this using one of several well-developed techniques; and inversion of the results from the transform domain to obtain the response. In this chapter, we develop the ordinary differential systems for acoustic, isotropic and anisotropic, elastic media. The important ray expansion is then introduced, to expand this into propagators for each continuous layer of the model. Three techniques that can be used to solve the ordinary differential equations when the layers are heterogeneous, are described: the *WKBJ asymptotic expansion*, the *WKBJ iterative solution* or *Bremmer series*, and the *Langer asymptotic expansion*. These methods are useful to describe various canonical solutions. However, for realistic media, a combination of methods might well be required and it is often more realistic to resort to numerical methods to solve the ordinary differential equations.

In the next three chapters, we discuss techniques that avoid the approximations of ray theory (Chapter 5) and can be used in stratified media, where the model parameters only vary as a function of one coordinate, z. As a result of this simplification, transform methods can be used to reduce the partial differential equations of motion and constitution to a system of ordinary differential equations. This is more easily solved than a partial differential equation. Initially, we consider models of homogeneous layers, and then allow for vertical heterogeneity.

7.1 One-dimensional differential systems

This chapter discusses the systems of ordinary differential equations for acoustic, isotropic and anisotropic media, and their solutions (Section 7.2). It should be read in conjunction with the next chapter, Chapter 8, which discusses techniques for inverting the temporal and spatial transforms to obtain the response or Green function in the time and space domain. The inverse transform technique employed depends on the model and method used to solve the ordinary differential equations. Not all sections in this chapter are pre-requisite for each inversion technique, e.g. only the two-dimensional acoustic system (Section 7.1.1) is needed in order to read the section on the Cagniard method (Section 8.1). Applying the same method to the elastic systems involves more complicated algebra, but not fundamentally different techniques. Appendix C.1, which describes the propagator matrix method, should also be read in conjunction with this chapter.

7.1.1 Acoustic waves in two dimensions

First we consider the simplest system, an acoustic medium in two dimensions. The equation of motion for the Green functions is equation (4.5.20) and the constitutive relation is equation (4.5.21), where we consider solutions in the x–z plane. Applying the Fourier transform with respect to time (3.1.1), the equations become expressions (4.5.32) and (4.5.33). As the model parameters, $\rho(z)$ and $\kappa(z)$, only depend on z, we can take the Fourier transform with respect to x, i.e. the transform (3.2.4). Derivatives with respect to x are replaced by a factor $\mathrm{i}\omega p$. Thus equations (4.5.32) and (4.5.33) become

$$-\mathrm{i}\omega\rho(z)\,\underline{\mathbf{v}} = -\begin{pmatrix} \mathrm{i}\omega p \\ \partial/\partial z \end{pmatrix} \underline{P} + \mathbf{I}\,\delta(z - z_{\mathrm{S}}) \tag{7.1.1}$$

$$-\mathrm{i}\omega \underline{P} = -\kappa(z)\begin{pmatrix} \mathrm{i}\omega p & \partial/\partial z \end{pmatrix} \underline{\mathbf{v}}, \tag{7.1.2}$$

where the arguments of the transformed variables, $\underline{P}$ and $\underline{\mathbf{v}}$, are (ω, p, z). The first component of equation (7.1.1) gives

$$\underline{v}_x = p\,\underline{P}/\rho(z), \tag{7.1.3}$$

except at the source depth, z_{S}. Substituting for $\underline{v}_x$ in equation (7.1.2), we can write the differential system as

$$\boxed{\frac{\mathrm{d}}{\mathrm{d}z}\underline{\mathbf{w}} = \mathrm{i}\omega\mathbf{A}\,\underline{\mathbf{w}} + \underline{\boldsymbol{\mathcal{F}}}\,\delta(z - z_{\mathrm{S}}),} \tag{7.1.4}$$

where

$$\underline{\mathbf{w}} = \begin{pmatrix} \underline{v}_z \\ -\underline{P} \end{pmatrix} \tag{7.1.5}$$

$$\mathbf{A} = \begin{pmatrix} 0 & p^2/\rho - 1/\kappa \\ -\rho & 0 \end{pmatrix} = \begin{pmatrix} 0 & -q_\alpha^2/\rho \\ -\rho & 0 \end{pmatrix} \tag{7.1.6}$$

$$\underline{\boldsymbol{\mathcal{F}}} = \begin{pmatrix} p/\rho & 0 \\ 0 & -1 \end{pmatrix}. \tag{7.1.7}$$

The two columns of the transformed source matrix, $\underline{\mathcal{F}}$ (7.1.7), give the two columns of the transformed Green function, $\underline{\mathbf{w}}$ (7.1.5), corresponding to the two components, f_x and f_z, of the unit, point force source. The vector (7.1.5), matrix (7.1.6) and variable q_α are analogous to those used for ray theory at an interface, cf. equations (6.1.1), (6.3.2) and (6.2.8). In the former case, the components of the vector are the amplitude coefficients – now they are transformed variables. As before, it is useful to find the eigen-solutions of the matrix $\mathbf{A}$, i.e. solutions of the equation

$$\mathbf{A}\mathbf{w} = p_z \mathbf{w}. \tag{7.1.8}$$

The eigenvalues are $p_z = \pm q_\alpha$, and we form a diagonal matrix of the eigenvalues

$$\mathbf{p}_z = \begin{pmatrix} q_\alpha & 0 \\ 0 & -q_\alpha \end{pmatrix}, \tag{7.1.9}$$

and a matrix with the eigenvectors as columns

$$\mathbf{W} = \frac{1}{\sqrt{2\rho\, q_\alpha}} \begin{pmatrix} q_\alpha & -q_\alpha \\ -\rho & -\rho \end{pmatrix}. \tag{7.1.10}$$

These satisfy equation (6.3.3).

7.1.2 A pressure line source

Expression (7.1.4) is the differential system for the transformed, *force* Green function of acoustic waves. Specific sources can be introduced by a suitable multiplicative factor (equation (4.5.34) as we are in the frequency domain). Often it is convenient to consider a pressure source, e.g. an explosion. Consider a cavity with sides L within which the pressure is given by $P_S(t)$. Then the pressure is

$$\begin{aligned} P(t, x, z) &= P_S(t) \Big\{ H(x + L/2) - H(x - L/2) \Big\} \\ &\qquad \Big\{ H(z - z_S + L/2) - H(z - z_S - L/2) \Big\} \\ &\to L^2 P_S(t)\, \delta(x)\, \delta(z - z_S), \end{aligned} \tag{7.1.11}$$

as $L \to 0$. This line source is illustrated in Figure 7.1. Letting $L^2 = A_S$, the cross-sectional area of the line pressure source, the force field is given by

$$\mathbf{f} = -\nabla P = -A_S P_S(t) \begin{pmatrix} \delta'(x)\, \delta(z - z_S) \\ \delta(x)\, \delta'(z - z_S) \end{pmatrix}. \tag{7.1.12}$$

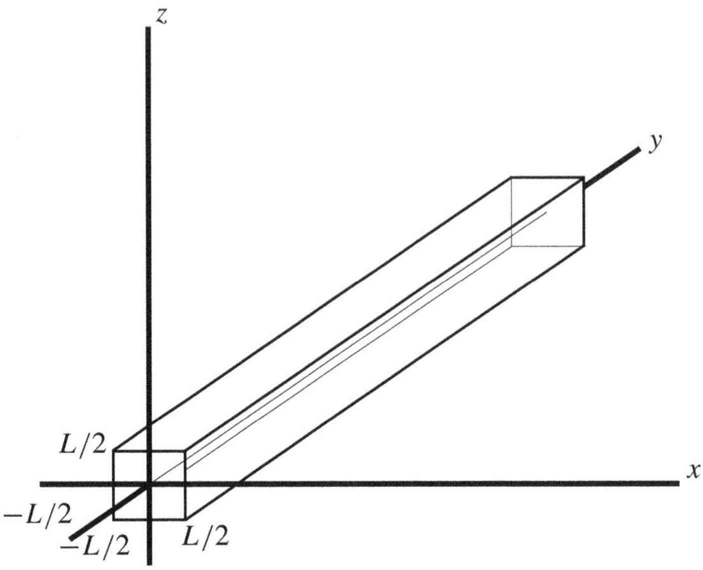

Fig. 7.1. A line cavity described by equation (7.1.11) within which the pressure is $P_S(t)$.

Transforming the equation of motion (4.5.33) becomes

$$-i\omega\rho(z)\mathbf{v} = -\begin{pmatrix} i\omega p \\ \partial/\partial z \end{pmatrix} P + A_S P_S(\omega) \begin{pmatrix} i\omega p^2 \delta(z - z_S)/\rho \\ \delta'(z - z_S) \end{pmatrix}, \quad (7.1.13)$$

cf. equation (7.1.1). Thus the multiplicative factor for the pressure source can be represented as a force term

$$\mathbf{f} = i\omega A_S P_S(\omega) \begin{pmatrix} p \\ \pm q_\alpha \end{pmatrix}, \quad (7.1.14)$$

which should post-multiply the Green function in the transformed domain, i.e. $\boldsymbol{\mathcal{F}}\mathbf{f}\,\delta(z - z_S)$ in equation (7.1.4) is equivalent to equation (7.1.13). In the time domain, a convolution with $-A_S P_S'(t)$ is introduced and the pressure source is represented by the 'force' operator

$$\mathbf{f} = -\begin{pmatrix} p \\ \pm q_\alpha \end{pmatrix} A_S P_S'(t) * . \quad (7.1.15)$$

For a moment tensor, pressure source (4.6.21), the force becomes

$$\mathbf{f} = \begin{pmatrix} p \\ \pm q_\alpha \end{pmatrix} M_S. \quad (7.1.16)$$

7.1.3 Acoustic waves in three dimensions

The results in Section 7.1.1 only require very minor modifications in three dimensions with a point source. We take the Fourier transform with respect to both horizontal coordinates, x_1 and x_2 (3.2.11). Equations (7.1.1) and (7.1.2) become

$$-\mathrm{i}\omega\rho(z)\underline{\mathbf{v}} = -\begin{pmatrix} \mathrm{i}\omega p_1 \\ \mathrm{i}\omega p_2 \\ \partial/\partial z \end{pmatrix} \underline{P} + \mathbf{I}\,\delta(z - z_{\mathrm{S}}) \tag{7.1.17}$$

$$-\mathrm{i}\omega\underline{P} = -\kappa(z)\begin{pmatrix} \mathrm{i}\omega p_1 & \mathrm{i}\omega p_2 & \partial/\partial z \end{pmatrix} \underline{\mathbf{v}}. \tag{7.1.18}$$

Using the first two components of equation (7.1.17) in equation (7.1.18), these equations can be rewritten as equation (7.1.4) where $\underline{\mathbf{w}}$ and $\mathbf{A}$ are still equations (7.1.5) and (7.1.6), but the transformed source matrix $\underline{\boldsymbol{\mathcal{F}}}$ becomes

$$\underline{\boldsymbol{\mathcal{F}}} = \begin{pmatrix} p_1/\rho & p_2/\rho & 0 \\ 0 & 0 & -1 \end{pmatrix}. \tag{7.1.19}$$

In the matrix $\mathbf{A}$ (7.1.6)

$$p^2 = p_1^2 + p_2^2. \tag{7.1.20}$$

It is convenient to define a two-dimensional vector of the transform slownesses

$$\mathsf{p} = \begin{pmatrix} p_1 \\ p_2 \end{pmatrix}. \tag{7.1.21}$$

As the differential system is identical in two and three dimensions, with only the source term differing, all the solutions will carry over. The acoustic differential system is 2×2, but in three dimensions, the force source and Green function have three components. Hence the matrix $\underline{\boldsymbol{\mathcal{F}}}$ is 2×3 with the three columns corresponding to the three components, f_x, f_y and f_z, of the unit, point force source.

7.1.4 Anisotropic waves in three dimensions

The anisotropic wave equation (4.5.45) and constitutive relation (4.5.46), in the frequency domain, are transformed with respect to the horizontal coordinates (Section 3.2). The equation of motion (4.5.45) becomes

$$-\mathrm{i}\omega\rho\,\underline{\mathbf{v}} = \mathrm{i}\omega p_\nu\,\underline{\mathbf{t}}_\nu + \frac{\partial \underline{\mathbf{t}}_3}{\partial x_3} + \mathbf{I}\,\delta(z - z_{\mathrm{S}}), \tag{7.1.22}$$

and the constitutive relation (4.5.46)

$$-\mathrm{i}\omega\, \underline{\mathbf{t}}_j = \mathrm{i}\omega\, p_\nu\, \mathbf{c}_{\nu j}\, \underline{\mathbf{v}} + \mathbf{c}_{j3}\, \frac{\partial \underline{\mathbf{v}}}{\partial x_3}, \tag{7.1.23}$$

where Greek subscripts are summed over 1 and 2. Equation (7.1.23) with $j = 3$ gives the ordinary differential equation for $\underline{\mathbf{v}}$

$$\frac{\mathrm{d}\underline{\mathbf{v}}}{\mathrm{d}x_3} = -\mathrm{i}\omega\, p_\nu\, \mathbf{c}_{33}^{-1}\, \mathbf{c}_{3\nu}\, \underline{\mathbf{v}} - \mathrm{i}\omega\, \mathbf{c}_{33}^{-1}\, \underline{\mathbf{t}}_3. \tag{7.1.24}$$

This can be substituted in equation (7.1.23) with $j = 1$ and 2, to find $\underline{\mathbf{t}}_\nu$ in terms of $\underline{\mathbf{v}}$ and $\underline{\mathbf{t}}_3$. Then substituting in equation (7.1.22) we obtain the differential equation for $\underline{\mathbf{t}}_3$. This and equation (7.1.24) can then be written compactly as

$$\frac{\mathrm{d}}{\mathrm{d}z}\mathbf{w} = \mathrm{i}\omega\mathbf{A}\mathbf{w} + \underline{\boldsymbol{\mathcal{F}}}\,\delta(z - z_{\mathrm{S}}), \tag{7.1.25}$$

where

$$\mathbf{w} = \begin{pmatrix} \underline{\mathbf{v}} \\ \underline{\mathbf{t}}_3 \end{pmatrix} \tag{7.1.26}$$

$$\underline{\boldsymbol{\mathcal{F}}} = \begin{pmatrix} \mathbf{0} \\ -\mathbf{I} \end{pmatrix}, \tag{7.1.27}$$

and the matrix $\mathbf{A}$ is defined exactly as in equation (6.3.14) with equations (6.3.15)–(6.3.17), except that the slowness is now the (complex) transform variable, p (Section 3.2). As this differential system (7.1.25) has exactly the same form as the acoustic system (7.1.4) (but is sixth-order rather than second), we have used the same notation. The transformed source matrix, $\underline{\boldsymbol{\mathcal{F}}}$ (7.1.27), is 6×3 with the three columns corresponding to the three components, f_x, f_y and f_z, of the unit, point force source.

7.1.5 Isotropic waves in three dimensions

In isotropic media we can take advantage of the simplified constitutive relations with definitions (4.4.55) and (4.4.56), to simplify the differential system (7.1.25). For instance

$$p_\nu\, \mathbf{c}_{3\nu} = \begin{pmatrix} 0 & 0 & \mu p_1 \\ 0 & 0 & \mu p_2 \\ \lambda p_1 & \lambda p_2 & 0 \end{pmatrix}. \tag{7.1.28}$$

However, to fully exploit the isotropy in the differential system, we must rotate the particle velocity $\underline{\mathbf{v}}$ and traction $\underline{\mathbf{t}}_3$ into coordinates aligned with the p vector. Thus if

$$p_1 = p\cos\chi \tag{7.1.29}$$

$$p_2 = p\sin\chi, \tag{7.1.30}$$

we consider particle velocity and traction components with respect to a system rotated by the angle χ. Thus, for instance,

$$\underline{\mathbf{v}}' = \begin{pmatrix} \cos\chi & \sin\chi & 0 \\ -\sin\chi & \cos\chi & 0 \\ 0 & 0 & 1 \end{pmatrix} \underline{\mathbf{v}} = \mathbf{R}\,\underline{\mathbf{v}}, \tag{7.1.31}$$

say, where $\mathbf{R}$ is the rotation matrix. This simplifies the differential system. For instance, applying to equation (7.1.28), we have

$$p_\nu\,\mathbf{R}\,\mathbf{c}_{3\nu}\,\mathbf{R}^{-1} = \begin{pmatrix} 0 & 0 & \mu p \\ 0 & 0 & 0 \\ \lambda p & 0 & 0 \end{pmatrix}, \tag{7.1.32}$$

where p is defined in equation (7.1.20). For the rotated variables, the matrix for the differential system (7.1.25) reduces to definition (6.3.47). The important property is that it separates into two systems, the second-order *SH* system

$$\mathbf{A}^{(25)\times(25)} = \begin{pmatrix} 0 & -1/\mu \\ \mu p^2 - \rho & 0 \end{pmatrix}, \tag{7.1.33}$$

and the fourth-order $P-SV$ system

$$\mathbf{A}^{(1346)\times(1346)} = \begin{pmatrix} 0 & -p & -1/\mu & 0 \\ -p\lambda/(\lambda+2\mu) & 0 & 0 & -1/(\lambda+2\mu) \\ \eta p^2 - \rho & 0 & 0 & -p\lambda/(\lambda+2\mu) \\ 0 & -\rho & -p & 0 \end{pmatrix}, \tag{7.1.34}$$

where η is defined in equation (6.3.48).

7.2 Solutions of one-dimensional systems

In this section we investigate solutions of the acoustic and elastic differential systems developed in the previous section, Section 7.1. We follow a general notation which will apply to the different systems: the second-order acoustic or isotropic *SH* systems ($m = 1$); the fourth-order isotropic $P-SV$ elastic system ($m = 2$); and the sixth-order anisotropic system ($m = 3$). We also consider solutions in two and three dimensions (where $\ell = 2$ or 3, respectively). General results and nomenclature for solutions of ordinary differential equations are discussed in Appendix C.1.

7.2.1 Waves in a homogeneous medium

The solution of a differential system such as (7.1.4) in a homogeneous medium is easily written in terms of the eigen-solutions of the matrix $\mathbf{A}$. The eigen-equations can be written

$$\mathbf{AW} = \mathbf{W}\mathbf{p}_z, \tag{7.2.1}$$

where $\mathbf{W}$ is a matrix with the eigenvectors as columns and $\mathbf{p}_z$ is the diagonal matrix of eigenvalues (cf. Section 6.3 and equation (6.3.3)). We generalize the definitions in Section 6.3 to apply for any (complex) slowness, but otherwise the results carry over. We order the eigenvalues as before, i.e. those propagating in the positive z direction first, ordered with increasing velocity, followed by those propagating in the negative z direction. It is straightforward to show that in a homogeneous medium ($\mathbf{A}$ constant), a fundamental matrix of differential system (7.1.4) is

$$\mathbf{F}(z) = \mathbf{W}e^{i\omega\mathbf{p}_z z}. \tag{7.2.2}$$

The columns of $\mathbf{F}$ are identified as up and down-going plane waves with slowness vectors

$$\mathbf{p} = \begin{pmatrix} \mathsf{p} \\ p_z \end{pmatrix}, \tag{7.2.3}$$

where p_z are the diagonal elements of matrix $\mathbf{p}_z$, the eigenvalues or vertical slownesses. A propagator matrix from z_0 is

$$\boxed{\mathbf{P}(z, z_0) = \mathbf{F}(z)\,\mathbf{F}^{-1}(z_0) = \mathbf{W}e^{i\omega\mathbf{p}_z(z-z_0)}\mathbf{W}^{-1},} \tag{7.2.4}$$

where the inverse matrix is simply

$$\mathbf{W}^{-1} = \begin{pmatrix} \acute{\mathbf{W}}^{\ddagger} \\ -\grave{\mathbf{W}}^{\ddagger} \end{pmatrix}, \tag{7.2.5}$$

using the notation of equation (6.3.32). The same orthonormality condition (6.3.33) applies in the transform domain, with the slownesses p_ν common to all waves. In an acoustic medium the propagator is simply

$$\boxed{\mathbf{P}(z, z_0) = \begin{pmatrix} \cos\omega q_\alpha(z-z_0) & -(iq_\alpha/\rho)\sin\omega q_\alpha(z-z_0) \\ -(i\rho/q_\alpha)\sin\omega q_\alpha(z-z_0) & \cos\omega q_\alpha(z-z_0) \end{pmatrix}.} \tag{7.2.6}$$

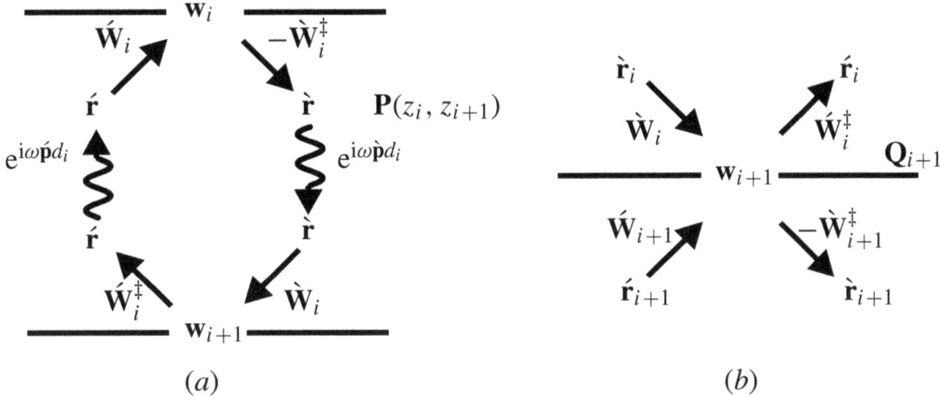

Fig. 7.2. An illustration of the 'layer' and 'interface' propagators: (*a*) the 'layer' propagator, such as the Haskell matrix (7.2.4), is constructed by decomposing the solution $\mathbf{w}$ into its eigenvectors using the matrix $\mathbf{W}^{-1}$; propagating across the layer using the matrix $\exp(\mathrm{i}\omega\mathbf{p}_z z)$; and recombining to give $\mathbf{w}$ using the matrix $\mathbf{W}$; (*b*) the 'interface' propagator, used in the Kennett ray expansion, e.g. reflection (7.2.57) connects the up and down-going waves, $\acute{\mathbf{r}}$ and $\grave{\mathbf{r}}$, through the reflection/transmission matrix, $\mathbf{Q}$ (6.3.22), of the interface.

The propagator matrix (7.2.4) is sometimes known as the *Haskell matrix* (after Haskell, 1953). The construction of the propagator is illustrated in Figure 7.2*a*: the solution $\mathbf{w}$ is decomposed into its eigenvectors using the matrix $\mathbf{W}^{-1}$; propagated across the layer using the matrix $\exp(\mathrm{i}\omega\mathbf{p}_z z)$; and recombined to give $\mathbf{w}$ using the matrix $\mathbf{W}$. The propagation part of the solution is illustrated in Figure 7.3*a*. Note that as $\mathrm{tr}(\mathbf{A}) = 0$, the Jacobi identity (C.1.16) gives $|\mathbf{P}(z, z_0)| = 1$, and the normalization used in $\mathbf{W}$ (7.1.10) makes $|\mathbf{F}(z)| = |\mathbf{W}| = (-1)^m$ for solution (7.2.2) when the differential system is $2m \times 2m$.

7.2.2 Direct waves

Knowing a fundamental matrix (7.2.2), we can easily obtain the solution (C.1.10) of the differential equation (7.1.4). It can be rewritten

$$\underline{\mathbf{w}}(z) = \mathbf{F}(z)\,\underline{\mathbf{r}}(z), \tag{7.2.7}$$

where, for arbitrary z_0,

$$\underline{\mathbf{r}}(z) = \underline{\mathbf{r}}(z_0) + \int_{z_0}^{z} \mathbf{F}^{-1}(\zeta)\,\underline{\boldsymbol{\mathcal{F}}}(\zeta)\,\delta(\zeta - z_S)\,\mathrm{d}\zeta. \tag{7.2.8}$$

The components of the matrix $\underline{\mathbf{r}}$ are the amplitudes of the up and down-going waves in the fundamental matrix $\mathbf{F}$ for each source component. It is $2m \times \ell$. The unknown component vector $\underline{\mathbf{r}}(z_0)$ is determined by applying boundary conditions

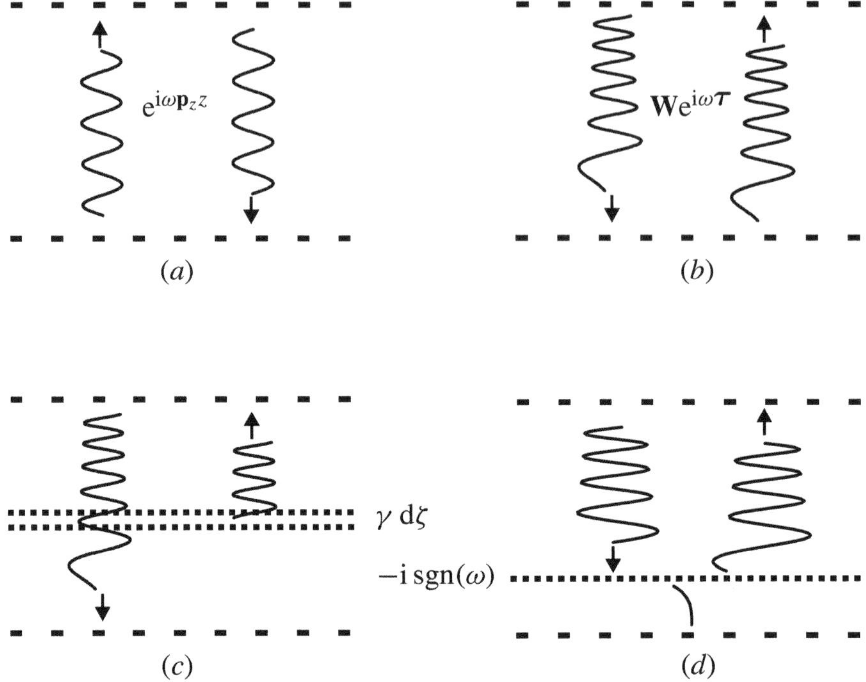

Fig. 7.3. An illustration of the different types of solutions of the differential system (7.1.4): (*a*) in a homogeneous layer, the propagation term is uniform (7.2.4); (*b*) in an inhomogeneous layer, the propagation term of the WKBJ asymptotic solution (7.2.104) changes wavelength and amplitude; (*c*) in the WKBJ iterative solution (7.2.125), the WKBJ solutions couple due to 'reflectors' $\gamma\, d\zeta$; and (*d*), the Langer propagating waves (7.2.159) couple at the turning point $z_\alpha(p)$.

on the solution. Let us take $z_0 = z_S - 0$, i.e. infinitesimally (or anywhere) below the source. Then only down-going waves can exist. We must have

$$\underline{\mathbf{r}}(z_S - 0) = \begin{pmatrix} \mathbf{0} \\ \grave{\underline{\mathbf{r}}} \end{pmatrix}. \tag{7.2.9}$$

Integrating through the source, we obtain

$$\underline{\mathbf{r}}(z_S + 0) = \underline{\mathbf{r}}(z_S - 0) + \int_{z_S - 0}^{z_S + 0} \mathbf{F}^{-1}(\zeta)\, \underline{\boldsymbol{\mathcal{F}}}(\zeta)\, \delta(\zeta - z_S)\, d\zeta = \begin{pmatrix} \acute{\underline{\mathbf{r}}} \\ \mathbf{0} \end{pmatrix}, \tag{7.2.10}$$

where only up-going waves exist above the source. Substituting for $\mathbf{F}$ and $\underline{\boldsymbol{\mathcal{F}}}$, and evaluating the trivial, delta function integral, we obtain

$$\begin{pmatrix} \acute{\underline{\mathbf{r}}} \\ \mathbf{0} \end{pmatrix} = \begin{pmatrix} \mathbf{0} \\ \grave{\underline{\mathbf{r}}} \end{pmatrix} + e^{-i\omega \mathbf{p}_z z_S} \begin{pmatrix} \acute{\mathbf{W}}^{\ddagger} \\ -\grave{\mathbf{W}}^{\ddagger} \end{pmatrix} \underline{\boldsymbol{\mathcal{F}}}. \tag{7.2.11}$$

Note that $\acute{\mathbf{W}}^{\ddagger}$ and $\grave{\mathbf{W}}^{\ddagger}$ are $m \times 2m$ and $\underline{\mathcal{F}}$ is $2m \times \ell$ – the ℓ columns in $\underline{\mathcal{F}}$ are for the different source components in the Green function. The sub-matrices $\underline{\acute{\mathbf{r}}}$ and $\underline{\grave{\mathbf{r}}}$ are $m \times \ell$.

Combining with the fundamental matrix (7.2.7), we obtain

$$\underline{\mathbf{w}} = \acute{\mathbf{W}} e^{i\omega \acute{\mathbf{p}}_z (z - z_S)} \acute{\mathbf{W}}^{\ddagger} \underline{\mathcal{F}} \quad \text{if} \quad z > z_S \tag{7.2.12}$$

$$= \grave{\mathbf{W}} e^{i\omega \grave{\mathbf{p}}_z (z - z_S)} \grave{\mathbf{W}}^{\ddagger} \underline{\mathcal{F}} \quad \text{if} \quad z < z_S, \tag{7.2.13}$$

where we have decomposed the eigenvalue matrix into up and down-going waves

$$\mathbf{p}_z = \begin{pmatrix} \acute{\mathbf{p}}_z & \mathbf{0} \\ \mathbf{0} & \grave{\mathbf{p}}_z \end{pmatrix}. \tag{7.2.14}$$

In these expressions we have maintained a general notation that can be used to describe general sources, waves and receivers. It is simple to interpret. Reading from right to left in expressions (7.2.12) and (7.2.13), $\underline{\mathcal{F}}$ converts the source force into the source excitation of the field variables, $\acute{\mathbf{W}}^{\ddagger}$ or $\grave{\mathbf{W}}^{\ddagger}$ resolves these into the amplitudes of basic waves, the exponential term propagates these to z, and $\acute{\mathbf{W}}$ and $\grave{\mathbf{W}}$ recombine these to give the field variables. It is then relatively straightforward to modify each part of these expressions for more complicated problems. Combining the inverse matrix (7.2.5) with the source matrix (7.1.27) using equations (6.3.32) and (6.3.25), we find both

$$\acute{\mathbf{W}}^{\ddagger} \underline{\mathcal{F}} = \mathbf{W}_{11}^{\mathrm{T}} = \acute{\mathbf{G}}^{\mathrm{T}} \tag{7.2.15}$$

$$\grave{\mathbf{W}}^{\ddagger} \underline{\mathcal{F}} = \mathbf{W}_{12}^{\mathrm{T}} = \grave{\mathbf{G}}^{\mathrm{T}}, \tag{7.2.16}$$

where the $\mathbf{G}$'s are the appropriate $\ell \times m$ matrices of the generalized energy-flux normalized polarization vectors, i.e. columns of $\mathbf{G}$ are from eigenvectors (6.3.29)

$$\mathbf{g} = \frac{1}{\sqrt{\pm 2\rho V_3}} \hat{\mathbf{g}} \tag{7.2.17}$$

(an identical result is obtained in the acoustic case less directly as $\mathbf{W}$ does not contain v_x explicitly, but is generated by the $\mathcal{F}_{11}$ element in matrix (7.1.7) with equation (7.1.3)). We sometimes refer to $\mathbf{g}$ as a *generalized polarization vector* to indicate that it is defined for any complex $\mathbf{p}$, not just a real ray value.

Using definitions (7.2.15) and (7.2.16) in equations (7.2.12) and (7.2.13), we find

$$\boxed{\underline{\mathbf{v}}_{\text{direct}} = \acute{\mathbf{G}} \, \acute{\boldsymbol{\Phi}} \, \acute{\mathbf{G}}^{\mathrm{T}} \quad \text{for} \quad z > z_S} \tag{7.2.18}$$

$$\boxed{\underline{\mathbf{v}}_{\text{direct}} = \grave{\mathbf{G}} \, \grave{\boldsymbol{\Phi}} \, \grave{\mathbf{G}}^{\mathrm{T}} \quad \text{for} \quad z < z_S,} \tag{7.2.19}$$

where the phase propagation matrices ($m \times m$) are

$$\acute{\boldsymbol{\Phi}} = \mathrm{e}^{\mathrm{i}\omega\acute{\mathbf{p}}_z(z-z_S)} \quad \text{for} \quad z > z_S \tag{7.2.20}$$

$$\grave{\boldsymbol{\Phi}} = \mathrm{e}^{\mathrm{i}\omega\grave{\mathbf{p}}_z(z-z_S)} \quad \text{for} \quad z < z_S. \tag{7.2.21}$$

Expressions (7.2.18) and (7.2.19) are the Green function dyadic for the direct waves. We have used a matrix notation so in elastic media it includes all three waves (in acoustic media $\boldsymbol{\Phi}$ is a scalar). We have included the subscript to indicate the ray type, i.e. a direct wave, to allow future generalizations. This expression for the transformed particle velocity Green function is $\ell \times \ell$. It clearly separates into a *source excitation* part, $\mathbf{G}^{\mathrm{T}}$, a directional propagation part, $\acute{\boldsymbol{\Phi}}$ or $\grave{\boldsymbol{\Phi}}$, and a *receiver conversion* part, $\mathbf{G}$ (cf. expression (5.4.35) in ray theory with $\mathbf{G} \to \mathbf{g}$ and $\boldsymbol{\Phi} \to \Phi$ for a single ray/wave).

In Chapter 6 we considered rays, whereas in this chapter we are considering wave solutions in the transformed domain, i.e. plane, spectral waves defined by (ω, p_1, p_2). Although the concepts are different, there are many similarities. In particular, the concept of a *ray expansion* applies to the wave solution. The wave solution can be expanded using reflection/transmission coefficients at interfaces. The only generalization is that the coefficients are used even if the slownesses are complex and the waves evanescent. With this generality, we use the terminology of rays and waves interchangeably when discussing the wave solution.

7.2.3 Reflected and transmitted waves

As we know a fundamental matrix in a homogeneous layer (7.2.2), or a propagator (7.2.4), it is relatively simple to introduce an interface into the model and consider reflected and transmitted waves. Suppose we introduce an interface at $z_2 < z_S$, so that medium 1 is for $z > z_2$, and medium 2 is for $z < z_2$ (Figure 7.4). We introduce the convention that the interface with the same index as a layer is at the top of the layer (cf. Section 0.1.6).

The chain rule for propagators (C.1.5) means that we can combine propagators for each layer

$$\mathbf{P}(z, z_3) = \mathbf{P}(z, z_2)\mathbf{P}(z_2, z_0), \tag{7.2.22}$$

where z_0 is arbitrary. Expressing the propagators as in equation (7.2.4), we can use the results in Section 6.3 to connect waves through the interface. To determine the solution of the differential system we could proceed from first principles using

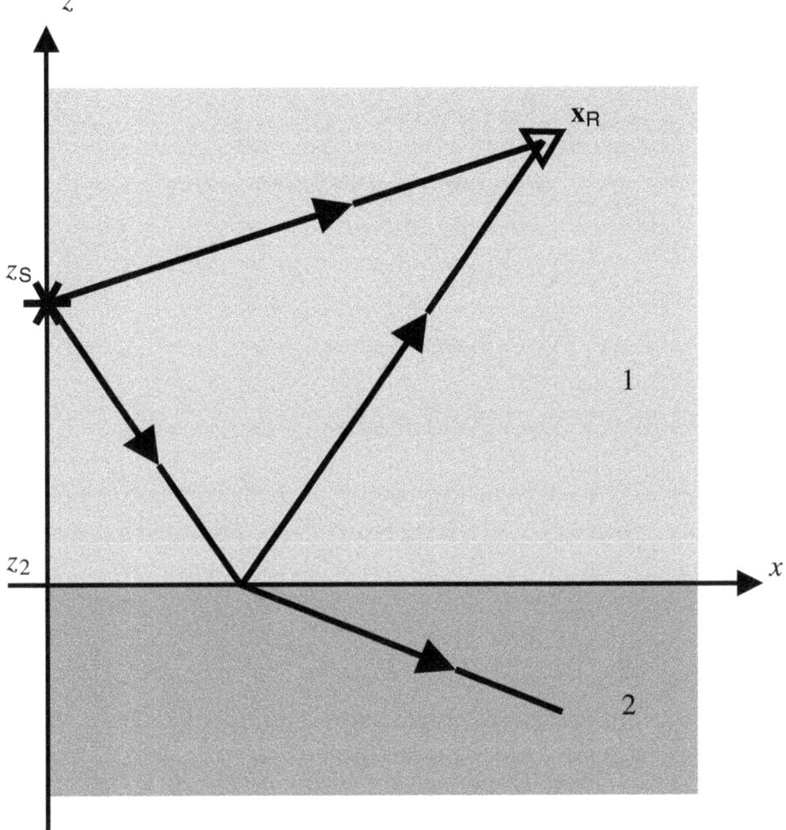

Fig. 7.4. An interface at z_2 below a source at z_S, with the direct waves from the source, and the reflected and transmitted waves indicated.

equation (C.1.9). However, it is simpler to construct the solution combining known results.

The source wave at the interface is result (7.2.13) with definition (7.2.16)

$$\mathbf{w}(z_2) = \grave{\mathbf{W}}_1 \mathrm{e}^{\mathrm{i}\omega\grave{\mathbf{p}}_{z1}(z_2 - z_S)} \grave{\mathbf{G}}_1^{\mathrm{T}}, \tag{7.2.23}$$

where we have added a subscript to indicate medium 1. Therefore for $z < z_2$ we take the solution

$$\mathbf{w}(z) = \grave{\mathbf{W}}_2 \mathrm{e}^{\mathrm{i}\omega\grave{\mathbf{p}}_{z2}(z - z_2)} \boldsymbol{\mathcal{T}}_{21} \mathrm{e}^{\mathrm{i}\omega\grave{\mathbf{p}}_{z1}(z_2 - z_S)} \grave{\mathbf{G}}_1^{\mathrm{T}}, \tag{7.2.24}$$

where $\mathcal{T}_{21}$ is the matrix of transmission coefficients (6.3.23). This has been designed to match (7.2.23) at the interface. For then

$$\mathbf{w}(z_2) = \mathbf{W}_2 \begin{pmatrix} \mathbf{0} \\ \mathcal{T}_{21} \end{pmatrix} \mathrm{e}^{\mathrm{i}\omega \grave{\mathbf{p}}_{z1}(z_2 - z_\mathrm{S})} \grave{\mathbf{G}}_1^{\mathrm{T}} \tag{7.2.25}$$

$$= \mathbf{W}_1 \begin{pmatrix} \mathcal{T}_{12} - \mathcal{T}_{11}\mathcal{T}_{21}^{-1}\mathcal{T}_{22} & \mathcal{T}_{11}\mathcal{T}_{21}^{-1} \\ -\mathcal{T}_{21}^{-1}\mathcal{T}_{22} & \mathcal{T}_{21}^{-1} \end{pmatrix} \begin{pmatrix} \mathbf{0} \\ \mathcal{T}_{21} \end{pmatrix}$$

$$\times \mathrm{e}^{\mathrm{i}\omega \grave{\mathbf{p}}_{z1}(z_2 - z_\mathrm{S})} \grave{\mathbf{G}}_1^{\mathrm{T}} \tag{7.2.26}$$

$$= \mathbf{W}_1 \begin{pmatrix} \mathcal{T}_{11} \\ \mathbf{I} \end{pmatrix} \mathrm{e}^{\mathrm{i}\omega \grave{\mathbf{p}}_{z1}(z_2 - z_\mathrm{S})} \grave{\mathbf{G}}_1^{\mathrm{T}} \tag{7.2.27}$$

$$= \left(\acute{\mathbf{W}}_1 \mathcal{T}_{11} + \grave{\mathbf{W}}_1 \right) \mathrm{e}^{\mathrm{i}\omega \grave{\mathbf{p}}_{z1}(z_2 - z_\mathrm{S})} \grave{\mathbf{G}}_1^{\mathrm{T}}, \tag{7.2.28}$$

using results (6.3.21). By design the second term in expression (7.2.28) corresponds to the source wave (7.2.23). Expression (7.2.27) contains the source waves (7.2.23) and the reflected waves. For $z_2 < z < z_\mathrm{S}$ the solution is

$$\mathbf{w}(z) = \mathbf{W}_1 \mathrm{e}^{\mathrm{i}\omega \mathbf{p}_{z1}(z - z_2)} \begin{pmatrix} \mathcal{T}_{11} \\ \mathbf{I} \end{pmatrix} \mathrm{e}^{\mathrm{i}\omega \grave{\mathbf{p}}_{z1}(z_2 - z_\mathrm{S})} \grave{\mathbf{G}}_1^{\mathrm{T}}. \tag{7.2.29}$$

From the direct waves (Section 7.2.2) we know the discontinuity in the solution at the source, $z = z_\mathrm{S}$, and for $z > z_\mathrm{S}$, the solution is

$$\mathbf{w}(z) = \acute{\mathbf{W}}_1 \left(\mathrm{e}^{\mathrm{i}\omega \grave{\mathbf{p}}_{z1}(z - z_\mathrm{S})} + \mathrm{e}^{\mathrm{i}\omega \acute{\mathbf{p}}_{z1}(z - z_2)} \mathcal{T}_{11} \mathrm{e}^{\mathrm{i}\omega \grave{\mathbf{p}}_{z1}(z_2 - z_\mathrm{S})} \right) \grave{\mathbf{G}}_1^{\mathrm{T}}, \tag{7.2.30}$$

i.e. the up-going direct and reflected waves. Thus the Green function dyadic for the reflected waves is

$$\boxed{\underline{\mathbf{Y}}_{\mathrm{reflect}} = \acute{\mathbf{G}}_1 \, \acute{\mathbf{\Phi}}_1(z - z_2) \mathcal{T}_{11} \, \grave{\mathbf{\Phi}}_1(z_2 - z_\mathrm{S}) \grave{\mathbf{G}}_1^{\mathrm{T}},} \tag{7.2.31}$$

and for the transmitted waves (7.2.24)

$$\boxed{\underline{\mathbf{Y}}_{\mathrm{transmit}} = \grave{\mathbf{G}}_2 \, \grave{\mathbf{\Phi}}_2(z - z_2) \mathcal{T}_{21} \, \grave{\mathbf{\Phi}}_1(z_2 - z_\mathrm{S}) \grave{\mathbf{G}}_1^{\mathrm{T}},} \tag{7.2.32}$$

where $\grave{\mathbf{\Phi}}(z) = \exp(\mathrm{i}\omega \grave{\mathbf{p}}_z z)$, etc. The matrix notation generates the complete set of reflections and transmissions – in anisotropic media there may be nine different waves.

7.2.4 The ray expansion

The technique used in Section 6.3 for computing the reflection/transmission coefficients from an interface can be extended to a stack of layers, and to the wave

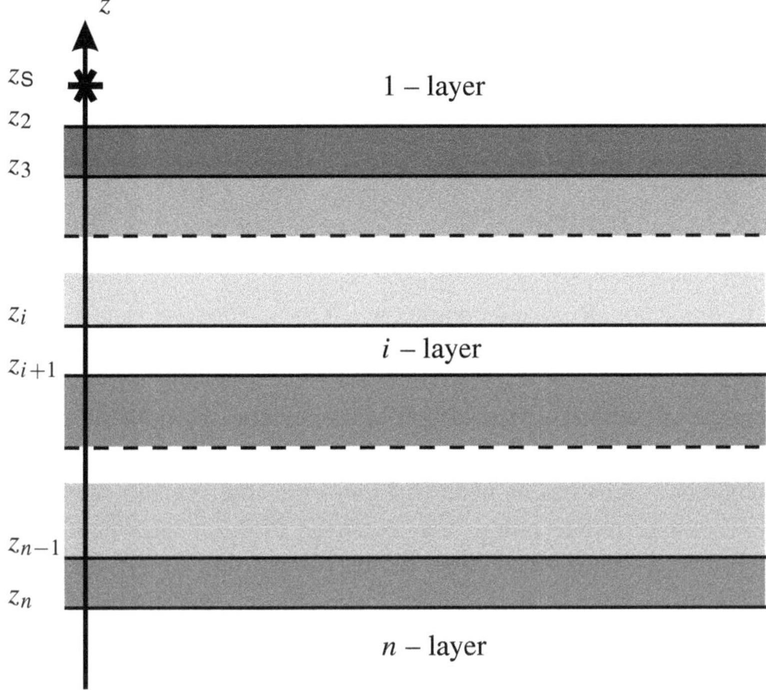

Fig. 7.5. A stack of n layers where the i-th layer has $z_i > z > z_{i+1}$ and the source lies at z_S. The first and n-th layers are half-spaces.

solution in the transform domain, i.e. plane waves defined by (ω, p_1, p_2). Consider a stack of n homogeneous layers where the i-th layer has $z_i > z > z_{i+1}$ (see Figure 7.5).

The propagator from z_n for the stack can be written

$$\mathbf{P}(z, z_n) = \mathbf{P}(z, z_2)\mathbf{P}(z_2, z_3)\ldots\mathbf{P}(z_{n-1}, z_n), \tag{7.2.33}$$

where for each layer we know the propagator (7.2.4). Thus the vector $\mathbf{w}$ at z_2 and z_n can be connected by

$$\mathbf{w}(z_2) = \mathbf{P}(z_2, z_3)\ldots\mathbf{P}(z_{n-1}, z_n)\mathbf{w}(z_n). \tag{7.2.34}$$

In order to present results in a concise manner, it is useful to modify the reflection/transmission coefficient notation slightly. The result must be independent of the coordinate direction so we avoid a notation which differentiates up and down-going waves. Consider a stack of layers between z_A and z_B. We represent reflections by $\mathbf{R}$ and transmissions by $\mathbf{T}$, and the reflecting/transmitting zone by a subscript, e.g. $\mathbf{R}_{BA}$. Thus Figure 6.6 is revised in Figure 7.6.

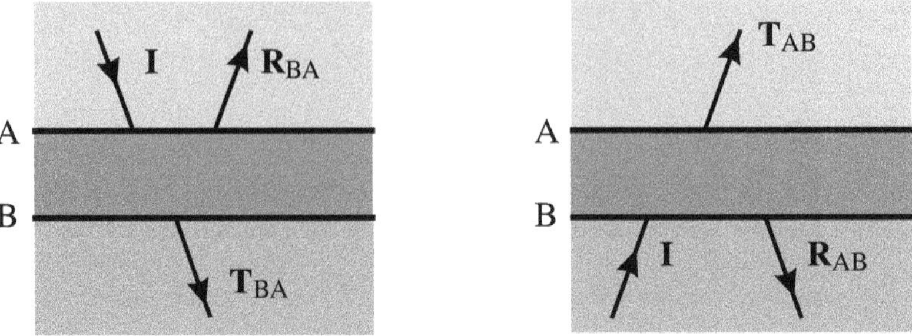

Fig. 7.6. Reflection, $\mathbf{R}_{\mathrm{BA}}$, and transmission, $\mathbf{T}_{\mathrm{BA}}$, coefficients from a stack A to B, and $\mathbf{R}_{\mathrm{AB}}$ and $\mathbf{T}_{\mathrm{AB}}$ from B to A.

Note that the symbol in the subscript now represents the reflection/transmission zone as seen from the incident ray, and not subscript indices representing the incident and generated media as used in Figure 6.6, etc. As the composite reflection/transmission formulae below, e.g. equation (7.2.52), are read from right to left, the subscript symbol should also be read from right to left, e.g. $\mathbf{R}_{\mathrm{BA}}$ are the reflection coefficients for an incident wave travelling in the direction from A to B. The matrix of all coefficients (6.3.23) would now be

$$\mathcal{T} = \begin{pmatrix} \mathbf{R}_{\mathrm{BA}} & \mathbf{T}_{\mathrm{AB}} \\ \mathbf{T}_{\mathrm{BA}} & \mathbf{R}_{\mathrm{AB}} \end{pmatrix}. \tag{7.2.35}$$

With this generalization, equation (6.3.18) becomes

$$\mathbf{W}_1 \begin{pmatrix} \mathbf{R}_{n2} & \mathbf{T}_{2n} \\ \mathbf{I} & \mathbf{0} \end{pmatrix} = \mathbf{P}(z_2, z_3) \ldots \mathbf{P}(z_{n-1}, z_n) \mathbf{W}_n \begin{pmatrix} \mathbf{0} & \mathbf{I} \\ \mathbf{T}_{n2} & \mathbf{R}_{2n} \end{pmatrix}, \tag{7.2.36}$$

for reflection/transmission experiments from the stack z_2 to z_n. Writing

$$\mathbf{W}_{2n} = \mathbf{P}(z_2, z_3) \ldots \mathbf{P}(z_{n-1}, z_n) \mathbf{W}_n, \tag{7.2.37}$$

propagating the eigenvectors at z_n to z_2, equation (7.2.36) can be solved as (6.3.19)

$$\mathcal{T} = \begin{pmatrix} \mathbf{R}_{n2} & \mathbf{T}_{2n} \\ \mathbf{T}_{n2} & \mathbf{R}_{2n} \end{pmatrix} = \begin{pmatrix} \mathbf{W}_1^{(.)\times(123)} & -\mathbf{W}_{2n}^{(.)\times(456)} \end{pmatrix}^{-1} \times \begin{pmatrix} -\mathbf{W}_1^{(.)\times(456)} & \mathbf{W}_{2n}^{(.)\times(123)} \end{pmatrix}. \tag{7.2.38}$$

Although this result is correct, there are two problems with it. Firstly it offers no physical insight into the signals within the solution. The expected rays are not evident. In fact, even for a very simple model, e.g. a single layer, expanding the denominator binomially, i.e. the determinant of the inverse matrix on the right-hand side, leads to terms with negative phase and wrong signs. Only when these cancel with the expansion of the numerator are we left with just terms that can be recognized as rays, i.e. positive phase corresponding to the propagation direction, and amplitudes corresponding to the product of the appropriate reflection/transmission coefficients. The second difficulty is that the accurate numerical evaluation of equation (7.2.38) is extremely difficult, and with finite-length arithmetic, breaks down at high frequencies. The problem arises when one or more slownesses are complex and the waves evanescent. Expression (7.2.38) will contain exponentially large and small terms. In evaluating the inverse matrix, differencing of terms is necessary. Again in the overall expression, the exponentially large terms must cancel as the actual waves decay in the direction of propagation. But numerically, the exponentially large terms dominate, and the significant small terms are lost in rounding errors. Various schemes have been developed to overcome this numerical problem. One of them is the ray expansion due to Kennett (1974, 1983), discussed in the next section.

Another advantage of Kennett's algorithm is the ease with which fluid layers can be included. Expression (7.2.36) breaks down if fluid layers are included, as at fluid–solid or fluid–fluid interfaces, the velocity–traction vector $\mathbf{w}$ need not be continuous, i.e. the tangential velocity may be discontinuous and the tangential traction must be zero (see Sections 4.3.2 and 6.5). It must be modified (and made more complicated) to include fluids. The results from Kennett's algorithm in the next section apply without modification to any mixture of fluid and solid layers, provided any coefficients that would generate shear waves in the fluid layers, and any propagation terms for shear waves in fluid layers, are ignored or numerically set to zero. In other words, any undefined terms can be ignored or set zero. The special interface (dis)continuity conditions are accommodated in the fluid–fluid or fluid–solid interface coefficients (Sections 6.3.1 and 6.5).

7.2.4.1 Kennett's ray expansion

We present Kennett's algorithm in a concise manner with our notation which is independent of coordinate direction. Consider two stacks of layers AB and BC and the combined stack AC (Figure 7.7). We assume that the media properties are continuous at B, i.e. combining the two stacks does not create a new interface at B. If an interface is required at B it must be included infinitesimally inside one stack.

The reflection/transmission experiments from each stack can be written

$$\mathbf{W}_{\mathrm{A}}\begin{pmatrix}\mathbf{R}_{\mathrm{BA}} & \mathbf{T}_{\mathrm{AB}}\\ \mathbf{I} & \mathbf{0}\end{pmatrix} = \mathbf{P}(z_{\mathrm{A}}, z_{\mathrm{B}})\mathbf{W}_{\mathrm{B}}\begin{pmatrix}\mathbf{0} & \mathbf{I}\\ \mathbf{T}_{\mathrm{BA}} & \mathbf{R}_{\mathrm{AB}}\end{pmatrix} \qquad (7.2.39)$$

$$\mathbf{W}_{\mathrm{B}}\begin{pmatrix}\mathbf{R}_{\mathrm{CB}} & \mathbf{T}_{\mathrm{BC}}\\ \mathbf{I} & \mathbf{0}\end{pmatrix} = \mathbf{P}(z_{\mathrm{B}}, z_{\mathrm{C}})\mathbf{W}_{\mathrm{C}}\begin{pmatrix}\mathbf{0} & \mathbf{I}\\ \mathbf{T}_{\mathrm{CB}} & \mathbf{R}_{\mathrm{BC}}\end{pmatrix}, \qquad (7.2.40)$$

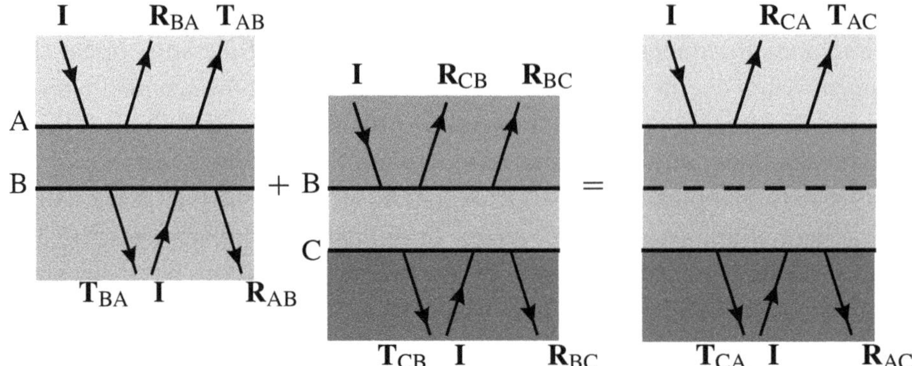

Fig. 7.7. Diagram of the reflection/transmission coefficients for two blocks AB and BC, combining to form a block AC.

and for the combined stack

$$\mathbf{W}_{\mathrm{A}}\begin{pmatrix}\mathbf{R}_{\mathrm{CA}} & \mathbf{T}_{\mathrm{AC}}\\ \mathbf{I} & \mathbf{0}\end{pmatrix} = \mathbf{P}(z_{\mathrm{A}}, z_{\mathrm{C}})\mathbf{W}_{\mathrm{C}}\begin{pmatrix}\mathbf{0} & \mathbf{I}\\ \mathbf{T}_{\mathrm{CA}} & \mathbf{R}_{\mathrm{AC}}\end{pmatrix}. \tag{7.2.41}$$

This can be expanded as

$$\mathbf{W}_{\mathrm{A}}\begin{pmatrix}\mathbf{R}_{\mathrm{CA}} & \mathbf{T}_{\mathrm{AC}}\\ \mathbf{I} & \mathbf{0}\end{pmatrix} = \mathbf{P}(z_{\mathrm{A}}, z_{\mathrm{B}})\mathbf{P}(z_{\mathrm{B}}, z_{\mathrm{C}})\mathbf{W}_{\mathrm{C}}\begin{pmatrix}\mathbf{0} & \mathbf{I}\\ \mathbf{T}_{\mathrm{CA}} & \mathbf{R}_{\mathrm{AC}}\end{pmatrix} \tag{7.2.42}$$

$$= \mathbf{P}(z_{\mathrm{A}}, z_{\mathrm{B}})\mathbf{W}_{\mathrm{B}} \times \begin{pmatrix}\mathbf{T}_{\mathrm{BC}} - \mathbf{R}_{\mathrm{CB}}\mathbf{T}_{\mathrm{CB}}^{-1}\mathbf{R}_{\mathrm{BC}} & \mathbf{R}_{\mathrm{CB}}\mathbf{T}_{\mathrm{CB}}^{-1}\\ -\mathbf{T}_{\mathrm{CB}}^{-1}\mathbf{R}_{\mathrm{BC}} & \mathbf{T}_{\mathrm{CB}}^{-1}\end{pmatrix}\begin{pmatrix}\mathbf{0} & \mathbf{I}\\ \mathbf{T}_{\mathrm{CA}} & \mathbf{R}_{\mathrm{AC}}\end{pmatrix} \tag{7.2.43}$$

$$= \mathbf{W}_{\mathrm{A}}\begin{pmatrix}\mathbf{T}_{\mathrm{AB}} - \mathbf{R}_{\mathrm{BA}}\mathbf{T}_{\mathrm{BA}}^{-1}\mathbf{R}_{\mathrm{AB}} & \mathbf{R}_{\mathrm{BA}}\mathbf{T}_{\mathrm{BA}}^{-1}\\ -\mathbf{T}_{\mathrm{BA}}^{-1}\mathbf{R}_{\mathrm{AB}} & \mathbf{T}_{\mathrm{BA}}^{-1}\end{pmatrix} \times \begin{pmatrix}\mathbf{T}_{\mathrm{BC}} - \mathbf{R}_{\mathrm{CB}}\mathbf{T}_{\mathrm{CB}}^{-1}\mathbf{R}_{\mathrm{BC}} & \mathbf{R}_{\mathrm{CB}}\mathbf{T}_{\mathrm{CB}}^{-1}\\ -\mathbf{T}_{\mathrm{CB}}^{-1}\mathbf{R}_{\mathrm{BC}} & \mathbf{T}_{\mathrm{CB}}^{-1}\end{pmatrix}\begin{pmatrix}\mathbf{0} & \mathbf{I}\\ \mathbf{T}_{\mathrm{CA}} & \mathbf{R}_{\mathrm{AC}}\end{pmatrix}, \tag{7.2.44}$$

where the chain rule (C.1.5) is used in equation (7.2.42), equations (7.2.40) and (6.3.21) are used in expression (7.2.43), and equations (7.2.39) and (6.3.21) are used in equation (7.2.44). Expanding the product of matrices, we obtain the iterative results

$$\boxed{\mathbf{T}_{\mathrm{CA}} = \mathbf{T}_{\mathrm{CB}}\left(\mathbf{I} - \mathbf{R}_{\mathrm{AB}}\mathbf{R}_{\mathrm{CB}}\right)^{-1}\mathbf{T}_{\mathrm{BA}}} \tag{7.2.45}$$

$$\boxed{\mathbf{R}_{\mathrm{CA}} = \mathbf{R}_{\mathrm{BA}} + \mathbf{T}_{\mathrm{AB}}\mathbf{R}_{\mathrm{CB}}\left(\mathbf{I} - \mathbf{R}_{\mathrm{AB}}\mathbf{R}_{\mathrm{CB}}\right)^{-1}\mathbf{T}_{\mathrm{BA}}.} \tag{7.2.46}$$

This result contains both equations (30) and (31) in Kennett (1974), i.e. adding a layer on either side of a stack. As no reference has been made to direction, this formula can be used for either direction, i.e. if A and C are interchanged. Note that

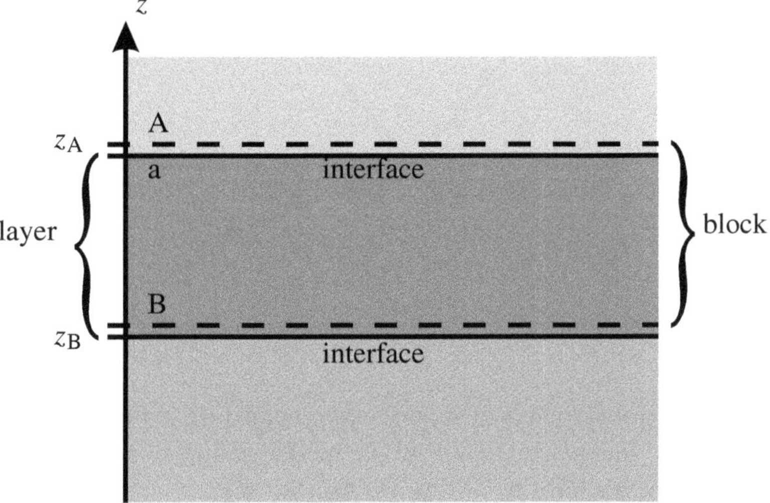

Fig. 7.8. Diagram of a single-layer block AB. The model has discontinuities at z_A and z_B. Assuming $z_A > z_B$, the homogeneous layer is for $z_A - 0 \geq z \geq z_B + 0$, but the block is $z_A + 0 \leq z \leq z_B + 0$, i.e. it includes the interface at z_A. Symbolically, we denote the layer by aB and the interface by Aa, i.e. Aa + aB = AB.

while it appears that the reverberation matrix changes if the direction is reversed, they are simply related:

$$(\mathbf{I} - \mathbf{R}_{AB}\mathbf{R}_{CB})^{-1}\,\mathbf{R}_{AB} = \mathbf{R}_{AB}\,(\mathbf{I} - \mathbf{R}_{CB}\mathbf{R}_{AB})^{-1}. \tag{7.2.47}$$

This algorithm, equations (7.2.45) and (7.2.46), applied iteratively in both directions, can be used to compute the reflection/transmission coefficients for any stack of layers. The starting point is the matrices for a single layer-interface block, which are simply formed from the interface reflection/transmission coefficients and the layer phase terms.

The building block for the above iteration scheme is the result for a block consisting of a single, homogeneous layer and an interface. Let us consider a block AB with an interface at A. A point infinitesimally inside the homogeneous layer at A is denoted by a, i.e. aB is the homogeneous layer, and Aa is the interface (Figure 7.8).

The above iterative results, equations (7.2.45) and (7.2.46), can be used to combine the layer and interface, i.e. symbolically, AB = Aa + aB. In the homogeneous layer there are no reflections and the transmissions are just the phase terms

$$\mathbf{R}_{Ba} = \mathbf{R}_{aB} = \mathbf{0} \tag{7.2.48}$$

$$\mathbf{T}_{Ba} = e^{i\omega\grave{\mathbf{p}}_z(z_B - z_A)} = \mathbf{\Phi}_{Ba} \tag{7.2.49}$$

$$\mathbf{T}_{aB} = e^{i\omega\acute{\mathbf{p}}_z(z_A - z_B)} = \mathbf{\Phi}_{aB}, \tag{7.2.50}$$

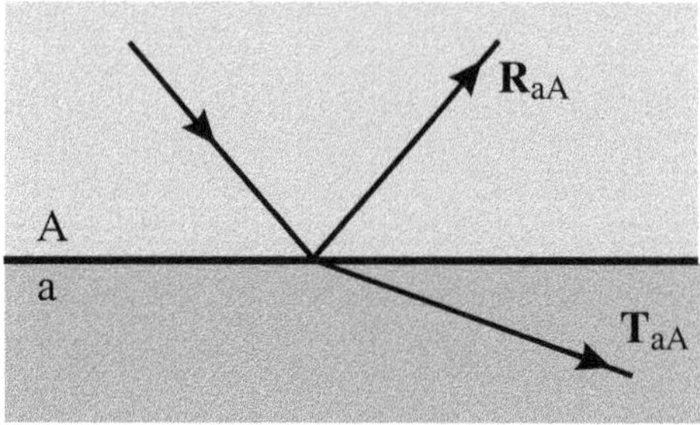

Fig. 7.9. Diagram of the reflection/transmission coefficients for an interface Aa.

where $\grave{\mathbf{p}}_z$ is the diagonal matrix of slownesses propagating in the direction AB and $\acute{\mathbf{p}}_z$ in direction BA. In acoustic and isotropic media, and in anisotropic media with up-down symmetry, we have $\grave{\mathbf{p}}_z = -\acute{\mathbf{p}}_z$ so $\boldsymbol{\Phi}_{\mathrm{aB}} = \boldsymbol{\Phi}_{\mathrm{Ba}}$.

At the interface, the coefficients can be computed using the techniques developed in Chapter 6. The coefficient matrix for the interface Aa is

$$\mathcal{T} = \begin{pmatrix} \mathbf{R}_{\mathrm{aA}} & \mathbf{T}_{\mathrm{Aa}} \\ \mathbf{T}_{\mathrm{aA}} & \mathbf{R}_{\mathrm{Aa}} \end{pmatrix}, \tag{7.2.51}$$

illustrated in Figure 7.9.

Using results (7.2.45) and (7.2.46) to combine the layer matrices (7.2.48), (7.2.49) and (7.2.50), and interface matrix (7.2.51), i.e. combinations Aa + aB and Ba + aA, the results for blocks AB and BA are

$$\mathbf{T}_{\mathrm{BA}} = \boldsymbol{\Phi}_{\mathrm{Ba}}\mathbf{T}_{\mathrm{aA}} \tag{7.2.52}$$

$$\mathbf{R}_{\mathrm{BA}} = \mathbf{R}_{\mathrm{aA}} \tag{7.2.53}$$

$$\mathbf{T}_{\mathrm{AB}} = \mathbf{T}_{\mathrm{Aa}}\,\boldsymbol{\Phi}_{\mathrm{aB}} \tag{7.2.54}$$

$$\mathbf{R}_{\mathrm{AB}} = \boldsymbol{\Phi}_{\mathrm{Ba}}\mathbf{R}_{\mathrm{Aa}}\,\boldsymbol{\Phi}_{\mathrm{aB}}. \tag{7.2.55}$$

These are illustrated in Figure 7.10.

These formulae, (7.2.52)–(7.2.55), contain the interface coefficients and propagation phases across the layer. Reading a formula from right to left corresponds to the propagation of the rays, e.g. $\mathbf{T}_{\mathrm{AB}}$ (7.2.54) first propagates across the layer from B to a, $\boldsymbol{\Phi}_{\mathrm{aB}}$, and then is transmitted through the interface from a to A, $\mathbf{T}_{\mathrm{Aa}}$. It is an important feature of the algebra that all terms in these expressions correspond to the propagation of the physical rays, and have the correct signs and direction of propagation, e.g. $\boldsymbol{\Phi}^{-1}$ does not appear.

Starting with equations (7.2.52)–(7.2.55), and applying results (7.2.45) and (7.2.46) iteratively for each layer, we can build up the total reflection/transmission coefficients. At each stage we just add the expected extra reflections and

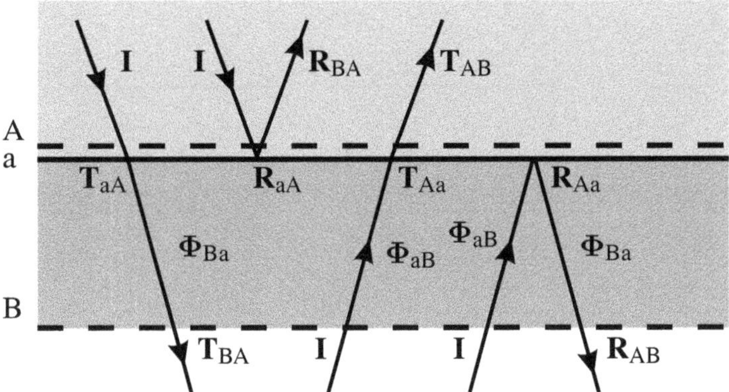

Fig. 7.10. Diagram of the reflection/transmission coefficients for a block AB: illustrated from left to right in the diagram are the four equations (7.2.52)–(7.2.55).

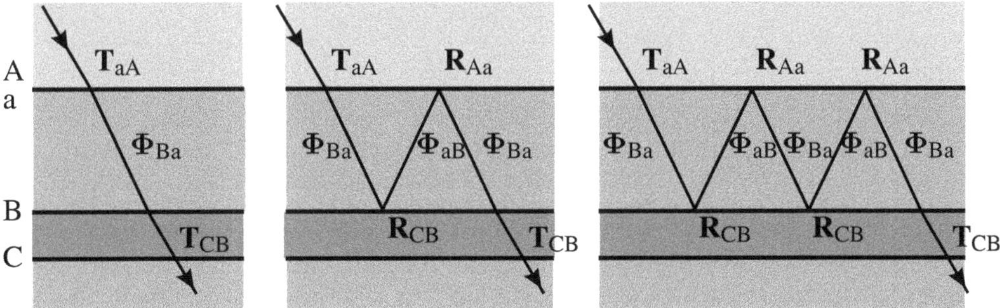

Fig. 7.11. Diagram of the first three signals making up the expansion (7.2.56) of $\mathbf{T}_{CA}$.

reverberations. Consider equation (7.2.45)

$$\begin{aligned}
\mathbf{T}_{CA} &= \mathbf{T}_{CB}\left(\mathbf{I}-\mathbf{R}_{AB}\mathbf{R}_{CB}\right)^{-1}\mathbf{T}_{BA} \\
&= \mathbf{T}_{CB}\mathbf{T}_{BA}+\mathbf{T}_{CB}\mathbf{R}_{AB}\mathbf{R}_{CB}\mathbf{T}_{BA}+\mathbf{T}_{CB}\left(\mathbf{R}_{AB}\mathbf{R}_{CB}\right)^{2}\mathbf{T}_{BA}+\dots \\
&= \mathbf{T}_{CB}\,\boldsymbol{\Phi}_{Ba}\mathbf{T}_{aA}+\mathbf{T}_{CB}\left(\boldsymbol{\Phi}_{Ba}\mathbf{R}_{Aa}\,\boldsymbol{\Phi}_{aB}\mathbf{R}_{CB}\right)\,\boldsymbol{\Phi}_{Ba}\mathbf{T}_{aA} \\
&\quad +\mathbf{T}_{CB}\left(\boldsymbol{\Phi}_{Ba}\mathbf{R}_{Aa}\,\boldsymbol{\Phi}_{aB}\mathbf{R}_{CB}\right)^{2}\,\boldsymbol{\Phi}_{Ba}\mathbf{T}_{aA}+\dots. \qquad (7.2.56)
\end{aligned}$$

These signals are illustrated in Figure 7.11.

From (7.2.46)

$$\begin{aligned}
\mathbf{R}_{CA} &= \mathbf{R}_{BA}+\mathbf{T}_{AB}\mathbf{R}_{CB}\left(\mathbf{I}-\mathbf{R}_{AB}\mathbf{R}_{CB}\right)^{-1}\mathbf{T}_{BA} \\
&= \mathbf{R}_{BA}+\mathbf{T}_{AB}\mathbf{R}_{CB}\mathbf{T}_{BA}+\mathbf{T}_{AB}\mathbf{R}_{CB}\mathbf{R}_{AB}\mathbf{R}_{CB}\mathbf{T}_{BA}+\dots \\
&= \mathbf{R}_{aA}+\mathbf{T}_{Aa}\,\boldsymbol{\Phi}_{aB}\mathbf{R}_{CB}\,\boldsymbol{\Phi}_{Ba}\mathbf{T}_{aA} \\
&\quad +\mathbf{T}_{Aa}\,\boldsymbol{\Phi}_{aB}\mathbf{R}_{CB}\left(\boldsymbol{\Phi}_{Ba}\mathbf{R}_{Aa}\,\boldsymbol{\Phi}_{aB}\mathbf{R}_{CB}\right)\,\boldsymbol{\Phi}_{Ba}\mathbf{T}_{aA}+\dots, \qquad (7.2.57)
\end{aligned}$$

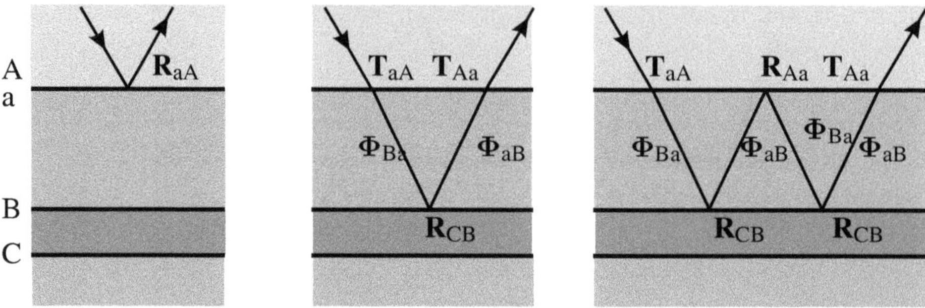

Fig. 7.12. Diagram of the first three signals making up the expansion (7.2.57) of $\mathbf{R}_{CA}$.

and again Figure 7.12 illustrates these signals. The matrix

$$(\mathbf{I} - \mathbf{R}_{AB}\mathbf{R}_{CB})^{-1} = (\mathbf{I} - \mathbf{\Phi}_{Ba}\mathbf{R}_{Aa}\,\mathbf{\Phi}_{aB}\mathbf{R}_{CB})^{-1} = \sum_{n=0}^{\infty}(\mathbf{\Phi}_{Ba}\mathbf{R}_{Aa}\,\mathbf{\Phi}_{aB}\mathbf{R}_{CB})^{n}\,, \tag{7.2.58}$$

is called the *reverberation matrix* and represents reverberations in the layer AB. The binomial expansion represents the series of reverberating rays. Each factor $(\mathbf{\Phi}_{Ba}\mathbf{R}_{Aa}\,\mathbf{\Phi}_{aB}\mathbf{R}_{CB})$ introduces one more set of reverberations, i.e. rays with an extra up and down-going segment in the layer (in a general anisotropic layer, an extra multiplicity of nine rays). The Kennett algorithm illustrates how the complete response is composed of reverberations, and also provides a mechanism for controlling the reverberations included in the solution through truncating the binomial expansion of the reverberation matrix (or through artificially setting some reflection coefficients to zero).

7.2.4.2 Source and receiver rays

Now we consider a complete model AB with a source S and receiver R included (Figure 7.13). Note that again we make no up/down distinction – we just label A and B so the order is ASRB. A and B can be at infinity, i.e. the model can be terminated with half-spaces. The model is considered as three blocks, AS, SR and RB.

At any position in the model, we can decompose the solution into the waves propagating towards A, $\mathbf{a}$, and towards B, $\mathbf{b}$. Components of the $m \times 1$ vectors $\mathbf{a}$ and $\mathbf{b}$ are the amplitudes of the qP and qS waves. Thus at the receiver, we have waves $\mathbf{a}_R$ and $\mathbf{b}_R$. The source radiates waves $[\mathbf{a}_S]$ towards A and $[\mathbf{b}_S]$ towards B, i.e. the saltus or discontinuity in $\mathbf{a}_S$ and $\mathbf{b}_S$. $[\mathbf{a}_S]$ and $[\mathbf{b}_S]$ are the vectors $\underline{\acute{\mathbf{r}}}$ and $\underline{\grave{\mathbf{r}}}$ used in (7.2.11) – the correspondence depends on whether A or B is in the positive x_3 direction. For simplicity, we indicate the waves (just) on the A side of S as $\mathbf{a}_{SA}$

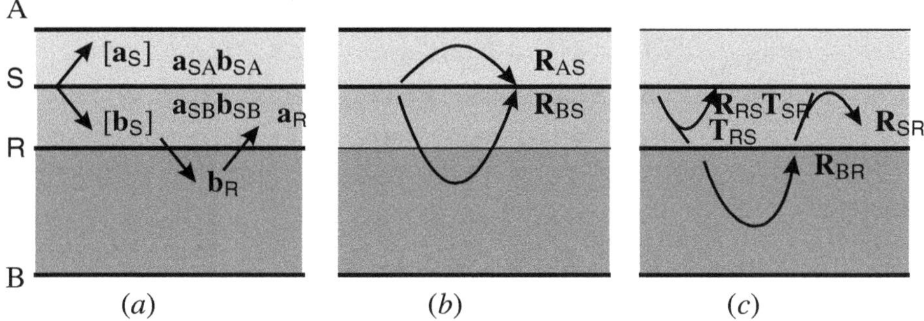

Fig. 7.13. Diagram of: (*a*) the model AB, source S and receiver R; (*b*) reflection/transmission coefficients needed for the source; (*c*) reflection/transmission coefficients needed for the receiver.

and $\mathbf{b}_{SA}$, and similarly on the B side (Figure 7.13*a*). Thus

$$\mathbf{a}_{SA} = \mathbf{a}_{SB} + [\mathbf{a}_S] \tag{7.2.59}$$

$$\mathbf{b}_{SB} = \mathbf{b}_{SA} + [\mathbf{b}_S], \tag{7.2.60}$$

expressing the discontinuity introduced by the source (by definition of the model blocks, the model is continuous at S, so no other interface discontinuity needs be considered). We need to know $\mathbf{a}_{SB}$ and $\mathbf{b}_{SB}$ in order to find the response at the receiver. The vectors in (7.2.59) and (7.2.60) are also connected by the reflections from AS and SB (Figure 7.13*b*), i.e.

$$\mathbf{b}_{SA} = \mathbf{R}_{AS}\mathbf{a}_{SA} \tag{7.2.61}$$

$$\mathbf{a}_{SB} = \mathbf{R}_{BS}\mathbf{b}_{SB}. \tag{7.2.62}$$

Solving these equations, (7.2.59)–(7.2.62), we obtain

$$\mathbf{b}_{SB} = (\mathbf{I} - \mathbf{R}_{AS}\mathbf{R}_{BS})^{-1} (\mathbf{R}_{AS}[\mathbf{a}_S] + [\mathbf{b}_S]), \tag{7.2.63}$$

and together with (7.2.62) this gives the required components.

To find the components at the receiver, $\mathbf{a}_R$ and $\mathbf{b}_R$, we need the standard formula connecting ray amplitudes across a block. From

$$\mathbf{b}_R = \mathbf{T}_{RS}\mathbf{b}_{SB} + \mathbf{R}_{SR}\mathbf{a}_R \tag{7.2.64}$$

$$\mathbf{a}_{SB} = \mathbf{T}_{SR}\mathbf{a}_R + \mathbf{R}_{RS}\mathbf{b}_{SB}, \tag{7.2.65}$$

(Figure 7.13*c*), we obtain

$$\begin{pmatrix} \mathbf{a}_R \\ \mathbf{b}_R \end{pmatrix} = \begin{pmatrix} \mathbf{T}_{SR}^{-1} & -\mathbf{T}_{SR}^{-1}\mathbf{R}_{RS} \\ \mathbf{R}_{SR}\mathbf{T}_{SR}^{-1} & \mathbf{T}_{RS} - \mathbf{R}_{SR}\mathbf{T}_{SR}^{-1}\mathbf{R}_{RS} \end{pmatrix} \begin{pmatrix} \mathbf{a}_{SB} \\ \mathbf{b}_{SB} \end{pmatrix}. \tag{7.2.66}$$

Combining with (7.2.62) and (7.2.63), we have

$$\begin{pmatrix} \mathbf{a}_\mathsf{R} \\ \mathbf{b}_\mathsf{R} \end{pmatrix} = \begin{pmatrix} \mathbf{T}_{\mathsf{SR}}^{-1} & -\mathbf{T}_{\mathsf{SR}}^{-1}\mathbf{R}_{\mathsf{RS}} \\ \mathbf{R}_{\mathsf{SR}}\mathbf{T}_{\mathsf{SR}}^{-1} & \mathbf{T}_{\mathsf{RS}} - \mathbf{R}_{\mathsf{SR}}\mathbf{T}_{\mathsf{SR}}^{-1}\mathbf{R}_{\mathsf{RS}} \end{pmatrix} \begin{pmatrix} \mathbf{R}_{\mathsf{BS}} \\ \mathbf{I} \end{pmatrix} \times (\mathbf{I} - \mathbf{R}_{\mathsf{AS}}\mathbf{R}_{\mathsf{BS}})^{-1} \quad (\mathbf{R}_{\mathsf{AS}}[\mathbf{a}_\mathsf{S}] + [\mathbf{b}_\mathsf{S}]) . \tag{7.2.67}$$

Substituting (7.2.46)

$$\mathbf{R}_{\mathsf{BS}} = \mathbf{R}_{\mathsf{RS}} + \mathbf{T}_{\mathsf{SR}}\mathbf{R}_{\mathsf{BR}} (\mathbf{I} - \mathbf{R}_{\mathsf{SR}}\mathbf{R}_{\mathsf{BR}})^{-1} \mathbf{T}_{\mathsf{RS}}, \tag{7.2.68}$$

the leading factor for $\mathbf{b}_\mathsf{R}$ reduces, i.e.

$$\mathbf{R}_{\mathsf{SR}}\mathbf{T}_{\mathsf{SR}}^{-1}\mathbf{R}_{\mathsf{BS}} + \mathbf{T}_{\mathsf{RS}} - \mathbf{R}_{\mathsf{SR}}\mathbf{T}_{\mathsf{SR}}^{-1}\mathbf{R}_{\mathsf{RS}} = (\mathbf{I} - \mathbf{R}_{\mathsf{SR}}\mathbf{R}_{\mathsf{BR}})^{-1} \mathbf{T}_{\mathsf{RS}}. \tag{7.2.69}$$

Thus using $\mathbf{a}_\mathsf{R} = \mathbf{R}_{\mathsf{BR}}\mathbf{b}_\mathsf{R}$, we obtain

$$\boxed{\begin{pmatrix} \mathbf{a}_\mathsf{R} \\ \mathbf{b}_\mathsf{R} \end{pmatrix} = \begin{pmatrix} \mathbf{R}_{\mathsf{BR}} \\ \mathbf{I} \end{pmatrix} (\mathbf{I} - \mathbf{R}_{\mathsf{SR}}\mathbf{R}_{\mathsf{BR}})^{-1} \mathbf{T}_{\mathsf{RS}} (\mathbf{I} - \mathbf{R}_{\mathsf{AS}}\mathbf{R}_{\mathsf{BS}})^{-1} \times (\mathbf{R}_{\mathsf{AS}}[\mathbf{a}_\mathsf{S}] + [\mathbf{b}_\mathsf{S}]) .} \tag{7.2.70}$$

The last two factors can be computed once for all receiver depths, but the first three factors must be recomputed for each receiver depth. The amplitudes of the waves $\mathbf{a}_\mathsf{R}$ and $\mathbf{b}_\mathsf{R}$ are then combined with appropriate polarizations to give the field at R. The reverberation factors in (7.2.70) can be expanded to generate reverberations above and below the source (the second inverse matrix), and above and below the receiver (the first inverse matrix). Thus we obtain

$$\boxed{\mathbf{v}_\mathsf{R} = \mathbf{G}_\mathsf{R} \begin{pmatrix} \mathbf{R}_{\mathsf{BR}} \\ \mathbf{I} \end{pmatrix} \sum_{n=0}^{\infty} (\mathbf{R}_{\mathsf{SR}}\mathbf{R}_{\mathsf{BR}})^n \, \mathbf{T}_{\mathsf{RS}} \sum_{m=0}^{\infty} (\mathbf{R}_{\mathsf{AS}}\mathbf{R}_{\mathsf{BS}})^m \, (\mathbf{R}_{\mathsf{AS}}[\mathbf{a}_\mathsf{S}] + [\mathbf{b}_\mathsf{S}]) ,} \tag{7.2.71}$$

where the polarizations in the matrix $\mathbf{G}_\mathsf{R}$ are ordered to correspond to the directions **a** and **b**. In this expression (7.2.71) for the complete response, the reflection and transmission matrices for parts of the model, $\mathbf{R}_{\mathsf{SR}}$, $\mathbf{R}_{\mathsf{BR}}$, $\mathbf{R}_{\mathsf{AS}}$, $\mathbf{R}_{\mathsf{BS}}$ and $\mathbf{T}_{\mathsf{RS}}$, can be calculated by repeated application of the Kennett algorithm (7.2.45) and (7.2.46), to stacks of layers. The complete response can then be expanded algebraically as a summation of rays.

Although we have presented the Kennett algorithm as an algebraic method of obtaining the ray expansion, we reiterate that it is also a powerful numerical method of obtaining the complete response of a layered model. As already discussed in relation to equation (7.2.38), direct numerical calculation of the transformed response from propagator matrices can be unstable, and the Kennett algorithm provides a useful solution. The numerical difficulties and an alternative algorithm are discussed below (Section 7.2.8).

Constructing the solution using the 'layer' propagator, the Haskell matrix (7.2.4), relies on the continuity of the solution $\mathbf{w}$ at interfaces and describes the propagation across a layer by decomposing the solution into its eigen-solutions. Constructing the solution using the Kennett ray expansion algorithm relies on decomposing the continuity of the solution $\mathbf{w}$ at an interface into connections between the eigen-solutions using the reflection/transmission coefficients in the interface matrix, $\mathbf{Q}$ (6.3.22). This is illustrated in Figure 7.2.

Before leaving Kennett's algorithm, we reiterate that it can be used without modification even if fluid layers exist by simply setting any coefficients that would generate shear waves in the fluid layers, and any propagation terms for shear waves in fluid layers, zero. The special interface conditions are accommodated in the interface coefficients (Sections 6.3.1 and 6.5).

7.2.4.3 The complete, generalized ray response

In the previous two sections, we have seen how the reflection and transmission coefficients of a layer stack can be developed iteratively, and expanded into a series where each term represents a ray, i.e. a product of reflection/transmission coefficients and phase terms, correctly sequenced to correspond to a physically realizable ray path (in a realizable ray path, the incident and generated wave types of the coefficients must match the preceding and following phase terms, both in type and layer). We have also seen how these can be combined to give the complete response with a source and receiver, connecting the waves excited at the source with the waves at the receiver (7.2.70). Thus starting with an interface and a homogeneous layer, we have the results (7.2.52)–(7.2.55) for a block. Two blocks can be combined to include all reverberations, equations (7.2.56) and (7.2.57), and this procedure can be continued, iteratively to build up a layer stack.

If the response is completely expanded, so all the terms are layer propagation or interface reflection/transmission coefficients, i.e. no composite stack or block terms remain, then certain simple rules must apply for valid, physically realizable ray paths. Propagation and interface terms must alternate. Within the sequence, the subscripts of all pairs must satisfy a few rules. Valid pairs are

$$\mathbf{R}_{\mathrm{yY}}\,\boldsymbol{\Phi}_{\mathrm{Yx}} \quad \text{or} \quad \mathbf{R}_{\mathrm{Yy}}\,\boldsymbol{\Phi}_{\mathrm{yX}} \tag{7.2.72}$$

$$\boldsymbol{\Phi}_{\mathrm{xY}}\mathbf{R}_{\mathrm{yY}} \quad \text{or} \quad \boldsymbol{\Phi}_{\mathrm{Xy}}\mathbf{R}_{\mathrm{Yy}} \tag{7.2.73}$$

$$\mathbf{T}_{\mathrm{yY}}\,\boldsymbol{\Phi}_{\mathrm{Yx}} \quad \text{or} \quad \mathbf{T}_{\mathrm{Yy}}\,\boldsymbol{\Phi}_{\mathrm{yX}} \tag{7.2.74}$$

$$\boldsymbol{\Phi}_{\mathrm{Zy}}\mathbf{T}_{\mathrm{yY}} \quad \text{or} \quad \boldsymbol{\Phi}_{\mathrm{zY}}\mathbf{T}_{\mathrm{Yy}}, \tag{7.2.75}$$

where the interfaces are order Xx, Yy and Zz. Notice the second symbol in the reflection subscript must match the neighbouring propagation subscript symbol, while the first symbol in the reflection subscript must form an interface with the second. For transmission, the neighbouring subscript symbols in the transmission and propagation must match, while the other subscript symbol of the transmission must form an interface. It is straightforward to see that the expansions in results (7.2.56) and (7.2.57) satisfy these rules.

We have used a matrix notation to group all the coefficients with the incident and generated rays propagating in the same direction together with the corresponding phase terms. In turn, the matrix notation can be expanded to give all the individual rays, in general 3^{n+1} rays in anisotropic media, for a term containing n reflection/transmission coefficients. Substituting in the complete response (7.2.70), with a similar expansion of the reverberation terms, we obtain an expansion of the complete response in terms which all represent physically realizable rays. Conversely, any physically realizable ray is contained in this expansion. We refer to this as *the ray expansion*.

While the existence of the ray expansion is, perhaps, intuitively obvious, the proof is non-trivial. Other methods of obtaining the response of a multi-layered medium, e.g. expression (7.2.38), lead to the ratio of a complicated numerator and denominator (from the determinant of the inverse of the matrix). When these are each expanded binomially, many terms are obtained which do not represent rays – they contain coefficients and phases with the wrong sign, etc. – and only when both are combined, do these cancel. The paper by Cisternas, Betancourt and Leiva (1973), where another method of obtaining the ray expansion was developed, contains an example. Spencer (1960) introduced the terms *generalized rays* and *generalized reflection and transmission coefficients* to describe the expansion of the transform-domain solution in a multi-layered media (although he did not have the algebraic tools of the Kennett ray expansion to rigorously obtain the expansion).

The Kennett algorithm for the ray expansion serves three purposes: it establishes that the complete response can be written as a ray expansion; it provides a numerically stable method for computing the response of a stack of layers; and it allows reverberations in the solution to be controlled. The difficulties of evaluating expressions such as result (7.2.38) numerically have already been discussed. In contrast, the Kennett algorithm only involves terms that represent propagation of rays. Rays are only summed and the cancellation of exponentially large terms is not required. The phase terms may be evanescent, but as they are only included in the propagation direction, they will be exponentially small not large. Exponential underflow will only occur when the corresponding ray is physically small.

The ray expansion allows ray methods (those that only apply to individual rays) to be used, and the total response to be evaluated by summation. We refer to each term in the complete ray expansion of the transformed response as a *generalized ray*, completely analogous to the rays of asymptotic ray theory (Chapter 5 and 6) but defined for complex slowness, **p**. In principle, the ray expansion is straightforward – in practice, it is difficult to enumerate or evaluate all the rays in the infinite series. In practice, the ray expansion is restricted to a finite number of rays, either by restricting the time window of interest (each ray will be causal, so later arrivals need not be included in an earlier window), or by applying an amplitude cut-off criterion (generally, multiples and reverberations will decay rapidly). Alternatively, for many purposes of interpretation and processing, only a restricted number of rays are needed. Nevertheless, there are situations where an infinite number of rays arrive within a finite time window, and the complete response requires the summation of this infinite sequence. In these circumstances, the ray expansion, while valid, may not be appropriate.

7.2.4.4 *The complete, unconverted ray expansion*

Although in principle the ray expansion is straightforward, it is difficult to enumerate all the rays in the infinite series explicitly or to write the expansion in a compact form. However, for unconverted rays, e.g. acoustic or *SH* rays, Hron (1971, 1972) has shown how this can be done.

Consider a stack of n layers. Using our convention, the ℓ-th layer lies between the ℓ-th and $(\ell+1)$-th interfaces, i.e. $z_\ell > z > z_{\ell+1}$. The first and n-th layers are half-spaces. In the ℓ-th layer, the ray has $2n_\ell$ segments, n_ℓ travelling upwards and the same number downwards. We assume that the source and receiver lie in the first layer at $z = z_S = z_R$, so $n_1 = 1$. We consider only reflections from below the source and receiver – composite models can be obtained using the result (7.2.71). All rays with the same values for the set of numbers n_ℓ have the same phase or temporal properties. We call these *kinematic analogues*. Figure 7.14 illustrates three rays which are kinematic analogues with $n_1 = 1$, $n_2 = 2$ and $n_3 = 2$.

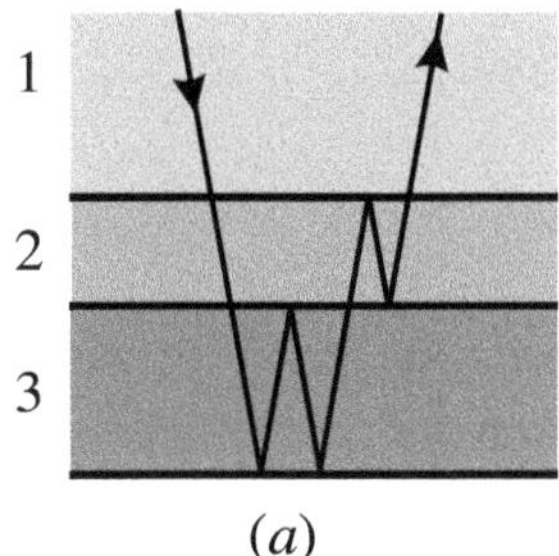

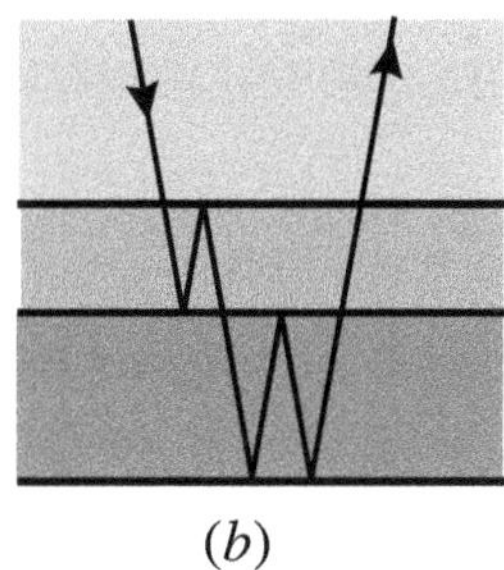

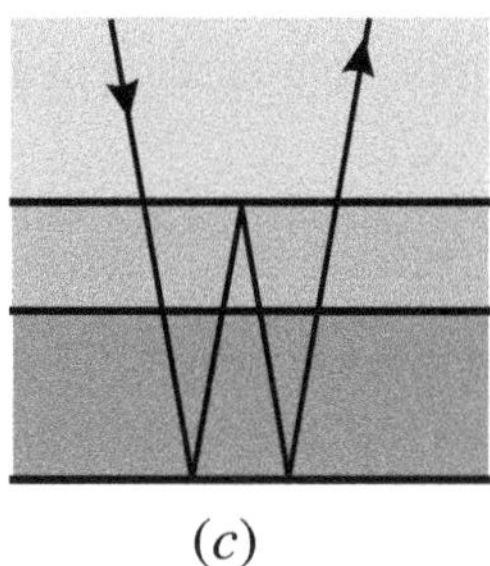

Fig. 7.14. Three rays which are kinematic analogues, with $n_1 = 1$, $n_2 = 2$ and $n_3 = 2$ ($2n_\ell$ is the number of segments in a layer). Rays (*a*) and (*b*) are dynamic analogues with $m_2 = 0$, $m_3 = 1$ and $m_4 = 2$, but (*c*) is not, with $m_2 = 0$, $m_3 = 0$ and $m_4 = 2$ (m_ℓ is the number of reflections from above an interface).

The number of reflections from above the ℓ-th interface is denoted by m_ℓ. From this we can deduce the number of transmissions through the ℓ-th interface, $\nu_\ell = n_{\ell-1} - m_\ell$ in each direction, and the number of reflections from below $n_\ell - \nu_\ell$. All rays with the same values for the set of numbers n_ℓ and m_ℓ have the same amplitude properties, i.e. the same reflection/transmission coefficients. We call these *dynamic analogues*. Figure 7.14 illustrates that of the three kinematic analogues, two are dynamic analogues (with $m_2 = 0$, $m_3 = 1$ and $m_4 = 2$), and one is not ($m_2 = 0$, $m_3 = 0$ and $m_4 = 2$). It is straightforward to see that m_2 is always zero, and that the range of m_ℓ is $\max(0, n_{\ell-1} - n_\ell) \le m_\ell \le n_{\ell-1} - 1$.

Using combinatorial theory, Hron (1971, 1972) has shown how the number of dynamic analogues can be calculated. The complete ray expansion can then be written (Chapman, 1977)

$$\mathcal{T}_{11} = \sum_{L=1}^{n-1} \sum_{\substack{n_1 = 1 \\ n_2 = 1 \\ \vdots \\ n_L = 1}}^{\substack{1 \\ \infty \\ \vdots \\ \infty}} e^{2 i \omega \sum_{\ell=1}^{L} n_\ell d_\ell q_\ell} \sum_{\substack{m_2 = 0 \\ m_3 = \max(0, n_2 - n_3) \\ \vdots \\ m_L = \max(0, n_{L-1} - n_L)}}^{\substack{0 \\ n_2 - 1 \\ \vdots \\ n_{L-1} - 1}}$$

$$\times {}_{L+1}\mathcal{T}_{11}^{n_L} \prod_{\ell=2}^{L} \left\{ C_{m_\ell}^{n_{\ell-1}} \, C_{\nu_\ell - 1}^{n_\ell - 1} \, {}_\ell\mathcal{T}_{11}^{m_\ell} \left({}_\ell\mathcal{T}_{12} \, {}_\ell\mathcal{T}_{21} \right)^{\nu_\ell} \, {}_\ell\mathcal{T}_{22}^{n_\ell - \nu_\ell} \right\}. \quad (7.2.76)$$

The symbol C_m^n is the standard binomial coefficient. The symbol ${}_\ell\mathcal{T}$ is used to denote the coefficients at the ℓ-th interface, and d_ℓ and q_ℓ are the thickness and vertical slowness, respectively, in the ℓ-th layer.

Unfortunately, a similar compact expression is not possible for converted rays, although Hron (1971, 1972) has obtained a similar expression for enumerating 'simply converted rays', i.e. rays with one converted segment.

7.2.4.5 An infinitesimal layer

If a layer is very thin, the ray expansion may not be appropriate as many reverberations will arrive in a short time window. It is instructive to consider an infinitesimal layer. The following results can either be considered as trivial, or fundamental and very important (or both)!

Consider a thin layer AB, embedded in a homogeneous medium, i.e. the interfaces Aa and Bb are identical, except for their orientation (Figure 7.15). For generality, we consider an anisotropic medium. The transmission coefficient through

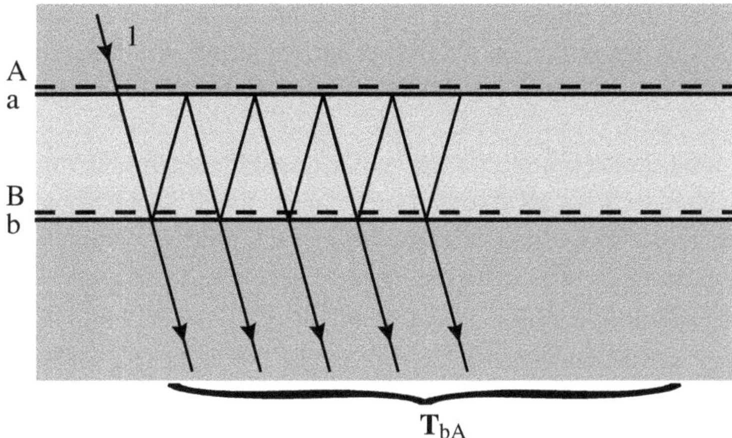

Fig. 7.15. A thin layer AB embedded in a homogeneous medium with the transmission, $\mathbf{T}_{\mathrm{bA}}$.

the layer including all reverberations is (equation (7.2.56) with C → b)

$$\mathbf{T}_{\mathrm{bA}} = \mathbf{T}_{\mathrm{bB}}\left(\mathbf{I} - \mathbf{R}_{\mathrm{AB}}\mathbf{R}_{\mathrm{bB}}\right)^{-1}\mathbf{T}_{\mathrm{BA}} \tag{7.2.77}$$

$$= \mathbf{T}_{\mathrm{bB}}\left(\mathbf{I} - \boldsymbol{\Phi}_{\mathrm{Ba}}\mathbf{R}_{\mathrm{Aa}}\,\boldsymbol{\Phi}_{\mathrm{aB}}\mathbf{R}_{\mathrm{bB}}\right)^{-1}\boldsymbol{\Phi}_{\mathrm{Ba}}\mathbf{T}_{\mathrm{aA}} \tag{7.2.78}$$

$$\longrightarrow \mathbf{T}_{\mathrm{bB}}\left(\mathbf{I} - \mathbf{R}_{\mathrm{Aa}}\mathbf{R}_{\mathrm{bB}}\right)^{-1}\mathbf{T}_{\mathrm{aA}}, \tag{7.2.79}$$

as the thickness of the layer tends to zero, for a fixed frequency.

Let us define the reflection/transmission coefficients, $\boldsymbol{\mathcal{T}}$, for the interface Aa with the half-space as medium 1 and the thin layer as medium 2. The interface Bb is identical but with the directions reversed including the slowness components, p_ν, parallel to the interface. Thus in expression (7.2.79) the coefficients map as

$$\mathbf{T}_{\mathrm{bB}} \longrightarrow \boldsymbol{\mathcal{T}}_{12}(-p_1, -p_2) \tag{7.2.80}$$

$$\mathbf{R}_{\mathrm{Aa}} \longrightarrow \boldsymbol{\mathcal{T}}_{22}(p_1, p_2) \tag{7.2.81}$$

$$\mathbf{R}_{\mathrm{bB}} \longrightarrow \boldsymbol{\mathcal{T}}_{22}(-p_1, -p_2) \tag{7.2.82}$$

$$\mathbf{T}_{\mathrm{Aa}} \longrightarrow \boldsymbol{\mathcal{T}}_{21}(p_1, p_2). \tag{7.2.83}$$

Thus the limit (7.2.79) becomes

$$\mathbf{T}_{\mathrm{bA}} \longrightarrow \mathbf{T}_{\mathrm{bB}}\left(\mathbf{I} - \mathbf{R}_{\mathrm{Aa}}\mathbf{R}_{\mathrm{bB}}\right)^{-1}\mathbf{T}_{\mathrm{aA}} \tag{7.2.84}$$

$$= \boldsymbol{\mathcal{T}}_{12}(-p_1, -p_2)\left(\mathbf{I} - \boldsymbol{\mathcal{T}}_{22}(p_1, p_2)\boldsymbol{\mathcal{T}}_{22}(-p_1, -p_2)\right)^{-1} \times \boldsymbol{\mathcal{T}}_{21}(p_1, p_2) \tag{7.2.85}$$

$$= \boldsymbol{\mathcal{T}}_{12}(-p_1, -p_2)\left(\mathbf{I} - \boldsymbol{\mathcal{T}}_{22}^{\mathrm{T}}(-p_1, -p_2)\boldsymbol{\mathcal{T}}_{22}(-p_1, -p_2)\right)^{-1} \times \boldsymbol{\mathcal{T}}_{12}^{\mathrm{T}}(-p_1, -p_2) \tag{7.2.86}$$

$$= \mathbf{I}. \tag{7.2.87}$$

The conversion from equation (7.2.84) to result (7.2.85) uses the mappings (7.2.80)–(7.2.83), from equation (7.2.85) to result (7.2.86) uses the reciprocity relationship (6.3.42), and equation (7.2.86) reduces to the identity matrix (7.2.87) using result (6.3.46).

Thus the total transmission through a thin layer, if the layer is thin compared with the wavelength, is unity. In contrast, the transmitted rays without reverberations, i.e. the zeroth-order transmission, are just the product of two transmission coefficient matrices, $\mathcal{T}_{12}\mathcal{T}_{12}^{\mathrm{T}}$, which in general is certainly not unity. If the waves are propagating in the layer, the elements in this product are less than unity, i.e. some energy is converted into different rays, but if the waves are evanescent, they may be greater than unity. It is a simple exercise to establish these facts using the simple acoustic coefficients (6.3.7) and (6.3.8), e.g. example (6.3.13). Although having the zeroth-order transmission greater than unity appears non-intuitive, it violates no physical law, as when the multiple waves are evanescent, their arrival times coincide, and the ray expansion makes no sense. All evanescent multiples must be considered together. Although the first term (the zeroth-order transmission) may be greater than unity, the multiples change sign, and the sum is unity reducing to result (7.2.87).

This result (7.2.87) can either be regarded as trivial and physically intuitively obvious, or of some importance.

Algebraically, the general proof for anisotropic media is hardly trivial as it depends on Kennett's ray expansion, the reflection/transmission coefficient reciprocity (6.3.42), and the inter-relationship (6.3.46).

Physically obvious, it may be. A low-frequency wave, where the wavelength is long compared with the layer thickness, should not 'see' the thin layer. The transmitted wave should not be sensitive to this small feature. While this argument is right, it is easy to carry it too far. Suppose the layer were a fluid. The above result no longer applies and

$$\mathbf{T}_{\mathrm{bA}} \not\rightarrow \mathbf{I}. \tag{7.2.88}$$

The proof goes wrong because $\mathbf{w}$ is no longer continuous at the interfaces. The shear eigenvectors degenerate. The layer response can still be modelled using result (7.2.78) provided the shear wave propagation terms are set to zero in $\mathbf{\Phi}$. A trivial example, shear waves normal to a thin fluid layer in an isotropic medium, immediately disproves result (7.2.87) and establishes result (7.2.88), as the transmitted shear wave will be zero whatever the layer thickness. If we try to take the limit of a solid becoming a fluid in result (7.2.87), we have two incompatible limits – the low-frequency requirement on the layer thickness is incompatible with the zero limit of the shear velocity, making the limiting wavelength zero. Thus even if the frequency is low, a thin fluid layer has an effect on the waves.

The result (7.2.87) is certainly important. It means that seismic waveform modelling, for finite frequency waves, is a robust process and a meaningful activity. Small variations in the model have a small effect on the seismic waveforms. If this were not true, every small detail in the model would be needed to obtain useful results. The real Earth undoubtedly contains thin layers as most well logs or geological sections clearly reveal. It also means that the complete response for a receiver located in or near to a thin layer is robust. It does not depend on the exact location of the receiver relative to the layer. The displacement for the complete response is continuous through the interfaces and layer. This is not true if only limited rays, e.g. the zeroth-order transmission, are used. Then the displacement inside the layer can be significantly different from that outside. Again this statement may seem trivial and intuitively obvious, but the author has frequently heard incorrect statements about the effect of placing a receiver in a slow or fast thin layer.

Numerical algorithms that use the Kennett ray expansion (Section 7.2.4.1) to calculate the response of a stack of layers frequently contain a switch to turn off reverberations, i.e. to replace the reverberation matrix (7.2.58) by the identity matrix, $\mathbf{I}$. If the layers are thin compared with the wavelength, as they must be for large values of the horizontal slowness, then this is very dangerous and numerical algorithms can become non-robust and physically meaningless, as the cancellation of reverberations is not modelled.

Overall, we would argue that result (7.2.87) is significant and worth remembering!

7.2.5 The WKBJ asymptotic expansion

In homogeneous layers, we have seen that a fundamental solution of the differential system (7.1.4) can be written as equation (7.2.2) where the acoustic eigenvector matrix $\mathbf{W}$ is given by equation (7.1.10) and its inverse by equation (6.3.32), and the diagonal eigenvalue matrix by equation (7.1.9). Similar systems exist for the isotropic and anisotropic elastic systems. We have discussed how the complete response of a stack of homogeneous layers can be obtained using the propagator matrices, equation (7.2.38) with result (7.2.37), or the Kennett ray expansion with repeated applications of equations (7.2.45) and (7.2.46). In this and the following sections, we discuss how the solution in a homogeneous layer can be generalized to an inhomogeneous layer.

In an inhomogeneous layer, in some circumstances we can generalize the homogenous-layer solution using the WKBJ asymptotic expansion. The method is widely used for second-order wave equations. Coddington and Levinson (1955, pp. 174–178) and Wasow (1965, Theorem 26.3) have considered the asymptotic solutions of general ordinary differential equations of the form (7.1.4). Richards (1971) and Chapman (1973) have considered the isotropic elastic cases. Garmany (1988*a*) has extended this to anisotropic media. The WKBJ ansatz is

$$\boxed{\mathbf{F}(z) = \mathbf{W}(z) \left\{ \sum_{m=0}^{\infty} \frac{\mathbf{W}^{(m)}(z)}{(-\mathrm{i}\omega)^m} \right\} \mathrm{e}^{\mathrm{i}\omega \boldsymbol{\tau}(z)}.} \tag{7.2.89}$$

The form of the ansatz is chosen for the same reasons as the ray theory ansatz (cf. Section 5.1.1). We then substitute this into the differential system (7.1.4) and show that we can find the unknown matrices, $\mathbf{W}^{(m)}$ and $\boldsymbol{\tau}$, in a consistent manner. Substituting in system (7.1.4), the coefficient of $(-\mathrm{i}\omega)^{-m} \exp(\mathrm{i}\omega\boldsymbol{\tau}(z))$ is

$$\mathbf{W}'\,\mathbf{W}^{(m)} + \mathbf{W}\,\mathbf{W}^{(m)\prime} - \mathbf{W}\,\mathbf{W}^{(m+1)}\;\boldsymbol{\tau}' = -\mathbf{A}\,\mathbf{W}\,\mathbf{W}^{(n+1)}, \tag{7.2.90}$$

with $\mathbf{W}^{(-1)} = \mathbf{0}$, and the prime indicates differentiation with respect to z. Taking $m = -1$, we have

$$\mathbf{W}\,\mathbf{W}^{(0)}\;\boldsymbol{\tau}' = \mathbf{A}\,\mathbf{W}\,\mathbf{W}^{(0)}. \tag{7.2.91}$$

Let us try $\mathbf{W}^{(0)} = \mathbf{I}$, which later we will check is consistent with the other equations. Then

$$\boldsymbol{\tau}' = \mathbf{W}^{-1}\,\mathbf{A}\,\mathbf{W} = \mathbf{p}_z, \tag{7.2.92}$$

using (7.2.1). Thus

$$\boxed{\boldsymbol{\tau}(z) = \int^z \mathbf{p}_z(\zeta)\,\mathrm{d}\zeta,} \tag{7.2.93}$$

an easily predicted result. Rearranging (7.2.90) we have

$$\mathbf{W}^{(m)\prime} - \mathbf{C}\,\mathbf{W}^{(m)} = \left[\mathbf{W}^{(m+1)}, \mathbf{p}_z\right], \tag{7.2.94}$$

where the commutator has been defined in equation (6.7.13), and

$$\boxed{\mathbf{C} = -\mathbf{W}^{-1}\,\mathbf{W}',} \tag{7.2.95}$$

cf. the differential form of the perturbation matrix (6.7.7). For the acoustic system

$$\boxed{\mathbf{C}(z) = \begin{pmatrix} 0 & \gamma_A \\ \gamma_A & 0 \end{pmatrix},} \tag{7.2.96}$$

with

$$\gamma_A = \frac{q_\alpha'}{2q_\alpha} - \frac{\rho'}{2\rho} = \frac{\partial}{\partial z} \ln\left(\frac{q_\alpha}{\rho}\right)^{1/2}, \tag{7.2.97}$$

cf. equations (6.7.22) and (6.7.23). Later we will derive the matrix $\mathbf{C}$ for isotropic and anisotropic elastic waves.

Whatever $\mathbf{W}^{(m+1)}$, the right-hand side of equation (7.2.94) has zero diagonal elements. Thus $\mathbf{W}^{(0)} = \mathbf{I}$ satisfies (7.2.94) provided

$$\mathbf{C} = -\left[\mathbf{W}^{(1)}, \mathbf{p}_z\right]. \qquad (7.2.98)$$

This can be solved for the off-diagonal elements of $\mathbf{W}^{(1)}$, i.e.

$$W_{12}^{(1)} = -W_{21}^{(1)} = \frac{\gamma_A}{2q_\alpha}. \qquad (7.2.99)$$

This equation imposes no restrictions on the diagonal elements of $\mathbf{W}^{(1)}$. These are found from the $m = 1$ equation (7.2.94). Then

$$W_{11}^{(1)\prime} = -W_{22}^{(1)\prime} = -\frac{\gamma_A^2}{2q_\alpha}, \qquad (7.2.100)$$

so

$$W_{11}^{(1)} = -W_{22}^{(1)} = -\int^z \frac{\gamma_A^2}{2q_\alpha}\,\mathrm{d}\zeta. \qquad (7.2.101)$$

This procedure can be continued, the equation (7.2.94) being solved for the off-diagonal elements of $\mathbf{W}^{(m+1)}$ and the diagonal elements of $\mathbf{W}^{(m)}$.

Thus in principle we can solve for all the matrices in the WKBJ asymptotic expansion (7.2.89). In practice, only the zeroth-order

$$\mathbf{F}(z) \simeq \mathbf{W}(z)\,\mathrm{e}^{\mathrm{i}\omega \boldsymbol{\mathcal{T}}(z)}, \qquad (7.2.102)$$

or the first-order

$$\mathbf{F}(z) \simeq \mathbf{W}(z)\left\{\mathbf{I} + \frac{\mathbf{W}^{(1)}(z)}{-\mathrm{i}\omega}\right\}\mathrm{e}^{\mathrm{i}\omega \boldsymbol{\mathcal{T}}(z)}, \qquad (7.2.103)$$

approximations are usually used. The zeroth-order propagator is

$$\mathbf{P}(z, z_0) \simeq \mathbf{W}(z)\mathrm{e}^{\mathrm{i}\omega(\boldsymbol{\mathcal{T}}(z) - \boldsymbol{\mathcal{T}}(z_0))}\mathbf{W}^{-1}(z_0), \qquad (7.2.104)$$

which only differs from the homogeneous propagator (7.2.4) in that the matrix $\mathbf{W}$ differs at the two depths, and the propagator term contains the integral to the varying slowness matrix $\mathbf{p}_z$. This solution is illustrated in Figure 7.3*b*.

The zeroth-order WKBJ approximation can be used for direct waves, reflections and transmissions in inhomogeneous layers. The previous results for homogeneous layers remain valid provided we replace terms such as $\sum q\,d$ by $\int q\,\mathrm{d}\zeta$. The approximation breaks down when q is small or zero, a problem we will return to later (Section 7.2.7). We can also use the WKBJ expansion to study reflections from higher-order discontinuities in the model (see below, e.g. equation (7.2.118)).

7.2.5.1 Elastic waves

With the exception of the results specifically for the acoustic system (equations (7.2.96), (7.2.97), (7.2.99), (7.2.100) and (7.2.101)), the above results apply to the elastic systems.

For the isotropic systems, the matrix (7.2.95) can be evaluated algebraically without difficulty. For the *SH* system (7.1.33), and the eigenvectors (6.3.52), the matrix is

$$\boxed{\mathbf{C}^{(25)\times(25)} = \begin{pmatrix} 0 & \gamma_H \\ \gamma_H & 0 \end{pmatrix},} \qquad (7.2.105)$$

where

$$\gamma_H = \frac{q_\beta'}{2q_\beta} + \frac{\mu'}{2\mu} = \frac{\partial}{\partial z} \ln \left(\mu q_\beta\right)^{1/2}, \qquad (7.2.106)$$

cf. perturbation equations (6.7.24) and (6.7.25). For the $P - SV$ system (7.1.34) with the eigenvectors (6.3.51) and (6.3.53), we obtain

$$\boxed{\mathbf{C}^{(1346)\times(1346)} = \begin{pmatrix} 0 & \gamma_T & -\gamma_S & \gamma_R \\ -\gamma_T & 0 & \gamma_R & \gamma_P \\ -\gamma_S & \gamma_R & 0 & \gamma_T \\ \gamma_R & \gamma_P & -\gamma_T & 0 \end{pmatrix},} \qquad (7.2.107)$$

where

$$\boxed{\begin{aligned} \gamma_P &= q_\alpha'/2q_\alpha - \rho'/2\rho + 2\beta^2 p^2 \mu'/\mu \\ \gamma_S &= q_\beta'/2q_\beta - \rho'/2\rho + 2\beta^2 p^2 \mu'/\mu \\ \gamma_R &= p(q_\alpha q_\beta)^{-1/2} \left(\beta^2 (p^2 - 2q_\alpha q_\beta)\mu'/\mu - \rho'/2\rho\right) \\ \gamma_T &= p(q_\alpha q_\beta)^{-1/2} \left(\beta^2 (p^2 + 2q_\alpha q_\beta)\mu'/\mu - \rho'/2\rho\right), \end{aligned}} \qquad (7.2.108)$$

cf. perturbation equations (6.7.26) and (6.7.27). The subscripts suggest the type of interaction between the wave types: γ_P is the interaction between the *P* waves travelling in opposite directions; γ_S is the interaction between the *SV* waves travelling in opposite directions (the sign purely depends on the definition of the eigenvectors (6.3.53) and which components change sign with direction); γ_R is the interaction between *P* and *SV* waves, and *vice versa*, propagating in opposite directions, i.e. reflective interactions; and, γ_T is the interaction between *P* and *SV* waves, and *vice versa*, propagating in the same direction, i.e. transmissive interactions. All apply in either direction.

In anisotropic media, the eigenvectors (6.3.25) must usually be found numerically as the solutions of the sixth-order eigen-equation (7.2.1). The inverse eigenvector matrix required in expression (7.2.95) can be obtained simply from equation (6.3.28), but **C** must be found numerically. However, we can avoid numerical differentiation of the eigenvectors, and replace it by model gradients, normally known explicitly from the model parameterization.

Differentiating the eigen-equation (7.2.1) and rearranging using the definition (7.2.95), we obtain

$$\mathbf{W}^{-1}\,\mathbf{A}'\,\mathbf{W} - \mathbf{p}_z' = [\,\mathbf{p}_z\,,\ \mathbf{C}\,]\,. \qquad (7.2.109)$$

This equation is analogous to the perturbation equation (6.7.12) and the following results are equivalent to results (6.7.15), (6.7.16) and (6.7.17). As $\mathbf{p}_z$ is diagonal, the right-hand side of equation (7.2.109) has zero diagonal elements so

$$\mathbf{p}_z' = \mathrm{diag}(\mathbf{W}^{-1}\mathbf{A}'\,\mathbf{W}) \qquad (7.2.110)$$

$$[\,\mathbf{p}_z\,,\ \mathbf{C}\,] = \text{off-diag}(\mathbf{W}^{-1}\mathbf{A}'\mathbf{W}), \qquad (7.2.111)$$

with the obvious notation for diagonal and off-diagonal elements. Thus

$$C_{ij} = \frac{(\mathbf{W}^{-1}\mathbf{A}'\,\mathbf{W})_{ij}}{q_i - q_j}, \qquad (7.2.112)$$

where q_i is the i-th diagonal element in $\mathbf{p}_z$. Thus the elements of **C** can be calculated using the derivatives of the model parameters in $\mathbf{A}'$. This result has been obtained by Chapman and Shearer (1989) and Frazer and Fryer (1989).

Using result (6.3.28) for $\mathbf{W}^{-1}$ in equation (7.2.112), it is straightforward to show that $\mathbf{W}^{\mathrm{T}}\mathbf{I}_2\mathbf{A}'\,\mathbf{W}$ is symmetric as $\mathbf{I}_2\mathbf{A}$ is symmetric. With $\mathbf{K} = \mathbf{I}_3$, this means that if **C** is divided into $m \times m$ sub-matrices, then the diagonal blocks are anti-symmetric, and the off-diagonal blocks are symmetrically related. The isotropic result (7.2.107) agrees with this symmetry. The diagonal elements of the matrix **C** are always zero.

7.2.5.2 Reflection from a second-order discontinuity

Let us consider a second-order discontinuity where the material properties are continuous, but the gradients are discontinuous. Equation (6.3.18) was used to solve for the reflection/transmission coefficients using the eigenvectors. At a second-order discontinuity, the eigenvectors are continuous, $\mathbf{W}_1(z_2) = \mathbf{W}_2(z_2)$, and so using the zeroth-order WKBJ approximation (7.2.102) we find that the matrix **Q** related to the reflection/transmission coefficients (6.3.22) is the identity

$$\mathbf{Q} \simeq \mathbf{W}_1(z_2)^{-1}\mathbf{W}_2(z_2) = \mathbf{I}, \qquad (7.2.113)$$

and with definition (0.1.5), the coefficient matrix reduces to

$$\mathcal{T} = \mathbf{I}_2. \tag{7.2.114}$$

Using the first-order WKBJ approximation (7.2.103), however, we find

$$\mathbf{Q} \simeq \left\{\mathbf{I} + \frac{\mathbf{W}_1^{(1)}(z_2)}{-\mathrm{i}\omega}\right\}^{-1} \mathbf{W}_1(z_2)^{-1}\mathbf{W}_2(z_2)\left\{\mathbf{I} + \frac{\mathbf{W}_2^{(1)}(z_2)}{-\mathrm{i}\omega}\right\} \tag{7.2.115}$$

$$\simeq \mathbf{I} - \frac{\left[\mathbf{W}^{(1)}(z_2)\right]}{-\mathrm{i}\omega}, \tag{7.2.116}$$

to first order in $1/\omega$. The saltus of $\mathbf{W}^{(1)}$ is

$$\left[\mathbf{W}^{(1)}(z_2)\right] = \mathbf{W}_1^{(1)}(z_2) - \mathbf{W}_2^{(1)}(z_2), \tag{7.2.117}$$

where we have taken the reference depth as z_2 so the integrals in the diagonal elements are zero. For the acoustic coefficients, equation (7.2.116) gives

$$\boxed{\mathcal{T}_{11} = \mathcal{T}_{22} = \frac{1}{\mathrm{i}\omega}\frac{[\gamma_A(z_2)]}{2q_\alpha}} \tag{7.2.118}$$

$$\mathcal{T}_{12} = \mathcal{T}_{21} = 1. \tag{7.2.119}$$

These reflection coefficients (7.2.118) from a second-order discontinuity can be used just as the coefficients from a first-order discontinuity, an interface – the factor $-\left[\gamma_A(z_2)\right]/2q_\alpha$ is used as the coefficient, and an integration with respect to time is introduced due to the factor $1/(-\mathrm{i}\omega)$. Thus we expect a reflection from a gradient discontinuity that is the integral of the incident pulse.

Suppose now that the discontinuity is smoothed out, so there is no gradient or higher-order discontinuity but just rapid changes. Intuitively we still expect a low-frequency reflection from the high gradient zone. However, the WKBJ asymptotic expansion completely fails to model the reflection from this zone. There is no coupling between the WKBJ solutions travelling in opposite directions. The WKBJ asymptotic expansion only models the distortion of the pulse as it propagates through the heterogeneity. However large the gradients and however many terms are taken in the asymptotic expansion, the expected low-frequency, partial reflections are never contained in the WKBJ asymptotic expansion – the so-called *WKB paradox* (Gray, 1982). To model reflections from heterogeneities, we must use the WKBJ iterative solution, which is developed in the next section.

7.2.6 The WKBJ iterative solution – the Bremmer series

The solution of the WKBJ paradox is well known. Instead of writing the solution as an asymptotic series in inverse powers of frequency (7.2.89), we write the solution as a combination of the zeroth-order WKBJ approximations, i.e.

$$\underline{\mathbf{w}}(z) = \mathbf{W}(z)\, \mathrm{e}^{\mathrm{i}\omega \boldsymbol{\mathcal{T}}(z)}\, \underline{\mathbf{r}}(z), \tag{7.2.120}$$

generalizing equation (7.2.7). The components of $\underline{\mathbf{r}}$ are the amplitudes of the up and down-going solutions. Substituting in the differential system (7.1.4), $\underline{\mathbf{r}}$ satisfies

$$\frac{\mathrm{d}}{\mathrm{d}z}\underline{\mathbf{r}} = \mathrm{e}^{-\mathrm{i}\omega \boldsymbol{\mathcal{T}}(z)}\, \mathbf{C}(z)\, \mathrm{e}^{\mathrm{i}\omega \boldsymbol{\mathcal{T}}(z)}\, \underline{\mathbf{r}} + \mathrm{e}^{-\mathrm{i}\omega \boldsymbol{\mathcal{T}}(z)}\, \mathbf{W}^{-1}(z)\, \underline{\boldsymbol{\mathcal{F}}}\, \delta(z - z_{\mathrm{S}}), \tag{7.2.121}$$

where $\mathbf{C}$ has been defined before in equation (7.2.95). This equation is often called the *coupling equation* as it represents coupling between components of $\underline{\mathbf{r}}$, i.e. between the zeroth-order WKBJ solutions.

In a homogeneous medium, $\mathbf{C} = \mathbf{0}$, and $\underline{\mathbf{r}}$ is constant. Normally the elements of $\mathbf{C}$ are small and the zeroth-order approximation is good. Therefore, we solve equation (7.2.121) iteratively using

$$\frac{\mathrm{d}}{\mathrm{d}z}\underline{\mathbf{r}}^{(k+1)} = \mathrm{e}^{-\mathrm{i}\omega \boldsymbol{\mathcal{T}}(z)}\, \mathbf{C}(z)\, \mathrm{e}^{\mathrm{i}\omega \boldsymbol{\mathcal{T}}(z)}\, \underline{\mathbf{r}}^{(k)} + \mathrm{e}^{-\mathrm{i}\omega \boldsymbol{\mathcal{T}}(z)}\, \mathbf{W}^{-1}(z)\, \underline{\boldsymbol{\mathcal{F}}}\, \delta(z - z_{\mathrm{S}}), \tag{7.2.122}$$

where $\underline{\mathbf{r}}^{(k)}$ is the k-th iteration. For the zeroth iteration we take

$$\frac{\mathrm{d}}{\mathrm{d}z}\underline{\mathbf{r}}^{(0)} = \mathrm{e}^{-\mathrm{i}\omega \boldsymbol{\mathcal{T}}(z)}\, \mathbf{W}^{-1}(z)\, \underline{\boldsymbol{\mathcal{F}}}\, \delta(z - z_{\mathrm{S}}), \tag{7.2.123}$$

and

$$\underline{\mathbf{r}}^{(0)}(z) = \underline{\mathbf{r}}^{(0)}(z_0) + \int_{z_0}^{z} \mathrm{e}^{-\mathrm{i}\omega \boldsymbol{\mathcal{T}}(\zeta)}\, \mathbf{W}^{-1}(\zeta)\, \underline{\boldsymbol{\mathcal{F}}}\, \delta(\zeta - z_{\mathrm{S}})\, \mathrm{d}\zeta, \tag{7.2.124}$$

generalizing result (7.2.8). As in Section 7.2.2, source radiation boundary conditions are applied so the up-going components of $\underline{\mathbf{r}}$ are zero below the source, and the down-going components are zero above the source. Substituting $\underline{\mathbf{r}}^{(0)}(z)$ in equation (7.2.120) will be equivalent to the zeroth-order WKBJ approximation and will model the direct waves. The solution of equation (7.2.122) is then

$$\boxed{\underline{\mathbf{r}}^{(k+1)}(z) = \underline{\mathbf{r}}^{(0)}(z) + \int^{z} \exp\left(-\mathrm{i}\omega \boldsymbol{\tau}(\zeta)\right) \mathbf{C}(\zeta) \exp\left(\mathrm{i}\omega \boldsymbol{\tau}(\zeta)\right) \underline{\mathbf{r}}^{(k)}(\zeta)\, \mathrm{d}\zeta,} \tag{7.2.125}$$

where again appropriate boundary conditions must be applied.

The convergence of the coupling series (7.2.125) is easily established if the elements of $\mathbf{C}$ are bounded. Provided the eigenvalues $\mathbf{p}_z$ remain real, i.e. provided no turning points or total reflections occur, the plane-wave response is causal. Thus for

$\mathrm{Im}(\omega) > 0$, the response is analytic and can be analytically continued in the upper ω plane (Section 3.1.1). It is sufficient to prove convergence along any line where $\mathrm{Im}(\omega)$ is a positive constant. The phase terms in the integral (7.2.125) always occur in the direction of propagation. Thus if we introduce a positive imaginary part to ω, the integrand decays exponentially. Whatever the magnitude of terms in $\mathbf{C}$, provided they are bounded, the positive imaginary part of ω can always be chosen large enough to reduce the magnitude of the integrand to ensure convergence. If the solution converges for $\mathrm{Im}(\omega) > 0$, it remains valid for $\mathrm{Im}(\omega) = 0$ as the result is causal. This proves that the iterative series converges. More mathematical details can be found in the papers by Verweij and de Hoop (1990) and de Hoop (1990).

If the coupling coefficients are unbounded (singular), then difficulties arise. Two important cases exist: a turning point and an interface. In the first case, the coupling coefficient, γ, has a simple pole. This can be solved using an asymptotic method, the Langer asymptotic expansion, which we discuss in the next section, Section 7.2.7. Alternatively, it can be shown that provided the pole is handled correctly, the coupling solution can still be used in the Cagniard method. This is discussed in more detail in Chapter 8, Section 8.3.2. Finally, at an interface, the coupling coefficient, γ has a delta-function singularity which coincides with a discontinuity in the amplitude coefficients, $\underline{\mathbf{r}}$. If an interface exists, then the discontinuity in the amplitude coefficients is easily found in terms of the reflection/transmission coefficients. It is not necessary to use the coupling equation through an interface.

At an interface, $z = z_j$, the amplitude coefficients are related by

$$\mathbf{W}_{j-1}\,\underline{\mathbf{r}}(z_j + 0) = \mathbf{w}(z_j) = \mathbf{W}_j\,\underline{\mathbf{r}}(z_j - 0), \tag{7.2.126}$$

equivalent to equation (6.3.18). Thus the saltus of $\underline{\mathbf{r}}$ is

$$\left[\underline{\mathbf{r}}\right] = \underline{\mathbf{r}}(z_j + 0) - \underline{\mathbf{r}}(z_j - 0) = (\mathbf{Q}_j - \mathbf{I})\,\underline{\mathbf{r}}(z_j - 0), \tag{7.2.127}$$

where $\mathbf{Q}_j$ is matrix (6.3.22) defined for the j-th interface. Thus formally, the coupling equation (7.2.121) can be generalized to

$$\begin{aligned}\frac{\mathrm{d}}{\mathrm{d}z}\,\underline{\mathbf{r}} = \mathrm{e}^{-\mathrm{i}\omega\boldsymbol{\mathcal{T}}(z)}\,\mathbf{C}(z)\,\mathrm{e}^{\mathrm{i}\omega\boldsymbol{\mathcal{T}}(z)}\,\underline{\mathbf{r}} + \sum_j(\mathbf{Q}_j - \mathbf{I})\,\delta(z - z_j)\\ + \mathrm{e}^{-\mathrm{i}\omega\boldsymbol{\mathcal{T}}(z)}\,\mathbf{W}^{-1}\,\underline{\boldsymbol{\mathcal{F}}}\,\delta(z - z_S),\end{aligned} \tag{7.2.128}$$

where the interface is excluded from $\mathbf{C}$. Effectively, it is not necessary to use the coupling equation through an interface. It is replaced by the reflection/transmission coefficient system. However, if a gradient zone exists, the coupling equation must be used. If the gradient zone is thin (a 'thin' interface), i.e. restricted to a zone narrow compared with the minimum significant wavelength, then convergence may be a problem. As the zone becomes narrower, we should approach the interface limit.

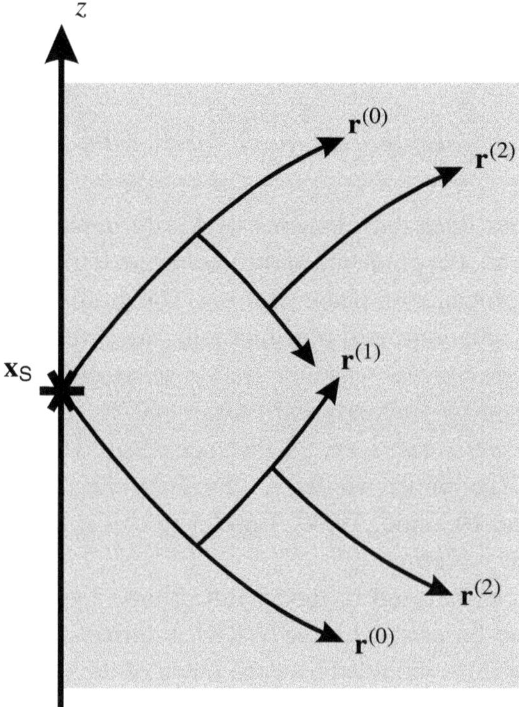

Fig. 7.16. The first three iterations of the WKBJ iterative solution including the direct rays and the first- and second-order reflections from the gradients.

To investigate this process, it is of interest to consider the coupling equation in the interface limit. In Section 9.1.2 we consider this convergence for a thin, acoustic interface.

Each iteration adds 'reflections' from the gradients. The terms $\gamma_A \mathrm{d}\zeta$ (from $\mathbf{C}\,\mathrm{d}\zeta$ for the acoustic case) behave just as reflection coefficients from the depth element $\mathrm{d}\zeta$ (Figure 7.16). For from equation (6.3.7)

$$\mathcal{T}_{11} = -\mathcal{T}_{22} = \frac{\rho_2 q_{\alpha\,1} - \rho_1 q_{\alpha\,2}}{\rho_2 q_{\alpha\,1} + \rho_1 q_{\alpha\,2}} \simeq -\frac{\delta(q_\alpha/\rho)}{2q_\alpha/\rho} \simeq -\gamma_A\,\delta\zeta. \tag{7.2.129}$$

The term γ_A is sometimes called the *differential reflection coefficient*. The coupling between the WKBJ approximations in the WKBJ iterative solution is illustrated in Figure 7.3*c*.

This concept can be made explicit by considering a reflecting region between z_1 and z_2. The reflection/transmission experiment can be written as (7.2.36)

$$\mathbf{W}_1 \begin{pmatrix} \boldsymbol{\mathcal{T}}_{11} & \boldsymbol{\mathcal{T}}_{12} \\ \mathbf{I} & \mathbf{0} \end{pmatrix} = \mathbf{P}(z_1, z_2)\mathbf{W}_2 \begin{pmatrix} \mathbf{0} & \mathbf{I} \\ \boldsymbol{\mathcal{T}}_{21} & \boldsymbol{\mathcal{T}}_{22} \end{pmatrix}. \tag{7.2.130}$$

As $z_1 \to z_2$, $\mathbf{P} \to \mathbf{I}$, $\boldsymbol{\mathcal{T}}_{11} \to \mathbf{0}$, $\boldsymbol{\mathcal{T}}_{12} \to \mathbf{I}$, $\boldsymbol{\mathcal{T}}_{21} \to \mathbf{I}$ and $\boldsymbol{\mathcal{T}}_{22} \to \mathbf{0}$. Differentiating (7.2.130) with respect to z_1 and using $\mathbf{P}' = \mathrm{i}\omega\mathbf{A}\mathbf{P}$, at $z_1 = z_2$ we obtain

$$\frac{\mathrm{d}}{\mathrm{d}z}\begin{pmatrix} \boldsymbol{\mathcal{T}}_{12} & \boldsymbol{\mathcal{T}}_{11} \\ -\boldsymbol{\mathcal{T}}_{22} & -\boldsymbol{\mathcal{T}}_{21} \end{pmatrix} = \mathbf{C} + \mathrm{i}\omega\mathbf{p}_z. \qquad (7.2.131)$$

Remembering that the diagonal elements of $\mathbf{C}$ are zero, e.g. results (7.2.96), (7.2.105), (7.2.107), etc., the gradients of the unconverted transmission coefficients are just given by the propagation phase changes. The gradients of the other coefficients, the reflections and converted transmissions, are given by the elements of $\mathbf{C}$ (apart from signs). This concept was first used in seismics by Scholte (1962), and using the above method by Richards and Frasier (1976). The iterative solution is often called the *Bremmer series* (see, for instance, Gray, 1982, 1983) after earlier work on radio waves (Bremmer, 1949*a*, p. 159; Bremmer, 1949*b*; Budden, 1966, Ch. 18; Clemmow and Heading, 1954). I prefer to call it the *WKBJ iterative* (as opposed to asymptotic) *solution*.

Any method that can be used to model reflections, e.g. the Cagniard method (Section 8.1), can also be used with the WKBJ iterative solution (Section 8.3.1) provided a depth integral is included over the range of the reflector depths.

7.2.7 The Langer asymptotic expansion

If a turning point exists, i.e. the vertical slowness is zero, both the WKBJ asymptotic expansion (Section 7.2.5) and the WKBJ iterative solution (Section 7.2.6) break down. The eigenvectors are no longer independent and the elements of $\mathbf{C}$ (7.2.112) are singular. The WKBJ asymptotic expansion breaks down because the first-order terms are infinite. The integrands of the WKBJ iterative solution (7.2.125) contain a singularity (γ_A has a pole). While it is possible to evaluate these integrals with the singularities properly, it is complicated (Chapman, 1974*a*, 1976). A simpler approach is to generalize the asymptotic expansion (Chapman, 1974*b*).

Let us consider the simple acoustic system (7.1.6). First we transform the dependent variables to separate the density and slowness behaviour. Defining

$$\boxed{\mathbf{L} = (\rho p)^{-1/2}\begin{pmatrix} p & 0 \\ 0 & \rho \end{pmatrix},} \qquad (7.2.132)$$

we transform $\mathbf{w}$ as

$$\mathbf{w} = \mathbf{L}\,\mathbf{y}. \qquad (7.2.133)$$

The differential system (7.1.4) (ignoring the source term), becomes

$$\frac{\mathrm{d}\mathbf{y}}{\mathrm{d}z} = \mathrm{i}\omega\,\mathbf{B}\,\mathbf{y} + \frac{\rho'}{2\rho}\,\mathbf{I}_3, \tag{7.2.134}$$

where $\mathbf{I}_3$ is defined in equation (0.1.5). The matrix $\mathbf{B}$ is

$$\mathbf{B} = \mathbf{L}^{-1}\,\mathbf{A}\,\mathbf{L} = \begin{pmatrix} 0 & -q_\alpha^2/p \\ -p & 0 \end{pmatrix}. \tag{7.2.135}$$

By design, the first term on the right-hand side of equation (7.2.134) contains the slowness behaviour, and the second term the density behaviour. The first term is asymptotically more significant due to the frequency factor. If we ignore the second term in equation (7.2.134), which will be small, $\mathrm{O}(1/\omega)$, at high frequencies, the differential equation reduces to

$$y_2'' + \omega^2\, q_\alpha^2\, y_2 = 0. \tag{7.2.136}$$

If the squared vertical slowness can be approximated by a linear function near the turning point $z_\alpha(p)$, where $p\,\alpha(z_\alpha(p)) = 1$ (cf. Section 2.3.1), then the differential equation can be reduced to a standard form. Thus the vertical squared slowness is approximately

$$q_\alpha^2 \simeq \left(-\frac{2\alpha'}{\alpha^3}\right)(z - z_\alpha(p)), \tag{7.2.137}$$

where in this first-order Taylor expansion, the velocity and its gradient are taken at the turning point. Substituting in equation (7.2.136), and changing the independent variable to

$$x = -\left(-\frac{2\omega^2\alpha'}{\alpha^3}\right)^{1/3}(z - z_\alpha), \tag{7.2.138}$$

we obtain the Stokes equation (Abramowitz and Stegun, 1965, §10.4.1)

$$y'' - x\,y = 0. \tag{7.2.139}$$

The solutions of the Stokes equation are the Airy functions (Abramowitz and Stegun, 1965, §10.4). Note that if z is positive upwards, $\alpha' < 0$, so the independent variable, x (7.2.138), in the Stokes equation is measured in the opposite direction to z (Figure 7.17). The Airy functions have negative argument above the turning point and positive below it.

As the squared vertical slowness is not exactly linear, we use Langer's (1937) approach to generalize by 'stretching' the z coordinate. We define a new depth

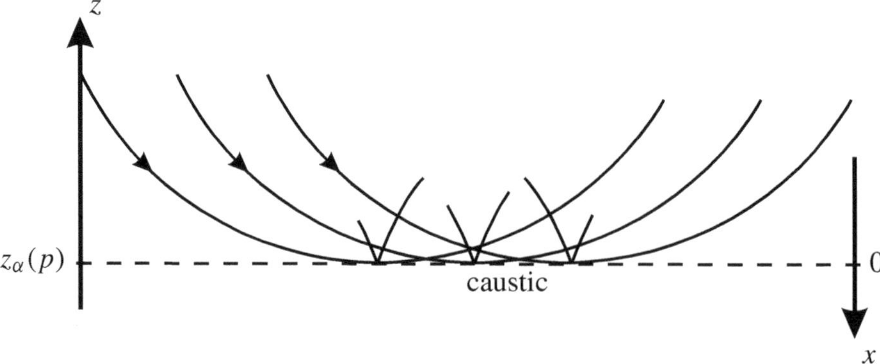

Fig. 7.17. Waves with constant horizontal slowness p at a turning point, $z_\alpha(p)$, forming a horizontal caustic. Parts of the cusped wavefronts are shown.

variable

$$\xi(z) = \left(\frac{3\omega\tau_\alpha}{2}\right)^{2/3}, \tag{7.2.140}$$

where

$$\tau_\alpha(p, z) = \int_{z_\alpha(p)}^{z} q_\alpha(p, \zeta)\,\mathrm{d}\zeta. \tag{7.2.141}$$

Near the turning point, we have

$$\tau_\alpha(p, z) \simeq \frac{2}{3}\left(-\frac{2\alpha'}{\alpha^3}\right)^{1/2}(z - z_\alpha(p))^{3/2} \tag{7.2.142}$$

$$\xi(z) \simeq \left(-\frac{2\omega^2\alpha'}{\alpha^3}\right)^{1/3}(z - z_\alpha(p))\,. \tag{7.2.143}$$

For simplicity, we have assumed $\omega > 0$ in (7.2.140) and (7.2.143). For $z > z_\alpha(p)$, τ and ξ are positive real, and for $z < z_\alpha(p)$, ξ is negative real and τ is negative imaginary. Airy functions with $-\xi$ as argument then approximately satisfy equation (7.2.136).

We now set up the solution in matrix notation. Consider the matrix

$$\boxed{\boldsymbol{\mathcal{A}} = \left(-\frac{2\pi\mathrm{i}}{\omega p\xi'}\right)^{1/2}\begin{pmatrix} \mathrm{i}\xi' Aj'(-\xi) & \mathrm{i}\xi' Bj'(-\xi) \\ -\omega p Aj(-\xi) & -\omega p Bj(-\xi)\end{pmatrix}.} \tag{7.2.144}$$

We introduce the notation Aj and Bj for linear combinations of the standard Airy functions, i.e. general solutions of Stokes equation (7.2.139). In different regions, we will use alternative combinations which represent travelling or evanescent solutions, and will be asymptotically equivalent to the WKBJ expansions. The matrix

$\boldsymbol{\mathcal{A}}$ has been designed so it approximately satisfies equation (7.2.134). It exactly satisfies

$$\boldsymbol{\mathcal{A}}' = \mathrm{i}\omega \mathbf{B} \boldsymbol{\mathcal{A}} - \frac{\xi''}{2\xi'} \mathbf{I}_3 \boldsymbol{\mathcal{A}}. \tag{7.2.145}$$

For negative arguments, the Airy functions are standing waves. The leading terms in the asymptotic expansions are given in Appendix D.2, equations (D.2.4) and (D.2.5). The asymptotic forms for the derivatives of the Airy functions corresponding to equations (D.2.4) and (D.2.5) are also useful (Abramowitz and Stegun, 1965, §10.4.62 and §10.4.67). We take linear combinations of the standard Airy functions to create travelling-wave solutions

$$Aj(-\xi) = \frac{1}{2}(\mathrm{i}\, Ai(-\xi) + Bi(-\xi)) \tag{7.2.146}$$

$$Bj(-\xi) = \frac{1}{2}(Ai(-\xi) + \mathrm{i}\, Bi(-\xi)), \tag{7.2.147}$$

which asymptotically reduce to

$$Aj(-\xi) \simeq \frac{1}{2\pi^{1/2}\xi^{1/4}}\, \mathrm{e}^{\mathrm{i}\zeta + \mathrm{i}\pi/4} \tag{7.2.148}$$

$$Bj(-\xi) \simeq \frac{1}{2\pi^{1/2}\xi^{1/4}}\, \mathrm{e}^{-\mathrm{i}\zeta + \mathrm{i}\pi/4}, \tag{7.2.149}$$

where

$$\zeta = \omega\tau_\alpha. \tag{7.2.150}$$

The common factors have been introduced in $\boldsymbol{\mathcal{A}}$ so that its inverse is simply

$$\boldsymbol{\mathcal{A}}^{-1} = \begin{pmatrix} -\mathcal{A}_{22} & \mathcal{A}_{12} \\ \mathcal{A}_{21} & -\mathcal{A}_{11} \end{pmatrix} \tag{7.2.151}$$

($|\boldsymbol{\mathcal{A}}|^{-1} = -1$, cf. $|\mathbf{W}| = -1$ and equation (7.2.5)). This result requires the Wronskian of the Airy functions (Abramowitz and Stegun, 1965, §10.4.10)

$$Ai(x)Bi'(x) - Ai'(x)Bi(x) = \frac{1}{\pi}. \tag{7.2.152}$$

Thus substituting $\zeta = \omega\tau_\alpha$ in equations (7.2.148) and (7.2.149), the first column of $\boldsymbol{\mathcal{A}}$ (7.2.144) reduces to the WKBJ asymptotic solution propagating in the positive direction, and the second column to the solution propagating in the negative direction, when $\xi \gg 1$. We have arranged the constant factors so that with the 'Airy' functions (7.2.146) and (7.2.147), we have simply

$$\mathbf{L}(z)\, \boldsymbol{\mathcal{A}}^{(\text{travelling})}(z) \to \mathbf{W}(z)\, \mathrm{e}^{\mathrm{i}\omega \boldsymbol{\mathcal{T}}(p,z)}, \tag{7.2.153}$$

where we have used $\xi^{1/2}\xi' = \omega q_\alpha$ (the '(travelling)' superscript indicates the choice (7.2.146) and (7.2.147)).

Alternatively, below the turning point, we need the Airy functions with positive arguments which are evanescent, (D.2.7) and (D.2.8). We choose

$$Aj(-\xi) = \frac{1}{2}\,Bi(-\xi) \tag{7.2.154}$$

$$Bj(-\xi) = Ai(-\xi), \tag{7.2.155}$$

which will be asymptotically equivalent to the WKBJ solutions in evanescent regions (using equations (D.2.7) and (D.2.8), and for the gradients, Abramowitz and Stegun, 1965, §10.4.61 and §10.4.66). Again the factors are arranged so equation (7.2.151) is still satisfied, and the columns of $\boldsymbol{\mathcal{A}}$ represent waves evanescent in the positive and negative directions, i.e.

$$\mathbf{L}(z)\,\boldsymbol{\mathcal{A}}^{\text{(evanescent)}}(z) \to \mathbf{W}(z)\,\mathrm{e}^{\mathrm{i}\omega\boldsymbol{\mathcal{T}}(p,z)}, \tag{7.2.156}$$

where the elements of $\boldsymbol{\mathcal{T}}$ are imaginary (the '(evanescent)' superscript indicates the choice (7.2.154) and (7.2.155)).

Having designed the matrix $\boldsymbol{\mathcal{A}}$ to satisfy the differential equation (7.2.145), we can write the solution as an asymptotic expansion

$$\mathbf{F}(z) = \mathbf{L}(z)\left\{\sum_{m=0}^{\infty}\frac{\mathbf{L}^{(m)}(z)}{(-\mathrm{i}\omega)^m}\right\}\boldsymbol{\mathcal{A}}(z). \tag{7.2.157}$$

We call this the *Langer asymptotic expansion*. Substituting in the differential system (7.1.4), we obtain from the coefficient of ω^{-m}

$$\mathbf{L}^{(m)\prime} - \frac{\rho'}{2\rho}\mathbf{I}_3\,\mathbf{L}^{(m)} - \frac{\xi''}{2\xi'}\mathbf{L}^{(m)}\,\mathbf{I}_3 = \left[\mathbf{B}\,,\,\mathbf{L}^{(m+1)}\right]. \tag{7.2.158}$$

These expressions are solved for $\mathbf{L}^{(m)}$ by a procedure similar to the WKBJ method, but somewhat more complicated. Wasow (1965), Chapman (1974*b*) and Woodhouse (1978) have given more details. For $m = -1$, the left-hand side is zero and $\mathbf{L}^{0)} = \mathbf{I}$ is a possible solution. By design of the matrix $\boldsymbol{\mathcal{A}}$, it can be confirmed from the $m = 0$ expression that this is consistent.

The zeroth-order Langer approximation

$$\boxed{\mathbf{F}(z) = \mathbf{L}(z)\,\boldsymbol{\mathcal{A}}(z),} \tag{7.2.159}$$

is the useful result. It is important to know that higher-order terms can be found in order to establish the validity of the asymptotic expansion, but in practice they are rarely used and we resort to numerical methods if a complete solution is needed. The zeroth-order Langer approximation (7.2.159), however, provides the crucial result connecting the down-going travelling wave with the up-going wave. In the travelling wave region ($\xi \gg 1$) it is most convenient to use $\boldsymbol{\mathcal{A}}^{(\text{travelling})}$ in equation (7.2.159), and in the evanescent region ($\xi \ll 1$), $\boldsymbol{\mathcal{A}}^{(\text{evanescent})}$. Near the turning point, neither form is preferred as the two waves couple. Of course $\boldsymbol{\mathcal{A}}^{(\text{travelling})}$ and $\boldsymbol{\mathcal{A}}^{(\text{evanescent})}$ are not equal but are linearly related

$$\boldsymbol{\mathcal{A}}^{(\text{evanescent})} = \boldsymbol{\mathcal{A}}^{(\text{travelling})} \begin{pmatrix} 1/2 & -\mathrm{i} \\ -\mathrm{i}/2 & 1 \end{pmatrix}. \tag{7.2.160}$$

If there are no sources or interfaces below the turning point, the solution must be evanescent decaying away from the turning point, i.e. the second column of $\boldsymbol{\mathcal{A}}^{(\text{evanescent})}$. Then we must have

$$\boldsymbol{\mathcal{A}}^{(\text{evanescent})} \begin{pmatrix} 0 \\ 1 \end{pmatrix} = \boldsymbol{\mathcal{A}}^{(\text{travelling})} \begin{pmatrix} -\mathrm{i} \\ 1 \end{pmatrix}. \tag{7.2.161}$$

The second column of $\boldsymbol{\mathcal{A}}^{(\text{travelling})}$ is the wave in-going (down-going) to the turning point, and the first the out-going (up-going) wave. The coupling in the Langer solution of the WKBJ travelling waves through the turning point is illustrated in Figure 7.3*d*. Thus the ratio of the out-going to in-going waves, the reflection coefficient, is

$$\mathrm{e}^{2\mathrm{i}\zeta - \mathrm{i}\pi/2} = \mathrm{e}^{2\mathrm{i}\omega\tau_\alpha - \mathrm{i}\pi/2}. \tag{7.2.162}$$

The phase $2\tau_\alpha$ is just as expected for the WKBJ wave travelling down to the turning point and up again, but the extra factor $-\mathrm{i}$ is interesting and important. Connecting through the turning point introduces the phase shift of $-\mathrm{i}\pi/2$.

Note that we have restricted our discussion of the Airy functions and the Langer asymptotic expansion to $\omega > 0$ (to avoid handling the awkward fractional powers). Using the symmetry that must exist in the response (3.1.9), we can now generalize this phase shift to

$$\boxed{\mathrm{e}^{-\mathrm{i}\,\mathrm{sgn}(\omega)\pi/2}.} \tag{7.2.163}$$

In the transform (ω, p) domain, the turning point is a caustic (Figure 7.17). Result (7.2.163) is a special case of the KMAH index introduced into the ray Green function (5.2.70). Rays propagating through an Airy caustic suffer a phase shift (7.2.163) corresponding to a Hilbert transform in the time domain (we avoid calling this a phase advance or retardation, as the Hilbert transform is a zero-phase operation). The phase shift follows immediately from the asymptotic form of the Airy function (D.2.4) where the out-going phase term $\exp(\mathrm{i}\zeta)$ has a factor $-\mathrm{i}$ relative to the in-going phase $\exp(-\mathrm{i}\zeta)$. We can also deduce that if the phase and group directions are opposite, as occasionally happens in anisotropic media and for surface waves, then the identification of the in-going and out-going waves is reversed, and the phase shift is $+\mathrm{i}$.

7.2.7.1 Elastic waves

The Langer asymptotic expansion for elastic waves can be obtained in a similar fashion. Wasow (1965, Theorem 25.1, 25.2 and 29.2) has established that the procedure is possible for differential systems of the form (7.1.4). Chapman (1974*b*) and Woodhouse (1978) have applied the method to isotropic elastic media. Kennett and Illingworth (1981) have used it in combination with the Kennett layer-matrix algorithm (Section 7.2.4.1) to obtain the propagator in piecewise smooth models. Garmany (1988*b*) has investigated the anisotropic case. We only consider a special case here – an anisotropic medium with up–down symmetry on a plane of symmetry, so the $qP - qSV$ system separates. The important example is transverse isotropic media (Section 4.4.4) with a vertical axis of symmetry (TIV), but some more general media are covered, e.g. symmetry planes in cubic media.

The *SH* system (7.1.33) is very similar to the acoustic system. Defining a transformation (7.2.133) with

$$\mathbf{L} = (\mu p)^{-1/2} \begin{pmatrix} \mu p & 0 \\ 0 & 1 \end{pmatrix}, \tag{7.2.164}$$

we obtain the differential system (cf. (7.2.134))

$$\frac{\mathrm{d}\mathbf{y}}{\mathrm{d}z} = \mathrm{i}\omega \mathbf{B}\,\mathbf{y} - \frac{\mu'}{2\mu}\mathbf{I}_3, \tag{7.2.165}$$

where

$$\mathbf{B} = \begin{pmatrix} 0 & -q_\beta^2/p \\ -p & 0 \end{pmatrix}. \tag{7.2.166}$$

Provided we substitute the shear velocity instead of the acoustic velocity, the zeroth-order Langer approximation (7.2.159) still applies.

The $P - SV$ system is necessarily more complicated. Chapman (1974*b*) and Woodhouse (1978) have given the transformations for the fourth-order isotropic system (7.1.34). Here we follow a more general procedure which extends the results to anisotropic media with up–down symmetry on planes of symmetry. Without loss in generality, we arrange the coordinates so that the symmetry plane is the x_1–x_3 plane and take $p_2 = 0$. Then the sixth-order differential system (7.1.25) separates into fourth and second-order systems. The fourth-order system, the first, third, fourth and sixth rows and columns, describes waves with the polarization in the plane, and the second-order system, the second and fifth rows and columns, has the polarization normal to the plane. We follow the normal convention of referring to these as the $qP - qSV$ and qSH systems. Henceforth, we only consider the $qP-qSV$ system – the vectors $\mathbf{v}$ and $\mathbf{t}_3$ are reduced to two components, and the matrix $\mathbf{A}$ is 4×4.

With up–down symmetry, the eigenvalues of the matrix $\mathbf{A}$ must exist in positive and negative pairs. We denote these by $\pm q_V$ and $\pm q_P$ for the qSV and qP waves. We use subscripts V and P here rather than β and α to distinguish from the purely isotropic results. For clarity, we write the matrix of eigenvectors as

$$\mathbf{W} = \begin{pmatrix} \acute{\mathbf{w}}^V & \acute{\mathbf{w}}^P & \grave{\mathbf{w}}^V & \grave{\mathbf{w}}^P \end{pmatrix}, \tag{7.2.167}$$

where the accent indicates the propagation direction, and the superscript the wave type. The elements of the up and down-going eigenvectors only differ by sign and we must have

$$\mathbf{W} = \begin{pmatrix} g_1^V & g_1^P & g_1^V & g_1^P \\ g_3^V & g_3^P & -g_3^V & -g_3^P \\ \sigma_{13}^V & \sigma_{13}^P & -\sigma_{13}^V & -\sigma_{13}^P \\ \sigma_{33}^V & \sigma_{33}^P & \sigma_{33}^V & \sigma_{33}^P \end{pmatrix}. \tag{7.2.168}$$

With these specializations, we can proceed to decompose the matrix $\mathbf{A}$. With the eigenvectors, the matrix $\mathbf{A}$ can be diagonalized (cf. equation (7.2.1))

$$\mathbf{p}_z = \mathbf{W}^{-1}\mathbf{A}\mathbf{W} \tag{7.2.169}$$

The algebra using the eigenvector matrix (7.2.167), $\mathbf{W}$, is not trivial, especially in anisotropic media, as all the elements are non-zero, and when eigenvalues are equal, degenerate. We require an alternative transformation of the form (7.2.133) that block-diagonalizes the matrix $\mathbf{A}$ (cf. equation (7.2.135)) and remains valid when the eigenvalues are degenerate, i.e.

$$\mathbf{L}^{-1}\mathbf{A}\mathbf{L} = \mathbf{B} = \begin{pmatrix} \mathbf{B}_V & \mathbf{0} \\ \mathbf{0} & \mathbf{B}_P \end{pmatrix} = \mathbf{B}_V \oplus \mathbf{B}_P = \mathbf{L}^{-1}\mathbf{W}\mathbf{p}_z\mathbf{W}^{-1}\mathbf{L}, \tag{7.2.170}$$

where the sub-matrices $\mathbf{B}_V$ and $\mathbf{B}_P$ are of the form (7.2.135) (we use the symbol $\oplus$ to indicate a matrix formed from sub-matrices on the diagonal). The matrix $\mathbf{L}$, valid for any eigen-system (7.2.168), is

$$\mathbf{L} = \frac{1}{2}\mathbf{W}\begin{pmatrix} n_1 & n_2 & 0 & 0 \\ 0 & 0 & n_3 & n_4 \\ n_1 & -n_2 & 0 & 0 \\ 0 & 0 & -n_3 & n_4 \end{pmatrix} \tag{7.2.171}$$

$$= \frac{1}{2}\left(n_1\left(\acute{\mathbf{w}}^V + \grave{\mathbf{w}}^V\right) \quad n_2\left(\acute{\mathbf{w}}^V - \grave{\mathbf{w}}^V\right) \quad n_3\left(\acute{\mathbf{w}}^P - \grave{\mathbf{w}}^P\right) \quad n_4\left(\acute{\mathbf{w}}^P + \grave{\mathbf{w}}^P\right)\right) \tag{7.2.172}$$

The elements of $\mathbf{L}$ are obtained from linear combinations of the columns of $\mathbf{W}$ with revised normalization. For the moment we just denote the re-normalization by $n_i/2$ – the factor of a half is introduced into equation (7.2.172) in order to simplify the elements. Using the orthonormal relationships (6.3.33), the inverse matrix $\mathbf{L}^{-1}$ can be obtained as

$$\mathbf{L}^{-1} = \begin{pmatrix} (\acute{\mathbf{w}}^V - \grave{\mathbf{w}}^V)^{\ddagger} \big/ n_1 \\ (\acute{\mathbf{w}}^V + \grave{\mathbf{w}}^V)^{\ddagger} \big/ n_2 \\ (\acute{\mathbf{w}}^P + \grave{\mathbf{w}}^P)^{\ddagger} \big/ n_3 \\ (\acute{\mathbf{w}}^P - \grave{\mathbf{w}}^P)^{\ddagger} \big/ n_4 \end{pmatrix}. \tag{7.2.173}$$

The matrices needed in equation (7.2.170) are

$$\mathbf{W}^{-1}\mathbf{L} = \begin{pmatrix} \acute{\mathbf{w}}^{V\ddagger} \\ \acute{\mathbf{w}}^{P\ddagger} \\ -\grave{\mathbf{w}}^{V\ddagger} \\ -\grave{\mathbf{w}}^{P\ddagger} \end{pmatrix} \mathbf{L} = \frac{1}{2} \begin{pmatrix} 1 & 1 & 0 & 0 \\ 0 & 0 & 1 & 1 \\ 1 & -1 & 0 & 0 \\ 0 & 0 & -1 & 1 \end{pmatrix} \begin{pmatrix} n_1 & 0 & 0 & 0 \\ 0 & n_2 & 0 & 0 \\ 0 & 0 & n_3 & 0 \\ 0 & 0 & 0 & n_4 \end{pmatrix}, \tag{7.2.174}$$

and

$$\mathbf{L}^{-1}\mathbf{W} = \begin{pmatrix} n_1^{-1} & 0 & 0 & 0 \\ 0 & n_2^{-1} & 0 & 0 \\ 0 & 0 & n_3^{-1} & 0 \\ 0 & 0 & 0 & n_4^{-1} \end{pmatrix} \begin{pmatrix} 1 & 0 & 1 & 0 \\ 1 & 0 & -1 & 0 \\ 0 & 1 & 0 & -1 \\ 0 & 1 & 0 & 1 \end{pmatrix}. \tag{7.2.175}$$

Substituting these expressions in equation (7.2.170) and simplifying we find

$$\mathbf{B}_\nu = \begin{pmatrix} 0 & q_\nu/g_\nu \\ g_\nu q_\nu & 0 \end{pmatrix}, \tag{7.2.176}$$

where ν is V or P (no summation over ν), with

$$g_V = \frac{n_1}{n_2} \tag{7.2.177}$$

$$g_P = \frac{n_3}{n_4}. \tag{7.2.178}$$

Thus the matrix $\mathbf{L}$ (7.2.172) block-diagonalizes the differential system.

We can now investigate the elements of the matrix $\mathbf{L}$ and its inverse $\mathbf{L}^{-1}$ in more detail. Using equation (7.2.168), we have

$$\mathbf{L} = \begin{pmatrix} n_1 g_1^V & 0 & 0 & n_4 g_1^P \\ 0 & n_2 g_3^V & n_3 g_3^P & 0 \\ 0 & n_2 \sigma_{13}^V & n_3 \sigma_{13}^P & 0 \\ n_1 \sigma_{33}^V & 0 & 0 & n_4 \sigma_{33}^P \end{pmatrix}. \tag{7.2.179}$$

The inverse matrix $\mathbf{L}^{-1}$ can be obtained by two methods. First from the definition (7.2.173), it is

$$\mathbf{L}^{-1} = -2 \begin{pmatrix} \sigma_{13}^V/n_1 & 0 & 0 & g_3^V/n_1 \\ 0 & \sigma_{33}^V/n_2 & g_1^V/n_2 & 0 \\ 0 & \sigma_{33}^P/n_3 & g_1^P/n_3 & 0 \\ \sigma_{13}^P/n_4 & 0 & 0 & g_3^P/n_4 \end{pmatrix}. \tag{7.2.180}$$

Alternatively, the matrix (7.2.179) is effectively two 2×2 blocks, and so can be inverted easily. It is

$$\mathbf{L}^{-1} = \begin{pmatrix} n_4' \sigma_{33}^P & 0 & 0 & -n_4' g_1^P \\ 0 & n_3' \sigma_{13}^P & -n_3' g_3^P & 0 \\ 0 & -n_2' \sigma_{13}^V & n_2' g_3^V & 0 \\ -n_1' \sigma_{33}^V & 0 & 0 & n_1' g_1^V \end{pmatrix}, \tag{7.2.181}$$

where the factors n_i' are the corresponding n_i divided by the appropriate determinant of a 2×2 sub-matrix. The connections are

$$n_1 n_4' = n_1' n_4 = \left(g_1^V \sigma_{33}^P - g_1^P \sigma_{33}^V \right)^{-1} \tag{7.2.182}$$

$$n_2 n_3' = n_2' n_3 = \left(g_3^V \sigma_{13}^P - g_3^P \sigma_{13}^V \right)^{-1}. \tag{7.2.183}$$

The orthonormality relation (6.3.33) can be applied to definition (7.2.168), and together with a comparison of equations (7.2.180) and (7.2.181) gives

$$\frac{n_1 n_4'}{2} = \frac{n_1' n_4}{2} = -\frac{2}{n_2 n_3'} = -\frac{2}{n_2' n_3} = -\frac{\sigma_{13}^V}{\sigma_{33}^P} = \frac{g_3^V}{g_1^P} = \frac{\sigma_{13}^P}{\sigma_{33}^V} = -\frac{g_3^P}{g_1^V}. \tag{7.2.184}$$

While the equality of all these expressions can be shown explicitly in isotropic media, the results are non-trivial in anisotropic media and would be tedious to prove explicitly. The equalities are useful as they allow factors to be removed which further simplify the matrix $\mathbf{L}$ and its inverse.

Using the equalities (7.2.184), we define

$$Z_V = \frac{\sigma_{33}^V}{g_1^V} = -\frac{\sigma_{13}^P}{g_3^P} \tag{7.2.185}$$

$$Z_P = \frac{\sigma_{33}^P}{g_1^P} = -\frac{\sigma_{13}^V}{g_3^V}. \tag{7.2.186}$$

(We use the notation Z as these quantities have the dimensions of impedance. They could be called *cross-impedances* as they are the ratio of orthogonal components of stress and velocity. We take the subscript from the first definition of each. The normalization (6.3.33) is equivalent to the relationships $2(Z_P - Z_V)g_1^V g_3^V = 2(Z_V - Z_P)g_1^P g_3^P = 1$.) The matrix $\mathbf{L}$ (7.2.179) can then be factored as

$$\mathbf{L} = \begin{pmatrix} 1 & 0 & 0 & -1 \\ 0 & 1 & 1 & 0 \\ 0 & -Z_P & -Z_V & 0 \\ Z_V & 0 & 0 & -Z_P \end{pmatrix} \begin{pmatrix} n_1 g_1^V & 0 & 0 & 0 \\ 0 & n_2 g_3^V & 0 & 0 \\ 0 & 0 & n_3 g_3^P & 0 \\ 0 & 0 & 0 & -n_4 g_1^P \end{pmatrix}$$
$$= \begin{pmatrix} 0 & -1 & 0 & 0 \\ 1 & 0 & 0 & -1 \\ -Z_P & 0 & 0 & Z_V \\ 0 & 0 & 1 & 0 \end{pmatrix} \begin{pmatrix} 0 & 1 & 0 & 0 \\ -1 & 0 & 0 & 1 \\ Z_V & 0 & 0 & -Z_P \\ 0 & 0 & -1 & 0 \end{pmatrix}$$
$$\times \begin{pmatrix} n_1 g_1^V & 0 & 0 & 0 \\ 0 & n_2 g_3^V & 0 & 0 \\ 0 & 0 & n_3 g_3^P & 0 \\ 0 & 0 & 0 & -n_4 g_1^P \end{pmatrix} \tag{7.2.187}$$

(the minus sign is introduced into the trailing normalization matrix anticipating the isotropic results and to simplify later results). Similarly, the inverse matrix $\mathbf{L}^{-1}$ (7.2.181) can be written as

$$\mathbf{L}^{-1} = \begin{pmatrix} -n'_4 g_1^P & 0 & 0 & 0 \\ 0 & -n'_3 g_3^P & 0 & 0 \\ 0 & 0 & -n'_2 g_3^V & 0 \\ 0 & 0 & 0 & n'_1 g_1^V \end{pmatrix} \begin{pmatrix} -Z_P & 0 & 0 & 1 \\ 0 & Z_V & 1 & 0 \\ 0 & -Z_P & -1 & 0 \\ -Z_V & 0 & 0 & 1 \end{pmatrix}$$

$$= \begin{pmatrix} -n_4' g_1^P & 0 & 0 & 0 \\ 0 & -n_3' g_3^P & 0 & 0 \\ 0 & 0 & -n_2' g_3^V & 0 \\ 0 & 0 & 0 & n_1' g_1^V \end{pmatrix}$$

$$\times \begin{pmatrix} 0 & Z_P & 1 & 0 \\ 1 & 0 & 0 & 0 \\ 0 & 0 & 0 & -1 \\ 0 & Z_V & 1 & 0 \end{pmatrix} \begin{pmatrix} 0 & Z_V & 1 & 0 \\ -1 & 0 & 0 & 0 \\ 0 & 0 & 0 & 1 \\ 0 & Z_P & 1 & 0 \end{pmatrix}. \tag{7.2.188}$$

The simplicity and repetitiveness of these expression is remarkable enough in isotropic media. In anisotropic media, it is truly unexpected.

The choice of the normalization factors n_i is arbitrary but for symmetry, the natural choice is

$$n_1 = n_1' = -\frac{2}{n_2} = \frac{2}{n_2'} = \left(-2\,\frac{g_3^V}{g_1^V}\right)^{1/2} \tag{7.2.189}$$

$$n_3 = -n_3' = -\frac{2}{n_4} = -\frac{2}{n_4'} = \left(2\,\frac{g_1^P}{g_3^P}\right)^{1/2} \tag{7.2.190}$$

(because of the up–down symmetry, the expressions in brackets are always positive for travelling waves). In isotropic media, these normalizations reduce to $(2p/q_\alpha)^{1/2}$ and $\left(2p/q_\beta\right)^{1/2}$, respectively, and expression (7.2.184) equals $-\left(q_\alpha q_\beta\right)^{1/2}$. Then we obtain for definitions (7.2.177) and (7.2.178)

$$g_V = \frac{g_3^V}{g_1^V} \tag{7.2.191}$$

$$g_P = -\frac{g_1^P}{g_3^P} \tag{7.2.192}$$

(both negative – for travelling waves, these are the tangent or cotangent of the polarization angle), which in isotropic media reduce to $-p/q_\beta$ and $-p/q_\alpha$, respectively. The cross-impedances (7.2.185) and (7.2.186) in isotropic media are

$$Z_V = 2\mu p \tag{7.2.193}$$

$$Z_P = 2\mu p - \rho/p. \tag{7.2.194}$$

We notice that for a horizontally travelling wave, the corresponding factor g_ν is singular, but in the block matrix $\mathbf{B}_\nu$ (7.2.176), one element (B_{12}) is zero and the other (B_{21}) remains finite. The trailing normalization matrix in $\mathbf{L}$ (7.2.187) simplifies to

$$\left(-2g_1^V g_3^V\right)^{1/2} \mathbf{I} \oplus \left(2g_1^P g_3^P\right)^{1/2} \mathbf{I} = s_V \mathbf{I} \oplus s_P \mathbf{I}, \tag{7.2.195}$$

say, and the leading normalization matrix in $\mathbf{L}^{-1}$ (7.2.188) to

$$\left(2g_1^P g_3^P\right)^{1/2} \mathbf{I} \oplus \left(-2g_1^V g_3^V\right)^{1/2} \mathbf{I} = s_P \mathbf{I} \oplus s_V \mathbf{I}. \tag{7.2.196}$$

These matrices just scale the solutions. In isotropic media the two scaling factors are equal, i.e. $s_V = s_P = (p/\rho)^{1/2}$, and normalization matrices (7.2.195) and (7.2.196) reduce to $(p/\rho)^{1/2}\,\mathbf{I}$.

Applying the transformation (7.2.133) with the matrix (7.2.187) to the differential system (7.1.4) (ignoring the source term) with equation (7.1.34), we obtain

$$\frac{\mathrm{d}\mathbf{y}}{\mathrm{d}z} = \mathrm{i}\omega\, \mathbf{B}\, \mathbf{y} - \mathbf{L}^{-1}\, \mathbf{L}', \tag{7.2.197}$$

where $\mathbf{B}$ has been given by equation (7.2.170). In isotropic media, the matrix $\mathbf{L}$ reduces to

$$\boxed{\mathbf{L} = (\rho\, p)^{-1/2} \begin{pmatrix} p & 0 & 0 & -p \\ 0 & p & p & 0 \\ 0 & \mu\,\Omega & -2\mu\, p^2 & 0 \\ 2\mu\, p^2 & 0 & 0 & \mu\,\Omega \end{pmatrix},} \tag{7.2.198}$$

and the second, lower-order term in equation (7.2.197) is

$$\mathbf{L}^{-1}\,\mathbf{L}' = \frac{2\mu' p^2}{\rho} \begin{pmatrix} 1 & 0 & 0 & -1 \\ 0 & -1 & -1 & 0 \\ 0 & 1 & 1 & 0 \\ 1 & 0 & 0 & -1 \end{pmatrix} + \frac{\rho'}{2\rho} \begin{pmatrix} -1 & 0 & 0 & 2 \\ 0 & 1 & 0 & 0 \\ 0 & 2 & -1 & 0 \\ 0 & 0 & 0 & 1 \end{pmatrix}. \tag{7.2.199}$$

Ignoring this term, the zeroth-order Langer approximation is

$$\boxed{\mathbf{F}(z) = \mathbf{L}(z)\left(\boldsymbol{\mathcal{A}}_V(z) \oplus \boldsymbol{\mathcal{A}}_P(z)\right)} \tag{7.2.200}$$

(cf. result (7.2.159)), with the obvious definitions for the matrices $\boldsymbol{\mathcal{A}}$ (7.2.144) with subscripts V and P.

Just as the WKBJ asymptotic expansion failed to model reflections from velocity gradients, the Langer asymptotic expansion does not model fully reflections from gradients. Wasow (1965), Chapman (1974*b*), and Woodhouse (1978) describe how to solve for higher-order terms but these are not very useful. In numerical solutions for the propagator in piecewise smooth, layered media (e.g. Kennett and Illingworth, 1981), normally only the zeroth-order term is used.

A Langer iterative solution (Chapman, 1981; Thomson and Chapman, 1984) can be formulated using the zeroth-order Langer approximation (7.2.159) just as in the WKBJ iterative solution in Section 7.2.6. Again, in practice, numerical solutions are normally preferred. Although these analytic extensions of the propagator to inhomogeneous layers – the WKBJ asymptotic expansion (Section 7.2.5), the WKBJ iterative solution (Section 7.2.6) and the Langer asymptotic expansion (Section 7.2.7) – are useful for approximate solutions and analysing canonical problems, their use in general applications is limited. When accurate numerical solutions are required, it is normally simpler to solve the ordinary differential system (7.1.25) numerically, or to approximate the model by thin, homogeneous layers. The accuracy of asymptotic or iterative schemes can be difficult to determine or control.

Although the Langer decomposition (7.2.170) was developed in order to obtain a solution (7.2.200) in a heterogeneous layer with a turning point, e.g. when the vertical slowness can be approximated by expression (7.2.137), and importantly to obtain the reflection coefficient (7.2.162), it is also useful in homogeneous media. It results in a particularly simple algorithm as so many elements in the matrices (7.2.187) and (7.2.188) are zero, unity or equal. In a homogeneous layer, it leads to an alternative for the propagator (7.2.4), i.e.

$$\mathbf{P}(z_{\mathrm{A}}, z_{\mathrm{B}}) = \mathbf{L}\,\mathbf{X}\,\mathbf{L}^{-1} = \mathbf{L}(\mathbf{X}_V \oplus \mathbf{X}_P)\,\mathbf{L}^{-1}, \tag{7.2.201}$$

where

$$\mathbf{X}_\nu = \begin{pmatrix} \cos(\omega q_\nu d) & \mathrm{i}/g_\nu \sin(\omega q_\nu d) \\ \mathrm{i}g_\nu \sin(\omega q_\nu d) & \cos(\omega q_\nu d) \end{pmatrix}, \tag{7.2.202}$$

with $d = z_{\mathrm{A}} - z_{\mathrm{B}}$, the layer thickness. $\mathbf{X}_V$ and $\mathbf{X}_P$ are defined with the appropriate vertical slowness, q_V or q_P, and g_V or g_P, respectively. As the separate waves propagate independently, the scalings (7.2.195) and (7.2.196) can be combined, i.e.

$$(s_V\mathbf{I} \oplus s_P\mathbf{I})\,(\mathbf{X}_V \oplus \mathbf{X}_P)\,(s_P\mathbf{I} \oplus s_V\mathbf{I}) = s_V s_P\,(\mathbf{X}_V \oplus \mathbf{X}_P)\,. \tag{7.2.203}$$

The scaling factors, $s_V s_P$, can be accumulated separately or, as the reflection coefficients only involve ratios, ignored. The propagation term in the propagator (7.2.201) can be factored as

$$\mathbf{X}_V \oplus \mathbf{X}_P = (\mathbf{X}_V \oplus \mathbf{I})\,(\mathbf{I} \oplus \mathbf{X}_P)\,. \tag{7.2.204}$$

In the next section, we demonstrate how using the propagator (7.2.201) with expansions (7.2.187), (7.2.188) and (7.2.204), a particularly simple and a robust algorithm is obtained for the reflection coefficients from a stack of layers.

7.2.8 Second-order minors

In Section 7.2.4.1, we have described Kennett's algorithm for solving the layer matrix problem, i.e. finding the reflection/transmission coefficients from a stack of homogeneous layers.

The elastodynamic response of a stack of plane layers to a plane, spectral wave is of fundamental importance in seismology. The propagator matrix solution is commonly called the *Haskell matrix* method after the classic publication on surface waves (Haskell, 1953 – it is sometimes called the *Thomson–Haskell* method after the earlier publication by Thomson, 1950, although that contained an error). The Haskell matrix method, and variants thereof, have been used for many studies of the reflectivity of waves, and surface or guided waves, in stacks of layers. They can be used to generalize the interface reflection/transmission coefficients used in ray theory, to model approximately the frequency-dependent effect of a layered structure at the reflector, or to model the complete response of a plane layered structure to an impulsive, point source.

Unfortunately for elastic waves the propagator method is numerically unstable at high frequencies if waves are evanescent in some layers. The dominant evanescent behaviour of the P waves compared with SV waves, for a given frequency and wave slowness parallel to the layers, causes loss of numerical precision in any fixed word-length calculation. Essentially, rounding errors in the SV wave solutions grow exponentially as the P wave, and their independence is lost. The differential system (7.1.25) is said to be stiff. Several solutions to this problem have been published. The independence of the solutions can be maintained by re-orthogonalizing the solutions at each interface, a method introduced by Pitteway (1965) in a related problem for radio waves. A similar approach has been used in seismology by Chapman and Phinney (1972) and Wang (1999). The original matrix system can be replaced by a second-order minor system, the so-called Δ-matrix method (Thrower, 1965; Dunkin, 1965), in which only the exponentially dominant solution is required. A faster version, the reduced Δ-matrix method (in which the sixth-order system is reduced to a fifth-order system), was developed later (Watson, 1970). This method was widely used in seismology before the ray expansion method was developed by Kennett (1974, 1983). In Section 7.2.4.1, we have described Kennett's algorithm for solving the layer matrix problem, i.e. finding the reflection/transmission coefficients from a stack of homogeneous layers. In this method, rather than finding the propagator matrix for the wavefields, the 'propagator' for the required reflection/transmission coefficients is found directly. All terms calculated represent rays in the ray expansion in the layer stack. Only exponentially small (not large) terms arise as rays must decay in their propagation direction. The Kennett algorithm is now very widely used as it is numerically robust, but also partly because it allows control of the ray expansion and partly because it extends without difficulty to anisotropic and fluid media.

Several other methods have been developed for solving the boundary conditions of a stack of layers. Knopoff (1964) developed an alternative matrix decomposition of the complete system of equations that is numerically robust and faster than the reduced Δ-matrix system (see Schwab and Knopoff, 1972, for a review of the *Knopoff* method and other publications – the method is sometimes called the *Schwab–Knopoff* method). Later Abo-Zena (1979) developed another related approach. Buchen and Ben-Hador (1996) have published a useful comparison of all the above methods.

For most purposes, any of the above methods is adequate. With improvements in computer performance, efficiency is less of an issue than it was when the algorithms were developed. Probably the most widely used algorithm, Kennett's method, is certainly not the most efficient. The calculation of the P and SV phase factors across each layer with trigonometrical functions, complex if attenuation is included, is required by all algorithms and is a significant fraction of the total cost so savings elsewhere are less significant. Nevertheless, the algebra and computer code for these algorithms are not trivial. The matrix elements in the Δ-matrix and Knopoff's methods are many and varied (see, for instance, example code in Schwab and Knopoff, 1972). The matrix elements in Kennett's method are reflection/transmission coefficients and these are algebraically complicated (see, for instance, equations (6.3.60)–(6.3.62), for the isotropic case). They have been published by many authors starting with Knott (1899) and Zoeppritz (1919) in different forms depending on the coordinate systems and basic waves, but as Kennett, Kerry and Woodhouse (1978) have commented, 'these results have in many cases been marred by minor errors and misprints'.

Here we describe yet another algorithm. It is based on the Langer block-diagonal decomposition of the differential system (Section 7.2.7), and the second-order minor method. It is numerically robust and efficient but we make no claim that it is better than any of the other algorithms. However, its implementation is certainly significantly simpler and is ideal for implementation in high-level languages such as Matlab, Java or C++. Although the theory of the Langer block-diagonal decomposition and second-order minors is well known, the algorithm had only been described in proceedings of a school (Woodhouse, 1980) and only for isotropic media. Recently it has been extended to anisotropic media by Chapman (2003). The Langer decomposition has already been developed for anisotropic media with up–down symmetry on planes of symmetry (Section 7.2.7), e.g. transverse isotropic media (Sections 4.4.4 and 6.3.4) with a vertical axis of symmetry (TIV), and the results in this section apply to the same media. Schoenberg and Protazio (1992) have considered the reflection coefficients in such media, exploiting the up–down symmetry.

The propagator (7.2.201) is perfectly straightforward to compute but will suffer from exactly the same numerical problems as the Haskell matrix (7.2.4). To solve this problem, we use the second-order minors to find the reflection/transmssion coefficients we need (Thrower, 1965; Dunkin, 1965). Combined with the Langer decomposition (7.2.170), this leads to a computationally simple algorithm. The algorithm is simple in the sense that it is easy to program and requires minimal computing operations. First we discuss the second-order minor algebra for the reflection/transmission coefficients, and then apply it to the Langer decomposition.

Second-order minors are formed from elements of a matrix by forming determinants of pairs of rows and columns. We will be concerned with the fourth-order $qP - qSV$ system. The second-order minors of two four-dimensional vectors, $\mathbf{y}^{(1)}$ and $\mathbf{y}^{(2)}$, can be arranged as a six-dimensional vector. We write this as

$$\{\mathbf{y}^{(1)}, \mathbf{y}^{(2)}\} = \begin{pmatrix} y_1^{(1)} y_2^{(2)} - y_2^{(1)} y_1^{(2)} \\ y_1^{(1)} y_3^{(2)} - y_3^{(1)} y_1^{(2)} \\ y_1^{(1)} y_4^{(2)} - y_4^{(1)} y_1^{(2)} \\ y_2^{(1)} y_3^{(2)} - y_3^{(1)} y_2^{(2)} \\ y_2^{(1)} y_4^{(2)} - y_4^{(1)} y_2^{(2)} \\ y_3^{(1)} y_4^{(2)} - y_4^{(1)} y_3^{(2)} \end{pmatrix}, \tag{7.2.205}$$

where $\{\ldots\}$ is the operation of forming the six-dimensional vector of second-order minors. By convention, we order the pair of indices (i, j) as k, where

$$k = 1 \leftrightarrow (i, j) = (1, 2) \tag{7.2.206}$$
$$k = 2 \leftrightarrow (i, j) = (1, 3) \tag{7.2.207}$$
$$k = 3 \leftrightarrow (i, j) = (1, 4) \tag{7.2.208}$$
$$k = 4 \leftrightarrow (i, j) = (2, 3) \tag{7.2.209}$$
$$k = 5 \leftrightarrow (i, j) = (2, 4) \tag{7.2.210}$$
$$k = 6 \leftrightarrow (i, j) = (3, 4). \tag{7.2.211}$$

Thus $\{\mathbf{y}^{(1)}, \mathbf{y}^{(2)}\}_k = y_i^{(1)} y_j^{(2)} - y_j^{(1)} y_i^{(2)}$. Similarly, we form a 6×6 matrix of second-order minors from a 4×4 matrix, i.e. $\{\mathbf{X}\}_{kn} = X_{il} X_{jm} - X_{jl} X_{im}$, where $k \leftrightarrow (i, j)$ and $n \leftrightarrow (l, m)$. Matrix $\{\mathbf{X}\}$ is known as the *compound matrix* of $\mathbf{X}$. These matrices of second-order minors satisfy the algebra (the Binet–Cauchy formula, Gantmacher 1959, Vol 1, p. 9)

$$\{\mathbf{X}\}\{\mathbf{Y}\} = \{\mathbf{XY}\} \tag{7.2.212}$$

$$\{\mathbf{X}\}\{\mathbf{y}^{(1)}, \mathbf{y}^{(2)}\} = \{\mathbf{Xy}^{(1)}, \mathbf{Xy}^{(2)}\}. \tag{7.2.213}$$

The second-order minors are useful because the expressions for the $qP - qSV$ reflection/transmission coefficients can be expressed in terms of second-order minors of the propagator matrix. As the only solution required is dominant, computing the second-order minors directly, we can retain numerical accuracy. Computing them by combining two vectors introduces numerical difficulties as exponentially large terms may be subtracted.

Now we must show how the solution for the $qP - qSV$ reflection coefficients can be written in terms of second-order minors. We consider the same situation as in Section 7.2.4, i.e. a stack of n homogeneous layers (Figure 7.5), with a source and receiver in the first layer (half-space). For simplicity, we calculate the reflection coefficients at the interface, $z_\mathsf{S} = z_\mathsf{R} = z_2$ – arbitrary source and receiver depths just introduce extra phase terms. The two reflection experiments with incident qSV and qP waves can be written as the first columns in equation (7.2.36) – we assume the reduced fourth-order $qSV - qP$ system where the second qSH rows/columns are omitted from the matrices. Let us write

$$\overline{\mathbf{P}}(z_2, z_n) = \mathbf{W}_1^{-1}\mathbf{P}(z_2, z_3) \ldots \mathbf{P}(z_{n-1}, z_n)\mathbf{W}_n. \tag{7.2.214}$$

Taking second-order minors of equation (7.2.36) and using (7.2.213), it reduces to

$$\begin{pmatrix} \mathcal{T}_{11}\mathcal{T}_{33} - \mathcal{T}_{13}\mathcal{T}_{31} \\ -\mathcal{T}_{13} \\ \mathcal{T}_{11} \\ -\mathcal{T}_{33} \\ \mathcal{T}_{31} \\ 1 \end{pmatrix} = \{\overline{\mathbf{P}}\} \begin{pmatrix} 0 \\ 0 \\ 0 \\ 0 \\ 0 \\ \mathcal{T}_{41}\mathcal{T}_{63} - \mathcal{T}_{61}\mathcal{T}_{43} \end{pmatrix}. \tag{7.2.215}$$

The sixth row can be solved for $\mathcal{T}_{41}\mathcal{T}_{63} - \mathcal{T}_{61}\mathcal{T}_{43}$, which can then be eliminated from the other rows, giving

$$\mathbf{R}_{n\,2} = \{\overline{\mathbf{P}}\}_{66}^{-1} \begin{pmatrix} \{\overline{\mathbf{P}}\}_{36} & -\{\overline{\mathbf{P}}\}_{26} \\ \{\overline{\mathbf{P}}\}_{56} & -\{\overline{\mathbf{P}}\}_{46} \end{pmatrix}. \tag{7.2.216}$$

Thus the $qP - qSV$ reflection coefficients can be computed simply from second-order minors of the matrices in equation (7.2.214). Note that only one solution, the sixth column, of the second-order minors, $\{\overline{\mathbf{P}}\}$, is required. This is the dominant solution, which combines the third and fourth columns, i.e. the down-going solutions of $\mathbf{W}_n$.

Result (7.2.216) applies however the propagator is obtained. With the Langer decomposition (7.2.201), it is particularly simple as the matrices (7.2.187), (7.2.188) and (7.2.204) have been factorized, and the individual matrices contain many zero, unit or equal elements. Thus the main term from the stack of layers is

$$\mathbf{Z} = \{\mathbf{X}_1\}\{\mathbf{L}_1^{-1}\}\left(\prod_{j=2}^{n-1}\{\mathbf{L}_j\}\{\mathbf{X}_j\}\{\mathbf{L}_j^{-1}\}\right)\{\mathbf{L}_n\}, \tag{7.2.217}$$

where

$$\{\overline{\mathbf{P}}\} = \{\mathbf{W}_1^{-1}\mathbf{L}_1\}\mathbf{Z}\{\mathbf{L}_n^{-1}\mathbf{W}_n\}. \tag{7.2.218}$$

Only the sixth column of $\{\mathbf{L}_n^{-1}\mathbf{W}_n\}$ is required, $\mathbf{x}_n$, say, which can be obtained from equation (7.2.175) with definitions (7.2.189)–(7.2.192)

$$\mathbf{x}_n = \frac{1}{n_1 n_2 n_3 n_4}\begin{pmatrix} 0 \\ -n_2 n_4 \\ n_2 n_3 \\ n_1 n_4 \\ -n_1 n_3 \\ 0 \end{pmatrix} = -\frac{1}{2(g_V g_P)^{1/2}}\begin{pmatrix} 0 \\ 1 \\ g_P \\ g_V \\ g_V g_P \\ 0 \end{pmatrix} \tag{7.2.219}$$

(the trailing scalings (7.2.195) can be ignored as they only introduce a common factor $(s_V s_P)_n$). Similarly using equation (7.2.174), we can expand $\{\mathbf{W}_1^{-1}\mathbf{L}_1\}$. Then equation (7.2.216) can be rewritten as

$$\begin{pmatrix} \mathcal{T}_{11} \\ \mathcal{T}_{31} \\ \mathcal{T}_{13} \\ \mathcal{T}_{33} \end{pmatrix} = \frac{\begin{pmatrix} 0 & g_V g_P & g_V & -g_P & -1 & 0 \\ 0 & 0 & 0 & 0 & 0 & 2(g_V g_P)^{1/2}\frac{s_V}{s_P} \\ 2(g_V g_P)^{1/2}\frac{s_P}{s_V} & 0 & 0 & 0 & 0 & 0 \\ 0 & -g_V g_P & g_V & -g_P & 1 & 0 \end{pmatrix}_1 \mathbf{Z}\mathbf{x}_n}{\begin{pmatrix} 0 & g_V g_P & g_V & g_P & 1 & 0 \end{pmatrix}_1 \mathbf{Z}\mathbf{x}_n}, \tag{7.2.220}$$

and the leading scaling (7.2.196) from $\{\mathbf{L}_1^{-1}\}$ in equation (7.2.217) has been included in result (7.2.220). Even for a single interface ($n = 2$), this reduces to a simple algorithm for the interface coefficients with

$$\mathbf{Z} = \{\mathbf{L}_1^{-1}\}\{\mathbf{L}_2\}. \tag{7.2.221}$$

With expression (7.2.221) substituted in result (7.2.220) we obtain results equal to the standard interface coefficients (e.g. equations (6.3.60)–(6.3.62) for the isotropic coefficients) but with a relatively simple algorithm.

With the Langer block-diagonal decomposition (7.2.201) and factorization (7.2.187), (7.2.188) and (7.2.204), the second-order minors are particularly simple to compute. Ignoring the trailing diagonal matrix in expression (7.2.187), which simply forms a diagonal matrix of second-order minors and reduces to the simple scaling (7.2.203), we have

$$\{\mathbf{L}\} = \begin{pmatrix} 1 & 0 & 0 & 0 & 1 & 0 \\ -Z_P & 0 & 0 & 0 & -Z_V & 0 \\ 0 & 0 & 0 & -1 & 0 & 0 \\ 0 & 0 & Z_V - Z_P & 0 & 0 & 0 \\ 0 & 1 & 0 & 0 & 0 & 1 \\ 0 & -Z_P & 0 & 0 & 0 & -Z_V \end{pmatrix}$$
$$\times \begin{pmatrix} 1 & 0 & 0 & 0 & 1 & 0 \\ -Z_V & 0 & 0 & 0 & -Z_P & 0 \\ 0 & 0 & 0 & -1 & 0 & 0 \\ 0 & 0 & Z_P - Z_V & 0 & 0 & 0 \\ 0 & 1 & 0 & 0 & 0 & 1 \\ 0 & -Z_V & 0 & 0 & 0 & -Z_P \end{pmatrix}. \tag{7.2.222}$$

Multiplication by the matrix $\{\mathbf{L}\}$ can be performed by the repeated applications of the sub-matrix

$$\begin{pmatrix} 1 & 1 \\ -Z_P & -Z_V \end{pmatrix}, \tag{7.2.223}$$

with the appropriate row and column indices. The inverse matrix is from expression (7.2.188), again ignoring the leading diagonal matrix

$$\{\mathbf{L}^{-1}\} = \begin{pmatrix} -Z_P & -1 & 0 & 0 & 0 & 0 \\ 0 & 0 & 0 & 0 & -Z_P & -1 \\ 0 & 0 & 0 & Z_P - Z_V & 0 & 0 \\ 0 & 0 & -1 & 0 & 0 & 0 \\ Z_V & 1 & 0 & 0 & 0 & 0 \\ 0 & 0 & 0 & 0 & Z_V & 1 \end{pmatrix}$$

$$\times \begin{pmatrix} Z_V & 1 & 0 & 0 & 0 & 0 \\ 0 & 0 & 0 & 0 & Z_V & 1 \\ 0 & 0 & 0 & Z_V - Z_P & 0 & 0 \\ 0 & 0 & -1 & 0 & 0 & 0 \\ -Z_P & -1 & 0 & 0 & 0 & 0 \\ 0 & 0 & 0 & 0 & -Z_P & -1 \end{pmatrix}. \tag{7.2.224}$$

Again multiplication by the matrix $\{\mathbf{L}^{-1}\}$ can be performed by repeated applications of the sub-matrix

$$\begin{pmatrix} Z_V & 1 \\ -Z_P & -1 \end{pmatrix}, \tag{7.2.225}$$

with the appropriate row and column indices.

The second-order minors of the propagation terms (7.2.204) are

$$\{\mathbf{X}_V \oplus \mathbf{I}\} = \begin{pmatrix} 1 & 0 & 0 & 0 & 0 & 0 \\ 0 & c & 0 & -is/g & 0 & 0 \\ 0 & 0 & c & 0 & -is/g & 0 \\ 0 & -igs & 0 & c & 0 & 0 \\ 0 & 0 & -igs & 0 & c & 0 \\ 0 & 0 & 0 & 0 & 0 & 1 \end{pmatrix} \tag{7.2.226}$$

$$\{\mathbf{I} \oplus \mathbf{X}_P\} = \begin{pmatrix} 1 & 0 & 0 & 0 & 0 & 0 \\ 0 & c & -is/g & 0 & 0 & 0 \\ 0 & -igs & c & 0 & 0 & 0 \\ 0 & 0 & 0 & c & -is/g & 0 \\ 0 & 0 & 0 & -igs & c & 0 \\ 0 & 0 & 0 & 0 & 0 & 1 \end{pmatrix}, \tag{7.2.227}$$

where c, s and g are used as shorthand for the appropriate cosine, sine and g_ν functions. These matrices contain the matrix (7.2.202) repeatedly, which can be coded as one function.

The algorithm is so straightforward that a Matlab program has been included in Chapman (2003) for the isotropic case. Multiplications of the six-dimensional vector by the matrices of second-order minors in (7.2.217) is performed using the factorizations (7.2.222), (7.2.224), (7.2.226) and (7.2.227). The operations on the third and fourth elements reduce to multiplying by $\pm\rho/p$. Thus if we define the functions

$$\ell_1(x_i, x_j) = x_i + x_j \tag{7.2.228}$$

$$\ell_2(x_i, x_j) = -Z_V x_i - Z_P x_j \tag{7.2.229}$$

$$\ell_3(x_i) = (Z_V - Z_P) x_i \tag{7.2.230}$$

$$\ell_4(x_i, x_j) = Z_V x_i + x_j \tag{7.2.231}$$

$$\ell_5(x_i, x_j) = -Z_P x_i - x_j, \tag{7.2.232}$$

we can very simply evaluate the products of the matrices (7.2.222) and (7.2.224) with a six-dimensional vector $\mathbf{x}$, i.e.

$$\{\mathbf{L}\}\mathbf{x} = \begin{pmatrix} \ell_1\left(\ell_1(x_2, x_6), \ell_1(x_1, x_5)\right) \\ \ell_2\left(\ell_1(x_2, x_6), \ell_1(x_1, x_5)\right) \\ \ell_3(x_3) \\ -\ell_3(x_4) \\ \ell_1\left(\ell_2(x_2, x_6), \ell_2(x_1, x_5)\right) \\ \ell_2\left(\ell_2(x_2, x_6), \ell_2(x_1, x_5)\right) \end{pmatrix} \tag{7.2.233}$$

$$\{\mathbf{L}^{-1}\}\mathbf{x} = \begin{pmatrix} \ell_5\left(\ell_4(x_1, x_2), \ell_4(x_5, x_6)\right) \\ \ell_5\left(\ell_5(x_1, x_2), \ell_5(x_5, x_6)\right) \\ \ell_3(x_3) \\ -\ell_3(x_4) \\ \ell_4\left(\ell_4(x_1, x_2), \ell_4(x_5, x_6)\right) \\ \ell_4\left(\ell_5(x_1, x_2), \ell_5(x_5, x_6)\right) \end{pmatrix}. \tag{7.2.234}$$

If we define a function

$$X(x_i, x_j, g) = cx_i - \mathrm{i}gsx_j, \tag{7.2.235}$$

then

$$\{\mathbf{X}_\beta \oplus \mathbf{I}\}\mathbf{x} = \begin{pmatrix} x_1 \\ X(x_2, x_4, g_V^{-1}) \\ X(x_3, x_5, g_V^{-1}) \\ X(x_4, x_2, g_V) \\ X(x_5, x_3, g_V) \\ x_6 \end{pmatrix} \tag{7.2.236}$$

$$\{\mathbf{I} \oplus \mathbf{X}_\alpha\}\mathbf{x} = \begin{pmatrix} x_1 \\ X(x_2, x_3, g_P^{-1}) \\ X(x_3, x_2, g_P) \\ X(x_4, x_5, g_P^{-1}) \\ X(x_5, x_4, g_P) \\ x_6 \end{pmatrix}. \tag{7.2.237}$$

Clearly because of their repetitive nature, these operations will be extremely simple and efficient to code and compute.

A complete algorithm would need to handle some special cases: $p = 0$, $\mu = 0$ or $q = 0$. The latter only requires the reduction

$$\mathbf{X}_\nu = \begin{pmatrix} 1 & 0 \\ -\mathrm{i}p\omega d & 1 \end{pmatrix}, \tag{7.2.238}$$

for the corresponding matrix $\mathbf{X}_\nu$ (7.2.202). The former two cases require consideration of different systems. If $p = 0$, the $P - SV$ waves separate and if $\mu = 0$ the system reduces to the acoustic system. However, as the numerical algorithm is robust, it is easier to handle these cases by replacing p or μ by numerically small quantities. At high frequencies it may be necessary to rescale the six-vector solution at intermediate depths in order to avoid numerical overflow (the final result (7.2.220) only contains ratios, so removing this exponential growth has no effect). Similarly it may be necessary to sub-divide layers in order to prevent overflow in a single layer.

Exercises

7.1 Show that the same differential systems (7.1.4) with the definitions (7.1.6), (7.1.33) and (7.1.34) for acoustic or isotropic elastic media, apply in cylindrical coordinates, except that the components of the vector $\mathbf{w}$ are transformed cylindrical components (see Exercise 4.11).

7.2 Investigate how the elastic matrix (7.1.34) is modified if buoyancy forces due to gravity are included. This result would be relevant in extremely soft sediments. How are the velocities and eigenvectors modified? A useful publication is Gilbert (1967).

7.3 Investigate how the differential system (7.1.4) behaves for elastic waves, i.e. result (7.1.34), as the shear modulus, μ, tends to zero. How is this compatible with the fluid limit, when the tangential displacement is discontinuous at interfaces? Gilbert (1998) has discussed numerical schemes for solving the differential system as $\mu \to 0$.

7.4 Use the propagator method to write down the solution in the transform domain for a homogeneous layer over a homogeneous half-space, without using the ray expansion method, i.e. using the Haskell matrix (7.2.4). Show that the complete response can then be expanded into terms that can be identified as the reverberating rays in the layer. First do this for SH waves (relatively simple), and then for $P - SV$ waves (algebraically messy). The advantages of applying Kennett's ray expansion method to the propagator should now be obvious!

7.5 *Further reading:* A differential system similar to the ordinary differential equation (7.1.4) applies in a stratified sphere when the elastic parameters are a function of the spherical radius. The horizontal derivatives are removed from the partial differential equations by the generalized Legendre transform (see Takeuchi and Saito, 1972, for the derivation of the differential system). A fundamental difference compared with the cartesian system is that the matrix $\mathbf{A}$ depends explicitly on the radius, r, so even in a homogeneous layer, a fundamental solution cannot be written as equation (7.2.2). The fundamental solution can be written in terms of spherical Bessel functions (see Phinney and Alexander, 1966, for the solutions). Although the fundamental matrix is more complicated, the matrix still has symplectic symmetries and the inverse fundamental matrix needed to form the propagator is known without explicitly inverting the fundamental matrix (first given by Teng, 1970, although without using the symplectic symmetries).

7.6 *Programming exercise:* Confirm numerically that the results from the algorithm given in Section 7.2.8 applied to a single interface, agree with those in programming Exercise 6.3 in Chapter 6 (Matlab code is given in Chapman, 2003).

7.7 *Programming exercise:* Generalize the code used in Exercise 7.6 (from Chapman, 2003) to TIV media by writing a new `EigenFactors` routine. Confirm that numerical results agree with those in programming Exercise 6.3 in Chapter 6.

7.8 Investigate the symplectic symmetry proportion of the differential system $d\underline{\mathbf{w}} = i\omega\mathbf{A}\,\underline{\mathbf{w}}$ with matrices (7.1.6), (6.3.15)–(6.3.17), (7.1.33) and (7.1.34) (and those in Exercises 7.2 and 7.5). Obtain an expression for the inverse of the propagator using a symplectic transform (cf. Section 6.3.2.1).

8

Inverse transforms for stratified media

Having obtained the transformed response of a stratified medium, i.e. in the spectral (frequency), plane-wave (wavenumber) domain, it is necessary to invert the transforms, to obtain the impulsive, point-source response. An elegant, exact technique, the *Cagniard–de Hoop–Pekeris method*, which can be used in models with homogeneous layers, or with the WKBJ iterative solution in stratified layers, and an approximate method, the *WKBJ seismogram method*, are developed in this chapter. For realistic models, these methods are often impractical and numerical methods are necessary. We describe the techniques necessary for the numerical, spectral method.

In this chapter, we investigate different methods of obtaining the Green function from the transformed response. The first problem of this type solved in seismology is now know as Lamb's problem after the classic paper by Lamb (1904). Lamb investigated the excitation of seismic waves in a homogenous half-space due to a point force source on the surface. He used what we would call asymptotic methods in the spectral domain (see Section 8.5), and explained the excitation of head and Rayleigh waves. As the same mathematical techniques can be used whether the source is on the surface or buried, we now refer to the problem of exciting waves in a homogeneous half-space due to any source as Lamb's problem. The problem was investigated in much greater detail by Lapwood (1949) and Garvin (1956), but it was not until the papers of Pekeris (1955*a, b*) that complete, exact solutions were obtained. Cagniard (1939) solved the more general problem of reflection and transmission at an interface between two homogeneous half-spaces using a similar method, but it was not until the translation (Cagniard, 1962) and the modification and simplification by de Hoop (1960) that the exact method became well known and popular. It is a variant of this method that we will discuss. The classic methods can be found in textbooks by Ewing, Jardetsky and Press (1957) and Båth (1968).

In the last chapter, Chapter 7, various methods of finding the transformed response for different signals were investigated. A variety of inverse transform methods are also available – exact analytic methods, approximate and numerical.

Not every inverse technique can be used with every transformed response, but in general the same inverse technique can be used for many different signals. To avoid repetition, we have tried to introduce a general notation in Chapter 7. The Green function consists of a dyadic of the generalized polarizations at the source and receiver, with propagation terms consisting of phase terms and reflection/transmission coefficients between. We have used a matrix notation for the Green dyadic, e.g. expressions (7.2.18) and (7.2.19) for the direct waves, expressions (7.2.31) and (7.2.32) for reflected and transmitted waves. Using the ray expansions (7.2.56) and (7.2.57), these can be generalized to any rays with any number of reflections, transmissions and reverberations in a stack of layers. In this chapter we often consider the rays individually, e.g. expand expressions (7.2.18) and (7.2.19) into three rays, and expressions (7.2.31) and (7.2.32) into nine rays. Mathematically the response is linear, and the individual rays can be solved separately and summed. Physically, we are often interested in a small subset of all the rays as the source may not generate all wave types (e.g. an explosion in isotropic media only generates P waves); the receiver may not be sensitive to all waves (or some waves may be removed by pre-processing); some reflection or transmission coefficients or products thereof may be sufficiently small that the waves can be neglected; and many waves may arrive outside the time window of interest.

8.0.8.1 Inverse transformations

In two dimensions, the impulse response is given by the inverse spectral and slowness Fourier transforms, (3.1.2) and (3.2.10)

$$\boxed{\underline{\mathbf{v}}(t, \mathbf{x}_\mathrm{R}) = \frac{1}{4\pi^2} \int_B |\omega| \int_{-\infty}^{\infty} \underline{\mathrm{v}}(\omega, p, z_\mathrm{R})\, \mathrm{e}^{\mathrm{i}\omega(p x_\mathrm{R} - t)} \mathrm{d}p\, \mathrm{d}\omega.} \tag{8.0.1}$$

In three dimensions the result from the inverse transform (3.2.15) is

$$\boxed{\underline{\mathbf{v}}(t, \mathbf{x}_\mathrm{R}) = \frac{1}{8\pi^3} \int_B \omega^2 \iint_{-\infty}^{\infty} \underline{\mathrm{v}}(\omega, \mathrm{p}, z_\mathrm{R})\, \mathrm{e}^{\mathrm{i}\omega(\mathrm{p}\cdot \mathrm{x}_\mathrm{R} - t)} \mathrm{dp}\, \mathrm{d}\omega.} \tag{8.0.2}$$

This chapter concerns evaluating these integrals analytically, numerically or approximately. Most of the results can be applied in acoustic, isotropic and anisotropic elastic media. However, for analytic methods, e.g. the Cagniard method (Section 8.1), this is complicated and we have not included the generality of anisotropy. Restricting results to isotropic media means that the propagation is axially symmetric and any asymmetry only arises through the source and receiver.

The inverse transformations, (8.0.1) and (8.0.2), are linear, so it is often useful to use the ray expansion to expand the transformed response into generalized rays.

Thus formally, we have

$$\underline{\mathbf{v}}(\omega, \mathsf{p}, z_\mathsf{R}) = \sum_{\text{rays}} \underline{\mathbf{v}}_{\text{ray}}(\omega, \mathsf{p}, z_\mathsf{R}), \tag{8.0.3}$$

summing the rays in the ray expansion (Section 7.2.4). The subscript 'ray' is used to enumerate the 'rays' in the expansion. The impulse response can be written

$$\underline{\mathbf{v}}(t, \mathbf{x}_\mathsf{R}) = \sum_{\text{rays}} \underline{\mathbf{v}}_{\text{ray}}(t, \mathbf{x}_\mathsf{R}), \tag{8.0.4}$$

where, for instance,

$$\boxed{\underline{\mathbf{v}}_{\text{ray}}(t, \mathbf{x}_\mathsf{R}) = \frac{1}{4\pi^2} \int_B |\omega| \int_{-\infty}^{\infty} \underline{\mathbf{v}}_{\text{ray}}(\omega, p, z_\mathsf{R})\, \mathrm{e}^{\mathrm{i}\omega(p x_\mathsf{R} - t)} \mathrm{d}p\, \mathrm{d}\omega,} \tag{8.0.5}$$

in two dimensions.

8.0.8.2 Generalized ray response

For simplicity, we introduce the notation

$$\boxed{\begin{aligned}\underline{\mathbf{v}}_{\text{ray}}(\omega, \mathsf{p}, z_\mathsf{R}) &= \mathbf{g}_\mathsf{R}(\mathsf{p})\, \mathcal{P}_{\text{ray}}(\omega, \mathsf{p}, z_\mathsf{R})\, \mathbf{g}_\mathsf{S}^{\mathrm{T}}(\mathsf{p}) \\ &= \underline{\boldsymbol{\mathcal{G}}}_{\text{ray}}(\mathsf{p})\, \mathrm{e}^{\mathrm{i}\omega\tau_{\text{ray}}(\mathsf{p}, z_\mathsf{R})},\end{aligned}} \tag{8.0.6}$$

for the transformed response of a single generalized ray in a layered medium. The spatial transform variable, p, is the horizontal slowness(es). In two dimensions $\mathsf{p} = p$, while in three dimensions $\mathsf{p} = (p_1, p_2)$. The propagator term in the transform domain is

$$\boxed{\mathcal{P}_{\text{ray}}(\omega, \mathsf{p}, z_\mathsf{R}) = \left(\prod_{\text{ray}} \mathcal{T}_{ij}(\mathsf{p})\right) \mathrm{e}^{\mathrm{i}\omega\tau_{\text{ray}}(\mathsf{p}, z_\mathsf{R})}.} \tag{8.0.7}$$

The frequency-independent part of the dyadic for the ray consists of the source and receiver polarizations and a product of reflection/transmission coefficients for the ray

$$\boxed{\underline{\boldsymbol{\mathcal{G}}}_{\text{ray}}(\mathsf{p}) = \left(\prod_{\text{ray}} \mathcal{T}_{ij}(\mathsf{p})\right) \mathbf{g}_\mathsf{R}(\mathsf{p})\, \mathbf{g}_\mathsf{S}^{\mathrm{T}}(\mathsf{p}).} \tag{8.0.8}$$

The phase function consists of the sum of vertical slowness times vertical segment lengths

$$\boxed{\tau_{\text{ray}}(\mathsf{p}, z_\mathsf{R}) = \sum_{\text{ray}} q_j d_j.} \tag{8.0.9}$$

It is difficult to write an explicit general notation for the product and sum in expressions (8.0.8) and (8.0.9) but completely straightforward to compute for specific generalized rays.

Expressions (8.0.6) and (8.0.8) are easily modified to generalize the source and/or receiver. For instance, for a receiver on an interface, such as the free surface of the Earth, the polarization, $\mathbf{g}_\mathsf{R}$, must be replaced by the interface polarization conversions (Section 6.6). Thus

$$\underline{\mathbf{v}}_{\mathrm{ray}}(\omega, \mathsf{p}, z_\mathsf{R}) = \mathbf{h}_{\mathrm{ray}}(\mathsf{p})\, \mathcal{P}_{\mathrm{ray}}(\omega, \mathsf{p}, z_\mathsf{R})\, \mathbf{g}_\mathsf{S}^{\mathrm{T}}(\mathsf{p}), \tag{8.0.10}$$

where the vector $\mathbf{h}_{\mathrm{ray}}$ is the appropriate interface polarization, e.g. one of results (6.6.10), (6.6.11) or (6.6.12) for a free surface. Similarly, the source term can be generalized, for example, to a point, moment tensor (4.6.17)

$$\mathbf{v}_{\mathrm{ray}}(\omega, \mathsf{p}, z_\mathsf{R}) = \mathbf{h}_{\mathrm{ray}}(\mathsf{p})\, \mathcal{P}_{\mathrm{ray}}(\omega, \mathsf{p}, z_\mathsf{R})\, \mathbf{g}_\mathsf{S}^{\mathrm{T}}(\mathsf{p})\, \mathbf{M}_\mathsf{S}(\mathsf{p})\, \mathbf{p}_\mathsf{S}(\mathsf{p}). \tag{8.0.11}$$

Symbolically, we can write a generalized ray as

$$\boxed{\mathbf{v}_{\mathrm{ray}}(\omega, \mathsf{p}, z_\mathsf{R}) = \boldsymbol{\mathcal{H}}_\mathsf{R}(\mathsf{p})\, \mathcal{P}_{\mathrm{ray}}(\omega, \mathsf{p}, z_\mathsf{R})\, \mathcal{M}_\mathsf{S}(\mathsf{p}),} \tag{8.0.12}$$

where $\mathcal{M}_\mathsf{S}$ represents the alternative source terms, e.g. a point force, $\mathcal{M}_\mathsf{S} = \mathbf{g}_\mathsf{S}^{\mathrm{T}}\, \mathbf{f}_\mathsf{S}$, a moment tensor $\mathcal{M}_\mathsf{S} = \mathbf{g}_\mathsf{S}^{\mathrm{T}}\, \mathbf{M}_\mathsf{S}\, \mathbf{p}_\mathsf{S}$, etc., and $\boldsymbol{\mathcal{H}}_\mathsf{R}$ alternative receiver types, e.g. $\boldsymbol{\mathcal{H}}_\mathsf{R} = \mathbf{g}_{\mathrm{ray}}$ for the particle velocity in the medium, $\boldsymbol{\mathcal{H}}_\mathsf{R} = \mathbf{h}_{\mathrm{ray}}$ for the particle velocity on an interface, etc.

8.1 Cagniard method in two dimensions

The Cagniard–de Hoop–Pekeris method is a particularly elegant method for evaluating, exactly and analytically, the inverse transforms of (some) functions of the form of expression (8.0.6). These include the direct ray (7.2.18) and (7.2.19), and any reflections and transmissions in the ray expansion in a medium of homogeneous layers. It can be used approximately for similar signals in inhomogeneous layers given by the WKBJ approximation (7.2.89), but not for rays with turning points. It can be extended to include all signals in inhomogeneous layers using the WKBJ iterative solution, (7.2.120) and (7.2.125), but at the expense of multiple depth integrals.

The method was developed by Cagniard (1939, 1960) and Pekeris (1955*a*, *b*), with an important modification by de Hoop (1960), and is sometimes called the *Cagniard–de Hoop–Pekeris* method. Many variations have appeared in the literature including the application to Lamb's problem in the textbook Fung (1965). For brevity, we refer to any variant of the method simply as the *Cagniard method*, and present only one version which is relatively simple and straightforward.

First we investigate the Cagniard method in two dimensions, i.e. a line source in three dimensions, and then extend it to three dimensions, i.e. a point source. We restrict our discussion to acoustic and isotropic elastic media. It is relatively simple to extend the method to transversely isotropic media, with a vertical axis of symmetry (TIV – van der Hijden, 1987), i.e. we still have axial symmetry in the model, but with general anisotropy, the method is very complicated.

There are many versions of the Cagniard method in the literature, although they all depend on the fundamental fact that both $\underline{\mathcal{G}}$ and τ in expression (8.0.6) are independent of frequency. The major difference depends on whether the integrals are analysed for real frequency (as in the inverse Fourier and Laplace transforms) or imaginary frequency (as in the forward Laplace transform). In the latter case, we rely on the analytic behaviour of the integrand in the positive, imaginary ω half-space (i.e. causality). Here we take the frequency real as this is required when the integrand is only known approximately (see Section 3.1).

Taking the inverse spatial and temporal transforms (3.2.10) and (3.1.2) with expression (8.0.6), we obtain

$$\boxed{\underline{\mathbf{v}}_{\mathrm{ray}}(t,\mathbf{x}_{\mathrm{R}}) = \frac{1}{4\pi^2}\int_B |\omega| \int_{-\infty}^{\infty} \underline{\mathcal{G}}_{\mathrm{ray}}(p)\mathrm{e}^{\mathrm{i}\omega(\widetilde{T}_{\mathrm{ray}}(p,\mathbf{x}_{\mathrm{R}})-t)}\mathrm{d}p\,\mathrm{d}\omega,} \tag{8.1.1}$$

where

$$\boxed{\begin{aligned}\widetilde{T}_{\mathrm{ray}}(p,\mathbf{x}_{\mathrm{R}}) &= p\,x_{\mathrm{R}} + \tau_{\mathrm{ray}}(p,z_{\mathrm{R}}) \\ &= T_{\mathrm{ray}}(p,z_{\mathrm{R}}) + p\,(x_{\mathrm{R}} - X_{\mathrm{ray}}(p,z_{\mathrm{R}})).\end{aligned}} \tag{8.1.2}$$

In this expression, and throughout this chapter, we use the same notation as, for instance, in Chapter 2, for the ray functions – slowness components p and q, time $T(p,z_{\mathrm{R}})$, range $X(p,z_{\mathrm{R}})$, and intercept time $\tau(p,z_{\mathrm{R}})$. The definitions of the functions are identical but they are generalized to complex slowness.

8.1.1 The Cagniard contour

In order to evaluate the integrals in expression (8.1.1), it is important to investigate the behaviour of the phase function, $\widetilde{T}_{\mathrm{ray}}(p,\mathbf{x}_{\mathrm{R}})$. In particular we need to know the contours in the complex p plane where $\widetilde{T}_{\mathrm{ray}}$ is real or imaginary. For real frequency, this allows us to determine where the integrand is oscillatory, or exponentially large or small.

8.1.1.1 Cagniard contour for uni-velocity waves

For unconverted waves with only one velocity, e.g. the direct ray or unconverted reflection, it is possible to find the inverse function $p(\widetilde{T}_{\mathrm{ray}})$ analytically. The forward function is of the form

$$\widetilde{T}_{\mathrm{ray}}(p, \mathbf{x}_{\mathrm{R}}) = p x_{\mathrm{R}} + q d, \tag{8.1.3}$$

where

$$\begin{aligned} d &= |z_{\mathrm{R}} - z_{\mathrm{S}}|, \text{ direct ray}, & (8.1.4)\\ &= z_{\mathrm{R}} + z_{\mathrm{S}} - 2z_1, \text{ reflected ray}. & (8.1.5) \end{aligned}$$

Examples of the direct ray are expressions (7.2.18) and (7.2.19). The vertical slowness is

$$q = \left(1/c^2 - p^2\right)^{1/2}, \tag{8.1.6}$$

where c is the wave velocity, i.e. α or β. Substituting definition (8.1.6) in expression (8.1.3), subtracting $p x_{\mathrm{R}}$ from both sides and squaring, the equation is quadratic in p. This is easily solved as

$$p = \frac{\widetilde{T}_{\mathrm{ray}}}{R} \sin\theta \pm \left(\frac{1}{c^2} - \frac{\widetilde{T}^2_{\mathrm{ray}}}{R^2} \right)^{1/2} \cos\theta, \tag{8.1.7}$$

where

$$R^2 = x_{\mathrm{R}}^2 + d^2, \tag{8.1.8}$$

and

$$\begin{aligned} \sin\theta &= x_{\mathrm{R}}/R & (8.1.9)\\ \cos\theta &= d/R. & (8.1.10) \end{aligned}$$

Obviously, R and θ have a simple physical interpretation: R is the ray length from source to receiver; and θ is the angle the ray makes with the vertical. With the specific analytic form for the inverse function $p(\widetilde{T}_{\mathrm{ray}})$, it is straightforward to investigate the $p - \widetilde{T}_{\mathrm{ray}}$ mapping. Obviously for p real and $-1/c < p < 1/c$, $\widetilde{T}_{\mathrm{ray}}$ is also real. For $\widetilde{T}_{\mathrm{ray}} > R/c$, p will be complex, i.e.

$$\boxed{p = \frac{\widetilde{T}_{\mathrm{ray}}}{R} \sin\theta \pm \mathrm{i} \left(\frac{\widetilde{T}^2_{\mathrm{ray}}}{R^2} - \frac{1}{c^2} \right)^{1/2} \cos\theta,} \tag{8.1.11}$$

As $\widetilde{T}_{\mathrm{ray}} \to \infty$,

$$p \to \frac{\widetilde{T}_{\mathrm{ray}}}{R} \sin\theta \pm \mathrm{i} \frac{\widetilde{T}_{\mathrm{ray}}}{R} \cos\theta, \tag{8.1.12}$$

which is asymptotic to the lines

$$\operatorname{Arg}(p) = \pm\Big(\frac{\pi}{2} - \theta\Big). \tag{8.1.13}$$

Rather than continue with the analysis for a uni-velocity wave, we generalize the discussion to include any converted reflection or transmission.

8.1.1.2 Cagniard contour for general waves

For any reflected or transmitted ray, definition (8.1.2) holds with

$$\tau_{\text{ray}}(p, z_{\text{R}}) = \sum_{\text{ray}} q_j d_j, \tag{8.1.14}$$

where d_j is the vertical distance (normally a layer thickness except for the source and receiver segments) propagated with vertical slowness q_j. The summation index j enumerates all the segments of the ray. The phase, $\widetilde{T}_{\text{ray}}$, is real and positive for $0 < p < 1/\max(c_j)$, where $\max(c_j)$ is the maximum velocity on the ray, and at the origin

$$\widetilde{T}_{\text{ray}}(0, \mathbf{x}_{\text{R}}) = \sum_{\text{ray}} d_j/c_j, \tag{8.1.15}$$

the vertical travel time. The gradient of the phase is

$$\frac{\partial \widetilde{T}_{\text{ray}}}{\partial p} = x_{\text{R}} - \sum_j \frac{p\, d_j}{q_j}, \tag{8.1.16}$$

and at the origin this is positive and equal to x_{R}. At $p = 1/\max(c_j)$, it is singular with $\partial \widetilde{T}_{\text{ray}}/\partial p \to -\infty$, and as it is continuous, there must be a saddle point in the range $0 < p < 1/\max(c_j)$, where

$$\frac{\partial \widetilde{T}_{\text{ray}}}{\partial p} = 0. \tag{8.1.17}$$

This occurs when

$$X_{\text{ray}}(p, z_{\text{R}}) = x_{\text{R}}, \tag{8.1.18}$$

i.e. at $p = p_{\text{ray}}$ say, where

$$X_{\text{ray}}(p, z_{\text{R}}) = \sum_j \frac{p\, d_j}{q_j}. \tag{8.1.19}$$

This equation is analogous to the range integral, cf. expressions (2.3.4), (2.3.7) or (5.7.5), and, of course, its solution is the horizontal slowness or ray parameter of the geometrical ray, i.e.

$$p_{\mathrm{ray}} = \frac{\sin\theta_j}{c_j} \quad \text{and} \quad q_j = \frac{\cos\theta_j}{c_j}. \tag{8.1.20}$$

The second derivative of the phase is given by

$$\frac{\partial^2 \widetilde{T}_{\mathrm{ray}}}{\partial p^2} = -\frac{\mathrm{d}X_{\mathrm{ray}}}{\mathrm{d}p} = -\sum_j \frac{d_j}{c_j^2 q_j^3}, \tag{8.1.21}$$

cf. expression (2.3.12), and at the saddle point this is negative (in fact for all $0 < p < 1/\max(c_j)$). This establishes that the steepest-descent path over the saddle point is at $-\mathrm{sgn}(\omega)\pi/4$ to the real axis (Figure 8.1). At the saddle point, the phase is

$$\begin{aligned}
\widetilde{T}_{\mathrm{ray}}(p_{\mathrm{ray}}, \mathbf{x}_{\mathrm{R}}) &= p_{\mathrm{ray}} x_{\mathrm{R}} + \tau_{\mathrm{ray}}(p_{\mathrm{ray}}, z_{\mathrm{R}}) \\
&= p_{\mathrm{ray}} X_{\mathrm{ray}}(p_{\mathrm{ray}}, z_{\mathrm{R}}) + \tau_{\mathrm{ray}}(p_{\mathrm{ray}}, z_{\mathrm{R}}) \\
&= \sum_j \frac{d_j}{c_j^2 q_j} \\
&= T_{\mathrm{ray}}(p_{\mathrm{ray}}, z_{\mathrm{R}}),
\end{aligned} \tag{8.1.22}$$

say, where $T_{\mathrm{ray}}(p, z_{\mathrm{R}})$ is the generalized travel-time function (cf. expressions (2.3.5), (2.3.8) and (5.7.6)).

When $|p| \to \infty$, we can approximate (8.1.2) by

$$\widetilde{T}_{\mathrm{ray}}(p, \mathbf{x}_{\mathrm{R}}) = p x_{\mathrm{R}} \pm \mathrm{i} p \sum_j d_j, \tag{8.1.23}$$

and thus $\widetilde{T}_{\mathrm{ray}}$ is real and tends to infinity on lines asymptotic to

$$\mathrm{Arg}(p) = \pm\left(\frac{\pi}{2} - \tilde{\theta}\right), \tag{8.1.24}$$

where

$$\tilde{\theta} = \tan^{-1}\left(x_{\mathrm{R}} \Big/ \sum_j d_j\right), \tag{8.1.25}$$

cf. equation (8.1.13).

8.1.2 The inverse transforms

If we can change the p contour of integration in the integral (8.1.1) to make $\widetilde{T}_{\mathrm{ray}}$ real, then the inverse transforms are readily evaluated. From the above analysis,

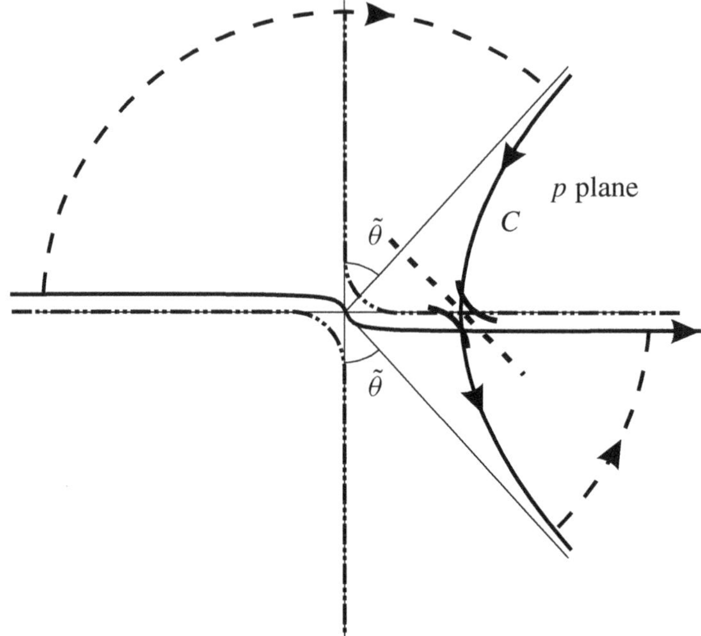

Fig. 8.1. The original p contour (see Figure 3.4) and its distortion to the Cagniard contour C when $\omega > 0$. The orientation of the saddle point, and therefore the positions of the valleys of the integrand, are indicated. For $\omega < 0$, the diagram is reflected in the real axis. The asymptotes defined by the angle $\tilde{\theta}$ (8.1.25) are indicated. The integrand is exponentially small along the contours indicated with dashed lines.

it is easily established that this can be achieved. We must remember that the arrangement of branch cuts, and the original p contour depend on the sign of the frequency (Figure 3.4). For positive frequencies, the exponential $\exp(\mathrm{i}\omega\widetilde{T}_{\mathrm{ray}})$ is exponentially small in the sectors between $\mathrm{Arg}(p) = 0$ and $-(\pi/2 - \tilde{\theta})$, and between $\mathrm{Arg}(p) = -\pi$ and $+(\pi/2 - \tilde{\theta})$. The contour can be distorted from the real p axis, to the contour C defined by $\mathrm{Im}(\widetilde{T}_{\mathrm{ray}}) = 0$ (see Figure 8.1) (for the uni-velocity case, this is given explicitly by expression (8.1.11)). This contour is asymptotic to the lines (8.1.24), is symmetrical about the real axis and passes through the real axis at the saddle point p_{ray} (defined by equation (8.1.18)) (Figure 8.1). It is sometimes called the *Cagniard contour*. Note that the portion in the first quadrant is distorted onto the lower Riemann sheet (where the branch cuts are defined by $\mathrm{Im}(q_j) = 0$, and the upper, physical sheet has $\mathrm{Im}(q_j) > 0$). For negative frequencies, the arrangement of the original and distorted contours is the mirror image in the real p axis (Figures 3.4 and 8.1). The opposing directions of the contour for positive and negative frequencies combine with the factor $|\omega|$ in the integral (8.1.1) to give

$$\underline{\mathbf{v}}_{\text{ray}}(t, \mathbf{x}_\text{R}) = \frac{1}{4\pi^2} \int_B \omega \int_C \underline{\mathcal{G}}_{\text{ray}}(p) \mathrm{e}^{\mathrm{i}\omega(\widetilde{T}_{\text{ray}}(p,\mathbf{x}_\text{R})-t)} \mathrm{d}p \, \mathrm{d}\omega, \qquad (8.1.26)$$

where C is defined in the direction of the distorted contour for positive frequencies (Figure 8.1), i.e. from $p = \infty \, \mathrm{e}^{+\mathrm{i}(\pi/2-\tilde{\theta})}$ to $p = \infty \, \mathrm{e}^{-\mathrm{i}(\pi/2-\tilde{\theta})}$.

We have assumed that in distorting the contour no singularities of the other part of the integrand, $\underline{\mathcal{G}}$, are encountered. For the direct wave, when $\underline{\mathcal{G}}$ contains the source excitation and receiver conversion functions (7.2.18), this is easily established. For general reflections or transmissions, $\underline{\mathcal{G}}$ may have branch cuts in the range $0 < p < p_{\text{ray}}$. We will investigate this in more detail below (see Section 9.1.3 on head waves). The function $\underline{\mathcal{G}}$ may also have poles, but these either lie on the real axis for $p > 1/\max(c_j)$, or on a lower Riemann sheet in a region not encountered by distorting the contour to C. We will investigate these poles in more detail later (see Section 9.1.5). For the moment, we will just assume that no singularities are encountered and the distortion is possible.

Changing the order of integration in integrals (8.1.26), and removing a factor $-\mathrm{i}\omega$ from the integrand, we obtain

$$\underline{\mathbf{v}}_{\text{ray}}(t, \mathbf{x}_\text{R}) = \frac{1}{4\pi^2} \frac{\mathrm{d}}{\mathrm{d}t} \int_C \int_B \mathrm{i} \underline{\mathcal{G}}_{\text{ray}}(p) \, \mathrm{e}^{\mathrm{i}\omega(\widetilde{T}_{\text{ray}}(p,\mathbf{x}_\text{R})-t)} \mathrm{d}\omega \, \mathrm{d}p \qquad (8.1.27)$$

$$= \frac{1}{2\pi} \frac{\mathrm{d}}{\mathrm{d}t} \int_C \mathrm{i} \underline{\mathcal{G}}_{\text{ray}}(p) \, \delta(\widetilde{T}_{\text{ray}}(p, \mathbf{x}_\text{R}) - t) \, \mathrm{d}p, \qquad (8.1.28)$$

as $\widetilde{T}_{\text{ray}}$ is real on the contour C. As $T(p_{\text{ray}}, z_\text{R}) < t < \infty$ on the Cagniard contour, for $t < T(p_{\text{ray}}, z_\text{R})$ there are no solutions of $t = \widetilde{T}_{\text{ray}}$ and the response is zero (causality). For $t > T(p_{\text{ray}}, z_\text{R})$ we have two contributing points where $t = \widetilde{T}_{\text{ray}}(p, \mathbf{x}_\text{R})$ at complex conjugate points on the Cagniard contour. Adding both solutions, we obtain twice the real part, but with the factor of i in the integrand, we obtain

$$\underline{\mathbf{v}}_{\text{ray}}(t, \mathbf{x}_\text{R}) = -\frac{1}{\pi} \frac{\mathrm{d}}{\mathrm{d}t} \operatorname{Im} \int_{T(p_{\text{ray}}, z_\text{R})}^{\infty} \underline{\mathcal{G}}_{\text{ray}}(p) \, \delta(\widetilde{T}_{\text{ray}} - t) \, \frac{\partial p}{\partial \widetilde{T}_{\text{ray}}} \, \mathrm{d}\widetilde{T}_{\text{ray}} \qquad (8.1.29)$$

$$= -\frac{1}{\pi} \frac{\mathrm{d}}{\mathrm{d}t} \operatorname{Im} \left(\underline{\mathcal{G}}_{\text{ray}}(p) \, \frac{\partial p}{\partial \widetilde{T}_{\text{ray}}} \right)_{p=p(t,\mathbf{x}_\text{R})}, \qquad (8.1.30)$$

or

$$\boxed{\underline{\mathbf{u}}_{\text{ray}}(t, \mathbf{x}_\text{R}) = -\frac{1}{\pi} \operatorname{Im} \left(\underline{\mathcal{G}}_{\text{ray}}(p) \, \frac{\partial p}{\partial \widetilde{T}_{\text{ray}}} \right)_{p=p(t,\mathbf{x}_\text{R})}.} \qquad (8.1.31)$$

The final bracket is evaluated at the point on the Cagniard contour that solves $t = \widetilde{T}_{\rm ray}(p, \mathbf{x}_{\rm R})$ in the fourth quadrant. The final result (8.1.31) contains no integrals – the two inverse transforms have effectively cancelled. It is worthwhile checking that the units of the final results are consistent. In two dimensions, unit($\underline{\mathbf{u}}$) = [M^{-1}LT] where we have unit($\underline{\boldsymbol{\mathcal{G}}}$) = unit($\mathbf{g}\,\mathbf{g}^{\rm T}$) = [M^{-1}L^{2}T].

8.1.3 The exact direct wave from a line explosion

Expression (8.1.31) is extremely simple to evaluate. In general, for any source/receiver and any combination of reflections and transmissions, it is necessary to do this numerically, but it is still straightforward and efficient. For the direct ray due to an explosion, the algebra is simple enough to obtain exact analytic results.

Using the source (7.1.15) for the direct ray due to a line explosion, expression (8.1.31) becomes

$$\mathbf{u}_P(t, \mathbf{x}_{\rm R}) = \frac{A_{\rm S} P_{\rm S}'(t)}{2\pi\rho\alpha^2} * \operatorname{Im}\left(\frac{1}{q_\alpha}\begin{pmatrix} p \\ \pm q_\alpha \end{pmatrix}\frac{\partial p}{\partial \widetilde{T}_P}\right)_{p=p(t,\mathbf{x}_{\rm R})}, \tag{8.1.32}$$

where only a P wave is excited and the velocity is the P wave velocity, $c = \alpha$.

For a uni-velocity wave, we can find a simple expression for q_α on the Cagniard contour

$$q_\alpha = (t - p x_{\rm R})/d \tag{8.1.33}$$

$$= \frac{\widetilde{T}_P}{R}\cos\theta + {\rm i}\left(\frac{\widetilde{T}_P^2}{R^2} - \frac{1}{\alpha^2}\right)^{1/2}\sin\theta, \tag{8.1.34}$$

corresponding to (8.1.11) in the fourth quadrant. Differentiating expression (8.1.11), we obtain simply

$$\frac{\partial p}{\partial \widetilde{T}_P} = -\frac{{\rm i} q_\alpha}{(\widetilde{T}_P^2 - R^2/\alpha^2)^{1/2}}. \tag{8.1.35}$$

Substituting in (8.1.32), the particle displacement due to a pressure line source is

$$\boxed{\mathbf{u}_P(t, \mathbf{x}_{\rm R}) = \frac{A_{\rm S}}{2\pi\rho\alpha^2 R}\begin{pmatrix} \sin\theta \\ \pm\cos\theta \end{pmatrix} P_{\rm S}'(t) * \frac{t H(t - R/\alpha)}{(t^2 - R^2/\alpha^2)^{1/2}}.} \tag{8.1.36}$$

As a check, we can confirm that unit($\mathbf{u}$) = [L] as unit($P_{\rm S}$) = [ML^{-1}T^{-2}] and the convolution cancels the time derivative. The result can also be compared with the two-dimensional Green dyadic (4.5.84). Taking the acoustic limit ($\beta \to 0$) and differentiating the Green dyadic with respect to the source position to obtain force

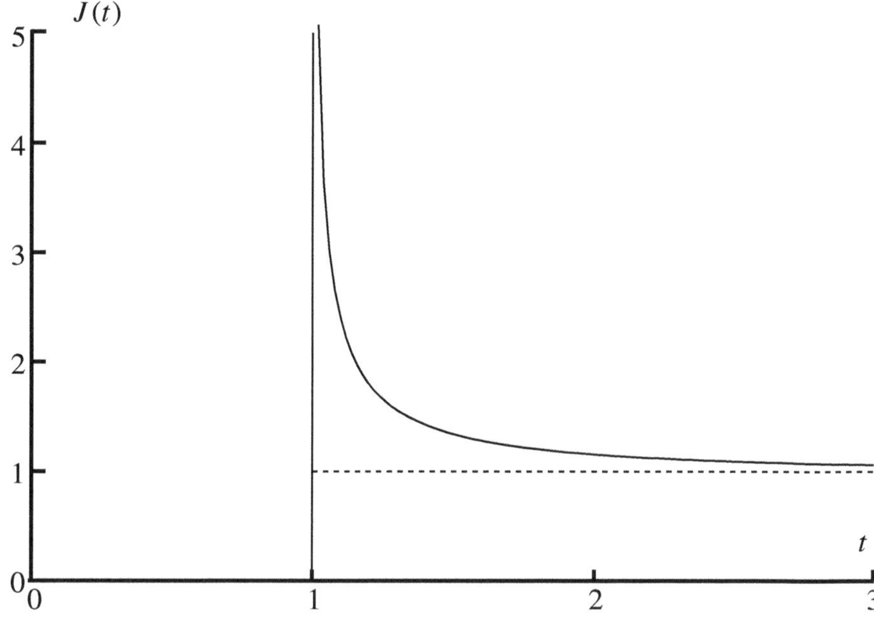

Fig. 8.2. The standard, two-dimensional, pressure Green function, $J(t)$ (8.1.37).

couples, we can obtain the exact solution for a pressure source from the Green dyadic (4.5.84) confirming result (8.1.36) (Exercise 8.3).

Expression (8.1.36) can be written in terms of a standard function

$$J(t) = \frac{tH(t-1)}{(t^2-1)^{1/2}}, \tag{8.1.37}$$

so the result becomes

$$\mathbf{u}_P(t, \mathbf{x}_\mathrm{R}) = \frac{A_\mathrm{S}}{2\pi\rho\alpha^2 R} \begin{pmatrix} \sin\theta \\ \pm\cos\theta \end{pmatrix} P'_\mathrm{S}(t) * J(\alpha t/R). \tag{8.1.38}$$

This two-dimensional, pressure Green function (8.1.37) is illustrated in Figure 8.2.

Although the response (8.1.38) can be written at all ranges in terms of one standard function (8.1.37), the significance of the different features varies with range. The standard function has an inverse square-root singularity at $t = 1$ and a unit asymptote as $t \to \infty$, i.e.

$$J(t) \simeq 2^{-1/2}\lambda(t-1) \qquad \text{for} \quad t \gtrsim 1 \tag{8.1.39}$$

$$\to 1 \qquad \text{as} \quad t \to \infty, \tag{8.1.40}$$

where $\lambda(t)$ is defined in (B.2.1). If the source pressure is a step function, $P_\mathrm{S}H(t)$, then the response consists of an inverse square-root singularity at $t = R/\alpha$, and a

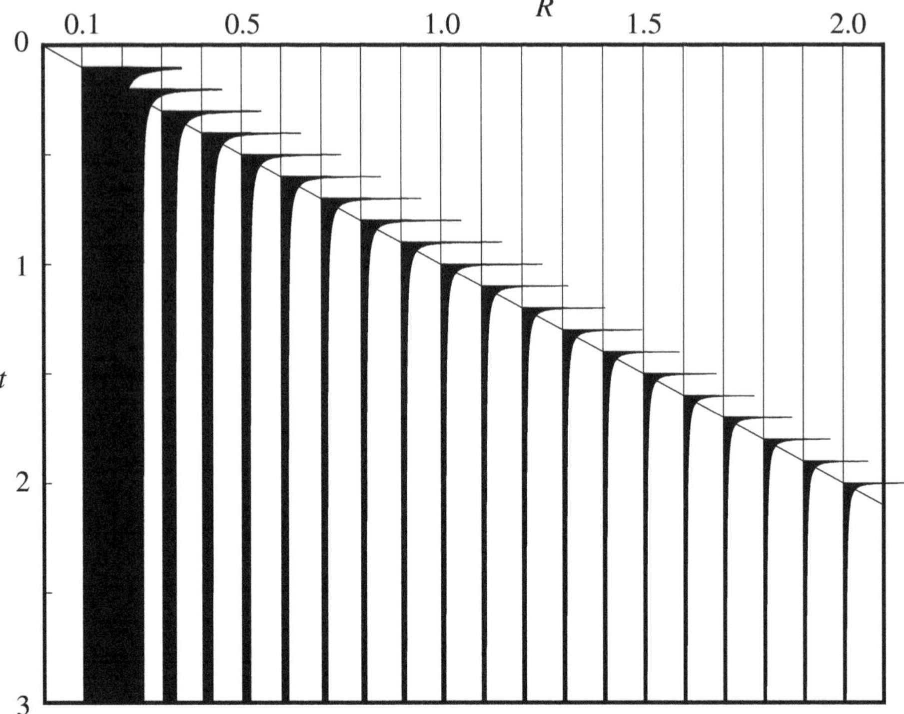

Fig. 8.3. The particle displacement for the direct pulse due to a line, pressure, step function source, $P_S H(t)$, illustrated for a sequence of ranges, R (units are scaled so $\alpha = 1$).

static deformation tail (Figure 8.3). Thus for $t \simeq R/\alpha$, the signal is approximately

$$\boxed{\mathbf{u}_P(t, \mathbf{x}_R) \simeq \frac{A_S P_S}{2^{3/2}\pi\rho\,\alpha^{5/2}R^{1/2}} \begin{pmatrix} \sin\theta \\ \pm\cos\theta \end{pmatrix} \lambda(t - R/\alpha).} \tag{8.1.41}$$

This is known as the *first-motion approximation.* At long times the deformation is

$$\mathbf{u}_P(t, \mathbf{x}_R) \simeq \frac{A_S P_S}{2\pi\rho\alpha^2 R} \begin{pmatrix} \sin\theta \\ \pm\cos\theta \end{pmatrix}. \tag{8.1.42}$$

The important properties of this solution (8.1.36) are that it is causal and longitudinal. The first motion decays as $R^{-1/2}$, i.e. cylindrical spreading. The pulse shape for a pressure step is $(t - R/\alpha)^{-1/2}$, the tail being due to the two-dimensional wave propagation (see Figure 4.16). This is known as the *far-field term.* Anticipating the simpler, point-source result (8.2.70), the displacement due to a step pressure, point source is, to first order, a delta function. Considering a line source as being made up of point sources, the tail $\lambda(t)$ is due to out-of-plane sources

(Figure 4.16 and Exercise 4.7). The decay is due to the extra propagation distance, obliquity and time delay from the out-of-plane positions. Finally, the static deformation (8.1.42) decays more rapidly, R^{-1}, and is known as the *near-field term*. Its significance clearly decreases as range increases in Figure 8.3.

8.2 Cagniard method in three dimensions

In three dimensions we use the two-dimensional spatial Fourier transform (3.2.14) with inverse (3.2.15). The form of the transformed solution is similar to that in two dimensions, and the Cagniard method can be applied to the same problems, i.e. the direct wave, reflections and transmissions in layered media. Thus in three dimensions, expressions (8.1.1) and (8.1.2) are replaced by

$$\boxed{\underline{\mathbf{v}}_{\text{ray}}(t, \mathbf{x}_\text{R}) = \frac{1}{8\pi^3} \int_B \omega^2 \iint_{-\infty}^{\infty} \underline{\mathcal{G}}_{\text{ray}}(\mathsf{p}) e^{i\omega(\widetilde{T}_{\text{ray}}(\mathsf{p}, \mathbf{x}_\text{R}) - t)} \mathrm{d}\mathsf{p}\, \mathrm{d}\omega,} \tag{8.2.1}$$

where

$$\boxed{\widetilde{T}_{\text{ray}}(\mathsf{p}, \mathbf{x}_\text{R}) = \mathsf{p} \cdot \mathsf{x}_\text{R} + \tau_{\text{ray}}(p, z_\text{R}),} \tag{8.2.2}$$

with $\mathsf{p} = (\,p_1\ p_2\,)^{\mathrm{T}}$ and $\mathsf{x}_\text{R} = (\,x_1\ x_2\,)^{\mathrm{T}}$, and $p = |\mathsf{p}|$. For future reference, we note that in three dimensions, unit($\underline{\mathbf{v}}$) = [M^{-1}]. Again we restrict the discussion to acoustic and isotropic elastic media, so the propagation terms are axially symmetric and asymmetry only arises through the source and receiver.

8.2.1 The inverse transforms

There are various methods of inverting the transforms (8.2.1). Without going into details, we should comment that although the final expressions appear to be significantly different, they must, of course, be equal. They all contain an integral in the complex p plane, and their equivalence depends on distorting one contour into another. We follow a technique which most closely follows the two-dimensional solution.

In expression (8.2.1) we let

$$p_1 = p \cos\phi \tag{8.2.3}$$

$$p_2 = p \sin\phi, \tag{8.2.4}$$

and

$$x_1 = x_\text{R} \cos\chi_\text{R} \tag{8.2.5}$$

$$x_2 = x_\text{R} \sin\chi_\text{R}, \tag{8.2.6}$$

i.e. (p, ϕ) and $(x_{\mathsf{R}}, \chi_{\mathsf{R}})$ are the polar components of p and x_{R}, respectively. Substituting in (8.2.1) and (8.2.2), we obtain

$$\underline{\mathbf{v}}_{\text{ray}}(t, \mathbf{x}_{\mathsf{R}}) = \frac{1}{8\pi^3} \int_B \omega^2 \int_0^\infty \int_0^{2\pi} \underline{\boldsymbol{\mathcal{G}}}_{\text{ray}}(p, \phi) \times \mathrm{e}^{\mathrm{i}\omega(p x_{\mathsf{R}} \cos(\phi - \chi_{\mathsf{R}}) + \tau_{\text{ray}}(p, z_{\mathsf{R}}) - t)}\, p \,\mathrm{d}\phi\, \mathrm{d}p\, \mathrm{d}\omega. \tag{8.2.7}$$

8.2.1.1 Point, pressure source

The dependence of $\underline{\boldsymbol{\mathcal{G}}}_{\text{ray}}(p, \phi)$ on ϕ complicates the solution but only arises through the source and receiver terms. The propagation term – the phase $\tau_{\text{ray}}(p, z_{\mathsf{R}})$ and any reflection/transmission coefficients in $\underline{\boldsymbol{\mathcal{G}}}_{\text{ray}}(p, \phi)$ – only depends on p for isotropic media considered here. For definiteness, let us consider the direct wave from a point, pressure source (an explosion as in expressions (4.6.22) and (7.1.16))

$$\boldsymbol{\mathcal{G}}_P(p, \phi) = -\frac{\mathrm{i}\omega V_{\mathsf{S}} P_{\mathsf{S}}(\omega)}{2\rho\, q_\alpha \alpha^2} \begin{pmatrix} p\cos\phi \\ p \sin\phi \\ \pm q_\alpha \end{pmatrix}. \tag{8.2.8}$$

We convert the angle variable to

$$\phi' = \phi - \chi_{\mathsf{R}}, \tag{8.2.9}$$

i.e. the angle between the p and x vectors. Then

$$\mathbf{v}_P(t, \mathbf{x}_{\mathsf{R}}) = \frac{V_{\mathsf{S}} P''_{\mathsf{S}}(t)}{16\pi^3 \rho\, \alpha^2} * \int_B \mathrm{i}\,\omega \int_0^\infty \int_0^{2\pi} \frac{p}{q_\alpha} \begin{pmatrix} p\cos(\phi' + \chi_{\mathsf{R}}) \\ p\sin(\phi' + \chi_{\mathsf{R}}) \\ \pm q_\alpha \end{pmatrix} \times \mathrm{e}^{\mathrm{i}\omega(p x_{\mathsf{R}} \cos\phi' + \tau_{\text{ray}}(p, z_{\mathsf{R}}) - t)} \mathrm{d}\phi'\, \mathrm{d}p\, \mathrm{d}\omega \tag{8.2.10}$$

$$= \frac{V_{\mathsf{S}} P''_{\mathsf{S}}(t)}{16\pi^3 \rho\, \alpha^2} * \int_B \mathrm{i}\,\omega \int_0^\infty \int_0^{2\pi} \frac{p}{q_\alpha} \begin{pmatrix} p\cos\chi_{\mathsf{R}} \cos\phi' \\ p\sin\chi_{\mathsf{R}}\cos\phi' \\ \pm q_\alpha \end{pmatrix} \times \mathrm{e}^{\mathrm{i}\omega(p x_{\mathsf{R}} \cos\phi' + \tau_{\text{ray}}(p, z_{\mathsf{R}}) - t)} \mathrm{d}\phi'\, \mathrm{d}p\, \mathrm{d}\omega, \tag{8.2.11}$$

as the terms involving $\sin\phi'$ make no overall contribution to the integral. The integrals for the components v_1 and v_2 only differ by the directivity factors $\cos\chi_{\mathsf{R}}$ and $\sin\chi_{\mathsf{R}}$. For brevity, let us assume $\chi_{\mathsf{R}} = 0$ ($x_2 = 0$). As the problem is axially symmetric, we can rotate the axes so the receiver lies on $x_2 = 0$. For simplicity, we assume $x_2 = 0$ without changing the notation to indicate the rotation. Thus

$$\mathbf{v}_P(t, \mathbf{x}_\mathrm{R}) = \frac{V_\mathrm{S} P_\mathrm{S}''(t)}{16\pi^3 \rho\, \alpha^2} * \int_B \mathrm{i}\,\omega \int_0^\infty \int_0^{2\pi} \frac{p}{q_\alpha} \begin{pmatrix} p\cos\phi' \\ 0 \\ \pm q_\alpha \end{pmatrix} \times \mathrm{e}^{\mathrm{i}\omega(p x_\mathrm{R} \cos\phi' + \tau_P(p, z_\mathrm{R}) - t)} \mathrm{d}\phi'\, \mathrm{d}p\, \mathrm{d}\omega. \quad (8.2.12)$$

The solution in the original coordinate system with x_2 and χ_R non-zero can then be obtained by rotating this solution.

The ϕ' integral is easily evaluated using the standard Parseval's integral representation of the Bessel functions (Abramowitz and Stegun, 1965, §9.1.21)

$$J_n(z) = \frac{\mathrm{i}^{-n}}{\pi} \int_0^\pi \mathrm{e}^{\mathrm{i}z\cos\theta} \cos(n\theta)\, \mathrm{d}\theta. \quad (8.2.13)$$

Using this in expression (8.2.12), we obtain

$$\mathbf{v}_P(t, \mathbf{x}_\mathrm{R}) = \frac{V_\mathrm{S} P_\mathrm{S}''(t)}{8\pi^2 \rho\, \alpha^2} * \int_B \mathrm{i}\,\omega \int_0^\infty \frac{p}{q_\alpha} \begin{pmatrix} \mathrm{i} p J_1(\omega p x_\mathrm{R}) \\ 0 \\ \pm\, q_\alpha J_0(\omega p x_\mathrm{R}) \end{pmatrix} \times \mathrm{e}^{\mathrm{i}\omega(\tau_P(p, z_\mathrm{R}) - t)} \mathrm{d}p\, \mathrm{d}\omega. \quad (8.2.14)$$

We could have obtained this result using the Fourier–Bessel transform (Section 3.3) on the wave equations, but that would have required transforming the wave equations with derivatives in cylindrical polar coordinates. The above technique is probably simpler, particularly when dealing with the vector equations of elasticity.

As we have noted in Chapter 3, Bessel integrals can be rewritten with Hankel functions (equation (3.3.5) becomes (3.3.7)). Thus expression (8.2.14) becomes

$$\mathbf{v}_P(t, \mathbf{x}_\mathrm{R}) = \frac{V_\mathrm{S} P_\mathrm{S}''(t)}{16\pi^2 \rho\, \alpha^2} * \int_B \mathrm{i}\,|\omega| \int_{-\infty}^\infty \frac{p}{q_\alpha} \begin{pmatrix} \mathrm{i} p H_1^{(1)}(\omega p x_\mathrm{R}) \\ 0 \\ \pm\, q_\alpha H_0^{(1)}(\omega p x_\mathrm{R}) \end{pmatrix} \times \mathrm{e}^{\mathrm{i}\omega(\tau_P(p, z_\mathrm{R}) - t)}\, \mathrm{d}p\, \mathrm{d}\omega. \quad (8.2.15)$$

The motivation for rewriting the integrals in this form is that with the asymptotic form for the Hankel functions (3.3.8), the integrals have essentially the same form as the two-dimensional inverse transforms (8.1.1).

The inverse Fourier transforms of the Hankel transforms are known (Appendix B.4). The factor $\exp(\mathrm{i}\omega\tau(p, z_\mathrm{R}))$ causes a time shift. We proceed as in two dimensions and change the order of integration and deform the p contour from the real axis to the Cagniard contour (Section 8.1.2). Using the symmetry of the integrand in the real p axis, we obtain for the particle displacement

$$\mathbf{u}_P(t,\mathbf{x}_R) = -\frac{V_S P_S''(t)}{2\pi^2\rho\,\alpha^2} * \operatorname{Im}\int_{C_-} \frac{p}{q_\alpha}\begin{pmatrix} p\left(\frac{t-\tau_P(p,z_R)}{p x_R}\right) \\ 0 \\ \pm q_\alpha \end{pmatrix} \times \frac{H(t-\widetilde{T}_P(p,\mathbf{x}_R))}{\sqrt{(t-\tau_P(p,z_R))^2 - p^2 x_R^2}}\,\mathrm{d}p, \tag{8.2.16}$$

where the integral is along Cagniard contour in the fourth quadrant. This is the exact result for the direct wave due to a point explosion. Compared with the two-dimensional result (8.1.31) we still have an integral. However, it is of finite extent, beginning at $p = \sin\theta/\alpha$ where the Cagniard contour leaves the real axis, and ending at the point where $\widetilde{T}_P(p,\mathbf{x}_R) = t$ (the point which gives the two-dimensional result). Thus three infinite integrals, the inverse transforms, have been replaced by one, non-oscillatory, finite integral

$$\boxed{\mathbf{u}_P(t,\mathbf{x}_R) = -\frac{V_S P_S''(t)}{2\pi^2\rho\,\alpha^2} * \int_{R/\alpha}^{t} \operatorname{Im}\left(\frac{p}{q_\alpha}\begin{pmatrix} p\left(\frac{t-\tau_P(p,z_R)}{p x_R}\right) \\ 0 \\ \pm q_\alpha \end{pmatrix} \times \frac{1}{\sqrt{(t-\tau_P(p,z_R))^2 - p^2 x_R^2}}\frac{\partial p}{\partial \widetilde{T}_P}\right)\mathrm{d}\widetilde{T}.} \tag{8.2.17}$$

Again it is sensible to confirm that unit($\mathbf{u}$) = [L].

8.2.1.2 An alternative method without Bessel or Hankel functions

An alternative method of reducing the triple integral (8.2.12) to the solution (8.2.17) is attractive as it does not involve or require knowledge of Bessel or Hankel functions. Starting with equation (8.2.12), we halve the angular range and double the range of the slowness integral while changing the order of integration

$$\mathbf{u}_P(t,\mathbf{x}_R) = -\frac{V_S P_S''(t)}{16\pi^3\rho\,\alpha^2} * \int_0^{\pi}\int_{-\infty}^{\infty}\int_B \frac{p}{q_\alpha}\begin{pmatrix} p\cos\phi' \\ 0 \\ \pm q_\alpha \end{pmatrix} \times e^{\mathrm{i}\omega(p x_R\cos\phi' + \tau_P(p,z_R) - t)}\mathrm{d}\omega\,\mathrm{d}p\,\mathrm{d}\phi'. \tag{8.2.18}$$

For each angle, ϕ', we distort the slowness contour from the real axis to the Cagniard contour defined by $\operatorname{Im}\left(\widetilde{T}_P\right) = 0$, where

$$\widetilde{T}_P(p,\mathbf{x}_R,\phi') = p\,x_R\cos\phi' + \tau_P(p,z_R), \tag{8.2.19}$$

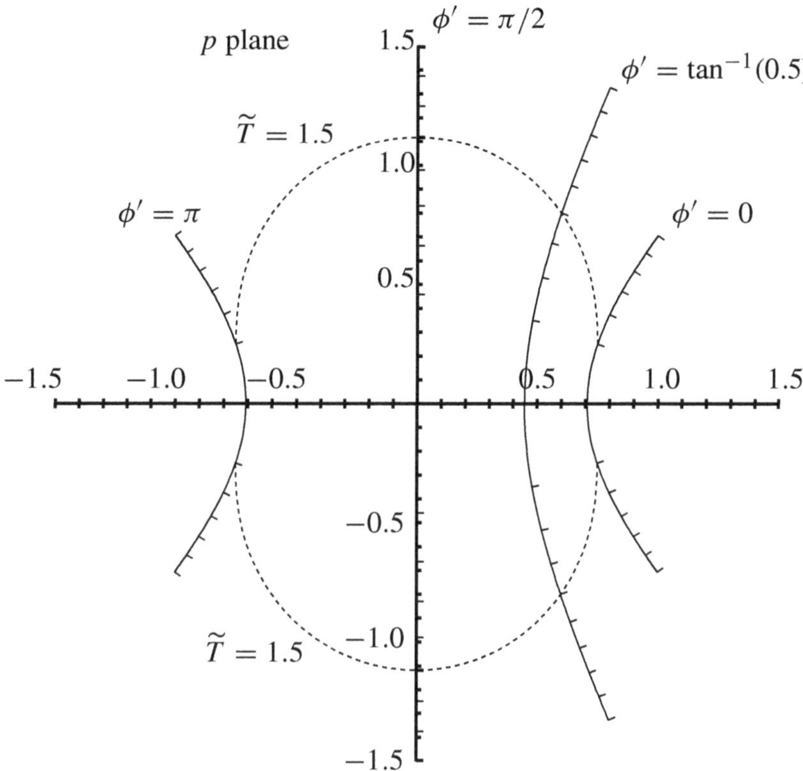

Fig. 8.4. Cagniard contours for $\phi' = 0$, $\pi/2$ and π, and an intermediate value $0 < \phi' = \tan^{-1}(0.5) < \pi/2$. The velocity, depth range and range are scaled to unity, $\alpha = 1$, $|z_S - z_R| = 1$ and $x_R = 1$. The tick marks on the Cagniard contours correspond to times $\widetilde{T} = 1$ to 2 in steps of 0.1. A ϕ' isochron for fixed time, $\widetilde{T} = 1.5$, is also shown with a dashed line.

cf. equation (8.1.3) – we add the extra argument ϕ' to the functions $\widetilde{T}_P$ and the inverse function defining the Cagniard contours, $p = p(t, \mathbf{x}_R, \phi')$. Each contour for fixed angle, ϕ', is analogous to the two-dimensional situation (Figure 8.1), but as ϕ' varies from 0 to π, the contour moves from the first and fourth quadrants to the second and third, e.g. for $\phi' = \pi/2$, the Cagniard contour is the imaginary axis (Figure 8.4).

With the slowness contour running along the Cagniard contour, which we denote by $C(\phi')$ as it depends on the angle ϕ', the Fourier transform with respect to frequency can be evaluated

$$\mathbf{u}_P(t, \mathbf{x}_R) = -\frac{V_S P_S''(t)}{8\pi^2 \rho\, \alpha^2} * \mathrm{Re} \int_0^{\pi} \int_{C(\phi')} \frac{p}{q_\alpha} \begin{pmatrix} p \cos\phi' \\ 0 \\ \pm q_\alpha \end{pmatrix}$$

$$\times\, \delta\left(t - \widetilde{T}_P(p, \mathbf{x}_R, \phi')\right) \mathrm{d}p\, \mathrm{d}\phi'. \tag{8.2.20}$$

Note the reversal in the direction of $C(\phi')$ depending on the sign of the frequency produces the required conjugate values of slowness (cf. equations (3.1.9) and (3.1.10)).

The slowness integral can now be evaluated from the contributions at the points on the Cagniard contours where

$$\widetilde{T}_P(p, \mathbf{x}_\mathrm{R}, \phi') = t. \tag{8.2.21}$$

This gives

$$\mathbf{u}_P(t, \mathbf{x}_\mathrm{R}) = -\frac{V_\mathrm{S} P_\mathrm{S}''(t)}{4\pi^2 \rho\, \alpha^2} * \int_0^\pi \mathrm{Re}\left(\frac{p}{q_\alpha} \begin{pmatrix} p\cos\phi' \\ 0 \\ \pm q_\alpha \end{pmatrix} \left(\frac{\partial p}{\partial \widetilde{T}_P} \right)_{\phi'} \right)_{p=p(t,\mathbf{x}_\mathrm{R},\phi')} \mathrm{d}\phi', \tag{8.2.22}$$

where for a given t and ϕ', equation (8.2.21) defines two slowness values, and by symmetry we will just use the values below the real axis with $\mathrm{Im}(p) \leq 0$. As ϕ' varies it defines a contour in the complex slowness plane, illustrated in Figure 8.4 and 8.5. We can convert this into a complex slowness integral, i.e.

$$\mathbf{u}_P(t, \mathbf{x}_\mathrm{R}) = -\frac{V_\mathrm{S} P_\mathrm{S}''(t)}{4\pi^2 \rho\, \alpha^2} * \mathrm{Re} \int_{p(t,\mathbf{x}_\mathrm{R},0)}^{p(t,\mathbf{x}_\mathrm{R},\pi)} \frac{p}{q_\alpha} \begin{pmatrix} p\cos\phi' \\ 0 \\ \pm q_\alpha \end{pmatrix} \left(\frac{\partial p}{\partial \widetilde{T}_P} \right)_{\phi'} \left(\frac{\partial \phi'}{\partial p} \right)_t \mathrm{d}p \tag{8.2.23}$$

$$= \frac{V_\mathrm{S} P_\mathrm{S}''(t)}{4\pi^2 \rho\, \alpha^2} * \mathrm{Re} \int_{p(t,\mathbf{x}_\mathrm{R},0)}^{p(t,\mathbf{x}_\mathrm{R},\pi)} \frac{p}{q_\alpha} \begin{pmatrix} p\cos\phi' \\ 0 \\ \pm q_\alpha \end{pmatrix} \left(\frac{\partial \phi'}{\partial \widetilde{T}_P} \right)_p \mathrm{d}p \tag{8.2.24}$$

$$= -\frac{V_\mathrm{S} P_\mathrm{S}''(t)}{4\pi^2 \rho\, \alpha^2} * \mathrm{Re} \int_{p(t,\mathbf{x}_\mathrm{R},0)}^{p(t,\mathbf{x}_\mathrm{R},\pi)} \frac{p}{q_\alpha} \begin{pmatrix} p \left(\frac{t-\tau_P(p,z_\mathrm{R})}{p x_\mathrm{R}} \right) \\ 0 \\ \pm q_\alpha \end{pmatrix} \times \frac{\mathrm{d}p}{\sqrt{p^2 x_\mathrm{R}^2 - (t - \tau_P(p, z_\mathrm{R}))^2}}. \tag{8.2.25}$$

To obtain expression (8.2.23), we change the variable of integration from ϕ' to p; in equation (8.2.24) we combine the partial derivatives using the chain rule; and in result (8.2.25), substitute for the partial derivatives and trigonometrical functions of ϕ' using definition (8.2.19).

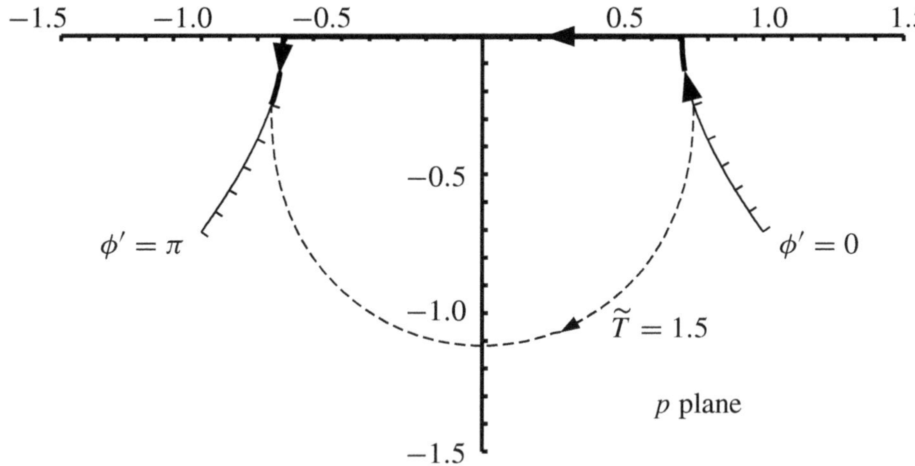

Fig. 8.5. The contour with t constant and ϕ' varying from 0 to π, and the distorted contour along the Cagniard contours $p(t, \mathbf{x}_{\mathrm{R}}, 0)$ and $p(t, \mathbf{x}_{\mathrm{R}}, \pi)$, and the real axis. The parameters are as in Figure 8.4.

The complex contour from $p(t, \mathbf{x}_{\mathrm{R}}, 0)$ to $p(t, \mathbf{x}_{\mathrm{R}}, \pi)$ can now be distorted to run along Cagniard contours for $\phi' = 0$ and π, and along the real axis (Figure 8.5). We can reduce the contour to the fourth quadrant by symmetry and obtain

$$\mathbf{u}_P(t, \mathbf{x}_{\mathrm{R}}) = \frac{V_{\mathrm{S}} P''_{\mathrm{S}}(t)}{2\pi^2 \rho\, \alpha^2} * \operatorname{Re} \int_0^{p(t,\mathbf{x}_{\mathrm{R}},0)} \frac{p}{q_\alpha} \begin{pmatrix} p \left(\frac{t - \tau_P(p, z_{\mathrm{R}})}{p x_{\mathrm{R}}} \right) \\ 0 \\ \pm q_\alpha \end{pmatrix}$$

$$\times \frac{\mathrm{d}p}{\sqrt{p^2 x_{\mathrm{R}}^2 - (t - \tau_P(p, z_{\mathrm{R}}))^2}} \tag{8.2.26}$$

$$= -\frac{V_{\mathrm{S}} P''_{\mathrm{S}}(t)}{2\pi^2 \rho\, \alpha^2} * \int_{R/\alpha}^{t} \operatorname{Im} \left(\frac{p}{q_\alpha} \begin{pmatrix} p \left(\frac{t - \tau_P(p, z_{\mathrm{R}})}{p x_{\mathrm{R}}} \right) \\ 0 \\ \pm q_\alpha \end{pmatrix} \right.$$

$$\left. \times \frac{1}{\sqrt{(t - \tau_P(p, z_{\mathrm{R}}))^2 - p^2 x_{\mathrm{R}}^2}} \left(\frac{\partial p}{\partial \widetilde{T}_P} \right)_{\phi'=0} \right) \mathrm{d}\widetilde{T}, \tag{8.2.27}$$

in agreement with expression (8.2.17).

Having obtained the exact solution for a point explosion by two methods, (8.2.17) and (8.2.27), we can now generalize the solution. It is straightforward to extend result (8.2.17) to more general waves, including generalized reflection and/or transmission coefficients in the integrand. For some source/receiver components it is necessary to use the second-order Bessel and Hankel functions. We outline this in the next section.

8.2.1.3 Generalized rays, sources and receivers

The three-dimensional Cagniard solution (8.2.17) is relatively simple as we have only considered a point, pressure source in an isotropic medium so the solution is axially symmetric and contains no reflection or transmission coefficients. It is straightforward to generalize this result to point sources without axial symmetry, e.g. the point source of stress glut of Sections 4.5.5 and 4.6, for generalized rays with reflection and/or transmission coefficients, and for receivers at interfaces, etc. We consider only isotropic media so any asymmetry arises only through the source and receiver terms. The results can be extended to TIV media without difficulty (van der Hijden, 1987) as the problem is still axially symmetric. The generalization to any anisotropy is complicated as asymmetry enters through the reflection and transmission coefficients, and the phase term, as well as the source and receiver. The method is sufficiently complicated that numerical, spectral techniques are normally employed. We do not consider the Cagniard method for anisotropic media here.

Spencer (1960) introduced the term generalized rays and investigated the Cagniard solution in a layered medium. Abramovici (1978) extended the Cagniard method to model various sources. Many authors used the method in layered models, first for a single layer over a half-space (Pekeris, Alterman and Abramovici, 1963, for *SH* waves, and Pekeris, Alterman, Abramovici and Jarosh, 1965, Abramovici, 1970, Abramovici and Gal-Ezer, 1978, and Abramovici, Kanasewich and Kelamis, 1982, for *P* and *SV* waves and explosive, vertical and horizontal force sources). The multi-layered problem has been investigated by Abramovici (1984) and Drijkoningen and Fokkema (1987).

For the direct ray considered so far, the propagation term (8.0.7) is simply $\mathcal{P}_{\mathrm{ray}}(\omega, \mathsf{p}, z_{\mathsf{R}}) = \exp(\mathrm{i}\omega q |z_{\mathsf{R}} - z_{\mathsf{S}}|)$. For a generalized ray, the propagation term contains a product of the appropriate reflection and transmission coefficients and a phase term (8.0.9) containing the depth integral of the vertical slowness. If the medium is isotropic, the coefficients and phase only depend on the magnitude of the horizontal slowness, $p = |\mathsf{p}|$, not its vector components, i.e.

$$\mathcal{P}_{\mathrm{ray}}(\omega, \mathsf{p}, z_{\mathsf{R}}) = \left(\prod_{\mathrm{ray}} \mathcal{T}_{ij}(p) \right) \mathrm{e}^{\mathrm{i}\omega\tau_{\mathrm{ray}}(p, z_{\mathsf{R}})}. \tag{8.2.28}$$

For a point force, the source term depends on the source ray type. The source term is $\mathbf{g}^{\mathrm{T}}\mathbf{f}_{\mathsf{S}}$ (4.5.77). If it is a P wave, the term is given by

$$\mathcal{M}_{\mathsf{S}}(\mathsf{p}) = \mathbf{g}_P^{\mathrm{T}}\, \mathbf{f}_{\mathsf{S}} = \alpha_{\mathsf{S}} w_3 \left(p_1\, f_x + p_2\, f_y \pm q_\alpha\, f_z \right), \tag{8.2.29}$$

where w_3 is given by definition (6.3.57) and terms are evaluated at the source depth, $z = z_S$. For an *SV* wave source

$$\mathcal{M}_S(\mathsf{p}) = \mathbf{g}_V^T \mathbf{f}_S = \beta_S w_1 \left(q_\beta (p_1 f_x + p_2 f_y)/p \pm p f_z\right), \qquad (8.2.30)$$

and for an *SH* wave

$$\mathcal{M}_S = \mathbf{g}_H^T \mathbf{f}_S = \beta_S w_2 \left(p_1 f_y - p_2 f_x\right)/p, \qquad (8.2.31)$$

where w_1 and w_2 are given by definitions (6.3.55) and (6.3.56), respectively.

For a point stress-glut, the source term is $\mathbf{g}^T \mathbf{M}_S \mathbf{p}$ (4.6.17). For a *P* wave this gives

$$\mathcal{M}_S(\mathsf{p}) = \mathbf{g}_P^T \mathbf{M}_S \mathbf{p} = \alpha_S w_3 \Big(M_{xx} p_1^2 + M_{yy} p_2^2 + M_{zz} q_\alpha^2 + 2M_{xy} p_1 p_2 \pm 2(M_{zx} p_1 + M_{yz} p_2) q_\alpha\Big). \qquad (8.2.32)$$

Similarly for an *SV* source

$$\mathcal{M}_S(\mathsf{p}) = \mathbf{g}_V^T \mathbf{M}_S \mathbf{p} = \frac{\beta_S w_1}{p}\Big((M_{xx} p_1^2 + M_{yy} p_2^2 - M_{zz} p^2 + 2M_{xy} p_1 p_2) q_\beta \pm (M_{zx} p_1 + M_{yz} p_2)\Omega\Big), \qquad (8.2.33)$$

where Ω is defined in equation (6.3.54), and for an *SH* source

$$\mathcal{M}_S(\mathsf{p}) = \mathbf{g}_H^T \mathbf{M}_S \mathbf{p} = \frac{\beta_S w_2}{p}\Big((M_{yy} p_1^2 - M_{xx}) p_1 p_2 + M_{xy}(p_1^2 - p_2^2) \pm (M_{yz} p_1 - M_{zx} p_2) q_\beta\Big). \qquad (8.2.34)$$

Mathematically, the important feature of the possible source terms, (8.2.29)–(8.2.34), is that they can all be expressed as

$$\mathcal{M}_S(\mathsf{p}) = \mathcal{M}_{mn}(p) p_1^m p_2^n, \qquad (8.2.35)$$

where the coefficients, $\mathcal{M}_{mn}$, are only functions of the horizontal slowness magnitude, p, and the components, p_1 and p_2, only enter through polynominal terms with $0 \le m + n \le 2$.

The receiver terms depend on the ray type. For a *P* wave, the polarization (6.3.53) generalizes to

$$\mathcal{H}_R(\mathsf{p}) = \mathbf{g}_P = w_3 \begin{pmatrix} p_1 \\ p_2 \\ \pm q_\alpha \end{pmatrix}, \qquad (8.2.36)$$

for an *SV* wave (6.3.51) to

$$\mathcal{H}_{\mathrm{R}}(\mathrm{p}) = \mathbf{g}_V = w_1 \begin{pmatrix} q_\beta p_1/p \\ q_\beta p_2/p \\ \mp p \end{pmatrix}, \tag{8.2.37}$$

and for an *SH* wave (6.3.52)

$$\mathcal{H}_{\mathrm{R}}(\mathrm{p}) = \mathbf{g}_H = w_2 \begin{pmatrix} -p_2/p \\ p_1/p \\ 0 \end{pmatrix}. \tag{8.2.38}$$

If a receiver is at an interface, these are replaced by the conversion coefficients, i.e. expressions (6.6.10), (6.6.11) and (6.6.12)

$$\mathcal{H}_{\mathrm{R}}(\mathrm{p}) = \mathbf{h}_P = \frac{1}{w_3 \mu \Delta^{\mathrm{Rayleigh}}} \begin{pmatrix} 2p_1 q_\beta \\ 2p_2 q_\beta \\ \pm \Omega \end{pmatrix} \tag{8.2.39}$$

$$\mathcal{H}_{\mathrm{R}}(\mathrm{p}) = \mathbf{h}_V = \frac{1}{w_1 \mu \Delta^{\mathrm{Rayleigh}}} \begin{pmatrix} \Omega p_1/p \\ \Omega p_2/p \\ \mp 2\mu q_\alpha \end{pmatrix} \tag{8.2.40}$$

$$\mathcal{H}_{\mathrm{R}}(\mathrm{p}) = \mathbf{h}_H = 2w_2 \begin{pmatrix} -p_2/p \\ p_1/p \\ 0 \end{pmatrix}. \tag{8.2.41}$$

Note that for *P* or *SV* waves these can be written as

$$\mathcal{H}_{\mathrm{R}}(\mathrm{p}) = \begin{pmatrix} h_1(p)\, p_1 \\ h_1(p)\, p_2 \\ h_3(p) \end{pmatrix}, \tag{8.2.42}$$

where the functions $h_1(p)$ and $h_3(p)$ take various forms (the important feature being that they depend only on the horizontal slowness magnitude, p), and for *SH* waves as

$$\mathcal{H}_{\mathrm{R}}(\mathrm{p}) = \begin{pmatrix} -h_2(p)\, p_2 \\ h_2(p)\, p_1 \\ 0 \end{pmatrix}. \tag{8.2.43}$$

Combining the appropriate source term, from the expressions (8.2.29)–(8.2.34), with the propagation term (8.2.28) and the appropriate receiver term, from expressions (8.2.36)–(8.2.41), we obtain

$$\mathbf{v}_{\mathrm{ray}}(\omega, \mathrm{p}, z_{\mathrm{R}}) = \mathcal{H}_{\mathrm{R}}(\mathrm{p})\, \mathcal{P}_{\mathrm{ray}}(\omega, \mathrm{p}, z_{\mathrm{R}}) \mathcal{M}_{\mathrm{S}}(\mathrm{p}) = \mathcal{G}_{\mathrm{ray}}(\mathrm{p}) \mathrm{e}^{\mathrm{i}\omega\, \tau_{\mathrm{ray}}(p, z_{\mathrm{R}})}. \tag{8.2.44}$$

For P and SV receiver waves, we have

$$\boldsymbol{\mathcal{G}}_{\rm ray}(\mathsf{p}) = \left(\prod_{\rm ray} \mathcal{T}_{ij}(p)\right) \mathcal{M}_{mn}(p) p_1^m p_2^n \begin{pmatrix} h_1(p)\, p_1 \\ h_1(p)\, p_2 \\ h_3(p) \end{pmatrix}, \tag{8.2.45}$$

and for SH receiver waves

$$\boldsymbol{\mathcal{G}}_{\rm ray}(\mathsf{p}) = \left(\prod_{\rm ray} \mathcal{T}_{ij}(p)\right) \mathcal{M}_{mn}(p) p_1^m p_2^n \begin{pmatrix} -h_2(p)\, p_2 \\ h_2(p)\, p_1 \\ 0 \end{pmatrix}, \tag{8.2.46}$$

with summation over m and n. The important features of expressions (8.2.45) and (8.2.46) are that the horizontal slowness components only enter through the simple polynomial terms. The more complicated functions, $\mathcal{T}_{ij}(p)$, $\mathcal{M}_{mn}(p)$, $h_k(p)$ and $\tau_{\rm ray}(p, z_{\rm R})$, only depend on the magnitude of the horizontal slowness, $p = (p_1^2 + p_2^2)^{1/2}$. This simplicity only exists for isotropic media or media with a vertical symmetry axis, e.g. TIV. In more general anisotropic media, all the more general functions depend on p_1 and p_2 in a complicated manner and simple analytic methods are impossible. Normally only numerical solutions are possible.

For simplicity, let us consider only one component of expression (8.2.45) or (8.2.46) and write it

$$\mathcal{G}_{mn}(p)\, p_1^m p_2^n, \tag{8.2.47}$$

where now $0 \le m + n \le 3$. Substituting in (8.2.1), we obtain

$$u_{mn}(t, \mathbf{x}_{\rm R}) = \frac{\rm i}{8\pi^3} \int_B \omega \iint_{-\infty}^{\infty} \mathcal{G}_{mn}(p)\, p_1^m p_2^n\, {\rm e}^{{\rm i}\omega(\widetilde{\mathcal{T}}_{\rm ray}(\mathsf{p},\mathbf{x}_{\rm R})-t)} {\rm d}\mathsf{p}\, {\rm d}\omega, \tag{8.2.48}$$

for a typical term for the displacement (no summation over m and n). Applying the transforms (8.2.3)–(8.2.6), equivalent to expression (8.2.7) we have

$$u_{mn}(t, \mathbf{x}_{\rm R}) = \frac{\rm i}{8\pi^3} \int_B \omega \int_0^\infty \int_0^{2\pi} \mathcal{G}_{mn}(p)\, p^{m+n+1} \cos^m\phi \sin^n\phi \\ \times {\rm e}^{{\rm i}\omega(p x_{\rm R} \cos(\phi-\chi_{\rm R}) + \tau_{\rm ray}(p, z_{\rm R}) - t)} {\rm d}\phi\, {\rm d}p\, {\rm d}\omega. \tag{8.2.49}$$

For simplicity, as the model is axially symmetric let us assume $\chi_{\rm R} = 0$. This implies that $x_2 = 0$, i.e. the coordinate system must have been rotated so the receiver lies in the x_1–x_3 plane. This in turn implies that the source components in expressions (8.2.29)–(8.2.34) are in this rotated system. Defining the rotation matrix, $\mathbf{R}$ (7.1.31), the coordinates in the rotated coordinated system, $\mathbf{x}'$, are given in terms of the original system by

$$\mathbf{x}' = \mathbf{R}\,\mathbf{x}. \tag{8.2.50}$$

Obviously $\tan\chi_\mathsf{R} = x_2/x_1$. The force components used in the source terms, (8.2.29)–(8.2.31), are similarly rotated

$$\mathbf{f}'_\mathsf{S} = \mathbf{R}\,\mathbf{f}_\mathsf{S}, \tag{8.2.51}$$

and the moment tensor components in expressions (8.2.32)–(8.2.34) are

$$\mathbf{M}'_\mathsf{S} = \mathbf{R}^\mathrm{T}\,\mathbf{M}_\mathsf{S}\,\mathbf{R}. \tag{8.2.52}$$

Using $\mathbf{f}'_\mathsf{S}$ and $\mathbf{M}'_\mathsf{S}$ in expressions (8.2.29)–(8.2.34), the inverse transforms give $\mathbf{v}'_\mathrm{ray}$, and the response in the original coordinates is obtained from $\mathbf{v}_\mathrm{ray} = \mathbf{R}^\mathrm{T}\,\mathbf{v}'_\mathrm{ray}$.

In the integrals (8.2.49), any terms with odd powers of $\sin\phi$, i.e. $n = 1$ or 3, are zero by symmetry. Terms with $n = 2$ can simply be reduced to $n = 0$ using $\sin^2\phi = 1 - \cos^2\phi$. Thus only integrals of the form

$$u(t, \mathbf{x}_\mathsf{R}) = \frac{\mathrm{i}}{8\pi^3}\int_B \omega \int_0^\infty \int_0^{2\pi} \mathcal{G}(p)\,p^{m+1}\cos^m\phi \\ \times \mathrm{e}^{\mathrm{i}\omega(p x_\mathsf{R}\cos\phi + \tau_\mathrm{ray}(p, z_\mathsf{R}) - t)}\mathrm{d}\phi\,\mathrm{d}p\,\mathrm{d}\omega, \tag{8.2.53}$$

are required, where $\mathcal{G}(p) = \mathcal{G}_{m\,0}(p)$ or $\mathcal{G}(p) = \mathcal{G}_{m-2\,2}(p)$.

The related integrals with $\cos^m\phi$ replaced by $\cos m\phi$ can be evaluated as above. The ϕ integral is evaluated to give a Bessel function (8.2.13). The integral over positive slownesses (3.3.5) can be converted into an integral with a Hankel function along the complete slowness axis (3.3.7), which can be distorted to the Cagniard contour. Then the Fourier transform can be inverted using the inverse Fourier transform of the Hankel function (B.4.14). Thus

$$\frac{\mathrm{i}}{8\pi^3}\int_B \omega \int_0^\infty \int_0^{2\pi} \mathcal{G}(p)\,p^{m+1}\cos m\phi\,\mathrm{e}^{\mathrm{i}\omega(p x_\mathsf{R}\cos\phi + \tau_\mathrm{ray}(p, z_\mathsf{R}) - t)}\mathrm{d}\phi\,\mathrm{d}p\,\mathrm{d}\omega \tag{8.2.54}$$

$$= \frac{\mathrm{i}^{m+1}}{4\pi^2}\int_B \omega \int_0^\infty \mathcal{G}(p)\,p^{m+1} J_m(\omega p x_\mathsf{R})\,\mathrm{e}^{\mathrm{i}\omega(\tau_\mathrm{ray}(p, z_\mathsf{R}) - t)}\mathrm{d}p\,\mathrm{d}\omega \tag{8.2.55}$$

$$= \frac{\mathrm{i}^{m+1}}{8\pi^2}\int_B |\omega| \int_{-\infty}^\infty \mathcal{G}(p)\,p^{m+1} H_m^{(1)}(\omega p x_\mathsf{R})\,\mathrm{e}^{\mathrm{i}\omega(\tau_\mathrm{ray}(p, z_\mathsf{R}) - t)}\mathrm{d}p\,\mathrm{d}\omega \tag{8.2.56}$$

$$= \frac{1}{\pi^2}\frac{\mathrm{d}}{\mathrm{d}t}\,\mathrm{Im}\int_C \mathcal{G}(p)\,p^{m+1}\frac{T_m\left((t - \tau_\mathrm{ray}(p, z_\mathsf{R}))/p x_\mathsf{R}\right)}{\sqrt{(t - \tau_\mathrm{ray}(p, z_\mathsf{R}))^2 - p^2 x_\mathsf{R}^2}}\,\mathrm{d}p. \tag{8.2.57}$$

Noting that as the original integral (8.2.54) and the Chebyshev polynomials $T_m(x)$

(B.4.2) contain $\cos m\phi$, we can deduce that integral (8.2.53) is

$$u(t, \mathbf{x}_\mathsf{R}) = \frac{\mathrm{i}}{8\pi^3} \int_B \omega \int_0^\infty \int_0^{2\pi} \mathcal{G}(p)\, p^{m+1} \cos^m \phi \times \mathrm{e}^{\mathrm{i}\omega(p x_\mathsf{R} \cos\phi + \tau_{\mathrm{ray}}(p, z_\mathsf{R}) - t)} \mathrm{d}\phi\, \mathrm{d}p\, \mathrm{d}\omega \tag{8.2.58}$$

$$= \frac{1}{\pi^2} \frac{\mathrm{d}}{\mathrm{d}t} \operatorname{Im} \int_C \mathcal{G}(p)\, p^{m+1} \frac{\left((t - \tau_{\mathrm{ray}}(p, z_\mathsf{R}))/p x_\mathsf{R}\right)^m}{\sqrt{(t - \tau_{\mathrm{ray}}(p, z_\mathsf{R}))^2 - p^2 x_\mathsf{R}^2}} \mathrm{d}p. \tag{8.2.59}$$

With this result, all terms for a general source and receiver, (8.2.45) or (8.2.46), can be evaluated.

The above technique is useful as the evaluation of the ϕ integral, to obtain a result (8.2.55) equivalent to a Fourier–Bessel transform (3.3.5) can be used for a general response when the ray expansion is not possible. Generally, rather than (8.2.55), we obtain integrals of the form

$$\frac{\mathrm{i}^{m+1}}{4\pi^2} \int_B \omega \int_0^\infty \mathcal{G}(\omega, p, z_\mathsf{R})\, p^{m+1}\, J_m(\omega p x_\mathsf{R})\, \mathrm{e}^{-\mathrm{i}\omega t} \mathrm{d}p\, \mathrm{d}\omega, \tag{8.2.60}$$

which can be evaluated numerically (see Section 8.5.2). Expressions like this can also be obtained using the Fourier–Bessel transform (Section 3.3) directly, rather than via the two-dimensional spatial Fourier transform. This requires developing vector Fourier–Bessel transforms for the equations of motion and the constitutive relation, a complication that is unnecessary for our purposes (see Exercise 7.1).

Another advantage of the above procedure is that the intermediate result (8.2.56) with the Hankel function is analogous to the Fourier transform in the two-dimensional solution (8.1.1).

Again an alternative procedure that requires no knowledge of Bessel or Hankel functions (e.g. results (8.2.13) and (B.4.14)) is even simpler and more attractive. The procedure is identical to equations (8.2.18)–(8.2.27) (Section 8.2.1.2), and Figures 8.4 and 8.5, except we start with the triple integral (8.2.58). Proceeding just as before we obtain result (8.2.59) without using Bessel or Hankel functions, or results (8.2.13) and (B.4.14).

8.2.2 The far-field approximation

The integrand in result (8.2.17) has an integrable singularity at its upper limit, $\widetilde{T} = t$. We can approximate the integral by expanding about this point. The important term is the denominator

$$\sqrt{(t-\tau_P(p,z_{\mathrm{R}}))^2-p^2x_{\mathrm{R}}^2}=\sqrt{(t-\widetilde{T}_P)(t-\widetilde{T}_P+2px_{\mathrm{R}})} \qquad (8.2.61)$$

$$\simeq\sqrt{(t-\widetilde{T}_P)(2px_{\mathrm{R}})}. \qquad (8.2.62)$$

Substituting in (8.2.17), and making a similar expansion for the first component, u_1, we have

$$\mathbf{u}_P(t,\mathbf{x}_{\mathrm{R}})\simeq-\frac{V_{\mathrm{S}}P_{\mathrm{S}}''(t)}{2^{3/2}\pi^2\rho\,\alpha^2x_{\mathrm{R}}^{1/2}} * \int_{R/\alpha}^{t}\mathrm{Im}\left(\frac{p^{1/2}}{q_\alpha}\begin{pmatrix}p\\0\\\pm q_\alpha\end{pmatrix}\frac{1}{\sqrt{t-\widetilde{T}_P}}\frac{\partial p}{\partial\widetilde{T}_P}\right)_{p=p_P(\widetilde{T},\mathbf{x}_{\mathrm{R}})}\mathrm{d}\widetilde{T} \qquad (8.2.63)$$

$$\simeq-\frac{V_{\mathrm{S}}P_{\mathrm{S}}''(t)}{2^{3/2}\pi^2\rho\,\alpha^2x_{\mathrm{R}}^{1/2}} * \lambda(t) * \mathrm{Im}\left(\frac{p^{1/2}}{q_\alpha}\begin{pmatrix}p\\0\\\pm q_\alpha\end{pmatrix}\frac{\partial p}{\partial\widetilde{T}_P}\right)_{p=p_P(t,\mathbf{x}_{\mathrm{R}})}, \qquad (8.2.64)$$

where the integral in result (8.2.64) is recognized as a convolution with $\lambda(t)$ (B.2.1). This is known as the *far-field approximation* as it requires

$$px_{\mathrm{R}}\gg\Delta t. \qquad (8.2.65)$$

Unfortunately it is not valid when x_{R} is small, e.g. a VSP experiment. As well as making an expansion in the time domain, it can be obtained by substituting the asymptotic form for the Hankel functions (3.3.8) in transforms (8.2.15) and using the two-dimensional inversion technique (Section 8.1.2).

8.2.2.1 Two to three dimensions

To convert from the two-dimensional result (8.1.31) to the far-field approximation for the three-dimensional result, it is necessary to replace

$$\boxed{A_{\mathrm{S}}\to\frac{V_{\mathrm{S}}}{2^{1/2}\pi x_{\mathrm{R}}^{1/2}}\frac{\mathrm{d}}{\mathrm{d}t}\lambda(t) * p^{1/2}.} \qquad (8.2.66)$$

Here we have obtained the extra terms from a comparison of the far-field approximation for the direct wave from an explosion. However, the result applies more generally and can be obtained from the asymptotic forms of Hankel functions in the inverse transforms (cf. equation (8.2.15)) or by evaluating the transverse slowness integral by the second-order saddle point method (using result (5.7.15) with result (D.1.11)). This conversion factor is also obtained by comparing two and three-dimensional ray theory, e.g. the dyadic results (5.2.82) and (5.2.71), respectively. The time dependence converts as

$$\frac{1}{\pi}\frac{\mathrm{d}}{\mathrm{d}t}\lambda(t) * \Lambda(t-T) = \Delta(t-T), \tag{8.2.67}$$

and the factor $(p/x_{\mathrm{R}})^{-1/2}$ converts to the spreading in the third dimension (5.7.15), the difference between the spreading functions, $\mathcal{S}^{(2)}$ and $\mathcal{S}^{(3)}$, in (5.2.82) and (5.2.71), e.g. definitions (5.2.80) and (5.2.67). The exact two-dimensional result (8.1.31) is replaced by the far-field, three-dimensional result

$$\boxed{\underline{\mathbf{u}}_{\mathrm{ray}}(t,\mathbf{x}_{\mathrm{R}}) \simeq -\frac{1}{\pi^2(2x_{\mathrm{R}})^{1/2}}\frac{\mathrm{d}}{\mathrm{d}t}\lambda(t) * \mathrm{Im}\left(p^{1/2}\underline{\boldsymbol{\mathcal{G}}}_{\mathrm{ray}}(p)\frac{\partial p}{\partial \widetilde{T}_{\mathrm{ray}}}\right)_{p=p_{\mathrm{ray}}(t,\mathbf{x}_{\mathrm{R}})}.} \tag{8.2.68}$$

In Figure 8.6 we have illustrated the far-field approximation (8.2.68) compared with the exact result (8.2.17) (and see equation (8.2.70), below). These are for an explosive source with time function $P_{\mathrm{S}}(t) = P_{\mathrm{S}} t H(t)$ so the displacement response has the form $H(t)$. It can be seen that except at very short ranges, the far-field approximation is very satisfactory.

Converting wave propagation in two dimensions to three dimensions is sometimes referred to as *2.5D wave propagation*. The term becomes more meaningful when the model is two dimensional, e.g. parameters vary as a function of x and z, but the waves propagate in a three-dimensional model from a point source, i.e. the two-dimensional model is extruded in the y direction. We return to this in Chapter 10, when methods valid in laterally heterogeneous models are considered. In this section, the model only has a one-dimensional variation, and the conversion (8.2.66) from two to three-dimensional propagation is equivalent to the far-field approximation.

Helmberger (1968) and Wiggins and Helmberger (1974) exploited the efficiency of the far-field approximation to compute synthetic seismograms in realistic, stratified Earth models (8.2.68). These were among the first publications to successfully model real data in complicated models, and use synthetic seismograms to improve interpretations (Helmberger, 1968, for a crustal model and Wiggins and Helmberger, 1974, for a mantle model), although Roever and Vining (1959*a*,*b*) and Strick (1959) had followed a similar approach to model laboratory data.

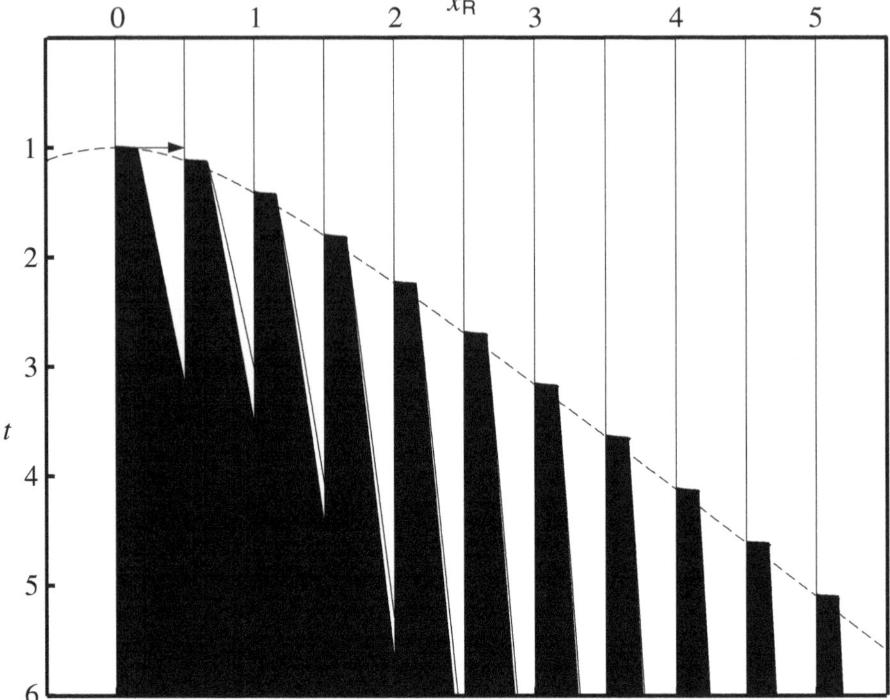

Fig. 8.6. The vertical component, u_z, of the displacement response due to an explosive source with time function $P_S(t) = P_S t H(t)$. The variable area plot is the exact solution (8.2.17) and equation (8.2.70), and the line plot is the far-field approximation (8.2.68). The parameters are normalized so the velocity is unity, $\alpha = 1$, the depth difference unity, $|z_R - z_S| = 1$, and the range varies from $x_R = 0$ to 5 in increments of 0.5. At $x_R = 0$, the far-field approximation is singular indicated by the arrow. The displacement is multiplied by the squared ray length, $R^2 = x_R^2 + (z_R - z_S)^2$, to remove the range and angular dependence of the first motion.

8.2.3 The exact direct wave from an explosion

For the direct wave, the exact solution can be obtained without approximation nor numerical methods. From result (8.2.16), we have

$$\begin{aligned}
&\operatorname{Im} \int_{\sin\theta/\alpha}^{p(t)} \frac{p}{q_\alpha} \begin{pmatrix} (t - q_\alpha d)/x_R \\ 0 \\ \pm q_\alpha \end{pmatrix} \frac{\mathrm{d}p}{\left((t - q_\alpha d)^2 - p^2 x_R^2\right)^{1/2}} \\
&= -\operatorname{Im} \int_{\cos\theta/\alpha}^{q_\alpha(t)} \begin{pmatrix} (t - q_\alpha d)/x_R \\ 0 \\ \pm q_\alpha \end{pmatrix} \frac{\mathrm{d}q_\alpha}{\left(R^2 q_\alpha^2 - 2t q_\alpha d + t^2 - x_R^2/\alpha^2\right)^{1/2}},
\end{aligned} \tag{8.2.69}$$

by a simple change of variable ($p\,\mathrm{d}p = -q_\alpha \mathrm{d}q_\alpha$). In the quadratic in the denominator, the coefficient combination '$b^2 - 4ac$' is

$$4z_{\mathrm{R}}^2 t^2 - 4R^2(t^2 - x_{\mathrm{R}}^2/\alpha^2) = -4x_{\mathrm{R}}^2(t^2 - R^2/\alpha^2) < 0,$$

so we can use the standard result from Abramowitz and Stegun (1965, §3.3.39)

$$\int_{\cos\theta/\alpha}^{q_\alpha(t)} \frac{q_\alpha \mathrm{d}q_\alpha}{\left(R^2 q_\alpha^2 - 2tq_\alpha d + t^2 - x_{\mathrm{R}}^2/\alpha^2\right)^{1/2}}$$
$$= \frac{1}{R^2}\left(R^2 q_\alpha^2 - 2tq_\alpha d + t^2 - x_{\mathrm{R}}^2/\alpha^2\right)^{1/2}\Big|_{\cos\theta/\alpha}^{q_\alpha(t)}$$
$$+ \frac{td}{R^2}\int_{\cos\theta/\alpha}^{q_\alpha(t)} \frac{\mathrm{d}q_\alpha}{\left(R^2 q_\alpha^2 - 2tq_\alpha d + t^2 - x_{\mathrm{R}}^2/\alpha^2\right)^{1/2}}.$$

At the lower limit, the first term is real, and at the upper limit zero, so only the integral need be retained. Thus (8.2.69) reduces to

$$-\frac{t}{R}\begin{pmatrix} \sin\theta \\ 0 \\ \pm\cos\theta \end{pmatrix} \operatorname{Im} \int_{\cos\theta/\alpha}^{q_\alpha(t)} \frac{\mathrm{d}q_\alpha}{\left(R^2 q_\alpha^2 - 2tq_\alpha d + t^2 - x_{\mathrm{R}}^2/\alpha^2\right)^{1/2}}.$$

The integral reduces to $\pi/2R$ using Abramowitz and Stegun (1965, §3.3.34)

$$\operatorname{Im} \int_{\cos\theta/\alpha}^{q_\alpha(t)} \frac{\mathrm{d}q_\alpha}{\left(R^2 q_\alpha^2 - 2tq_\alpha d + t^2 - x_{\mathrm{R}}^2/\alpha^2\right)^{1/2}}$$
$$= \operatorname{Im}\left(\frac{1}{R}\sinh^{-1}\frac{R^2 q_\alpha - td}{x_{\mathrm{R}}(t^2 - R^2/\alpha^2)^{1/2}}\right)\Big|_{\cos\theta/\alpha}^{q_\alpha(t)}$$
$$= \operatorname{Im}\left(\frac{1}{R}\sinh^{-1}\frac{R\,q_\alpha(t) - t\cos\theta}{\sin\theta\,(t^2 - R^2/\alpha^2)^{1/2}}\right)$$
$$= \operatorname{Im}\left(\frac{1}{R}\sinh^{-1}\frac{\mathrm{i}(t^2 - R^2/\alpha^2)^{1/2}\sin\theta}{\sin\theta\,(t^2 - R^2/\alpha^2)^{1/2}}\right) = \frac{\pi}{2R},$$

where at the lower limit the expression is real and doesn't contribute, and we have used result (8.1.34) for $q_\alpha(t)$. Using these results in expression (8.2.69) and hence in expression (8.2.16), we obtain

$$\begin{aligned} \mathbf{u}_P(t, \mathbf{x}_\mathrm{R}) \\ &= \frac{V_\mathrm{S}}{4\pi\rho\,\alpha^2 R^2}\begin{pmatrix}\sin\theta\\0\\\pm\cos\theta\end{pmatrix} P''_\mathrm{S}(t) * tH(t - R/\alpha) \\ &= \frac{V_\mathrm{S}}{4\pi\rho\,\alpha^3 R}\begin{pmatrix}\sin\theta\\0\\\pm\cos\theta\end{pmatrix}\left(P'_\mathrm{S}(t - R/\alpha) + \alpha P_\mathrm{S}(t - R/\alpha)/R\right). \end{aligned} \tag{8.2.70}$$

Substituting in the constitutive relation (4.4.3), we can obtain the exact result for the pressure

$$P_P(t, \mathbf{x}_\mathrm{R}) = \frac{V_\mathrm{S}}{4\pi\rho\,\alpha^2 R} P''_\mathrm{S}(t - R/\alpha), \tag{8.2.71}$$

where we have expressed the polarization as $\hat{\mathbf{g}} = \mathbf{R}/R$ and used $\nabla \cdot \mathbf{R} = 3$ and $\nabla R = \mathbf{R}/R$ in the simplification.

The result (8.2.70) can be compared with the point source, Green dyadic obtained in Chapter 4. Taking the derivative of the force dyadic (4.5.71), to obtain force couples and forming a point pressure source, we obtain exactly result (8.2.70) (see Exercise 4.9). Thus if the source is a step function in pressure, $P_\mathrm{S}(t) = P_\mathrm{S} H(t)$ (cf. the moment tensor source $M_\mathrm{S} = V_\mathrm{S} P_\mathrm{S}$ and the result (4.6.22)), the particle displacement consists of a delta function, far-field term and a step function, near-field term. The far-field term is the time derivative of the pressure source. It decays as R^{-1}, i.e. spherical spreading. The dependence on material properties is $1/\rho\,\alpha^3$ (a strong dependence on the velocity). The near-field term decays as R^{-2}. It is the static deformation due to the step function in pressure, which causes the medium to permanently deform outwards. Although the three-dimensional Green function (8.2.70) is not very complicated, for the sake of completeness and comparison with the two-dimensional Green function (Figure 8.3), it is illustrated in Figure 8.7.

8.3 Cagniard method in stratified media

The application of the Cagniard method to invert the transforms in expression (8.0.2) depends on three features: the separation of the transformed response into a frequency-independent coefficient, $\underline{\mathcal{G}}_\mathrm{ray}(p)$ (8.0.8), a phase that depends linearly on frequency (8.0.6), and the behaviour of the slowness integral (8.0.9) which allows the contour of integration to be distorted as required. It can be applied to any ray in a model of homogeneous layers. For reflections and transmissions in models

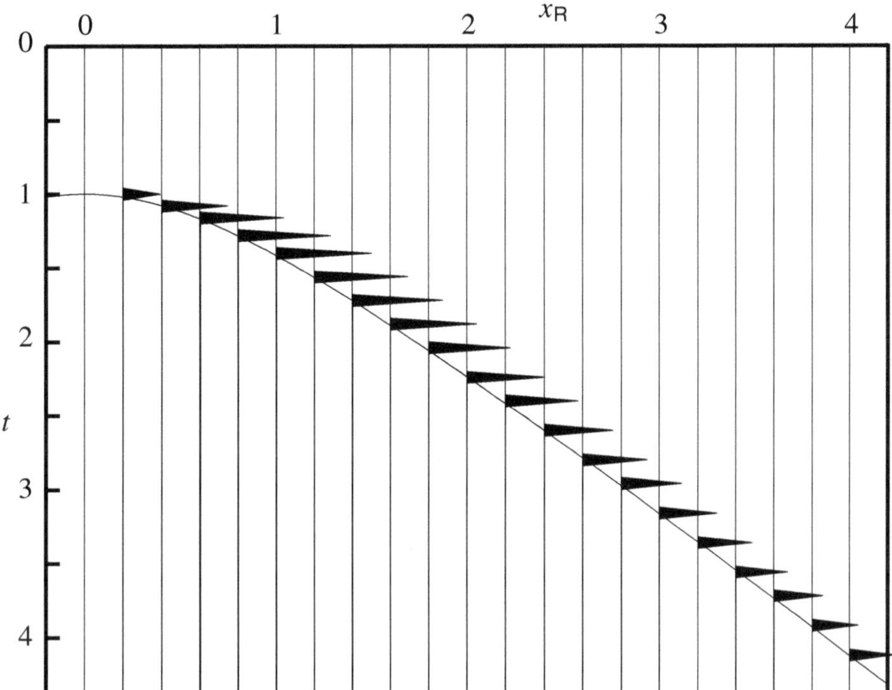

Fig. 8.7. The three-dimensional response (8.2.70) for a pressure source $P_S(t) = P_S H(t)$. Compared with Figure 8.3 for the two-dimensional Green function, the seismograms are for the horizontal displacement, u_x, at fixed depth, $|z_S - z_R| = 1$, and ranges from $x_R = 0$ to 4, so the amplitude is zero at zero range, peaks at $x_R = 1$ and then decays. The velocity is taken as unity, $\alpha = 1$, so the time axis is from $t = 0$ to 4.4. The delta function is plotted as a triangular pulse with $\Delta t = 0.04$. The near-field term is barely visible with these parameters and scale.

with inhomogeneous layers, the applications are limited. If we assume the zeroth-order WKBJ approximation (7.2.102), the above conditions are still satisfied. In the neighbourhood of the saddle point corresponding to the ray, the behaviour is similar to that for homogeneous layers. Unfortunately, this is not sufficient, as the WKBJ approximation is not valid for all slowness values. If the layers are inhomogeneous, for some larger slownesses, rays will turn rather than be reflected, and the WKBJ approximation will break down (as the vertical slowness q will be zero at a turning point). Thus the Cagniard method can only be used in a non-rigorous manner. We can even include higher-order terms in the WKBJ expansion, e.g. the first-order approximation (7.2.103), by a simple time integration, but as discussed before these are not very useful. They do not model the reflections from gradients – the WKB paradox (Gray, 1982, Section 7.2.5) – and they break down at turning points.

If a turning point exists and the reflected signal can be modelled with the approximate phase shift (7.2.163), then again the first two requirements for the Cagniard method are satisfied: frequency only appears as a linear phase term. However, saddle points corresponding to turning rays behave differently from reflections, and again we cannot distort the contour and use the Cagniard method. If the turning points are near interfaces, source or receiver then we must use the Langer approximation (7.2.159), and the first conditions for the Cagniard method are violated (the frequency appears inside the Airy functions).

8.3.1 The WKBJ iterative solution

The Cagniard method can, however, be applied to the WKBJ iterative solution (Section 7.2.6), and with the expense of extra depth integrals (as in iteration (7.2.125)), we can obtain the response of layered, inhomogeneous media. For simplicity, we consider the far-field approximation for the three-dimensional response. Let us consider the first-order iteration for the reflected signal from a velocity gradient. Generalizing expression (8.2.68), the response is

$$\begin{aligned}\underline{\mathbf{u}}_{\text{ray}}(t, \mathbf{x}_{\text{R}}) \simeq & -\frac{1}{\pi^2 (2x_{\text{R}})^{1/2}} \frac{\mathrm{d}}{\mathrm{d}t} \lambda(t) \\ & * \operatorname{Im} \int \left(p^{1/2} \underline{\boldsymbol{\mathcal{G}}}_{\text{ray}}(p) \, \gamma_P(p, z) \frac{\partial p}{\partial \widetilde{T}_{\text{ray}}} \right)_{p=p(t, \mathbf{x}_{\text{R}}, z)} \mathrm{d}z, \qquad (8.3.1)\end{aligned}$$

where the differential reflection coefficient for P waves, $\gamma_P(p, z)$, is defined in (7.2.108) and

$$\widetilde{T}_{\text{ray}}(p, \mathbf{x}_{\text{R}}, z) = p \, x_{\text{R}} + \int_z^{z_{\text{S}}} + \int_z^{z_{\text{R}}} q_\alpha(p, \zeta) \, \mathrm{d}\zeta, \qquad (8.3.2)$$

where the 'reflector' is at a depth z. The slowness substituted in the integrand, $p = p(t, \mathbf{x}_{\text{R}}, z)$, solves the equation

$$\widetilde{T}_{\text{ray}}(p, \mathbf{x}_{\text{R}}, z) = t. \qquad (8.3.3)$$

Expression (8.3.1) models the signals reflected by the velocity gradient. The reflection coefficient from a depth element is $\gamma_P \, dz$. The depth integral could be approximated by replacing the gradient by thin homogeneous layers. Then the integral is replaced by the sum of first-order reflections from the small interfaces. Often this is a convenient method of evaluating expression (8.3.1) numerically.

For a fixed depth, z, the equation (8.3.3) defines a Cagniard contour with the standard properties – for t less than the geometrical arrival time, the contour is on the real slowness axis; at the geometrical arrival time the contour leaves the real axis; as time increases the contour is asymptotic to the lines given by expression (8.1.25) where

$$\tan\tilde{\theta} = x_{\mathrm{R}}/(z_{\mathrm{S}} + z_{\mathrm{R}} - 2z). \tag{8.3.4}$$

For different depths, the geometrical arrival time varies. It can be found by solving

$$X_{\mathrm{ray}}(p, z_{\mathrm{R}}, z) = x_{\mathrm{R}}, \tag{8.3.5}$$

for the geometrical ray parameter, where

$$X_{\mathrm{ray}}(p, z_{\mathrm{R}}, z) = -\frac{\partial \tau_{\mathrm{ray}}(p, z_{\mathrm{R}}, z)}{\partial p} = \int_{z}^{z_{\mathrm{S}}} + \int_{z}^{z_{\mathrm{R}}} \frac{p\,\mathrm{d}\zeta}{q_{\alpha}(p, \zeta)}. \tag{8.3.6}$$

The geometrical arrival is then given by

$$T_{\mathrm{ray}}(p, z_{\mathrm{R}}, z) = \int_{z}^{z_{\mathrm{S}}} + \int_{z}^{z_{\mathrm{R}}} \frac{\mathrm{d}\zeta}{\alpha^2(\zeta) q_{\alpha}(p, \zeta)}. \tag{8.3.7}$$

These results generalize expressions (2.3.7) and (2.3.8) for reflections from depth z.

8.3.2 The isochrons and a turning point

The exact form of the Cagniard contours (8.3.3) depends on the details of the velocity depth function, $\alpha(z)$. In Figure 8.8, the situation is illustrated when the velocity function simply increases smoothly with depth. As the depth increases (z decreases), the saddle point exists for a smaller p value and the contour is closer to the imaginary axis ($\tilde{\theta}$ in expression (8.3.4) is smaller). The geometrical arrival time increases so for a fixed time, t, there will be a depth below which the time is before the arrival time and the Cagniard contour is on the real axis. Points on the real axis do not contribute to the integral (8.3.1) as the integrand is real, so the depth integral has an effective maximum depth $z = z_{\min}(t, \mathbf{x}_{\mathrm{R}})$ which solves

$$T_{\mathrm{ray}}(p, z_{\mathrm{R}}, z) = t. \tag{8.3.8}$$

As the depth decreases, the situation is somewhat more complicated. For a given range, a turning ray may exist where equation (8.3.5) is solved when

$$p = 1/\alpha(z) \tag{8.3.9}$$

(see Figure 8.9). The depth that solves equations (8.3.5) and (8.3.9) simultaneously is denoted by

$$z = z_{\mathrm{ray}}(\mathbf{x}_{\mathrm{R}}), \tag{8.3.10}$$

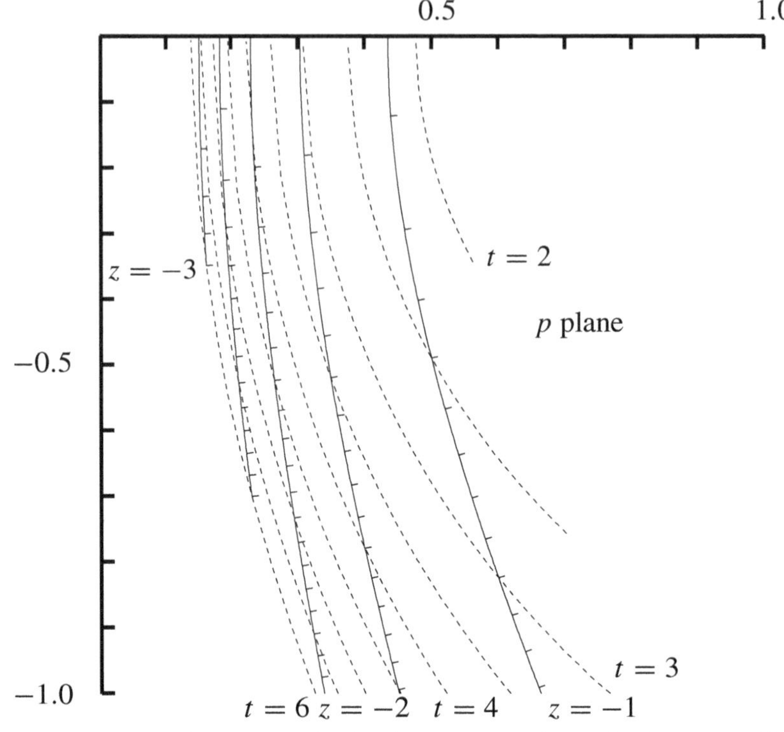

Fig. 8.8. Cagniard contours which solve equation (8.3.3) for fixed z are illustrated for a range of depths in a model where the velocity increases smoothly with depth. A simple linear function, $\alpha = 1 - z/10$, is used. The range is unity, $x_R = 1$, and the depths vary from $z = -1$ to -3 in increments of -0.5. The tick marks on the Cagniard contours are at time intervals of 0.1. Also indicated are isochrons which solve equation (8.3.3) for fixed time, t, and varying depth, z. Isochrons with $t = 2$ to 6 in increments of 0.5 are shown with dashed lines.

the maximum depth of the turning ray through $\mathbf{x}_R$. The corresponding geometrical ray parameter is $p_{ray} = 1/\alpha(z_{ray})$. For depths less than this depth (8.3.10), no geometrical rays exist. Nevertheless a Cagniard contour still exists. For $z > z_{ray}$, the contour leaves the real axis at $p = 1/\alpha(z)$. The time at this point is

$$\begin{aligned} t &= \widetilde{T}_{ray}(1/\alpha(z), \mathbf{x}_R, z) \\ &= T_{ray}(1/\alpha(z), z_R, z) + \left(x_R - X_{ray}(1/\alpha(z), z_R, z) \right) \Big/ \alpha(z), \end{aligned} \tag{8.3.11}$$

which is the time of a turning ray plus a segment propagating horizontally for the extra horizontal distance, $x_R - X_{ray}(1/\alpha(z), z_R, z)$, with a velocity $\alpha(z)$. This 'ray' path is illustrated in Figure 8.9.

As the depth decreases (z increases), the time (8.3.11) increases. Thus for any given time, t, there will be a depth that solves (8.3.11). This defines an effective

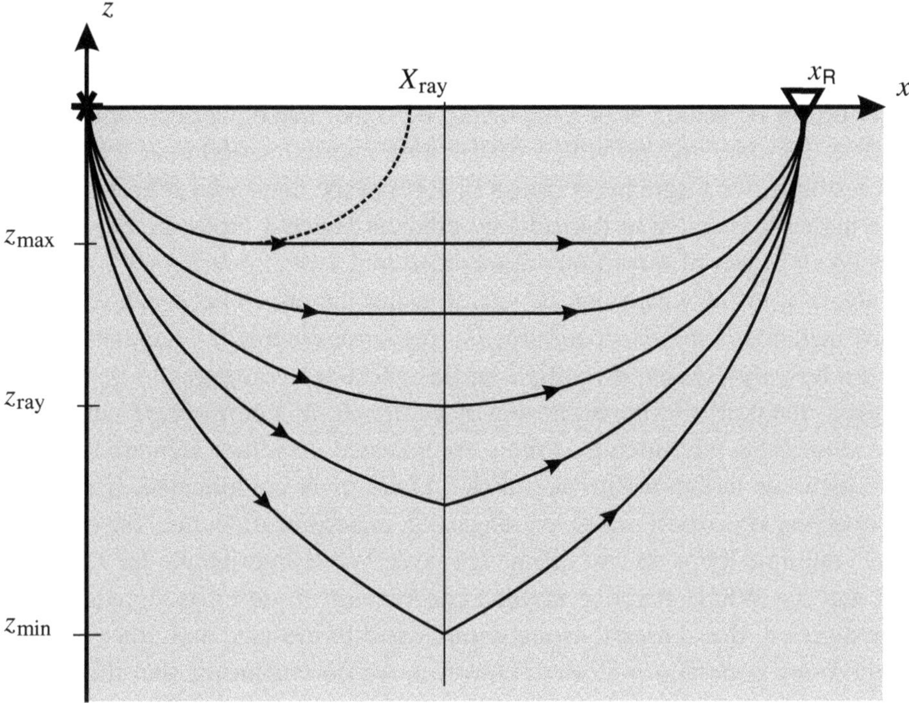

Fig. 8.9. Ray paths contributing to the depth integral (8.3.1). A turning ray for $z > z_{ray}$ is illustrated with a dashed line – displaced by the length of the horizontal section, $x_R - X_{ray}$ (see expression (8.3.11)).

minimum depth for the integral

$$z = z_{max}(t, \mathbf{x}_R). \tag{8.3.12}$$

Thus the response (8.3.1) can be written

$$\underline{\mathbf{u}}_{ray}(t, \mathbf{x}_R) = -\frac{1}{\pi^2 (2x_R)^{1/2}} \frac{d}{dt} \lambda(t) \\ * \text{Im} \int_{z_{min}(t,\mathbf{x}_R)}^{z_{max}(t,\mathbf{x}_R)} \left(p^{1/2} \underline{\boldsymbol{\mathcal{G}}}_{ray}(p)\, \gamma_P(p, z) \frac{\partial p}{\partial \widetilde{T}_{ray}} \right)_{p=p(t,\mathbf{x}_R,z)} dz, \tag{8.3.13}$$

where for a given time t, the depth integral is over a finite range. The integral is along an isochron defined by expression (8.3.2) for fixed t and varying z.

The point where the Cagniard contour leaves the real axis for $z > z_{ray}$ is not a saddle point. At this point, the differential coefficient, γ_P, is singular (in the limit of infinitesimal thin layers, a branch cut and the saddle point coalesce). A detailed analysis of the behaviour near this point, and near the point p_{ray}, is beyond the scope of this text, but has been made in Appendix B of Chapman (1976). Although γ_P is singular, the depth integral can be evaluated. Suitable changes of variable are described in Appendix B of Chapman (1976) that make the depth integration numerically feasible, and provide a first-motion approximation near the turning ray. Interestingly, the Cagniard response (8.3.13) of the first-order reflections does not give the turning signal in the velocity gradient but $\pi/3$ times the turning ray. Only when all orders of reflection are included in a series of terms with multiple depth integrals, all of which can be solved using the above approach using the Cagniard method, is the exact turning ray response obtained. Chapman (1976) has shown how this series of multiple-order reflections converges to the correct amplitude as the power series expansion of $2\sin(\pi/6) = 1$ converges to unity.

If the velocity–depth function is more complicated, then the Cagniard contours and the isochrons in the depth integral (8.3.1) are more complicated. If multiple arrivals (turning rays) exist at a given range, e.g. in a triplication, then the isochron may have multiple loops on and off the real axis. While in principle the Cagniard method and the WKBJ iterative method can be used to describe signals in any velocity function, the expressions are complicated to evaluate and other numerical methods are generally preferred. However, we do emphasize that this is just a practical limitation. Although it has often been stated that the WKBJ iterative method (the Bremmer series) breaks down when a turning point exists, and cannot be used, this is *not true*. Together with the Cagniard method, the response from the WKBJ iterative method can be evaluated even when the differential coefficient, γ_P, is singular. This was demonstrated in Chapman (1976) but has not been exploited elsewhere. We do not include numerical examples here.

8.4 Real slowness methods

The Cagniard method, Section 8.1 (two dimensions) and Section 8.2 (three dimensions), is a particularly elegant method for obtaining the impulse response with a line or point source. The essential feature of the solution is that the inverse Fourier transform of the integrand with respect to frequency is known exactly. The intermediate result, the impulse response for a fixed slowness, p, is straightforward. In two dimensions, the slowness integral can be evaluated exactly while in three dimensions it is straightforward to approximate analytically or evaluate numerically, being of finite range and non-oscillatory. As the intermediate result is the impulse response in the slowness domain, we call it a *slowness method*.

In this section, we consider other slowness methods that can be used when the requirements of the Cagniard method are not met. As discussed in Section 3.4.2, the temporal and spatial inverse Fourier transforms can be reduced to a Radon transform. Applying (3.4.15) to the two-dimensional result (8.0.1), we obtain

$$\underline{\mathbf{u}}_{\text{ray}}(t, \mathbf{x}_{\text{R}}) = -\frac{1}{2\pi} \int_{-\infty}^{\infty} \bar{\underline{\mathbf{v}}}_{\text{ray}}(t - px_{\text{R}}, p, z_{\text{R}})\, \mathrm{d}p. \tag{8.4.1}$$

Apart from the Hilbert transform, the integral is the slant stack of the impulse velocity response $\underline{\mathbf{v}}_{\text{ray}}(t, p, z_{\text{R}})$. This result is very instructive. As $\underline{\mathbf{v}}_{\text{ray}}(t, p, z_{\text{R}})$ is normally localized along tau-p curves, it clearly illustrates the formation of the seismogram. In the next section we exploit this analytically with the WKBJ approximation. More generally, the method can be used for numerical calculations. However, when $\underline{\mathbf{v}}_{\text{ray}}(\omega, p, z_{\text{R}})$ is evaluated numerically for the full response of a realistic multi-layered model, there is little advantage in expression (8.4.1) over a *spectral method* (Section 8.5). With both methods, the most significant numerical problem is to obtain adequate slowness sampling to avoid aliasing in the slowness integral. The spectral method using the Filon method (Section 8.5.2.3) handles this as well as any, as it interpolates the amplitude and phase separately.

8.4.1 The WKBJ seismogram

If we can approximate the transformed response as in expression (8.0.6), or as a sum of similar terms, then the slant stack (8.4.1) is straightforward. This is true for signals that can be modelled using the WKBJ asymptotic approximation (Section 7.2.5), with reflections, transmissions (Sections 7.2.3 and 7.2.4) and turning points (Section 7.2.7), and including extra depth integrals, with the WKBJ iterative solution (Section 7.2.6).

The inverse spectral Fourier transform of expression (8.0.6) is trivial, particularly if we use the delta function generalized with a complex argument (B.1.9). Thus

$$\underline{\mathbf{v}}_{\text{ray}}(t, p, z_{\text{R}}) = \operatorname{Re}\left(\underline{\mathcal{G}}_{\text{ray}}(p)\, \Delta\left(t - \tau_{\text{ray}}(p, z_{\text{R}})\right)\right). \tag{8.4.2}$$

Substituting in expression (8.4.1), we obtain

$$\underline{\mathbf{u}}_{\text{ray}}(t, \mathbf{x}_{\text{R}}) = -\frac{1}{2\pi} \operatorname{Im} \int_{-\infty}^{\infty} \underline{\mathcal{G}}_{\text{ray}}(p)\, \Delta\left(t - \widetilde{T}_{\text{ray}}(p, \mathbf{x}_{\text{R}})\right) \mathrm{d}p. \tag{8.4.3}$$

Apart from assuming that expression (8.0.6) is valid, which may be an approximation, this result is exact. We now approximate further by restricting the integral to the range where τ_{ray} is real. Thus

$$\underline{\mathbf{u}}_{\text{ray}}(t, \mathbf{x}_{\text{R}}) \simeq -\frac{1}{2\pi} \operatorname{Im}\left(\Delta(t) * \int_{\operatorname{Im}(\tau_{\text{ray}})=0} \underline{\mathcal{G}}_{\text{ray}}(p)\, \delta\left(t - \widetilde{T}_{\text{ray}}(p, \mathbf{x}_{\text{R}})\right) \mathrm{d}p\right), \tag{8.4.4}$$

where we have taken the analytic delta function outside to simplify evaluating the integral. The integral is restricted to the slowness range where τ_{ray} is real so the argument of the delta function in the integral is real. Thus the integral can be evaluated exactly

$$\boxed{\underline{\mathbf{u}}_{\text{ray}}(t, \mathbf{x}_{\text{R}}) \simeq -\frac{1}{2\pi} \operatorname{Im}\left(\Delta(t) * \sum_{\widetilde{T}_{\text{ray}}(p,\mathbf{x}_{\text{R}})=t} \frac{\underline{\boldsymbol{\mathcal{G}}}_{\text{ray}}(p)}{|\partial \widetilde{T}_{\text{ray}}/\partial p|} \right),} \tag{8.4.5}$$

where the summation is over solutions of the equation

$$\widetilde{T}_{\text{ray}}(p, \mathbf{x}_{\text{R}}) = t. \tag{8.4.6}$$

Note that we have the modulus of the gradient as whatever its sign, the contribution from the delta function in integral (8.4.4) is positive. This expression is similar to the Cagniard solution (8.1.31) but with several important differences. Expression (8.4.5) is approximate as it is inevitable that for some slownesses the phase τ_{ray} will be complex, whereas for homogeneous layers the Cagniard solution is exact. Expression (8.4.5) only uses real slownesses whereas the Cagniard solution uses complex slownesses. Expression (8.4.5) can be used for the same signals as the Cagniard method, but it can also be used more generally. When it is used for the same signals as the Cagniard method, some of the slownesses used in expression (8.4.5) may contribute to the Cagniard solution, but others may not.

The two-dimensional WKBJ seismogram (8.4.5) is easily converted into the far-field approximation for the three-dimensional response using the operation (8.2.66). The result is

$$\boxed{\underline{\mathbf{u}}_{\text{ray}}(t, \mathbf{x}_{\text{R}}) \simeq -\frac{1}{2^{3/2}\pi^2 x_{\text{R}}^{1/2}} \frac{\mathrm{d}}{\mathrm{d}t} \operatorname{Im}\left(\Lambda(t) * \sum_{\widetilde{T}_{\text{ray}}(p,\mathbf{x}_{\text{R}})=t} \frac{p^{1/2}\underline{\boldsymbol{\mathcal{G}}}_{\text{ray}}(p)}{|\partial \widetilde{T}_{\text{ray}}/\partial p|} \right).} \tag{8.4.7}$$

In Exercise 8.8, the relationship of the real slowness methods to the Radon transform (Section 3.4.2) is discussed. Chapman (1978) has developed an exact three-dimensional real slowness method (related to the Radon transform (3.4.23) which can be reduced to the far-field approximation (8.4.7) (cf. equation (3.4.28)).

Expressions (8.4.5) and (8.4.7) are known as the *WKBJ seismogram*, as given the WKBJ approximation (8.0.6) with real τ_{ray}, there are no further approximations in the two-dimensional inverse transforms. In the three-dimensional WKBJ seismogram (8.4.7), the additional far-field approximation has been made. Any complex function $\underline{\boldsymbol{\mathcal{G}}}_{\text{ray}}$ is allowed, and provided τ_{ray} is real, its form is arbitrary. The errors from neglecting the part of the integral (8.4.3) where the delay time τ_{ray} is complex are not normally very important as then the transformed waveform

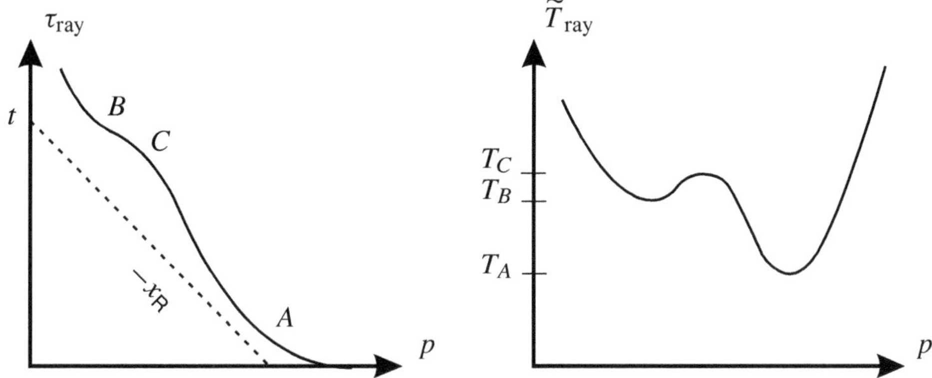

Fig. 8.10. The τ_{ray} and $\widetilde{T}_{\text{ray}}$ functions for a turning ray with a caustic. The range x_{R} is within the triplication with three arrivals at times T_A, T_B and T_C. A line with gradient $-x_{\mathrm{R}}$ and time $t < T_A$ is indicated in the $\tau(p)$ plot.

is evanescent. The important error is from the end-point just where τ_{ray} becomes complex (Thomson and Chapman, 1986).

The construction of the seismogram using expression (8.4.5) is extremely straightforward. In Figure 8.10, typical τ_{ray} and $\widetilde{T}_{\text{ray}}$ functions are shown for a turning ray with a triplication (cf. Section 2.4.4). The range x_{R} is within the triplication so there are three points where the lines $\tau = t - px_{\mathrm{R}}$ are tangent to τ_{ray}. At these points, the gradient is

$$\frac{\partial \tau_{\text{ray}}}{\partial p} = -X_{\text{ray}}(p, z_{\mathrm{R}}) = -x_{\mathrm{R}}, \tag{8.4.8}$$

where τ_{ray} and X_{ray} are defined as in integrals (2.3.15) and (2.3.7). Solving this equation, the slownesses are the geometrical values where $X(p_{\text{ray}}, z_{\mathrm{R}}) = x_{\mathrm{R}}$. The $\widetilde{T}_{\text{ray}}$ function is stationary at these points with its value equal to the geometrical travel times, i.e.

$$\begin{aligned}\widetilde{T}_{\text{ray}}(p, \mathbf{x}_{\mathrm{R}}) &= px_{\mathrm{R}} + \tau_{\text{ray}}(p, z_{\mathrm{R}})\\ &= T_{\text{ray}}(p, z_{\mathrm{R}}) + p\left(x_{\mathrm{R}} - X_{\text{ray}}(p, z_{\mathrm{R}})\right)\\ &= T_{\text{ray}}(p_{\text{ray}}, z_{\mathrm{R}}) \quad \text{when} \quad p = p_{\text{ray}}\end{aligned} \tag{8.4.9}$$

$$\frac{\partial \widetilde{T}_{\text{ray}}}{\partial p} = x_{\mathrm{R}} - X_{\text{ray}}(p, z_{\mathrm{R}}) = 0 \quad \text{when} \quad p = p_{\text{ray}}, \tag{8.4.10}$$

where the arrival time function, T_{ray}, is defined as in integral (2.3.8).

In the example (Figure 8.10), the function $\widetilde{T}_{\text{ray}}$ has three stationary points corresponding to three geometrical arrivals. The waveform is determined by solving $\widetilde{T}_{\text{ray}} = t$ and summing the magnitudes of the inverse gradient. At the geometrical

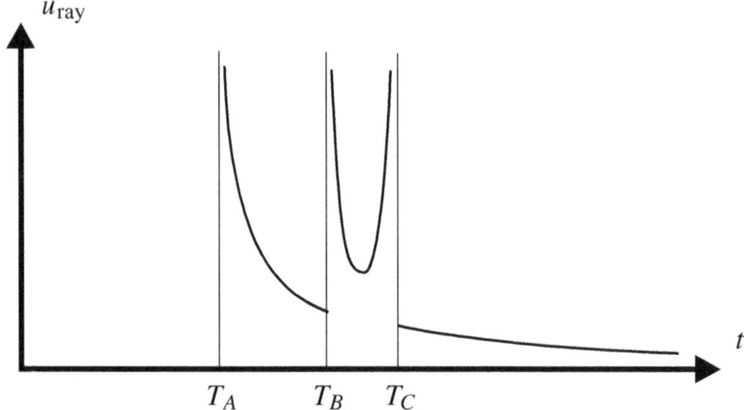

Fig. 8.11. The waveform corresponding to Figure 8.10, with three geometrical arrivals T_A, T_B and T_C.

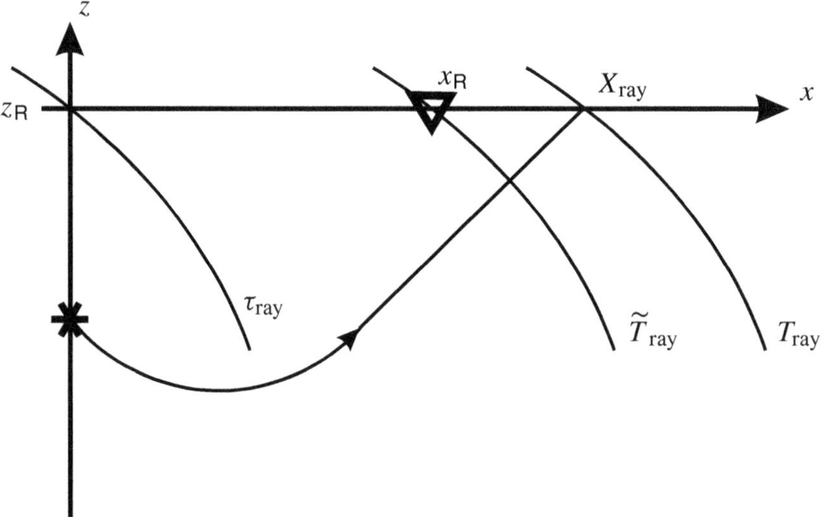

Fig. 8.12. Plane waves with propagation time τ_{ray}, T_{ray} and $\widetilde{T}_{ray}$ at $x = 0$, X_{ray} and arbitrary x_R, respectively.

arrivals, the waveform is singular; before the first arrival, there are no solutions of $\widetilde{T}_{ray} = t$; immediately after the first arrival $t > T_A$, there are two solutions; immediately after the second $t > T_B$, four; and after the third $t > T_C$, just two again. In Figure 8.11, we sketch the waveform corresponding to the situation in Figure 8.10.

The construction also has a simple physical interpretation. In Figure 8.12 we illustrate wavefronts with a constant slowness, p (plane wavefronts in a homogeneous region). Consider a wavefront through the source. The delay time,

$\tau_{\text{ray}}(p, z_R)$, is the propagation time for this wavefront when it passes through the receiver depth, z_R, at $x = 0$, after reflection or turning in the model. At a range, $X_{\text{ray}}(p, z_R)$, i.e. the geometrical range for a ray with the wavefront's slowness, the propagation time is $T_{\text{ray}}(p, z_R) = \tau_{\text{ray}}(p, z_R) + pX_{\text{ray}}(p, z_R)$, i.e. the geometrical travel-time for the ray. Consider an extra increment of range $\mathrm{d}x$. The extra distance propagated by the wavefront is $\mathrm{d}x \sin\theta = c\, p\, \mathrm{d}x$ (cf. equation (2.3.9) and Figure 2.10). Thus the extra propagation time is $p\,\mathrm{d}x$. Thus to any location, the propagation time is

$$\widetilde{T}_{\text{ray}}(p, \mathbf{x}_R) = \tau_{\text{ray}}(p, z_R) + p\, x_R \tag{8.4.11}$$

$$= T_{\text{ray}}(p, z_R) + p(x_R - X_{\text{ray}}(p, z_R)). \tag{8.4.12}$$

Given that $\widetilde{T}_{\text{ray}}$ is the propagation time of a wavefront with slowness p, the slowness integral then represents the summation of these waves to simulate a line source. The wavefronts then contribute to the impulse response at the propagation time $\widetilde{T}_{\text{ray}} = t$. The factor $|\partial \widetilde{T}_{\text{ray}}/\partial p|$ arises from the density of plane waves in the slowness-to-time mapping.

Figure 8.12 illustrates the extra time $p(x_R - X_{\text{ray}})$ for the wavefront with slowness p propagating from the geometrical range X_{ray} to the receiver x_R (negative in this diagram). Notice that in order to make this argument, the solution must be decomposed into wavefronts with constant slowness. Sometimes a similar intuitive but less rigorous argument is made for rays without decomposing into constant-slowness wavefronts (cf. Figure 2.10).

8.4.2 Numerical band-limited WKBJ seismograms

The impulse response, expression (8.4.5), is analytic but contains singularities. Numerically these are difficult to handle when we compute a discrete time series – suppose the sampled point $t = T_0 + n\Delta T$ exactly equals T_{ray} or is so close that numerical overflow occurs. If we try to digitize the impulse response, we have lost the game (temporal aliasing). As in an instrument, we must band-limit *before* digitization – in an instrument this means while the signal is still analogue; with the impulse response, while the result is still analytic, not numerical.

The simplest way to band-limit the response is to use a boxcar filter. We define a normalized, dimensionless boxcar

$$B(t) = \frac{1}{2}\Big(H(t+1) - H(t-1)\Big). \tag{8.4.13}$$

Then $B(t/\Delta t)/\Delta t$ is a filter with a zero at the Nyquist frequency $\omega_N = \pi/\Delta t$. The spectrum is the function $\operatorname{sinc}(\pi\omega/\omega_N)$. The filter is illustrated in Figure 8.13

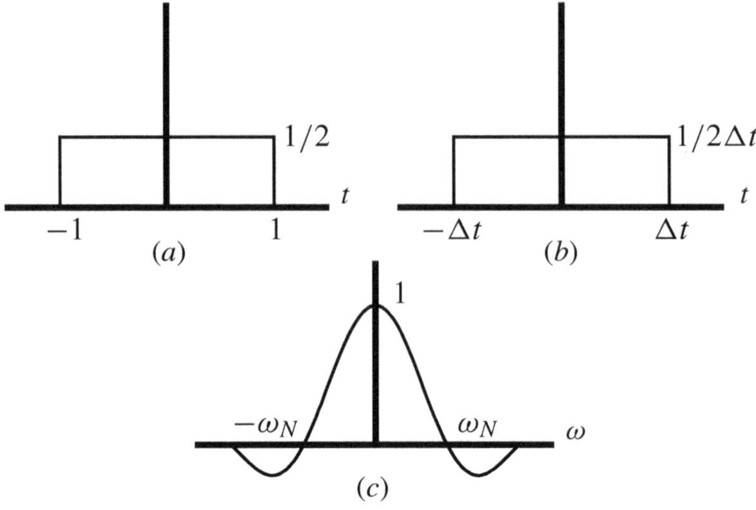

Fig. 8.13. The dimensionless boxcar (*a*), $B(t)$; the smoothing filter (*b*), $B(t/\Delta t)/\Delta t$; and, its spectrum (*c*), $\mathrm{sinc}(\pi\omega/\omega_N)$.

Applying this filter to expression (8.4.4), we obtain

$$\frac{1}{\Delta t}B\left(\frac{t}{\Delta t}\right) * \underline{\mathbf{u}}_{\mathrm{ray}}(t,\mathbf{x}_{\mathrm{R}})$$
$$\simeq -\frac{1}{2\pi\,\Delta t}\,\mathrm{Im}\left(\Delta(t) * \int_{\mathrm{Im}(\tau_{\mathrm{ray}})=0} \underline{\mathcal{G}}_{\mathrm{ray}}(p)\, B\left(\frac{t-\widetilde{T}_{\mathrm{ray}}(p,\mathbf{x}_{\mathrm{R}})}{\Delta t}\right)\mathrm{d}p\right),$$

or

$$\boxed{\frac{1}{\Delta t}B\left(\frac{t}{\Delta t}\right) * \underline{\mathbf{u}}_{\mathrm{ray}}(t,\mathbf{x}_{\mathrm{R}}) \simeq -\frac{1}{4\pi\,\Delta t}\,\mathrm{Im}\left(\Delta(t) * \int_{\widetilde{T}_{\mathrm{ray}}=t\pm\Delta t} \underline{\mathcal{G}}_{\mathrm{ray}}(p)\,\mathrm{d}p\right).} \tag{8.4.14}$$

The integral is over (narrow) bands of slowness defined by

$$\widetilde{T}_{\mathrm{ray}}(p,\mathbf{x}_{\mathrm{R}}) = t \pm \Delta t, \tag{8.4.15}$$

illustrated in Figure 8.14. A similar expression is obtained for the three-dimensional result (8.4.7).

Effectively we are integrating result (8.4.5) with respect to t $(=\widetilde{T}_{\mathrm{ray}})$ and replacing it by an integral over slowness, p. Note the result is now non-singular. Singularities are replaced by large values. It appears that we have lost the advantage that the integrals disappear in the final result, as expression (8.4.14) contains an integral. However, these integrals are over very small slowness ranges. Normally

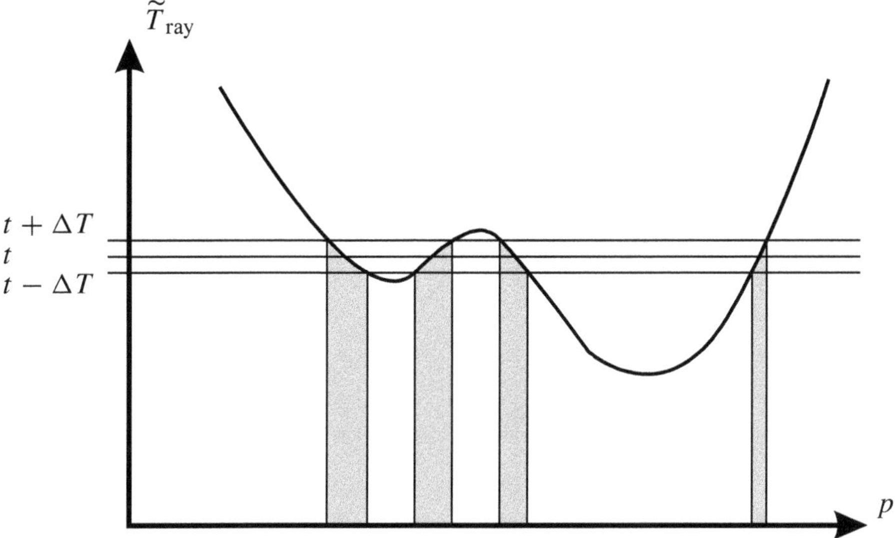

Fig. 8.14. Bands of slowness defined by $\widetilde{T}_{\text{ray}} = t \pm \Delta t$ (8.4.15).

$\underline{\mathcal{G}}_{\text{ray}}(p)$ varies slowly and smoothly with slowness p (an exception is at a branch cut causing a head wave discussed in Section 9.1.3 – poles are not in the range $\text{Im}(\tau_{\text{ray}}) = 0$) so $\int \underline{\mathcal{G}}_{\text{ray}} \mathrm{d}p$ can be easily evaluated by the trapezoidal rule.

The crude boxcar filter is used to obtain this simple result. A better filter could be used but with the extra expense of evaluating these narrow integrals. In practice it is probably simpler to initially over-sample the time series and then to apply further smoothing to the time series before decimating.

Not only do we require a discretized time series, i.e. $t = T_0 + n\Delta t$, but the ray results are also sampled discretely, i.e. the functions $\widetilde{T}_{\text{ray}}$ and $\underline{\mathcal{G}}_{\text{ray}}$ are only known for discrete rays. This is illustrated in Figure 8.15 where linear interpolation of the $\widetilde{T}_{\text{ray}}$ function between the sampled rays is assumed. With the band-limiting, errors due to interpolating (linearly) are small. Results are robust with respect to numerical errors and approximations, and small changes in the model. In contrast, the impulse result (8.4.5) is unstable. A small change in the model may create singularities. In the band-limited result (8.4.14), such changes or errors are not significant. We don't even need to be aware of the presence of singularities. The value of stabilizing results and calculating directly band-limited results is a very important feature of the WKBJ seismogram algorithm. Its significance can't be over-stated.

The algorithm to compute WKBJ seismograms is extremely simple. In outline it is

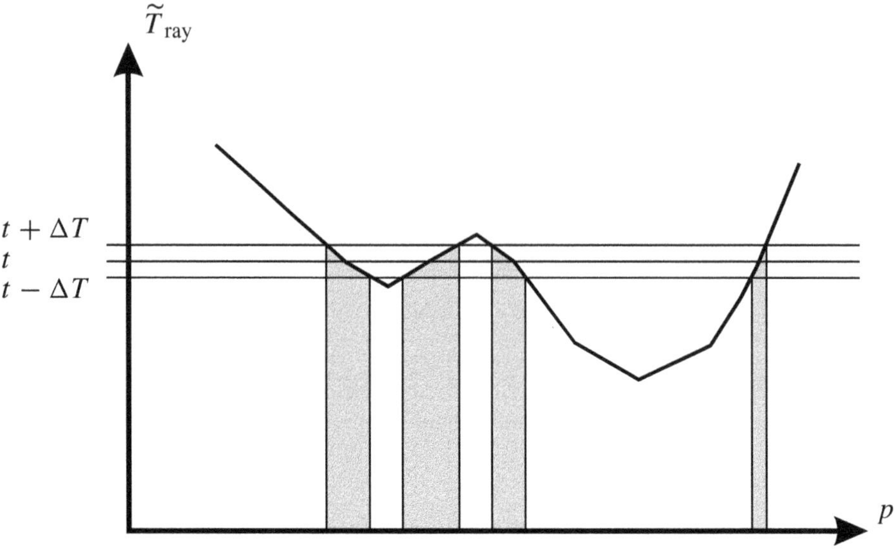

Fig. 8.15. Bands of slowness defined by $\widetilde{T}_{ray} = t \pm \Delta t$ (8.4.15) when the $\widetilde{T}_{ray}$ function is linearly interpolated between sampled rays.

- `for each slowness interval:`
  - `label values at minimum and maximum of interval with` A `and` B`, respectively, e.g.` $\widetilde{T}_A$ `and` $\widetilde{T}_B$`;`
  - `initialize` $n = \text{int}((\widetilde{T}_A - T_0)/\Delta t) + 1$ `for the first included discrete time point;`
  - `initialize` $\underline{\boldsymbol{\mathcal{G}}}_{n-1} = \mathbf{0}$ `and` $\Delta p = p_B - p_A$`;`
  - `loop over` $t = T_0 + n\Delta t$ `values;`
    - `integral to` t `is` $\underline{\boldsymbol{\mathcal{G}}}_n = f\,\Delta p\,((2-f)\,\underline{\boldsymbol{\mathcal{G}}}_A + f\,\underline{\boldsymbol{\mathcal{G}}}_B)/2$ `where` $f = (t - \widetilde{T}_A)/(\widetilde{T}_B - \widetilde{T}_A)$`;`
    - `add` $\underline{\boldsymbol{\mathcal{G}}}_n - \underline{\boldsymbol{\mathcal{G}}}_{n-1}$ `to points` $n-1$ `and` n`;`
    - `if` $t < \widetilde{T}_B$`, increment` n `and repeat loop;`
    - `end loop;`
  - `end interval;`

Apart from some constant factors, i.e. $1/2$, $1/2\Delta t$, etc., and convolutions, this is the complete algorithm.

The WKBJ seismogram algorithm is efficient and robust to compute. Suppose we trace n_p rays and require n_t points in the seismogram. To form the seismogram we can step through each slowness interval in order, determining which times it contributes to and adding the contributions to the seismogram. Thus the number of computations is

$$an_t + bn_p, \tag{8.4.16}$$

where a and b are operation counts. An alternative algorithm where we step through each time in order, searching all slowness intervals for contributions would take

$$c n_t n_p, \tag{8.4.17}$$

operations. Normally the operation count (8.4.16) is significantly less than the count (8.4.17). Spectral methods discussed in Section 8.5 have an operation count like expression (8.4.17).

8.4.3 Example – *PKP*, *PKiKP* and *PKIKP*

As an example of the WKBJ seismogram method, we consider the core rays used as the ray tracing example in Chapter 2 (Figures 2.31–2.34). Using the algorithm described in the previous section (equation (8.4.14) together with the two-to-three dimension transformation (8.2.66)), the seismograms in Figure 8.16 are calculated.

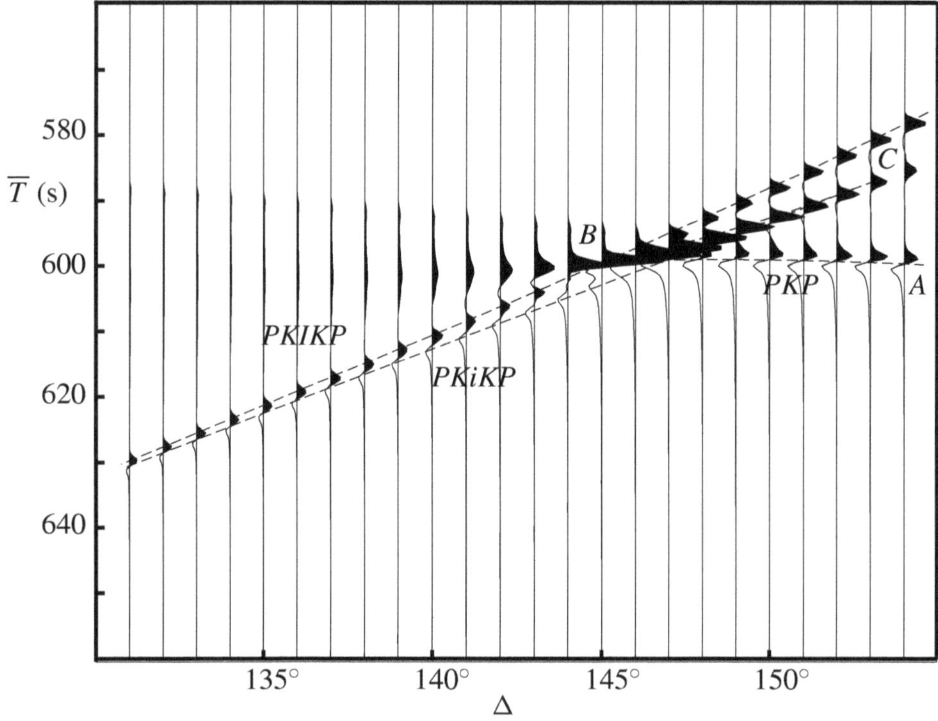

Fig. 8.16. Waveforms for the core rays *PKP*, *PKiKP* and *PKIKP* in the model 1066B (Figure 2.31). The reduced travel time, $\overline{T}$ (2.5.48), is plotted with a dashed line (as in Figure 2.34). The reducing slowness is $u_0 = 4$ s/°. The same points are marked as in Figures 2.33 and 2.34. The source is a smoothed, pressure step.

The numerical ray-tracing results used have been illustrated in Figures 2.31–2.34 and details of the model parameters can be found in Gilbert and Dziewonski (1975) and Figure 2.31. The seismograms have been calculated for the angular range $\Delta_R = 131^\circ$ to 154° to illustrate the caustic at about 144.6°. They are for a source which is a step function of pressure, subject to the smoothing on the algorithm (8.4.14) with $\Delta t = 0.5$ s. In addition, the convolution operator is a smoothed version of the analytic operator, $(d/dt)\Lambda(t) *$, and for efficiency is approximated by a rational approximation (Chapman, Chu Jen-Yi and Lyness, 1988). The plot is superimposed on the reduced travel time (Figure 2.34). The waveforms of interest are the Hilbert transform on the branch AB, the interference near the caustic B and the non-geometrical signal beyond the caustic (at $\Delta < 144.6^\circ$). The canonical form of the wavefunction near the caustic is analysed in the next chapter (Section 9.2.7).

8.5 Spectral methods

In the previous sections, we have used slowness methods to invert the transformed solutions to obtain the impulse response. The methods have been called slowness methods as the inverse Fourier transform with respect to frequency is evaluated first, giving intermediate results in the slowness domain. We have concentrated on these slowness methods as, for 'ray' signals, the results are particularly simple and elegant. Nevertheless, the traditional method, shall we say classical method of solution, is to evaluate the inverse transform with respect to slowness first and obtain the spectrum as the intermediate result. This corresponds to the order of the forward transforms, where the wave equation is normally transformed to the frequency domain first and the intermediate result is the spectrum. We shall refer to such as *spectral methods*. Unlike slowness methods, where fairly general results can be obtained for 'ray' signals, the analytic results using the spectral method are normally asymptotic. While there is a large body of mathematics dealing with these asymptotic methods, the slowness methods are normally easier to evaluate numerically and more generally valid. We therefore only investigate the lowest-order asymptotic results which, in general, are equivalent to the first-motion approximations of the slowness methods. For a more general signal, e.g. the complete response of a multi-layered, heterogeneous medium, we must evaluate the inverse transforms numerically. In this case, either a slowness or spectral method is possible. In practice, the spectral approach is normally preferred and the numerical slowness integral is performed using the Filon method (Section 8.5.2.3).

The most widely used method is known as the *reflectivity method*, introduced by Fuchs and Müller (1971). In the original reflectivity method, the reflective response of a stack of layers was calculated by performing the inverse spatial transform as an integral over real angles of incidence. The method has since been generalized to include the complete response of a multi-layered, heterogeneous medium with, for instance, the free surface (including the free surface complicates the numerical integral as the integrand contains more significant singularities). The slowness integral must be extended beyond real angles to obtain the complete response. The term reflectivity method is now used more generally to refer any spectral method where the contour of integration is still basically along the real axis (see Section 8.5.2). An alternative method is to distort the contour of the slowness integral significantly so that it approximates the steepest descent path for the signals being studied. This method has been called *full-wave theory* (a slightly confusing name as many other methods contain the full-wave response in the solution, and the special feature of this method was the choice of contour in the inverse transforms not the solutions of the differential system). It was used by Chapman and Phinney (1972), Choy (1977), Cormier and Richards (1977), Aki and Richards (1980, 2002, Chapter 9), etc. It is now less popular than the reflectivity method as the contour must be designed for the particular signals being studied and ideally is also a function of the receiver position. We have restricted the discussion in Section 8.5.2 on numerical issues of the spectral method to a generalization of the reflectivity method.

In this section we outline the spectral method and describe the Filon method for numerically evaluating the slowness integral accurately and efficiently. In Section 9.3, we describe the leading asymptotic term for various 'ray' signals using the spectral method.

8.5.1 Slowness integrals

8.5.1.1 Two dimensions

In two dimensions, the spectral result is given by the inverse Fourier transform with respect to horizontal slowness (3.2.10), i.e.

$$\underline{\mathbf{v}}(\omega, \mathbf{x}_{\mathrm{R}}) = \frac{|\omega|}{2\pi} \int_{-\infty}^{\infty} \underline{\mathbf{v}}(\omega, p, z_{\mathrm{R}})\, \mathrm{e}^{\mathrm{i}\omega p x_{\mathrm{R}}}\, \mathrm{d}p. \tag{8.5.1}$$

If the transformed response is a 'ray' signal (8.0.6), then the spectrum is

$$\underline{\mathbf{v}}_{\mathrm{ray}}(\omega, \mathbf{x}_{\mathrm{R}}) = \frac{|\omega|}{2\pi} \int_{-\infty}^{\infty} \underline{\mathcal{G}}_{\mathrm{ray}}(p)\, \mathrm{e}^{\mathrm{i}\omega \widetilde{T}_{\mathrm{ray}}(p, \mathbf{x}_{\mathrm{R}})}\, \mathrm{d}p, \tag{8.5.2}$$

where $\widetilde{T}_{\mathrm{ray}}$ is defined in equation (8.1.2). If the response is a sum of ray signals, then the terms can be integrated separately and summed afterwards. At high frequencies, these integrals are highly oscillatory and difficult to evaluate numerically.

8.5.1.2 Three dimensions

In three dimensions, the result requires a two-dimensional inverse slowness transform (3.2.15), i.e.

$$\underline{\mathbf{v}}(\omega, \mathbf{x}_R) = \frac{\omega^2}{4\pi^2} \iint_{-\infty}^{\infty} \underline{\mathbf{v}}(\omega, \mathsf{p}, z_R)\, e^{i\omega \mathsf{p}\cdot\mathsf{x}_R}\, d\mathsf{p} \tag{8.5.3}$$

$$= \frac{\omega^2}{4\pi^2} \iint_{-\infty}^{\infty} \underline{\mathcal{G}}_{\text{ray}}(\mathsf{p})\, e^{i\omega \widetilde{T}_{\text{ray}}(\mathsf{p},\mathbf{x}_R)}\, d\mathsf{p}, \tag{8.5.4}$$

for a 'ray' response. Recall that the sans serif font is used for vectors in the two-dimensional, horizontal sub-space, e.g. equation (3.2.13).

The phase function is

$$\widetilde{T}_{\text{ray}}(\mathsf{p}, \mathbf{x}_R) = \tau_{\text{ray}}(\mathsf{p}, z_R) + \mathsf{p}\cdot\mathsf{x}_R \tag{8.5.5}$$

$$= T_{\text{ray}}(\mathsf{p}, z_R) + \mathsf{p}\cdot\left(\mathsf{x}_R - \mathsf{X}_{\text{ray}}(\mathsf{p}, z_R)\right). \tag{8.5.6}$$

Again the inverse integrals (8.5.4) are difficult to evaluate numerically.

8.5.2 Numerical methods

In many circumstances the analytic methods we have investigated, together with the approximate results in Chapter 9, are adequate to describe seismic signals of interest. However, in complicated, realistic models the methods may be inadequate and we will have to revert to numerical methods. For instance, if the transformed response for a multi-layered or inhomogeneous medium is found using a layer-matrix approach, e.g. expression (7.2.71) without the ray expansion, possibly including results for inhomogeneous layers for the propagators – Sections 7.2.5, 7.2.6 and 7.2.7 – then numerical methods will be necessary to perform the inverse transforms. This does not negate the use of approximate methods, as even when numerical solutions are feasible and available, approximate methods are useful to predict the behaviour of the integrals and signals.

In wave modelling, and processing, integrals of the form

$$\int_{-\infty}^{\infty} e^{\pm iap}\mathbf{K}(a, p)\, dq \tag{8.5.7}$$

$$\int_{0}^{\infty} J_n(ap)\mathbf{K}(a, p)\, dq \tag{8.5.8}$$

$$\int_{-\infty}^{\infty} H_n^{(1 \text{ or } 2)}(ap)\mathbf{K}(a, p)\, dq, \tag{8.5.9}$$

frequently occur, i.e. Fourier or Fourier–Bessel type integrals, e.g. equations (8.0.1), (8.2.14) and (8.2.15). The Hankel integrals (3.3.7) are obtained from the

Bessel integrals (3.3.5). Asymptotically, these Hankel integrals have the form of the Fourier integral (3.2.10) using the asymptotic result (3.3.8). The singularity at the origin is excluded by taking the principal value.

We have separated a known, oscillatory part (exponential, Bessel or Hankel function) from a kernel, $\mathbf{K}(a, p)$. For $ap \gg 1$, the former is highly oscillatory while the latter is assumed to vary less rapidly. The parameter a is typically $x_\mathrm{R}\omega$ where ω is frequency and x_R the range. At large frequencies and/or range, numerical difficulties arise evaluating the integrals, even though the kernel may be well sampled, as the oscillating function may have many cycles in each sampling interval. The Filon method (Filon, 1929) is designed to solve this problem. Frazer and Gettrust (1984) and Mallick and Frazer (1987) have developed the use of the Filon method in slowness integrals, and discussed many of the topics in this section.

Fourier and Fourier–Bessel integrals also occur in other wave-modelling and processing applications. The techniques described in this section are applicable and most can be used whether the kernel is calculated numerically or from data.

In the integrals (8.5.7), (8.5.8) and (8.5.9), the independent variable, q, and the argument, p, are related. In many circumstances, they are identical

$$p = q. \tag{8.5.10}$$

In other applications, the variable q is introduced as a change of variable to simplify the kernel, e.g.

$$\int \frac{p}{q} \ldots \mathrm{d}p = -\int \ldots \mathrm{d}q, \tag{8.5.11}$$

where

$$p^2 + q^2 = \frac{1}{c^2}. \tag{8.5.12}$$

The branch-point singularity in the p integral at $p = 1/c$, $q = 0$, is removed in the q integral. A p integral along the real axis is replaced by a q integral along the real and imaginary q axes, e.g.

$$\int_0^\infty \frac{p}{q} \ldots \mathrm{d}p = -\int_{1/c}^{0} \ldots \mathrm{d}q - \int_0^{\mathrm{i}\infty} \ldots \mathrm{d}q. \tag{8.5.13}$$

In other cases, the p–q mapping may just be a non-linear relationship to improve the sampling of the kernel. We assume that there is a one-to-one relationship between p and q, and that both values are given at the sampled points.

The order of the Bessel function is usually small ($n = 0$, 1 or 2). We assume n is not negative, as results for negative orders can always be obtained simply from the positive order.

With this introduction, we now investigate numerical issues that arise evaluating integrals of the form (8.5.7), (8.5.8) or (8.5.9), and the inverse Fourier transform (3.1.2). The difficulties arise because infinite integrals are approximated by discrete forms over a finite range. The Fourier integral (3.1.2) is normally approximated by a discrete Fourier series and evaluated as a fast Fourier transform (FFT). The slowness integral is approximated by a low-order, quadrature method.

If the fast Fourier transform uses n_ω frequencies, and the slowness integral n_p slownesses, then the cost of evaluating the integrand is going to be proportional to the product $n_\omega n_p$ (cf. equations (8.4.16) and (8.4.17)). This normally dominates the overall cost and the cost of evaluating the slowness quadrature and the FFT is less significant. Nevertheless, we may require the integral for several kernel functions, i.e. **K** is a vector/tensor. Performing all integrals simultaneously may again lead to some computational savings. Therefore, it is desirable to reduce the number of frequencies and slownesses as much as possible while retaining sufficient accuracy and avoiding numerical problems. We do not consider methods of estimating the accuracy of the numerical integration. We assume that the kernel has been sampled sufficiently that the Filon method and the FFT will be accurate. We do not consider adaptive methods of altering the sampling. It is normally sufficient and simple to test that results are accurate by comparison with greater sampling, and with this experience, use optimal values.

Typically we require the integral for many values of the parameter a, e.g. for many frequencies and ranges. Computationally it may be advantageous if these are evenly distributed, e.g.

$$a_j = j x_{\mathsf{R}} \Delta\omega, \tag{8.5.14}$$

with $j = 0$ to $n_\omega = n_t/2$, where $\Delta\omega = 2\pi/T$ and $T = n_t \Delta t$ for the Fourier transform of a discrete time series of n_t points, time T long. The Nyquist frequency is $\omega_{n_\omega} = \pi/\Delta t$.

The numerical issues with the FFT are simpler and well known, so we consider them first.

8.5.2.1 *The frequency integral*

The slowness integral is normally followed by a Fourier integral (3.1.2). It is approximated by a discrete Fourier series, performed numerically as a fast Fourier transform (FFT). The infinite frequency range is restricted by the Nyquist frequency which can be set from the bandwidth of the source and receiver functions. It is normally convenient to include the source and receiver spectral response in the kernel in order to band-limit the result. The Nyquist frequency determines the time step, $\Delta t = \pi/\omega_{n_\omega}$, and the frequency step depends on the the time window required, $\Delta\omega = 2\pi/T$. The number of frequencies and time samples are then determined from $n_t = 2n_\omega = T/\Delta t$. Given the bandwidth of the source and receiver, there are two options to reduce the number of frequencies at which the kernel must be calculated: if the source or receiver is band-limited, we may be able to take the kernel as zero outside the band, particularly at low frequencies; and we may be able to reduce the time window, T, required. The former is straightforward to apply; the latter may introduce difficulties due to wrap-around.

The length of the time series can be reduced by shifting the time origin. In the frequency domain, we introduce a phase shift $\exp(-\mathrm{i}\omega t_{\text{shift}})$, so $t = t_{\text{shift}}$ is the effective origin in the inverse FFT. If t_{shift} is chosen before the first arrival, and it is normally easy to estimate a suitable value, e.g. a conservative estimate is $t_{\text{shift}} = x_{\mathsf{R}}/c_{\max}$, then no signals will be lost (the result is still causal) and the time window can be reduced accordingly.

It is well known that, using the discrete Fourier series instead of a Fourier integral, signals that arrive for times $t > T$ (or $t > T + t_{\text{shift}}$ if the origin is shifted) are wrapped into the time window $0 < t < T$ at $t' = t - nT$ where n is an integer. If the time window is short, or the response contains strong reverberations, then the wrapped signals can be very significant. In order to avoid wrap-round problems in the discrete transform, it is useful to introduce a small, imaginary part to the frequency. This is equivalent to multiplying by $\exp(-\mathrm{Im}(\omega)t)$ in the time domain:

$$f'(t) = f(t)\mathrm{e}^{-\lambda t} = \frac{1}{2\pi}\int f(\omega)\mathrm{e}^{(-\mathrm{i}\omega-\lambda)t}\mathrm{d}\omega \tag{8.5.15}$$

$$= \frac{1}{2\pi}\int f(\omega' + \mathrm{i}\lambda)\mathrm{e}^{-\mathrm{i}\omega' t}\mathrm{d}\omega', \tag{8.5.16}$$

where $\omega' = \omega - \mathrm{i}\lambda$. The exponential decay, $\exp(-\lambda t)$, suppresses late arrivals and reduces wrap-round when using the discrete Fourier transform. Expression (8.5.16) is the integral approximated by the inverse discrete Fourier transform with ω' real. In it, the response, $f(\omega)$, is evaluated with a complex frequency with positive imaginary part, λ. Having evaluated the spectrum with a small positive, imaginary part of frequency (the spectrum is smoother in the positive, imaginary frequency plane as the response is causal – Section 3.1.1), we evaluate the inverse discrete transform (normally using a fast Fourier transform) with real frequency. In the time domain we can correct by multiplying $f'(t)$ by $\exp(\lambda t)$ which corrects the true signals but 'under-corrects' wrapped signals, i.e. arrivals for $t > T$ wrapped are decayed by $\exp(-\lambda t)$ but only amplified by $\exp(+\lambda t')$, a net suppression of $\exp(-n\lambda T)$. The decay factor, $\lambda = \mathrm{Im}(\omega)$, should be chosen to suppress wrapped signals significantly, while not reducing true signals so much that they are lost in the noise, i.e. $\epsilon = \exp(-\lambda T)$ might be 0.01. Then

$$\lambda = -\frac{1}{T}\ln\epsilon. \tag{8.5.17}$$

The disadvantages of this technique are:

- Post-multiplying be $\exp(\lambda t)$ will amplify numerical noise. While theoretically no signals are lost, numerically some noise will be amplified and may swamp the signals. The number ϵ should not be too small.
- The technique does not work if used in combination with time-windowing *after* the first arrival. If the time shift, t_{shift}, is greater than the first arrival time, as it might be if we were studying S waves, then the earlier arrivals, which will also wrap-round, will be amplified in $f'(t)$ and again in the post-multiplication.
- Related to the previous problem, but perhaps more subtle, acausal numerical noise will also be amplified in $f'(t)$. Theoretically, the response is, of course, causal but numerical errors in its evaluation can create acausal noise. This will be amplified in $f'(t)$ and again in the post-multiplication. Care must be taken to minimize the acausal noise, if the technique is to be used. We shall return to this point below when discussing the contour of integration.

We conclude that it is useful to compute the kernel with a small, positive imaginary part given by expression (8.5.17).

8.5.2.2 The slowness integral

The evaluation of the slowness integral, (8.5.7), (8.5.8) or (8.5.9), presents a number of numerical difficulties: the integrand is oscillatory and may contain singularities; and the range of the integral is infinite or semi-infinite. Numerically, the integral is evaluated for a finite range. We do not consider methods of evaluating a (semi-)infinite integral explicitly. Therefore it is required that the integrand decays sufficiently within the finite range, either naturally or by distorting the contour of integration or by introducing appropriate weighting functions.

The following discussion of the slowness integral is divided into four subsections:

- The Filon method is introduced to handle the oscillatory behaviour of the integrand.
- Small changes in the contour of integration are discussed to avoid aliasing problems from discrete sampling around singularities.
- Large changes in the contour of integration are discussed in order to improve the convergence of the integral.
- The use and design of weighting functions are discussed in order to avoid changing the contour of integration.

The discussion is simplest for the Fourier integral (8.5.7), and each sub-section contains minor additional comments for the Bessel (8.5.8) or Hankel (8.5.9) integrals.

8.5.2.3 Filon method

The Filon method allows the exact evaluation of integrals of the form

$$\int_x^y f(p)\mathrm{e}^{g(p)}\mathrm{d}p, \tag{8.5.18}$$

when $f(p)$ and $g(p)$ are linear in the interval (x, y). $f(p)$ and $g(p)$ may be complex and $f'(p)$ and $g'(p)$ of any magnitude. Typically $g(p)$ is imaginary and $g'(p)$ large, and the integrand is highly oscillatory.

Evaluating integral (8.5.18) exactly when $f(p)$ and $g(p)$ are linear gives (Mallick and Frazer, 1987)

$$\int_x^y f(p)\mathrm{e}^{g(p)}\mathrm{d}p = \frac{\Delta p}{\Delta g}\left(\Delta(f\mathrm{e}^g) - \frac{\Delta f\,\Delta(\mathrm{e}^g)}{\Delta g}\right), \tag{8.5.19}$$

where

$$\Delta p = y - x \tag{8.5.20}$$

$$\Delta f = f(y) - f(x)\text{ , etc.} \tag{8.5.21}$$

If $\Delta g = 0$ this expression breaks down and if it is small, numerical instabilities will occur. A rearrangement leads to

$$\int_x^y f(p)\mathrm{e}^{g(p)}\mathrm{d}p = \frac{\Delta p}{2}\mathrm{e}^{g(x)}\left(f(x) + f(y) + \Delta g(1+\delta)f(y) - \delta\Delta f\right), \tag{8.5.22}$$

where

$$\delta = \frac{2(\mathrm{e}^{\Delta g} - 1 - \Delta g)}{\Delta g^2} - 1 \tag{8.5.23}$$

$$= 2\sum_{n=1}^{\infty}\frac{\Delta g^n}{(n+2)!}. \tag{8.5.24}$$

Note that if $\Delta g \to 0$, $\delta \to 0$ and this reduces to the trapezoidal rule

$$\int_x^y f(p)\mathrm{e}^{g(p)}\mathrm{d}p = \frac{\Delta p}{2}\mathrm{e}^{g(x)}\left(f(x) + f(y)\right), \tag{8.5.25}$$

which is accurate if $f(p)$ is approximately linear. The Taylor expansion (8.5.24) should be used in result (8.5.22) when $|\Delta g| \ll 1$.

The Filon rule, (8.5.22), can be used to evaluate integrals of the form (8.5.7). Integrals of the form (8.5.8) or (8.5.9) have to be modified into the standard form (8.5.18). The Bessel integral (8.5.8) is split into two Hankel integrals

$$\int J_n(ap)\mathbf{K}(a, p)\,\mathrm{d}q = \frac{1}{2}\int\left(H_n^{(1)}(ap) + H_n^{(2)}(ap)\right)\mathbf{K}(a, p)\,\mathrm{d}q. \tag{8.5.26}$$

This separation is only used when the functions are oscillatory. Near the origin, we evaluate the Bessel integral using the trapezoidal rule for the complete integrand. We do not consider Hankel integrals near the origin due to the singularity. The Bessel integral is divided at a point where $\mathrm{Re}(ap) = x_c \;\; (= 1, \text{say})$. In the oscillatory region, the Hankel integrals (8.5.9) or (8.5.26) are reduced to the Fourier integrals

$$\int H_n^{(1 \text{ or } 2)}(ap)\mathbf{K}(a, p)\,\mathrm{d}q = \int \mathrm{e}^{\pm iap}\mathbf{K}'(a, p)\,\mathrm{d}q, \qquad (8.5.27)$$

where

$$\mathbf{K}'(a, p) = \left(\mathrm{e}^{\mp iap} H_n^{(1 \text{ or } 2)}(ap)\right)\mathbf{K}(a, p). \qquad (8.5.28)$$

From the asymptotic form of the Hankel function, the extra factor included in the kernel is non-oscillatory. Standard methods can be used to evaluate the Bessel and Hankel functions (e.g. Press, Flannery, Teulolsky and Vetterling, 1986, Section 6.4).

8.5.2.4 Slowness contour to avoid singularities

In this sub-section we discuss a small distortion of the contour from the real axis in order to avoid singularities and evaluate the integral numerically. In the next sub-section, we discuss more general distortions of the contour into the complex p plane in order to improve the convergence of the (infinite) integral.

Although the contour of integration of the Fourier integral (8.5.7) is usually written as along the real axis, singularities often exist on this contour and must be avoided. Branch points exist where wave slownesses are zero and poles exist corresponding to surface and interface waves. For the integral with the positive exponent (that is $\exp(+iap)$ with a, i.e. frequency and range, positive), the integral should be distorted to be infinitesimally below/above these singularities on the positive/negative axis (see Figure 3.4). For simplicity, we restrict our discussion to this case, e.g. positive exponent $\exp(iap)$, positive (real) frequency ω, and positive range x, so the most significant part of the integral (where the saddle points of geometrical rays exist) is the positive p axis. Results for the opposite signs can be obtained by symmetry. We emphasize that in all the following discussion, $\omega > 0$ and $x \geq 0$ are assumed.

Provided no singularities are crossed, using a complex contour should not alter the analytic integral (provided the contour ends are equivalent). However, it can significantly alter the numerical behaviour and result. This is because we avoid poles, normally on the real axis, and sample singularities better. The poles may or may not be physically significant. In the latter case they are a term of the form $\exp(-c)/(p - p_c)$ where c is large enough that the contribution is very small (an example is interface, Stoneley waves where the solution decays exponentially away from the interface). Numerically, poles will cause problems whether significant or not (slowness aliasing), as the result will be non-robust depending on how close an integration point is to the pole. Making the contour have a small, negative imaginary p value will reduce these problems. The singularity will be smeared out and the simple quadrature rule will become accurate and less sensitive to the sampling points. The exact value of $\mathrm{Im}(p)$, or shape of the contour, should not be important except that too large a value will introduce significant exponential behaviour into various terms in the integrand causing other numerical problems. Certainly a negative value for $\mathrm{Im}(p)$ comparable with the sampling step is sensible, as then any singularity which is small compared with the sampling, i.e. when $\exp(-c)/\Delta p$ is numerically small, will be missed completely and results will not depend on the sampling relative to p_c. If the contour is nearer the real axis, then the integral will be aliased and its value will depend on the positions of the samples relative to the singularities.

The contour for the Fourier–Bessel integral (8.5.8) is just the positive real axis, again infinitesimally below the singularities. This integral can be converted into the Hankel integral (8.5.9) with the same contour as the Fourier integral (Figure 3.4) corresponding to the first kind of Hankel function, $H^{(1)}(ap)$ – the integral with the second kind of Hankel function, $H^{(2)}(ap)$, has the same contour as the Fourier integral with a negative exponent, $\exp(-iap)$. The added difficulty with the Hankel integrals (8.5.9) is the singularity at the origin, $p = 0$, which should be evaluated as a principal value (the first kind of Hankel function has a branch cut along the negative real axis). Analytically, this is not a fundamental problem as the singular part cancels from the symmetry of the integrand (the Bessel integral (8.5.8) is not singular). Numerically it can be achieved provided the algorithm respects this symmetry, e.g. sampling points must be symmetric, and avoids a small region around the origin (alternatively, a symmetrical portion of the Hankel integral around the origin can be evaluated as a Bessel integral).

8.5.2.5 Slowness contour to improve convergence

We now discuss convergence of the integrals (8.5.7), (8.5.8) and (8.5.9) and the possible use of complex p contours.

Although the integrals have an infinite range, convergence for large p is rapid due to the decay of $\mathbf{K}$. For p larger than the largest slowness in the model, all the vertical slownesses become imaginary and the wavefunctions decay evanescently. Asymptotically, we must have $\mathbf{K} \sim O(\exp(-\omega|\mathrm{Re}(p)|d))$, where d is the minimum vertical distance travelled, e.g. the vertical distance from the source to receiver. Only if the direct ray is included and the source and receiver are at the

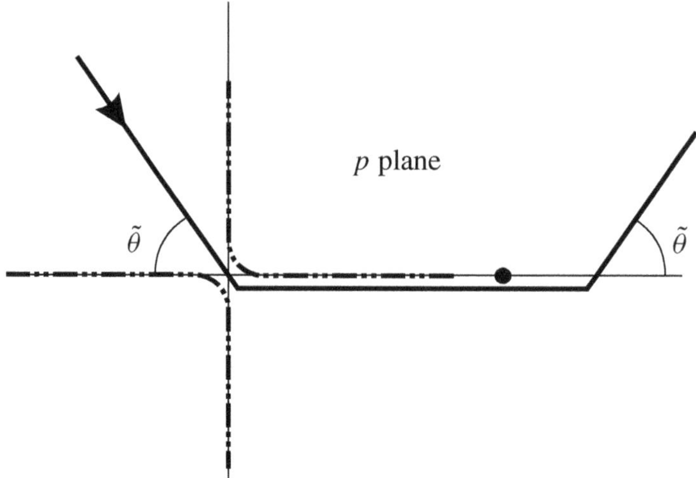

Fig. 8.17. The integration contour for the Fourier integral (8.5.7) distorted to decay rapidly. The complex parts of the contour make angles $\tilde{\theta} = \tan^{-1}(x_R/d)$ with the real axis.

same depth, does **K** not decay – the Fourier integral is purely oscillatory and the Bessel integral only decays as $O(p^{-1/2})$.

The convergence of the integral can be improved by distorting the contour. Provided no singularities are crossed, the value of the integral is not altered. If the ray expansion is applied to the integrand, then it consists of a summation of rays. The 'full-wave' method (Aki and Richards, 1980, 2002, Chapter 9) uses a contour distorted to follow, at least approximately, the steepest-descent path over the saddle points. Note that part of this contour will pass through the branch cut, off the physical Riemann sheet defined by $\text{Im}(q) > 0$. For large $|p|$, however, the behaviour is different. For $\text{Re}(p) \gg 0$, the integrand behaves as $\exp(-\omega p(d - \text{i}x_R))$, and the maximum decay occurs at $\text{Arg}(p) = \tan^{-1}(x_R/d)$ (as x_R increases, the $\omega p x_R$ term dominates, and a positive imaginary part of p causes exponential decay). When $\text{Re}(p) \ll 0$, the integrand behaves as $\exp(\omega p(d + \text{i}x_R))$, and $\text{Arg}(p) = \pi - \tan^{-1}(x_R/d)$. Thus a contour of integration with optimal decay as $|p|$ increases is illustrated in Figure 8.17 (Frazer and Gettrust, 1984).

For $\text{Re}(p) > 0$, the contour in Figure 8.17 has been kept infinitesimally below the real axis to a moderate slowness value so that:

- It passes through all the ray saddle points on the real axis, where the integrand is most significant.
- It avoids being distorted onto the lower Riemann sheet around the branch point (messy).
- It avoids picking up residues of poles on the real axis (if poles exist for large slownesses, their residues should be included but may not be significant due to the large slowness).

For Re$(p) < 0$, it is immediately distorted into the second quadrant as there are no saddle points on the negative real axis. On the negative real axis, the integrand is highly oscillatory with no stationary points. This is difficult to evaluate numerically yet contributes little, and distorting the contour takes advantage of the rapid decay.

There are a number of disadvantages of using a contour like Figure 8.17:

- The optimum contour depends on the range, x_{R}. The kernel **K** is independent of range, and so there is a significant computational advantage if the same contour, i.e. same p values, can be used for all ranges.
- For the Fourier–Bessel integral (8.5.8), we would have to convert it into a Hankel integral (8.5.9). There is no point in distorting the contour for the Bessel integral off the real axis, as it contains one part that grows exponentially and one that decays. In order to evaluate a Hankel integral with a contour like Figure 8.17, we must avoid the singularity at the origin and obtain the principal value. This can be done by making the contour locally symmetric about the origin (Figure 8.18*a*). Unfortunately the size of the singularity, and the necessary size of perturbation to the contour, is $p_c \sim O(1/\omega x_{\mathrm{R}})$, again range (and frequency) dependent.
- Care must be taken to find singularities on the real axis.

While none of these difficulties are fatal, they mitigate against using such a contour.

Before leaving these complex contours, let us consider the contour shown in Figure 8.18*a* for the Hankel integral. The negative part can be reflected through the origin, and evaluated with the second kind of Hankel function (Figure 8.18*b*). For the contour on the real axis ($p < p_c$), the two parts combine to give the Bessel function. Effectively for the rest of the integral, we have split the Bessel function into its two parts, and distorted the contour differently. We shall return to this

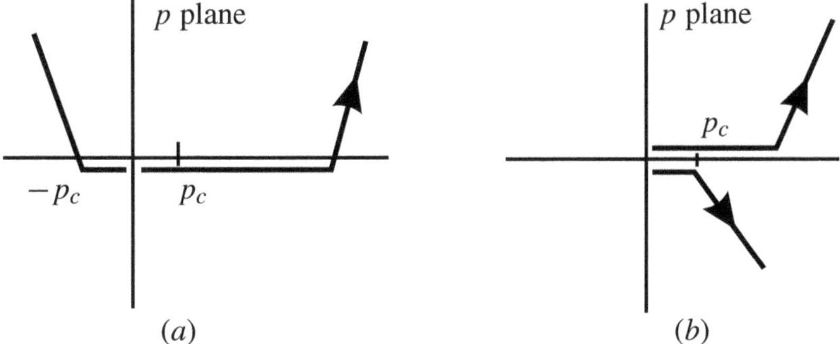

Fig. 8.18. The integration contour for the Hankel integral (8.5.9) distorted to handle the singularity at the origin: (*a*) the contour is symmetrical for $-p_c < p < p_c$ and an infinitesimal portion at the origin in omitted; (*b*) the contour in (*a*) is reflected through the origin.

concept – handling the two parts of the Bessel function in (8.5.8) differently – below.

8.5.2.6 Weighting functions

Although it would be nice to evaluate the integral numerically in a manner that optimizes the analytic convergence, and followed from the analytic behaviour, this is not necessary. It is known that the tails of the integral contribute little to its total value, either because the slowness is too large to propagate, or because negatively travelling waves are not important. While we cannot eliminate these parts of the integral completely, it is pointless to expend too much energy computing them accurately when they are known to be of minor importance. If the integral is abruptly terminated, the end-point will cause errors and arrivals with the slowness of the end-point, often in the time window of interest. The contribution from terminating a decaying, oscillatory integral will be $\sim O(1/a)$, and may be much larger than the significant contribution from the integral. A standard, numerical method of terminating any integral is to include a weighting function to increase the rate of convergence. If the weighting function decreases smoothly towards the end-point, the numerical error is much reduced and is insensitive to the exact location of the end-point. Thus we replace the integrals (8.5.7), (8.5.8) and (8.5.9) by

$$\int_{-\infty}^{\infty} \mathrm{e}^{\pm iap} \mathbf{K}(a, p) w(a, p) \, \mathrm{d}q \tag{8.5.29}$$

$$\int_{0}^{\infty} J_n(ap) \mathbf{K}(a, p) w(a, p) \, \mathrm{d}q \tag{8.5.30}$$

$$\int_{-\infty}^{\infty} H_n^{(1 \text{ or } 2)}(ap) \mathbf{K}(a, p) w(a, p) \, \mathrm{d}q, \tag{8.5.31}$$

where the weighting function $w(a, p) = 1$ for the significant part of the integral, e.g. p less than slownesses in the model, and $w(a, p) \to 0$ as $|p| \to \infty$. The integrals can be terminated when $w(a, p)$ is numerically insignificant. Note that while we were reluctant to choose a contour that depended on range (and/or frequency), as that would require recomputing $\mathbf{K}$, the weighting function can be dependent on range and frequency (through a), as it will be inexpensive to compute.

Thus rather than distort the contour, and use different contours for different frequencies and ranges, we prefer to use the same contour and improve the convergence using a weighting function.

As was mentioned above, the convergence criteria for the two parts of the Bessel function differ (Figure 8.18*b*). This can be handled by weighting the two Hankel functions differently:

$$\int_0^\infty J_n(ap)\mathbf{K}(a,p)\,\mathrm{d}q$$
$$= \frac{1}{2}\int_0^\infty \left(w^{(1)}(a,p)H_n^{(1)}(ap) + w^{(2)}(a,p)H_n^{(2)}(ap)\right)\mathbf{K}(a,p)\,\mathrm{d}q. \tag{8.5.32}$$

It is necessary that

$$w^{(1)}(a,p) = w^{(2)}(a,p) \quad \text{for} \quad 0 \leq p < p_c, \tag{8.5.33}$$

to handle the singularity at $p = 0$ correctly (so that for $0 \leq p < p_c$ we can use the Bessel function) (normally the weights in this range would be unity, but zero is possible if this part of the integral is not important). For $p > p_c$, $w^{(2)}(a,p)$ would normally decay more rapidly than $w^{(1)}(a,p)$. We shall return below to designing these weighting functions, but first let us complete the description of the contour. The advantages of using a contour with a small, negative imaginary p part (to avoid singularities on the real axis) and positive, imaginary frequency (to avoid wrap-around problems) have been mentioned above.

The purpose of $\mathrm{Im}(p) < 0$ is to avoid singularities on the real axis, and to smooth the integrand avoiding aliasing problems with very narrow singularities (which are not significant, anyway) due to undersampling. It has been suggested in the literature that the contour should be distorted to $\mathrm{Arg}(p) = -\phi$ (where ϕ is some small, positive angle). However, from our discussion above it is necessary that $\phi < \tan^{-1}(d/x_\mathrm{R})$ to obtain convergence. There seems to be no particular advantage to this choice over the simpler $\mathrm{Im}(p) = \text{constant}$.

The purpose of $\mathrm{Im}(\omega) > 0$ is to decay later arrivals that will wrap around in the finite Fourier transform. The value of the slowness p integral, the spectrum, is smoothed by making the frequency complex (this is in contrast to the purpose of $\mathrm{Im}(p)$ which does not alter the value of the integral, just makes it easier to evaluate). The value of $\mathrm{Im}(\omega)$ must be taken constant, λ, so it can be compensated by multiplying by $\exp(\lambda t)$ in the time domain. Mallick and Frazer (1987) have suggested that the p contour should be distorted into the fourth quadrant, so that ωp remains real (as in the original integral with respect to k, the wavenumber). An advantage is that the contour is moved away from the singularities on the real axis, smoothing the integrand and hence the integral. This means that $\mathrm{Arg}(p) = -\tan^{-1}(\lambda/\mathrm{Re}(\omega))$, a function of frequency, which would be extremely inefficient to handle numerically. However, for realistic values of $\mathrm{Im}(\omega)$, the distortion is unnecessary anyway. If ϵ is the decay required in the time series (length T), then (8.5.17)

$$\lambda = -\frac{1}{T}\ln\epsilon = -\frac{\Delta\omega}{2\pi}\ln\epsilon, \tag{8.5.34}$$

where $\Delta\omega$ is the frequency sampling. Thus for $\epsilon = 0.01$, this gives $\lambda = 0.73\Delta\omega$ and for $\omega = j\Delta\omega$, $\text{Arg}(p) = -\tan^{-1}(0.73/j)$. Except when j is very small, this is small. For efficiency and simplicity, as λ is small compared with ω except at the lowest frequencies, we do not alter the contour for $\text{Im}(\omega)$.

We now return to the design of the weight functions, $w(a, p)$. The integral must always be terminated at some finite limit, $p_{\max}$, say. In order to avoid end-point errors from this limit, the end of the integral is tapered to zero over 20%, say, of the range using a smooth function, e.g. a Gaussian

$$w(a, p) = 1 \qquad \text{for} \quad \text{Re}(p/p_{\max}) < f \tag{8.5.35}$$

$$= \frac{e^{-g^2\text{Re}(p/p_{\max}-f)^2/(1-f)^2} - e^{-g^2}}{1 - e^{-g^2}} \qquad \text{for} \quad f < \text{Re}(p/p_{\max}) < 1, \tag{8.5.36}$$

with $f = 0.8$ and $g = 3$, say, and the extra factors are included to force $w(a, p_{\max}) = 0$. Normally, the lower limit of the Fourier–Bessel integral (8.5.30) is zero, and $w(a, 0) = 1$ with no taper. However, if a non-zero, lower limit is used (which might be relevant if all the ranges are large), then a similar taper can be applied there. A similar taper should be applied at the lower, negative limit of the Fourier integral (8.5.29).

The weighting of the second kind of Hankel function (8.5.32), or for $p < 0$ in the Fourier integral (8.5.29), is important, especially if an imaginary part of the frequency is used to reduce wrap-around. Numerical noise from this part of the integral will contribute at negative (acausal) times (this is easily seen as the phase becomes negative). These signals will be *amplified* by the imaginary part of the frequency, will wrap into the time window of interest and be amplified again when multiplied by $\exp(\lambda t)$. Particularly troublesome is any error from the taper of the integral at $\pm p_{\max}$. This will suffer maximum amplification and may wrap-round anywhere in the time window. We emphasize that this part of the exact integral does not contribute acausal signals – the response must be causal. But noise in the numerical evaluation of the integral can easily be acausal. It is therefore essential that numerical noise is reduced as much as possible from this part of the integral. However, it contributes little to the complete integral, so we do not want to expend too much effort on reducing the noise. Nevertheless, it cannot be ignored completely as for small ranges the saddle points are near the origin and, particularly at low frequencies, spread over negative p values (at zero range, all saddle points are symmetrical about the origin). We therefore design a weighting function, $w^{(2)}(a, p)$, for the second kind of Hankel function (8.5.32), or $w(a, p)$

for $p < 0$ in the Fourier integral (8.5.29), so it is wide enough to include the significant contributions from the saddle points, but narrow enough to cut off the integrand and avoid acausal noise. For the Fourier–Bessel integral (8.5.32), we also impose a lower limit on the width of the weighting function, so that the weight is unity when the argument of the Bessel function is less than a fixed value. This avoids numerical difficulties in splitting the integral into Hankel functions, when their behaviour is singular, and neither the Filon nor trapezium integration rule would be accurate. Let us call the width of the weighting function p_{w} such that for $0 < |p| < p_{\mathrm{w}}$, the weighting function is unity. Below, we refer to the first and second parts of the integral, indicating the two kinds of Hankel functions in integral (8.5.32) or the positive and negative slownesses in integral (8.5.29). The width p_{w} must be sufficient to include the significant contributions from the saddle points in the second part of the integral.

First, let us just comment that wrap-round signals from the second part of the integral, with or without amplification, are easily recognized by their negative slowness, i.e. negative slope in the (t, x) domain. Sometimes they can just be ignored, but sometimes, particularly with amplification, they can be very troublesome.

In order to design the weighting function we need to estimate the width necessary to include the significant features of the integral, i.e. saddle points near the origin that contribute to the second part of the integral. Within this width, the weighting function should be unity. Beyond this we allow the weight to decay to zero, smoothly. Note that we are not interested in a detailed analysis of the integrand – rather we just need an estimate that can be computed efficiently. Provided the weighting is only a smooth cutoff, errors will be small and not serious. At zero range $(x = 0)$, all the saddle points are at the origin $(p = 0)$, equally in the first and second parts of the integral. The width of the weighting function, p_{w}, should be large enough to cover the widest saddle point. As the range increases, the saddle points move away from the origin at different rates and separate. In order to estimate the required width of the weighting function we have to consider both the width and position of the saddle points (the effective width of the saddle point in the *second* part of the integral is the difference of the width and position).

The width of a ray saddle point is

$$k_1 \left(\omega \frac{\partial x}{\partial p} \right)^{-1/2}, \tag{8.5.37}$$

where k_1 is a small constant (at the width of the saddle point, the phase has changed by $\exp(\mathrm{i}k_1^2/2)$ so $k_1 \sim 4$ will include a few cycles). Near the origin, the geometrical spreading function is given by

$$\frac{\partial x}{\partial p} \simeq \frac{x}{p} \tag{8.5.38}$$

(l'Hôpital's rule), and the range is approximately

$$x = \int \frac{p}{q} \mathrm{d}z \simeq p \int c \, \mathrm{d}z = p \int c^2 \mathrm{d}T = p\bar{c^2}t \tag{8.5.39}$$

(q is the vertical ray slowness and $\bar{c^2}$ is the mean square velocity with respect to time – result (2.5.2)). To obtain the width of the weighting function p_w, we subtract the position of the saddle point p from the width of the saddle point (8.5.37). Using approximations (8.5.38) and (8.5.39) in expression (8.5.37), we obtain

$$p_\mathrm{w} = \frac{k_1}{\left(\omega \bar{c^2} t\right)^{1/2}} - \frac{x}{\bar{c^2} t}, \tag{8.5.40}$$

where the arrival time t will vary from the first arrival, t_{min}, to the maximum time of interest, $t_{\mathrm{max}} = T$. The mean square velocity, $\bar{c^2}$, also varies with time and arrival type, but for simplicity we treat it as constant (compared with the other terms in expression (8.5.40)). To set the weighting function we need to analyse the function p_w (8.5.40). We need to consider the maximum significant value of expression (8.5.40) in the range of interest. This may be at the limits $t = t_{\mathrm{min}}$ or t_{max}, or at a stationary point, or may be zero (as we are not interested in negative values when the saddle points do not contribute significantly to the second part of the integral). The function p_w has a positive, maximum value at

$$t_\mathrm{w} = \frac{4\omega x^2}{k_1^2 \bar{c^2}}, \tag{8.5.41}$$

and a zero at $t = t_\mathrm{w}/4$. For $t < t_\mathrm{w}/4$, the saddle point has moved further from the origin than its width, and the negative values are not significant (the saddle width is entirely in the first part of the integral – the effective width is zero).

Consider the dimensionless parameter

$$x' = \frac{x}{k_1} \left(\frac{\omega}{\bar{c^2} t_{\mathrm{max}}} \right)^{1/2}, \tag{8.5.42}$$

and the dimensionless width

$$p'_\mathrm{w} = p_\mathrm{w} \frac{\left(\omega \bar{c^2} t_{\mathrm{min}}\right)^{1/2}}{k_1}. \tag{8.5.43}$$

Equation (8.5.40) reduces to dimensionless form

$$\frac{p'_\mathrm{w}}{a} = \frac{1}{t'^{1/2}} - \frac{x'}{t'}, \tag{8.5.44}$$

(as plotted in Figure 8.19) where $t' = t/t_{\mathrm{max}}$ and $a = (t_{\mathrm{min}}/t_{\mathrm{max}})^{1/2}$, and the stationary point (8.5.41) is at $t'_\mathrm{w} = 4x'^2$. The definition of the width can be divided in

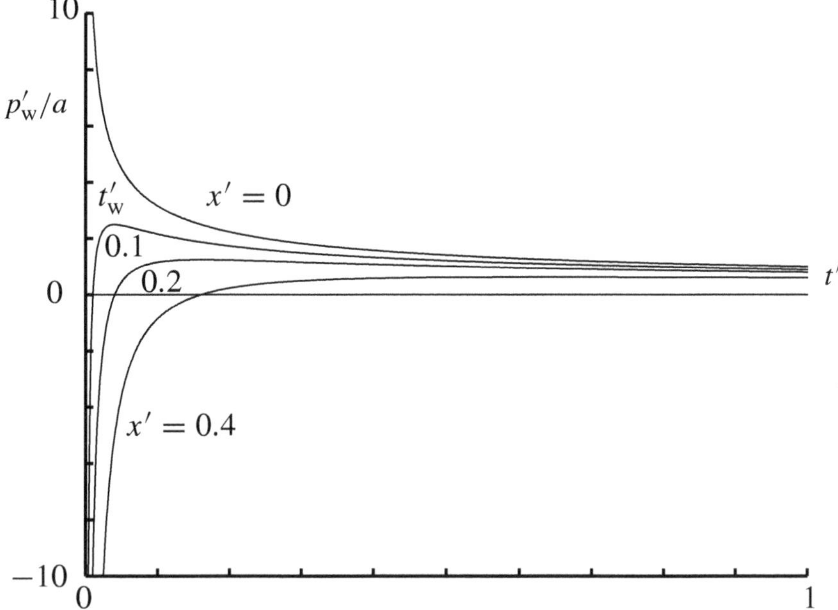

Fig. 8.19. The effective, dimensionless saddle point width, $p'_{\rm w}/a$, plotted against dimensionless time, t' (8.5.44). Four curves are plotted for $x' = 0$, 0.1, 0.2 and 0.4. The maximum at $t' = t'_{\rm w}$ is indicated on the curve for $x' = 0.1$. Only the part of the curve between $x'^2 < t' < 1$ is significant ($p'_{\rm w}$ positive).

four ranges (Figure 8.20)

$$p'_{\rm w} = 1 - x'/a \quad \text{for} \quad 0 < x' < a/2 \tag{8.5.45}$$

$$= a/4x' \quad \text{for} \quad a/2 < x' < 1/2 \tag{8.5.46}$$

$$= a(1 - x') \quad \text{for} \quad 1/2 < x' < 1 \tag{8.5.47}$$

$$= 0 \quad \text{for} \quad x' > 1, \tag{8.5.48}$$

where the linear regions (8.5.45) and (8.5.47) arise from $t = t_{\rm min}$ and $t_{\rm max}$ ($t' = a^2$ and 1 in equation (8.5.44)), respectively, and the hyperbola (8.5.46) from result (8.5.44) at $t' = 4x'^2$, the maximum (8.5.41). For $x' > 1$, the non-negative constraint applies, i.e. the saddle is further from the origin than its width.

In addition, we impose the limit that the argument of the Bessel function must not be less than k_2 outside the width of the weighting function, i.e. $p_{\rm w}$ must be the greater of results (8.5.45), (8.5.46), (8.5.47) or (8.5.48), and

$$p_{\rm w} = \frac{k_2}{\omega x} \quad \text{or} \quad p'_{\rm w} = \frac{ak_2}{4k_c x'}, \tag{8.5.49}$$

where $k_c = k_1^2/4$. Note that this is a hyperbola like result (8.5.46). If $k_2 > k_1^2/4 = k_c$, then the argument limit (8.5.49) always applies; if $k_2 < k_1^2/4 = k_c$, then the

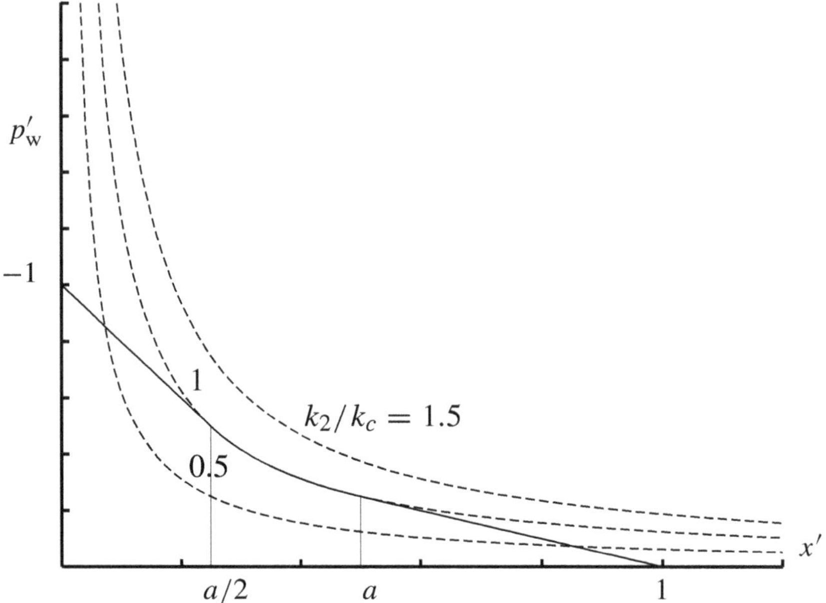

Fig. 8.20. The dimensionless width, p'_w (8.5.43), plotted against the dimensionless parameter, x' (8.5.42). The function (8.5.45), (8.5.46), (8.5.47) and (8.5.48) is the solid line for $a = 0.5$. Function (8.5.49) is the dotted line with $k_2 = k_c = k_1^2/4$, $k_2 = 1.5k_c$ and $k_2 = 0.5k_c$. In the first two cases, the argument limit (8.5.49) applies everywhere; in the last case, the saddle-point limit applies in mid-ranges.

saddle-point limit will apply in mid-range, but the argument limit (8.5.49) will always apply at small or large enough values of x'. We have already suggested that a value of $k_1 \sim 4$ is reasonable. The matching value of k_2 is 4, which is also reasonable. Therefore, for simplicity we just use width (8.5.49). Although the argument about saddle points and their positions was important, it has had no effect on the final result! In addition, we always impose a taper when $\mathrm{Re}(p/p_{\max}) > f$ (8.5.36). The same width can be used in the Fourier integral (8.5.29).

This algorithm is used for numerical examples in the next chapter.

Exercises

8.1 Prove result (8.1.35).

8.2 The results for the direct rays from a line source in a homogeneous medium, e.g. (4.5.84) and (8.1.36), can be obtained exploiting the cylindrical symmetry about the line source. Using the Fourier–Bessel transforms (Section 3.3) to obtain the result as an inverse Fourier transform of a Hankel function (Appendix B.4).

8.3 Result (4.5.84) for the two-dimensional force Green function can be compared with the Cagniard solution (8.1.36) for the two-dimensional pressure Green function. By differentiating result (4.5.84) to form a force dipole, and summing two orthogonal dipoles to form a pressure source, show that the two results are equivalent – remember the extra time differential in result (8.1.36) (it is fairly easy to confirm the equivalence of the leading term – rather tedious to do for the complete expression).

8.4 Exercise 7.4 in Chapter 7 developed the solution for a homogeneous layer over a half-space without the ray expansion. For *SH* waves, investigate the conditions under which the transformed response has poles in the complex p plane. Assume that the shear velocity of the half-space is greater than that of the layer. Sketch the dispersion behaviour of these poles, first as a plot of ω *v.* $(k = \omega p)$, and then as the phase velocity $(c = p^{-1})$ *v.* ω. Using the method of residues, find the spectral response, and then using the method of stationary phase, approximate the time response. Show that these poles give rise to a dispersed signal, where the low frequencies arrive first with the frequency increasing with time. Later the high-frequency signals are superimposed with the frequency decreasing with time. Finally, the two frequencies converge and a large amplitude, known as the *Airy phase*, is observed. To find the waveform of this signal, you will need the third-order saddle-point results in Appendix D.2. These dispersed waves are known as *Love waves*. They are *locked modes* generated by the constructive interference of waves in the surface layer that are totally reflected at the interface.

How does the amplitude of the Airy phase behave compared with the dispersed part of the Love wave? Why?

This problem was described by Knopoff (1958).

How is the solution altered if the velocity in the half-space is lower than that of the layer?

The dispersion of Love waves can be investigated analytically or numerically, but the equivalent waves for the $P - SV$ system will probably need numerical solutions. Show that dispersed waves exist for the $P - SV$ system (assume the shear velocity in the half-space is greater than that of the layer). Show that the lowest-order wave behaves like the Rayleigh wave (see Chapter 9) – at low frequencies it has the behaviour of the Rayleigh wave velocity of the half-space and at high frequencies of the layer. Higher-order waves are dispersed between the shear velocities in the half-space and layer. Again Airy phases exist. This problem was described by Spencer (1965).

8.5 Show that in a fluid layer over a fluid half-space, dispersive waves exist described by very similar mathematics to the Love waves – acoustic waveguide modes. This problem has been described in the classic paper by Pekeris (1948).

If the fluid layers are replaced by solids with very low shear velocities, what happens to the acoustic modes? Their velocities are outside the range of the shear velocities. Therefore, the waves are not totally reflected at the interface and some shear energy leaks into the half-space. Hence the waves are called *leaky modes*. What happens to the singularities of the response? But if the shear velocities are very low, the conversion to shear energy is very small, and the leaky mode may propagate significant distances.

8.6 The two-dimensional Cagniard result (8.1.31) of the far-field approximation for the three-dimensional result (8.2.68) contains singularities. Discuss how the calculations should be performed so they are numerically robust to aliasing problems (cf. the WKBJ seismogram method, Section 8.4.2 – or perform the convolution with the impulse, $\lambda(t)$, in a manner that does not suffer from aliasing problems).

The results of the integrals in the exact, three-dimensional Cagniard method (8.2.17) and (8.2.59) are not singular, but the integrand has singularities. Discuss numerical methods for evaluating these integrals.

8.7 *Further reading:* So many variants of the Cagniard method have appeared in the literature, that it is probably confusing to study them. However a particularly elegant alternative has been published by Burridge (1968).

8.8 *Further reading:* The two-dimensional, real slowness inverse method (8.4.1) depends on the inverse spectral Fourier transform of $\exp(i\omega px)$ being the Dirac delta function $\delta(t - px)$ (cf. the Radon transform (3.4.15)). With the WKBJ approximation the inverse slowness integral (8.4.4) is trivial to evaluate. In the text we have only developed the far-field approximation for the three-dimensional, real slowness inverse method. Chapman (1978) has shown how using the inverse spectral Fourier transform of the Bessel function, (B.4.7), an exact three-dimensional, real slowness inverse method can be developed, where the slowness integral contains a convolution with the inverse Fourier transform of the Bessel function (cf. the Radon transform (3.4.23)). Approximating the singularities of this function, we can obtain the far-field approximation where the three-dimensional result can be obtained by including some extra factors and a convolution (8.2.66) (cf. Radon transform (3.4.19) and (3.4.28)).

8.9 *Further reading:* In Exercise 7.5 of Chapter 7, the spherical system was discussed. Performing the inverse transforms at high frequencies is an interesting problem. The summation of modes is converted into an integral using the Watson transform (Watson, 1918). Nussenzveig (1965) has given a thorough analysis for scattering by an impenetrable sphere. In addition to the normal ray expansion at interfaces, we need the so-called *rainbow* or *Debye expansion* to expand rays that pass through the centre of the sphere. This has been used extensively in electromagnetic theory by van der Pol and Bremmer (1937*a,b*) and in seismology by Scholte (1956). A review is contained in Chapman and Phinney (1972). The application of the slowness method to the spherical system is discussed by Chapman (1978, 1979).

9

Canonical signals

Many seismic signals – *direct rays, partial and total reflections, transmissions, head waves, interface waves, tunnelling waves, caustics* and *Fresnel shadows* – can be described by the inverse transform methods developed in the last chapter. First-motion approximations to the Cagniard or WKBJ methods, or asymptotic spectral methods can be used to investigate these signals. In addition the spectral method can be used for *deep shadow* signals.

In this chapter, we investigate approximate solutions for many canonical signals, e.g. direct rays, partial and total reflections, transmissions, head waves, interface waves, tunnelling waves, caustics, Fresnel and deep shadows. Unfortunately, often these canonical signals do not occur in isolation and for realistic, complicated models it is necessary to use numerical methods as described in the last chapter (Section 8.5.2) for the complete response. With modern computers, these numerical computations are entirely practical and are preferred for forward modelling. Nevertheless, it is worth investigating approximations for canonical signals as it aids an understanding of wave propagation, and for interpreting the results of modelling and real data. While it is rare that the approximate results described here are used for complete, forward modelling, they are frequently employed to check more complete methods, and as an aid to interpretation or in an inverse problem, when complete modelling is too expensive.

The nomenclature for signals, particularly non-geometrical, is not standardized. In this chapter, we denote rays in multi-layered media with a sequence of letters with a subscript for each ray segment, e.g. P_1S_2 is a ray with a P ray segment in layer 1 and an S segment in layer 2 (therefore the source must be in layer 1, receiver in layer 2 and it must have a converted transmission at the interface, $z = z_2$). Head-wave segments are indicated by a lowercase letter, e.g. $P_1p_2S_2$ is the head wave associated with the above ray, with the P velocity from layer 2. Interface waves are indicated by an overbar, e.g. $\bar{S}$ is the Rayleigh wave. The evanescent segment of a tunnelling ray is indicated by a star, e.g. $P_1 P_2^*$ is a ray that tunnels to a receiver in the fast layer 2.

9.1 First-motion approximations using the Cagniard method

In this section, we describe the signals that can be modelled using the Cagniard method and can easily be approximated. If the transformed particle velocity is given by equation (8.0.6), with expressions (8.0.8), (8.0.9) and (8.1.2), then the inverse transforms (8.1.1) can be evaluated exactly to give result (8.1.31). For simplicity and because the Earth is three dimensional, we present these results for the far-field approximation in three dimensions (8.2.68), rather than in two dimensions (8.1.31). In this section we investigate first-motion approximations for this result.

9.1.1 Direct waves and partial reflections from interfaces

In general, it is not possible to solve $p = p(t, \mathbf{x}_{\mathrm{R}})$ explicitly and to substitute in result (8.2.68), and we must resort to numerical calculations. However, it is straightforward to obtain the first-motion approximation for the geometrical arrival. The saddle point is at $p = p_{\mathrm{ray}}$, which solves equation (8.1.18). To lowest order, approximations for the phase and its gradient near the saddle point are

$$\begin{aligned} \widetilde{T}_{\mathrm{ray}}(p, \mathbf{x}_{\mathrm{R}}) &\simeq T_{\mathrm{ray}}(p_{\mathrm{ray}}, z_{\mathrm{R}}) + \frac{1}{2}\frac{\partial^2 \widetilde{T}_{\mathrm{ray}}}{\partial p^2}(p - p_{\mathrm{ray}})^2 \\ &= T_{\mathrm{ray}} - \frac{1}{2}\frac{\mathrm{d}X_{\mathrm{ray}}}{\mathrm{d}p}(p - p_{\mathrm{ray}})^2 \end{aligned} \tag{9.1.1}$$

$$\frac{\partial \widetilde{T}_{\mathrm{ray}}}{\partial p} = x_{\mathrm{R}} - X_{\mathrm{ray}} \simeq -\frac{\mathrm{d}X_{\mathrm{ray}}}{\mathrm{d}p}(p - p_{\mathrm{ray}}). \tag{9.1.2}$$

Setting $\widetilde{T}_{\mathrm{ray}} = t$, in expansion (9.1.1) and solving for $p(t, \mathbf{x}_{\mathrm{R}})$, we have

$$p(t, \mathbf{x}_{\mathrm{R}}) - p_{\mathrm{ray}} \simeq \pm \mathrm{i} \sqrt{\frac{2\left(t - T_{\mathrm{ray}}\right)}{\mathrm{d}X_{\mathrm{ray}}/\mathrm{d}p}}, \tag{9.1.3}$$

and obtain

$$\frac{\partial \widetilde{T}_{\mathrm{ray}}}{\partial p} = \mathrm{i} \sqrt{2 \frac{\mathrm{d}X_{\mathrm{ray}}}{\mathrm{d}p} \left(t - T_{\mathrm{ray}}\right)}, \tag{9.1.4}$$

in the fourth p quadrant. Thus the first-motion approximation to result (8.2.68) is

$$\boxed{\underline{\mathbf{u}}_{\mathrm{direct}}(t, \mathbf{x}_{\mathrm{R}}) \simeq \frac{1}{2\pi} \left(\frac{p_{\mathrm{ray}}}{x_{\mathrm{R}}\left(\mathrm{d}X_{\mathrm{ray}}/\mathrm{d}p\right)} \right)^{1/2} \underline{\mathcal{G}}_{\mathrm{ray}}(p_{\mathrm{ray}})\, \delta(t - T_{\mathrm{ray}}),} \tag{9.1.5}$$

where we have used the simplification $\lambda(t) * \lambda(t) = \pi H(t)$. The geometrical arrival due to the line pressure step source has the form $(t - T_{\rm ray})^{-1/2}$ and due to a point pressure step source is $\delta(t - T_{\rm ray})$. This result (9.1.5) is equivalent to the geometrical ray approximation. We have unit($\underline{\boldsymbol{\mathcal{G}}}$) = [M^{-1}L^{2}T] and expression (9.1.5) agrees with unit($\underline{\mathbf{u}}$) = [M^{-1}T].

Having considered the direct wave, it is straightforward to extend the result to a partially reflected signal. The transformed, reflected acoustic signal is given by

$$\underline{\mathbf{v}}_{P_1P_1}(\omega, p, z_{\rm R}) = \underline{\boldsymbol{\mathcal{G}}}_{P_1P_1}(p)e^{i\omega\tau_{P_1P_1}(p,z_{\rm R})}, \tag{9.1.6}$$

with

$$\underline{\boldsymbol{\mathcal{G}}}_{P_1P_1} = \mathcal{T}_{11}(p)\,\mathbf{g}_{P_1}\mathbf{g}_{P_1}^{\rm T} \tag{9.1.7}$$

$$\tau_{P_1P_1}(p, z_{\rm R}) = q_{\alpha 1}(z_{\rm R} + z_{\rm S} - 2z_2) = q_{\alpha 1}d, \tag{9.1.8}$$

where $\mathcal{T}_{11}$ is the generalized reflection coefficient (6.3.7). We introduce subscripts 1 and 2 for the media $z > z_2$ and $z < z_2$, respectively. The complete response is a sum of the direct and reflected waves (7.2.30), but as the problem is linear, we can invert these separately and sum the results. The form of expressions (9.1.6) and (9.1.7) is exactly as before, and we can proceed with the same technique. The only complication is the reflection coefficient, $\mathcal{T}_{11}$, which introduces other branch points at $p = \pm 1/\alpha_2$. The Cagniard contour leaves the real axis at $p = \sin\theta_1/\alpha_1$. If $\alpha_1 > \alpha_2$, the reflection coefficient, $\mathcal{T}_{11}$, is always real at the saddle point. If $\alpha_2 > \alpha_1$, then provided $\theta_1 < \sin^{-1}(\alpha_1/\alpha_2)$ so $p < 1/\alpha_2$, the reflection coefficient, $\mathcal{T}_{11}$, is real. Then expression (9.1.5) still applies and describes the first-motion approximation of a partial reflection. We consider the case when $\theta_1 > \sin^{-1}(\alpha_1/\alpha_2)$ so $p > 1/\alpha_2$ below (Sections 9.1.3 and 9.1.4).

9.1.2 Partial reflections in stratified media

In Section 8.3.1, the Cagniard method was extended to stratified media using the WKBJ iterative solution (Section 7.2.6). The result (8.3.1) modelled the signal as a depth integral of differential reflectors. For fixed depth, the response is obtained by the Cagniard method evaluating the integrand along the Cagniard contour. For fixed time, the signal is obtained by integrating along a contour that connects equal time points on the Cagniard contours. The first-motion approximation of this result (8.3.1) is easily obtained, and illustrates how velocity gradients cause low-frequency, partial reflections.

For fixed depth, the important contribution comes from the neighbourhood of the saddle point. The arrival time corresponding to the saddle point varies as the depth varies. The complete result (8.3.1) is approximated by its behaviour near the saddle points on the real slowness axis. For each depth, z, in the depth integral, the same approximations (9.1.1)–(9.1.4) are used. Thus in the integrand of expression (8.3.1), we approximate the singular term by (cf. equation (9.1.4))

$$\left(\frac{\partial \widetilde{T}_{\mathrm{ray}}}{\partial p}\right)_z = \mathrm{i}\sqrt{2\left(\frac{\partial X_{\mathrm{ray}}}{\partial p}\right)_z (t - T_{\mathrm{ray}})}, \tag{9.1.9}$$

where $X_{\mathrm{ray}}(p, z_{\mathrm{R}}, z)$ and $T_{\mathrm{ray}}(p, z_{\mathrm{R}}, z)$ are given by equations (8.3.6) and (8.3.7), respectively, and

$$\left(\frac{\partial X_{\mathrm{ray}}}{\partial p}(p, z_{\mathrm{R}}, z)\right)_z = \int_z^{z_{\mathrm{S}}} + \int_z^{z_{\mathrm{R}}} \frac{\mathrm{d}\zeta}{\alpha^2(\zeta)\, q_\alpha^3(p, \zeta)}. \tag{9.1.10}$$

For each depth, z, we can solve equation (8.3.5) for the geometrical ray parameter corresponding to a reflection from that depth, i.e.

$$p = p_{\mathrm{ray}}(\mathbf{x}_{\mathrm{R}}, z), \tag{9.1.11}$$

solves $X_{\mathrm{ray}}(p, z_{\mathrm{R}}, z) = x_{\mathrm{R}}$. From this we can find the geometrical arrival time

$$T_{\mathrm{ray}} = T_{\mathrm{ray}}(p_{\mathrm{ray}}, z_{\mathrm{R}}, z) = \widetilde{T}_{\mathrm{ray}}(p_{\mathrm{ray}}, \mathbf{x}_{\mathrm{R}}, z). \tag{9.1.12}$$

Substituting approximation (9.1.9) in expression (8.3.1), we obtain the first-motion approximation

$$\underline{\mathbf{u}}_{\mathrm{stratified}}(t, \mathbf{x}_{\mathrm{R}}) \simeq \frac{1}{2\pi x_{\mathrm{R}}^{1/2}} \int \left(\frac{p^{1/2} \underline{\boldsymbol{\mathcal{G}}}_{\mathrm{ray}}(p)\, \gamma_P(p, z)}{\left(\partial X_{\mathrm{ray}}/\partial p\right)_z} \,\delta\left(t - T_{\mathrm{ray}}(p, z_{\mathrm{R}}, z)\right)\right)_{p = p_{\mathrm{ray}}(\mathbf{x}_{\mathrm{R}}, z)} \mathrm{d}z. \tag{9.1.13}$$

The range of the depth integral is restricted to depths for which the saddle-point approximations are valid, and for regions where the coupling coefficient, $\gamma_P(p, z)$, is significant. For times that are appropriate to reflections from the significant regions of $\gamma_P(p, z)$, the delta function will be non-zero, i.e. for a given time t, there will be a depth z which solves

$$T_{\mathrm{ray}}(p, z_{\mathrm{R}}, z)\big|_{p = p_{\mathrm{ray}}(\mathbf{x}_{\mathrm{R}}, z)} = t. \tag{9.1.14}$$

In order to calculate the depth integral, we need the derivative

$$\left.\frac{\mathrm{d}T_{\mathrm{ray}}}{\mathrm{d}z}\right|_{p = p_{\mathrm{ray}}(\mathbf{x}_{\mathrm{R}}, z)} = \left(\frac{\partial T_{\mathrm{ray}}}{\partial z}\right)_p + \left(\frac{\partial T_{\mathrm{ray}}}{\partial p}\right)_z \frac{\mathrm{d}p_{\mathrm{ray}}}{\mathrm{d}z}, \tag{9.1.15}$$

where the final term arises as $p_{\rm ray}$ changes with the depth z. This derivative can be obtained by differentiating the equation for $p_{\rm ray}$, i.e. equation (8.3.5)

$$0 = \left(\frac{\partial X_{\rm ray}}{\partial p}\right)_z \frac{{\rm d}p_{\rm ray}}{{\rm d}z} + \left(\frac{\partial X_{\rm ray}}{\partial z}\right)_p . \tag{9.1.16}$$

Combining (9.1.16) with (9.1.15), we obtain the simple result

$$\left.\frac{{\rm d}T_{\rm ray}}{{\rm d}z}\right|_{p=p_{\rm ray}(\mathbf{x}_{\rm R},z)} = -2q_\alpha(p,z), \tag{9.1.17}$$

where we have used the partial derivatives of expressions (8.3.6) and (8.3.7)

$$\left(\frac{\partial T_{\rm ray}}{\partial z}\right)_p = -\frac{2}{\alpha^2(z)\, q_\alpha(p,z)} \tag{9.1.18}$$

$$\left(\frac{\partial X_{\rm ray}}{\partial z}\right)_p = -\frac{2p}{q_\alpha(p,z)} \tag{9.1.19}$$

$$\left(\frac{\partial T_{\rm ray}}{\partial p}\right)_z = p\left(\frac{\partial X_{\rm ray}}{\partial p}\right)_z . \tag{9.1.20}$$

Using this derivative (9.1.17), the integral of the first-motion approximation to result (8.3.5), equation (9.1.13), can be evaluated to give

$$\boxed{\underline{\mathbf{u}}_{\rm stratified}(t,\mathbf{x}_{\rm R}) \simeq \frac{1}{4\pi x_{\rm R}^{1/2}} \left(\frac{p^{1/2}\underline{\boldsymbol{\mathcal{G}}}_{\rm ray}(p)\,\gamma_P(p,z)}{\left(\partial X_{\rm ray}/\partial p\right)_z\, q_\alpha(p,z)}\right)_{\substack{T_{\rm ray}(p,z_{\rm R},z)=t\\ X_{\rm ray}(p,z_{\rm R},z)=x}} .} \tag{9.1.21}$$

For a given time, t, we simultaneously solve

$$T_{\rm ray}(p,z_{\rm R},z) = t \tag{9.1.22}$$

$$X_{\rm ray}(p,z_{\rm R},z) = x_{\rm R}, \tag{9.1.23}$$

for the reflection depth, z, and the geometrical ray parameter, $p = p_{\rm ray}$, and substitute the appropriate values in the expression on the right-hand side of expression (9.1.21). Through the simultaneous solutions of equations (9.1.22) and (9.1.23), the solutions for p and z, and hence the right-hand side of expression (9.1.21), are functions of time, t.

In this expression (9.1.21) all the terms vary with depth. However, in many circumstances, the support of γ_P is restricted to a small range of depths, e.g. a narrow transition between two homogeneous layers. In these circumstances, only γ_P varies rapidly, involving local depth derivatives of material properties, and all the other factors are approximately constant. Then we can simplify expression (9.1.21) to

$$\underline{\mathbf{u}}_{\text{stratified}}(t, \mathbf{x}_\mathsf{R}) \simeq \frac{1}{4\pi x_\mathsf{R}^{1/2}} \left(\frac{p^{1/2} \underline{\boldsymbol{\mathcal{G}}}_{\text{ray}}(p)}{(\partial X_{\text{ray}}/\partial p)_z \, q_\alpha(p, z)} \right)_{\substack{p = \bar{p} \\ z = \bar{z}}} \times \Big(\gamma_P(p, z) \Big)_{\substack{T_{\text{ray}}(p, z_\mathsf{R}, z) = t \\ X_{\text{ray}}(p, z_\mathsf{R}, z) = x_\mathsf{R}}}, \tag{9.1.24}$$

where $\bar{p}$ and $\bar{z}$ are mean values appropriate to the transition zone, i.e.

$$x_\mathsf{R} = X_{\text{ray}}(\bar{p}, z_\mathsf{R}, \bar{z}). \tag{9.1.25}$$

The first term is taken constant with the mean values, whereas the solutions of equations (9.1.22) and (9.1.23) are used in the second term giving the time variation with depth. Normally the change in the geometrical ray parameter, p_{ray}, through the reflecting zone is small and we can use the mean slowness value in the second term. Then the approximate response is

$$\underline{\mathbf{u}}_{\text{stratified}}(t, \mathbf{x}_\mathsf{R}) \simeq \frac{1}{4\pi x_\mathsf{R}^{1/2}} \left(\frac{p^{1/2} \underline{\boldsymbol{\mathcal{G}}}_{\text{ray}}(p)}{(\partial X_{\text{ray}}/\partial p)_z \, q_\alpha(p, z)} \right)_{\substack{p = \bar{p} \\ z = \bar{z}}} \times \Big(\gamma_P(\bar{p}, z) \Big)_{T_{\text{ray}}(\bar{p}, z_\mathsf{R}, z) + \bar{p}(x_\mathsf{R} - X_{\text{ray}}(\bar{p}, z_\mathsf{R}, z)) = t}, \tag{9.1.26}$$

where we have used the plane-wave approximation (i.e. slowness $p = \bar{p}$ is constant) for the time.

This expression (9.1.26) for the partial reflections from a heterogeneous zone is sometimes called the *convolution model* as the response for a general source function is obtained by a convolution with the time function

$$\Big(\gamma_P(\bar{p}, z) \Big)_{T_{\text{ray}}(\bar{p}, z_\mathsf{R}, z) + \bar{p}(x_\mathsf{R} - X_{\text{ray}}(\bar{p}, z_\mathsf{R}, z)) = t}, \tag{9.1.27}$$

which maps depth functions to time, where locally in the heterogeneous zone we have the plane-wave approximation (i.e. slowness $p = \bar{p}$ is constant). Roughly, the time function is proportional to depth derivatives of the model properties. This is in agreement with asymptotic ray theory, where we saw that discontinuities in the model caused impulsive (delta function) reflections (9.1.5), whereas discontinuities in the model gradient caused time-integrated (step function) reflections (e.g. result (7.2.118) for the reflection coefficient).

9.1.2.1 The 'thin' interface limit

As discussed above, the reflection from a thin heterogeneous zone can be modelled using the WKBJ iterative solution. The reflected signal maps the coupling coefficient, γ_P, into the time domain. The reflected signal is frequency dependent and depends on this mapping. If the zone is thick, then the reflections will be lower frequency. If the heterogeneous zone is thin, then the reflection will be more delta-like. It is interesting to ask how this signal approximates the reflection from an interface, as the zone becomes infinitesimally thin (Thomson and Chapman, 1984). As the zone becomes thinner, we might expect that multiple reflections in the transition zone would become important. Although we will only investigate the behaviour of the acoustic, plane wave solution – a very simple model – the results are instructive for more general scattering problems, e.g. the generalized Born scattering method in three dimensions (Chapter 10, Section 10.3).

In Section 7.2.6, we showed that provided the coupling coefficient, γ_P, is bounded, the series (7.2.125) converges. Throughout this section we assume the plane-wave approximation, i.e. slowness $p = \bar{p}$ is constant, and for brevity, we omit it from our expressions. Let us define the admittance

$$Y(z) = \frac{q_\alpha(\bar{p}, z)}{\rho(z)}. \tag{9.1.28}$$

Then the coupling coefficient (7.2.97) can be written as

$$\gamma_A(z) = \frac{\mathrm{d}}{\mathrm{d}z}\Gamma(z), \tag{9.1.29}$$

where

$$\Gamma(z) = \ln Y^{1/2}. \tag{9.1.30}$$

In the limit of a thin layer we need the saltus of Γ across the layer, i.e.

$$[\Gamma] = \Gamma(\bar{z}_+) - \Gamma(\bar{z}_-), \tag{9.1.31}$$

where in general we indicate properties just above the thin layer by a subscript $_+$ and just below by a subscript $_-$. The first iteration (7.2.125) for the reflected wave at $\bar{z}_+$ is

$$r_1^{(1)} = \int_{-\infty}^{\bar{z}_+} \frac{\mathrm{d}}{\mathrm{d}z_1}\Gamma(z_1)\,\mathrm{d}z_1 = \int_{\Gamma_-}^{\Gamma_+} d\Gamma_1 = [\Gamma]. \tag{9.1.32}$$

We have assumed that the zone is thin enough that the phase difference across it is not important. In making the change of variable from z_1 to Γ_1, we have assumed that Γ is monotonic, but the result holds even if this is not true.

Obviously this is an approximation for the reflection coefficient from a discontinuity (6.3.7) as

$$[\Gamma] = \frac{1}{2}\ln\frac{Y_+}{Y_-} \simeq \frac{Y_+ - Y_-}{2Y_-} \simeq \mathcal{T}_{11}, \tag{9.1.33}$$

using the expansion for $\ln(1 + x)$.

The second-order iteration generates down-going waves and does not contribute to the reflection coefficient. The third-order iteration contributes to the reflection coefficient. It is

$$r_1^{(3)} = -\int_{\bar{z}_-}^{\bar{z}_+} \frac{d\Gamma_1}{dz_1} \int_{z_1}^{\bar{z}_+} \frac{d\Gamma_2}{dz_2} \int_{\bar{z}_-}^{z_2} \frac{d\Gamma_3}{dz_3} dz_3\, dz_2\, dz_1$$

$$= -\int_{\Gamma_-}^{\Gamma_+} d\Gamma_1 \int_{\Gamma_1}^{\Gamma_+} d\Gamma_2 \int_{\Gamma_-}^{\Gamma_2} d\Gamma_3 \tag{9.1.34}$$

$$= -\frac{1}{3}\,[\Gamma]^3, \tag{9.1.35}$$

where the final result is obtained by simple algebra. Proceeding in a similar manner, the fifth-order iteration is

$$r_1^{(5)} = \frac{2}{15}\,[\Gamma]^5, \tag{9.1.36}$$

after some algebra. We reiterate that the approximations, (9.1.32), (9.1.35) and (9.1.36), assume that the reflecting zone is thin enough that the phase terms in the depth integrals (7.2.125) do not vary significantly in the integrals and can be treated as constant, i.e. excluded from the integrals (9.1.32) or (9.1.34).

This procedure can be continued indefinitely and we obtain

$$r_1 = [\Gamma] - \frac{1}{3}\,[\Gamma]^3 + \frac{2}{15}\,[\Gamma]^5 - \cdots$$

$$= \tanh\,[\Gamma] = \frac{e^{2[\Gamma]} - 1}{e^{2[\Gamma]} + 1} = \frac{Y_+ - Y_-}{Y_+ + Y_-} = \mathcal{T}_{11}, \tag{9.1.37}$$

exactly the reflection coefficient (6.3.7) from an interface. In Appendix A.1 it is shown that the general odd-multiple integral of the form (9.1.34) is a term in the expansion of the hyperbolic tangent function (A.1.24) as used in expression (9.1.37).

The even-order iterations give the transmitted wave. Without going into details we obtain

$$r_2 = 1 - \frac{1}{2}\,[\Gamma]^2 + \frac{5}{24}\,[\Gamma]^4 - \cdots$$

$$= \operatorname{sech}\,[\Gamma] = \frac{2}{e^{2[\Gamma]} + 1} = \frac{2(Y_+Y_-)^{1/2}}{Y_+ + Y_-} = \mathcal{T}_{21}, \tag{9.1.38}$$

agreeing with the interface transmission coefficient (6.3.8). Again in Appendix A.1 it is shown that the general even-multiple integral of the form (9.1.34) is a term in the expansion of the hyperbolic secant function (A.1.19).

Thus provided the singularity of γ_A is handled correctly, the coupling integral can be evaluated for thin heterogeneous zones and converges to the correct reflection or transmission coefficient. The rate of convergence is easily established from the expansions of the hyperbolic tangent or secant functions. Although we have not generalized this method to the elastic case, as the coupling coefficients are more complicated and cannot be written as a perfect differential (cf. result (9.1.29)), it indicates the rate of convergence that might be expected.

Although this result is simple and straightforward, it is slightly more subtle than it first appears. Although the iteration series were recognized as expansions of the hyperbolic tangent and secant, it is known that these series only converge for

$$||[\Gamma]|| < \frac{\pi}{2} \tag{9.1.39}$$

(due to the poles at $\pm i\pi/2$ of the trigonometrical functions). If the heterogeneous zone is thin, but not infinitesimal, this seems to contradict the proof in Section 7.2.6, that the series (7.2.125) converges provided the coupling coefficient, γ_A, is bounded. However, we have made the assumption that the phase across the thin layer can be neglected. If the layer has finite thickness, this is not true. Higher-order iterations will be spread out in the time domain over an interval $n\,[\tau]$, where $[\tau]$ is the phase across the layer, i.e.

$$[\tau] = \int_{\bar{z}_-}^{\bar{z}_+} q_\alpha(\bar{p}, z)\,\mathrm{d}z, \tag{9.1.40}$$

and n is the iteration order. If we are calculating the band-limited response of a thin layer, where $\pi/\Delta t$ is the Nyquist frequency, and Δt the time sampling interval, then for iterations with $n\,[\tau] < \Delta t$, the reflected signals lie within one sample interval and the iterations converge as the terms in the expansion of the hyperbolic tangent or secant functions. For n large, the iterations converge as the signals are spread over $n\,[\tau]\,/\Delta t$ samples.

9.1.3 Head waves

Having considered the direct wave and partial reflections, it is straightforward to extend the result to a totally reflected signal. The transformed, reflected acoustic signal is given by expression (9.1.6) including the interface reflection coefficient (9.1.7). An interesting phenomenon occurs if the second medium is faster than the first ($\alpha_2 > \alpha_1$). As the Cagniard contour leaves the real p axis at $p = \sin\theta_1/\alpha_1$, this may be to the right of the branch point $p = 1/\alpha_2$. In this case, when $\theta_1 > \sin^{-1}(\alpha_1/\alpha_2)$, the Cagniard contour has to loop around the branch point (Figure 9.1). Expression (8.2.68) remains valid for the response, provided the part of the Cagniard contour on the real axis is included in $p(t, \mathbf{x}_\mathrm{R})$. For the part of the Cagniard contour on the real axis between $p = 1/\alpha_2$ and $p = \sin\theta_1/\alpha_1$, the phase function, $\widetilde{T}_{P_1P_1}$, is real (as required), but the reflection coefficient, $\mathcal{T}_{11}$, is complex.

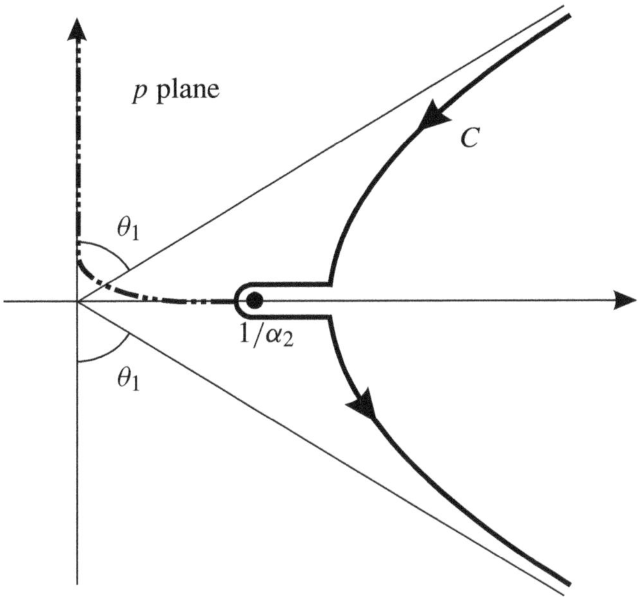

Fig. 9.1. The Cagniard contour C looping around the branch point at $p = 1/\alpha_2$, when $\theta_1 > \sin^{-1}(\alpha_1/\alpha_2)$. The dot-dashed line indicates the branch cut running from the branch point along the real axis to $p = 0$ and then along the positive imaginary axis to $p = \mathrm{i}\infty$.

Expression (8.2.68) remains causal, but instead of starting at the geometrical arrival, $T_{P_1P_1}(\sin\theta_1/\alpha_1)$, it starts where $p = 1/\alpha_2$ at

$$\begin{aligned} \widetilde{T}_{P_1P_1}(1/\alpha_2, \mathbf{x}_\mathrm{R}) &= \frac{x_\mathrm{R}}{\alpha_2} + \left(\frac{1}{\alpha_1^2} - \frac{1}{\alpha_2^2}\right)^{1/2} d \\ &= T_{P_1P_1}(1/\alpha_2, z_\mathrm{R}) + \frac{\ell_{P_1p_2P_1}(\mathbf{x}_\mathrm{R})}{\alpha_2} \\ &= T_{P_1p_2P_1}(\mathbf{x}_\mathrm{R}), \end{aligned} \tag{9.1.41}$$

say, where

$$\ell_{P_1p_2P_1}(\mathbf{x}_\mathrm{R}) = x_\mathrm{R} - X_{P_1P_1}(1/\alpha_2, z_\mathrm{R}). \tag{9.1.42}$$

From Chapter 2, we recognize (9.1.41) as being the arrival time of the head wave, and expression (9.1.42) as being the length of the head-wave segment. As $\theta_1 > \sin^{-1}(\alpha_1/\alpha_2)$, $\ell_{P_1p_2P_1} > 0$. In the ray notation $P_1p_2P_1$, the lowercase indicates the head-wave segment in the second medium.

We can now investigate the response near the head-wave arrival time (9.1.41). Near the branch cut we have

$$\widetilde{T}_{P_1P_1}(p,\mathbf{x}_R) \simeq T_{P_1p_2P_1}(\mathbf{x}_R) + \frac{\partial \widetilde{T}_{P_1P_1}(1/\alpha_2,\mathbf{x}_R)}{\partial p}(p-1/\alpha_2), \quad (9.1.43)$$

where

$$\frac{\partial \widetilde{T}_{P_1P_1}(1/\alpha_2,\mathbf{x}_R)}{\partial p} = x_R - X_{P_1P_1}(1/\alpha_2, z_R) = \ell_{P_1p_2P_1}. \quad (9.1.44)$$

The reflection coefficient is

$$\mathcal{T}_{11}(p) \simeq \mathcal{T}_{11}(1/\alpha_2) + \frac{\partial \mathcal{T}_{11}}{\partial q_{\alpha 2}} q_{\alpha 2} = 1 - \frac{2\rho_1}{\rho_2(1/\alpha_1^2 - 1/\alpha_2^2)^{1/2}} q_{\alpha 2}, \quad (9.1.45)$$

with

$$q_{\alpha 2} \simeq \mathrm{i}\left(\frac{2}{\alpha_2}\right)^{1/2}(p-1/\alpha_2)^{1/2}, \quad (9.1.46)$$

below the branch cut in the fourth p quadrant. Setting $\widetilde{T}_{P_1P_1}(p,\mathbf{x}_R) = t$ in approximation (9.1.43), we can solve for $p(t,\mathbf{x}_R)$ around the branch point, and substitute in expression (9.1.45) using approximation (9.1.46), to obtain the first-motion approximation for result (8.2.68)

$$\underline{\mathbf{u}}_{P_1p_2P_1}(t,\mathbf{x}_R) \simeq \mathbf{g}_{P_1}\mathbf{g}_{P_1}^{\mathrm{T}} \frac{\rho_1}{\pi\rho_2(1/\alpha_1^2 - 1/\alpha_2^2)^{1/2}\alpha_2 \ell_{P_1p_2P_1}^{3/2} x_R^{1/2}} H\left(t - T_{P_1p_2P_1}\right). \quad (9.1.47)$$

Notice that this has the form of the integral of the direct pulse, and decays as $\ell_{P_1p_2P_1}^{-3/2} x_R^{-1/2}$. It will be lower frequency and decay rapidly with range. We can confirm that the result (9.1.47) agrees with unit$(\underline{\mathbf{u}}_{P_1p_2P_1}) = [\mathrm{M}^{-1}\mathrm{T}]$.

9.1.3.1 General result for head waves

In general $\underline{\mathcal{G}}_{\mathrm{ray}}$ may contain the product of many reflection/transmission coefficients. These will contain vertical slownesses for all media contributing to the coefficients. If a vertical slowness is also present in the intercept function, τ_{ray}, then it will not cause a head wave. Vertical slownesses that are not in the intercept function, τ_{ray}, but are in the reflection/transmission coefficients in $\underline{\mathcal{G}}_{\mathrm{ray}}$, may cause head waves. Any velocities that are larger than the maximum velocity in the intercept function will cause head waves if the range is beyond the corresponding critical ray. In elastic models with S segments, P head waves can exist for transmissions, as well as P and S head waves for reflections. It is left to Exercise 9.1 to investigate the possibilities at various interfaces. Suppose one such velocity is c_n (in an elastic model it may be a P or S velocity). Expanding near the corresponding branch point at $p = p_n = 1/c_n$, we have

$$p(t, \mathbf{x}_\mathrm{R}) \simeq p_n + \frac{t - \widetilde{T}_{\mathrm{ray}}(p_n, \mathbf{x}_\mathrm{R})}{x_\mathrm{R} - X_{\mathrm{ray}}(p_n, z_\mathrm{R})} = p_n + \frac{t - T_n(\mathbf{x}_\mathrm{R})}{\ell_n(\mathbf{x}_\mathrm{R})}, \tag{9.1.48}$$

and

$$\underline{\mathcal{G}}_{\mathrm{ray}}(p) \simeq \underline{\mathcal{G}}_{\mathrm{ray}}(1/c_n) + \frac{\partial \underline{\mathcal{G}}_{\mathrm{ray}}(p_n)}{\partial q_n} q_n, \tag{9.1.49}$$

where q_n is the corresponding vertical slowness,

$$\ell_n(\mathbf{x}_\mathrm{R}) = x_\mathrm{R} - X_{\mathrm{ray}}(p_n, z_\mathrm{R}), \tag{9.1.50}$$

the length of the head wave, and

$$T_n(\mathbf{x}_\mathrm{R}) = \widetilde{T}_{\mathrm{ray}}(p_n, \mathbf{x}_\mathrm{R}), \tag{9.1.51}$$

its arrival time. Combining these results, the first-motion approximation for this head wave in (8.2.68) is

$$\boxed{\underline{\mathbf{u}}_{\mathrm{head}}(t, \mathbf{x}_\mathrm{R}) \simeq -\frac{1}{2\pi c_n \ell_n^{3/2} x_\mathrm{R}^{1/2}} \mathrm{Re}\left(\frac{\partial \underline{\mathcal{G}}_{\mathrm{ray}}(p_n)}{\partial q_n} H(t - T_n)\right).} \tag{9.1.52}$$

Notice that if the head wave is not the first arrival, then $\partial \underline{\mathcal{G}}_{\mathrm{ray}}(p_n)/\partial q_n$ may be complex and the simple head-wave pulse (9.1.52) is phase shifted. $H(t - T_n)$ is interpreted as an analytic time series, the integral of $\Delta(t - T_n)$. Again we can check that unit($\underline{\mathbf{u}}_{\mathrm{head}}$) = $[\mathrm{M}^{-1}\mathrm{T}]$.

It is also interesting to note that if a head wave exists for a reverberation, where a reflection coefficient enters $\underline{\mathcal{G}}_{\mathrm{ray}}$ more than once, then the corresponding head wave contains this multiplicity (see Figure 9.2). A head wave can exist at either reflection point, but not at both simultaneously as the plane wavefront from the first head wave does not generate a head wave at the second reflection. A head wave is only generated when a curved wavefront intersects an interface (see Figure 2.7).

As an illustration of the head wave and its approximation, we consider a reflection in a simple acoustic model (Figure 9.3). As the head wave is only generated at larger ranges, we have used the far-field approximation for the three-dimensional response (cf. Figure 8.6). The source is an explosive, point source with time function $P_\mathrm{S}(t) = P_\mathrm{S} t H(t)$. The first-motion approximation for the head wave (9.1.47)

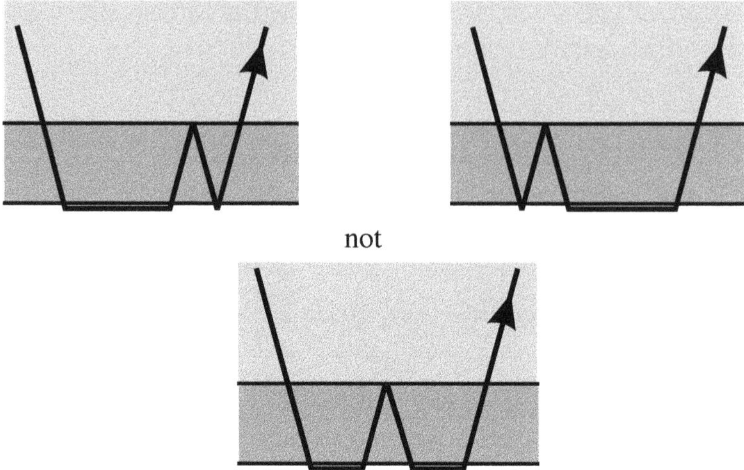

Fig. 9.2. Head waves from a ray reflecting twice – head waves can be generated at either reflection point (upper diagrams) but not at both (lower diagram).

is of the same form. More details of the response at one range ($x_R = 6$) are shown in Figure 9.4 with the first-motion approximation (9.1.47) plotted as a dashed line.

9.1.4 Total reflections

When a head wave exists, the reflection coefficient is complex at the saddle point. Therefore the first-motion result (9.1.5) must be generalized. Near the saddle point, expansion (9.1.1) still applies, but expression (9.1.3) must be generalized and also used on the real axis

$$p - p_{\mathrm{ray}} \simeq -\sqrt{\frac{2\left(T_{\mathrm{ray}} - t\right)}{\mathrm{d}X_{\mathrm{ray}}/\mathrm{d}p}}, \tag{9.1.53}$$

where expression (9.1.4) becomes

$$\frac{\partial \widetilde{T}_{\mathrm{ray}}}{\partial p} = \sqrt{2\frac{\mathrm{d}X_{\mathrm{ray}}}{\mathrm{d}p}\left(T_{\mathrm{ray}} - t\right)}. \tag{9.1.54}$$

This gives the first motion for the geometrical arrival as

$$\boxed{\underline{\mathbf{u}}_{\mathrm{total}}(t, \mathbf{x}_R) \simeq \frac{1}{2\pi}\left(\frac{p_{\mathrm{ray}}}{x_R\left(\mathrm{d}X_{\mathrm{ray}}/\mathrm{d}p\right)}\right)^{1/2} \mathrm{Re}\left(\underline{\mathcal{G}}_{\mathrm{ray}}(p_{\mathrm{ray}})\Delta(t - T_{\mathrm{ray}})\right),} \tag{9.1.55}$$

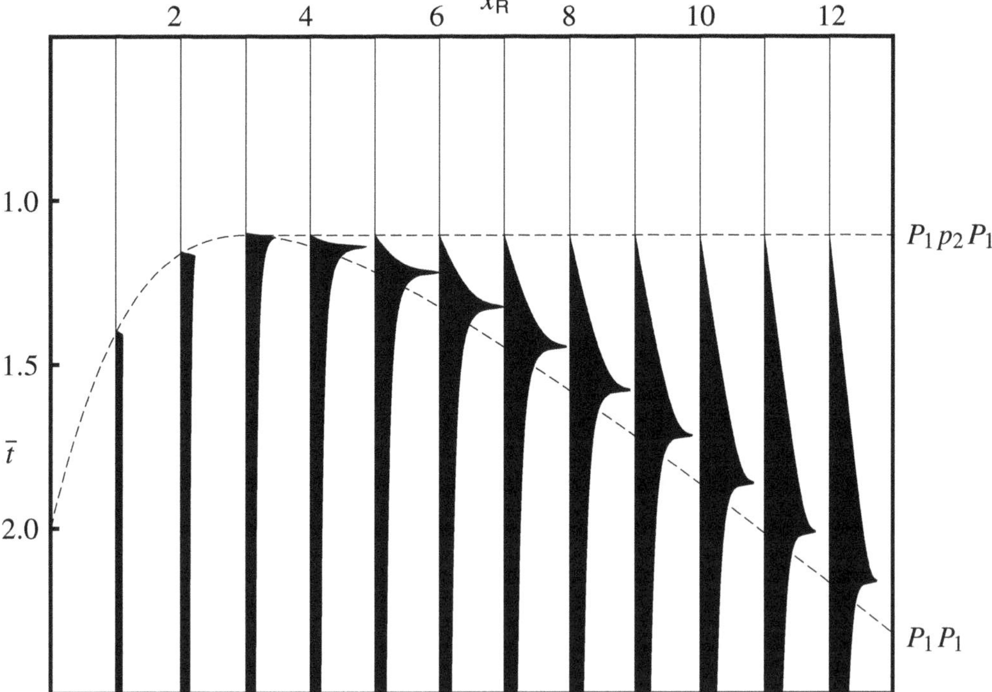

Fig. 9.3. The reflections in an acoustic model with two homogeneous half-spaces. The model parameters are normalized so $\alpha_1 = 1$, $\rho_1 = 1$, $\alpha_2 = 1.2$ and $\rho_2 = 1.1$. The source and receiver are unit distance from the interface, i.e. $|z_S - z_2| = |z_R - z_2| = 1$, and the ranges are from $x_R = 1$ to 12 in unit steps. The time axis is the reduced time $\bar{t} = t - x_R/\alpha_2$. The vertical component of displacement, u_z, for an explosive, point source with time function $P_S(t) = P_S t H(t)$ is plotted multiplied by the range x_R. The arrival times of the reflection P_1P_1 and head wave $P_1p_2P_1$ are plotted with a dashed line.

where the analytic time series $\Delta(t)$ is defined in expression (B.1.8). The complex reflection coefficient causes part of the signal to be the Hilbert transform of the direct wave. Thus expression (9.1.55) can be expanded as

$$\underline{\mathbf{u}}_{\text{total}}(t, \mathbf{x}_R) \simeq \frac{1}{2\pi}\left(\frac{p_{\text{ray}}}{x_R\,(\mathrm{d}X_{\text{ray}}/\mathrm{d}p)}\right)^{1/2} \times \left(\operatorname{Re}\left(\underline{\boldsymbol{\mathcal{G}}}_{\text{ray}}(p_{\text{ray}})\right)\delta(t - T_{\text{ray}}) - \operatorname{Im}\left(\underline{\boldsymbol{\mathcal{G}}}_{\text{ray}}(p_{\text{ray}})\right)\bar{\delta}(t - T_{\text{ray}})\right), \tag{9.1.56}$$

where $\operatorname{Re}\left(\underline{\boldsymbol{\mathcal{G}}}_{\text{ray}}(p_{\text{ray}})\right)$ causes the in-phase signal, and $-\operatorname{Im}\left(\underline{\boldsymbol{\mathcal{G}}}_{\text{ray}}(p_{\text{ray}})\right)$ the Hilbert transformed signal.

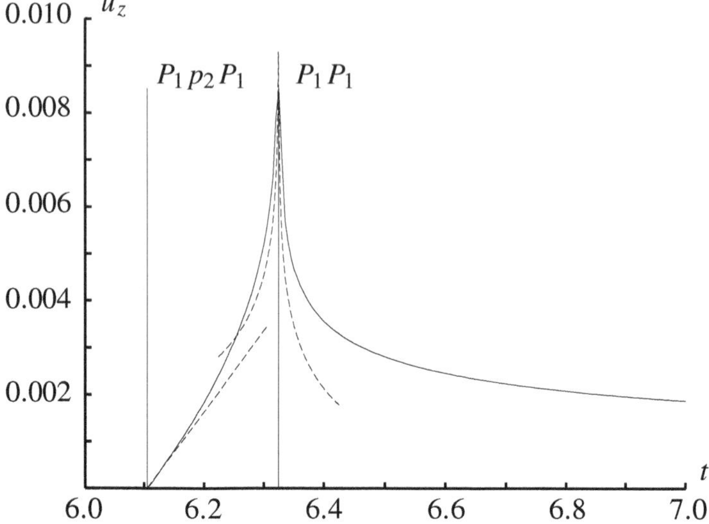

Fig. 9.4. The exact and approximate head wave and total reflection at $x_{\mathrm{R}} = 6$. The model parameters are as in Figure 9.3. The head-wave approximation (9.1.47) and total reflection approximation (9.1.56) for the source time function $P_{\mathrm{S}}(t) = P_{\mathrm{S}} t H(t)$ are plotted as dashed lines – the head-wave has the same form as the source, the in-phase part of the total reflection is of the form $H(t)$ and the out-phase signal $-\ln(|t|)/\pi$. The arrival times of the head wave, $P_1 p_2 P_1$, and the total reflection, $P_1 P_1$, are marked with thin vertical lines.

Figure 9.3 illustrates total reflections from a simple acoustic interface, and Figure 9.4 compares the complete far-field response with the first-motion approximation (9.1.56). Note that the latter is not a very good approximation as the reflection coefficient varies fairly rapidly near the saddle point.

9.1.5 Interface waves

Having investigated the signals due to saddle points and branch cuts, in this section we investigate the signals generated by poles of $\underline{\boldsymbol{\mathcal{G}}}_{\mathrm{ray}}$.

9.1.5.1 Acoustic waves

The reflection/transmission coefficients for acoustic waves have the denominator

$$\Delta = \rho_2 q_{\alpha 1} + \rho_1 q_{\alpha_2}. \tag{9.1.57}$$

Because the signs of $q_{\alpha 1}$ and $q_{\alpha 2}$ are the same on the physical Riemann sheet, it is impossible for Δ to be zero or for the coefficient to have a pole. Poles do exist on the lower Riemann sheets, $(+-)$ or $(-+)$, indicating the signs of the roots q_1 and q_2, when

$$p = \left(\frac{\rho_2^2/\alpha_1^2 - \rho_1^2/\alpha_2^2}{\rho_2^2 - \rho_1^2} \right)^{1/2}, \tag{9.1.58}$$

either on the real or imaginary axes. At a free, acoustic surface there are no poles. Because the poles are nowhere near the Cagniard contour, they do not cause arrivals directly.

9.1.5.2 *Elastic waves at a free surface – the Rayleigh wave*

At a free surface in an isotropic, elastic medium (Section 6.4.3), the reflection coefficients have the denominator (6.4.6)

$$\boxed{\Delta^{PV} = 4p^2 q_\alpha q_\beta + \Omega^2,} \tag{9.1.59}$$

where we have dropped the medium 2 subscript. For $p > 1/\beta > 1/\alpha$, both q_α and q_β are imaginary and there is a possibility that $\Delta^{PV} = 0$ on the physical Riemann sheet.

Letting $y = p^2\beta^2$ and $u = \beta^2/\alpha^2$ in the equation $\Delta^{PV} = 0$, it can be reduced to the cubic polynomial

$$f(y) = (1-u)y^3 + (u - 3/2)y^2 + y/2 - 1/16 = 0. \tag{9.1.60}$$

Notice that by squaring $q_\alpha q_\beta$ we have included roots that may be on lower Riemann sheets. The y^4 term cancels in (9.1.60) so we only have three roots. Although u is a convenient dimensionless parameter to use in equation (9.1.60), Poisson's ratio is an alternative commonly used to characterize elastic media. Poisson's ration, ν, is defined in terms of the ratio of the lateral contraction to longitudinal extension of a wire (Fung, 1965, p. 129 and Exercise 4.1*c*). It is related to our parameter by

$$\nu = \frac{1-2u}{2(1-u)} \quad \text{or} \quad u = \frac{1-2\nu}{2(1-\nu)}. \tag{9.1.61}$$

The parameter u is in the range $0 < u < 1/2$ corresponding to Poisson's ratio, ν, varying from 1/2 to 0. Clearly $f(1) < 0$ and $f(y) > 0$ as $y \to \infty$, so a root must exist for $1 < y < \infty$. This root is with $p > 1/\beta > 1/\alpha$ so both q_α and q_β are imaginary so it will satisfy the original equation. This root is called the *Rayleigh wave* and we denote its velocity, $1/p$, by γ. Let us solve equation (9.1.60) explicitly for three values of u.

For **Poisson's ratio** $\nu = \mathbf{0}$, $u = 1/2$ and (9.1.60) becomes

$$y^3 - 2y^2 + y - \frac{1}{8} = \left(y - \frac{3}{4} - \frac{\sqrt{5}}{4}\right)\left(y - \frac{3}{4} + \frac{\sqrt{5}}{4}\right)\left(y - \frac{1}{2}\right) = 0. \tag{9.1.62}$$

The root we are immediately interested in is $y = 3/4 + \sqrt{5}/4$, which converts to $p = 1.1441/\beta$ for the slowness, or

$$\gamma = 0.87403\beta, \tag{9.1.63}$$

for the velocity.

For **Poisson's ratio** $\nu = \mathbf{1/4}$, a Poisson solid with $\lambda = \mu$, $u = 1/3$ and equation (9.1.60) becomes

$$\frac{2}{3}y^3 - \left(\frac{1}{3} - \frac{3}{2}\right)y^2 + \frac{1}{2}y - \frac{1}{16} = \frac{2}{3}\left(y - \frac{3}{4} - \frac{\sqrt{3}}{4}\right)\left(y - \frac{3}{4} + \frac{\sqrt{3}}{4}\right)\left(y - \frac{1}{4}\right) = 0. \tag{9.1.64}$$

The root we are immediately interested in is $y = 3/4 + \sqrt{3}/4$, which converts to $p = 1.0877/\beta$ for the slowness, or

$$\gamma = 0.91940\beta, \tag{9.1.65}$$

for the velocity.

Finally for **Poisson's ratio** $\nu = \mathbf{1/2}$, a fluid or incompressible solid, $u = 0$ and

$$y^3 - \frac{3}{2}y^2 + \frac{1}{2}y - \frac{1}{16} = (y - 1.09575)\left(y^2 - 0.40425y + 0.057039\right) = 0 \tag{9.1.66}$$

(the numerical factorization is approximate). The root we are immediately interested in is $y = 1.09575$, which converts to $p = 1.0468/\beta$ for the slowness, or

$$\gamma = 0.95531\beta, \tag{9.1.67}$$

for the velocity.

The monotonic behaviour of γ ($\mathrm{d}\gamma/\mathrm{d}\nu > 0$) is easily confirmed (numerically it is easy to find the root as it is close to $y = 1$) (Figure 9.5).

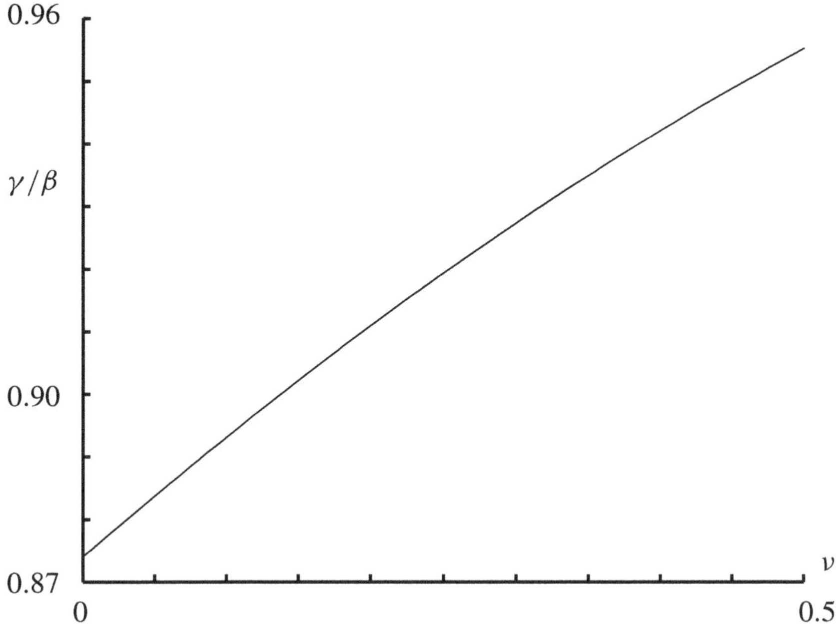

Fig. 9.5. The behaviour of the Rayleigh wave velocity, γ, as a function of the Poisson's ratio, ν.

If we consider the *SS* reflected signal, the far-field, three-dimensional Cagniard solution will be (results (8.2.68) and (8.0.8))

$$\underline{\mathbf{u}}_{SS}(t, \mathbf{x}_R) \simeq -\frac{1}{\pi^2 (2x_R)^{1/2}} \frac{\mathrm{d}}{\mathrm{d}t} \lambda(t) * \mathrm{Im} \left(p^{1/2} \mathcal{T}_{44}(p) \, \mathbf{g}_S \, \mathbf{g}_S^{\mathrm{T}} \frac{\partial p}{\partial \widetilde{T}_{SS}} \right)_{p=p(t,\mathbf{x}_R)} . \tag{9.1.68}$$

Cagniard contours for the *SS* reflected signal are illustrated in Figure 9.6.

We have already investigated the totally reflected *SS* reflection due to the saddle point at $p = \sin\theta/\beta$ between the branch cuts at $p = 1/\alpha$ and $1/\beta$. The branch point at $p = 1/\alpha$ causes the head wave *SpS*. We have now shown that $\mathcal{T}_{44}(p)$ has a pole at $p = 1/\gamma$ which may be near the contour at some time after the reflected arrival. This pole will influence the value of result (9.1.68). Let us approximate the terms by expanding about the pole. As $\mathcal{T}_{44}$ is singular, we expand its inverse

$$\mathcal{T}_{44}^{-1}(p) \simeq \mathcal{T}_{44}^{-1}(1/\gamma) + (p - 1/\gamma) \frac{\partial \mathcal{T}_{44}^{-1}(1/\gamma)}{\partial p}, \tag{9.1.69}$$

to first-order, and by definition the constant term is zero. We expand $\widetilde{T}_{SS}$ about its (complex) value at $p = 1/\gamma$ and on the contour we must have

$$t \simeq \widetilde{T}_{SS}(1/\gamma, \mathbf{x}) + \frac{\partial \widetilde{T}_{SS}(1/\gamma, \mathbf{x})}{\partial p} (p - 1/\gamma). \tag{9.1.70}$$

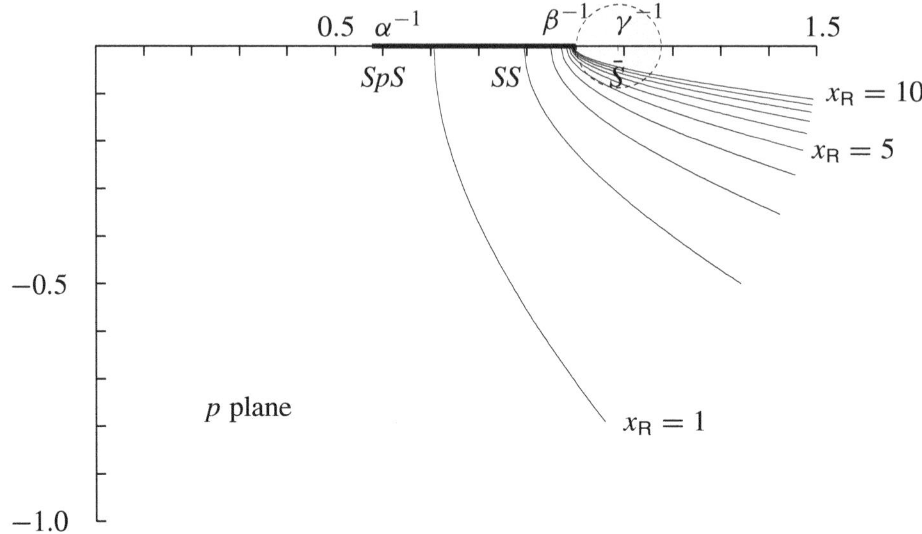

Fig. 9.6. The Cagniard contours for the SS reflected signals. The velocities are scaled so $\beta = 1$, and the results are for a Poisson solid, $\nu = 1/4$, with $\alpha = \sqrt{3}$. The Rayleigh wave velocity, γ, is given by equation (9.1.65). The vertical propagation distance is unity, $d = 1$, and contours are shown for horizontal ranges $x_R = 1$ to 10 in unit increments. Contours for $x_R > 5$ intersect the circle of convergence around the Rayleigh pole.

Thus on the contour near the pole, we have

$$\underline{\mathbf{u}}_{SS}(t, \mathbf{x}_R)$$

$$\simeq -\frac{1}{\pi^2 (2x_R)^{1/2}} \frac{\mathrm{d}}{\mathrm{d}t} \lambda(t) * \mathrm{Im}\left(p^{1/2} \mathcal{T}_{44}(p)\, \mathbf{g}_S\, \mathbf{g}_S^{\mathrm{T}} \frac{\partial p}{\partial \widetilde{T}_{SS}} \right)_{p=p(t,\mathbf{x}_R)} \tag{9.1.71}$$

$$\simeq -\frac{1}{\pi^2 (2x_R)^{1/2}} \frac{\mathrm{d}}{\mathrm{d}t} \lambda(t) * \mathrm{Im}\left(\frac{p^{1/2} \mathbf{g}_S\, \mathbf{g}_S^{\mathrm{T}}}{(p - 1/\gamma)\, \partial \mathcal{T}_{44}^{-1}/\partial p} \frac{\partial p}{\partial \widetilde{T}_{SS}} \right)_{p=p(t,\mathbf{x}_R)} \tag{9.1.72}$$

$$\simeq -\frac{1}{\pi^2 (2x_R)^{1/2}} \frac{\mathrm{d}}{\mathrm{d}t} \lambda(t) * \mathrm{Im}\left(\frac{p^{1/2} \mathbf{g}_S\, \mathbf{g}_S^{\mathrm{T}}}{\partial \mathcal{T}_{44}^{-1}/\partial p} \frac{1}{t - \widetilde{T}_{SS}(1/\gamma, \mathbf{x}_R)} \right) \tag{9.1.73}$$

Now

$$\widetilde{T}_{SS}(1/\gamma, \mathbf{x}_R) = \frac{x_R}{\gamma} + \mathrm{i} \left(\frac{1}{\gamma^2} - \frac{1}{\beta^2} \right)^{1/2} (2z_2 - z_S - z_R), \tag{9.1.74}$$

which is complex and $\mathbf{g}_S \mathbf{g}_S^{\mathrm{T}}$ will have some terms imaginary (q_β) and some terms real (p). The response will be

$$\boxed{\underline{\mathbf{u}}_{\bar{S}}(t, \mathbf{x}_R) \simeq -\frac{1}{\pi(2\gamma x_R)^{1/2}} \frac{\partial p}{\partial \mathcal{T}_{44}^{-1}} \frac{\mathrm{d}}{\mathrm{d}t}\lambda(t) * \mathrm{Im}\big(\Delta\big(t - \widetilde{T}_{SS}(1/\gamma, \mathbf{x}_R)\big)\, \mathbf{g}_S\, \mathbf{g}_S^{\mathrm{T}}\big),}$$

(9.1.75)

where we have used the symbol $\bar{S}$ for the Rayleigh pulse. The pulse shape in two dimensions, the term Im(...), is $t/(t^2+a^2)$ or $a/(t^2+a^2)$ centred on $t = x_R/\gamma$ and we have used the analytic Dirac delta function generalized for complex argument (B.1.8). In two dimensions, there is no decay with range as the wave does not spread out – it just propagates horizontally. In three dimensions, the decay with range is $x_R^{-1/2}$, i.e. cylindrical spreading. The pulse decays with depth as the imaginary, second term in expression (9.1.74) increases in magnitude. The convolution in (9.1.75) can be simplified as we have

$$\frac{\mathrm{d}}{\mathrm{d}t}\lambda(t) * \mathrm{Im}\left(\frac{b}{t-a}\right) = \frac{\mathrm{d}}{\mathrm{d}t}\mathrm{Re}\left(\frac{b}{(t-a)^{1/2}}\right) \tag{9.1.76}$$

(with a and b complex). Although it appears trivial, this result is most easily proved though the Fourier transforms in Appendix B.1.

The Rayleigh pulse is only generated as a distinct pulse when the Cagniard contour is close to the pole. The pole is close to the branch point $p = 1/\beta$, so the expansion about the pole is only valid within this radius. The Cagniard contour is given by expression (8.1.7) and substituting $t = x_R/\gamma$, the relevant point on the contour is

$$p = \frac{\sin^2\theta}{\gamma} - \mathrm{i}\left(\frac{\sin^2\theta}{\gamma} - \frac{1}{\beta^2}\right)^{1/2} \cos\theta. \tag{9.1.77}$$

The distance from the Rayleigh pole is then

$$\left|p - \frac{1}{\gamma}\right| = \frac{\cos\theta}{\gamma}\left(1 - \frac{\gamma^2}{\beta^2}\right)^{1/2}. \tag{9.1.78}$$

We require that this satisfies

$$\left|p - \frac{1}{\gamma}\right| < \frac{1}{\gamma} - \frac{1}{\beta}, \tag{9.1.79}$$

for the Rayleigh wave to be developed as a separate arrival. With result (9.1.78) in condition (9.1.79), this becomes

$$\cos\theta < \sqrt{\frac{\beta-\gamma}{\beta+\gamma}}, \tag{9.1.80}$$

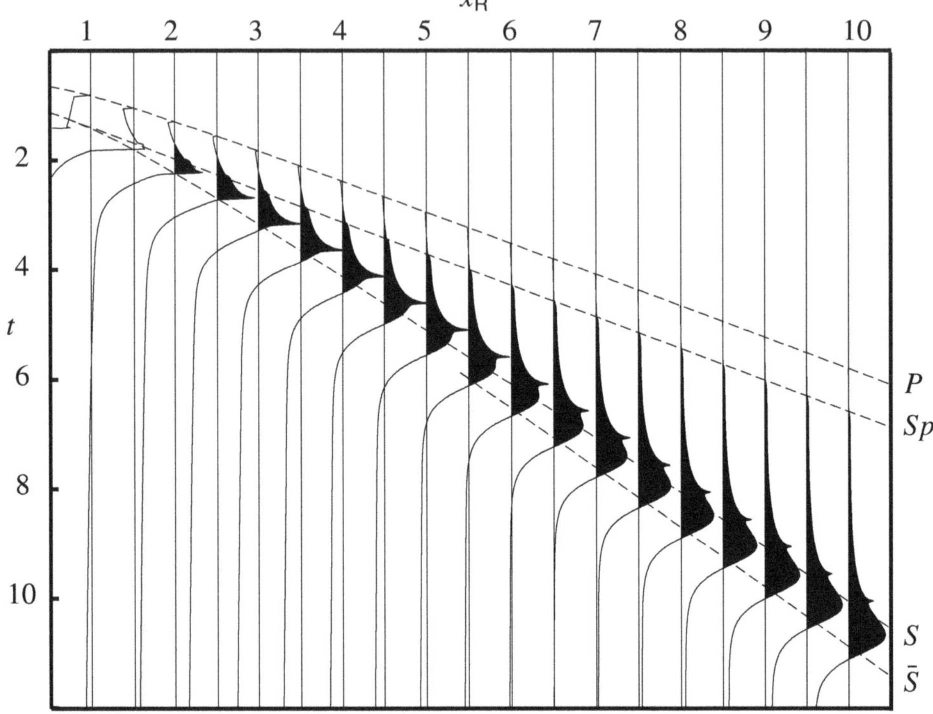

Fig. 9.7. The vertical surface ($z_R = z_2$) displacement u_z due to point, force source at unit depth, $z_2 - z_R = 1$, below a free surface. The source time function is of the form $f_S(t) = f_S H(t)$. The half-space is a Poisson solid, $\nu = 1/4$. The velocity is normalized with $\beta = 1$ so $\alpha = \sqrt{3}$. The ranges are $x_R = 1$ to 10 with unit steps. The arrival times of the direct *P* and *S* waves, the head wave *Sp* and the Rayleigh wave at x_R/γ are illustrated.

or

$$\tan\theta > \sqrt{\frac{2\beta}{\beta - \gamma}} \simeq 5. \tag{9.1.81}$$

The circle of convergence (9.1.79) is illustrated in Figure 9.6. Thus roughly, the range must be five or more times the source plus receiver depths for a separate Rayleigh wave to be developed. This is illustrated in Figure 9.7 for a vertical, force source with time function $f_S(t) = f_S H(t)$. The vertical displacement at the surface, $z_R = z_2$, is plotted and the direct pulse has the same form, $H(t)$. The source is at unit depth, $z_2 - z_S = 1$, and the *P* and *S* waves are included with the appropriate free surface conversion coefficients, (6.6.12) and (6.6.10), respectively. The seismograms have been scaled by $\sqrt{x_R}$ so the Rayleigh wave has approximately constant amplitude. The development of the Rayleigh wave for $x_R > 5$ is obvious

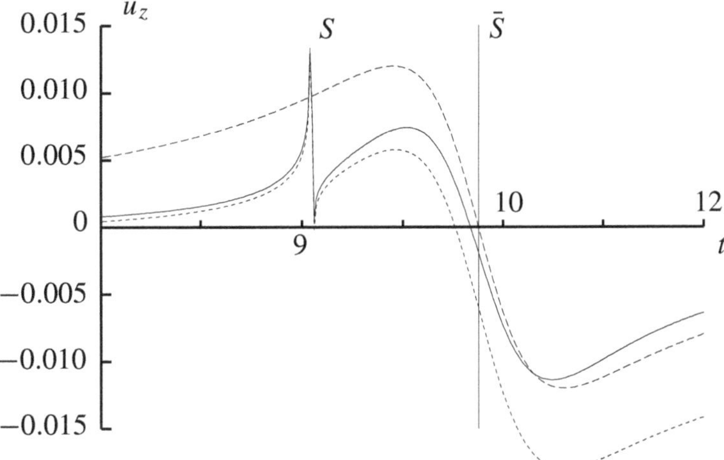

Fig. 9.8. As Figure 9.7 for the range $x_R = 10$ except that the three-dimensional source is of the form $f_S(t) = f_S\lambda(t)$. The total response is shown with a solid line, the Rayleigh wave approximation (9.1.75) with a long-dashed line. As only the *S* wave contributes to the Rayleigh wave (the Cagniard contour for the *P* wave is far from the Rayleigh pole), the *S* wave contribution is shown with a short-dashed line (the *P* wave contribution only shifts and distorts the response slightly at this time). The arrival time of the Rayleigh wave x_R/γ is marked with a thin vertical line.

and dominates the response. The direct *P* wave is barely visible on this scale as the receiver is almost on a radiation node of the source (Figure 4.14). The head wave *Sp* and direct *S* are visible.

Details of the response at $x_R = 10$ together with the Rayleigh wave approximation (9.1.75) are shown in Figure 9.8. For simplicity, this is the two-dimensional response, i.e. the form of the source is $f_S(t) = f_S\delta(t)$ in two dimensions or $f_S(t) = f_S\lambda(t)$ in three dimensions. It is clear that the Rayleigh wave term (9.1.75) well approximates the total response.

9.1.5.3 The leaking Rayleigh wave

In discussing the Rayleigh wave, we have ignored two of the roots of the cubic (9.1.60) and only analysed the root with $y > 1$. We must now investigate whether the other two roots are important.

For Poisson's ratio $\nu = 0$, $u = 1/2$ and the other roots are

$$y = \frac{3}{4} - \frac{\sqrt{5}}{4} \quad \text{and} \quad \frac{1}{2}. \tag{9.1.82}$$

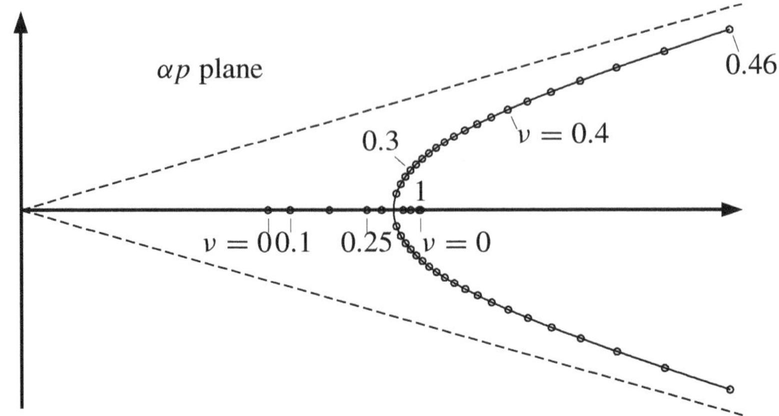

Fig. 9.9. The $\bar{P}$ roots of the Rayleigh equation plotted in the complex αp plane for $\nu = 0$ to $1/2$. The positions of the poles are plotted for $\nu = 0$, 0.1, 0.2, 0.25 and then at intervals of 0.01 to 0.46. The branch point $\alpha p = 1$ coincides with the upper pole when $\nu = 0$ (9.1.83).

It is best to normalize the root with respect to α, i.e.

$$\alpha p = 0.61803 \quad \text{and} \quad 1. \tag{9.1.83}$$

For Poisson's ratio $\nu = 1/4$, $u = 1/3$ and the other roots are

$$y = \frac{3}{4} - \frac{\sqrt{3}}{4} \quad \text{and} \quad \frac{1}{4}. \tag{9.1.84}$$

Again

$$\alpha p = 0.97517 \quad \text{and} \quad 0.86603. \tag{9.1.85}$$

For Poisson's ratio $\nu = 1/2$, $u = 0$ and the other roots are

$$y = 0.20212 \pm \mathrm{i}\,0.12722 = 0.23883\,\mathrm{e}^{\mathrm{i}\,0.56178}. \tag{9.1.86}$$

Thus

$$\beta p = 0.46670\,\mathrm{e}^{\mathrm{i}\,0.28088}, \tag{9.1.87}$$

and as $\alpha \to \infty$ when $\nu \to 1/2$, the roots are asymptotic to the line $\mathrm{Arg}(\alpha p) = \pm 0.28088$.

The complete behaviour of these roots as a function of ν is illustrated in Figure 9.9.

Clearly when the roots are real for $\alpha p < 1$, they cannot satisfy the Rayleigh denominator (9.1.59) unless the slownesses are taken on non-physical Riemann sheets. Nevertheless, the poles are near the branch point $p = 1/\alpha$ and when $\nu \simeq 0.45$, the poles are near the contour between $p = 1/\alpha$ and $1/\beta$. This causes a pulse which has been called $\bar{P}$ (Gilbert and Laster, 1962) and been discussed in more detail by Chapman (1972). It has been observed and analysed by Roth and Holliger (2000).

When the range is beyond critical, so the Cagniard contour lies along the real axis for $p > 1/\alpha$, the nearest singularity may be the $\bar{P}$ pole on the lower Riemann sheet through the branch cut. At this time, after the *SpS* head wave but before the geometrical arrival *SS*, the $\bar{P}$ pulse arrives and is described by an expression like result (9.1.73) but for the $\bar{P}$ pole, i.e.

$$\boxed{\underline{\mathbf{u}}_{\bar{P}}(t, \mathbf{x}_{\mathrm{R}}) \simeq -\frac{1}{\pi^2 (2x_{\mathrm{R}})^{1/2}} \frac{\mathrm{d}}{\mathrm{d}t} \lambda(t) * \operatorname{Im}\left(\frac{p_{\bar{P}}^{1/2} \mathbf{g}_S \mathbf{g}_S^{\mathrm{T}}}{\partial \mathcal{T}_{44}^{-1}/\partial p} \frac{1}{t - \widetilde{T}_{SS}(p_{\bar{P}}, \mathbf{x}_{\mathrm{R}})} \right).} \tag{9.1.88}$$

This pulse is only important when $\nu \simeq 0.45$, which corresponds to a soft sediment. This is illustrated in Figures 9.10 and 9.11. Apart from changing the Poisson ratio to $\nu = 0.45$, Figure 9.10 is as Figure 9.7. As the shear wave velocity is normalized ($\beta = 1$), the P velocity is now much greater ($\alpha \simeq 3.316$). The approximate arrival time of the leaking $\bar{P}$ pulse, $\operatorname{Re}\left(\widetilde{T}_{SS}(p_{\bar{P}}, \mathbf{x}_{\mathrm{R}})\right)$, is also shown. The pulse is broad and not large, but arrives between the head wave *Sp* and the direct *S* pulse. The leaking $\bar{P}$ pulse approximation (9.1.88) is illustrated in Figure 9.11.

In general, the reflection coefficient for any interface may have poles (zeros of the denominator). When the velocities of the two media are similar, these may be real poles analogous to the Rayleigh pole (they are called Stoneley waves on a solid–solid interface and Scholte waves on a fluid–solid interface). Often the poles are in complex positions on non-physical Riemann sheets but lie near branch cuts on the real axis. The mathematical methods to analyse these signals are identical to the above. Phinney (1961) investigated general interface waves and obtained results similar to ours. Strick (1959) and Roever and Vining (1959*a*, *b*) investigated pseudo-Rayleigh waves at a fluid–solid interface. The early papers on interface waves, e.g. Stoneley (1924), Scholte (1947), etc. were concerned with the existence of real interface waves. Outside a certain range of parameters, these waves did not exist. However, the above analysis of the leaking Rayleigh wave shows that there will not be a sudden cut-off at the limit of the real interface waves. Beyond the limit, the poles become complex and while the imaginary part is small the behaviour is very similar.

In general, the three-dimensional response due to a pole $p = p_{\mathrm{pole}}$ is

$$\boxed{\underline{\mathbf{u}}_{\mathrm{pole}}(t, \mathbf{x}_{\mathrm{R}}) \simeq -\frac{1}{\pi^2 (2x_{\mathrm{R}})^{1/2}} \frac{\mathrm{d}}{\mathrm{d}t} \lambda(t) * \operatorname{Im}\left(\frac{p^{1/2} \left((p - p_{\mathrm{pole}}) \underline{\boldsymbol{\mathcal{G}}}_{\mathrm{ray}} \right)}{t - \widetilde{T}_{\mathrm{ray}}} \right)_{p = p_{\mathrm{pole}}},} \tag{9.1.89}$$

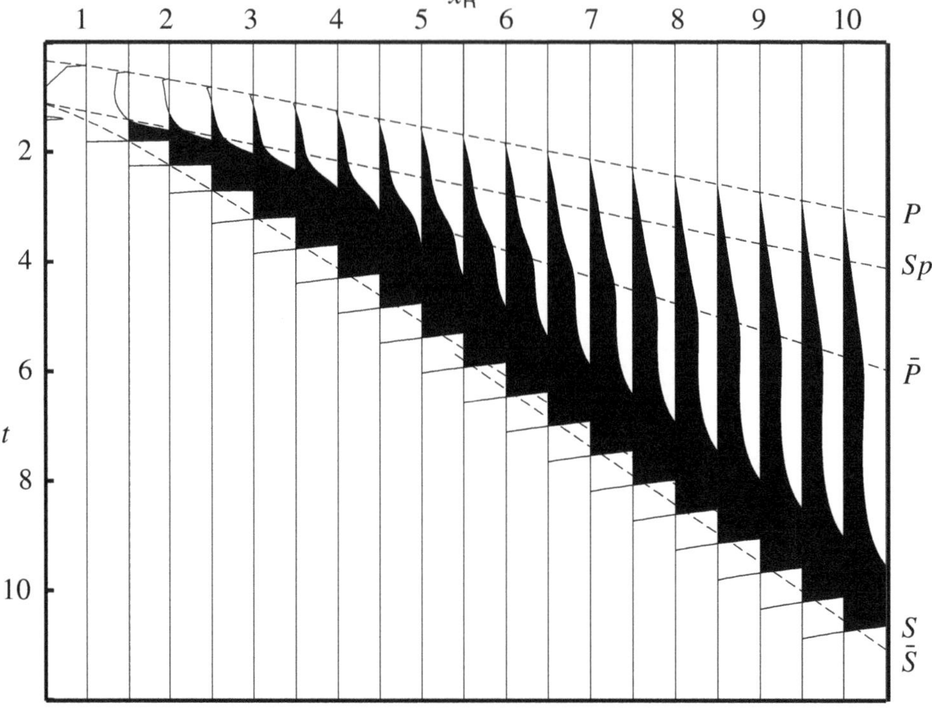

Fig. 9.10. As Figure 9.7 except that the Poisson's ratio has been increased from 0.25 to 0.45. The amplitude has been multiplied by a factor of three so the leaking $\bar{P}$ pulse is visible. In addition to the arrivals shown in Figure 9.7, the approximate arrival time of the leaking $\bar{P}$ pulse, $\mathrm{Re}\left(\widetilde{T}_{SS}(p_{\bar{P}}, \mathbf{x}_{\mathrm{R}})\right)$, is also shown with a dashed line.

where the combination $((p - p_{\mathrm{pole}})\underline{\mathcal{G}}_{\mathrm{ray}})$ is non-singular at the pole, i.e. if $\underline{\mathcal{G}}_{\mathrm{ray}}(p) \simeq \underline{\widetilde{\mathcal{G}}}_{\mathrm{pole}}(p)/g(p)$ where $g(p_{\mathrm{pole}})$ is zero and $\underline{\widetilde{\mathcal{G}}}_{\mathrm{pole}}(p)$ is well behaved in the neighbourhood of the pole, then

$$\left((p - p_{\mathrm{pole}})\underline{\mathcal{G}}_{\mathrm{ray}}\right)\Big|_{p=p_{\mathrm{pole}}} = \frac{\underline{\widetilde{\mathcal{G}}}_{\mathrm{pole}}(p_{\mathrm{pole}})}{g'(p_{\mathrm{pole}})}. \tag{9.1.90}$$

In expression (9.1.89), $\widetilde{T}_{\mathrm{ray}}$ is complex. For a real interface wave, e.g. Rayleigh wave, $\underline{\mathcal{G}}_{\mathrm{ray}}$ is real, while for a leaking wave, it is complex. This result (9.1.89) applies whether the pole is real or complex, provided it lies close to the Cagniard contour. This generally means that the range must be large so the contour lies close to the real axis. If complex, the pole will lie on a non-physical Riemann sheet but again it may lie close to the Cagniard contour through a branch cut.

Again a simple check shows $\mathrm{unit}(\underline{\mathbf{u}}_{\mathrm{pole}}) = [\mathrm{M}^{-1}\mathrm{T}]$.

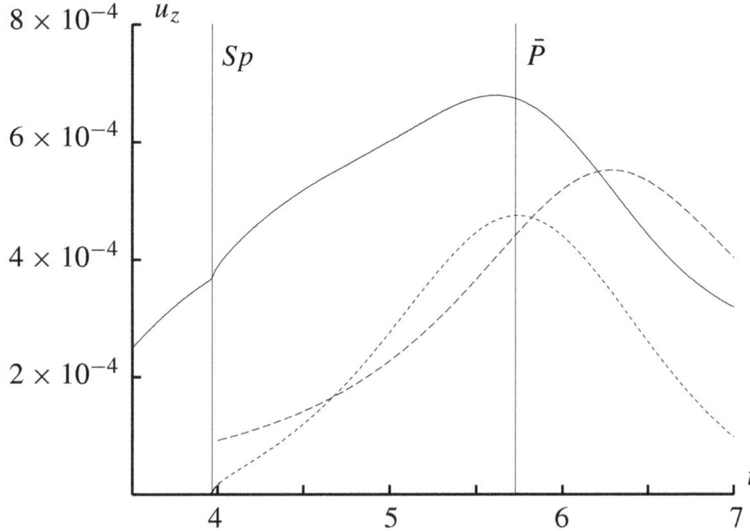

Fig. 9.11. As Figure 9.10 for the range $x_R = 10$ except that the three-dimensional source is of the form $f_S(t) = f_S\lambda(t)$. The total response is shown with a solid line, the leaking $\bar{P}$ approximation (9.1.88) with a long-dashed line. As only the *S* wave contributes to the leaking (the Cagniard contour for the *P* wave is far from the leaking pole), the *S* wave contribution is shown with a short-dashed line (the *P* wave contribution only shifts and distorts the response slightly at this time). The arrival time of the leaking $\bar{P}$ pulse, $\mathrm{Re}\left(\widetilde{T}_{SS}(p_{\bar{P}}, \mathbf{x}_R)\right)$, is also shown with a thin vertical line. The head wave *Sp* and its arrival time are visible.

9.1.6 Tunnelling waves

Tunnelling waves are important in situations where a wave can propagate evanescently through a thin high-velocity region. Various publications have discussed such waves in different situations but unfortunately there is no agreement about notation or nomenclature. First we describe the different situations in which tunnelling is important, mentioning the different nomenclature that has been used. The mathematics is similar for all cases so we use the same nomenclature in all cases (but unfortunately this makes it different from published papers).

Consider a simple acoustic transmission through an interface from a high to low velocity (Figure 9.12). If the source is near the interface, at large ranges the ray P_1P_2 from the source is almost parallel to the interface (p is slightly less than $1/\alpha_1$), and is refracted into the second medium at slightly less than the critical angle. In addition, a tunnelling wave exists that is evanescent between the source and interface, and then propagates to the receiver. We denote this by $P_1^*P_2$ where the asterisk indicates an evanescent portion. The slowness of this wave must be at $p \simeq \sin\theta/\alpha_2$ where $\tan\theta = x_R/(z_2 - z_R)$.

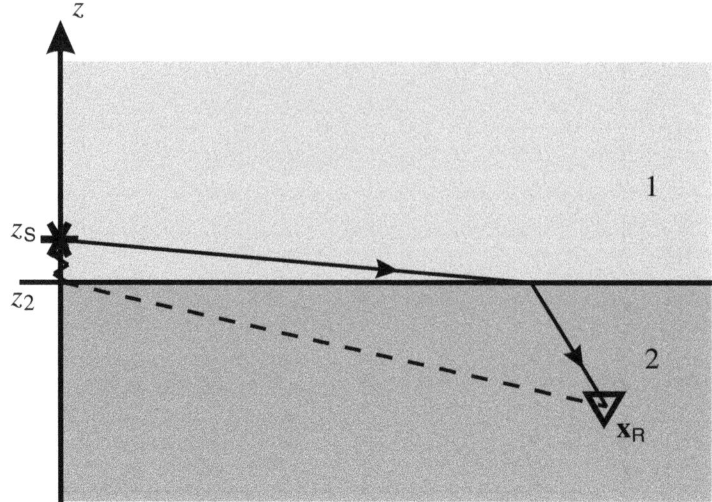

Fig. 9.12. Rays for a transmitted acoustic wave where $\alpha_1 > \alpha_2$ and $z_S - z_2$ is small.

Tunnelling also occurs in the reciprocal situation (as it must as reciprocity applies to the complete wavefield not just the geometrical rays). A receiver lies slightly below an interface in a higher velocity medium (Figure 9.13). An important example of this is for a buried receiver in the seafloor. Stephen and Bolmer (1985) studied this and called it a *direct wave root*. We would call it $P_1 P_2^*$. Another example would be a buried receiver near the Earth's surface with an air wave.

In an elastic half-space, the faster velocity can arise from the P velocity compared with the S velocity. If the source is located near a free surface, an evanescent P wave can excite an S wave (Figure 9.14). Hron and Mikhailenko (1981) have called this the S^* wave which they noticed in numerical modelling. In our notation, we prefer to call it the P^*S wave to emphasize that the evanescent portion of the ray is a P wave. Various other authors (Daley and Hron, 1983; Gutowski, Hron, Wagner and Treitel, 1984; Kim and Behrens, 1986) have studied the S^* wave.

Finally, the transmission through a thin, high-velocity layer can also exhibit tunnelling (Figure 9.15). We denote the geometrical ray that refracts along the thin layer as $P_1 P_2 P_3$ and the tunnelling wave as $P_1 P_2^* P_3$. If the thin layer becomes infinitesimally thin, the tunnelling signal must become the direct ray (and the refraction and reverberations in the thin layer must disappear). Hong and Helmberger (1977), Helmberger and Hadley (1981) and Fuchs and Schulz (1976) have encountered these signals. They are clearly very important in ray tracing between boreholes with velocity models from logs which normally exhibit many thin high-velocity layers.

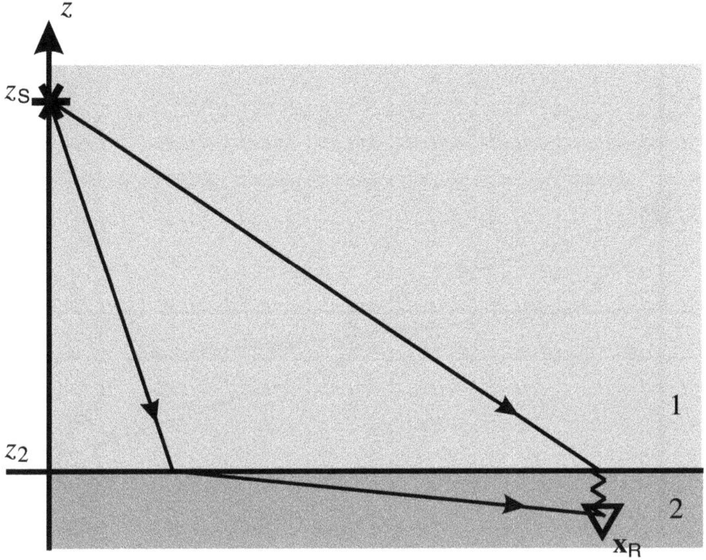

Fig. 9.13. Rays for a transmitted acoustic wave where $\alpha_2 > \alpha_1$ and $z_2 - z_{\mathrm{R}}$ is small.

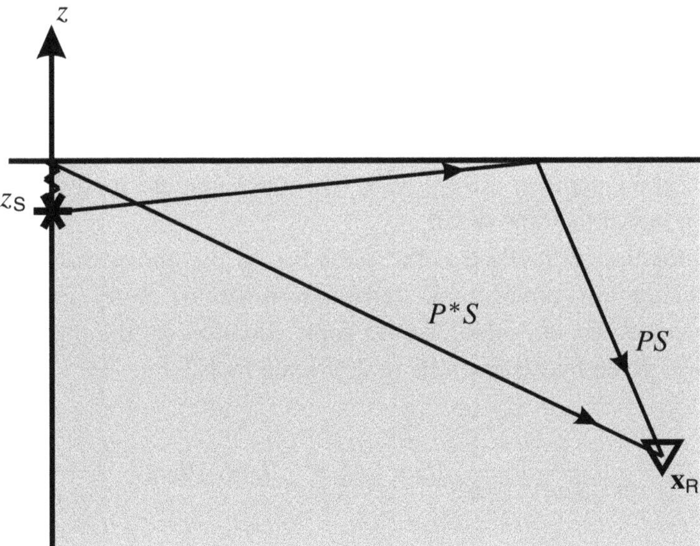

Fig. 9.14. Rays for a reflected *PS* wave with the source near the interface.

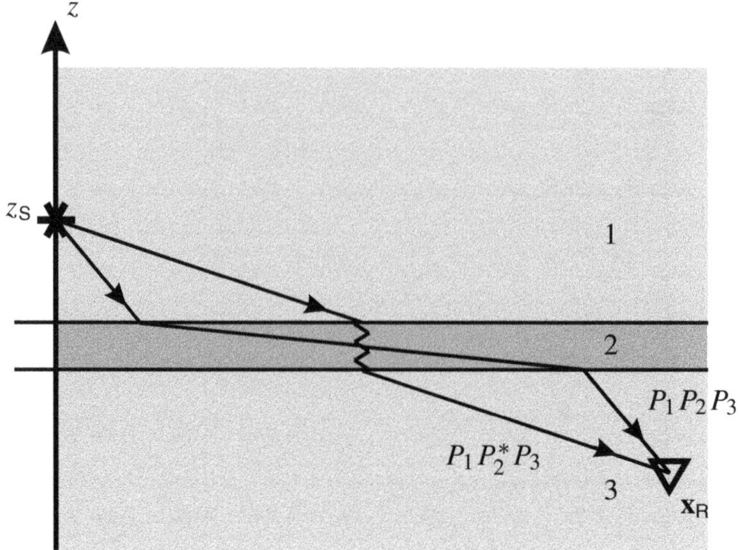

Fig. 9.15. Rays for a transmitted $P_1P_2P_3$ wave through a thin, high-velocity layer ($\alpha_2 > \alpha_1 = \alpha_3$) including the 'tunnelling' ray $P_1P_2^*P_3$.

The common feature of all these examples is the existence of a thin 'layer' (or ray segment) with higher velocity. The complete response, including the tunnelling signal, can be described by one 'ray' using the Cagniard solution. The thin-layer problem in the final example, Figure 9.15, is more complicated as multiple reverberations occur in the layer. With the exception of the authors for the last example, the Cagniard solution has not been used. Hong and Helmberger (1977) noticed the distortion of the Cagniard contour, but did not analyse the signal in detail. Later, Drijkoningen and Chapman (1988) have used the Cagniard method and analysed the tunnelling signal in some detail.

As the mathematical method is the same for all the above examples, we consider the first situation – an acoustic transmission with $\alpha_1 \gg \alpha_2$ (Figure 9.12). For simplicity we consider only the far-field approximation for the three-dimensional Green function given by expressions (8.2.68) and (8.1.2)

$$\underline{\mathbf{u}}_{P_1P_2}(t,\mathbf{x}_\mathrm{R}) \simeq -\frac{1}{\pi^2(2x_\mathrm{R})^{1/2}}\frac{\mathrm{d}}{\mathrm{d}t}\lambda(t) * \mathrm{Im}\left(p^{1/2}\mathcal{T}_{21}(p)\,\mathbf{g}_2\,\mathbf{g}_1^\mathrm{T}\frac{\partial p}{\partial \widetilde{T}_{P_1P_2}}\right)_{p=p(t,\mathbf{x}_\mathrm{R})}, \tag{9.1.91}$$

where

$$\widetilde{T}_{P_1P_2}(p,\mathbf{x}_\mathrm{R}) = px_\mathrm{R} + q_{\alpha\,1}(z_\mathrm{S} - z_2) + q_{\alpha\,2}(z_2 - z_\mathrm{R}). \tag{9.1.92}$$

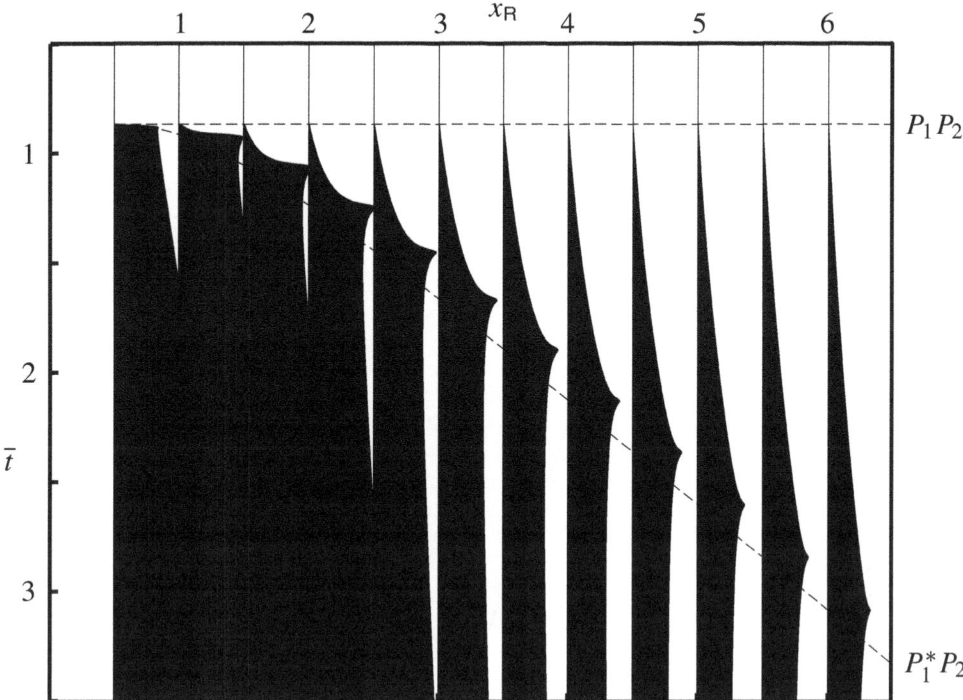

Fig. 9.16. The response in the acoustic model illustrated in Figure 9.12 to an explosive, point source with time function of the form $P_S(t) = P_S t H(t)$. The 'thick' layer has normalized parameters $\alpha_2 = 1$ and $d_2 = 1$ and the 'thin' layer has parameters $\alpha_1 = 2$ and $d_1 = 0.01$. The vertical displacement for ranges from $x_R = 0.5$ to 6 in increments of 0.5 are shown. The time axis is the reduced time $\bar{t} = t - x_R/\alpha_1$. The time of the geometrical arrival, $T_{P_1 P_2}$, and the tunnelling wave, R/α_2, are shown with dashed lines. The response are multiplied by the range, x_R.

For simplicity, as we are not interested specifically in the source radiation pattern, we specialize this to an explosive source (cf. (8.1.32))

$$\mathbf{u}_{P_1 P_2}(t, \mathbf{x}_R) \simeq \frac{V_0 P_0''(t)}{\pi^2 (2x_R)^{1/2} \alpha_1^2} * \lambda(t) * \text{Im}\left(\frac{p^{1/2}}{\rho_2 q_{\alpha 1} + \rho_1 q_{\alpha 2}} \begin{pmatrix} p \\ -q_{\alpha 2} \end{pmatrix} \frac{\partial p}{\partial \widetilde{T}_{P_1 P_2}} \right)_{p = p(t, \mathbf{x}_R)}. \quad (9.1.93)$$

Results for this expression for the vertical displacement are illustrated in Figure 9.16 for a source with time function of the form $P_S(t) = P_S t H(t)$. Note that the 'geometrical' arrival decays rapidly with range (the amplitudes have been multiplied by range x_R) and have the approximate form of a head wave, i.e. form

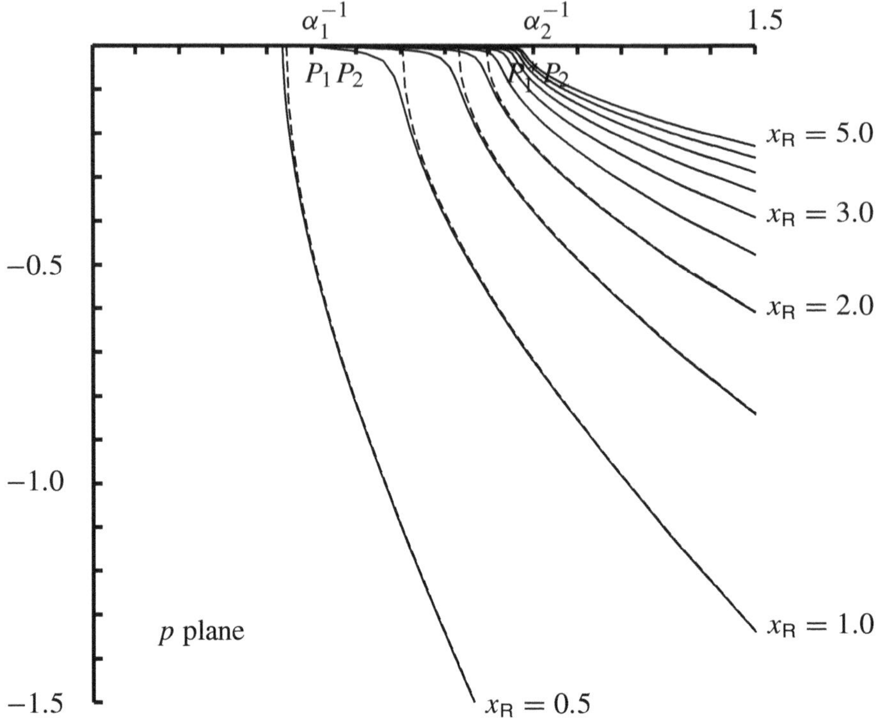

Fig. 9.17. Cagniard contours for the P_1P_2 wave. The model has parameters as in Figure 9.16. Contours for ranges from $x_R = 0.5$ to 5 in increments of 0.5 are shown. Up to $x_R = 2$, the contours with $\alpha_1 = \alpha_2$ are illustrated with a dash line.

$tH(t)$. The tunnelling wave has the approximate form of a direct wave with a phase shift due to the complex transmission coefficient, i.e. the in-phase part has the form $H(t)$ and the out-phase part the form $-\ln(|t|)/\pi$.

Cagniard contours for the P_1P_2 wave are illustrated in Figure 9.17. The contour must leave the real axis for $p < 1/\alpha_1$ and be asymptotic to the line

$$\text{Arg}(p) = \tan^{-1}\left(\frac{x_R}{z_S - z_R}\right) - \frac{\pi}{2} = \theta_0 - \frac{\pi}{2}, \quad \text{say.} \tag{9.1.94}$$

In between the saddle point and the asymptote, it turns and runs close to the real axis before turning rapidly at approximately $p = \sin\theta_0/\alpha_2$. Although it is difficult to see this analytically, by considering the two extreme cases we can see the rough behaviour. If $z_2 \to z_R$, we have the simple behaviour of a direct ray. The contour leaves the axis at $p = \sin\theta_0/\alpha_1$ and curves smoothly to the asymptote. As $z_2 \to z_S$, we have a head wave in the first medium due the branch point at $p = 1/\alpha_1$. The contour runs along the real axis to $p = \sin\theta_0/\alpha_2$, at which point the contour leaves the real axis and again curves smoothly to the asymptote. If $z_S - z_2$ is small (compared with $z_S - z_R$), the contour must be close to this case, but leave the axis for $p < 1/\alpha_1$, as illustrated in Figure 9.17.

Two interesting features exist for the solution (9.1.93) with the contour illustrated in Figure 9.17: near $p = 1/\alpha_1$, the geometrical arrival occurs, but soon afterwards a 'head-wave' like feature will occur as the contour passes close to the branch point; and near $p = \sin\theta_0/\alpha_2$, a signal will be caused by the rapid change in $\partial p/\partial \widetilde{T}$. First let us investigate the geometrical ray/head-wave combination. Let us define

$$p = \frac{1}{\alpha_1} + \xi. \tag{9.1.95}$$

Setting $\widetilde{T} = t$, and expanding we obtain

$$t \simeq T_{p_1 P_2}(\mathbf{x}_R) + \ell_{p_1 P_2}(\mathbf{x}_R)\,\xi, \tag{9.1.96}$$

where

$$T_{p_1 P_2}(\mathbf{x}_R) = \frac{x_R}{\alpha_1} + (z_2 - z_R)\left(\frac{1}{\alpha_2^2} - \frac{1}{\alpha_1^2}\right)^{1/2}, \tag{9.1.97}$$

is the arrival time of a head wave $p_1 P_2$ for a source on the interface, and

$$\ell_{p_1 P_2}(\mathbf{x}_R) = x_R - (z_2 - z_R)\left(\frac{\alpha_1^2}{\alpha_2^2} - 1\right)^{1/2}, \tag{9.1.98}$$

is the length of this head wave in the first medium. In expression (9.1.96), we have neglected a term in $\xi^{1/2}(z_S - z_2)$ as the source is close to the interface which means that we must have

$$|\xi| > \frac{2}{\alpha_1}\left(\frac{z_S - z_2}{\ell_{p_1 P_2}}\right)^2, \tag{9.1.99}$$

i.e. the expansion is not valid too close to the branch point. Expression (9.1.96) can be solved for ξ.

In expression (9.1.93) we expand in terms of ξ where the important term is $q_{\alpha\,1}$ – other terms are treated as constant. Thus

$$\frac{\partial \widetilde{T}_{P_1 P_2}}{\partial p} \simeq \ell_{p_1 P_2} - \frac{z_S - z_2}{(-2\xi\alpha_1)^{1/2}} \tag{9.1.100}$$

$$\rho_1 q_{\alpha\,2} + \rho_2 q_{\alpha\,1} \simeq \rho_1 \frac{\cos\theta_c}{\alpha_2} + \rho_2\left(-\frac{2\xi}{\alpha_1}\right)^{1/2}, \tag{9.1.101}$$

where θ_c is the critical angle ($\sin\theta_c = \alpha_2/\alpha_1$). Substituting in (9.1.93), taking the leading terms in the reciprocals of (9.1.100) and (9.1.101), and substituting for ξ

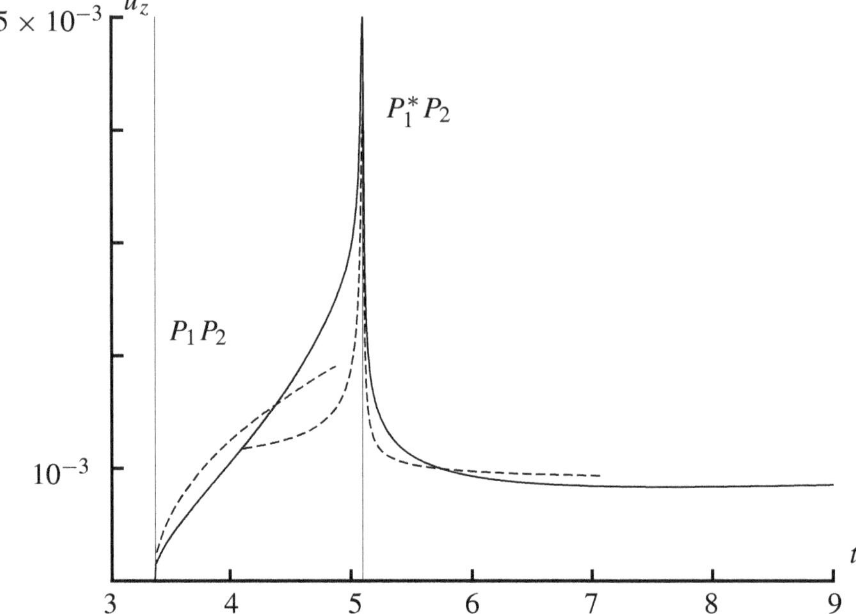

Fig. 9.18. The waveform P_1P_2 illustrating the distorted geometrical pulse and the large, smooth tunnelling pulse. As Figure 9.16 except that the source function is of the form $P_S(t) = P_S\mu(t)$ (defined in (9.2.30)), i.e. effectively the two-dimensional response with source of the form $P_S(t) = P_S H(t)$. The model parameters are as in Figure 9.16, with $\rho_1 = 1.5$ and $\rho_2 = 1$. The range is $x_R = 5$. The geometrical arrival time is $T_{P_1P_2} \simeq 3.3660$ and the tunnelling arrival $P_1^* P_2$ is at approximately $t = R/\alpha_1 \simeq 5.0990$. The arrival times are indicated with thin vertical lines. The approximations (9.1.102) and (9.1.115) are illustrated with dashed lines. The D.C. level of the latter is shifted to correspond to the large time asymptote. The singularity at $T_{P_1P_2}$ is not resolved at this scale.

from (9.1.96), we obtain in three dimensions

$$\mathbf{u}_{P_1P_2}(t, \mathbf{x}_R) \simeq \frac{V_0 P_0'(t)}{2\pi\alpha_1^3 \rho_1 \cos\theta_c \, \ell_{P_1P_2}^{3/2} \, x_R^{1/2}} \begin{pmatrix} \sin\theta_c \\ -\cos\theta_c \end{pmatrix} * \left((z_S - z_2)\, \delta\left(t - T_{P_1P_2}\right) + \frac{\rho_2\alpha_2}{\rho_1\cos\theta_c} H\left(t - T_{P_1P_2}\right) \right). \tag{9.1.102}$$

The first term is the geometrical arrival, and the second term, a head-wave 'like' arrival (but it is not the first arrival). Details near the geometrical arrival may not be correct due to the inadequacies of our expansions, but the general form is correct. The waveform (9.1.102) is illustrated in Figure 9.18. For simplicity, this is effectively the two-dimensional response for a source of the form $P_S(t) = P_S H(t)$.

In three dimensions, this corresponds to a source of the form $P_S(t) = P_S\mu(t)$ ($\mu(t)$ is defined in equation (9.2.30)). A simple check shows (9.1.102) agrees with unit($\mathbf{u}$) = [L].

Near $p = \sin\theta_2/\alpha_2$, where the rapid change in $\partial p/\partial \widetilde{T}$ will cause a signal, we expand using

$$p = \frac{\sin\theta_2}{\alpha_2} + \epsilon, \tag{9.1.103}$$

where $\tan\theta_2 = x_R/(z_2 - z_R)$. Defining values of $-\mathrm{i}\,q_{\alpha\,1}$ and $q_{\alpha\,2}$ at $p = \sin\theta_2/\alpha_2$, i.e.

$$q_1 = \left(\frac{\sin^2\theta_2}{\alpha_2^2} - \frac{1}{\alpha_1^2}\right)^{1/2} \tag{9.1.104}$$

$$q_2 = \frac{\cos\theta_2}{\alpha_2}, \tag{9.1.105}$$

the 'time' (9.1.92) is

$$\widetilde{T}_{P_1P_1}\left(\frac{\sin\theta_2}{\alpha_2}, \mathbf{x}_R\right) = \frac{R}{\alpha_2} + \mathrm{i}\,q_1(z_S - z_2), \tag{9.1.106}$$

where R is the ray length of $P_1^* P_2$ in the second medium (Figure 9.12), i.e. $R^2 = x_R^2 + (z_2 - z_R)^2$. In the first derivative

$$\frac{\partial \widetilde{T}_{P_1P_1}}{\partial p} = x_R - \frac{p}{q_{\alpha\,1}}(z_S - z_2) - \frac{p}{q_{\alpha\,2}}(z_2 - z_R), \tag{9.1.107}$$

the first and last terms cancel, and the second term is small (due to $z_S - z_2$). Assuming the second derivative

$$\frac{\partial^2 \widetilde{T}_{P_1P_1}}{\partial p^2} = -\frac{1}{\alpha_1^2 q_{\alpha\,1}^3}(z_S - z_2) - \frac{1}{\alpha_2^2 q_{\alpha\,2}^3}(z_2 - z_R) \tag{9.1.108}$$

$$\simeq -\frac{1}{\alpha_2^2 q_2^3}(z_2 - z_R), \tag{9.1.109}$$

dominates, the small perturbation in equation (9.1.103) is

$$\epsilon = -\mathrm{i}\,\frac{2^{1/2}\alpha_2 q_2^{3/2}}{(z_2 - z_R)^{1/2}}\left(t - \frac{R}{\alpha_2} - \mathrm{i}q_1(z_S - z_2)\right)^{1/2} \tag{9.1.110}$$

$$= -\mathrm{i}\,\frac{\alpha_2 q_2^{3/2}}{(z_2 - z_R)^{1/2}}(u - \mathrm{i}v), \tag{9.1.111}$$

say, where

$$u = \left(\left\{ \left(t - \frac{R}{\alpha_2} \right)^2 + q_1^2 (z_S - z_2)^2 \right\}^{1/2} + t - \frac{R}{\alpha_2} \right)^{1/2} \qquad (9.1.112)$$

$$v = \left(\left\{ \left(t - \frac{R}{\alpha_2} \right)^2 + q_1^2 (z_S - z_2)^2 \right\}^{1/2} - t + \frac{R}{\alpha_2} \right)^{1/2} . \qquad (9.1.113)$$

The derivative is then

$$\frac{\partial \widetilde{T}_{P_1 P_1}}{\partial p} = \mathrm{i} \frac{(z_2 - z_R)^{1/2}}{\alpha_2 q_2^{3/2}} (u - \mathrm{i} v), \qquad (9.1.114)$$

and substituting in result (9.1.93), treating all other terms as constant, we obtain

$$\mathbf{u}_{P_1^* P_2}(t, \mathbf{x}_R) \simeq - \frac{V_0 P_0''(t) q_2^2}{2^{1/2} \pi^2 \alpha_1^2 (z_2 - z_R)} \frac{1}{\rho_1^2 q_2^2 + \rho_2^2 q_1^2} \begin{pmatrix} \sin\theta_2 \\ -\cos\theta_2 \end{pmatrix} * \lambda(t) * \frac{\rho_1 q_2 u + \rho_2 q_1 v}{u^2 + v^2} . \qquad (9.1.115)$$

Although the final time function appears to be rather messy, it can be reduced to a simple convolution. Removing the time shift R/α_2, the functions $u/(u^2 + v^2)$ and $v/(u^2 + v^2)$ are $2^{-1/2}$ times the standard functions (B.3.2) (with $a = q_1(z_S - z_2)$). Via their Fourier transforms (B.3.2), these can be seen to be equivalent to the convolutions (B.3.3) and (B.3.4). Thus (9.1.115) can be rewritten

$$\mathbf{u}_{P_1^* P_2}(t, \mathbf{x}_R) \simeq - \frac{V_0 P_0'(t) q_2^2}{2\pi \alpha_1^2 (z_2 - z_R)} * \mathrm{Re} \left(\frac{\Delta\left(t - \widetilde{T}_{P_1 P_1}(\sin\theta_2/\alpha_2, \mathbf{x}_R)\right)}{\rho_1 q_2 + \mathrm{i} \rho_2 q_1} \right), \qquad (9.1.116)$$

where we have used the delta function with complex argument (B.1.9). Thus the tunnelling pulse $P_1^* P_2$ is like the source pulse convolved with $a/(t^2 + a^2)$, a result we might have expected intuitively. It includes the in-phase, $\delta(t)$, and Hilbert transform, $\bar{\delta}(t)$, parts as the transmission coefficient is complex (as $\theta_2 > \theta_c$). As $z_S - z_2$ increases, it decays, and as it decreases towards zero, it becomes more delta-like. The complete waveform of the $P_1 P_2$ signal is illustrated in Figure 9.18. Expression (9.1.116) agrees with unit($\mathbf{u}$) = [L].

Another interesting aspect of a thin layer (Figure 9.15) is how the reverberations cancel at large ranges. We consider this in more detail in Section 9.3.4.1.

9.1.6.1 General result for tunnelling waves

We consider a general result for tunnelling waves where the thin layers of total thickness, d_{max}, exist with high velocity, c_{max}. We do not consider a possible generalization where tunnelling occurs in layers with different high velocities. In the general result (8.2.68), we consider the case when the phase function (8.1.2) can be written

$$\widetilde{T}_{ray}(p, \mathbf{x}_R) = px_R + \tau_{ray}(p, z_R) + q_{max}d_{max}, \tag{9.1.117}$$

where

$$q_{max} = \left(\frac{1}{c_{max}^2} - p^2\right)^{1/2}, \tag{9.1.118}$$

is the vertical slowness for layers with the maximum velocity, c_{max}, and their thickness, d_{max}, is small compared with the other layers. Thus as $|p| \to \infty$

$$\tau_{ray}(p, z_R) \to \pm i p d, \tag{9.1.119}$$

where $d_{max} \ll d$.

Near the branch point $p = 1/c_{max}$, the geometrical arrival exists. We expand using

$$p = \frac{1}{c_{max}} + \xi, \tag{9.1.120}$$

so

$$q_{max} = \left(-\frac{2\xi}{c_{max}}\right)^{1/2}, \tag{9.1.121}$$

and

$$t \simeq \widetilde{T}_{max}(\mathbf{x}_R) + \ell_{max}(\mathbf{x}_R)\xi, \tag{9.1.122}$$

where

$$\widetilde{T}_{max}(\mathbf{x}_R) = \frac{x_R}{c_{max}} + \tau_{ray}\left(c_{max}^{-1}, z_R\right) \tag{9.1.123}$$

$$\ell_{max}(\mathbf{x}_R) = x_R - X_{ray}\left(c_{max}^{-1}, z_R\right). \tag{9.1.124}$$

Expression (9.1.123) for $\widetilde{T}_{max}(\mathbf{x}_R)$ is the arrival time of the 'ray' with slowness $p = 1/c_{max}$ and 'head-wave' segments of length ℓ_{max}. The range function is $X_{ray}(p, z_R) = -d\tau_{ray}/dp$. With the approximations

$$\frac{\partial \widetilde{T}_{max}}{\partial p} \simeq \ell_{max} + d_{max} \left(-2\xi c_{max}\right)^{-1/2} \tag{9.1.125}$$

$$\underline{\boldsymbol{\mathcal{G}}}_{ray}(p) \simeq \underline{\boldsymbol{\mathcal{G}}}_{ray}\left(c_{max}^{-1}\right) + \frac{\partial \underline{\boldsymbol{\mathcal{G}}}_{ray}}{\partial q_{max}}\left(-\frac{2\xi}{c_{max}}\right)^{1/2}, \tag{9.1.126}$$

substituted in expression (8.2.68), we obtain the approximation in three dimensions

$$\underline{\mathbf{u}}_{ray}(t, \mathbf{x}_R) \simeq -\frac{1}{2\pi c_{max} \ell_{max}^{3/2} x_R^{1/2}} \times \mathrm{Re}\left(\frac{\partial \underline{\boldsymbol{\mathcal{G}}}_{ray}}{\partial q_{max}} H\left(t - \widetilde{T}_{max}\right) + \underline{\boldsymbol{\mathcal{G}}}_{ray}\left(c_{max}^{-1}\right) d_{max} \delta\left(t - \widetilde{T}_{max}\right)\right), \tag{9.1.127}$$

where expression (9.1.122) has been used to convert from ξ to t. This expression describes the geometrical arrival immediately followed by a 'head-wave' like arrival.

To find the tunnelling signal, we need to solve for the 'ray' in the non-geometrical layers, i.e. $x_R = X_{ray}(p_{ray})$. We then expand

$$p = p_{ray} + \epsilon, \tag{9.1.128}$$

but we can take

$$q_{max} = \mathrm{i}\left(p_{ray}^2 - \frac{1}{c_{max}^2}\right)^{1/2} = \mathrm{i}\, q_{ray}, \tag{9.1.129}$$

say, a constant. At $p = p_{ray}$, we have

$$\widetilde{T}_{ray}(p_{ray}, \mathbf{x}_R) \simeq T_{ray}(p_{ray}, z_R) + q_{max} \mathrm{d}_{max} \tag{9.1.130}$$

$$\frac{\partial \widetilde{T}_{ray}}{\partial p} = x_R - X_{ray}(p, z_R) - \frac{p}{q_{max}} \mathrm{d}_{max} \simeq 0 \tag{9.1.131}$$

$$\frac{\partial^2 \widetilde{T}_{ray}}{\partial p^2} \simeq -\frac{\mathrm{d}X_{ray}}{\mathrm{d}p}. \tag{9.1.132}$$

Thus

$$t \simeq T_{ray}(\mathbf{x}_R) + \mathrm{i}\, q_{ray} d_{max} - \frac{1}{2}\frac{\mathrm{d}X_{ray}}{\mathrm{d}p}\epsilon^2, \tag{9.1.133}$$

which can be solved for ϵ

$$\epsilon \simeq -\mathrm{i}\left(\frac{\mathrm{d}X_{\mathrm{ray}}}{\mathrm{d}p}\right)^{-1/2}(u - \mathrm{i}v), \tag{9.1.134}$$

where

$$\begin{matrix} u \\ v \end{matrix} = \left(\left\{\left(t - T_{\mathrm{ray}}\right)^2 + q_{\mathrm{ray}}^2 d_{\mathrm{max}}^2\right\}^{1/2} \pm t \mp T_{\mathrm{ray}}\right)^{1/2}. \tag{9.1.135}$$

Differentiating expression (9.1.133) and substituting in result (8.2.68), with approximation (9.1.130) we obtain in three dimensions

$$\boxed{\underline{\mathbf{u}}_{\mathrm{tunnel}}(t, \mathbf{x}_{\mathrm{R}}) \simeq \frac{1}{2\pi}\left(\frac{p_{\mathrm{ray}}}{x_{\mathrm{R}}\left(\mathrm{d}X_{\mathrm{ray}}/\mathrm{d}p\right)}\right)^{1/2} \mathrm{Re}\left(\underline{\boldsymbol{\mathcal{G}}}_{\mathrm{ray}}(p_{\mathrm{ray}})\Delta\left(t - \widetilde{T}_{\mathrm{ray}}\right)\right).} \tag{9.1.136}$$

This describes the tunnelling signal where both $\underline{\boldsymbol{\mathcal{G}}}_{\mathrm{ray}}(p_{\mathrm{ray}})$ and $\widetilde{T}_{\mathrm{ray}} = T_{\mathrm{ray}} + \mathrm{i}\, q_{\mathrm{ray}} d_{\mathrm{max}}$ are complex.

This analysis of the tunnelling wave applies to all the examples illustrated in Figures 9.12, 9.13, 9.14 and 9.15. However, in the thin-layer case (Figure 9.15), the situation is more complicated due to reverberations within the layer. At small ranges a series of multiple reverberations within the layer will exist. At large ranges, these will still exist but will become evanescent. Because the reverberatory part is evanescent, the reverberations will all be centred on the same arrival time and the signals will combine. If the decay per reverberation is significant (a frequency-dependent criterion depending on the magnitude of $\omega q_{\mathrm{ray}} d_{\mathrm{max}}$), then the leading term (9.1.136) will be a good approximation. If the layer is thin and the decay small, then it will be necessary to consider the reverberation series. This is investigated further in Section 9.3.4.1.

9.2 First-motion approximations for WKBJ seismograms

In Chapter 8 we introduced the WKBJ seismogram method (Section 8.4.1). We considered an example (Figures 8.10 and 8.11), where three singularities exist in the waveform corresponding to the three geometrical arrivals. It is straightforward to find approximations near these times. We approximate the $\widetilde{T}_{\mathrm{ray}}$ by a second-order Taylor expansion. The first derivative is zero at the geometrical arrival (8.4.10). Thus

$$\widetilde{T}_{\mathrm{ray}}(p, \mathbf{x}_{\mathrm{R}}) \simeq T_{\mathrm{ray}}(p_{\mathrm{ray}}) - \frac{1}{2}\frac{\mathrm{d}X_{\mathrm{ray}}}{\mathrm{d}p}(p - p_{\mathrm{ray}})^2. \tag{9.2.1}$$

Although the results are superficially similar to those in the previous section using the Cagniard method, there are two important differences: the slowness used is always real; and the saddle points can have either orientation ($\mathrm{d}X_{\mathrm{ray}}/\mathrm{d}p > 0$ or < 0). The method remains valid for turning rays, which we now consider.

9.2.1 An arrival on a forward branch

Consider the example in Figures 8.10 and 8.11. If $\mathrm{d}X_{\mathrm{ray}}/\mathrm{d}p < 0$, as at $p = p_A$ and p_B, the function has a minimum at the geometrical arrival time (T_A or T_B), and

$$p \simeq p_{\mathrm{ray}} \pm \sqrt{-\frac{2(t - T_{\mathrm{ray}})}{\mathrm{d}X_{\mathrm{ray}}/\mathrm{d}p}}, \tag{9.2.2}$$

when $t = \widetilde{T}_{\mathrm{ray}}$. The gradient can be approximated as

$$\frac{\partial \widetilde{T}_{\mathrm{ray}}}{\partial p} \simeq \sqrt{-2(\mathrm{d}X_{\mathrm{ray}}/\mathrm{d}p)(t - T_{\mathrm{ray}})} \quad \text{for} \quad t > T_{\mathrm{ray}}, \tag{9.2.3}$$

and expression (8.4.7) by

$$\underline{\mathbf{u}}_{\mathrm{ray}}(t, \mathbf{x}_{\mathrm{R}}) \simeq -\frac{1}{2\pi}\left(-\frac{p_{\mathrm{ray}}}{x_{\mathrm{R}}\,(\mathrm{d}X_{\mathrm{ray}}/\mathrm{d}p)}\right)^{1/2} \mathrm{Im}\left(\underline{\mathcal{G}}_{\mathrm{ray}}(p_{\mathrm{ray}})\,\Delta(t - T_{\mathrm{ray}})\right). \tag{9.2.4}$$

This is a common situation for turning rays, with $\underline{\mathcal{G}}_{\mathrm{ray}}$ imaginary due to the turning point (cf. equation (7.2.163)). Then

$$\boxed{\underline{\mathbf{u}}_{\mathrm{ray}}(t, \mathbf{x}_{\mathrm{R}}) \simeq -\frac{1}{2\pi}\left(-\frac{p_{\mathrm{ray}}}{x_{\mathrm{R}}\,(\mathrm{d}X_{\mathrm{ray}}/\mathrm{d}p)}\right)^{1/2} \mathrm{Im}\left(\underline{\mathcal{G}}_{\mathrm{ray}}(p_{\mathrm{ray}})\right)\,\delta(t - T_{\mathrm{ray}}).} \tag{9.2.5}$$

Note that this has the same form as the direct ray (8.2.70) or (9.1.5).

9.2.2 A reversed branch arrival

Sometimes $\mathrm{d}X_{\mathrm{ray}}/\mathrm{d}p > 0$ for turning rays, as at $p = p_C$ in Figure 8.10 (Section 2.4.4 and equation (2.3.26)) so for $t < T_{\mathrm{ray}}$ on the real axis

$$p \simeq p_{\mathrm{ray}} \pm \sqrt{\frac{2(T_{\mathrm{ray}} - t)}{\mathrm{d}X_{\mathrm{ray}}/\mathrm{d}p}}. \tag{9.2.6}$$

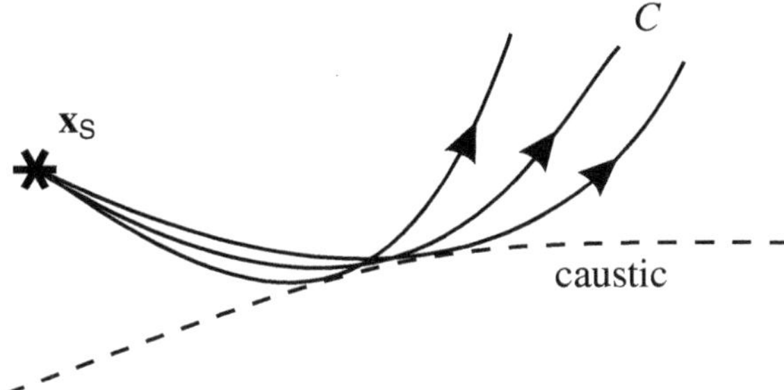

Fig. 9.19. Rays corresponding to the situation in Figure 8.10, where ray C has touched a caustic.

Turning rays with $\mathrm{d}X_{\mathrm{ray}}/\mathrm{d}p > 0$ are illustrated in Figure 9.19. The gradient near the saddle point is approximately

$$\frac{\partial \widetilde{T}_{\mathrm{ray}}}{\partial p} \simeq \sqrt{2(\mathrm{d}X_{\mathrm{ray}}/\mathrm{d}p)(T_{\mathrm{ray}} - t)} \quad \text{for} \quad t > T_{\mathrm{ray}}, \tag{9.2.7}$$

and we obtain

$$\boxed{\underline{\mathbf{u}}_{\mathrm{ray}}(t, \mathbf{x}_{\mathrm{R}}) \simeq \frac{1}{2\pi} \left(\frac{p_{\mathrm{ray}}}{x_{\mathrm{R}}\,(\mathrm{d}X_{\mathrm{ray}}/\mathrm{d}p)} \right)^{1/2} \mathrm{Re}\left(\underline{\mathcal{G}}_{\mathrm{ray}}(p_{\mathrm{ray}})\, \Delta(t - T_{\mathrm{ray}}) \right).} \tag{9.2.8}$$

For a simple turning ray this reduces to

$$\underline{\mathbf{u}}_{\mathrm{ray}}(t, \mathbf{x}_{\mathrm{R}}) \simeq -\frac{1}{2\pi} \left(\frac{p_{\mathrm{ray}}}{x_{\mathrm{R}}\,(\mathrm{d}X_{\mathrm{ray}}/\mathrm{d}p)} \right)^{1/2} \mathrm{Im}\left(\underline{\mathcal{G}}_{\mathrm{ray}}(p_{\mathrm{ray}}) \right) \bar{\delta}(t - T_{\mathrm{ray}}). \tag{9.2.9}$$

This is the Hilbert transform of the direct ray. In the language of equation (5.2.70), the KMAH index for this geometrical ray is $\sigma_{\mathrm{ray}} = 1$ as the ray has touched one caustic (Figure 9.19).

9.2.3 An arrival with two turning points

In the above situation we have assumed that $\underline{\mathcal{G}}_{\mathrm{ray}}$ is imaginary, as for a single turning ray. The caustic exists because the velocity gradient is large enough that $\mathrm{d}X_{\mathrm{ray}}/\mathrm{d}p > 0$ (Section 2.4.4, Figure 2.22). However, caustics don't really need very special conditions. Consider a ray with a double bounce, as illustrated in Figure 9.20. Normally $\underline{\mathcal{G}}_{\mathrm{ray}}$ will be real, containing a factor $(-\mathrm{i}\ \mathrm{sgn}(\omega))^2 = -1$, and if the velocity gradient is uniform, $\mathrm{d}X_{\mathrm{ray}}/\mathrm{d}p < 0$ (just double the value for a

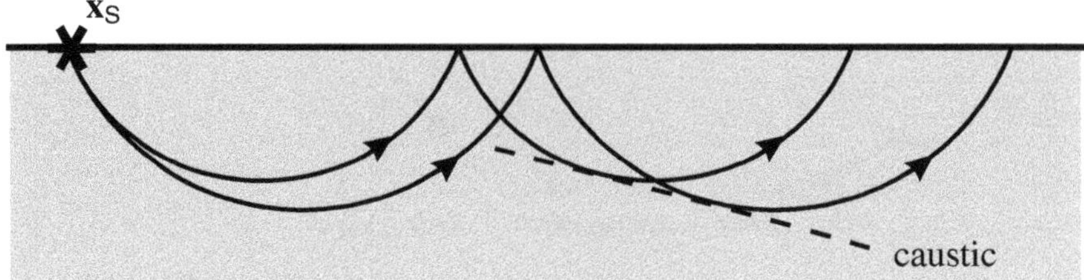

Fig. 9.20. Rays with two turning points forming a caustic.

single-bounce ray). Thus the first-motion approximation will be

$$\underline{\mathbf{u}}_{\text{ray}}(t, \mathbf{x}_R) \simeq -\frac{1}{2\pi^2} \frac{\mathrm{d}}{\mathrm{d}t} \operatorname{Im}\left(\Lambda(t) * \left(-\frac{p_{\text{ray}}}{x_R\,(\mathrm{d}X_{\text{ray}}/\mathrm{d}p)}\right)^{1/2} \frac{\underline{\boldsymbol{\mathcal{G}}}_{\text{ray}}(p)}{\sqrt{(t - T_{\text{ray}})}}\right) \tag{9.2.10}$$

$$\simeq -\frac{1}{2\pi} \left(-\frac{p_{\text{ray}}}{x_R\,(\mathrm{d}X_{\text{ray}}/\mathrm{d}p)}\right)^{1/2} \operatorname{Re}\left(\underline{\boldsymbol{\mathcal{G}}}_{\text{ray}}(p_{\text{ray}})\right) \bar{\delta}(t - T_{\text{ray}}), \tag{9.2.11}$$

i.e. a Hilbert transform as the ray has touched a caustic (see Figure 9.20).

9.2.3.1 Aside – a min–max phase

Signals that are Hilbert transformed are normally more difficult to pick. The direct pulse is impulsive, but its Hilbert transform is emergent. The Hilbert transform of a causal pulse is always acausal (but this does not make the complete signal acausal as the Hilbert transform arises from the first-motion approximation which is only valid in a small time window). An alternative argument that does not require knowledge of the Hilbert transform indicates why a double-bounce ray will be difficult to pick. The double-bounce ray is an example of a *min–max* signal. For a normal reflection, the ray is a *min–min* path. If the reflection point is perturbed so the reflection no longer satisfies Snell's law, then the travel time is increased. This is most easily seen geometrically, without algebra, by considering the image point of the source (see Figure 9.21). The path, a straight line, from the image point is lengthened when the reflection point is perturbed. This is true whatever direction the perturbation of the reflection point, either in the ray plane or normal to it. Therefore the path is called min–min. In contrast, if the reflection point of a double-bounce ray is perturbed in the ray plane, the travel time is decreased. This is most easily seen by considering a large perturbation where the reflection point approaches the source or receiver. Then the travel time is reduced to the single-bounce ray which is less than the double-bounce ray. As the travel time is still increased for a perturbation normal to the ray plane, the ray is known as min–max. As it is min–max, the exact analytic response has a Hilbert

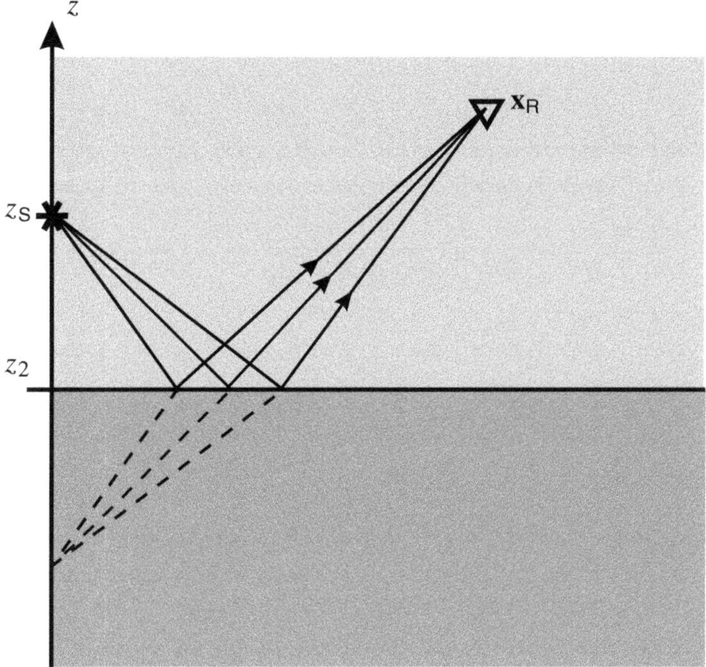

Fig. 9.21. A reflected ray path, equivalent to the path from the image point, lengthens when the reflection point is perturbed (in any direction).

transform (9.2.11). Another feature of a min–max phase is that if the interface at the reflector is imperfect, as in the real Earth it must be, then reflections from points other than the mid-point will arrive earlier. The observed signal will be emergent. This is a well-known phenomenon for signals such as *PP* in the whole Earth.

9.2.4 A reflection

For a reflection $\mathrm{d}X_{\mathrm{ray}}/\mathrm{d}p > 0$ (always), so $\widetilde{T}_{\mathrm{ray}}$ is always maximum. Thus

$$\boxed{\underline{\mathbf{u}}_{\mathrm{ray}}(t, \mathbf{x}_{\mathrm{R}}) \simeq -\frac{1}{2\pi^2}\frac{\mathrm{d}}{\mathrm{d}t}\,\mathrm{Im}\left(\Lambda(t) * \frac{p^{1/2}\underline{\boldsymbol{\mathcal{G}}}_{\mathrm{ray}}(p)}{\sqrt{x_{\mathrm{R}}(\mathrm{d}X_{\mathrm{ray}}/\mathrm{d}p)(T_{\mathrm{ray}} - t)}}\right).} \tag{9.2.12}$$

$\underline{\boldsymbol{\mathcal{G}}}_{\mathrm{ray}}$ may be complex if we have a total reflection. For the real part we obtain

$$\underline{\mathbf{u}}_{\mathrm{ray}}(t, \mathbf{x}_{\mathrm{R}}) \simeq -\frac{1}{2\pi^2}\frac{\mathrm{d}}{\mathrm{d}t}\,\bar{\lambda}(t) * \frac{\mathrm{Re}\left(p^{1/2}\underline{\boldsymbol{\mathcal{G}}}_{\mathrm{ray}}(p)\right)}{\sqrt{x_{\mathrm{R}}\,(\mathrm{d}X_{\mathrm{ray}}/\mathrm{d}p)(T_{\mathrm{ray}} - t)}} \tag{9.2.13}$$

$$\simeq \frac{1}{2\pi}\,\frac{\mathrm{Re}\left(p_{\mathrm{ray}}^{1/2}\,\underline{\boldsymbol{\mathcal{G}}}_{\mathrm{ray}}(p_{\mathrm{ray}})\right)}{\sqrt{x_{\mathrm{R}}\,(\mathrm{d}X_{\mathrm{ray}}/\mathrm{d}p)}}\,\delta(t - T_{\mathrm{ray}}). \tag{9.2.14}$$

Note the two Hilbert transforms (both acausal) combine to give the causal partial reflection – awkward but correct. For the imaginary part

$$\underline{\mathbf{u}}_{\text{ray}}(t, \mathbf{x}_{\mathrm{R}}) \simeq -\frac{1}{2\pi^2} \frac{\mathrm{d}}{\mathrm{d}t} \lambda(t) * \frac{\mathrm{Im}\left(p^{1/2} \underline{\mathcal{G}}_{\text{ray}}(p)\right)}{\sqrt{x_{\mathrm{R}}\, (\mathrm{d}X_{\text{ray}}/\mathrm{d}p)(T_{\text{ray}} - t)}} \tag{9.2.15}$$

$$\simeq -\frac{1}{2\pi} \frac{\mathrm{Im}\left(p_{\text{ray}}^{1/2} \underline{\mathcal{G}}_{\text{ray}}(p_{\text{ray}})\right)}{\sqrt{x_{\mathrm{R}}\, (\mathrm{d}X_{\text{ray}}/\mathrm{d}p)}} \bar{\delta}(t - T_{\text{ray}}). \tag{9.2.16}$$

Combining these results we obtain

$$\underline{\mathbf{u}}_{\text{ray}}(t, \mathbf{x}_{\mathrm{R}}) \simeq \frac{1}{2\pi} \left(\frac{p_{\text{ray}}}{x_{\mathrm{R}}(\mathrm{d}X_{\text{ray}}/\mathrm{d}p)}\right)^{1/2} \mathrm{Re}\left(\underline{\mathcal{G}}_{\text{ray}}(p_{\text{ray}})\, \Delta(t - T_{\text{ray}})\right), \tag{9.2.17}$$

in complete agreement with the expression (9.1.55). Thus the Cagniard and WKBJ methods both model reflections, but the construction and contributing slownesses are quite different.

9.2.5 A general first-motion approximation – the KMAH indices

Above we have derived the first-motion approximation for various signals. These can all be combined in one expression. Suppose the $\underline{\mathcal{G}}_{\text{ray}}$ is written as

$$\boxed{\underline{\mathcal{G}}_{\text{ray}}(p) = \left(\prod_{\text{ray}} \mathcal{T}_{ij}\right) \mathrm{e}^{-\mathrm{i}\frac{\pi}{2}\mathrm{sgn}(\omega)\tilde{\sigma}_{\text{ray}}(p,\mathcal{L}_n)}\, \mathbf{g}\,\mathbf{g}^{\mathrm{T}} = \underline{\tilde{\mathcal{G}}}_{\text{ray}}(p) \mathrm{e}^{-\mathrm{i}\frac{\pi}{2}\mathrm{sgn}(\omega)\tilde{\sigma}_{\text{ray}}(p,\mathcal{L}_n)},} \tag{9.2.18}$$

i.e. a product of reflection/transmission coefficients (that may be complex), the polarization dyadic and a phase factor due to turning points. The function $\tilde{\sigma}_{\text{ray}}(p, \mathcal{L}_n)$ counts the number of turning points. The notation emphasizes that this is a count of caustics in the transform, p, domain, i.e. an extension of the KMAH index to the p domain. Substituting in result (8.4.7) we have

$$\boxed{\begin{aligned}\underline{\mathbf{u}}_{\text{ray}}(t, \mathbf{x}_{\mathrm{R}}) &\simeq -\frac{1}{2^{3/2}\pi^2 x_{\mathrm{R}}^{1/2}} \\ &\times \frac{\mathrm{d}}{\mathrm{d}t} \mathrm{Im}\left(\Lambda(t - T_{\text{ray}}) * \sum_{\tilde{T}_{\text{ray}}(p,\mathbf{x}_{\mathrm{R}})=t} \frac{p^{1/2}\underline{\tilde{\mathcal{G}}}_{\text{ray}} \mathrm{e}^{-\mathrm{i}\pi\tilde{\sigma}_{\text{ray}}(p,\mathcal{L}_n)/2}}{|\partial\tilde{T}_{\text{ray}}/\partial p|}\right).\end{aligned}} \tag{9.2.19}$$

This is a general first-motion approximation for geometrical rays including direct rays, reflections, transmissions, turning rays, etc. with or without caustics. If $\mathrm{d}X_{\text{ray}}/\mathrm{d}p > 0$ for a geometrical ray, the first-motion approximation for the arrival (9.2.19) is

$$\underline{\mathbf{u}}_{\mathrm{ray}}(t, \mathbf{x}_{\mathrm{R}}) \simeq \frac{1}{2\pi} \operatorname{Re}\left(\frac{p_{\mathrm{ray}}^{1/2} \underline{\tilde{\mathcal{G}}}_{\mathrm{ray}} \mathrm{e}^{-\mathrm{i}\pi\tilde{\sigma}_{\mathrm{ray}}(p,\mathcal{L}_n)/2}}{\sqrt{x_{\mathrm{R}}\,(\mathrm{d}X_{\mathrm{ray}}/\mathrm{d}p)}} \Delta(t - T_{\mathrm{ray}}) \right). \quad (9.2.20)$$

The ray theory result (5.2.71) is

$$\underline{\mathbf{u}}_{\mathrm{ray}}(t, \mathbf{x}_{\mathrm{R}}) \simeq \frac{1}{2\pi} \operatorname{Re}\left(\frac{p_{\mathrm{ray}}^{1/2} \underline{\tilde{\mathcal{G}}}_{\mathrm{ray}} \mathrm{e}^{-\mathrm{i}\pi\sigma_{\mathrm{ray}}(\mathbf{x}_{\mathrm{R}},\mathcal{L}_n)/2}}{\sqrt{x_{\mathrm{R}}\,|\mathrm{d}X_{\mathrm{ray}}/\mathrm{d}p|}} \Delta(t - T_{\mathrm{ray}}) \right). \quad (9.2.21)$$

Thus, we must have $\sigma_{\mathrm{ray}}(\mathbf{x}_{\mathrm{R}}, \mathcal{L}_n) = \tilde{\sigma}_{\mathrm{ray}}(p, \mathcal{L}_n)$ for the connection between the KMAH indices in the spatial and transform domains. If $\mathrm{d}X_{\mathrm{ray}}/\mathrm{d}p < 0$

$$\underline{\mathbf{u}}_{\mathrm{ray}}(t, \mathbf{x}_{\mathrm{R}}) \simeq -\frac{1}{2\pi} \operatorname{Im}\left(\frac{p_{\mathrm{ray}}^{1/2} \underline{\tilde{\mathcal{G}}}_{\mathrm{ray}} \mathrm{e}^{-\mathrm{i}\pi\tilde{\sigma}_{\mathrm{ray}}(p,\mathcal{L}_n)/2}}{\sqrt{-x_{\mathrm{R}}\,(\mathrm{d}X_{\mathrm{ray}}/\mathrm{d}p)}} \Delta(t - T_{\mathrm{ray}}) \right) \quad (9.2.22)$$

$$\simeq \frac{1}{2\pi} \operatorname{Re}\left(\mathrm{i}\, \frac{p_{\mathrm{ray}}^{1/2} \underline{\tilde{\mathcal{G}}}_{\mathrm{ray}} \mathrm{e}^{-\mathrm{i}\pi\tilde{\sigma}_{\mathrm{ray}}(p,\mathcal{L}_n)/2}}{\sqrt{-x_{\mathrm{R}}\,(\mathrm{d}X_{\mathrm{ray}}/\mathrm{d}p)}} \Delta(t - T_{\mathrm{ray}}) \right), \quad (9.2.23)$$

and the connection between the KMAH indices in the slowness and spatial domains is $\sigma_{\mathrm{ray}}(\mathbf{x}_{\mathrm{R}}, \mathcal{L}_n) = \tilde{\sigma}_{\mathrm{ray}}(p, \mathcal{L}_n) - 1$. Combining these results we have in general

$$\boxed{\sigma_{\mathrm{ray}}(\mathbf{x}_{\mathrm{R}}, \mathcal{L}_n) = \tilde{\sigma}_{\mathrm{ray}}(p, \mathcal{L}_n) + \frac{1}{2}\left(\operatorname{sgn}\left(\frac{\mathrm{d}X_{\mathrm{ray}}}{\mathrm{d}p} \right) - 1 \right).} \quad (9.2.24)$$

As an example of these general results, let us consider a simple case of a turning ray with a caustic. The ray can be divided into three parts: before the turning point, I; after the turning point but before the caustic, II; and after the caustic, III. This is illustrated in Figure 9.22. The numerical values of the terms in expression (9.2.24) are given in Table 9.1.

We should emphasize that although the first-motion approximation to the WKBJ seismogram (9.2.17) is of interest as it confirms that the method models the arrivals correctly (and how), the exact, band-limited WKBJ result (8.4.14) is so easy to compute that it is normally used. Examples of this for the core rays on forward and reversed branches have already been shown in Figure 8.16.

9.2.6 Head waves

Above we have considered the WKBJ results for reflections, both partial and total (9.2.17). In Section 9.1.3 we considered the first-motion approximations for head

Table 9.1. *Numerical values of terms in expression (9.2.24) in the regions indicated in Figure 9.22*

	$\mathrm{sgn}(\mathrm{d}X_{\mathrm{ray}}/\mathrm{d}p)$	$\tilde{\sigma}_{\mathrm{ray}}$	σ_{ray}
I	1	0	0
II	−1	1	0
III	1	1	1

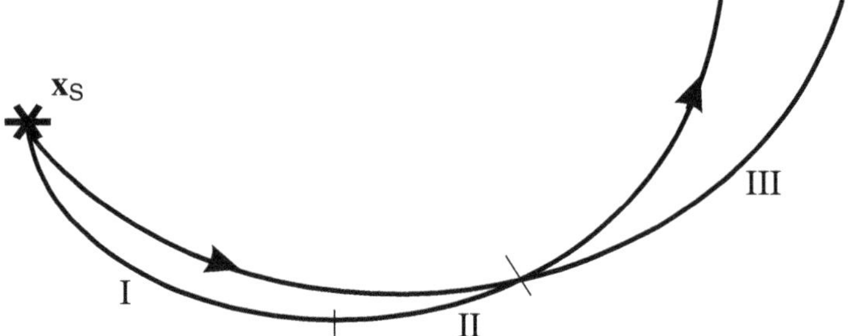

Fig. 9.22. A ray with a turning point and caustic illustrating the KMAH indices in the three regions: I – direct ray before the turning point; II – turning ray before caustic; and III – turning ray after caustic. The numerical results are illustrated in Table 9.1.

waves using the Cagniard method. It remains to show that the WKBJ method also gives the correct head-wave signal.

The function $\widetilde{T}_{\mathrm{ray}}$ always has a maximum for a reflection. The head-wave is caused by the behaviour of $\underline{\boldsymbol{\mathcal{G}}}_{\mathrm{ray}}$ at the critical point $p = p_n = 1/c_n$, say (cf. Section 9.1.3). The difference compared with the Cagniard method is that the branch cut contributes both before and after the critical range, $X_{\mathrm{ray}}(p_n, z_{\mathrm{R}})$ (Figure 9.23). At the critical point, the $\widetilde{T}_{\mathrm{ray}}$ function has slope

$$\ell_n(\mathbf{x}_{\mathrm{R}}) = \frac{\partial \widetilde{T}_{\mathrm{ray}}}{\partial p} = x_{\mathrm{R}} - X_{\mathrm{ray}}(p_n, z_{\mathrm{R}}), \tag{9.2.25}$$

the length of the head-wave segment. The time at the critical slowness is

$$\widetilde{T}_{\mathrm{ray}}(p_n, \mathbf{x}_{\mathrm{R}}) = T_{\mathrm{ray}}(p_n, z_{\mathrm{R}}) + \ell_n(\mathbf{x}_{\mathrm{R}})/c_n = T_n(\mathbf{x}_{\mathrm{R}}), \text{ say,} \tag{9.2.26}$$

the head-wave arrival time, and the slowness value that contributes at time t is

$$p(t, \mathbf{x}_{\mathrm{R}}) = p_n + \frac{t - T_n(\mathbf{x}_{\mathrm{R}})}{\ell_n(\mathbf{x}_{\mathrm{R}})}. \tag{9.2.27}$$

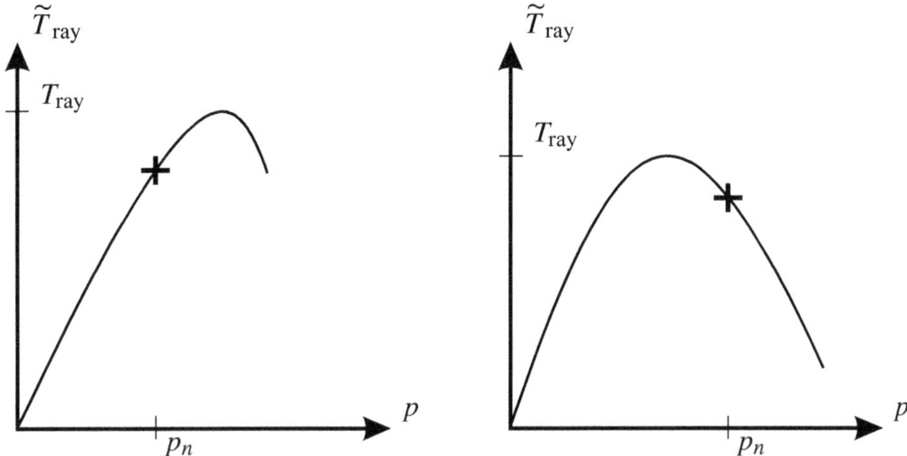

Fig. 9.23. The $\widetilde{T}_{\text{ray}}$ functions for a reflection with a critical point at $p_n = 1/c_n$, illustrated on either side of the critical point $X_{\text{ray}}(1/c_n, \mathbf{x}_{\text{R}})$.

This remains valid whether t is greater or less than $\widetilde{T}_{\text{ray}}(p_n, \mathbf{x}_{\text{R}})$ and whatever the sign of ℓ_n. For expression (9.1.49), we obtain

$$\underline{\mathcal{G}}_{\text{ray}}(p) \simeq \underline{\mathcal{G}}_{\text{ray}}(1/c_n) + \frac{\partial \underline{\mathcal{G}}_{\text{ray}}(1/c_n)}{\partial q_n} \left(\frac{2}{c_n}\right)^{1/2} \left(\frac{1}{c_n} - p\right)^{1/2}. \quad (9.2.28)$$

With approximations (9.2.27) and (9.2.28) near the branch point, the head-wave approximation is

$$\boxed{\underline{\mathbf{u}}_{\text{head}}(t, \mathbf{x}_{\text{R}}) \simeq -\frac{1}{2\pi^2 x_{\text{R}}^{1/2} c_n} \frac{\mathrm{d}}{\mathrm{d}t} \operatorname{Im} \left(\Lambda(t) * |\ell_n|^{-1} \frac{\partial \underline{\mathcal{G}}_{\text{ray}}(p_n)}{\partial q_n} \left(\frac{T_n - t}{\ell_n}\right)^{1/2} \right).} \quad (9.2.29)$$

We must consider two situations $\ell_n > 0$ and $\ell_n < 0$ and two contributions from $t > T_n$ and $t < T_n$ in each case. For simplicity, let us define the functions

$$\mu(t) = H(t)\, t^{1/2} \quad (9.2.30)$$

$$\bar{\mu}(t) = -H(-t)(-t)^{1/2}, \quad (9.2.31)$$

half the integrals of $\lambda(t)$ (B.2.1) and $\bar{\lambda}(t)$ (B.2.4), respectively.

For simplicity, let us assume that $\partial \underline{\mathcal{G}}_{\text{ray}}/\partial p$ is real. When $\ell_n > 0$, expression (9.2.29) becomes

$$\underline{\mathbf{u}}_{\text{head}}(t, \mathbf{x}_{\text{R}}) = -\frac{1}{2\pi^2 c_n x_{\text{R}}^{1/2}} |\ell_n|^{-3/2} \frac{\partial \underline{\mathcal{G}}_{\text{ray}}(p_n)}{\partial q_n} \times \frac{\mathrm{d}}{\mathrm{d}t} \operatorname{Im}\left(\Lambda\,(t - T_n) * \left(\mathrm{i}\,\mu(t) - \bar{\mu}(t)\right)\right) \tag{9.2.32}$$

$$= -\frac{1}{2\pi\, c_n\, \ell^{3/2}\, x_{\text{R}}^{1/2}} \frac{\partial \underline{\mathcal{G}}_{\text{ray}}(p_n)}{\partial q_n} H\,(t - T_n)\,. \tag{9.2.33}$$

This agrees with the Cagniard approximation (9.1.52). Notice that in expression (9.2.32), there is a contribution from before and after the slowness p_n. In the convolution we have $\lambda * \mu - \bar{\lambda} * \bar{\mu} = \pi H$, i.e. overall the two contributions are equal and double, despite the fact that one has two Hilbert transforms.

When $\ell_n < 0$, we have

$$\underline{\mathbf{u}}_{\text{head}}(t, \mathbf{x}_{\text{R}}) = -\frac{1}{2\pi^2 c_n x_{\text{R}}^{1/2}} |\ell_n|^{-3/2} \frac{\partial \underline{\mathcal{G}}_{\text{ray}}(1/c_n)}{\partial q_n} \times \frac{\mathrm{d}}{\mathrm{d}t} \operatorname{Im}\left(\Lambda\,(t - T_n) * \left(\mathrm{i}\,\bar{\mu}(t) - \mu(t)\right)\right) = 0, \tag{9.2.34}$$

where in the convolution we have $\lambda * \bar{\mu} - \bar{\lambda} * \mu = 0$, i.e. overall, the two contributions are equal and opposite, and cancel. There is no head-wave signal for $\ell_n < 0$. In the Cagniard method there is no head-wave signal because the contour does not touch the branch cut; in the WKBJ method there is no signal, because two equal and opposite contributions for $p < 1/c_n$ and $p > 1/c_n$ cancel.

In order to obtain numerically accurate head-wave results, either the doubling for the signal when $\ell_n > 0$, or cancellation when $\ell_n < 0$, it is necessary to sample the branch point adequately. This requires the discrete slowness points to be distributed symmetrically about the branch point, i.e.

$$p = 1/c_n \pm p_i, \tag{9.2.35}$$

with a greater density of points near the branch point. Results illustrating the success of this are illustrated in Figure 9.24. The model parameters are identical to those in Figure 9.3, but the calculations are performed with the WKBJ seismogram algorithm (8.4.14) together with the far-field, two-to-three dimensions conversion (8.2.66). For efficiency, the convolution operator (8.2.66) is performed using a smoothed form of the analytic operator and a rational approximation given by Chapman, Chu Jen-Yi and Lyness (1988). The results are virtually identical to those using the Cagniard method (Figure 9.3) except for some long-period drift

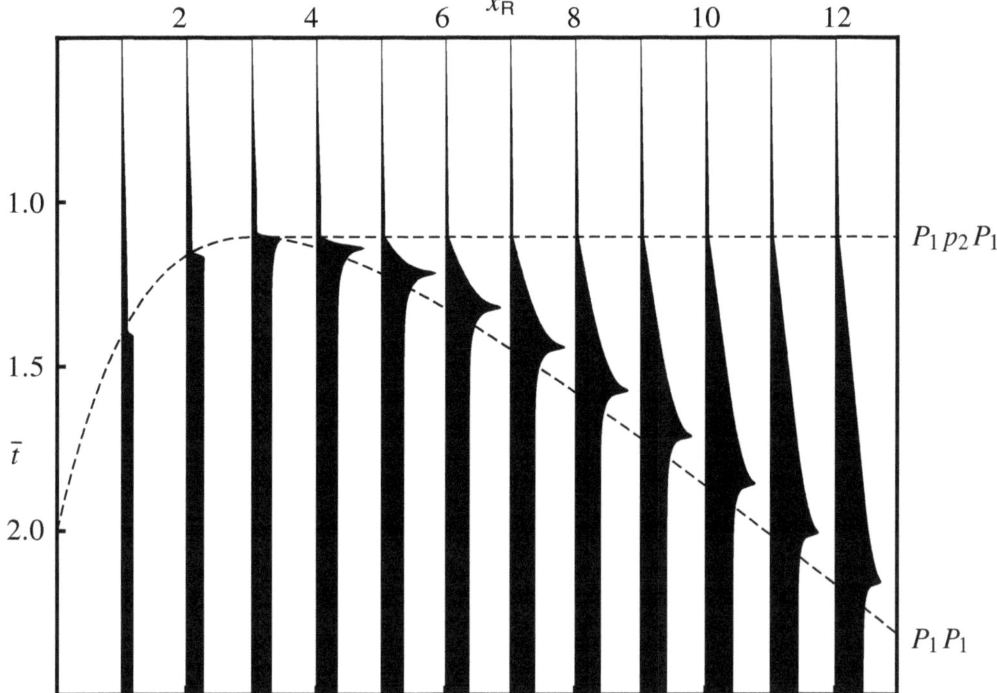

Fig. 9.24. The reflections in an acoustic model with two homogeneous half-spaces calculated using the WKBJ seismogram method (8.4.14). The head wave and total reflection are visible. The model parameters and other details are all as in Figure 9.3.

due to the acausal part in the signal construction (Figure 9.23) and the rational approximation. No error due to the acausal branch cut and equation (9.2.34) is visible. The WKBJ seismogram is faster to compute as it only depends on real ray-tracing results.

As an aside we might mention that the same doubling or cancellation at the branch points must also occur in the slowness integral of the spectral method (Section 8.5), as the same real slowness and inverse Fourier transforms are performed (numerically in the spectral integral, analytically in the WKBJ method). To obtain numerically accurate head-wave results with the spectral method it is necessary to distribute discrete points as described above. The success of this is illustrated below in Figure 9.31.

9.2.7 The Airy caustic

A major advantage of the WKBJ seismogram method is its ability to handle multiple arrivals forming caustics. The simplest form of caustic is an Airy caustic (so-called as its spectrum is given by an Airy function – see equation (9.3.53)), where two geometrical rays coalesce. This was first investigated in a seismic context by Jeffreys (1939) for the *PKP* caustic (Figure 8.16). Rather than approximate the phase function by a second-order Taylor expansion (9.2.1), we need a third-order expansion. We denote the inflection point by $p = p_A(z_\mathrm{R})$, where

$$\frac{\mathrm{d}X_{\mathrm{ray}}}{\mathrm{d}p}(p_A, z_\mathrm{R}) = 0. \tag{9.2.36}$$

The phase function can then be approximated by

$$\widetilde{T}_{\mathrm{ray}}(p, \mathbf{x}_\mathrm{R}) \simeq \widetilde{T}_A(\mathbf{x}_\mathrm{R}) + \ell_A(\mathbf{x}_\mathrm{R})\,(p - p_A(z_\mathrm{R})) - X_A''(\mathbf{x}_\mathrm{R})\,(p - p_A(z_\mathrm{R}))^3\,/6, \tag{9.2.37}$$

replacing (9.2.1). In this expression

$$\widetilde{T}_A(\mathbf{x}_\mathrm{R}) = \widetilde{T}_{\mathrm{ray}}(p_A, \mathbf{x}_\mathrm{R}) \tag{9.2.38}$$

$$= T_{\mathrm{ray}}(p_A, z_\mathrm{R}) + \ell_A(\mathbf{x}_\mathrm{R})p_A(z_\mathrm{R}) \tag{9.2.39}$$

$$X_A''(z_\mathrm{R}) = \frac{\mathrm{d}^2 X_{\mathrm{ray}}}{\mathrm{d}p^2}(p_A, z_\mathrm{R}) \tag{9.2.40}$$

$$\ell_A(\mathbf{x}_\mathrm{R}) = x_\mathrm{R} - X_{\mathrm{ray}}(p_A, z_\mathrm{R}). \tag{9.2.41}$$

For the sake of definiteness, we consider only the case where $X_A'' > 0$, and a simple turning point, i.e. $\underline{\boldsymbol{\mathcal{G}}}_{\mathrm{ray}}(p_A)$ is imaginary (cf. equation (9.2.5)). We illustrate the travel-time function, $T_{\mathrm{ray}}(X)$ and the delay function $\tau_{\mathrm{ray}}(p, z_\mathrm{R})$ in Figure 9.25.

To construct the seismogram, we need to consider three cases: the illuminated region, $\ell_A > 0$; the caustic, $\ell_A = 0$; and the shadow region, $\ell_A < 0$. These are illustrated in Figure 9.26.

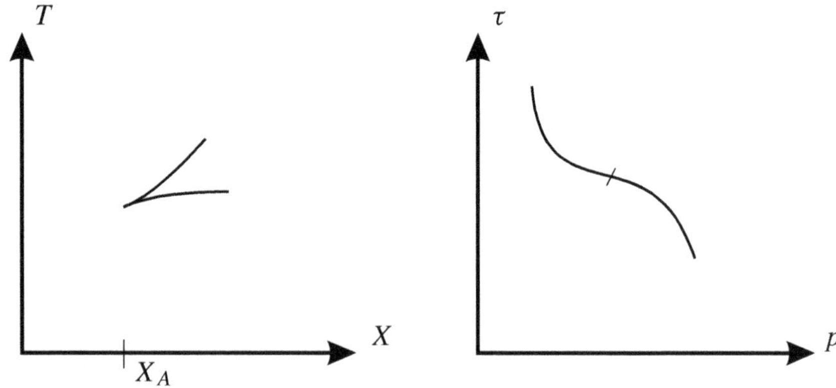

Fig. 9.25. The travel-time function, $T_{\mathrm{ray}}(X)$ and the delay function $\tau_{\mathrm{ray}}(p, z_\mathrm{R})$ for an Airy caustic with $X_A'' > 0$.

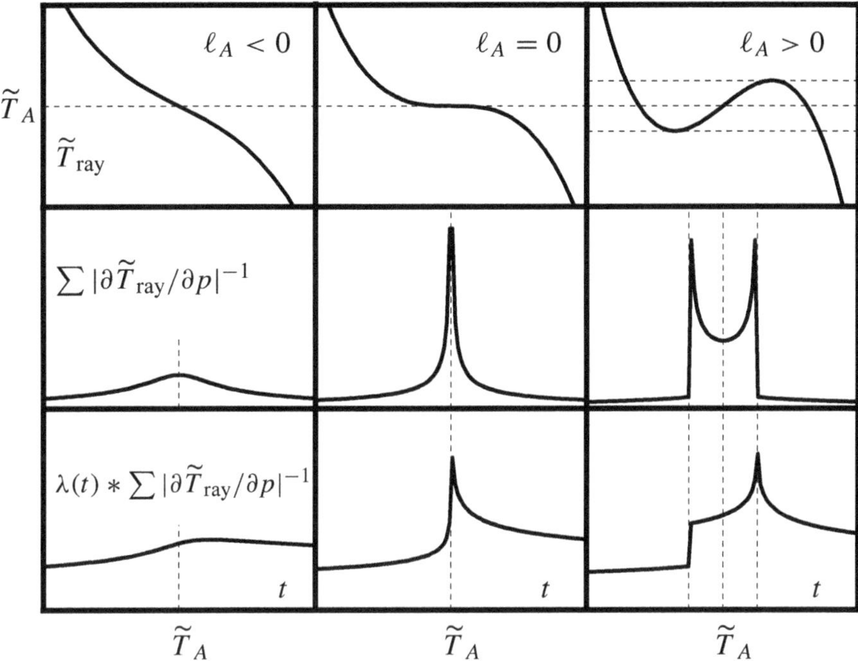

Fig. 9.26. The phase function $\widetilde{T}_{\text{ray}}$, $\sum |\partial \widetilde{T}_{\text{ray}}/\partial p|^{-1}$ and $\lambda(t) * \sum |\partial \widetilde{T}_{\text{ray}}/\partial p|^{-1}$ (in the three rows) for the three regions $\ell_A < 0$, $\ell_A = 0$ and $\ell_A > 0$ (in the three columns) of an Airy caustic.

Near the caustic the response is given by (8.4.7)

$$\underline{\mathbf{u}}_{\text{Airy}}(t, \mathbf{x}_{\text{R}}) \simeq -\frac{p_A^{1/2} \text{Im}\left(\underline{\mathcal{G}}_{\text{ray}}(p_A)\right)}{2^{3/2}\pi^2 x_{\text{R}}^{1/2}} \frac{\text{d}}{\text{d}t} \left(\lambda(t) * \sum_{\widetilde{T}_{\text{ray}}(p, \mathbf{x}_{\text{R}})=t} \frac{1}{|\partial \widetilde{T}_{\text{ray}}/\partial p|} \right), \tag{9.2.42}$$

where $\widetilde{T}_{\text{ray}}$ is approximated by expansion (9.2.37). Let us just investigate the time series

$$\lambda(t) * \sum_{t=\widetilde{T}_{\text{ray}}(p, \mathbf{x}_{\text{R}})} \frac{1}{|\partial \widetilde{T}_{\text{ray}}/\partial p|} = \int_{-\infty}^{\infty} \frac{1}{(t-t')^{1/2}} \sum_{t'=\widetilde{T}_{\text{ray}}(p, \mathbf{x}_{\text{R}})} \frac{1}{|\partial \widetilde{T}_{\text{ray}}/\partial p|} \, \text{d}t'. \tag{9.2.43}$$

Changing the variable of integration to p, the convolution integral can be rewritten

$$\int_{(\,)>0} \frac{\text{d}p}{\left(t - \widetilde{T}_A - \ell_A(p - p_A) + X_A''(p - p_A)^3/6\right)^{1/2}}, \tag{9.2.44}$$

where the integral is over all values of p for which the expression in the bracket is positive (so the square root is real).

Integrals of this form have been investigated by Burridge (1963*a,b*). Integral (9.2.44) can be reduced to a standard form, the function $C(t, y)$ which is defined and described in Appendix D.2. With the change of variable

$$p = p_A + 2(3/X''_A)^{1/3} z_{\mathrm{R}}, \tag{9.2.45}$$

integral (9.2.44) reduces to (D.2.11) and the response (9.2.42) can be written

$$\boxed{\begin{aligned}\underline{\mathbf{u}}_{\mathrm{Airy}}(t, \mathbf{x}_{\mathrm{R}}) \simeq & -\frac{1}{\pi^2}\left(\frac{3}{X''_A}\right)^{1/3}\left(\frac{p_A}{2x_{\mathrm{R}}}\right)^{1/2} \mathrm{Im}\left(\underline{\boldsymbol{\mathcal{G}}}_{\mathrm{ray}}(p_A)\right) \\ & \times \frac{\mathrm{d}}{\mathrm{d}t} C\left(t - \widetilde{T}_A, \frac{2\ell_A}{(9X''_A)^{1/3}}\right).\end{aligned}} \tag{9.2.46}$$

The dimensions of this expression check as in three dimensions unit($\underline{\boldsymbol{\mathcal{G}}}$) = $[\mathrm{M}^{-1}\mathrm{L}^2\mathrm{T}]$ and unit($\underline{\mathbf{u}}$) = $[\mathrm{M}^{-1}\mathrm{T}]$ as unit(C) = $[\mathrm{T}^{-1/6}]$ here. When $\ell_A \neq 0$, the change of variable

$$p = p_A + (8|\ell_A|/X''_A)^{1/2} z_{\mathrm{R}}, \tag{9.2.47}$$

reduces the integral (9.2.44) to $C(t, \pm 1)$. Thus overall we have

$$\begin{aligned}\underline{\mathbf{u}}_{\mathrm{Airy}}(t, \mathbf{x}_{\mathrm{R}}) \simeq & -\frac{1}{2\pi}\left(\frac{p_A}{x_{\mathrm{R}}}\right)^{1/2} \mathrm{Im}\left(\underline{\boldsymbol{\mathcal{G}}}_{\mathrm{ray}}(p_A)\right) \\ & \times \frac{2^{1/2}}{\pi}\left(\frac{3}{X''_A}\right)^{1/3} \frac{\mathrm{d}}{\mathrm{d}t} C\left(t - \widetilde{T}_A, 0\right) \quad \text{if} \quad \ell_A = 0, \qquad (9.2.48) \\ & \times \frac{3^{1/2}}{\pi}\left(\frac{2}{\ell_A X''_A}\right)^{1/4} \frac{\mathrm{d}}{\mathrm{d}t} C\left(\frac{3(t - \widetilde{T}_A)(X''_A)^{1/2}}{(2|\ell_A|)^{3/2}}, \mathrm{sgn}(\ell_A)\right), \\ & \quad \text{if} \quad \ell_A \neq 0. \qquad (9.2.49)\end{aligned}$$

The integrals of the waveforms (9.2.48) and (9.2.49) are illustrated in Figure 9.27.

To lowest order, the waves at and near an Airy caustic are described by these simple expressions (9.2.48) and (9.2.49). The waveforms can be derived from the 'standard' functions $C(t, 0)$ and $C(t, \pm 1)$ (Figure D.3) with suitable scaling in amplitude and time. Notice the geometrical arrivals in the illuminated region when $\ell_A > 0$. These are the singularities of $C(t, 1)$ at $t = \pm 1$ – the arrival at $t = -1$ is on the forward branch and at $t = +1$ is the Hilbert transform on the reversed

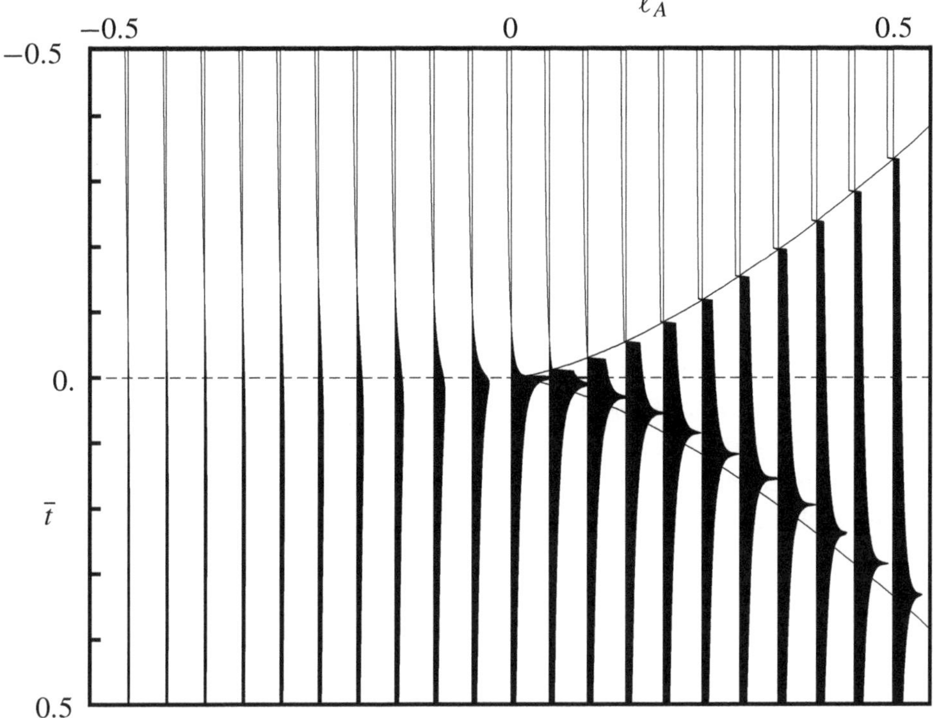

Fig. 9.27. The integrated waveforms near an Airy caustic are given by $C(t - \widetilde{T}_A, 2\ell_A(9X''_A)^{-1/3})$ (9.2.46). The time axis is the reduced travel time, $\bar{t} = t - \widetilde{T}_A$, and the range axis is the reduced range, $\ell_A = x_\mathrm{R} - X_A$, with $X''_A = 1$. To clarify the appearance of the plot, the D.C. level of the waveform at $\ell_A = -0.5$ has been subtracted from all waveforms.

branch. The singularity at $t = 0$ in $C(t, 0)$ is at the caustic. In the shadow, the 'arrival' is a diffracted, smooth pulse, $C(t, -1)$.

The function C, in expressions (9.2.48) and (9.2.49), is centred on $\widetilde{T}_A$ with geometrical arrivals at

$$t = \widetilde{T}_A \pm \frac{(2\ell_A)^{3/2}}{3(X''_A)^{1/2}} = \widetilde{T}_A \pm \Delta T_A, \tag{9.2.50}$$

when $\ell_A > 0$, corresponding to slownesses

$$p = p_A \pm \left(\frac{2\ell_A}{X''_A}\right)^{1/2} \tag{9.2.51}$$

(from the stationary points of the expansion (9.2.37)). The function in expression (9.2.49) simplifies to $C\left((t - \widetilde{T}_A)/\Delta T_A, \operatorname{sgn}(\ell_A)\right)$. At these geometrical rays

$$\frac{\partial^2 \widetilde{T}_\mathrm{ray}}{\partial p^2} = -\frac{\mathrm{d}X_\mathrm{ray}}{\mathrm{d}p} = \mp(2\ell_A X''_A)^{1/2}, \tag{9.2.52}$$

a factor that has been deliberately separated in expression (9.2.49) to show how the geometrical arrivals in it reduce to the standard result (9.2.5) using result (D.2.24).

Similar results apply for a caustic with $X''_A < 0$. In Section 9.3.5, we derive the equivalent spectral result. Higher-order terms involving amplitude variations between the two branches, or differences between the phase function $\widetilde{T}_{\text{ray}}$ and a cubic (9.2.37), can be derived in a similar manner. However, in general it is so simple to use the complete WKBJ expression (8.4.7) without further approximation that higher-order expansions are rarely worthwhile. Nevertheless, the lowest-order term investigated here is of interest as it indicates the behaviour to be expected at and near a caustic. It is interesting to note that in all situations the interfering and diffracted signals near an Airy caustic can be described by the simple functions $C(t, \pm 1)$ and $C(t, 0)$ scaled according to the time delay, $\widetilde{T}_A$, and the time shift, ΔT_A. We should emphasize that normally the WKBJ expression (8.4.14) is evaluated numerically without bothering with the first-motion approximation. An example of this is Figure 8.16 for the *PKP* core caustic.

9.2.8 The Fresnel shadow

If a velocity gradient exists above an interface, the rays turning in the gradient are terminated at the grazing ray and a reflection from the interface exists (Section 2.4.2). The wavefront of the turning rays is 'truncated' by the interface and the grazing ray (with $p = 1/c_1$) defines a shadow edge at $x = X(1/c_1, z_R)$. There is no special behaviour of the turning ray except it does not exist for $p < 1/c_1$. The reflection also terminates at the shadow edge $x = X(1/c_1, z_R)$ with $\mathrm{d}X/\mathrm{d}p \to +\infty$ as $p \to 1/c_1$ from below.

The behaviour near the shadow edge can be separated into two parts: the truncation of the turning wavefront at $p = 1/c_1$; and the interaction of the wave with the interface. In this section we describe the first part which can be investigated easily using the WKBJ seismogram method. We call this a *Fresnel shadow* as the spectrum is given by the Fresnel function (see equation (9.3.60) and Appendix D.3). The interaction of the wave with the interface is more complicated and requires the spectral method (Section 8.5).

Let us define

$$p_1 = 1/c_1 \tag{9.2.53}$$

$$X_1(z_R) = X_{\text{ray}}(1/c_1, z_R) \tag{9.2.54}$$

$$\ell_1(\mathbf{x}_R) = x_R - X_1(z_R) \tag{9.2.55}$$

$$T_1(z_R) = T_{\text{ray}}(1/c_1, z_R) \tag{9.2.56}$$

$$\widetilde{T}_1(\mathbf{x}_R) = T_1(z_R) + p_1\ell_1(\mathbf{x}_R). \tag{9.2.57}$$

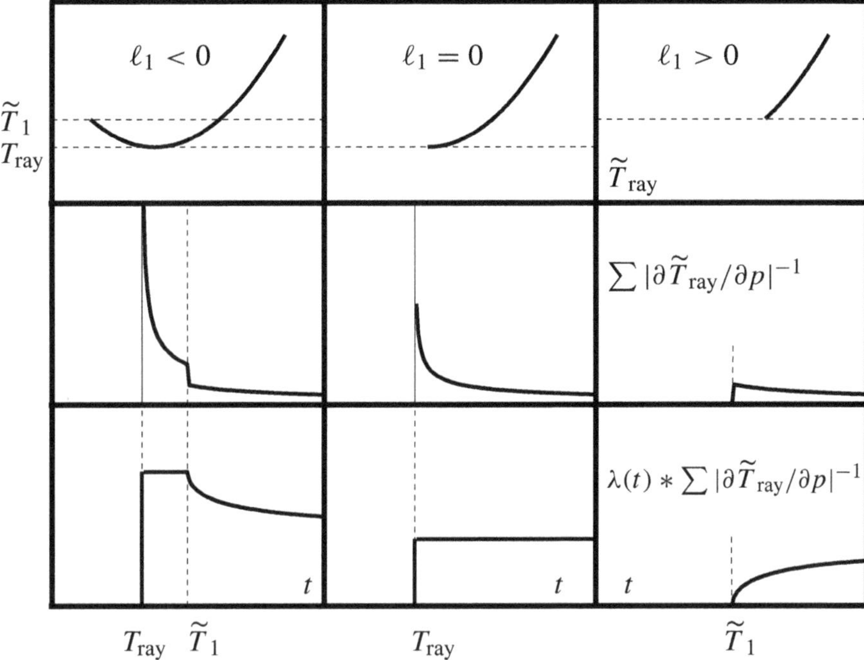

Fig. 9.28. The phase function $\widetilde{T}_{\text{ray}}$, $\sum |\partial \widetilde{T}_{\text{ray}}/\partial p|^{-1}$ and $\lambda(t) * \sum |\partial \widetilde{T}_{\text{ray}}/\partial p|^{-1}$ (in the three rows) for the three regions $\ell_1 < 0$, $\ell_1 = 0$ and $\ell_1 > 0$ (in the three columns) of a Fresnel shadow.

For simplicity, we assume a simple turning ray on a forward branch, so without the interface (or for $p \gg p_1$), expression (9.2.5) holds. Proceeding similarly when an interface exists, solutions of equation (9.2.2) only exist for $p > p_1$ (Figure 9.28). If $x_{\text{R}} < X_1(z_{\text{R}})$ ($\ell_1 < 0$), the stationary point exists at $p = p_{\text{ray}}$ with $t = T_{\text{ray}}$, but for $t > \widetilde{T}_1$ only one solution contributes to the seismogram. Thus we have

$$\underline{\mathbf{u}}_{\text{Fresnel}}(t, \mathbf{x}_{\text{R}}) \simeq -\frac{1}{2\pi} \left(-\frac{p_{\text{ray}}}{x_{\text{R}}\,(\mathrm{d}X_{\text{ray}}/\mathrm{d}p)} \right)^{1/2} \operatorname{Im}\left(\underline{\boldsymbol{\mathcal{G}}}_{\text{ray}}(p_{\text{ray}}) \right) \times \frac{1}{2\pi} \frac{\mathrm{d}}{\mathrm{d}t} \left(\lambda(t) * \left(2 - H(t - \widetilde{T}_1) \right) \lambda(t - T_{\text{ray}}) \right). \tag{9.2.58}$$

For $x_{\text{R}} > X_1(z_{\text{R}})$ ($\ell_1 > 0$), the stationary point would exist with $p_{\text{ray}} < p_1$ and we only have one solution contributing for $t > \widetilde{T}_1$. Thus

$$\underline{\mathbf{u}}_{\text{Fresnel}}(t, \mathbf{x}_{\text{R}}) \simeq -\frac{1}{2\pi} \left(-\frac{p_{\text{ray}}}{x_{\text{R}}\,(\mathrm{d}X_{\text{ray}}/\mathrm{d}p)} \right)^{1/2} \operatorname{Im}\left(\underline{\boldsymbol{\mathcal{G}}}_{\text{ray}}(p_{\text{ray}}) \right) \times \frac{1}{2\pi} \frac{\mathrm{d}}{\mathrm{d}t} \left(\lambda(t) * H(t - \widetilde{T}_1) \lambda(t - T_{\text{ray}}) \right). \tag{9.2.59}$$

The convolutions in results (9.2.58) and (9.2.59) can be simplified using

$$\lambda(t) * H(t-1)\,\lambda(t) = 2Fi(t), \tag{9.2.60}$$

where $Fi(t)$ is defined in equation (D.3.21). Expressions (9.2.58) and (9.2.59) can then be rewritten

$$\underline{\mathbf{u}}_{\text{Fresnel}}(t, \mathbf{x}_{\text{R}}) \simeq -\frac{1}{2\pi}\left(-\frac{p_{\text{ray}}}{x_{\text{R}}\,(\mathrm{d}X_{\text{ray}}/\mathrm{d}p)}\right)^{1/2} \operatorname{Im}\left(\underline{\mathcal{G}}_{\text{ray}}(p_{\text{ray}})\right) \times \frac{\mathrm{d}}{\mathrm{d}t} F\left(\frac{t-T_{\text{ray}}}{\widetilde{T}_1 - T_{\text{ray}}},\ \operatorname{sgn}(\ell_1)\right), \tag{9.2.61}$$

where the 'standard' Fresnel shadow function, $F(t, y)$, is defined in equations (D.3.25)–(D.3.27). The integrals of the waveforms (9.2.61) are illustrated in Figure 9.29.

A diffracted signal with slowness p_1 is generated. At the shadow edge X_1 the signal is reduced to a half the geometrical value (D.3.26). On both sides of the

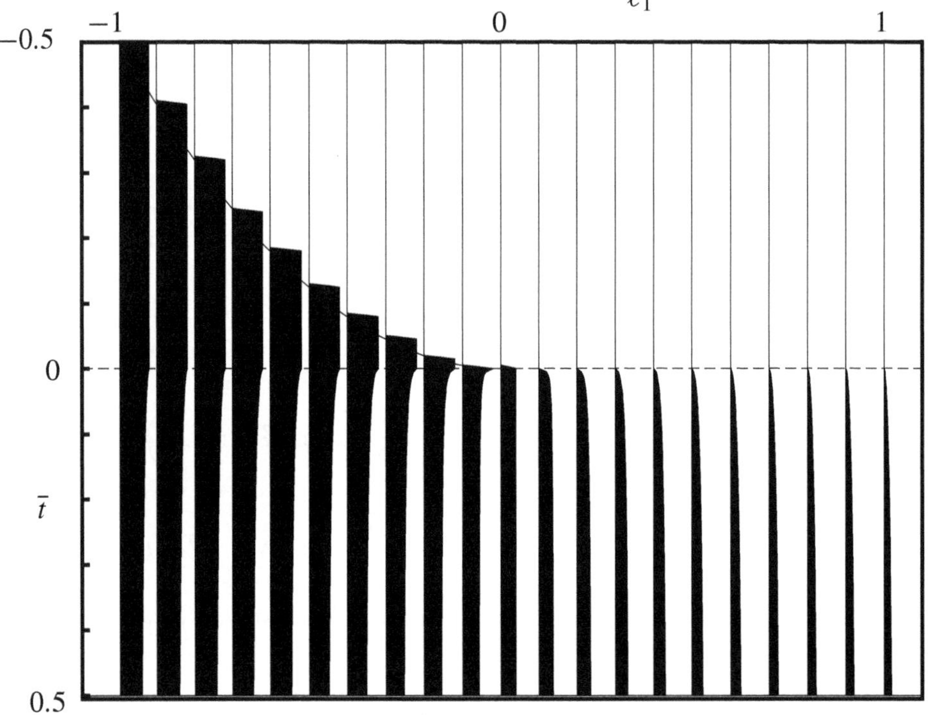

Fig. 9.29. The integrated waveforms near a Fresnel shadow are given by, $F\left((t-T_{\text{ray}})/(\widetilde{T}_1 - T_{\text{ray}}),\ \operatorname{sgn}(\ell_1)\right)$ (9.2.61). The time axis is the reduced travel time, $\bar{t} = t - \widetilde{T}_1$, and the range is the reduced range, $\ell_1 = x_{\text{R}} - X_1$, with $X_1' = -1$.

shadow edge, a diffracted arrival arrives at $\widetilde{T}_1$ with slowness p_1. The differential of $Fi(t)$ (D.3.21) is (cf. equation (D.3.18))

$$\dot{Fi}(t) = \frac{H(t-1)}{2t(t-1)^{1/2}}, \tag{9.2.62}$$

so compared with the geometrical ray $\delta(t - T_{\text{ray}})$, the diffracted pulse is

$$\simeq \pm \frac{1}{2\pi} \frac{1}{(t - \widetilde{T}_1)^{1/2}(\widetilde{T}_1 - T_{\text{ray}})^{1/2}}, \tag{9.2.63}$$

to first order, i.e. it is lower frequency $O(\omega^{-1/2})$. Near the end-point, p_1, we have assumed that the phase function is quadratic, i.e.

$$\widetilde{T} \simeq \widetilde{T}_1 + (p - p_1)\ell_1 - X_1'(p - p_1)^2/2. \tag{9.2.64}$$

Thus the stationary point for the geometrical ray is at

$$p_{\text{ray}} \simeq p_1 + \ell_1 / X_1', \tag{9.2.65}$$

and the geometrical arrival time is

$$T_{\text{ray}} \simeq \widetilde{T}_1 + \ell_1^2/2X_1' = \widetilde{T}_1 - \Delta T_1, \tag{9.2.66}$$

say. Thus the decay of the diffracted signal (9.2.63) from the shadow edge is

$$\left(\widetilde{T}_1 - T_{\text{ray}}\right)^{-1/2} = (-2X_1')^{1/2}/\ell_1. \tag{9.2.67}$$

It is interesting to note that in all situations, the diffracted signal near a Fresnel shadow can be described by the one function $Fi(t)$ scaled according to the time delay, T_{ray}, and the time shift, ΔT_1.

The Fresnel shadow result is a useful approximation to describe the behaviour at the shadow, in particular the decay to a half the geometrical amplitude at the shadow edge. It does, however, neglect the interaction of the wave with the interface. At the shadow edge, this is a small but non-negligible effect, and in the deep shadow it is very important. The description of these signals will be discussed later (Section 9.3.7).

9.3 Spectral methods

Asymptotic methods can be used to approximate the slowness integrals given in Section 8.5.1. In general, these results are equivalent to the first-motion approximations given in the previous section, Section 9.2. Normally the first-motion approximations to the slowness results are easier to obtain and do not require a knowledge of the special functions that often arise in the spectral domain. Nevertheless, we briefly summarize the results here for the sake of completeness and as the methods are the classical, more traditional approach. We only give the leading term in the asymptotic expansions.

9.3.1 Geometrical rays – method of stationary phase

9.3.1.1 Two dimensions

The integral (8.5.2) is highly oscillatory and can be approximated by the *method of stationary phase*. It has stationary points corresponding to geometrical rays, i.e.

$$\frac{\partial \widetilde{T}_{\mathrm{ray}}}{\partial p} = \frac{\partial \tau_{\mathrm{ray}}}{\partial p} + x_{\mathrm{R}} = 0, \tag{9.3.1}$$

when

$$x_{\mathrm{R}} = -\frac{\partial \tau_{\mathrm{ray}}}{\partial p} = X_{\mathrm{ray}}(p, z_{\mathrm{R}}), \tag{9.3.2}$$

at $p = p_{\mathrm{ray}}(\mathbf{x}_{\mathrm{R}})$, say. In general, multiple stationary points may exist. Expanding about a stationary point, its contribution can be evaluated by the second-order saddle-point method (D.1.11). The orientation of the saddle point depends on the second derivative,

$$\frac{\partial^2 \widetilde{T}_{\mathrm{ray}}}{\partial p^2} = -\frac{\mathrm{d}X_{\mathrm{ray}}}{\mathrm{d}p}, \tag{9.3.3}$$

and the sign of the frequency. Using $|\omega| = -\mathrm{i}\omega \exp(\mathrm{i}\,\mathrm{sgn}(\omega)\pi/2)$, we obtain

$$\boxed{\begin{aligned}\underline{\mathbf{u}}_{\mathrm{ray}}(\omega, \mathbf{x}_{\mathrm{R}}) \simeq \lambda(\omega) &\frac{\underline{\mathcal{G}}_{\mathrm{ray}}(p_{\mathrm{ray}})}{\pi\left(2\left|X'_{\mathrm{ray}}(p_{\mathrm{ray}}, z_{\mathrm{R}})\right|\right)^{1/2}} \\ &\times \mathrm{e}^{\mathrm{i}\omega\, T_{\mathrm{ray}}(p_{\mathrm{ray}}, z_{\mathrm{R}}) - i\frac{\pi}{4}\,\mathrm{sgn}(\omega)\left(\mathrm{sgn}\left(\frac{\mathrm{d}X_{\mathrm{ray}}}{\mathrm{d}p}\right)-1\right)}\end{aligned}} \tag{9.3.4}$$

($\lambda(\omega)$ denotes the spectrum of $\lambda(t)$, i.e. definition (B.2.2)). The inverse Fourier transform (3.1.2) of this spectrum agrees with the general first-motion approximation for the WKBJ seismogram (9.2.20) (see Appendix B.2). For a simple turning ray, $\underline{\mathcal{G}}_{\mathrm{ray}}$ is imaginary and $X'_{\mathrm{ray}} < 0$ on the forward branch, making $\underline{\mathbf{u}}_{\mathrm{ray}} \sim \lambda(\omega)$. On a reversed branch, $X'_{\mathrm{ray}} > 0$ and $\underline{\mathbf{u}}_{\mathrm{ray}} \sim -\mathrm{i}\,\mathrm{sgn}(\omega)\lambda(\omega)$.

9.3.1.2 Three dimensions

The integral (8.5.4) is highly oscillatory and can be approximated by the method of stationary phase in two dimensions. It has saddle-points when

$$\nabla_{\mathrm{p}} \widetilde{T}_{\mathrm{ray}}(\mathrm{p}, \mathbf{x}_{\mathrm{R}}) = \mathrm{x}_{\mathrm{R}} - \mathrm{X}_{\mathrm{ray}}(\mathrm{p}, z_{\mathrm{R}}) = 0, \tag{9.3.5}$$

at geometrical rays, i.e. when

$$\mathsf{x}_\mathsf{R} = \mathsf{X}_{\text{ray}}(\mathsf{p}, z_\mathsf{R}), \tag{9.3.6}$$

with $\mathsf{p} = \mathsf{p}_{\text{ray}}$, say (remember the sans serif font is used to indicate vectors in the two-dimensional space). Near a saddle point, the phase is approximated by the second-order behaviour

$$\widetilde{T}_{\text{ray}}(\mathsf{p}, \mathbf{x}_\mathsf{R}) \simeq T_{\text{ray}}(\mathsf{p}_{\text{ray}}, z_\mathsf{R}) + \frac{1}{2}(\mathsf{p} - \mathsf{p}_{\text{ray}})^{\mathrm{T}} \nabla_\mathsf{p} \left(\nabla_\mathsf{p} \widetilde{T}_{\text{ray}}\right)^{\mathrm{T}} (\mathsf{p} - \mathsf{p}_{\text{ray}}). \tag{9.3.7}$$

Using the general, multi-dimensional saddle point formula (D.1.18) with $m = 2$, the saddle-point approximation for (8.5.4) is

$$\boxed{\underline{\mathbf{u}}_{\text{ray}}(\omega, \mathbf{x}_\mathsf{R}) \simeq \frac{\underline{\boldsymbol{\mathcal{G}}}_{\text{ray}}(\mathsf{p}_{\text{ray}})}{2\pi \left\| \nabla_\mathsf{p} \mathsf{X}_{\text{ray}}^{\mathrm{T}} \right\|^{1/2}} \, \mathrm{e}^{\mathrm{i}\omega\, T_{\text{ray}}(\mathsf{p}_{\text{ray}}, z_\mathsf{R}) - \mathrm{i}\frac{\pi}{4}\,\mathrm{sgn}(\omega)\left(\mathrm{sgn}\left(\nabla_\mathsf{p} \mathsf{X}_{\text{ray}}^{\mathrm{T}}\right) - 2\right)}.} \tag{9.3.8}$$

The geometrical spreading matrix $\nabla_\mathsf{p} \mathsf{X}_{\text{ray}}^{\mathrm{T}}$ is 2×2.

9.3.1.3 Two to three dimensions

The three-dimensional result (8.5.4) with a double slowness integral can sometimes be approximated by a single slowness integral analogous to the two-dimensional result (8.5.2).

Writing $\mathsf{p} = (p_1, p_2)$ and $\mathsf{x} = (x_1, x_2)$, and assuming we have rotated the coordinate system so $x_2 = 0$, the slowness integral over the transverse slowness, p_2, has a stationary point when $p_2 = 0$. Thus we approximate the integral (8.5.4) by

$$\begin{aligned} \underline{\mathbf{v}}_{\text{ray}}(\omega, \mathbf{x}_\mathsf{R}) \simeq \frac{\omega^2}{4\pi^2} \int_{-\infty}^{\infty} & \underline{\boldsymbol{\mathcal{G}}}_{\text{ray}}(p_1)\, \mathrm{e}^{\mathrm{i}\omega \widetilde{T}_{\text{ray}}(p_1, \mathbf{x}_\mathsf{R})} \\ \times \int_{-\infty}^{\infty} & \mathrm{e}^{\mathrm{i}\omega(\partial^2 \widetilde{T}_{\text{ray}}/\partial p_2^2)\, p_2^2/2}\, \mathrm{d}p_2\, \mathrm{d}p_1, \end{aligned} \tag{9.3.9}$$

where $\underline{\boldsymbol{\mathcal{G}}}_{\text{ray}}(p_1)$ and $\widetilde{T}_{\text{ray}}(p_1, \mathbf{x})$ are the three-dimensional functions with $p_2 = 0$ and are identical to the two-dimensional functions in (8.5.2). Now from result (5.7.15), we have

$$\frac{\partial^2 \widetilde{T}_{\text{ray}}}{\partial p_2^2} = -\frac{\partial X_2}{\partial p_2} = -\frac{X_1}{p_1}. \tag{9.3.10}$$

Using this in integral (9.3.9) with the second-order saddle-point approximation

(D.1.11), we obtain

$$\underline{\mathbf{v}}_{\mathrm{ray}}(\omega, \mathbf{x}_{\mathrm{R}}) \simeq \frac{|\omega|}{2\pi} \int_{-\infty}^{\infty} \frac{1}{\pi} \left(\frac{p_1}{2X_1(p_1, z_{\mathrm{R}})} \right)^{1/2} \times (-\mathrm{i}\omega\,\lambda(\omega))\, \underline{\mathcal{G}}_{\mathrm{ray}}(p_1)\, \mathrm{e}^{\mathrm{i}\omega \widetilde{T}_{\mathrm{ray}}(p_1, \mathbf{x}_{\mathrm{R}})}\, \mathrm{d}p_1. \tag{9.3.11}$$

This expression (9.3.11) is exactly analogous to the two-dimensional integral (8.5.2) with the extra factor

$$\frac{1}{\pi} \left(\frac{p_1}{2X_1(p_1, z_{\mathrm{R}})} \right)^{1/2} \Big(-\mathrm{i}\omega\,\lambda(\omega) \Big). \tag{9.3.12}$$

If the main contribution comes from the geometrical ray, then $X_1(p_1, z_{\mathrm{R}})$ can be approximated by the range x, and this factor is equivalent to the conversion operation (8.2.66). In the three-dimensional expression (9.3.8), the spreading factor is (cf. result (5.7.16))

$$\left| \nabla_{\mathrm{p}} \mathbf{X}_{\mathrm{ray}}^{\mathrm{T}} \right| = \frac{X_1}{p_1} \frac{\mathrm{d}X_1}{\mathrm{d}p_1}, \tag{9.3.13}$$

and expression (9.3.8) is equal to (9.3.4) combined with (9.3.12)

$$\underline{\mathbf{u}}_{\mathrm{ray}}(\omega, \mathbf{x}_{\mathrm{R}}) = \frac{1}{2\pi} \left(\frac{p_{\mathrm{ray}}}{x_{\mathrm{R}} |X'_{\mathrm{ray}}|} \right)^{1/2} \underline{\mathcal{G}}_{\mathrm{ray}}\, \mathrm{e}^{\mathrm{i}\omega T_{\mathrm{ray}} - i\frac{\pi}{4}\,\mathrm{sgn}(\omega) \left(\mathrm{sgn}(X'_{\mathrm{ray}}) - 1\right)}. \tag{9.3.14}$$

It is worth commenting that these asymptotic, spectral results do not break down at zero range as $x_{\mathrm{R}} \to 0$. In expression (9.3.10)

$$\frac{X_1}{p_1} \to \frac{\partial X_1}{\partial p_1}, \tag{9.3.15}$$

by l'Hopital's rule. Necessarily, by symmetry, the 'widths' of the saddle point in p_1 and p_2 become equal at $x_{\mathrm{R}} = 0$. The spectral result becomes

$$\underline{\mathbf{u}}_{\mathrm{ray}}(\omega, \mathbf{x}_{\mathrm{R}}) = \frac{1}{2\pi \left| X'_{\mathrm{ray}} \right|} \underline{\mathcal{G}}_{\mathrm{ray}}\, \mathrm{e}^{\mathrm{i}\omega T_{\mathrm{ray}} - \mathrm{i}\frac{\pi}{4}\,\mathrm{sgn}(\omega) \left(\mathrm{sgn}(X'_{\mathrm{ray}}) - 1\right)}, \tag{9.3.16}$$

in agreement with the geometrical result (Section 5.7).

9.3.2 Head waves

Using the slowness method, either the Cagniard method or the WKBJ method, we have shown how a branch point in the transformed response causes a head wave, e.g. expression (9.1.52) or (9.2.29). In the spectral method, we reduce the slowness integral to a branch-line integral. The contour along the real axis is distorted

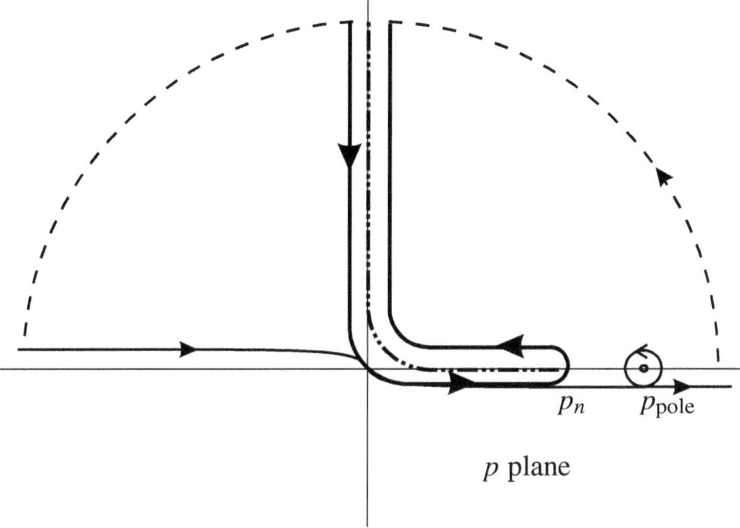

Fig. 9.30. The original p contour and the distorted contour around the branch cut and a pole. The dot-dashed line indicates the branch cut from the branch point at $p = p_n$ and along the positive imaginary axis from $p = 0$ to $p = i\infty$. The diagram is for $\omega > 0$ – for $\omega < 0$, it is reflected in the real axis.

into the first and second quadrants so it runs along either side of the branch cut $\text{Im}(\omega q_n) = 0$ (the branch cut is defined so $\text{Im}(\omega q_n) > 0$ on the physical Riemann sheet). The original contour runs infinitesimally below the positive p axis (Figure 3.4) for positive, real frequencies, and the distorted contour is illustrated in Figure 9.30. As well as the branch-line integral, distorting the contour may pick up residues of poles which we discuss below.

To evaluate the branch-line integral we consider the integral (8.5.2) or (9.3.11) with the approximation (9.1.49) near the branch point $p = p_n = 1/c_n$. The standard method to evaluate a branch-line integral is to change the variable of integration to q_n. We consider the leading term in the integral.

Using $p\,\mathrm{d}p = -q_n \mathrm{d}q_n$, we obtain from integral (8.5.2), with the significant varying part of expression (9.1.49)

$$\underline{\mathbf{v}}_{\text{ray}}(\omega, \mathbf{x}_{\text{R}}) \simeq -\frac{\omega}{2\pi} \int_{-\infty}^{\infty} \frac{\partial \underline{\boldsymbol{\mathcal{G}}}_{\text{ray}}(p_n)}{\partial q_n} \frac{q_n^2}{p} \mathrm{e}^{\mathrm{i}\omega \widetilde{T}_{\text{ray}}} dq_n, \tag{9.3.17}$$

near $p = p_n$ for $\omega > 0$. Below the branch cut $\text{Re}(q_n) > 0$, so the distorted integral (Figure 9.30) runs from $q_n = +\infty$ to $-\infty$. The phase in integral (9.3.17) is approximately

$$\widetilde{T}_{\text{ray}} \simeq T_n - \ell_n c_n q_n^2/2. \tag{9.3.18}$$

The branch-line integral has been converted into a saddle-point integral by the change of variable. This method has been developed by Lapwood (1949) and Budden (1961) to study head waves. More complicated spectral methods with poles near saddle points, e.g. to study the $\bar{P}$ pole, are known (Ott, 1943; van der Waerden, 1950), but we do not pursue them here. Substituting the approximation (9.3.18) in integral (9.3.17), and including the extra factors in expression (9.3.11) for the far-field approximation in three dimensions, we obtain (for $\omega > 0$)

$$\underline{\mathbf{v}}_{\text{ray}}(\omega, \mathbf{x}_{\text{R}}) \simeq \mathrm{i} \left(\frac{\omega}{2\pi}\right)^{3/2} \left(\frac{c_n}{x_{\text{R}}}\right)^{1/2} \frac{\partial \underline{\boldsymbol{\mathcal{G}}}_{\text{ray}}(p_n)}{\partial q_n} \mathrm{e}^{\mathrm{i}\omega T_n + \mathrm{i}\pi/4}$$
$$\times \int_{-\infty}^{\infty} q_n^2 \mathrm{e}^{-\mathrm{i}\omega \ell_n c_n q_n^2/2} dq_n. \tag{9.3.19}$$

Integrating by parts, this can be reduced to the standard second-order saddle-point integral (D.1.11) and we obtain

$$\underline{\mathbf{u}}_{\text{ray}}(\omega, \mathbf{x}_{\text{R}}) \simeq \frac{1}{\mathrm{i}\omega} \frac{1}{2\pi c_n \ell_n^{3/2} x_{\text{R}}^{1/2}} \frac{\partial \underline{\boldsymbol{\mathcal{G}}}_{\text{ray}}(p_n)}{\partial q_n} \mathrm{e}^{\mathrm{i}\omega T_n}, \tag{9.3.20}$$

the displacement spectrum. The inverse Fourier transform of this agrees with the first-motion approximation (9.1.52).

In evaluating the branch-line integral, we have ignored the behaviour except near the branch point. We have only considered the leading term in the expanded integrand. In general, there will also be a saddle point on the real slowness axis corresponding to the reflected arrival. At high frequency, the saddle point and branch cut can be treated as separate features except at the critical point. At low frequencies, at and near the critical point, they interact and must be considered together. Analytic methods exist for handling a saddle point and branch cut together, but we do not pursue that here. Instead, numerical methods would normally be used.

As a simple example of the numerical spectral method, we illustrate the same reflection and head wave as in Figures 9.3 (Cagniard method) and 9.24 (WKBJ seismogram method) in Figure 9.31. These are calculated using the numerical methods described in Section 8.5.2. The results differ somewhat due to the low-frequency drift of the results. The frequency Fourier integral is approximated by a finite Fourier transform and this results in some errors at low-frequencies. The differences are exaggerated by the low-frequency source, i.e. $P_{\text{S}}(t) = P_{\text{S}} t H(t)$, but the singularities at the head wave and reflection are virtually identical. For a realistic source pulse, the differences are inconsequential. This numerical result illustrates partial and total reflections (9.3.4) and head waves (9.3.20). To obtain this accurate result, the slowness values in the slowness integral must be placed symmetrically about the branch cut, as mentioned above with respect to the WKBJ algorithm (Section 9.2.6).

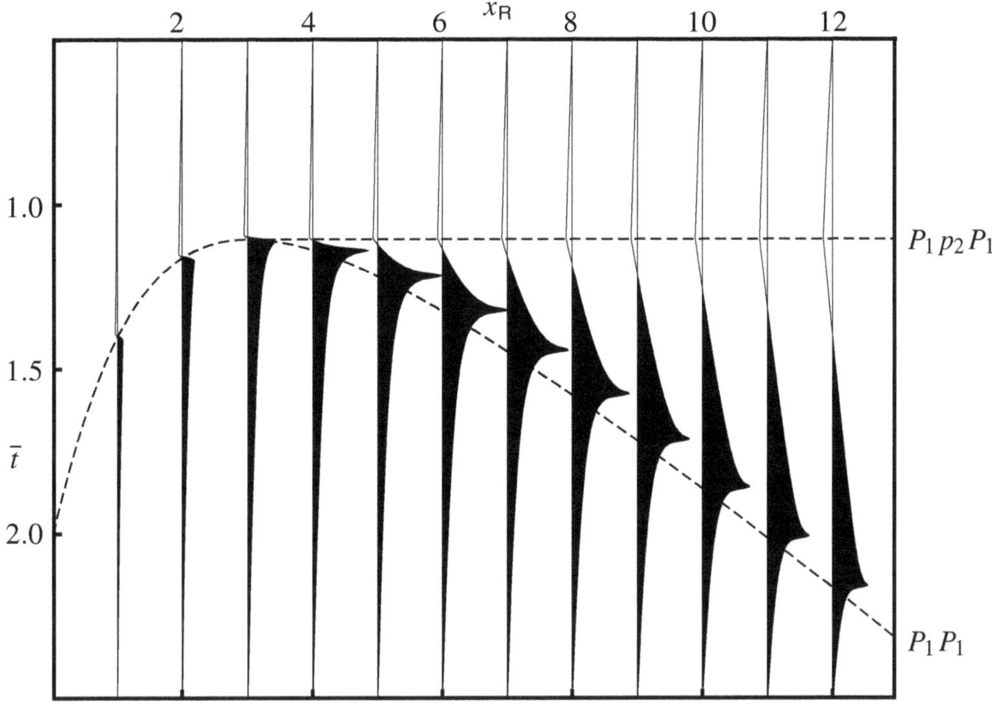

Fig. 9.31. The reflections in an acoustic model with two homogeneous half-spaces calculated using the numerical spectral method (Section 8.5.2). The head wave and total reflection are visible. The model parameters and other details are all as in Figure 9.3.

9.3.3 Interface waves

In the Cagniard method, interface waves are caused by the proximity of the Cagniard contour to poles of the response, e.g. expression (9.1.89). In the spectral method, the contour is distorted to pick up the residue of the poles, either on the real axis (Figure 9.30), or through the branch cut on a lower Riemannn sheet for leaking poles.

Near a pole, we approximate $\underline{\mathcal{G}}_{\text{ray}}(p)$ in integral (9.3.11) by (cf. equation (9.1.90))

$$\underline{\mathcal{G}}_{\text{ray}}(p) \simeq \frac{\underline{\widetilde{\mathcal{G}}}_{\text{pole}}(p_{\text{pole}})}{g'(p_{\text{pole}})(p - p_{\text{pole}})}. \tag{9.3.21}$$

The residue of this pole in integral (9.3.11) is then (for $\omega > 0$)

$$\underline{\mathbf{u}}_{\text{pole}}(\omega, \mathbf{x}_R) \simeq \frac{\mathrm{i}\omega}{\pi} \left(\frac{p_{\text{pole}}}{2x_R}\right)^{1/2} \frac{\underline{\widetilde{\mathcal{G}}}_{\text{pole}}(p_{\text{pole}})}{g'(p_{\text{pole}})} \, \mathrm{e}^{\mathrm{i}\omega \widetilde{T}_{\text{pole}}(p_{\text{pole}}, \mathbf{x}_R)}, \tag{9.3.22}$$

where $\widetilde{T}_{\text{ray}}(p_{\text{pole}}, \mathbf{x}_{\text{R}})$ is complex. Using result (B.1.9), this agrees with the time result (9.1.89) from the Cagniard method.

Using the Cagniard method, it is very straightforward to establish the range of validity of the interface wave approximation (9.1.89). It requires that the Cagniard contour should be near enough to the pole so that an approximation about the pole is valid. This places restrictions on the location of the receiver (and the material properties) and the time window.

With the spectral method, it is not as easy to establish or understand when the residue of the pole is a useful approximation. It requires that the residue be compared with the rest of the integral. In particular, leaking interface waves are difficult to understand in the spectral domain as, being on non-physical Riemann sheets, they grow exponentially away from the interface ($\widetilde{T}_{\text{ray}}(p_{\text{pole}}, \mathbf{x}_{\text{R}})$ contains a negative imaginary part). Exponential growth away from the interface seems to contradict energy conservation and the radiation condition. This has led to statements in the literature questioning the physical existence of leaking waves. However, the leaking interface approximation is not valid for all receiver positions nor all times. The complete wavefield does not violate these conditions. With the Cagniard method, the nature of the approximations and their range of validity are immediately apparent.

9.3.4 Tunnelling waves

As an example of a tunnelling wave, we consider the phase function (9.1.117). The slowness integral (9.3.11) has a saddle point when

$$\frac{\partial \widetilde{T}_{\text{ray}}}{\partial p} = x_{\text{R}} - X_{\text{ray}}(p, z_{\text{R}}) - \frac{p}{q_{\max}} d_{\max} = 0. \tag{9.3.23}$$

As $d_{\max}$ is small, this is approximately zero at $p = p_{\text{ray}}$, where

$$x_{\text{R}} = X_{\text{ray}}(p_{\text{ray}}, z_{\text{R}}), \tag{9.3.24}$$

i.e. the 'geometrical' ray solution in the slower layers. To find the exact position of the saddle point, we consider the expansion about this slowness. The second derivative is

$$\frac{\partial^2 \widetilde{T}_{\text{ray}}}{\partial p^2} = -\frac{\mathrm{d}X_{\text{ray}}(p_{\text{ray}}, z_{\text{R}})}{\mathrm{d}p} - \frac{\mathrm{i}\, d_{\max}}{c_{\max}^2 q_{\text{ray}}^3}. \tag{9.3.25}$$

At $p = p_{\text{ray}}$, $q_{\max} = \mathrm{i}\, q_{\text{ray}}$ is positive imaginary (9.1.129). Thus expanding about $p = p_{\text{ray}}$, we have

$$\frac{\partial \widetilde{T}_{\text{ray}}}{\partial p} \simeq \frac{\mathrm{i}\, p_{\text{ray}} d_{\max}}{q_{\text{ray}}} + \frac{\partial^2 \widetilde{T}_{\text{ray}}}{\partial p^2} (p - p_{\text{ray}}). \tag{9.3.26}$$

To first order in $d_{\max}$, the saddle point is at

$$p = p_{\text{ray}} + \frac{\mathrm{i}\, p_{\text{ray}} d_{\max}}{q_{\text{ray}} (\mathrm{d}X_{\text{ray}}/\mathrm{d}p)}, \tag{9.3.27}$$

and the phase is

$$\widetilde{T}_{\text{ray}}(p, \mathbf{x}_{\mathrm{R}}) = T_{\text{ray}} + \mathrm{i}\, q_{\text{ray}} d_{\max} \tag{9.3.28}$$

(the gradient and second derivative of $\widetilde{T}_{\text{ray}}$ at p_{ray} lead to corrections that are second order in $d_{\max}$ as the shift in slowness is first order (9.3.27)). Distorting the contour so it passes over the saddle point and evaluating integral (9.3.11) by the second-order saddle-point method, we obtain to lowest order

$$\boxed{\underline{\mathbf{u}}_{\text{tunnel}}(t, \mathbf{x}_{\mathrm{R}}) \simeq \frac{1}{2\pi} \left(\frac{p_{\text{ray}}}{x_{\mathrm{R}} X'_{\text{ray}}} \right)^{1/2} \underline{\boldsymbol{\mathcal{G}}}_{\text{ray}}(p_{\text{ray}})\, \mathrm{e}^{\mathrm{i}\omega T_{\text{ray}} - |\omega| q_{\text{ray}} d_{\max}},} \tag{9.3.29}$$

the spectral result corresponding to (9.1.136).

9.3.4.1 Thin-layer reverberations

As mentioned in Section 9.1.6, the situation is more complicated for a thin layer (Figure 9.15) due to the layer reverberations. This can be investigated most easily for the simple acoustic case. The complete transformed propagation term (8.0.7) for transmission through a thin acoustic layer is

$$\mathcal{P}(\omega, p, z_{\mathrm{R}}) = \mathcal{T}^2 \mathrm{e}^{\mathrm{i}\omega\tau + \mathrm{i}\omega q_2 d_2} \sum_{n=0}^{\infty} \mathcal{R}^{2n} \mathrm{e}^{2\mathrm{i}n\omega q_2 d_2} \tag{9.3.30}$$

$$= \mathcal{T}^2 \mathrm{e}^{\mathrm{i}\omega\tau + \mathrm{i}\omega q_2 d_2} \left(1 - \mathcal{R}^{2n} \mathrm{e}^{2\mathrm{i}\omega q_2 d_2}\right)^{-1}, \tag{9.3.31}$$

where

$$\tau = q_1 (z_{\mathrm{S}} - z_{\mathrm{R}} - d_2) \tag{9.3.32}$$

$$d_2 = z_1 - z_2, \tag{9.3.33}$$

and

$$\mathcal{R} = \frac{\rho_1 q_2 - \rho_2 q_1}{\rho_1 q_2 + \rho_2 q_1} \tag{9.3.34}$$

$$\mathcal{T} = \frac{2\sqrt{\rho_1 \rho_2 q_1 q_2}}{\rho_1 q_2 + \rho_2 q_1}, \tag{9.3.35}$$

from expressions (6.3.7) and (6.3.8).

If q_2 is real, then the series (9.3.30) converges as $\mathcal{R} < 1$. When q_2 is imaginary, $|\mathcal{R}| = 1$, but the series converges due to the evanescent decay of $\exp(-2|\omega q_2| d_2)$.

As $d_2 \to 0$, it is trivial to establish that $\mathcal{P}(\omega, p, z_R) \to \exp(i\omega\tau)$ because $1 - \mathcal{R}^2 = \mathcal{T}^2$. Physically, this is the expected result – that an infinitesimal thin layer will not affect the wave propagation – but it is important to realize that this is not true for the leading term in the ray expansion. In fact when q_2 is imaginary, the leading term in the ray expansion

$$\mathcal{P}(\omega, p, z) = \mathcal{T}^2 e^{i\omega\tau + i\omega q_2 d_2}, \tag{9.3.36}$$

can easily be amplified compared with the complete response. For

$$\mathcal{T}^2 = 1 - \mathcal{R}^2 = 1 - e^{-4i\,\mathrm{sgn}(\omega)\phi}, \tag{9.3.37}$$

so

$$|\mathcal{T}^2| = 2\sin 2\phi, \tag{9.3.38}$$

using the total reflection coefficient (6.3.11) with ϕ defined in equation (6.3.12) (strictly in these expressions, we should understand $\phi = \mathrm{sgn}(\omega)|\phi|$). If $\phi > \pi/12$, then $|\mathcal{T}^2| > 1$. This can lead to some unexpected results if the ray expansion is used in a model with many thin, high-velocity layers.

In the slowness domain, it is straightforward to invert the complete expression (9.3.30). In general, we obtain

$$\mathcal{P}(t, p, z_R) = \mathrm{Re}\left(\mathcal{T}^2 \sum_{n=0}^{\infty} \mathcal{R}^{2n} \Delta\left(t - \tau - (2n+1)q_2 d_2\right)\right). \tag{9.3.39}$$

If q_2 is real ($p < 1/\alpha_2$), this reduces to a decaying reverberation series of delta functions (Figure 9.32).

If q_2 is imaginary ($p > 1/\alpha_2$), we can substitute (9.3.37) in equation (9.3.30) to obtain

$$\mathcal{P}(\omega, p, z_R) = e^{i\omega\tau} \sin\psi \left(2i\,\mathrm{sgn}(\omega)\right) \sum_{n=0}^{\infty} e^{-(2n+1)(i\,\mathrm{sgn}(\omega)\psi + |\omega q_2| d_2)} \tag{9.3.40}$$

$$= e^{i\omega\tau} \sin\psi \; Tun(\omega|q_2|d_2, \psi), \tag{9.3.41}$$

where (6.3.12)

$$\psi = 2\phi = 2\tan^{-1}\left(\frac{\rho_1 |q_2|}{\rho_2 q_1}\right), \tag{9.3.42}$$

and

$$Tun(\omega, \psi) = 2i\,\mathrm{sgn}(\omega) \sum_{n=0}^{\infty} e^{-(2n+1)(i\,\mathrm{sgn}(\omega)\psi + |\omega|)} \tag{9.3.43}$$

$$= i\,\mathrm{sgn}(\omega)\,\mathrm{csch}(i\,\mathrm{sgn}(\omega)\psi + |\omega|). \tag{9.3.44}$$

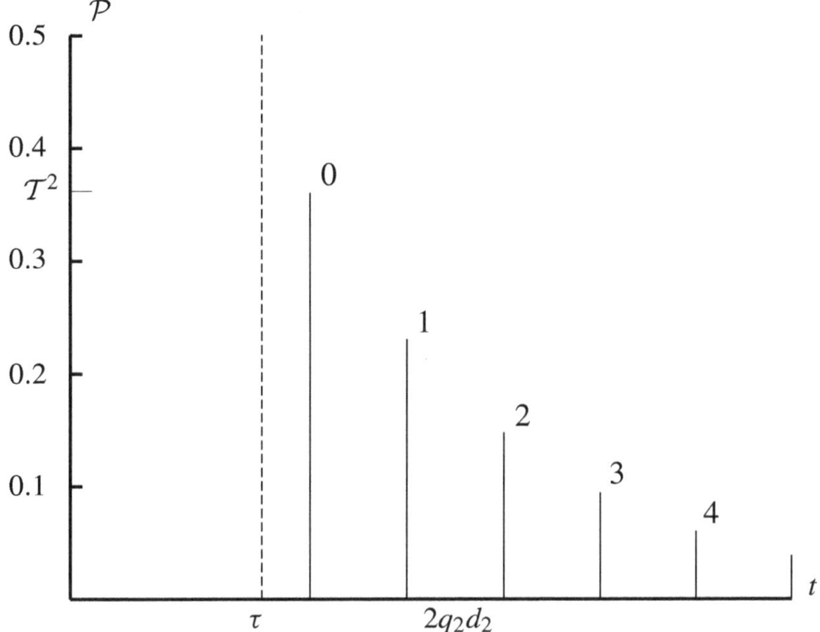

Fig. 9.32. The reverberating time series of delta functions from a thin, high-velocity layer with $p < 1/\alpha_2$, result (9.3.39), with $\mathcal{R} = 0.8$ (so $\mathcal{T}^2 = 0.36$).

The inverse transform of (9.3.41) can then be written

$$\mathcal{P}(t, p, z_{\mathrm{R}}) = \frac{1}{|q_2|d_2} \sin(2\phi)\, Tun\left(\frac{t-\tau}{|q_2|d_2}, 2\phi\right) \tag{9.3.45}$$

$$\to \delta(t-\tau) \qquad \text{as} \qquad d_2 \to 0, \tag{9.3.46}$$

where the inverse transforms of (9.3.43) and (9.3.44) can be written

$$Tun(t, \psi) = \frac{2}{\pi} \operatorname{Re} \sum_{n=0}^{\infty} \frac{\mathrm{e}^{-(2n+1)\mathrm{i}|\psi|}}{t-(2n+1)\mathrm{i}} \tag{9.3.47}$$

$$= \frac{\mathrm{e}^{-|\psi|t}}{1+\mathrm{e}^{-\pi t}}. \tag{9.3.48}$$

Expression (9.3.47) is the inverse Fourier transform of series (9.3.43) using result (B.1.9). The inverse Fourier transform (9.3.48) of expression (9.3.44) is obtained using Erdélyi, Magnus, Oberhettinger and Tricomi (1954, §1.9(1)) for the inverse cosine transform of the hyperbolic cosecant function. The variable of integration in the inverse transform is changed to $\omega' = \omega + \mathrm{i}|\psi|$, but the contour of integration can be restored to the real axis without encountering the singularities of the hyperbolic function on the imaginary axis. With definition (9.3.42), we are interested in

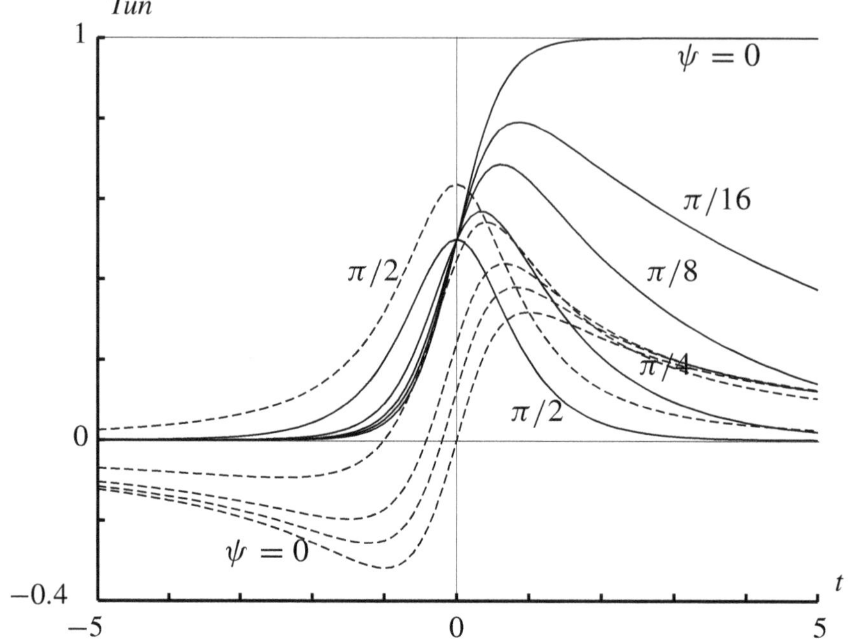

Fig. 9.33. The 'standard' function $Tun(t, \psi)$ used for the reverberating, tunnelling time series (solid lines), and the first term (9.3.50) in the reverberations (dashed lines).

the range $0 < \psi < \pi$ although as we have the symmetry

$$Tun(t, \pi - \psi) = Tun(-t, \psi), \tag{9.3.49}$$

we need only consider the range $0 < \psi < \pi/2$. Figure 9.33 illustrates the function (9.3.48) for several values of $\psi = 2\phi$ in this range. For $d_2 \to 0$, the standard time functions, (9.3.48) and Figure 9.33, are compressed and as $Tun(t, \psi) \to 0$ as $t \to \infty$, become delta-like (there is an awkward limit if $\psi = 0$, when $\mathcal{T} = 0$, $\mathcal{R} = 1$ and $\mathcal{P}(\omega, p, z_{\mathrm{R}}) = 0$ except when $d_2 \to 0$, when $\mathcal{P}(\omega, p, z_{\mathrm{R}}) \to 1$).

For comparison, we have also plotted the first term in the ray expansion for each case

$$Tun^{(0)}(t, \psi) = \frac{2}{\pi} \frac{t \cos \psi + \sin \psi}{t^2 + 1}, \tag{9.3.50}$$

with a dashed line in Figure 9.33. Of interest are the pulse differences between the complete reverberation and the first term when ψ is small, and the pulse attenuation in the complete reverberation when ψ is larger.

When reverberations are significant, it is normally easiest to calculate the response using the numerical spectra method (Section 8.5.2). Numerical calculations for the tunnelling example in Figure 9.16 (with no reverberations) simply confirm

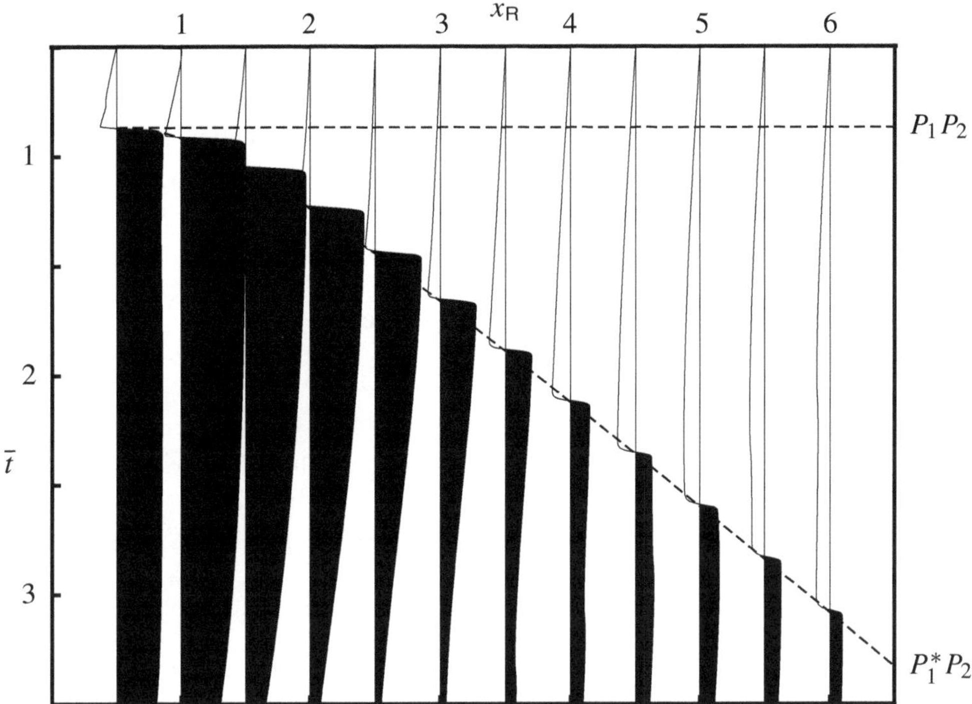

Fig. 9.34. The response in the acoustic model illustrated in Figure 9.15 to an explosive, point source with time function of the form $P_S(t) = P_S t H(t)$. The model parameters are identical to Figure 9.16, except thin-layer reverberations are included (see the text for model details).

the Cagniard results and for brevity are not included here. They only differ in the low-frequency drift typical of numerical spectral results (e.g. as Figure 9.31). Instead, we present numerical calculations for a situation as in Figure 9.15 where reverberations in a thin, high-velocity exist. The model parameters are just as in the tunnelling example (Figures 9.12 and 9.16) except the signal transmits through a thin layer, i.e. as in Figure 9.15. The source and receiver are in identical slow half-spaces (with the slow parameters from layer 2 in the tunnelling example). The vertical distances propagated in the two media are identical to Figure 9.12, i.e. $(z_S - z_2) + (z_3 - z_R) = 1$ and $z_2 - z_3 = 0.01$, with $\alpha_1 = \alpha_3 = 1$ and $\alpha_2 = 2$, but reverberations are now present. The full response is shown in Figure 9.34 calculated using the numerical methods in Section 8.5.2. Note the response is much simplified compared with Figure 9.16, and is virtually that of a direct wave without the high-velocity layer. The thin layer is invisible and with the reverberations the transmission through it is approximately unity (7.2.87). The calculations have not been extended to small ranges where the reverberation series (9.3.39) would be relevant.

9.3.5 The Airy caustic

When a geometrical arrival on a forward branch ($X' < 0$) and a reversed branch ($X' > 0$) coalesce, the second-order saddle-point method breaks down (result (9.3.4) is singular). The third-order saddle-point method (Section D.2) must be used. We assume that the phase function can be expanded as a cubic (9.2.37) and for simplicity, we consider the case with $\underline{\mathcal{G}}_{\text{ray}}(p_A)$ imaginary, $X''_A > 0$ and $\omega > 0$. The two-dimensional integral (8.5.2) is approximated by

$$\underline{\mathbf{u}}_{\text{Airy}}(\omega, \mathbf{x}_{\text{R}}) \simeq -\frac{1}{2\pi} \operatorname{Im}\left(\underline{\mathcal{G}}_{\text{ray}}(p_A)\right) \mathrm{e}^{\mathrm{i}\omega \widetilde{T}_A} \times \int_{-\infty}^{\infty} \mathrm{e}^{\mathrm{i}\omega \ell_A (p-p_A) - \mathrm{i}\omega X''_A (p-p_A)^3/6} \, \mathrm{d}p. \tag{9.3.51}$$

This can be expressed using the Airy function using result (D.2.3) so

$$\underline{\mathbf{u}}_{\text{Airy}}(\omega, \mathbf{x}_{\text{R}}) \simeq -\left(\frac{2}{\omega X''_A}\right)^{1/3} \operatorname{Im}\left(\underline{\mathcal{G}}_{\text{ray}}(p_A)\right) \mathrm{e}^{\mathrm{i}\omega \widetilde{T}_A} Ai\left(-\frac{2^{1/3}\omega^{2/3}\ell_A}{(X''_A)^{1/3}}\right). \tag{9.3.52}$$

Similar expressions can be obtained if $\underline{\mathcal{G}}_{\text{ray}}(p_A)$ is complex, $X''_A < 0$ or $\omega < 0$. Higher-order terms due to non-cubic behaviour of the phase, or variation in the amplitude, lead to an asymptotic series (Chester, Friedman and Ursell, 1957).

Equivalent expressions for the three-dimensional response are similar. Using the far-field conversion factor (9.3.12), we obtain

$$\boxed{\underline{\mathbf{u}}_{\text{Airy}}(\omega, \mathbf{x}_{\text{R}}) \simeq -\frac{p_A^{1/2}\omega^{1/6}}{2^{1/6}\pi^{1/2}(X_A)^{1/2}(X''_A)^{1/3}} \operatorname{Im}\left(\underline{\mathcal{G}}_{\text{ray}}(p_A)\right) \times \mathrm{e}^{\mathrm{i}\omega \widetilde{T}_A - \mathrm{i}\pi/4} Ai\left(-\frac{2^{1/3}\omega^{2/3}\ell_A}{(X''_A)^{1/3}}\right).} \tag{9.3.53}$$

The dimensions of this expression check as in two-dimensions, unit($\underline{\mathcal{G}}$) = $[\mathrm{M}^{-1}\mathrm{L}^2\mathrm{T}]$ and unit $(\underline{\mathbf{u}}) = [\mathrm{M}^{-1}\mathrm{T}^2]$.

Burridge (1963*a*) has shown that the inverse Fourier transform of the Airy function (as in expression (9.3.53)) leads to the time function $C(t, y)$ (D.2.12) which we describe in Appendix D.2. The equivalent results in the time domain have already been given in expression (9.2.46). Although classically the caustic was studied in the frequency domain (Jeffreys, 1939) – hence the name Airy caustic – it is simpler to study it in the time domain. The time function $C(t, y)$, although not elementary, is easy to calculate and apart from the known singularities, behaves smoothly. In contrast, the Airy function $Ai(x)$ is more complicated and needs different asymptotic or power series expansions in different regions. Nevertheless, it is useful to consider the asymptotic form of result (9.3.52) in the shadow, $\ell_A < 0$. The Airy function is evanescent and we can use the asymptotic result (D.2.7). The result is

$$\boxed{\begin{aligned}\underline{\mathbf{u}}_{\text{Airy}}(\omega,\mathbf{x}_{\text{R}}) \simeq &-\frac{p_A^{1/2}}{2^{5/4}\pi(X_A)^{1/2}(|\ell_A|X_A'')^{1/4}}\,\text{Im}\left(\underline{\mathcal{G}}_{\text{ray}}(p_A)\right)\\ &\times \text{e}^{\text{i}\omega\widetilde{T}_A-\text{i}\pi/4}\exp\left(-\frac{(2|\ell_A|)^{3/2}\omega}{3(X_A'')^{1/2}}\right).\end{aligned}} \tag{9.3.54}$$

The important feature of expression (9.3.54) is the exponential decay with frequency into the shadow. The inverse Fourier transform of expression (9.3.54) is given by result (B.1.4) and (B.1.5), which agrees approximately with the scaled time derivative of $C(t,-1)$ (cf. result (D.2.33)), i.e. expression (9.2.49).

Exactly at the caustic ($\ell_A = 0$), the spectrum (9.3.53) reduces to

$$\begin{aligned}\underline{\mathbf{u}}_{\text{Airy}}(\omega,\mathbf{x}_{\text{R}}) \simeq &-\frac{p_A^{1/2}\omega^{1/6}}{2^{1/6}3^{2/3}\pi^{1/2}\Gamma(2/3)(X_A)^{1/2}(X_A'')^{1/3}}\,\text{Im}\left(\underline{\mathcal{G}}_{\text{ray}}(p_A)\right)\\ &\times \text{e}^{\text{i}\omega\widetilde{T}_A-\text{i}\pi/4},\end{aligned} \tag{9.3.55}$$

using result (D.2.19). This is the Fourier transform of the time derivative of $C(t,0)$, i.e. expressions (D.2.20) and (D.2.21), and agrees with expression (9.2.46). Compared with the geometrical rays, we have a spectrum $O(\omega^{1/6})$, i.e. the focus at the caustic accentuates the high frequencies.

In the illuminated region ($\ell_A > 0$), we use the asymptotic approximation (D.2.4) to simplify the spectrum (9.3.53)

$$\begin{aligned}\underline{\mathbf{u}}_{\text{Airy}}(\omega,\mathbf{x}_{\text{R}}) \simeq &-\frac{p_A^{1/2}}{2^{1/4}\pi(X_A)^{1/2}(X_A'')^{1/4}\ell_A^{1/4}}\,\text{Im}\left(\underline{\mathcal{G}}_{\text{ray}}(p_A)\right)\\ &\times \text{e}^{\text{i}\omega\widetilde{T}_A-\text{i}\pi/4}\sin\left(\omega\frac{(2\ell_A)^{3/2}}{3(X_A'')^{1/2}}+\frac{\pi}{4}\right).\end{aligned} \tag{9.3.56}$$

Using the expressions (9.2.50) and (9.2.52) for the geometrical arrivals, the Airy function in (9.3.53) reduces to $Ai\left(-(3\omega\Delta\widetilde{T}_A/2)^{3/2}\right)$, and result (9.3.56) reduces to

$$\underline{\mathbf{u}}_{\text{Airy}}(\omega,\mathbf{x}_{\text{R}}) \simeq -\frac{p_A^{1/2}\text{e}^{\text{i}\omega\widetilde{T}_A}}{2\pi(X_A)^{1/2}|X_A'|^{1/2}}\,\text{Im}\left(\underline{\mathcal{G}}_{\text{ray}}(p_A)\right)\left(\text{e}^{-\text{i}\omega\Delta T_A} - i\,\text{e}^{\text{i}\omega\Delta T_A}\right), \tag{9.3.57}$$

agreeing with expression (9.3.14) for an arrival on a forward branch at $t = T_A - \Delta T_A$, and a Hilbert transformed pulse on the reversed branch at $t = T_A + \Delta T_A$.

This in turn agrees with the scaled time derivative of the function $C(t, 1)$ (expressions (D.2.23) and (D.2.24)), i.e. result (9.2.49).

9.3.6 The Fresnel shadow

The spectral result for a geometrical ray has been obtained using the second-order saddle-point method (9.3.8). If the ray is near an interface, then the method breaks down as the slowness integrand varies rapidly near the grazing value. The transformed response, $\underline{\mathbf{v}}(\omega, \mathsf{p}, z_{\mathsf{R}})$, is complicated because of the interaction of the wave with the interface. As a first approximation we can ignore the interaction with the interface and just take the turning point solution when the turning point is above the interface, i.e. in the two-dimensional expression (8.5.1) we have

$$\underline{\mathbf{v}}_{\mathrm{Airy}}(\omega, \mathsf{p}, z_{\mathsf{R}}) \simeq H(p - p_1)\underline{\boldsymbol{\mathcal{G}}}_{\mathrm{ray}}(p)\mathrm{e}^{\mathrm{i}\omega\widetilde{T}_{\mathrm{ray}}(p,\mathbf{x}_{\mathsf{R}})}, \tag{9.3.58}$$

where $p = p_1$ is the grazing slowness (cf. definition (9.2.53)). The slowness integral (8.5.2) then becomes

$$\underline{\mathbf{v}}_{\mathrm{Fresnel}}(\omega, \mathbf{x}_{\mathsf{R}}) = \frac{|\omega|}{2\pi}\int_{p_1}^{\infty} \underline{\boldsymbol{\mathcal{G}}}_{\mathrm{ray}}(p)\,\mathrm{e}^{\mathrm{i}\omega\widetilde{T}_{\mathrm{ray}}(p,\mathbf{x}_{\mathsf{R}})}\,\mathrm{d}p, \tag{9.3.59}$$

and can be approximated by an incomplete saddle point (Appendix D.3). Similar incomplete saddle-point integrals occur in the three-dimensional integrals.

The integral of an incomplete saddle can be approximated by special functions (Appendix D.3). The spectral results for a geometrical ray (e.g. results (9.3.14)) are modified by a factor

$$\boxed{\begin{aligned} Fr\!\left(\left(-\frac{\omega(\mathrm{d}X_{\mathrm{ray}}/\mathrm{d}p)}{\pi}\right)^{1/2}(p_1 - p_{\mathrm{ray}})\right) &\simeq Fr\!\left(\left(-\frac{\omega}{\pi(\mathrm{d}X_{\mathrm{ray}}/\mathrm{d}p)}\right)^{1/2}\ell_1\right) \\ &\simeq Fr\!\left((2\omega\Delta T_1)^{1/2}/\pi\right), \end{aligned}} \tag{9.3.60}$$

where the function $Fr(z)$ is defined in expression (D.3.4) or (D.3.7), and ℓ_1 is defined in equation (9.2.55). The inverse Fourier transform of the Fresnel function is known, results (D.3.22) and (D.3.24), and it is straightforward to show that the spectral result (9.3.60) is equivalent to the slowness result (9.2.61) with definition (9.2.62).

In the shadow, $\ell_1 > 0$, the asymptotic form for the spectrum (D.3.10) gives a factor

$$Fr\left(\left(-\frac{\omega}{\pi(\mathrm{d}X_{\mathrm{ray}}/\mathrm{d}p)}\right)^{1/2}\ell_1\right) \simeq \frac{1}{\ell_1}\left(-\frac{X'_{\mathrm{ray}}}{2\pi\omega}\right)^{1/2}\mathrm{e}^{-\mathrm{i}\omega\ell_1/2X'_1+\mathrm{i}\pi/4}. \tag{9.3.61}$$

Using expression (9.2.66), this can be rewritten

$$Fr\left(\left(-\frac{\omega}{\pi(\mathrm{d}X_{\mathrm{ray}}/\mathrm{d}p)}\right)^{1/2}\ell_1\right) \simeq \frac{1}{2\pi}\frac{1}{(\Delta T_1)^{1/2}}\left(\frac{\pi}{\omega}\right)^{1/2}\mathrm{e}^{\mathrm{i}\pi/4}\,\mathrm{e}^{\mathrm{i}\omega\,\Delta T_1} \tag{9.3.62}$$

$$\longleftrightarrow \frac{1}{2\pi}\frac{1}{(\Delta T_1)^{1/2}}\lambda(t-\Delta T_1). \tag{9.3.63}$$

This agrees with expression (9.2.63).

9.3.7 The deep shadow

The Fresnel shadow only describes the truncation of the turning ray. The interaction of the wave with the interface is more complicated. In order to investigate the behaviour in the shadow, we consider the simple case of a rigid interface in an acoustic medium. More realistic interfaces can be investigated using similar techniques, but the results are algebraically messy and add little insight. The simple problem studied here is instructive to describe the type of behaviour expected but, in general for realistic models, it is simpler to use numerical methods. The behaviour in the deep shadow has been described by Duwalo and Jacobs (1959), Gilbert (1960) and Knopoff and Gilbert (1961).

We consider a medium with a negative velocity gradient above an interface at $z = z_2$, i.e. $\alpha_1' < 0$. We assume that the zeroth-order term in the Langer asymptotic expansion (7.2.159) is a good approximation to the solution. Thus the wave solution is

$$\mathbf{w}(z) = \mathbf{F}(z)\mathbf{r} = \mathbf{L}(z)\boldsymbol{\mathcal{A}}^{(\mathrm{travelling})}(z)\mathbf{r}, \tag{9.3.64}$$

where we have used the travelling-wave form, (7.2.146) and (7.2.147), as these are needed at the source and receiver, i.e. the ratio of the components of the vector $\mathbf{r}$, r_1/r_2, gives the reflected wave relative to the incident wave. At the rigid interface, the displacement must be zero, so equation (7.1.5) gives

$$\mathbf{w}(z_2) = \begin{pmatrix} 0 \\ -\underline{P} \end{pmatrix} = \mathbf{L}(z_2)\boldsymbol{\mathcal{A}}^{(\mathrm{travelling})}(z_2)\mathbf{r}. \tag{9.3.65}$$

Thus using results (7.2.132) and (7.2.144), the ratio of the components of $\mathbf{r}$ is

$$\frac{r_1}{r_2} = -\frac{Bj'(-\xi_1)}{Aj'(-\xi_1)}, \tag{9.3.66}$$

where definition (7.2.140) gives

$$\xi_1 = \xi(z_2) = \left(\frac{3\omega\tau_1}{2}\right)^{2/3}, \tag{9.3.67}$$

and definition (7.2.141) gives

$$\tau_1 = \tau_\alpha(p, z_2) = \int_{z_\alpha(p)}^{z_2} q_\alpha(p, \zeta)\, d\zeta \tag{9.3.68}$$

(for simplicity, as we frequently have fractional powers of frequency, e.g. in equation (9.3.67), we assume $\omega > 0$ in this section, and use result (3.1.9) for negative frequencies). Using result (9.3.66), the propagation term in the transformed response (8.0.7) can be written

$$\boxed{\mathcal{P}_{\text{ray}}(\omega, p, z_R) = \left(\prod_{\text{ray}'} \mathcal{T}_{ij}\right)\left(-\frac{Bj'(-\xi_1)}{Aj'(-\xi_1)}\right) e^{i\omega\tau_{\text{ray}}(p, z_R)},} \tag{9.3.69}$$

where the product $\prod_{\text{ray}'}$ excludes the coefficient (9.3.66) from the interface at z_2 and the turning point at $z_\alpha(p)$. The delay-time function is typically

$$\tau_{\text{ray}}(p, z_R) = \tau_\alpha(p, z_S) + \tau_\alpha(p, z_R) = \int_{z_\alpha(p)}^{z_S} + \int_{z_\alpha(p)}^{z_R} q_\alpha(p, \zeta)\, d\zeta. \tag{9.3.70}$$

Using the appropriate travelling-wave definitions, (7.2.146) and (7.2.147), the 'coefficient' (9.3.66) can be written

$$-\frac{Bj'(-\xi_1)}{Aj'(-\xi_1)} = i\,\frac{Ai'(-\xi_1) + iBi'(-\xi_1)}{Ai'(-\xi_1) - iBi'(-\xi_1)} \tag{9.3.71}$$

$$= e^{-i\pi/6}\,\frac{Ai'(\xi_1 e^{i\pi/3})}{Ai'(\xi_1 e^{-i\pi/3})} \tag{9.3.72}$$

$$= i\, e^{2i\,\phi(\xi_1)}, \tag{9.3.73}$$

where expression (9.3.72) is obtained using the identity

$$Ai(ze^{\pm i\pi/3}) = \frac{1}{2}\, e^{\mp i\pi/3}\left(Ai(-z) \pm i\, Bi(-z)\right) \tag{9.3.74}$$

$$Ai'(ze^{\pm i\pi/3}) = \frac{1}{2}\, e^{\pm i\pi/3}\left(Ai'(-z) \pm i\, Bi'(-z)\right), \tag{9.3.75}$$

and in expression (9.3.73), the function $\phi(\xi)$ is defined as (Abramowitz and Stegun, 1965, §10.4.70)

$$\phi(\xi) = \tan^{-1}\left(\frac{Bi'(-\xi)}{Ai'(-\xi)}\right). \tag{9.3.76}$$

We denote the slowness of the grazing ray as $p_1 = 1/\alpha_1$ so that $z_\alpha(p_1) = z_2$. For $p > p_1$, the turning point is above the interface, $z_\alpha(p) > z_2$ and $\xi_1 < 0$ (τ_1 is negative, imaginary). The Airy functions in expression (9.3.71) are evanescent,

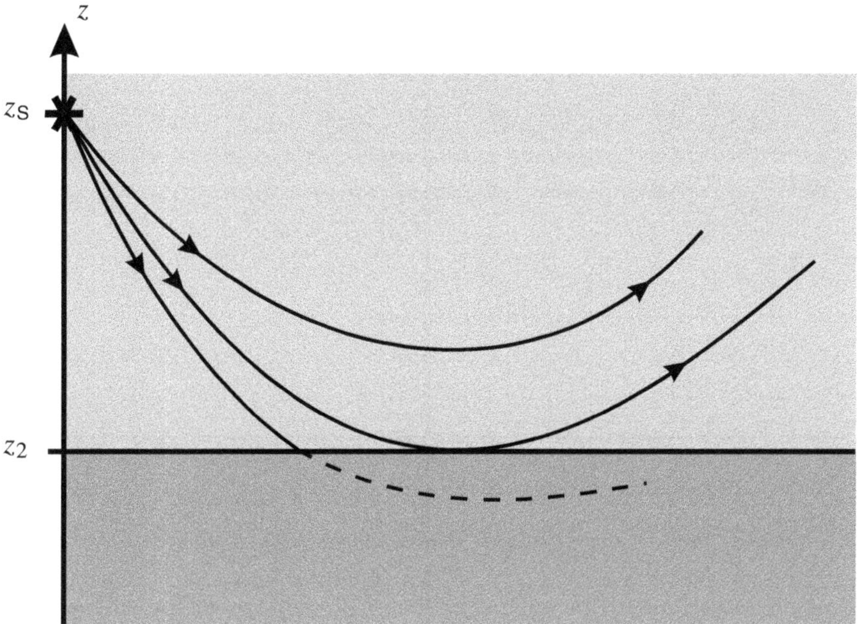

Fig. 9.35. Rays with turning points above, at and below an interface.

(D.2.7) and (D.2.8), and asymptotically expression (9.3.71) reduces to

$$-\frac{Bj'(-\xi_1)}{Aj'(-\xi_1)} \longrightarrow -\mathrm{i}, \tag{9.3.77}$$

when $\xi_1 \ll -1$. Physically, if the turning-point is well above the interface, then the wave is exponentially small at the interface and the interface has little influence (Figure 9.35). The limit (9.3.77) is equivalent to the standard turning-ray result (7.2.163).

When $p < p_1$, the turning point would be below the interface. We continue the velocity with a constant gradient below the interface in order to define a virtual turning point. Then $z_\alpha(p) < z_2$ and $\xi_1 > 0$. From the derivatives of the asymptotic forms (7.2.148) and (7.2.149), the 'coefficient' (9.3.71) reduces to

$$-\frac{Bj'(-\xi_1)}{Aj'(-\xi_1)} \longrightarrow \mathrm{e}^{-2\mathrm{i}\omega\,\tau_1}, \tag{9.3.78}$$

when $\xi_1 \gg 1$. Combining with expression (9.3.69), we obtain

$$\mathcal{P}_{\mathrm{ray}}(\omega,\, p,\, z_{\mathrm{R}}) = \left(\prod_{\mathrm{ray}'} \mathcal{T}_{ij}\right) \mathrm{e}^{\mathrm{i}\omega(\tau_{\mathrm{ray}}(p,z_{\mathrm{R}})-2\tau_1)}, \tag{9.3.79}$$

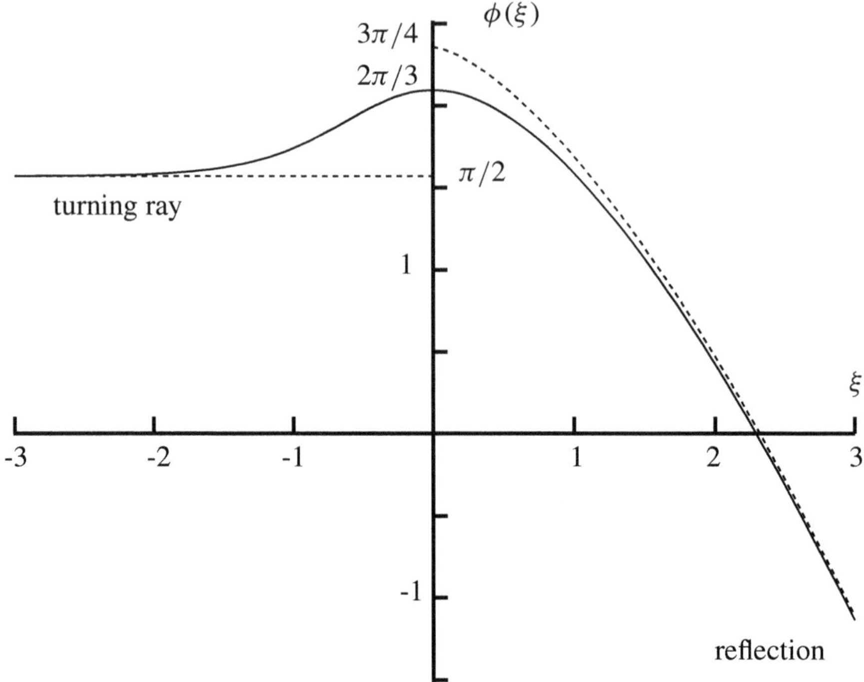

Fig. 9.36. The function $\phi(\xi)$ (9.3.76).

where with function (9.3.70)

$$\tau_{\text{ray}}(p, z_R) - 2\tau_1 = \int_{z_2}^{z_S} + \int_{z_2}^{z_R} q_\alpha(p, \zeta)\, d\zeta. \tag{9.3.80}$$

This corresponds to the signal reflected from the interface with a reflection coefficient $\mathcal{T}_{11} = 1$ (which can be obtained from result (6.3.7) by letting $\rho_2 \to \infty$ to simulate a rigid interface).

The function $\phi(\xi)$ (9.3.76) is plotted in Figure 9.36. The asymptotes are

$$\phi(\xi) \longrightarrow \frac{3\pi}{4} - \frac{2}{3}\xi^{3/2} = \frac{3\pi}{4} - \omega\tau \quad \text{if} \quad \xi \gg 1 \tag{9.3.81}$$

$$\longrightarrow \frac{\pi}{2} \quad \text{if} \quad \xi \ll -1. \tag{9.3.82}$$

The former (9.3.81) corresponds to a reflection, and the later (9.3.82) to the turning ray.

Using result (9.3.69) in the transformed response (8.0.6), we can obtain the response near the interface shadow. For convenience, we revise (8.0.8) and define

$$\underline{\widetilde{\mathcal{G}}}_{\mathrm{ray}}(p) = \left(\prod_{\mathrm{ray}'} \mathcal{T}_{ij} \right) \mathbf{g}\,\mathbf{g}_{\mathrm{S}}^{\mathrm{T}}. \tag{9.3.83}$$

Substituting in expression (8.0.2), and approximating the transverse slowness integral by the second-order saddle point method (9.3.11) to reduce to a single slowness integral, we obtain

$$\underline{\mathbf{v}}_{\mathrm{shadow}}(\omega, \mathbf{x}_{\mathrm{R}}) \simeq \frac{\omega^{3/2} p_1^{1/2} \mathrm{e}^{-5\mathrm{i}\pi/12} \underline{\widetilde{\mathcal{G}}}_{\mathrm{ray}}(p_1)}{2^{3/2}\pi^{3/2} X_1^{1/2}} \int_{-\infty}^{\infty} \frac{Ai'(\xi_1 \mathrm{e}^{\mathrm{i}\pi/3})}{Ai'(\xi_1 \mathrm{e}^{-\mathrm{i}\pi/3})} \mathrm{e}^{\mathrm{i}\omega \widetilde{T}_{\mathrm{ray}}(p, \mathbf{x}_{\mathrm{R}})}\, \mathrm{d}p, \tag{9.3.84}$$

where we have approximated some of the integrand by the grazing values.

The Airy function is oscillatory on the negative real axis (D.2.4). Its derivative has zeros at $-\sigma_j'$, say,

$$Ai'(-\sigma_j') = 0, \tag{9.3.85}$$

(the important value will be $\sigma_1' \simeq 1.01879$, Abramowitz and Stegun 1965, Table 10.13). Thus the integrand of expression (9.3.84) has poles when

$$\xi_1 \mathrm{e}^{-\mathrm{i}\pi/3} = -\sigma_j'. \tag{9.3.86}$$

Using definition (7.2.143), when $p \simeq p_1$ we have

$$\xi_1 \simeq \frac{2^{1/3}\omega^{2/3}\alpha_1}{(-\alpha_1')^{2/3}}(p_1 - p), \tag{9.3.87}$$

so the poles are at

$$p_j' \simeq p_1 + \omega^{-2/3} V_j \mathrm{e}^{\mathrm{i}\pi/3}, \tag{9.3.88}$$

where

$$V_j = \frac{(-\alpha_1')^{2/3}\sigma_j'}{2^{1/3}\alpha_1}. \tag{9.3.89}$$

The poles of illustrated in Figure 9.37. Near the poles, we have

$$\frac{Ai'(\xi_1 \mathrm{e}^{\mathrm{i}\pi/3})}{Ai'(\xi_1 \mathrm{e}^{-\mathrm{i}\pi/3})} \simeq -\mathrm{i}\, \frac{(-\alpha_1')^{2/3}}{2^{4/3}\omega^{2/3}\alpha_1} \frac{Bi'(-\sigma_j')}{\sigma_j' Ai(-\sigma_j')} \frac{1}{p - p_j'}, \tag{9.3.90}$$

where we have used results (9.3.75) and (9.3.85) to simplify the numerator, and results (7.2.139) and (9.3.87) to simplify the denominator.

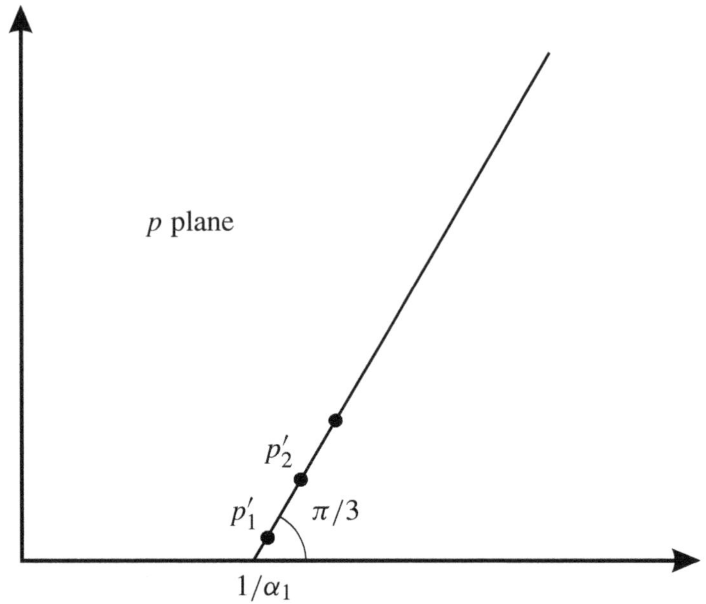

Fig. 9.37. The positions of the poles p'_j (9.3.88) of the integral (9.3.84).

The phase term in the integrand of expression (9.3.84) can be approximated by

$$\widetilde{T}_{\mathrm{ray}}(p, \mathbf{x}_{\mathrm{R}}) \simeq \widetilde{T}_{\mathrm{ray}}(p_1, \mathbf{x}_{\mathrm{R}}) + \left(\partial \widetilde{T}_{\mathrm{ray}}/\partial p\right)(p'_j - p_1) \tag{9.3.91}$$

$$= \widetilde{T}_1 + \omega^{-2/3} v_j \ell_1 \mathrm{e}^{\mathrm{i}\pi/3}, \tag{9.3.92}$$

at the poles, with

$$\widetilde{T}_1(\mathbf{x}_{\mathrm{R}}) = \widetilde{T}_{\mathrm{ray}}(p_1, \mathbf{x}_{\mathrm{R}}) = T_1(z_{\mathrm{R}}) + p_1 \ell_1(\mathbf{x}_{\mathrm{R}}) \tag{9.3.93}$$

$$T_1(z_{\mathrm{R}}) = \widetilde{T}_{\mathrm{ray}}(p_1, X_1) = T_{\mathrm{ray}}(p_1, z_{\mathrm{R}}) \tag{9.3.94}$$

$$X_1(z_{\mathrm{R}}) = X_{\mathrm{ray}}(p_1, z_{\mathrm{R}}) \tag{9.3.95}$$

$$\ell_1(\mathbf{x}_{\mathrm{R}}) = x_{\mathrm{R}} - X_1(z_{\mathrm{R}}). \tag{9.3.96}$$

The shadow edge is at X_1 with a geometrical travel time of T_1. The distance from the shadow edge is ℓ_1 and $\widetilde{T}_1$ is an arrival time with slowness p_1 extending from the shadow edge.

The contour of integration in integral (9.3.84) can be distorted into the upper p plane, picking up the residues of the poles at $p = p'_j$. Provided $\ell_1 > 0$, i.e. $x > X_1$, the contribution from these poles decreases exponentially as j and ℓ_1 increase (see result (9.3.97) below). We therefore approximate the integral by the residue of the first pole ($j = 1$), and using result (9.3.90) to evaluate the residue, obtain

$$\boxed{\begin{aligned}\underline{\mathbf{v}}_{\text{shadow}}(\omega, \mathbf{x}_{\text{R}}) \simeq\ & \frac{(-\alpha_1')^{2/3}\,\widetilde{\underline{\mathcal{G}}}_{\text{ray}}(p_1)}{2^{11/6}\alpha_1^{3/2}X_1^{1/2}}\,\frac{Bi'(-\sigma_j')}{\sigma_j' Ai(-\sigma_j')} \\ & \times \left(\omega^2\left(\frac{\pi}{\omega}\right)^{1/2} \mathrm{e}^{\mathrm{i}\omega\widetilde{T}_1+\mathrm{i}\pi/4}\right)\left(\omega^{-2/3}\mathrm{e}^{-\omega^{1/3}V_1\ell_1\mathrm{e}^{-\mathrm{i}\pi/6}-2\mathrm{i}\pi/3}\right),\end{aligned}} \tag{9.3.97}$$

where we have factored the frequency dependence for later convenience. This approximation is valid provided

$$\omega^{1/3}V_1\ell_1 \gg 1, \tag{9.3.98}$$

so the exponential term is small and the residue series is well approximated by the first term.

The important part of this spectrum is the exponential. The exponent is

$$\mathrm{i}\omega\widetilde{T}_1 - \omega^{1/3}V_1\ell_1\mathrm{e}^{-\mathrm{i}\pi/6} = i\omega\left(T_1 + p_1\ell_1 + \omega^{-2/3}V_1\ell_1/2\right) - \omega^{1/3}3^{1/2}V_1\ell_1/2. \tag{9.3.99}$$

The final term in equation (9.3.99) describes an exponential decay of the spectrum into the shadow where the decay constant is proportional to the cube root of the frequency. The first term in equation (9.3.99) describes the arrival time. The signal is delayed relative to an arrival with slowness p_1 extending from the shadow edge. The delay is inversely proportional to $\omega^{2/3}$, i.e. lower frequencies arrive later.

The inverse Fourier transform (3.1.2) of the final term in result (9.3.97) can be obtained by the change of variable

$$\omega^{1/3} = \frac{\mathrm{e}^{2\mathrm{i}\pi/3}}{(3t)^{1/3}}\,\xi, \tag{9.3.100}$$

when $\omega > 0$. Then the term reduces to a form investigated in Appendix D.2.3, i.e. equation (D.2.37). Using the function $Sh^{(3)}(t)$ defined in equation (D.2.47), we obtain

$$\boxed{\underline{\mathbf{u}}_{\text{shadow}}(t, \mathbf{x}_{\text{R}}) \simeq -\frac{3^{5/2}(-3\alpha_1')^{2/3}\,\widetilde{\underline{\mathcal{G}}}_{\text{ray}}(p_1)}{2^{11/6}\alpha_1^{3/2}X_1^{1/2}(V_1\ell_1)^{11/2}}\,\frac{Bi'(-\sigma_j')}{\sigma_j' Ai(-\sigma_j')}\,Sh^{(3)}\left(\frac{3(t-\widetilde{T}_1)}{(V_1\ell_1)^3}\right).} \tag{9.3.101}$$

The time function $Sh^{(3)}(t)$ is an emergent signal with a delayed peak (Figure D.6). The waveforms in the deep shadow are illustrated in Figure 9.38. As we progress deeper into the shadow the signal decays and becomes lower frequency.

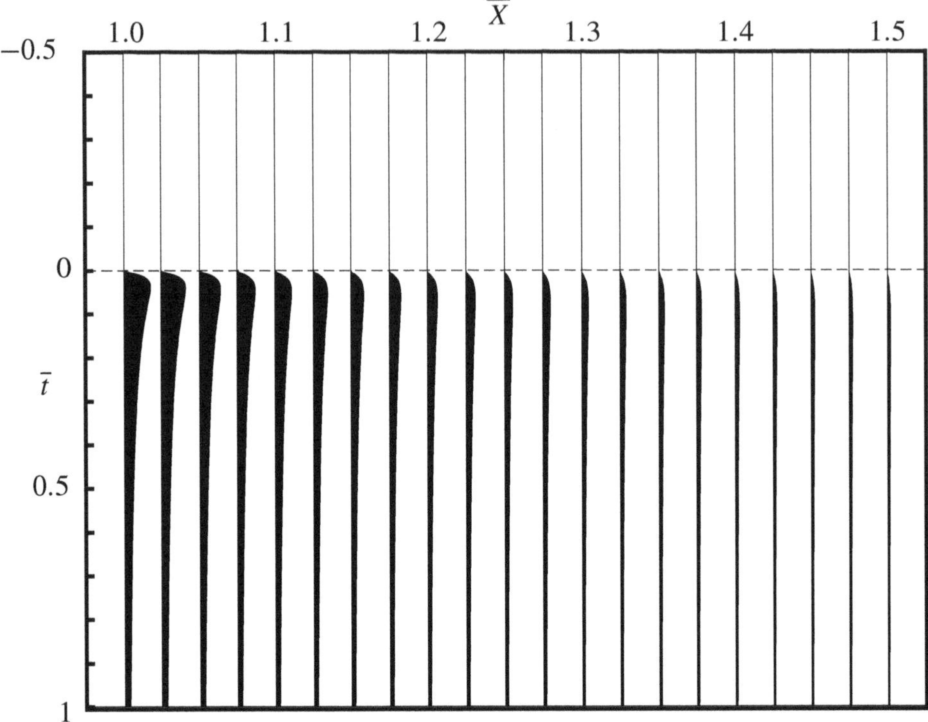

Fig. 9.38. The waveforms in the deep shadow given by approximation $(V_1\ell_1)^{-11/2}Sh^{(3)}\big(3(V_1\ell_1)^3(t-\widetilde{T}_1)\big)$ (9.3.101) with $V_1 = 1$. The time axis is the reduced travel time, $\bar{t} = t - \widetilde{T}_1$, and the range axis is the reduced range, $\overline{X} = \ell_1 = x_{\mathsf{R}} - X_1$.

The algebra in the derivation of expression (9.3.101) is sufficiently involved that is worth checking the units. We have unit$(V_1\ell_1) = [\mathrm{T}^{1/3}]$ and unit $\big(Sh^{(3)}(t/a)\big) = [\mathrm{T}]$ if unit$(a) = [\mathrm{T}]$. Then it is straightforward to confirm that unit$(\underline{\mathbf{u}}) = [\mathrm{M}^{-1}\mathrm{T}]$.

Finally, we illustrate the behaviour across a shadow edge in Figure 9.39. These have been calculated using the numerical methods of Section 8.5.2 with the propagation term (9.3.69) and the far-field approximation, i.e. the integral (9.3.84). The model is acoustic with a linear gradient and a rigid interface. The shadow edge is indicated and the Fresnel shadow behaviour near the edge (Section 9.2.8 and Figure 9.29), and the decay in the shadow (this section and Figure 9.38) is evident. A more detailed analysis of the different approximations and their ranges of validity for realistic interface conditions have been made in the early papers by Scholte (1956) and Duwalo and Jacobs (1959), the substantial theoretical papers by Nussenzveig (1965, 1969*a, b*) and Ansell (1978), and the numerical studies by Chapman and Phinney (1970, 1972).

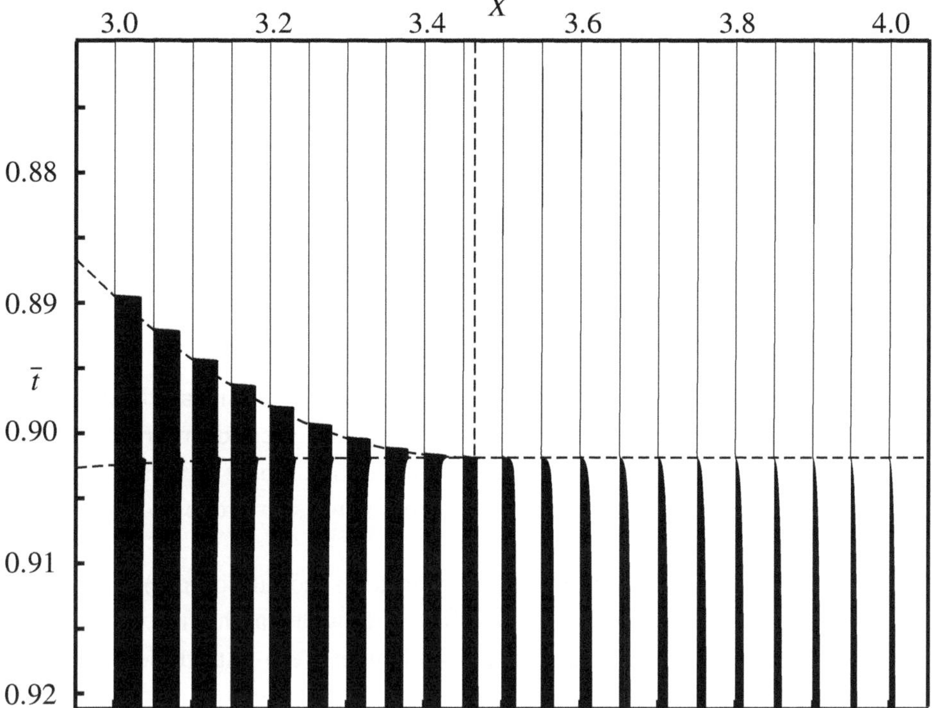

Fig. 9.39. The waveforms across a shadow. The model is normalized so the source and receiver are at zero depth, $z_S = z_R = 0$, and the velocity has a linear unit gradient, $\alpha = 1 - z$, to a rigid interface at $z = z_2 = -1$. The shadow slowness is $p_1 = 1/2$ and the shadow edge is at $X_1 = 2\sqrt{3}$ (indicated with a dashed line). The shadow time is $T_1 \simeq 2.634$ and the waveforms are plotted against the reduced travel time, $\bar{t} = t - x_R/2$. The reduced time $T_1 - p_1 X_1$ in the shadow, and the turning ray and reflection travel times in the illuminated region, are indicated with dashed lines.

Exercises

9.1 Head waves can exist on reflected or transmitted waves, with velocities from either side of the interface. There are four kinds of head waves: (1) a reflected wave with velocity from the transmitted medium; (2) a transmitted wave with velocity from the transmitted medium; (3) a transmitted wave with velocity from the incident medium; and (4) a reflected wave with velocity from the incident medium. Which head waves exists depends on the incident wave type, and the arrangement of the velocities. There are six cases for the velocities. In the following table, the six cases are listed together with the number of head waves possible for the two incident rays (in the first medium). Confirm these figures, identifying the kind and the ray notation of the possible head waves.

Case	Velocities	P	S
1	$\alpha_2 > \beta_2 > \alpha_1 > \beta_1$	5	6
2	$\alpha_2 > \alpha_1 > \beta_2 > \beta_1$	3	6
3	$\alpha_2 > \alpha_1 > \beta_1 > \beta_2$	3	5
4	$\alpha_1 > \alpha_2 > \beta_2 > \beta_1$	0	6
5	$\alpha_1 > \alpha_2 > \beta_1 > \beta_2$	0	5
6	$\alpha_1 > \beta_1 > \alpha_2 > \beta_2$	0	3

9.2 The first-motion approximation is based on first-order Taylor expansions about the geometrical arrival. Investigate the second-order terms for a point source in a homogeneous medium, and compare the results with the exact result. Do the second-order terms always improve the approximation?

9.3 Show that at a fixed frequency (e.g. in the spectral domain), the particle motion of a Rayleigh wave is an ellipse. Show that at the free surface it is a retrograde ellipse but that at depth it changes from retrograde to prograde (and at some depth it is vertical). Show that at the free surface the ratio of vertical to horizontal displacement varies from $|u_z/u_x| = 1.83924$ for Poisson's ratio $\nu = 1/2$, to 1.27201 for Poisson's ratio $\nu = 0$.

9.4 *Further reading:* The shadow results, Sections 9.3.6 and 9.3.7, have been extensively studied in a spherical Earth, where even with homogeneous layers, the spherical surfaces cast shadows. Early publications are Duwalo and Jacobs (1959), Gilbert (1960) and Knopoff and Gilbert (1961). Other papers are Phinney and Alexander (1966) and Chapman and Phinney (1972).

9.5 *Further reading:* The amplitudes of head waves were obtained using the Cagniard and WKBJ methods and an expansion about the branch points of the reflection/transmission coefficients (9.1.52). A completely different method is used in the textbook by Červený and Ravindra (1971). Show that the two methods agree.

10

Generalizations of ray theory

Ray theory often breaks down, and transform methods are only valid for stratified models. In this chapter, extensions of these methods are developed which bridge some of these gaps. The *Maslov method* combines the advantages of ray theory and the WKBJ seismogram method to provide a method valid at caustics in generally, heterogeneous models; *quasi-isotropic ray theory* extends ray theory to model the frequency-dependent coupling that exists between *qS* rays in heterogeneous, anisotropic media; *Born scattering theory* extends ray theory to model signals scattered by perturbations to a reference model, or heterogeneities where ray theory is inaccurate; and the *Kirchhoff surface integral method* extends ray theory to model reflections from non-planar interfaces. These methods are all computationally relatively inexpensive and straightforward to apply.

Unfortunately, difficulties never come alone, and in realistic, complex models, a combination of all these methods and more may be needed. Apart from numerical methods, such as finite-difference techniques which are extremely expensive for realistic, complex media at body-wave frequencies, no such comprehensive method has been developed. But enough for today – that is for tomorrow!

In previous chapters we have investigated asymptotic ray theory (Chapter 5), valid in three-dimensional, heterogeneous media but breaking down at singularities, and transform methods (Chapter 8), only valid in stratified media (a 'one-dimensional' structure but three-dimensional wave propagation) but which remain valid at the singularities of ray theory. In this chapter, we introduce some extensions of ray theory which contain some of the advantages of both methods: validity in general, heterogeneous media and at (some of) the singularities of ray theory, and efficient evaluation. In Section 10.1, we discuss Maslov asymptotic ray theory that combines the advantages of asymptotic ray theory with the WKBJ seismogram; in Section 10.2, we extend ray theory in anisotropic media so it remains valid in the isotropic limit; in Section 10.3, we develop a generalization of Born scattering theory to describe the signals scattered in regions where asymptotic ray theory breaks down; and finally in Section 10.4, we specialize the volume scattering of

Born theory to scattering by an interface to describe signals reflected by non-planar surfaces.

10.1 Maslov asymptotic ray theory

Maslov asymptotic ray theory extends the WKBJ seismogram method (Section 8.4.1) to heterogeneous media. This was introduced to seismology by Chapman and Drummond (1982) after theory by Maslov (1965, 1972). First we develop the result for a two-dimensional structure as this is virtually identical to the WKBJ seismogram. Then we extend this to three dimensions. Chapman and Keers (2002) have investigated computational details of the Maslov algorithm in three dimensions.

In the transform method, the slowness integral (8.5.2) can be approximated by the second-order saddle-point method to give the result (9.3.4) which is exactly equivalent to the ray approximation, i.e. the leading term in the asymptotic ray series (5.1.1). However, if the slowness integral (8.5.2) is evaluated exactly or numerically, it remains valid in regions where the ray approximation (9.3.4) is singular, e.g. at critical points, caustics, etc. This suggests that we should extend ray theory by representing the response as a 'slowness' integral

$$\underline{\mathbf{v}}(\omega, \mathbf{x}_{\mathsf{R}}) = \left(\frac{|\omega|}{2\pi}\right)^m \int \underline{\tilde{\mathbf{v}}}(\omega, \mathsf{q})\, e^{i\omega \widetilde{T}(\mathsf{q}, \mathbf{x}_{\mathsf{R}})}\, d\mathsf{q} \tag{10.1.1}$$

$$\simeq \left(\frac{|\omega|}{2\pi}\right)^m \int \underline{\tilde{\mathbf{v}}}^{(0)}(\mathsf{q})\, e^{i\omega \widetilde{T}(\mathsf{q}, \mathbf{x}_{\mathsf{R}})}\, d\mathsf{q}, \tag{10.1.2}$$

where $m = 1$ or 2 is the dimension of the integral. The factor $|\omega|^m$ is introduced as, with hindsight, it makes the approximation for $\underline{\tilde{\mathbf{v}}}^{(0)}$ in integral (10.1.2) independent of frequency. We will define below the functions $\underline{\tilde{\mathbf{v}}}^{(0)}$, $\widetilde{T}$ and the variable q. Just as integral (8.5.2) reduced to result (9.3.4) when the second-order saddle-point method was a good approximation, expression (10.1.2) must reduce to the ray approximation (5.1.1) when that is valid.

Expression (10.1.2) should only be considered as an ansatz, just as approximation (5.1.1) was an ansatz for ray theory. It doesn't matter that in heterogeneous media we cannot use transform theory to obtain this integral (although using the methods of pseudo-differential and Fourier integral operators, transform methods can be extended to heterogeneous media, e.g. Duistermaat, 1995). The rigorous transform methods possible in a stratified model suggest this ansatz, but the fact that we cannot transform the wave equation in a heterogeneous medium does not prevent us from using an integral representation as our ansatz. In the original Maslov theory, the variable q in integral (10.1.2) is the slowness on a surface through the receiver. This precise physical interpretation is not necessary and we consider the variable q as just any variable that parameterizes rays. With any parameterization, the integral (10.1.2) can be transformed into an integral with respect to the slowness at the receiver by a simple change of variable. The integrand will then contain the extra factor of the Jacobian for the mapping between the parameterization q and the slowness. This Jacobian can be absorbed into $\underline{\tilde{\mathbf{v}}}^{(0)}(\mathsf{q})$, so we can consider integral (10.1.2) generally with any parameterization.

10.1.1 Two dimensions

Let us first consider the case of two-dimensional wave propagation. The ray approximation is given by result (5.4.28)

$$\underline{\mathbf{v}}(\omega, \mathbf{x}_{\mathsf{R}}) \simeq f^{(2)}(\omega)\, \underline{\mathbf{v}}^{(0)}(\mathbf{x}_{\mathsf{R}})\, \mathrm{e}^{\mathrm{i}\omega T}. \tag{10.1.3}$$

In the two-dimensional case, q in integral (10.1.2) is a scalar. If q parameterizes the rays, i.e. the direction a ray leaves the source, we can generalize the earlier notation for the geometrical travel time and range functions to include q as an argument. Later we will be more specific about our choice of parameterization. We consider an arbitrary surface through the receiver $\mathbf{x}_{\mathsf{R}}$ (for simplicity and by analogy with the transform results, this is normally taken as a plane). We call this the *target surface*. Shooting the ray with parameter q, it hits the surface at $\mathbf{X}(\mathsf{q})$ with travel time, $T(\mathsf{q})$, and slowness $\mathbf{p}(\mathsf{q})$ (see Figure 10.1) (for brevity we have not included any notation for the surface or the ray path in these functions).

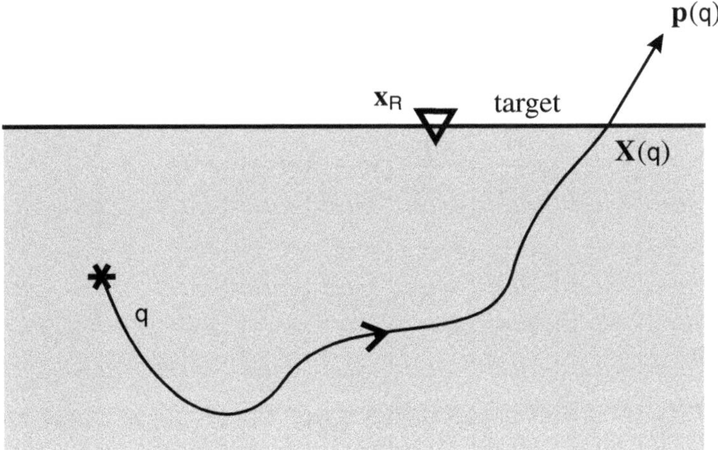

Fig. 10.1. A ray with 'parameter' q hitting the target surface through the receiver $\mathbf{x}_{\mathsf{R}}$ at $\mathbf{X}(\mathsf{q})$ with slowness $\mathbf{p}(\mathsf{q})$.

As normal transform theory cannot be used to reduce the wave equations to ordinary differential equations which can be solved, how are we to determine the functions $\tilde{\underline{\mathbf{v}}}^{(0)}$ and $\widetilde{T}$ in equation (10.1.2)? Obviously there is some arbitrariness, but they should agree with the WKBJ seismogram method when the model is stratified, and lead to agreement with ray theory in general models when it is valid. In fact, if the frequency is high enough, ray theory is valid almost everywhere and breaks down only in very narrow zones around the singular surfaces or points. This agreement is therefore a stringent requirement and defines the functions $\tilde{\underline{\mathbf{v}}}^{(0)}$ and $\widetilde{T}$ almost everywhere. The required continuity and smoothness of the solution (as it satisfies the wave equations) can then be used to define the functions $\tilde{\underline{\mathbf{v}}}^{(0)}$ and $\widetilde{T}$ everywhere.

The natural choice (but not unique – see Kendall and Thomson, 1993) for the phase function is

$$\boxed{\widetilde{T}(\mathsf{q}, \mathbf{x}_\mathsf{R}) = T(\mathsf{q}) + \mathbf{p}(\mathsf{q}) \cdot \left(\mathbf{x}_\mathsf{R} - \mathbf{X}(\mathsf{q})\right),} \tag{10.1.4}$$

where the dot product extracts the slowness component in the target surface, which we denote by $p_\parallel$, a scalar in the two-dimensional problem. Thus equation (10.1.4) reduces to

$$\widetilde{T}(\mathsf{q}, \mathbf{x}_\mathsf{R}) = T(\mathsf{q}) + p_\parallel \left(x_\parallel - X_\parallel(\mathsf{q})\right), \tag{10.1.5}$$

where $x_\parallel$ is a scalar parameter, not necessarily a cartesian component, that defines the receiver position on the target surface. Using the slowness definition (5.1.6), we have

$$\frac{\mathrm{d}T}{\mathrm{d}\mathsf{q}} = \mathbf{p}(\mathsf{q}) \cdot \frac{\mathrm{d}\mathbf{X}}{\mathrm{d}\mathsf{q}}, \tag{10.1.6}$$

so a saddle point of the phase function exists when

$$\frac{\partial \widetilde{T}}{\partial \mathsf{q}} = \frac{\mathrm{d}p_\parallel}{\mathrm{d}\mathsf{q}} \left(x_\parallel - X_\parallel(\mathsf{q})\right) = 0, \tag{10.1.7}$$

i.e. when

$$\mathbf{X}(\mathsf{q}_{\mathrm{ray}}) = \mathbf{x}_\mathsf{R}, \tag{10.1.8}$$

say, or when

$$\frac{\mathrm{d}p_\parallel}{\mathrm{d}\mathsf{q}} = 0. \tag{10.1.9}$$

We will return to the latter possibility later. The ray parameter $\mathsf{q}_{\mathrm{ray}}$ is the value that solves the two-point ray-tracing problem (10.1.8). At the saddle point, the phase function, $\widetilde{T}$ (10.1.4), obviously reduces to the two-point travel time, $T(\mathsf{q}_{\mathrm{ray}})$, i.e.

$$\widetilde{T}\left(\mathsf{q}_{\mathrm{ray}}, \mathbf{X}(\mathsf{q}_{\mathrm{ray}})\right) = T(\mathsf{q}_{\mathrm{ray}}), \tag{10.1.10}$$

and the second derivative is

$$\left.\frac{\partial^2 \widetilde{T}}{\partial \mathsf{q}^2}\right|_{\mathrm{ray}} = -\frac{\mathrm{d}p_{\|}}{\mathrm{d}\mathsf{q}}\frac{\mathrm{d}X_{\|}}{\mathrm{d}\mathsf{q}}. \tag{10.1.11}$$

The contribution to the integral (10.1.2) can be evaluated by the second-order saddle-point approximation (D.1.11). The result is

$$\underline{\mathbf{v}}(\omega, \mathbf{x}_{\mathrm{R}}) \simeq f^{(2)}(\omega) \frac{\mathrm{e}^{\mathrm{i}\omega T(\mathsf{q}_{\mathrm{ray}}) + \mathrm{i}\frac{\pi}{4}\,\mathrm{sgn}(\omega)\left(\mathrm{sgn}\left(\partial^2\widetilde{T}/\partial\mathsf{q}^2|_{\mathrm{ray}}\right)+1\right)}}{\left|\partial^2\widetilde{T}/\partial\mathsf{q}^2\right|_{\mathrm{ray}}^{1/2}} \underline{\tilde{\mathbf{v}}}^{(0)}(\mathsf{q}_{\mathrm{ray}}). \tag{10.1.12}$$

This must agree with the ray approximation (10.1.3), when the latter is valid. Thus

$$\boxed{\underline{\tilde{\mathbf{v}}}^{(0)}(\mathsf{q}) = \underline{\mathbf{v}}^{(0)}\left(\mathbf{X}(\mathsf{q})\right) \left|\frac{\mathrm{d}p_{\|}}{\mathrm{d}\mathsf{q}}\right|^{1/2} \left|\frac{\mathrm{d}X_{\|}}{\mathrm{d}\mathsf{q}}\right|^{1/2} \mathrm{e}^{\mathrm{i}\frac{\pi}{4}\,\mathrm{sgn}(\omega)\left(\mathrm{sgn}\left(\frac{\mathrm{d}p_{\|}}{\mathrm{d}\mathsf{q}}\frac{\mathrm{d}X_{\|}}{\mathrm{d}\mathsf{q}}\right)-1\right)}.} \tag{10.1.13}$$

Note in expression (10.1.12), the second derivative of the phase function, $\widetilde{T}$, is evaluated at $\mathsf{q} = \mathsf{q}_{\mathrm{ray}}$ which solves equation (10.1.8), and the right-hand side is a function of $\mathsf{q}_{\mathrm{ray}}$. In expression (10.1.13), the variable is q, and the range $\mathbf{x}_{\mathrm{R}}$ has been replaced by $\mathbf{X}(\mathsf{q})$. In regions where ray theory is valid, we have defined $\underline{\tilde{\mathbf{v}}}^{(0)}(\mathsf{q})$. Where ray theory breaks down, the function $\underline{\tilde{\mathbf{v}}}^{(0)}$ remains non-singular as the factor $|\mathrm{d}X_{\|}/\mathrm{d}\mathsf{q}|^{1/2}$ cancels with a similar factor in the denominator of the ray amplitude coefficient, $\underline{\mathbf{v}}^{(0)}(\mathbf{x}_{\mathrm{R}})$ (see result (5.7.12)). We will analyse this in more detail later.

Requiring that the Maslov ansatz (10.1.2) agrees with the ray approximation (10.1.3) has allowed us to define the functions in the integrand, $\underline{\tilde{\mathbf{v}}}^{(0)}$ (10.1.13) and $\widetilde{T}$ (10.1.4). This is adequate for our purposes. More rigorously, we can use the methods of pseudo-differential and Fourier integral operators to generalize the integrand to an asymptotic series (Duistermaat, 1995). However, for the leading term, this complication is unnecessary (we have emphasized that properly, approximation (10.1.2) is only the leading term in an asymptotic series by including the suffix on the function $\underline{\tilde{\mathbf{v}}}^{(0)}(\mathsf{q})$). Usually, only the leading term has been used in numerical computations.

The terms in the Maslov integral, $\underline{\tilde{\mathbf{v}}}^{(0)}$ (10.1.13) and $\widetilde{T}$ (10.1.4), can all be evaluated by normal ray shooting (cf. Chapter 5). We can now proceed to evaluate the integral using the same technique as for the WKBJ seismogram (Section 8.4.1). Applying the inverse Fourier transform (3.1.2) with respect to frequency to integral

(10.1.2), changing the order of integration, we obtain (cf. result (8.4.5))

$$\boxed{\underline{\mathbf{u}}(t, \mathbf{x}_{\mathrm{R}}) \simeq -\frac{1}{2\pi} \operatorname{Im} \left(\Delta(t) * \sum_{\widetilde{T}(\mathsf{q}, \mathbf{x}_{\mathrm{R}})=t} \frac{\underline{\tilde{\mathbf{v}}}^{(0)}(\mathsf{q})}{|\partial \widetilde{T}/\partial \mathsf{q}|} \right),} \tag{10.1.14}$$

where the summation is over solutions of the equation

$$\widetilde{T}(\mathsf{q}, \mathbf{x}_{\mathrm{R}}) = t. \tag{10.1.15}$$

We call result (10.1.14) the *Maslov seismogram.*

For three-dimensional wave propagation in a two-dimensional model, commonly called *2.5D wave propagation,* we can use the usual conversion factor (8.2.66). Expression (10.1.14) is replaced by

$$\underline{\mathbf{u}}(t, \mathbf{x}_{\mathrm{R}}) \simeq -\frac{1}{2^{3/2}\pi^2} \frac{\mathrm{d}}{\mathrm{d}t} \operatorname{Im} \left(\Lambda(t) * \sum_{\widetilde{T}(\mathsf{q}, \mathbf{x}_{\mathrm{R}})=t} \frac{\underline{\tilde{\mathbf{v}}}^{(0)}(\mathsf{q})(\partial X_2/\partial p_2)^{1/2}}{|\partial \widetilde{T}/\partial \mathsf{q}|} \right), \tag{10.1.16}$$

where X_2 and p_2 are the transverse range and slowness, respectively. The derivative is calculated using integral (5.7.13).

Note that although we have used the second-order saddle-point approximation to find $\underline{\tilde{\mathbf{v}}}^{(0)}(\mathsf{q})$ (at all values of q), no such approximation need be made in evaluating results (10.1.14) or (10.1.16). Although the second-order saddle-point approximation is used to define $\underline{\tilde{\mathbf{v}}}^{(0)}(\mathsf{q})$, it is used for *all* geometrical rays to define the function at *all* parameters q. Values at *all* parameters, not just the ray value, are then used in the Maslov seismogram (10.1.14). The response can be evaluated without the approximation that the phase function, $\widetilde{T}(\mathsf{q}, \mathbf{x}_{\mathrm{R}})$, is quadratic and the amplitude function, $\underline{\tilde{\mathbf{v}}}^{(0)}(\mathsf{q})$, is constant, requirements for the second-order saddle-point approximation.

10.1.1.1 Numerical band-limited Maslov seismograms

In order to obtain the numerical response, the impulse result (10.1.14) must be smoothed. Exactly the same band-limiting algorithm can be applied as was used for the WKBJ seismogram (cf. result (8.4.14))

$$\boxed{\frac{1}{\Delta t} B\left(\frac{t}{\Delta t}\right) * \mathbf{u}(t, \mathbf{x}_{\mathrm{R}}) \simeq -\frac{1}{4\pi\,\Delta t} \operatorname{Im} \left(\Delta(t) * \int_{\widetilde{T}=t\pm\Delta t} \underline{\tilde{\mathbf{v}}}^{(0)}(\mathsf{q})\, \mathrm{d}\mathsf{q} \right).} \tag{10.1.17}$$

This result provides an expression that is numerically robust and efficient to compute. As we have taken q to be a unique parameterization of the rays, e.g. the

Table 10.1. *The formation of geometrical arrivals with the two-dimensional Maslov algorithm (10.1.14).*

	$\mathrm{sgn}\left(\partial^2\widetilde{T}/\partial\mathrm{q}^2\right)$	$\underline{\tilde{\mathbf{v}}}^{(0)}$	$\sum_{\widetilde{T}=t}\lvert\partial\widetilde{T}/\partial\mathrm{q}\rvert^{-1}$	$\mathrm{Im}\left(\Delta(t)\ast\right.$	$\underline{\mathbf{u}}(t)$
$\mathrm{Re}\left(\underline{\mathbf{v}}^{(0)}\right)$	$+1$	$-\mathrm{i}$	$\lambda(t)$	$-\delta(t)\ast$	$\lambda(t)$
	-1	1	$\bar{\lambda}(t)$	$\bar{\delta}(t)\ast$	$\lambda(t)$
$\mathrm{Im}\left(\underline{\mathbf{v}}^{(0)}\right)$	$+1$	1	$\lambda(t)$	$\bar{\delta}(t)\ast$	$-\bar{\lambda}(t)$
	-1	i	$\bar{\lambda}(t)$	$\delta(t)\ast$	$-\bar{\lambda}(t)$

take-off angle, θ_S, at the source, exactly the same numerical algorithm can be used as for WKBJ seismograms. The range of integral (10.1.2) is over real rays that reach the target surface, and the integrand is single valued. Alternative variables that parameterize the rays are possible. If we had taken q to be the component of slowness in the target surface, $p_\parallel$, then the integrand of (10.1.2) may be multi-valued (see the discussion on pseudo-caustics below).

10.1.1.2 Geometrical arrivals in the Maslov seismogram

All geometrical arrivals are contained in the Maslov seismogram (10.1.14). The formation of the impulse response depends on the sign of the second derivative, $\partial^2\widetilde{T}/\partial\mathrm{q}^2$, and the phase of the Maslov amplitude coefficient, $\underline{\tilde{\mathbf{v}}}^{(0)}$. The latter depends on the phase of the geometrical amplitude, $\underline{\mathbf{v}}^{(0)}$, and again the sign of the second derivative, $\partial^2\widetilde{T}/\partial\mathrm{q}^2$, through equation (10.1.13). The sign of the second derivative, $\partial^2\widetilde{T}/\partial\mathrm{q}^2$, depends on the signs of the derivatives $\mathrm{d}p_\parallel/\mathrm{dq}$ and $\mathrm{d}X_\parallel/\mathrm{dq}$ but we need only consider the combination. Table 10.1 summarizes the form of the terms in expression (10.1.14) for each case, and shows that the results are consistent.

The first column lists the two cases of geometrical rays we need consider: the real and imaginary parts of the geometrical amplitude coefficient, $\underline{\mathbf{v}}^{(0)}$. For two-dimensional wave propagation, the real part leads to a pulse with the form of the function $\lambda(t)$ (B.2.1), and the imaginary part, its Hilbert transform, $\bar{\lambda}(t)$ (B.2.3), as indicated in the final column of Table 10.1. The second column lists the two cases of $\mathrm{sgn}\left(\partial^2\widetilde{T}/\partial\mathrm{q}^2\right)$ and the third column the resultant phase of the Maslov amplitude coefficient, $\underline{\tilde{\mathbf{v}}}^{(0)}$, from equation (10.1.13). The fourth column gives the pulse shape obtained from $\sum_{\widetilde{T}=t}\lvert\partial\widetilde{T}/\partial\mathrm{q}\rvert^{-1}$ in expression (10.1.14) – it again depends on $\mathrm{sgn}\left(\partial^2\widetilde{T}/\partial\mathrm{q}^2\right)$ as this controls whether the phase function, $\widetilde{T}$, is maximum or minimum at the geometrical ray. The fifth column indicates the function needed from the operation $\mathrm{Im}\left(\Delta(t)\ast\right.$ depending on the phase of the Maslov amplitude coefficient, $\underline{\tilde{\mathbf{v}}}^{(0)}$, in the third column. Finally, the sixth column combines the fourth

Table 10.2. *The formation of geometrical arrivals with the 2.5D Maslov algorithm (10.1.16).*

	$\mathrm{sgn}\,(\partial^2\tilde{T}/\partial \mathrm{q}^2)$	$\tilde{\underline{\mathbf{v}}}^{(0)}$	$\sum_{\tilde{T}=t} \lvert\partial\tilde{T}/\partial \mathrm{q}\rvert^{-1}$	$\mathrm{Im}\,(\Lambda(t)*$	$\underline{\mathbf{u}}(t)$
$\mathrm{Re}\,(\underline{\mathbf{v}}^{(0)})$	$+1$	$-\mathrm{i}$	$\lambda\,(t)$	$-\lambda(t)*$	$\delta(t)$
	-1	1	$\bar{\lambda}\,(t)$	$\bar{\lambda}(t)*$	$\delta(t)$
$\mathrm{Im}\,(\underline{\mathbf{v}}^{(0)})$	$+1$	1	$\lambda\,(t)$	$\bar{\lambda}(t)*$	$-\bar{\delta}(t)$
	-1	i	$\bar{\lambda}(t)$	$\lambda(t)*$	$-\bar{\delta}(t)$

and fifth columns, as in equation (10.1.14), to give the form of the displacement Green function, $\underline{\mathbf{u}}$. Despite the different entries in the fourth and fifth columns, the important result is that the entries in the sixth column are consistent with those predicted from the first column, i.e. geometrical ray theory and Maslov theory agree for all types of geometrical arrivals. Although overall there are only two pulses (column six in Table 10.1) depending on the phase of geometrical amplitude coefficient, $\underline{\mathbf{v}}^{(0)}$ (first column), the intermediate results (columns two to five) lead to four distinct cases.

The results for 2.5D wave propagation according to the Maslov method (10.1.16) are very similar, and are summarized in Table 10.2. Only the final two columns differ from Table 10.1.

10.1.1.3 Pseudo-caustics

The integrand (10.1.2) is stationary if equation (10.1.7) is satisfied. This occurs at the geometrical rays (10.1.8) and at points where equation (10.1.9) is satisfied, so called *pseudo-caustics*. We have investigated geometrical rays and must now investigate pseudo-caustics. Assuming the medium is homogeneous at the target surface, then condition (10.1.9) means that neighbouring rays are parallel (sometimes a pseudo-caustic is called a *telescopic point* as the rays are parallel). This is analogous to a caustic where neighbouring rays do not spread in the spatial, $\mathbf{x}$, domain. Here the rays do not spread in the slowness, $\mathbf{p}$, domain. Although the integrand is stationary, and the denominator in expression (10.1.14) is zero, the response does not have the geometrical ray singularity as the numerator in expression (10.1.13) is also zero. Suppose that the condition (10.1.9) is true at a parameter $\mathrm{q} = \mathrm{q}_{\mathrm{pseudo}}$. This is illustrated in Figure 10.2 where two pseudo-caustics exist when $p_{\parallel}(\mathrm{q})$ is stationary.

If we had used the target slowness, $p_{\parallel}$, as the ray parameterization, then $\mathrm{q} = p_{\parallel}$ and $\mathrm{d}p_{\parallel}/\mathrm{dq} = 1$ so the integrand contains no non-geometrical stationary points. However, the integrand of expression (10.1.2) is now multi-valued as multiple rays

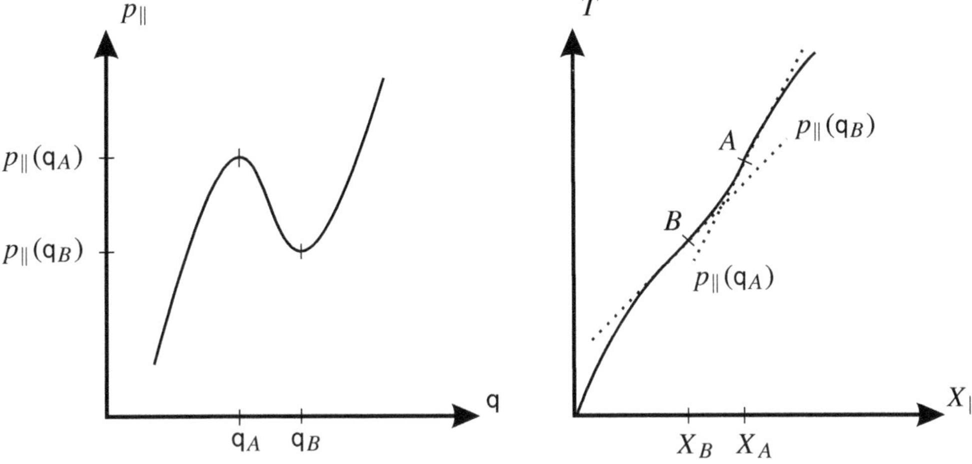

Fig. 10.2. The slowness function $p_{\parallel}(\mathsf{q})$ and travel time $T(X_{\parallel})$ with two pseudo-caustics, $\mathsf{q}_{\text{pseudo}} = \mathsf{q}_A$ or q_B. Note that for some values of $p_{\parallel}$, multiple rays, i.e. values of q, exist. In this example, $\mathsf{q}_A < \mathsf{q}_B$, but $p_{\parallel}(\mathsf{q}_A) > p_{\parallel}(\mathsf{q}_B)$ as indicated.

exist with the same slowness, $p_{\parallel}$ (see Figure 10.2), and the integral must be performed over all branches. The branches end and join at slownesses corresponding to the pseudo-caustics, $p_{\parallel} = \mathsf{q}_{\text{pseudo}}$. These end-points cause the pseudo-caustic arrivals. Frazer and Phinney (1980) have analysed these arrivals in some detail in the frequency domain – here we use the equivalent time-domain result (10.1.14).

Let us consider the situation illustrated in Figure 10.2. First we investigate the pulse shapes due to the pseudo-caustic. Two pseudo-caustics exist at $\mathsf{q} = \mathsf{q}_A$ and q_B where condition (10.1.9) is satisfied. At q_A, the slowness $p_{\parallel}$ is maximum and at q_B, minimum, i.e.

$$\frac{\mathrm{d}^2 p_{\parallel}}{\mathrm{d}\mathsf{q}^2}(\mathsf{q}_A) < 0 \quad \text{and} \quad \frac{\mathrm{d}^2 p_{\parallel}}{\mathrm{d}\mathsf{q}^2}(\mathsf{q}_B) > 0. \tag{10.1.18}$$

Let us assume that apart from the anomaly that causes the pseudo-caustics, the model has a simple velocity gradient so

$$\frac{\mathrm{d}X_{\parallel}}{\mathrm{d}\mathsf{q}} < 0, \tag{10.1.19}$$

and the arrivals are on a forward branch (assuming q increases with ray angle or horizontal slowness). The travel-time curve, $T(X_{\parallel})$, is illustrated in Figure 10.2. It has inflection points at $x_{\parallel} = X_{\parallel}(\mathsf{q}_A)$ and $X_{\parallel}(\mathsf{q}_B)$. The slope of this curve is, of course, the slowness, $p_{\parallel} = \mathrm{d}T/\mathrm{d}X_{\parallel}$. The phase function (10.1.4) depends on the receiver location. For $x_{\parallel} > X_A$ (where we use $X_A = X_{\parallel}(\mathsf{q}_A)$, etc. for brevity), the phase function is illustrated in Figure 10.3, as a function of q and $p_{\parallel}$.

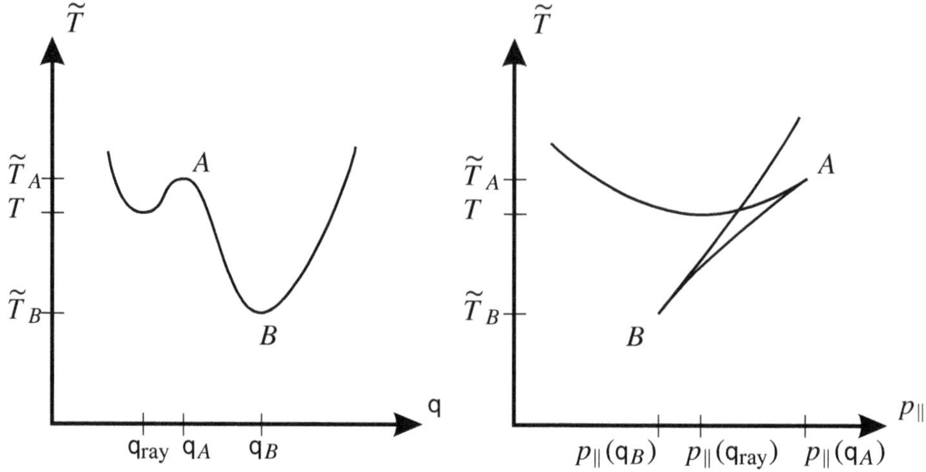

Fig. 10.3. The phase function, $\widetilde{T}$, for $x_\| > X_A$ corresponding to Figure 10.2, as a function of q and $p_\|$.

Both as a function of q and $p_\|$, a stationary point exists corresponding to the geometrical ray, T_{ray}. When the phase function (10.1.4) is considered as a function of parameter q, $\widetilde{T}(\mathsf{q})$, two more stationary points exist at $\widetilde{T}_A$ and $\widetilde{T}_B$, corresponding to the pseudo-caustics. Superficially, these are the same as geometrical arrivals. When the phase function (10.1.4) is considered as a function of the target slowness $p_\|$, $\widetilde{T}(p_\|)$ (Figure 10.3), these points become 'caustics', hence the name *pseudo-caustics*.

The second derivative of the phase function, $\widetilde{T}$, is

$$\frac{\partial^2 \widetilde{T}}{\partial \mathsf{q}^2} = \frac{\mathrm{d}^2 p_\|}{\mathrm{d}\mathsf{q}^2}\left(x_\| - X_\|(\mathsf{q})\right) - \frac{\mathrm{d}p_\|}{\mathrm{d}\mathsf{q}}\frac{\mathrm{d}X_\|}{\mathrm{d}\mathsf{q}}. \tag{10.1.20}$$

At a geometrical ray, $\mathsf{q} = \mathsf{q}_{\text{ray}}$, $x_\| = X_\|(\mathsf{q}_{\text{ray}})$, and

$$\frac{\partial^2 \widetilde{T}}{\partial \mathsf{q}^2} = -\frac{\mathrm{d}p_\|}{\mathrm{d}\mathsf{q}}\frac{\mathrm{d}X_\|}{\mathrm{d}\mathsf{q}}. \tag{10.1.21}$$

At a pseudo-caustic, $\mathsf{q} = \mathsf{q}_{\text{pseudo}}$, $\mathrm{d}p_\|/\mathrm{d}\mathsf{q} = 0$, and

$$\frac{\partial^2 \widetilde{T}}{\partial \mathsf{q}^2} = \frac{\mathrm{d}^2 p_\|}{\mathrm{d}\mathsf{q}^2}\left(x_\| - X_\|(\mathsf{q})\right). \tag{10.1.22}$$

The sign of this depends on both the range and the function $p_\|(\mathsf{q})$. In Figure 10.4, we illustrate the phase function $\widetilde{T}(\mathsf{q})$ for three different ranges in Figure 10.2. As the range, $x_\|$, decreases, the geometrical ray parameter, q_{ray}, increases from below the interval q_A to q_B, to within it and final above it. The stationary values of the

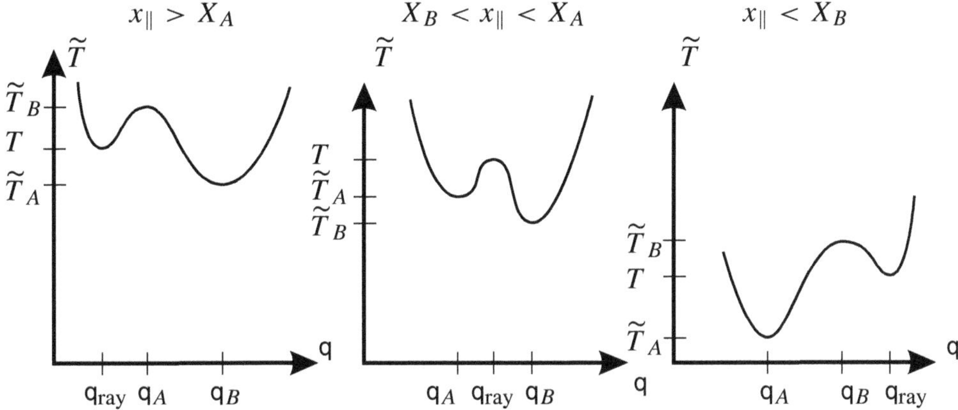

Fig. 10.4. The phase function, $\widetilde{T}(\mathsf{q}, \mathbf{x})$, for $x_\parallel > X_A$, $X_B < x_\parallel < X_A$ and $x_\parallel < X_B$ corresponding to Figure 10.2.

phase function, $\widetilde{T}(\mathsf{q})$, change roles in the different ranges. For $X_B < x_\parallel < X_A$, the geometrical arrival becomes a local maximum of $\widetilde{T}$ rather than a minimum as $\mathrm{d}p_\parallel/\mathrm{d}\mathsf{q} < 0$ in expression (10.1.21). However, whereas

$$e^{i\frac{\pi}{4}\operatorname{sgn}(\omega)\left(\operatorname{sgn}\left(\frac{\mathrm{d}p_\parallel}{\mathrm{d}\mathsf{q}}\frac{\mathrm{d}X_\parallel}{\mathrm{d}\mathsf{q}}\right)-1\right)} = -i\operatorname{sgn}(\omega), \tag{10.1.23}$$

for $x_\parallel > X_A$ or $x_\parallel < X_B$, for $X_B < x_\parallel < X_A$

$$e^{i\frac{\pi}{4}\operatorname{sgn}(\omega)\left(\operatorname{sgn}\left(\frac{\mathrm{d}p_\parallel}{\mathrm{d}\mathsf{q}}\frac{\mathrm{d}X_\parallel}{\mathrm{d}\mathsf{q}}\right)-1\right)} = 1. \tag{10.1.24}$$

Factor (10.1.23) is the same factor that arises from the turning point in transform methods (7.2.163). Combining these factors (10.1.23) and (10.1.24) with the minimum or maximum behaviour of $\widetilde{T}$ at the geometrical ray, we obtain the same geometrical signal, i.e. for $x_\parallel > X_A$ or $x_\parallel < X_B$, expression (10.1.14) contains $-\delta(t) * \lambda(t - T_{\mathrm{ray}}) = -\lambda(t - T_{\mathrm{ray}})$ – the first row in Table 10.1, and for $X_B < x_\parallel < X_A$, $\bar{\delta}(t) * \bar{\lambda}(t - T_{\mathrm{ray}}) = -\lambda(t - T_{\mathrm{ray}})$ – the second row in Table 10.1. Between the two pseudo-caustics, we again have the construction of two Hilbert transforms 'cancelling', and the geometrical arrival has the same form everywhere. In Figure 10.4, whether the pseudo-caustics are maximum or minimum depends on the location of $x_\parallel$. The signs of $x_\parallel - X_\parallel(\mathsf{q})$ in equation (10.1.22) change sign.

Near a pseudo-caustic, the phase function is stationary and normally

$$\widetilde{T}(\mathsf{q}, \mathbf{x}_\mathrm{R}) \simeq \widetilde{T}(\mathsf{q}_{\mathrm{pseudo}}, \mathbf{x}_\mathrm{R}) + \frac{1}{2}\frac{\partial^2 \widetilde{T}}{\partial \mathsf{q}^2}(\mathsf{q} - \mathsf{q}_{\mathrm{pseudo}})^2, \tag{10.1.25}$$

so the derivative is

$$\frac{\partial \widetilde{T}}{\partial \mathsf{q}} \simeq \frac{\partial^2 \widetilde{T}}{\partial \mathsf{q}^2}(\mathsf{q}-\mathsf{q}_{\text{pseudo}}). \tag{10.1.26}$$

Similarly, the slowness is stationary (Figure 10.2)

$$p_{\|} \simeq p_{\text{pseudo}} + \frac{1}{2}\frac{\mathrm{d}^2 p_{\|}}{\mathrm{d}\mathsf{q}^2}(\mathsf{q}-\mathsf{q}_{\text{pseudo}})^2, \tag{10.1.27}$$

and its gradient is

$$\frac{\mathrm{d}p_{\|}}{\mathrm{d}\mathsf{q}} \simeq \frac{\mathrm{d}^2 p_{\|}}{\mathrm{d}\mathsf{q}^2}(\mathsf{q}-\mathsf{q}_{\text{pseudo}}). \tag{10.1.28}$$

Let us consider the pseudo-caustic $\mathsf{q}_{\text{pseudo}} = \mathsf{q}_A$ so $p_{\|}$ is a local maximum. Then for $\mathsf{q} < \mathsf{q}_A$, $\mathrm{d}p_{\|}/\mathrm{d}\mathsf{q} > 0$ so in expression (10.1.13), result (10.1.23) holds. For $\mathsf{q} > \mathsf{q}_A$, $\mathrm{d}p_{\|}/\mathrm{d}\mathsf{q} < 0$ and result (10.1.24) holds. Substituting approximations (10.1.25), (10.1.26) and (10.1.28) in result (10.1.14) with results (10.1.23) and (10.1.24), we obtain

$$\underline{\mathbf{u}}(t,\mathbf{x}_{\mathsf{R}}) \simeq \left(\delta(t) - \bar{\delta}(t)\right) * \frac{\underline{\mathbf{v}}^{(0)}(\mathbf{x}_A)\,|\mathrm{d}X_{\|}/\mathrm{d}\mathsf{q}|^{1/2}}{2^{1/4}|\mathrm{d}^2 p_{\|}/\mathrm{d}\mathsf{q}^2|^{1/4}|x_{\|} - X_A|^{3/4}|\widetilde{T}_A - t|^{1/4}}. \tag{10.1.29}$$

In the frequency domain $t^{-1/4} \to O(\omega^{-3/4})$ compared with the geometrical arrivals $O(\omega^{-1/2})$ (in two dimensions) (cf. equation (64) in Frazer and Phinney, 1980). Expression (10.1.29) describes the arrivals due to the slowness of the pseudo-caustic. These arrivals are artifacts of the transform method – the signals are errors and are not correct. The exact details of expression (10.1.29) are not important, in as much as the predictions are in error, but certain basic properties allow the signals to be recognized. Compared with the geometrical arrival, the pseudo-caustic signal has an extra factor $O(\omega^{-1/4})$. Thus asymptotically at high frequency, the pseudo-caustic arrivals are less significant, and the asymptotic result is correct. The signal also decays as $|x_{\|} - X_A|^{-3/4}$ and will be less significant away from the pseudo-caustic. The pseudo-caustic artifacts are arrivals at a constant slowness, $\mathsf{q}_{\text{pseudo}}$. The arrival time is

$$t = \widetilde{T}(\mathsf{q}_{\text{pseudo}},\mathbf{x}_{\mathsf{R}}) = T(\mathsf{q}_{\text{pseudo}}) + \mathsf{q}_{\text{pseudo}}\left(x_{\|} - X_{\|}(\mathsf{q}_{\text{pseudo}})\right). \tag{10.1.30}$$

Expression (10.1.29) breaks down exactly at the pseudo-caustic due to the zero denominator. The pseudo-caustic and geometrical points combine and the second derivative (10.1.20) is zero. The phase function becomes cubic (see Figure 10.5).

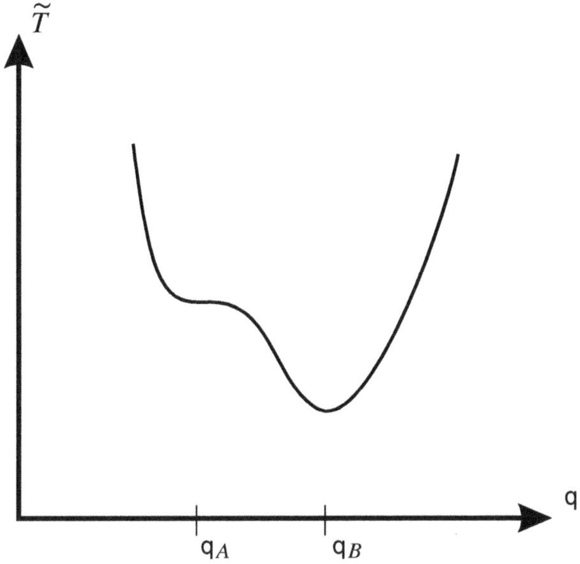

Fig. 10.5. The phase function $\widetilde{T}(\mathrm{q}, \mathbf{x}_\mathrm{R})$ at $x_\parallel = X_A$ corresponding to Figure 10.2.

The third derivative is

$$\frac{\partial^3 \widetilde{T}}{\partial \mathrm{q}^3} = \frac{\mathrm{d}^3 p_\parallel}{\mathrm{d}\mathrm{q}^3} \left(x_\parallel - X_\parallel(\mathrm{q})\right) - 2\frac{\mathrm{d}^2 p_\parallel}{\mathrm{d}\mathrm{q}^2}\frac{\mathrm{d}X_\parallel}{\mathrm{d}\mathrm{q}} - \frac{\mathrm{d}p_\parallel}{\mathrm{d}\mathrm{q}}\frac{\mathrm{d}^2 X_\parallel}{\mathrm{d}\mathrm{q}^2} \tag{10.1.31}$$

$$= -2\frac{\mathrm{d}^2 p_\parallel}{\mathrm{d}\mathrm{q}^2}\frac{\mathrm{d}X_\parallel}{\mathrm{d}\mathrm{q}} < 0, \tag{10.1.32}$$

in the example (Figure 10.5). The phase function is approximated by

$$\widetilde{T}(\mathrm{q}, \mathbf{x}) \simeq T_\mathrm{ray} + \frac{1}{6}\frac{\partial^3 \widetilde{T}}{\partial \mathrm{q}^3}(\mathrm{q} - \mathrm{q}_\mathrm{ray})^3, \tag{10.1.33}$$

and the derivative by

$$\frac{\partial \widetilde{T}}{\partial \mathrm{q}} \simeq \frac{1}{2}\frac{\partial^3 \widetilde{T}}{\partial \mathrm{q}^3}(\mathrm{q} - \mathrm{q}_\mathrm{ray})^2. \tag{10.1.34}$$

Substituting results (10.1.28), (10.1.33) and (10.1.34) in equation (10.1.14) with results (10.1.23), (10.1.24) and (10.1.32), we obtain

$$\underline{\mathbf{u}}(t, \mathbf{x}_\mathrm{R}) \simeq \left(\frac{2}{3}\right)^{1/2} \underline{\mathbf{v}}^{(0)}(\mathbf{x}_A)\, \lambda(t - T_\mathrm{ray}). \tag{10.1.35}$$

In this, the solution for $t = \widetilde{T}$ for $\mathrm{q} < \mathrm{q}_A$ contributes $-\mathrm{i}\,\lambda(t - T_\mathrm{ray})$, whereas the contribution from $\mathrm{q} > \mathrm{q}_A$ is $\bar{\lambda}(t - T_\mathrm{ray})$. These combine and double, again two Hilbert transforms 'cancelling'. This result is the correct geometrical arrival, but with an extra factor $(2/3)^{1/2}$, i.e. an amplitude error of about −18% (cf. equation (73) in Frazer and Phinney, 1980).

Pseudo-caustics are therefore a source of artifacts and amplitude errors in the Maslov seismogram (10.1.14). No ideal solution has been found. They are particularly irritating as, at pseudo-caustics, geometrical ray theory is a good approximation (as the wavefront is plane, the rays are parallel). It is, however, important to note that caustics and pseudo-caustics never coincide. This follows from Liouville's theorem (Section 5.2.2.4), as the volume mapping in phase space never reduces to zero. Maslov in his original papers suggested taking advantage of this and combining the two solutions, geometrical ray theory and Maslov seismograms. Smooth weighting functions switch between the two solutions, always summing to unity. At pseudo-caustics, the Maslov weighting function would be zero, and at caustics, the geometrical ray weighting function would be zero. As the weighting functions varied smoothly, the result is *asymptotically* correct everywhere. However, in practice, caustics and pseudo-caustics are often so numerous and close together, that the method is not practical.

10.1.2 Three dimensions

In three-dimensional wave propagation, we still take the ansätze (10.1.1) and (10.1.2), but the ray parameterization, the variable of integration q, is now two dimensional to define the ray direction at the source. We follow the notation used in Chapman and Keers (2002) who have investigated the algorithm for the Maslov method in three dimensions. The ray approximation is now result (5.4.28)

$$\underline{\mathbf{v}}(\omega, \mathbf{x}_\mathsf{R}) \simeq f^{(3)}(\omega)\, \underline{\mathbf{v}}^{(0)}(\mathbf{x}_\mathsf{R})\, e^{i\omega T}. \qquad (10.1.36)$$

The natural choice for the phase function is still expression (10.1.4). The integral (10.1.3) has a stationary-phase point when

$$\nabla_\mathsf{q} \widetilde{T} = \left(\nabla_\mathsf{q} \mathbf{p}^\mathrm{T}\right)(\mathbf{x}_\mathsf{R} - \mathbf{X}(\mathsf{q})) = 0, \qquad (10.1.37)$$

as by the definition of slowness

$$\nabla_\mathsf{q} T = \left(\nabla_\mathsf{q} \mathbf{X}^\mathrm{T}\right)\mathbf{p}. \qquad (10.1.38)$$

In our notation, ∇_q is a 2×1 vector, and the right-hand sides of equations (10.1.37) and (10.1.38) are 2×3 times 3×1 matrices.

A saddle point exists when

$$\mathbf{X}(\mathsf{q}_\mathrm{ray}) = \mathbf{x}_\mathsf{R}, \qquad (10.1.39)$$

i.e. when $\mathsf{q} = \mathsf{q}_{\text{ray}}$, the ray parameters that satisfy the two-point ray. At this point

$$\widetilde{T}(\mathsf{q}_{\text{ray}}, \mathbf{x}_{\mathsf{R}}) = T(\mathsf{q}_{\text{ray}}) = T_{\text{ray}}, \tag{10.1.40}$$

say. For the moment, we ignore points where $\nabla_{\mathsf{q}} \widetilde{T}$ is zero (10.1.37) without $\mathbf{X}(\mathsf{q}_{\text{ray}}) = \mathbf{x}_{\mathsf{R}}$ (10.1.39).

At the saddle point, the 2×2 matrix of second derivatives of $\widetilde{T}$ is

$$\nabla_{\mathsf{q}} \left(\nabla_{\mathsf{q}} \widetilde{T}\right)^{\mathrm{T}}\Big|_{\text{ray}} = -\left(\nabla_{\mathsf{q}} \mathbf{p}^{\mathrm{T}}\right) \left(\nabla_{\mathsf{q}} \mathbf{X}^{\mathrm{T}}\right)^{\mathrm{T}}, \tag{10.1.41}$$

where the right-hand side is 2×3 times 3×2 matrices (we should note that it is fortuitous that we can write these second derivatives using simple vector-matrix notation. Other than at the saddle point, we have

$$\frac{\partial \widetilde{T}}{\partial \mathsf{q}_\mu \mathsf{q}_\nu} = \frac{\partial p_i}{\partial \mathsf{q}_\mu \mathsf{q}_\nu} (x_i - X_i) - \frac{\partial p_i}{\partial \mathsf{q}_\nu} \frac{\partial X_i}{\partial \mathsf{q}_\mu}, \tag{10.1.42}$$

and the first term contains a third-order tensor (contracted with a first-order tensor). Expression (10.1.42) cannot be written simply in vector-matrix notation. Only at the saddle point (10.1.39), when the first term is zero, can expression (10.1.42) be written as equation (10.1.41) since the second term contains only two second-order tensors contracted). Using the second-order stationary-phase method (D.1.18), with $\delta\mathsf{q} = \mathsf{q} - \mathsf{q}_{\text{ray}}$, the integral (10.1.2) is approximately

$$\underline{\mathbf{v}}(\omega, \mathbf{x}_{\mathsf{R}}) \simeq \frac{\omega^2}{4\pi^2} \underline{\tilde{\mathbf{v}}}^{(0)}(\mathsf{q}_{\text{ray}})\, \mathrm{e}^{\mathrm{i}\omega T_{\text{ray}}} \int \mathrm{e}^{\mathrm{i}\omega\, \delta\mathsf{q}^{\mathrm{T}} \nabla_{\mathsf{q}} (\nabla_{\mathsf{q}} \widetilde{T})^{\mathrm{T}}|_{\text{ray}} \delta\mathsf{q}/2} \mathrm{d}\mathsf{q} \tag{10.1.43}$$

$$= \mathrm{i}\, f^{(3)}(\omega)\, \underline{\tilde{\mathbf{v}}}^{(0)}(\mathsf{q}_{\text{ray}})\, \mathrm{e}^{\mathrm{i}\omega T_{\text{ray}}} \frac{\mathrm{e}^{\mathrm{i}\frac{\pi}{4}\, \mathrm{sgn}\left(\omega\, \nabla_{\mathsf{q}} (\nabla_{\mathsf{q}} \widetilde{T})^{\mathrm{T}}|_{\text{ray}}\right)}}{\left\| \nabla_{\mathsf{q}} \left(\nabla_{\mathsf{q}} \widetilde{T}\right)^{\mathrm{T}} \right\|_{\text{ray}}^{1/2}}. \tag{10.1.44}$$

Comparing with result (10.1.36), we must have

$$\boxed{\underline{\tilde{\mathbf{v}}}^{(0)}(\mathsf{q}) = -\mathrm{i}\, \mathrm{sgn}(\omega)\, \underline{\mathbf{v}}^{(0)}(\mathbf{X}(\mathsf{q})) \left\| \nabla_{\mathsf{q}} \left(\nabla_{\mathsf{q}} \widetilde{T}\right)^{\mathrm{T}} \right\|_{\text{ray}}^{1/2} \mathrm{e}^{-\mathrm{i}\frac{\pi}{4}\, \mathrm{sgn}\left(\omega\, \nabla_{\mathsf{q}} (\nabla_{\mathsf{q}} \widetilde{T})^{\mathrm{T}}|_{\text{ray}}\right)}.} \tag{10.1.45}$$

Using expressions (10.1.4) and (10.1.45), we can evaluate the inverse transforms (10.1.2). Note in expression (10.1.44), the second derivative of $\widetilde{T}$ is evaluated at $\mathsf{q} = \mathsf{q}_{\text{ray}}$ which solves equation (10.1.39), and the right-hand side is a function of q_{ray}. In expression (10.1.45), the variable is q and the range $\mathbf{x}_{\mathsf{R}}$ has been replaced by $\mathbf{X}(\mathsf{q})$.

It is important to note that the derivatives in expression (10.1.41) are not simply fundamental solutions of the dynamic equations (5.2.29). The required derivatives are on the target surface, not the wavefront. In expression (10.1.41), the ray functions, $\widetilde{T}$, $\mathbf{p}$ and $\mathbf{X}$ are all defined on the target surface. The derivatives must be modified for the extra path length (as were the derivatives on an interface – Section 6.2.2). Thus

$$\left(\nabla_{\mathrm{q}}\mathbf{X}^{\mathrm{T}}\right)^{\mathrm{T}} = \left(\mathbf{I} - \frac{\dot{\mathbf{x}}\hat{\mathbf{n}}^{\mathrm{T}}}{\dot{\mathbf{x}}^{\mathrm{T}}\hat{\mathbf{n}}}\right)\mathbf{J}_{xp} = \boldsymbol{\Pi}_1\mathbf{J}_{xp}, \tag{10.1.46}$$

where $\hat{\mathbf{n}}$ is normal to the target surface, the matrix $\boldsymbol{\Pi}_1$ is defined in equation (6.2.13), and $\dot{\mathbf{x}} = \mathbf{V}$ is the ray velocity. Similarly the slowness differential is modified by the extra ray path and

$$\nabla_{\mathrm{q}}\mathbf{p}^{\mathrm{T}} = \left(\mathbf{J}_{pp} - \frac{\dot{\mathbf{p}}\hat{\mathbf{n}}^{\mathrm{T}}}{\dot{\mathbf{x}}^{\mathrm{T}}\hat{\mathbf{n}}}\mathbf{J}_{xp}\right)^{\mathrm{T}} = \left(\mathbf{J}_{pp} + \boldsymbol{\Pi}_2\mathbf{J}_{xp}\right)^{\mathrm{T}}, \tag{10.1.47}$$

where again the matrix $\boldsymbol{\Pi}_2$ is defined in equation (6.2.13), and $\dot{\mathbf{p}}$ is obtained from the relevant kinematic ray equation, (5.1.15), (5.1.27) or (5.3.21). Combining these results, (10.1.46) and (10.1.47), expression (10.1.41) becomes

$$\nabla_{\mathrm{q}}\left(\nabla_{\mathrm{q}}\widetilde{T}\right)^{\mathrm{T}}\Big|_{\mathrm{ray}} = -\left(\mathbf{J}_{pp} + \boldsymbol{\Pi}_2\mathbf{J}_{xp}\right)^{\mathrm{T}}\boldsymbol{\Pi}_1\mathbf{J}_{xp} \tag{10.1.48}$$

$$= -\mathbf{Q}_{\mathrm{S}}^{\mathrm{T}}\left(\mathbf{P}_{pp} + \boldsymbol{\Pi}_2\mathbf{P}_{xp}\right)^{\mathrm{T}}\boldsymbol{\Pi}_1\mathbf{P}_{xp}\mathbf{Q}_{\mathrm{S}}, \tag{10.1.49}$$

where the matrix

$$\mathbf{Q}_{\mathrm{S}} = \begin{pmatrix}\mathbf{q}_1 & \mathbf{q}_2\end{pmatrix} = \mathbf{J}_{\mathrm{S}}^{(456)\times(34)} \tag{10.1.50}$$

is the relevant part of the initial matrix (5.2.28).

To evaluate the response with the integral (10.1.2) ($m = 2$), we follow the WKBJ algorithm (Section 8.4.1), and apply the inverse Fourier frequency transform (3.1.2) first. We obtain

$$\underline{\mathbf{u}}(t, \mathbf{x}_{\mathrm{R}}) \simeq -\frac{1}{4\pi^2}\frac{\mathrm{d}}{\mathrm{d}t}\operatorname{Re}\iint \underline{\tilde{\mathbf{v}}}^{(0)}(\mathrm{q})\,\Delta\left(t - \widetilde{T}(\mathrm{q}, \mathbf{x}_{\mathrm{R}})\right)\mathrm{dq},$$

or

$$\boxed{\underline{\mathbf{u}}(t, \mathbf{x}_{\mathrm{R}}) \simeq -\frac{1}{4\pi^2}\frac{\mathrm{d}}{\mathrm{d}t}\operatorname{Re}\left(\Delta(t) * \int_{\widetilde{T}=t}\frac{\underline{\tilde{\mathbf{v}}}^{(0)}(\mathrm{q})}{\left|\nabla_{\mathrm{q}}\widetilde{T}\right|}\,\mathrm{d}q\right),} \tag{10.1.51}$$

where the integral is along the line in q space where $\widetilde{T}(\mathrm{q}, \mathbf{x}_{\mathrm{R}}) = t$. The form of this integral for a typical situation in a triplication is illustrated in Figure 10.6 (cf. Figure 8.10).

Note that although we have used the second-order saddle-point approximation to find $\underline{\tilde{\mathbf{v}}}^{(0)}(\mathrm{q})$ (at all values of q), no such approximation need be made in evaluating (10.1.51). The response can be evaluated without the approximation that the phase function, $\widetilde{T}(\mathrm{q}, \mathbf{x}_{\mathrm{R}})$, is quadratic and the amplitude function, $\underline{\tilde{\mathbf{v}}}^{(0)}(\mathrm{q})$, constant, requirements for the second-order, saddle-point approximation.

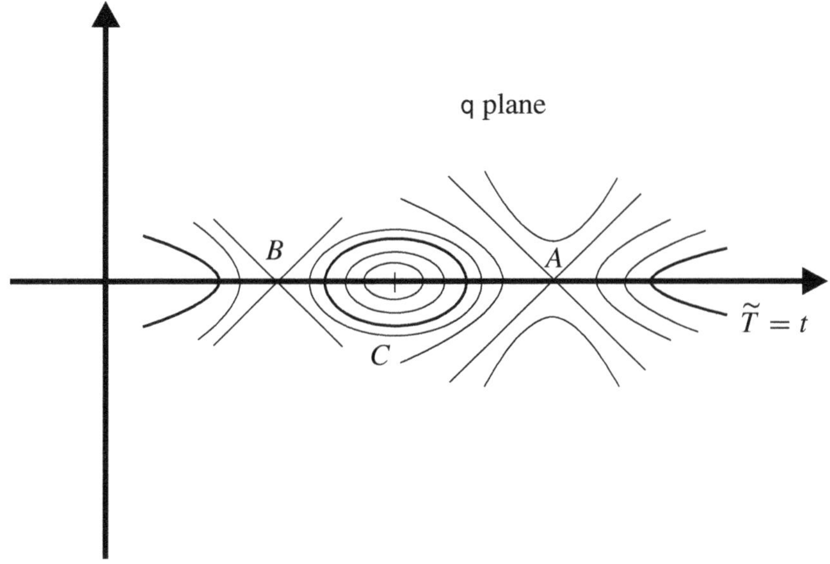

Fig. 10.6. The isochrons, $\widetilde{T} = t$, in expression (10.1.51) for a situation within a triplication similar to Figure 8.10. Contours of the function $\widetilde{T}_{\text{ray}}$ are illustrated and the isochrons for $T_A < T_B < t < T_C$ are indicated with heavier lines.

10.1.2.1 Numerical band-limited Maslov seismograms

In practice, as with the WKBJ algorithm, we have to smooth the analytic, impulse response (10.1.51) in order to avoid singularities. Using the boxcar (8.4.13) with width $2\Delta t$, we obtain from expression (10.1.51)

$$\boxed{\frac{1}{\Delta t} B\left(\frac{t}{\Delta t}\right) * \underline{\mathbf{u}}(t, \mathbf{x}_{\mathrm{R}}) = -\frac{1}{8\pi^2 \Delta t} \frac{\mathrm{d}}{\mathrm{d}t} \operatorname{Re}\left(\Delta(t) * \iint_{\widetilde{T}=t\pm\Delta t} \underline{\tilde{\mathbf{v}}}^{(0)}(\mathsf{q})\, \mathrm{d}\mathsf{q}\right),} \tag{10.1.52}$$

where the area integral is over strips defined by $\widetilde{T}(\mathsf{q}, \mathbf{x}_{\mathrm{R}}) = t \pm \Delta t$. The form of these integrals is illustrated in Figure 10.7.

Rays are shot to the target surface with different parameters, q. It is convenient to form triangular ray tubes in the two-dimensional q space. To test whether the density of rays is sufficient, we can test whether paraxial extrapolation from rays, with interpolation between the three rays, is sufficiently accurate within the tube. If not, more rays can be added to sub-divide the tube. Having obtained triangular ray tubes with enough rays to obtain accurate results anywhere on the target surface, we can perform the integration (10.1.52). Assuming linear interpolation of the phase function $\widetilde{T}$ within the tube, the form of the integral (10.1.52) is illustrated in Figure 10.8. The strip defined by $\widetilde{T}(\mathsf{q}, \mathbf{x}_{\mathrm{R}}) = t \pm \Delta t$ has parallel, straight sides in each triangular tube.

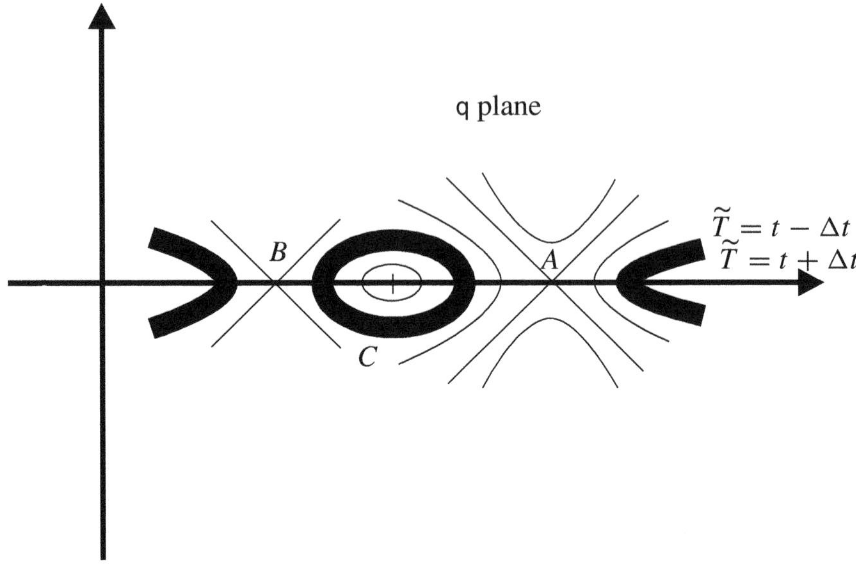

Fig. 10.7. The area of the integrals in expression (10.1.52) illustrated for the same situation as in Figure 10.6. The shaded area are strips defined by $\widetilde{T} = t \pm \Delta t$.

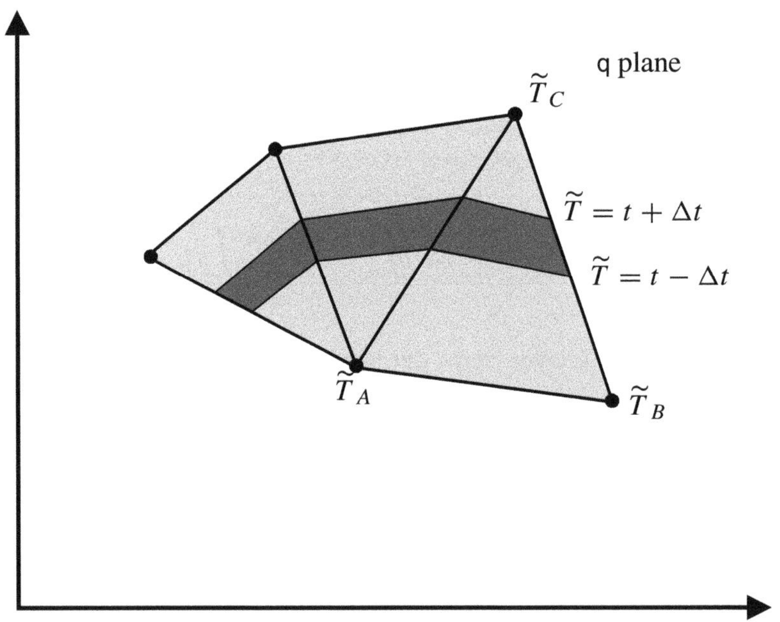

Fig. 10.8. The band-limited Maslov integral. The amplitude $\underline{\tilde{\mathbf{v}}}^{(0)}(\mathsf{q})$ is integrated over the strips defined by $\widetilde{T}(\mathsf{q}, \mathbf{x}_\mathsf{R}) = t \pm \Delta t$. Within the triangular ray tube (the rays are dots), the $\widetilde{T}$ function is linearly interpolated.

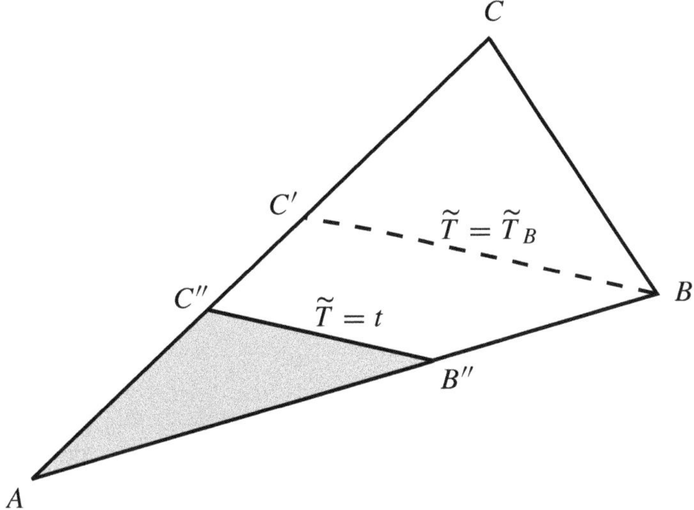

Fig. 10.9. A ray tube in the q plane. The phase at C' is the same as B. For a time between A and B, i.e. C'', the triangle $AB''C''$ is similar to ABC' and its area is easily calculated (10.1.53).

If the amplitude function $\underline{\tilde{\mathbf{v}}}^{(0)}(\mathrm{q})$ is also linearly interpolated, an extremely simple, efficient numerical algorithm results (Spencer, Chapman and Kragh, 1997). It is analogous to the WKBJ algorithm. Consider a ray tube (triangle) ABC (Figure 10.9), with phases $\widetilde{T}_A$, etc. and amplitudes, $\underline{\tilde{\mathbf{v}}}_A^{(0)}$, etc. Let us assume that the phases are ordered $\widetilde{T}_A < t < \widetilde{T}_B < \widetilde{T}_C$. The line $B''C''$ is where $\widetilde{T} = t$ and C' is where $\widetilde{T} = \widetilde{T}_B$ on the side AC (Figure 10.9).

As triangles $AB''C''$ and ABC' are similar, the area of the triangle $AB''C''$ is simply

$$\Delta_{AB''C''} = \left(\frac{t - \widetilde{T}_A}{\widetilde{T}_B - \widetilde{T}_A}\right)^2 \Delta_{ABC'} = \frac{(t - \widetilde{T}_A)^2}{(\widetilde{T}_B - \widetilde{T}_A)(\widetilde{T}_C - \widetilde{T}_A)} \Delta_{ABC}. \quad (10.1.53)$$

To obtain the integral of the amplitude $\underline{\tilde{\mathbf{v}}}^{(0)}$ over this triangle with linear interpolation, we just need the mean of the amplitude at the vertices. These are A, where the amplitude is $\underline{\tilde{\mathbf{v}}}_A^{(0)}$, and B'' and C'' with amplitudes

$$\underline{\tilde{\mathbf{v}}}_{B''}^{(0)} = \underline{\tilde{\mathbf{v}}}_A^{(0)} + \left(\underline{\tilde{\mathbf{v}}}_B^{(0)} - \underline{\tilde{\mathbf{v}}}_A^{(0)}\right)\left(\frac{t - \widetilde{T}_A}{\widetilde{T}_B - \widetilde{T}_A}\right) \quad (10.1.54)$$

$$\underline{\tilde{\mathbf{v}}}_{C''}^{(0)} = \underline{\tilde{\mathbf{v}}}_A^{(0)} + \left(\underline{\tilde{\mathbf{v}}}_C^{(0)} - \underline{\tilde{\mathbf{v}}}_A^{(0)}\right)\left(\frac{t - \widetilde{T}_A}{\widetilde{T}_C - \widetilde{T}_A}\right). \quad (10.1.55)$$

Thus

$$\iint_{t=\widetilde{T}_A}^{t} \underline{\tilde{\mathbf{v}}}^{(0)}(\mathrm{q})\,\mathrm{dq}$$

$$= \left\{ \underline{\tilde{\mathbf{v}}}_A^{(0)} + \frac{1}{3}\left(\frac{\underline{\tilde{\mathbf{v}}}_B^{(0)} - \underline{\tilde{\mathbf{v}}}_A^{(0)}}{\widetilde{T}_B - \widetilde{T}_A} + \frac{\underline{\tilde{\mathbf{v}}}_C^{(0)} - \underline{\tilde{\mathbf{v}}}_A^{(0)}}{\widetilde{T}_C - \widetilde{T}_A} \right)(t - \widetilde{T}_A) \right\} \Delta_{AB''C''} \tag{10.1.56}$$

$$= \left\{ \underline{\tilde{\mathbf{v}}}_A^{(0)} + \frac{1}{3}\left(\frac{\underline{\tilde{\mathbf{v}}}_B^{(0)} - \underline{\tilde{\mathbf{v}}}_A^{(0)}}{\widetilde{T}_B - \widetilde{T}_A} + \frac{\underline{\tilde{\mathbf{v}}}_C^{(0)} - \underline{\tilde{\mathbf{v}}}_A^{(0)}}{\widetilde{T}_C - \widetilde{T}_A} \right)(t - \widetilde{T}_A) \right\} \left(\frac{t - \widetilde{T}_A}{\widetilde{T}_B - \widetilde{T}_A} \right)^2 \Delta_{ABC'}. \tag{10.1.57}$$

Similar results hold for $t > \widetilde{T}_B$. Repeating at intervals of Δt, results for strips are obtained by subtraction. Only a few arithmetic operations are needed to compute (10.1.57) for each triangle $AB''C''$, as many factors are common across the triangle ABC'.

As with the WKBJ algorithm, this band-limiting algorithm, with linear interpolation, is crucial for obtaining an inexpensive, robust numerical result.

As an example of seismograms calculated using the Maslov algorithm, we consider the Valhall gas-cloud model (Section 5.1.4.1). This was used in Chapter 5 to illustrate ray tracing in a three-dimensional model. The model which simulates the Valhall gas cloud was described by Brandsberg-Dahl, de Hoop and Ursin (2003). Rays traced in this model are illustrated in Figure 5.5. Using the results of this ray tracing including ray directions out of the symmetry plane, Maslov seismograms have been calculated. The source is at $\mathbf{x}_\mathrm{S} = (4.68, 0, -1.5)$ km. The band-limited result (10.1.52), with the approximation (10.1.57) for the area integrals, has been used. The integrand has been calculated from expression (10.1.45). The results are shown in Figures 10.10 and 10.11. The profile in Figure 10.10 is on the symmetry plane $y_\mathrm{R} = 0$ km through the region of focusing ($x_\mathrm{R} = 4$ km to 5 km). The strong focusing is clearly visible together with the caustics, reversed branch arrivals and non-geometrical signals beyond the caustics (see Figure 5.5). The profile in Figure 10.11 is off the symmetry plane with $y_\mathrm{R} = 0.005$ km. The focusing is reduced but the multiple arrivals still exist.

The computational cost of computing these seismograms is small compared with the ray tracing. The algorithms used for Figures 10.10 and 10.11, and for Figures 10.26, 10.27 and 10.33 later in the chapter, which involve two- and three-dimensional integrals, are all very straightforward, e.g. using approximation (10.1.57) based on linear interpolation in triangular elements. The results contain some small signals due to numerical artifacts from the edges of integrals, etc. These could easily be reduced by smoothing the integrand at the edge of the integral domain, but for simplicity only the basic algorithms have been used here.

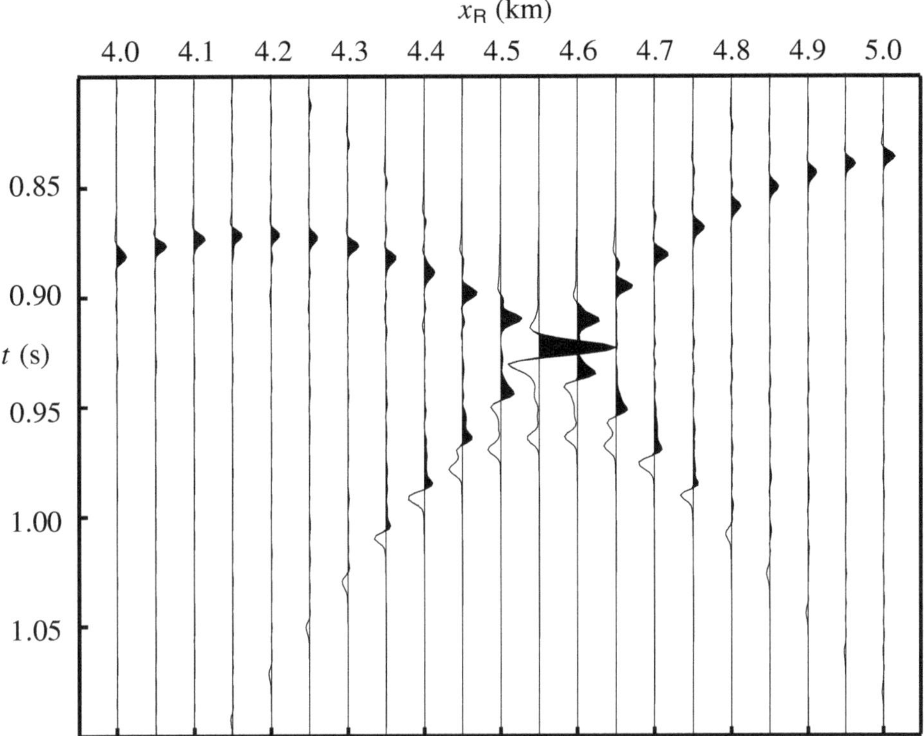

Fig. 10.10. Maslov seismograms in the Valhall model (Section 5.1.4.1). Profiles are for $y_R = 0$ km for $x_R = 4$ km to 5 km at intervals of 0.05 km. The seismograms run from $t = 0.8$ s to 1.1 s with a digitization interval of 0.0015 s. The results are smoothed by the boxcar integration (10.1.52) and by the smoothing used in the rational approximation used to approximate the convolution operator (Chapman, Chu Jen-Yi and Lyness, 1988).

10.1.2.2 Geometrical arrivals

Expression (10.1.51) is the *Maslov seismogram* in three dimensions. It must contain all the geometrical arrivals. The formation of these arrivals is summarized in Table 10.3. This table is similar to Table 10.1, except that we now have three options for $\operatorname{sgn}\left(\nabla_q(\nabla_q\widetilde{T})^{\mathrm{T}}\big|_{\mathrm{ray}}\right)$. The signature of the 2×2 matrix can be $+2$, 0 or -2. Examples of the last two values are illustrated in Figure 10.6.

Let us consider the first-motion approximation for the case when $\operatorname{sgn}\left(\nabla_q(\nabla_q\widetilde{T})^{\mathrm{T}}\big|_{\mathrm{ray}}\right) = -2$, i.e. the third and sixth rows in Table 10.3. In a stratified model, this occurs on reversed branches when

$$\frac{\partial^2\widetilde{T}}{\partial p_1^2} < 0 \quad \text{and} \quad \frac{\partial^2\widetilde{T}}{\partial p_2^2} < 0, \tag{10.1.58}$$

Table 10.3. *Contributions to different arrivals in the Maslov algorithm (10.1.51).*

	$\mathrm{sgn}\left(\nabla_q(\nabla_q\widetilde{T})^{\mathrm{T}}\big\rvert_{\mathrm{ray}}\right)$	$\underline{\tilde{\mathbf{v}}}^{(0)}$	$\int_{\widetilde{T}=t}\lvert\nabla_q\widetilde{T}\rvert^{-1}\mathrm{d}q$	$\mathrm{Re}\,(\Delta(t)*$	$\mathbf{u}(t)$
$\mathrm{Re}\,(\underline{\mathbf{v}}^{(0)})$	$+2$	-1	$H(t)$	$-\delta(t)*$	$\delta(t)$
	0	$-\mathrm{i}$	$\bar{H}(t)$	$\bar{\delta}(t)*$	$\delta(t)$
	-2	1	$-H(t)$	$-\delta(t)*$	$\delta(t)$
$\mathrm{Im}\,(\underline{\mathbf{v}}^{(0)})$	$+2$	$-\mathrm{i}$	$H(t)$	$\bar{\delta}(t)*$	$-\bar{\delta}(t)$
	0	1	$\bar{H}(t)$	$\delta(t)*$	$-\bar{\delta}(t)$
	-2	i	$-H(t)$	$-\bar{\delta}(t)*$	$-\bar{\delta}(t)$

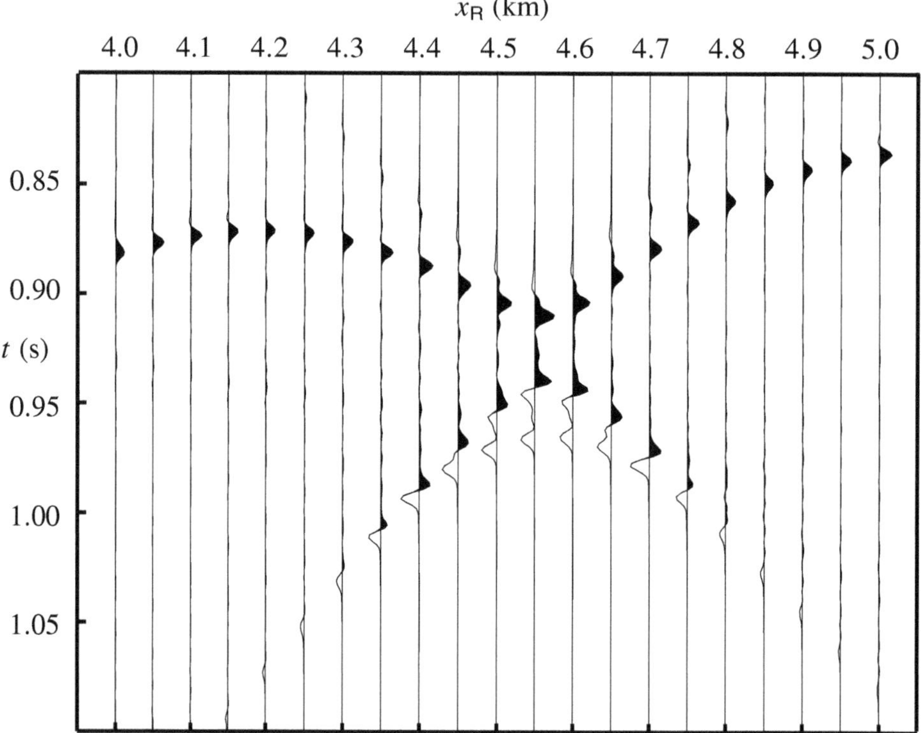

Fig. 10.11. As Figure 10.10 except the profile is for $y_{\mathrm{R}} = 0.005$ km. The seismograms are plotted to the same scale as in Figure 10.10.

e.g. for the arrival T_C in Figure 10.6. The phase function can be approximated by

$$\widetilde{T} \simeq T_C + \frac{1}{2}\frac{\partial^2\widetilde{T}}{\partial p_1^2}(p_1 - p_C)^2 + \frac{1}{2}\frac{\partial^2\widetilde{T}}{\partial p_2^2}\,p_2^2, \tag{10.1.59}$$

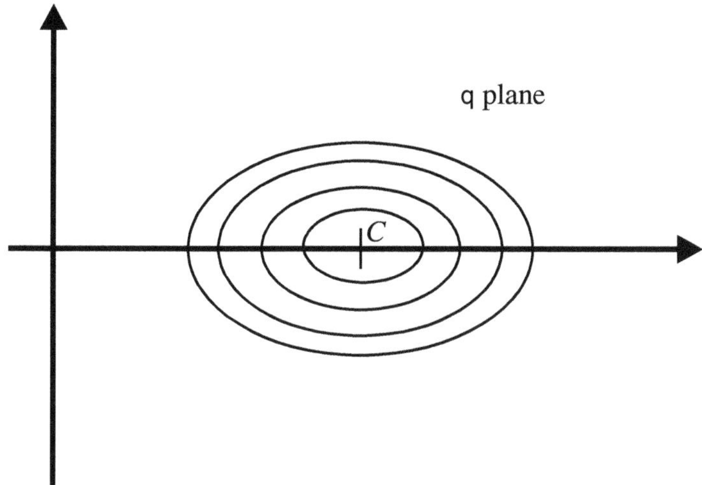

Fig. 10.12. The isochrons in expression (10.1.51) for a geometrical ray with $\operatorname{sgn}\left(\nabla_{\mathrm{q}}(\nabla_{\mathrm{q}}\widetilde{T})^{\mathrm{T}}\right) = -2$.

where the horizontal slowness p_1 in the propagation plane, and p_2 in the transverse direction, are components of q. For $t < T_C$, the integration line in expression (10.1.51) is an ellipse

$$2(T_C - t) \simeq \left(-\frac{\partial^2 \widetilde{T}}{\partial p_1^2}\right)(p_1 - p_C)^2 + \left(-\frac{\partial^2 \widetilde{T}}{\partial p_2^2}\right) p_2^2. \quad (10.1.60)$$

This is illustrated in Figure 10.12, an approximate detail of Figure 10.6. For $t > T_C$, no solutions of $\widetilde{T} = t$ exists and the integral is zero.

The gradient in expression (10.1.51) is

$$\left|\nabla_{\mathrm{q}}\widetilde{T}\right| \simeq \left(\left(\frac{\partial^2 \widetilde{T}}{\partial p_1^2}\right)^2 (p_1 - p_C)^2 + \left(\frac{\partial^2 \widetilde{T}}{\partial p_2^2}\right)^2 p_2^2\right)^{1/2}. \quad (10.1.61)$$

The length element is

$$\mathrm{d}q = \left(1 + \left(\frac{\mathrm{d}p_1}{\mathrm{d}p_2}\right)^2\right)^{1/2} \mathrm{d}p_2, \quad (10.1.62)$$

and with

$$\frac{\mathrm{d}p_1}{\mathrm{d}p_2} \simeq -\frac{p_1 - p_C}{p_2} \frac{\partial^2 \widetilde{T}/\partial p_1^2}{\partial^2 \widetilde{T}/\partial p_2^2}, \quad (10.1.63)$$

reduces to

$$\mathrm{d}q \simeq \left|\nabla_{\mathrm{q}}\widetilde{T}\right| \mathrm{d}p_2 \Big/ (p_1 - p_C)\frac{\partial^2 \widetilde{T}}{\partial p_1^2}. \tag{10.1.64}$$

Hence the integral in expression (10.1.51) reduces to

$$\int_{\widetilde{T}=t} \frac{\tilde{\underline{\mathbf{v}}}^{(0)}(\mathrm{q})}{\left|\nabla_{\mathrm{q}}\widetilde{T}\right|}\,\mathrm{d}q \simeq 4\int_0^{P_2} \frac{\tilde{\underline{\mathbf{v}}}^{(0)}(\mathrm{q})\,\mathrm{d}p_2}{(p_1 - p_C)\left(-\partial^2\widetilde{T}/\partial p_1^2\right)}, \tag{10.1.65}$$

where the upper limit P_2 is the semi-axis of the ellipse

$$P_2 \simeq \left(\frac{2(T_C - t)}{-\,\partial^2\widetilde{T}/\partial p_2^2}\right)^{1/2}. \tag{10.1.66}$$

Substituting from approximation (10.1.60) in the denominator of integral (10.1.65), the integral reduces to

$$\int_{\widetilde{T}=t} \frac{\tilde{\underline{\mathbf{v}}}^{(0)}(\mathrm{q})}{\left|\nabla_{\mathrm{q}}\widetilde{T}\right|}\,\mathrm{d}q \simeq 4\int_0^{P_2} \frac{\tilde{\underline{\mathbf{v}}}^{(0)}(\mathrm{q})\,\mathrm{d}p_2}{\left(P_2^2 - p_2^2\right)^{1/2}(-\partial^2\widetilde{T}/\partial p_1^2)^{1/2}(-\partial^2\widetilde{T}/\partial p_2^2)^{1/2}} \tag{10.1.67}$$

$$= 4\,\tilde{\underline{\mathbf{v}}}^{(0)}(\mathrm{q})\,\left\|\nabla_{\mathrm{q}}(\nabla_{\mathrm{q}})^{\mathrm{T}}\widetilde{T}\right\|_{\mathrm{ray}}^{-1/2}\,\sin^{-1}(p_2/P_2)\Big|_0^{P_2}$$

$$= 2\pi\,\tilde{\underline{\mathbf{v}}}^{(0)}(\mathrm{q})\,\left\|\nabla_{\mathrm{q}}(\nabla_{\mathrm{q}})^{\mathrm{T}}\widetilde{T}\right\|_{\mathrm{ray}}^{-1/2}\,H(T_C - t), \tag{10.1.68}$$

using integral (A.0.4). As $\mathrm{sgn}\left(\nabla_{\mathrm{q}}(\nabla_{\mathrm{q}}\widetilde{T})^{\mathrm{T}}\big|_{\mathrm{ray}}\right) = -2$, expression (10.1.45) simplifies to

$$\tilde{\underline{\mathbf{v}}}^{(0)}(\mathrm{q}) = \underline{\mathbf{v}}^{(0)}(\mathbf{x})\,\left\|\nabla_{\mathrm{q}}\left(\nabla_{\mathrm{q}}\widetilde{T}\right)^{\mathrm{T}}\right\|_{\mathrm{ray}}^{1/2}, \tag{10.1.69}$$

and substituting results (10.1.68) and (10.1.69) in expression (10.1.51), we obtain

$$\underline{\mathbf{u}}(t,\mathbf{x}_{\mathrm{R}}) = -\frac{1}{2\pi}\frac{\mathrm{d}}{\mathrm{d}t}\,\mathrm{Re}\left(\underline{\mathbf{v}}^{(0)}(\mathbf{x}_{\mathrm{R}})\,\Delta(t) * H(T_C - t)\right) = \frac{1}{2\pi}\mathrm{Re}\left(\underline{\mathbf{v}}^{(0)}(\mathbf{x}_{\mathrm{R}})\,\Delta(t - T_C)\right). \tag{10.1.70}$$

This result, the standard geometrical result (5.4.35), is summarized in the third and sixth rows of Table 10.3.

The case $\mathrm{sgn}\left(\nabla_{\mathrm{q}}(\nabla_{\mathrm{q}}\widetilde{T})^{\mathrm{T}}\big|_{\mathrm{ray}}\right) = +2$ is less common, and never occurs in stratified media. The first-motion approximation is obtained in a similar manner to the above, except for sign changes and the integral is non-zero for times after the geometrical arrival.

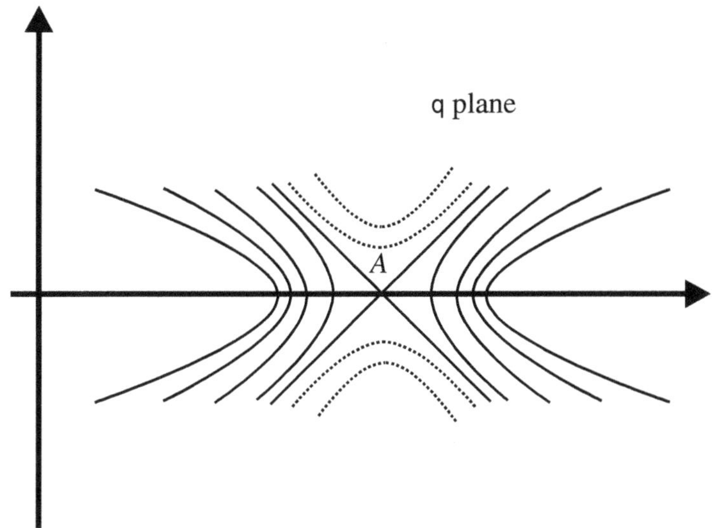

Fig. 10.13. The isochrons in expression (10.1.51) for a geometrical ray with $\operatorname{sgn}\left(\nabla_q(\nabla_q\widetilde{T})^T\big|_{ray}\right) = 0$. Isochrons for $t > T_A$ are shown with solid lines, and for $t < T_A$ with dotted lines.

The other case, when $\operatorname{sgn}\left(\nabla_q(\nabla_q\widetilde{T})^T\big|_{ray}\right) = 0$, which occurs on forward branches with

$$\frac{\partial^2\widetilde{T}}{\partial p_1^2} > 0 \quad \text{and} \quad \frac{\partial^2\widetilde{T}}{\partial p_2^2} < 0, \tag{10.1.71}$$

is slightly more complicated to evaluate. The integration lines are defined by

$$2(t - T_A) \simeq \left(\frac{\partial^2\widetilde{T}}{\partial p_1^2}\right)(p_1 - p_A)^2 - \left(-\frac{\partial^2\widetilde{T}}{\partial p_2^2}\right) p_2^2, \tag{10.1.72}$$

which describes hyperbolae both for $t < T_A$ and $t > T_A$. These are illustrated in Figure 10.13.

The algebra proceeds much as before until

$$\int_{\widetilde{T}=t} \frac{\underline{\tilde{\mathbf{v}}}^{(0)}(\mathrm{q})}{|\nabla_q\widetilde{T}|}\, dq \simeq 4\int_0^\infty \frac{\underline{\tilde{\mathbf{v}}}^{(0)}(\mathrm{q})\, dp_2}{\left(2(t - T_A) + \left(-\partial^2\widetilde{T}/\partial p_2^2\right) p_2^2\right)^{1/2} \left(\partial^2\widetilde{T}/\partial p_1^2\right)^{1/2}}. \tag{10.1.73}$$

The integral gives a $\sinh^{-1}$ function (A.0.2) but the definite integral is infinite. This is connected with the ambiguous D.C. level of the Hilbert transform of a delta function. The simplest way to proceed is to take the time derivative of this expression so

$$\frac{\mathrm{d}}{\mathrm{d}t}\int_{\widetilde{T}=t}\frac{\tilde{\underline{\mathbf{v}}}^{(0)}(\mathsf{q})}{\left|\nabla_{\mathsf{q}}\widetilde{T}\right|}\,dq$$

$$\simeq -4\int_0^\infty \frac{\tilde{\underline{\mathbf{v}}}^{(0)}(\mathsf{q})\,\mathrm{d}p_2}{\left(2(t-T_A)+\left(-\partial^2\widetilde{T}/\partial p_2^2\right)\,p_2^2\right)^{3/2}\left(\partial^2\widetilde{T}/\partial p_1^2\right)^{1/2}} \tag{10.1.74}$$

$$= -\left.\frac{2\,\tilde{\underline{\mathbf{v}}}^{(0)}(\mathsf{q})\,p_2}{(t-T_A)\left(2(t-T_A)+\left(-\partial^2\widetilde{T}/\partial p_2^2\right)\,p_2^2\right)^{1/2}\left(\partial^2\widetilde{T}/\partial p_1^2\right)^{1/2}}\right|_0^\infty$$

$$= -\frac{2\,\tilde{\underline{\mathbf{v}}}^{(0)}(\mathsf{q})}{(t-T_A)\left(-\partial^2\widetilde{T}/\partial p_2^2\right)^{1/2}\left(\partial^2\widetilde{T}/\partial p_1^2\right)^{1/2}}. \tag{10.1.75}$$

Simplifying result (10.1.45) and substituting in expression (10.1.75), the first-motion approximation to expression (10.1.51) becomes again

$$\underline{\mathbf{u}}(t,\mathbf{x}_{\mathsf{R}}) = \frac{1}{2\pi}\,\mathrm{Re}\left(\underline{\mathbf{v}}^{(0)}(\mathbf{x}_{\mathsf{R}})\,\Delta(t-T_A)\right), \tag{10.1.76}$$

where we have recognized $(t-T_A)^{-1} = \pi\,\bar{\delta}(t-T_A)$, (B.1.6), in result (10.1.75). These results are summarized in the second and fourth rows of Table 10.3.

10.1.2.3 Pseudo-caustics

In three dimensions, with a two-dimensional integral, the condition for pseudo-caustics (10.1.37) is somewhat more complicated than in two dimensions. The scalar equation (10.1.7) is replaced by the eigen-equation (10.1.37). Although the equation has been written as a 2×3 matrix times a 3×1 vector, the target restricts $\mathbf{X}(\mathsf{q})$ to a surface. In the target surface, we denote the slowness and range by $\mathsf{p}(\mathsf{q})$ and $\mathsf{X}(\mathsf{q})$. For simplicity, we normally consider a plane, so p and X are two-dimensional vectors. Then equation (10.1.37) can be rewritten as

$$\nabla_{\mathsf{q}}\widetilde{T} = \left(\nabla_{\mathsf{q}}\mathsf{p}^{\mathsf{T}}\right)\left(\mathsf{x}_{\mathsf{R}} - \mathsf{X}(\mathsf{q})\right) = 0. \tag{10.1.77}$$

The matrix $\left(\nabla_{\mathsf{q}}\mathsf{p}^{\mathsf{T}}\right)$ is 2×2, and the equation is an eigen-equation.

Apart from the geometrical ray condition (10.1.39), stationary points exist when the matrix $\left(\nabla_{\mathsf{q}}\mathsf{p}^{\mathsf{T}}\right)$ has a zero eigenvalue, with the eigenvector in the direction $(\mathsf{x}_{\mathsf{R}} - \mathsf{X}(\mathsf{q}))$. Let us consider the results in the two-dimensional q parameter space. By design, as q parameterizes the ray direction at the source, the functions $\mathsf{p}(\mathsf{q})$ and $\mathsf{X}(\mathsf{q})$ are single-valued in this space. Now consider the derivatives $\nabla_{\mathsf{q}}\mathsf{p}^{\mathsf{T}}$ and $\nabla_{\mathsf{q}}\mathsf{X}^{\mathsf{T}}$. In general, these 2×2 matrices will have rank 2, but there will be lines where the rank is reduced to 1. The lines where $\mathrm{rank}\left(\nabla_{\mathsf{q}}\mathsf{X}^{\mathsf{T}}\right) = 1$ are on caustic surfaces and are Airy caustics. In three dimensions, as the target surface is moved,

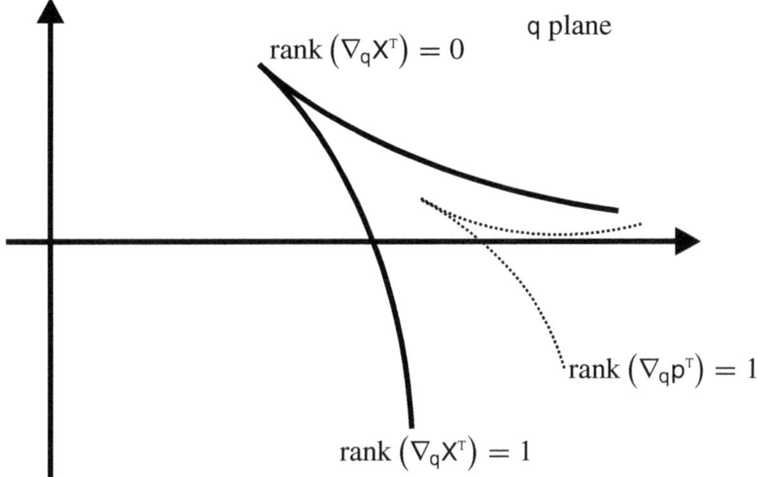

Fig. 10.14. Caustic and pseudo-caustic lines in q space.

lines where rank $(\nabla_q X^T) = 1$ on the target surface, map out a caustic surface. Occasionally, two such lines on the target surface will intersect and the rank of the matrix will be reduced to zero. Points with rank $(\nabla_q X^T) = 0$ form a Pearcey caustic (Pearcey, 1946). In three dimensions, as the target surface is moved, points where rank $(\nabla_q X^T) = 0$ on the target surface map out a caustic line. We have not explicitly investigated the Pearcey caustic, but the Maslov algorithms (10.1.14) and (10.1.51) remain valid and can be used to compute the complicated waveforms when three arrivals interfere (see Exercise 10.1).

Similarly, lines and points will exist in q space where rank $(\nabla_q p^T) = 1$ and 0, respectively. These are pseudo-caustics. Examples of caustic and pseudo-caustic lines are illustrated in Figure 10.14.

Although we expect lines where rank $(\nabla_q p^T) = 1$ to exist, only occasionally will the eigenvector corresponding to the zero eigenvalue be in the direction $x_R - X(q)$. Thus, in general, the stationary condition (10.1.77) will only be satisfied at a few points on these lines. In general, these points will depend on the range x_R and so will occur at different slownesses at different ranges. Constant slowness artifacts will not be generated across a seismic section.

The exception is on a two-dimensional symmetry plane in a three-dimensional model. The line rank $(\nabla_q p^T) = 1$ will be symmetrical about this symmetry plane and therefore perpendicular to it. For receivers on the symmetry plane, the stationary condition (10.1.77) will always be satisfied by a point on the plane. An example of this is a model with axial symmetry about the source (physically unlikely but feasible). Any line rank $(\nabla_q p^T) = 1$ will be a circle, with the zero eigenvector in

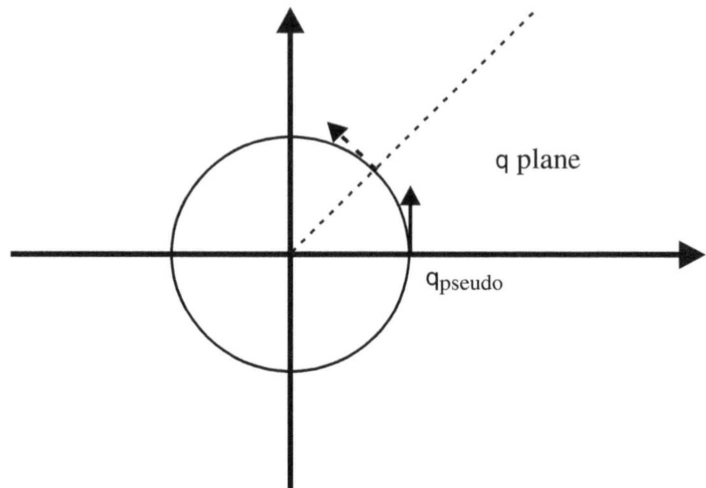

Fig. 10.15. Pseudo-caustic circle in q space for an axially symmetric situation. The direction of the eigenvector corresponding to the non-zero eigenvalue is shown – in the radial direction the eigenvalue is zero, i.e. towards the receiver. A pseudo-caustic point always exists on the circle in the receiver direction.

the radial direction, i.e. always in the direction $x_R - X(q)$. This is illustrated in Figure 10.15.

More realistically, a two-dimensional model extruded in the perpendicular cartesian direction will have a symmetry plane. Pseudo-caustics on this plane will always satisfy the stationarity condition (10.1.77) for receivers on the plane. The situation is similar to the axially symmetric example (Figure 10.15), except that the pseudo-caustic line will no longer be a circle but will still be perpendicular to the symmetry plane (Figure 10.16). The slowness at this pseudo-caustic will always cause a stationary point and a constant-slowness artifact for receivers on the symmetry plane. A similar situation will exist on any symmetry plane in a three-dimensional model. This situation is similar to the pseudo-caustics that exist in the two-dimensional solution (10.1.14), where effectively symmetry is always assumed in the third direction. However, in the three-dimensional Maslov solution (10.1.51), with a two-dimensional q integral, the stationary condition (10.1.77) is only satisfied at a point (e.g. Figures 10.15 and 10.16). We would therefore expect the pseudo-caustic artifact in the three-dimensional Maslov solution (10.1.51), to be less significant than in the two-dimensional solution.

In addition, there will be points in q space where $\operatorname{rank}\left(\nabla_q p^T\right) = 0$ (Figure 10.14). These will be independent of the receiver and so will cause constant-slowness pseudo-caustic arrivals.

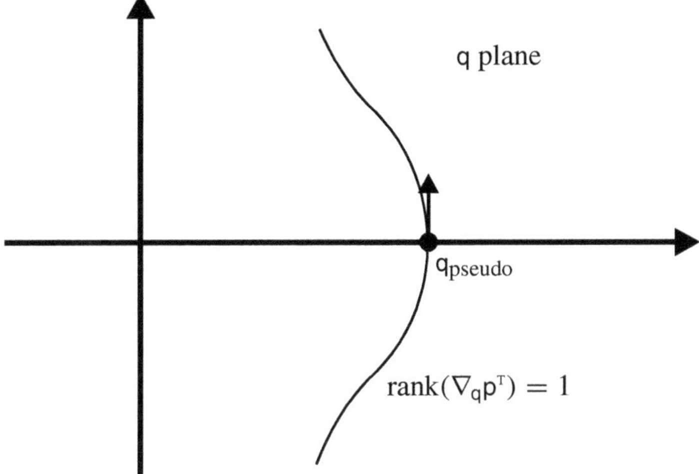

Fig. 10.16. A line where $\mathrm{rank}(\nabla_q p^T) = 1$ in a model with a symmetry plane. A pseudo-caustic point exists on the symmetry plane. For receivers on the symmetry plane, the receiver direction will correspond to the zero eigenvalue.

A complete analysis of the pseudo-caustic problem in three dimensions, with a two-dimensional parameter integral, has not been performed yet.

10.2 Quasi-isotropic ray theory

The development of anisotropic ray theory (Sections 5.3 and 5.4) assumed that the three ray types were non-degenerate. The isotropic ray theory (Sections 5.5 and 5.6) assumed that the *S* rays were always degenerate. In anisotropic media, in certain directions, the quasi-shear (*qS*) rays may degenerate and their velocities be equal. In these directions, degenerate theory must be used.

Near these degenerate directions, and in the case of weak, heterogeneous anisotropy, in all directions, standard anisotropic ray theory breaks down as the *qS* velocities are similar (G_N in definition (5.4.3) is small) and the additional terms (5.4.4) blow up (errors in asymptotic ray theory are bounded by the next term so the ray result is useless when the additional term is large). In homogeneous media, the two *qS* waves propagate independently as they do in anisotropic media when the velocities are significantly different (making the additional term small). They interact when these conditions fail as the time separation of the two waves is small compared with the pulse period. This occurs when $L\Delta V/V^2 < 2\pi/\omega$ where ΔV is the difference in *qS* velocities and L is a characteristic length of the heterogeneities. If we use anisotropic ray theory in weakly anisotropic media or near degeneracies, and assume the qS waves propagate independently, then absurd results can be obtained. Rapid changes in polarization are predicted for each independent ray which are clearly not allowed physically. This situation is illustrated in Figure 10.17, where a qS ray is traced through a simple TI, one-dimensional model with a linear gradient. The axis of symmetry is 15 degrees below the horizontal, slightly out of the plane of propagation. At the point A, the ray direction is close to this direction and the polarization changes rapidly. The ray direction is never exactly on the symmetry axis, and the polarization always changes continuously. The rate of change of the polarization is purely geometrical, depending on the proximity of the axial singularity, and does not depend on the frequency or wavelength as required by the wave equation. Thus the anisotropic ray result (Figure 10.17) is physically absurd.

To use ray methods at or near degeneracies and in weakly anisotropic media, we need the *quasi-isotropic (QI) ray theory* introduced by Zillmer, Kashtan and Gajewski (1998) and Pšenčík (1998) using a method by Kravtsov and Orlov (1990, p. 233). The effects of anisotropy and differences in the S velocities are modelled as perturbations to the isotropic solution by treating the anisotropic part of the model as another small parameter, in addition to $1/\omega$, in the asymptotic expansion. Before we develop quasi-isotropic ray theory, we need to outline the results of ray perturbation theory (Červený, 1982; Hanyga, 1982*b*; Červený and Jech, 1982; Jech and Pšenčík, 1989; Pšenčík and Gajewski, 1998: Zillmer, Kashtan and Gajewski, 1998).

10.2.1 Ray perturbation theory

We consider the general situation where ray results are known for a model described by elastic parameters c^0_{ijkl} and we require approximations for a different model described by parameters c_{ijkl}. We assume that ray theory is valid and that the model perturbation

$$\Delta c_{ijkl} = c_{ijkl} - c^0_{ijkl} \tag{10.2.1}$$

is small enough that first-order perturbation theory is useful. (We use the symbol Δ to distinguish perturbations due to changes in the model from perturbations to the rays in the original model, i.e. paraxial rays. These were indicated by d or δ, as in Section 5.2.2.)

10.2.1.1 Kinematic perturbations

The Christoffel equation (5.3.16) for the perturbed medium is

$$\left((c^0_I + \Delta c_I)^2 \mathbf{I} - \hat{\mathbf{\Gamma}}^0 - \Delta\,\hat{\mathbf{\Gamma}}\right)\left(\hat{\mathbf{g}}^0_I + \Delta\hat{\mathbf{g}}_I\right) = \mathbf{0}, \tag{10.2.2}$$

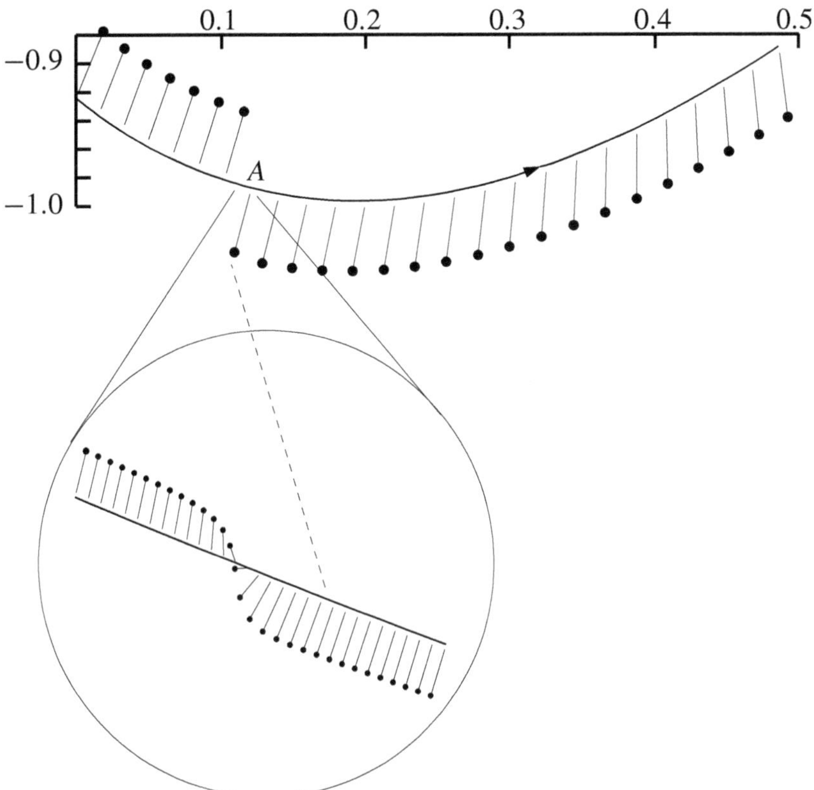

Fig. 10.17. The polarization on a qSV ray in a TI, one-dimensional model. The elastic parameters correspond to the Green Horn shale (Jones and Wang, 1981) scaled in the vertical direction to give a linear velocity model (Section 5.7.2.2). Therefore the ray path can be found from the slowness surface, although here it is found by numerically solving the ray equations (5.3.20) and (5.3.21) with constraint (5.7.1). The velocity is normalized so the shear velocities are unity on the symmetry axis, i.e. $C_{44}/\rho = 1$, at unit depth, $x_3 = -1$. With the velocity–depth law (5.7.48), the starting point is $\mathbf{x}_S = (0, 0, -\cos(\pi/8))$ and the initial slowness direction is $\hat{\mathbf{p}} = (\cos(\pi/8), 0, -\sin(\pi/8))$. The total travel time for the ray arc shown is $T = 0.5$ with the polarization plotted at $\Delta T = 0.02$ intervals. The enlarged portion is for $T = 0.130$ to 0.145 with the polarization plotted at $\Delta T = 0.0005$ intervals. The axis of symmetry is 15 degrees below the horizontal and 0.001 radians out of the plane of propagation. At the point A, the ray direction is close to this axis. Except near A, the ray polarization is very close to the propagation plane. The normalized polarization, $\hat{\mathbf{g}}_{qSV}$ is indicated by the lines ending in dots (amplitude changes due to geometrical spreading are not indicated). The enlarged plot shows that the polarization changes continuously near the point A. Incidentally it is also evident that near the end of the ray, when the direction is about 45 degrees from the symmetry axis, the polarization is significantly non-orthogonal to the ray (as is expected because the wavefront is distorted by a cusp, Figure 5.13).

where the superscript ^{0} indicates the unperturbed values and there is no summation over the ray type, I, throughout this section. Expanding and dropping the second-order terms, we obtain

$$\left(c_I^{0^2}\mathbf{I} - \hat{\boldsymbol{\Gamma}}^0\right)\Delta\hat{\mathbf{g}}_I + \left(2c_I^0\,\Delta c_I\mathbf{I} - \Delta\,\hat{\boldsymbol{\Gamma}}\right)\hat{\mathbf{g}}_I^0 = \mathbf{0}, \tag{10.2.3}$$

where, of course, the purely unperturbed terms satisfy the Christoffel equation

$$\left(c_I^{0^2}\mathbf{I} - \hat{\boldsymbol{\Gamma}}^0\right)\hat{\mathbf{g}}_I^0 = \mathbf{0}. \tag{10.2.4}$$

Pre-multiplying equation (10.2.3) by $\hat{\mathbf{g}}_I^{0^{\mathrm{T}}}$, the first term is zero from the transpose of equation (10.2.4), and we obtain

$$\Delta c_I = \frac{\hat{\mathbf{g}}_I^{0^{\mathrm{T}}}\,\Delta\,\hat{\boldsymbol{\Gamma}}\hat{\mathbf{g}}_I^0}{2c_I}, \tag{10.2.5}$$

for the perturbation of the phase velocity.

In general, the perturbation $\Delta\hat{\boldsymbol{\Gamma}}$ contains perturbations to the model and from the slowness direction

$$\Delta\hat{\boldsymbol{\Gamma}} = \hat{p}_j^0\,\hat{p}_k^0\,\Delta\mathbf{a}_{jk} + 2\hat{p}_j^0\,\Delta\hat{p}_k\,\mathbf{a}_{jk}^0. \tag{10.2.6}$$

Thus the numerator in expression (10.2.5) can be expanded as

$$\hat{\mathbf{g}}_I^{0\,\mathrm{T}}\,\Delta\hat{\boldsymbol{\Gamma}}\hat{\mathbf{g}}_I^0 = B_{II}^0 + 2c_I^0\,\Delta\hat{\mathbf{p}}\cdot\mathbf{V}^0, \tag{10.2.7}$$

where for future purposes it is convenient to define the symmetric matrix elements

$$B_{MN} = \hat{p}_j\,\hat{p}_k\,\hat{\mathbf{g}}_M^{\mathrm{T}}\,\Delta\mathbf{a}_{jk}\,\hat{\mathbf{g}}_N, \tag{10.2.8}$$

and we have used (5.3.20) for the definition of the ray velocity, $\mathbf{V}^0$.

Let us now consider the perturbation to the travel time. Writing this as

$$T = \int_{\mathcal{L}}\mathbf{p}\cdot\mathrm{d}\mathbf{x} = \int_{\mathcal{L}}\frac{\hat{\mathbf{p}}\cdot\mathrm{d}\mathbf{x}}{c}, \tag{10.2.9}$$

the travel-time perturbation is

$$\Delta T \simeq \int_{\mathcal{L}^0}\Delta\mathbf{p}\cdot\mathrm{d}\mathbf{x} \tag{10.2.10}$$

$$\simeq -\int_{\mathcal{L}^0}\frac{\Delta c}{c^2}\,\hat{\mathbf{p}}^0\cdot\mathrm{d}\mathbf{x} + \int_{\mathcal{L}^0}\frac{\Delta\hat{\mathbf{p}}\cdot\mathrm{d}\mathbf{x}}{c} \tag{10.2.11}$$

$$\simeq -\int_{\mathcal{L}^0}\frac{\Delta c}{c}\,\mathrm{d}T + \int_{\mathcal{L}^0}\frac{\Delta\hat{\mathbf{p}}\cdot\mathrm{d}\mathbf{x}}{c}. \tag{10.2.12}$$

Fermat's principle, that the error in calculating the travel time is second-order in errors in the ray path, has allowed us to calculate the first-order travel-time perturbation from perturbations on the unperturbed ray path, $\mathcal{L}^0$. Substituting the perturbed phase velocity (10.2.5), the final term in expression (10.2.7) cancels with the final term in result (10.2.12) and we obtain

$$\Delta T_I \simeq -\int_{\mathcal{L}^0} \frac{B_{II}}{2c_I^{0\,2}}\,\mathrm{d}T. \tag{10.2.13}$$

This result for the travel-time perturbation due to an anisotropic model perturbation (10.2.8) was derived by Červený (1982), Hanyga (1982*b*) and Červený and Jech (1982).

Although Fermat's principle allows the travel-time perturbation to be calculated without knowledge of the perturbed ray path, $\mathcal{L}^0 + \Delta\mathcal{L}$ (or the perturbed phase or ray directions), first-order perturbations of these occur. They can be calculated by perturbing the ray equation (Farra and Madariaga, 1987; Farra, 1989; Farra and Le Bégat, 1995). With the anisotropic ray equations, (5.3.20) and (5.3.21) written as equation (5.1.29), we obtain

$$\frac{\mathrm{d}\Delta\mathbf{y}}{\mathrm{d}T} = \mathbf{I}_1\nabla_{\mathbf{y}}(\nabla_{\mathbf{y}}H^0)^{\mathrm{T}}\Delta\mathbf{y} + \mathbf{I}_1\nabla_{\mathbf{y}}(\Delta H) = \mathbf{D}^0\Delta\mathbf{y} + \mathbf{I}_1\nabla_{\mathbf{y}}(\Delta H). \tag{10.2.14}$$

Compared with dynamic ray equations (5.2.19), the extra term, which acts as a 'source' term in the dynamic propagator, is due to perturbations to the Hamiltonian (5.3.18) on the unperturbed ray. The solution of this equation using result (C.1.9) is

$$\Delta\mathbf{y}(T) = \mathbf{P}^0(T, T_{\mathrm{S}})\,\Delta\mathbf{y}(T_{\mathrm{S}}) + \int_{T_{\mathrm{S}}}^{T} \mathbf{P}^0(T, T')\,\mathbf{I}_1\nabla_{\mathbf{y}}(\Delta H)(T')\,\mathrm{d}T'. \tag{10.2.15}$$

With $\Delta\mathbf{x}_{\mathrm{S}} = \Delta\mathbf{x}_{\mathrm{R}} = \mathbf{0}$, this equation can be solved for $\Delta\mathbf{p}_{\mathrm{S}}$, so giving the first-order perturbations to the ray path, $\Delta\mathbf{y}$, along the ray.

10.2.1.2 Polarization perturbations

The perturbation to the polarization can be obtained from equation (10.2.3). The eigenvectors of the Christoffel equation (5.3.16) are orthogonal, assuming the eigenvalues are non-degenerate, so the perturbation to the normalized polarization must be

$$\Delta\hat{\mathbf{g}}_I = g_{IJ}\,\hat{\mathbf{g}}_J^0 + g_{IK}\,\hat{\mathbf{g}}_K^0, \tag{10.2.16}$$

where I, J and K are a cyclic arrangement of the indices 1, 2 and 3 and there is no summation. Substituting this (10.2.16) in equation (10.2.3) and pre-multiplying by $\hat{\mathbf{g}}_J^{0^\mathrm{T}}$, we obtain

$$g_{IJ} = \frac{\hat{\mathbf{g}}_J^{0^\mathrm{T}}\, \Delta\hat{\mathbf{\Gamma}}\, \hat{\mathbf{g}}_I^0}{c_I^{0^2} - c_J^{0^2}}. \tag{10.2.17}$$

If the phase direction is fixed, this gives

$$\Delta\hat{\mathbf{g}}_I = \frac{B_{IJ}^0}{c_I^{0^2} - c_J^{0^2}}\, \hat{\mathbf{g}}_J^0 + \frac{B_{IK}^0}{c_I^{0^2} - c_K^{0^2}}\, \hat{\mathbf{g}}_K^0, \tag{10.2.18}$$

for the polarization perturbation. Clearly this breaks down if $c_I^0 = c_J^0$ or c_K^0, an important degenerate case as it applies to shear rays in isotropic media. In this case we must use degenerate perturbation theory (Jech and Pšenčík, 1989).

10.2.1.3 Degenerate polarizations

Let us consider the degenerate case when $c_1^0 = c_2^0 = c^0$. The corresponding polarizations are arbitrary except that they must be in the plane orthogonal to $\hat{\mathbf{g}}_3^0$. We choose any two mutually orthonormal vectors, $\hat{\mathbf{e}}_1^0$ and $\hat{\mathbf{e}}_2^0$. Then any polarizations can be written as

$$\hat{\mathbf{g}}_1^0 = \cos\phi\, \hat{\mathbf{e}}_1^0 + \sin\phi\, \hat{\mathbf{e}}_2^0 \tag{10.2.19}$$

$$\hat{\mathbf{g}}_2^0 = -\sin\phi\, \hat{\mathbf{e}}_1^0 + \cos\phi\, \hat{\mathbf{e}}_2^0, \tag{10.2.20}$$

where ϕ is a rotation angle in the plane orthogonal to $\hat{\mathbf{g}}_3^0$. Substituting in the perturbation equation (10.2.3), we can pre-multiply by $\hat{\mathbf{g}}_1^{0^\mathrm{T}}$ or $\hat{\mathbf{g}}_2^{0^\mathrm{T}}$ to obtain two simultaneous equations

$$\cos\phi\, (2c^0\Delta c - B_{11}) + \sin\phi\, B_{12} = 0 \tag{10.2.21}$$

$$-\sin\phi\, B_{21} + \cos\phi\, (2c^0\Delta c - B_{22}) = 0, \tag{10.2.22}$$

where again we have taken the phase direction, $\hat{\mathbf{p}}$, fixed and the elements $B_{\eta\nu}$ are defined using the vectors $\hat{\mathbf{e}}_\eta^0$. The condition for a solution of the simultaneous equations (10.2.21) and (10.2.22) requires that the eigenvalues, b_η, of the symmetric matrix

$$\mathbf{B} = \begin{pmatrix} B_{11} & B_{12} \\ B_{21} & B_{22} \end{pmatrix}, \tag{10.2.23}$$

equal the perturbation $2c^0\Delta c_\eta$, i.e.

$$2c^0\Delta c_\eta = b_\eta = \frac{1}{2}\left(B_{11} + B_{22} \mp B\right), \tag{10.2.24}$$

with

$$B = \left((B_{11} - B_{22})^2 + 4B_{12}^2\right)^{1/2}. \tag{10.2.25}$$

In equation (10.2.24), the negative sign corresponds to the index $\eta = 1$, the slower wave (following our convention of ordering the ray types with increasing velocity), and the positive sign to $\eta = 2$. The eigenvalues give the rotation matrix

$$\mathbf{\Phi} = \begin{pmatrix} \cos\phi & -\sin\phi \\ \sin\phi & \cos\phi \end{pmatrix}, \tag{10.2.26}$$

where

$$\cos\phi = \frac{1}{\sqrt{2}}\left(1 + \frac{B_{11} - B_{22}}{B}\right)^{1/2} \tag{10.2.27}$$

$$\sin\phi = \frac{\operatorname{sgn}(B_{12})}{\sqrt{2}}\left(1 - \frac{B_{11} - B_{22}}{B}\right)^{1/2}, \tag{10.2.28}$$

or equivalently

$$\tan 2\phi = \frac{2B_{12}}{B_{11} - B_{22}}. \tag{10.2.29}$$

The solutions for the angle ϕ and the polarizations (10.2.19) and (10.2.20) are independent of the magnitude of the perturbations $\Delta\mathbf{a}_{jk}$, and depend only on the ratios of the parameters, i.e. if we write $\Delta\mathbf{a}_{jk} = \epsilon\, \mathbf{a}^1_{jk}$, then the polarizations are independent of the parameter ϵ. Even an infinitesimal anisotropic perturbation ($\epsilon \to 0$) removes the degeneracy and defines the polarizations (assuming, of course, that the anisotropy does remove the degeneracy). We therefore refer to these as the *infinitesimal-anisotropy polarizations*. This feature is indicative of the breakdown of anisotropic ray theory in the isotropic limit discussed above – the isotropic polarizations depend on the ray history, i.e. the solution of differential equation (5.6.8), whereas the infinitesimal-anisotropy polarizations only depend on the local anisotropy, $\mathbf{a}^1_{jk}$. If the polarizations $\hat{\mathbf{e}}^0_\eta$ are chosen to be in the directions required by the (infinitesimal) anisotropy, i.e. by $\mathbf{a}^1_{jk}$, then $B_{12} = 0$ and $\phi = 0$ and equation (10.2.24) simplifies to

$$2c^0 \Delta c_\eta = b_\eta = B_{\eta\eta}, \tag{10.2.30}$$

as in the non-degenerate case (10.2.5).

Finally, we can find the perturbations to the polarizations caused by the first-order perturbation. Taking the polarizations $\hat{\mathbf{g}}^0_\eta$, equations (10.2.19) and (10.2.20), we can write the perturbation as in equation (10.2.16) (with $I = 1$, $J = 2$ and $K = 3$, say). Then taking equation (10.2.3), and pre-multiplying by $\hat{\mathbf{g}}_3^{0^{\mathrm{T}}}$ we obtain

g_{13} as before, results (10.2.17) and (10.2.18)

$$g_{13} = \frac{B_{13}^0}{c_1^{0^2} - c_3^{0^2}}. \tag{10.2.31}$$

To obtain g_{12}, we must retain second-order terms in equation (10.2.2) and pre-multiply by $\hat{\mathbf{g}}_2^{0^\mathrm{T}}$ to obtain

$$g_{12}\left(2c_1^0\Delta c_1 - B_{22}^0\right) - g_{13}B_{23}^0 = 0 \tag{10.2.32}$$

(this is second-order, all first-order terms being zero). Using result (10.2.24), this reduces to

$$g_{12} = \frac{B_{12}^0 B_{23}^0}{\left(B_{22}^0 - B_{11}^0\right)\left(c_1^{0^2} - c_3^{0^2}\right)}, \tag{10.2.33}$$

giving the first-order perturbation to the polarization provided $B_{11}^0 \neq B_{22}^0$, i.e. provided the anisotropy removes the degeneracy.

Having reviewed ray perturbation theory, we can now proceed to develop quasi-isotropic ray theory.

10.2.2 Quasi-isotropic ray equations

We consider a weakly anisotropic medium and factor the elastic parameters in an isotropic part (4.4.49)

$$c_{ijkl}^0 = \lambda\delta_{ij}\delta_{kl} + \mu\left(\delta_{ik}\delta_{jl} + \delta_{il}\delta_{jk}\right), \tag{10.2.34}$$

and a small anisotropic part, Δc_{ijkl} (10.2.1). We assume that Δc_{ijkl} is small and of order $1/\omega$ with a dimensionless condition describing the quasi-isotropic situation $(\omega\,\Delta c_{ijkl}L)/2\pi\rho V^3 \sim 1$. Substituting (10.2.1) in the constitutive relation (4.5.36), and using the same ray ansatz (5.3.1), we obtain a revised equation (5.3.3)

$$\mathbf{c}_{jk}^0\frac{\partial\mathbf{v}^{(m-1)}}{\partial x_k} - \mathrm{i}\omega\Delta\mathbf{c}_{jk}\frac{\partial\mathbf{v}^{(m-2)}}{\partial x_k} = \mathbf{t}_j^{(m)} + p_k\mathbf{c}_{jk}^0\mathbf{v}^{(m)} - \mathrm{i}\omega p_k\Delta\mathbf{c}_{jk}\mathbf{v}^{(m-1)}. \tag{10.2.35}$$

Note that the new terms in this equation are frequency dependent and have been included as $O(\omega\,\Delta c_{ijkl}L/2\pi\rho V^3) \sim 1$. Equation (5.3.2) remains valid, and as before $\mathbf{t}_j^{(m)}$ can be eliminated between equations (5.3.2) and (10.2.35) to obtain a revised version of equation (5.3.5)

$$\boldsymbol{\mathcal{N}}^0\left(\mathbf{v}^{(m)}\right) - \overline{\boldsymbol{\mathcal{M}}}^0\left(\mathbf{v}^{(m-1)}, \mathbf{t}_j^{(m-1)}\right) = \mathrm{i}\omega p_j\Delta\mathbf{c}_{jk}\left(p_k\mathbf{v}^{(m-1)} - \frac{\partial\mathbf{v}^{(m-2)}}{\partial x_k}\right). \tag{10.2.36}$$

The operators $\mathcal{N}^0$ and $\overline{\mathcal{M}}^0$ are in the isotropic model, e.g. equations (5.3.6) and (5.3.7).

10.2.2.1 The quasi-isotropic eikonal

The $m = 0$ terms are as before for the isotropic part of the model, c^0_{ijkl}, i.e. equation (10.2.36) reduces to equation (5.3.11) when $m = 0$. Thus the rays are traced in the isotropic part of the model (see Section 5.5). The P ray is longitudinal, i.e. $\hat{\mathbf{g}}_3 = \alpha\mathbf{p}$. The S rays are degenerate and the polarizations are in the plane perpendicular to the ray.

10.2.2.2 The quasi-isotropic transport equation

Letting $m = 1$ in expression (10.2.36) and proceeding as before, the transport equation (5.4.16) is modified to

$$\nabla \cdot \mathbf{N} = -\mathrm{i}\omega\,\mathbf{v}^{(0)\,\mathrm{T}} p_j p_k \Delta\mathbf{c}_{jk}\mathbf{v}^{(0)} \tag{10.2.37}$$

$$= -\frac{\mathrm{i}\omega\rho\, v^{(0)\,2} B_{II}}{c^2}, \tag{10.2.38}$$

(no summation over I) using definition (10.2.8) and assuming no density perturbation, where I defines the ray type.

10.2.2.3 P quasi-isotropic rays

As the isotropic eikonal applies, the polarization is still longitudinal, i.e. $\mathbf{v}^{(0)} = v_3^{(0)}\hat{\mathbf{g}}_3$ where $\hat{\mathbf{g}}_3 = \alpha\,\mathbf{p}$.

For P waves, equation (10.2.37) reduces to (cf. equation (5.4.9))

$$\frac{\mathrm{d}}{\mathrm{d}T}\ln\left(\rho\, v_3^{(0)\,2}\right) = -\nabla\cdot\mathbf{V} - \mathrm{i}\omega\,\frac{B_{33}}{\alpha^2}, \tag{10.2.39}$$

where B_{33} is defined in equation (10.2.8). Combining with equation (5.2.14), we obtain

$$\frac{\mathrm{d}}{\mathrm{d}T}\ln\left(\rho\,\alpha v_3^{(0)\,2} J\right) = -\mathrm{i}\omega\,\frac{B_{33}}{\alpha^2} \tag{10.2.40}$$

and the solution is

$$v_3^{(0)} = \frac{\text{constant}}{\sqrt{\rho\,\alpha|J|}}\exp\left(\mathrm{i}\omega\Delta T_3\right), \tag{10.2.41}$$

where

$$\Delta T_3 = -\int_{\mathcal{L}_n}\frac{B_{33}}{2\alpha^3}\,\mathrm{d}s, \tag{10.2.42}$$

is the travel-time shift due to the anisotropic perturbation (10.2.13).

The additional terms are evaluated as before, equation (5.4.4), including the extra QI term

$$v_\eta^{(1)} = \frac{1}{\rho G_\eta} \hat{\mathbf{g}}_\eta^{\mathrm{T}} \left(\overline{\boldsymbol{\mathcal{M}}}(\mathbf{v}^{(0)}, \mathbf{t}_j^{(0)}) + \mathrm{i}\omega p_j p_k \Delta \mathbf{c}_{jk} \mathbf{v}^{(0)} \right). \tag{10.2.43}$$

Specializing to the quasi-isotropic P wave ($\mathbf{v}^{0)} = v_3^{(0)} \hat{\mathbf{g}}_3$), we obtain the isotropic terms (5.6.12) plus additional QI terms

$$\begin{aligned} v_\eta^{(1)} = &-\hat{\mathbf{g}}_\eta^{\mathrm{T}} \left(\alpha \frac{\partial v_3^{(0)}}{\partial g_\eta} + \frac{v_3^{(0)}}{\alpha^2 - \beta^2} \left((\alpha^2 - \beta^2) \nabla \alpha - 4\alpha\beta \, \nabla \beta \right.\right. \\ &\left.\left. + \alpha(\alpha^2 - 2\beta^2) \nabla (\ln \rho) \right) \right) \\ &+ \frac{\mathrm{i}\omega\alpha^2 \hat{\mathbf{g}}_\nu^{\mathrm{T}}}{\rho(\beta^2 - \alpha^2)} \left(p_j \left(\mathbf{c}_{jk}^0 + \mathbf{c}_{kj}^0 \right) \mathbf{v}^{(0)} \frac{\partial(\Delta T_3)}{\partial x_k} + p_j p_k \Delta \mathbf{c}_{jk} \mathbf{v}^{(0)} \right). \end{aligned} \tag{10.2.44}$$

Using results (5.5.6) and (5.5.7), and definition (10.2.8), we obtain

$$\begin{aligned} v_\eta^{(1)} = &-\hat{\mathbf{g}}_\eta^{\mathrm{T}} \left(\alpha \frac{\partial v_3^{(0)}}{\partial g_\eta} + \frac{v_3^{(0)}}{\alpha^2 - \beta^2} \left((\alpha^2 - \beta^2) \nabla \alpha - 4\alpha\beta \, \nabla \beta \right.\right. \\ &\left.\left. + \alpha(\alpha^2 - 2\beta^2) \nabla (\ln \rho) \right) \right) - \mathrm{i}\omega v_3^{(0)} \left(\alpha \frac{\partial(\Delta T_3)}{\partial g_\eta} + \frac{B_{\eta 3}}{\alpha^2 - \beta^2} \right). \end{aligned} \tag{10.2.45}$$

The two extra QI terms were described by Pšenčík (1998). They are proportional to frequency, so overall their contribution is independent of frequency (and the perturbed polarization due to anisotropy remains linear in contrast to additional terms due to heterogeneity). The first QI term is the correction to the polarization due to perturbations of the wavefront. If the travel-time perturbation, ΔT_3, varies across the wavefront (with derivatives $\partial(\Delta T_3)/\partial g_\eta$), then the slowness (direction) $\mathbf{p}$ changes with a corresponding change in polarization. This is an alternative to using the dynamic propagator (10.2.15). The second term corrects the polarization for deviations from the slowness vector in the anisotropic medium. It agrees with the result of ray perturbation theory (10.2.18) (Jech and Pšenčík, 1989; Pšenčík and Gajewski, 1998).

10.2.2.4 S quasi-isotropic rays

For qS rays, the polarization is degenerate. The shear polarization can be written as a combination of any two orthogonal vectors in the transverse plane. Various choices suggest themselves, e.g. the isotropic shear polarizations, or the polarizations in the anisotropic medium, themselves perhaps estimated by perturbation theory. A final possibility is to use the normal and binormal vectors as a basis but given that the isotropic polarizations are known, this offers no advantages. Zillmer, Kashtan and Gajewski (1998) and Pšenčík (1998) have investigated these possibilities and we follow a similar general approach here, showing that using the isotropic shear polarizations or the infinitesimal-anisotropy polarizations, (10.2.19) and (10.2.20), the equations are analytically equivalent although the corresponding differential equations have different properties. Using the anisotropic polarizations is equivalent to Coates and Chapman (1990*b*) who investigated the coupling between qS waves using Born scattering theory (Section 10.3).

The shear-wave polarization can always be written

$$\mathbf{v}^{(0)} = v_\nu^{(0)} \hat{\mathbf{e}}_\nu^0, \tag{10.2.46}$$

where $\hat{\mathbf{e}}_\nu^0$ are two mutually orthogonal vectors in the plane orthogonal to the direction $\hat{\mathbf{g}}_3^0$ in the isotropic medium (cf. equations (10.2.19) and (10.2.20)). Substituting in equation (10.2.36) and pre-multiplying by $\hat{\mathbf{e}}_\eta^{\mathrm{T}}$, result (5.6.2) is modified to

$$\hat{\mathbf{e}}_\eta^{\mathrm{T}} \boldsymbol{\mathcal{M}}(v_\nu^{(0)} \hat{\mathbf{e}}_\nu) = -\frac{\mathrm{i}\omega\rho}{\beta^2} B_{\eta\nu} v_\nu^{(0)}, \tag{10.2.47}$$

where the elements $B_{\eta\nu}$ (10.2.8) are calculated using the vectors $\hat{\mathbf{e}}_\eta$. Consequently, equation (5.6.3) is modified to

$$\delta_{\nu\eta} \Big(2\mu \mathbf{p} \cdot \nabla v_\nu^{(0)} + v_\nu^{(0)} \mathbf{p} \cdot \nabla \mu + \mu v_\nu^{(0)} \nabla \cdot \mathbf{p} \Big) + 2\rho v_\nu^{(0)} \, \hat{\mathbf{e}}_\eta^{0\,\mathrm{T}} \frac{\mathrm{d}\hat{\mathbf{e}}_\nu^0}{\mathrm{d}T} = -\frac{\mathrm{i}\omega\rho}{\beta^2} B_{\eta\nu} v_\nu^{(0)}. \tag{10.2.48}$$

Knowing the solution of the isotropic equation (5.6.3), we make the substitution

$$\boxed{\check{v}_\nu^{(0)} = v_\nu^{(0)} \sqrt{\rho\beta |J|},} \tag{10.2.49}$$

to remove the amplitude behaviour of the geometrical ray approximation from the amplitude coefficient. The *GRA compensated amplitudes*, $\check{v}_\nu^{(0)}$, should be approximately constant. Equation (10.2.48) then becomes

$$\boxed{\frac{\mathrm{d}\check{\mathbf{v}}^{(0)}}{\mathrm{d}T} = \gamma \, \mathbf{I}_1 \check{\mathbf{v}}^{(0)} - \frac{\mathrm{i}\omega}{2\beta^2} \mathbf{B} \check{\mathbf{v}}^{(0)},} \tag{10.2.50}$$

where matrix $\mathbf{I}_1$ is defined in equation (0.1.5) and

$$\gamma = \hat{\mathbf{e}}_2^0 \cdot \frac{\mathrm{d}\hat{\mathbf{e}}_1^0}{\mathrm{d}T} = -\hat{\mathbf{e}}_1^0 \cdot \frac{\mathrm{d}\hat{\mathbf{e}}_2^0}{\mathrm{d}T}, \tag{10.2.51}$$

as $\hat{\mathbf{e}}_1^0 \cdot \hat{\mathbf{e}}_2^0 = 0$.

Various alternatives can be used to define the vectors $\hat{\mathbf{e}}_\nu^0$ and the corresponding coefficients $\check{\mathbf{v}}^{(0)}$. The two we consider here are the polarizations in the isotropic medium and the infinitesimal-anisotropy polarizations. A third alternative, which we do not consider, is to use the exact polarizations from the anisotropic model (as in Coates and Chapman, 1990*b*). As other aspects of the solution are only first order, it seems doubtful whether this is worth the extra expense. To distinguish the two choices we denote the isotropic polarizations by $\overline{\mathbf{g}}_\nu$, which are found by solving equations (5.6.8) in the isotropic model, and the infinitesimal-anisotropy polarizations by $\widetilde{\mathbf{g}}_\nu$, which are found from equations (10.2.19) and (10.2.20). The same overbar and tilde are use to denote terms evaluated with the corresponding vectors. The relationship between the two choices, illustrated in Figure 10.18, can be written

$$\widetilde{\mathbf{G}} = \overline{\mathbf{G}}\,\boldsymbol{\Phi}, \tag{10.2.52}$$

where the polarization vectors have been combined in a 3×2 matrix, e.g.

$$\widetilde{\mathbf{G}} = \begin{pmatrix} \widetilde{\mathbf{g}}_1 & \widetilde{\mathbf{g}}_2 \end{pmatrix}, \tag{10.2.53}$$

and equation (10.2.26) defines the rotation matrix, $\boldsymbol{\Phi}$.

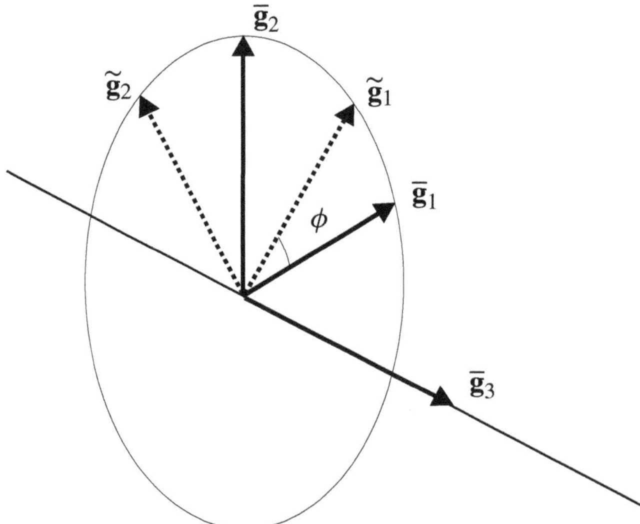

Fig. 10.18. The isotropic polarization vectors, $\overline{\mathbf{g}}_\nu$, and the infinitesimal-anisotropy polarization vectors, $\widetilde{\mathbf{g}}_\nu$, related by a rotation ϕ about the ray direction $\overline{\mathbf{g}}_3$.

First we consider using the infinitesimal-anisotropy polarizations, $\widetilde{\mathbf{g}}_\nu$. In differential equation (10.2.50), it is straightforward to show that

$$\widetilde{\gamma} = \frac{\mathrm{d}\phi}{\mathrm{d}T} = \dot{\phi}, \tag{10.2.54}$$

using the orthornomality of the vectors and result (5.6.8), and that matrix $\mathbf{B}$ is diagonal (10.2.30), i.e.

$$\widetilde{\mathbf{B}} = \begin{pmatrix} b_1 & 0 \\ 0 & b_2 \end{pmatrix} = \mathbf{b}. \tag{10.2.55}$$

Equation (10.2.50) becomes

$$\frac{\mathrm{d}\widetilde{\mathbf{v}}^{(0)}}{\mathrm{d}T} = \dot{\phi}\,\mathbf{I}_1\widetilde{\mathbf{v}}^{(0)} - \frac{\mathrm{i}\omega}{2\beta^2}\,\mathbf{b}\widetilde{\mathbf{v}}^{(0)}. \tag{10.2.56}$$

It is straightforward to understand the physical significance to the two terms on the right-hand side of this differential equation. The second term models the propagation difference of the quasi-shear rays compared with the isotropic shear ray and results in the *shear-ray splitting*. If the first term can be ignored, a fundamental matrix of system (10.2.56) is

$$\mathbf{\Lambda} = \mathrm{e}^{\mathrm{i}\omega\Delta\mathbf{T}}, \tag{10.2.57}$$

where

$$\Delta\mathbf{T} = -\int_0^T \frac{\mathbf{b}}{2\beta^2}\,\mathrm{d}T. \tag{10.2.58}$$

This diagonal matrix gives the time shifts, ΔT_ν, of the two quasi-shear rays (equations (10.2.13) and (10.2.30)). By convention, we have ordered the two quasi-shear rays so $\Delta T_1 > \Delta T_2$. This solution corresponds to the non-degenerate behaviour when the two quasi-shear rays propagate independently. On the other hand, if the first term on the right-hand side of equation (10.2.56) dominates and the second term can be ignored, then the two coefficients couple together due to the off-diagonal elements, $\pm\dot{\phi}$. This is called *quasi-shear ray coupling* (Coates and Chapman, 1990*b*). The same problem was investigated using transform methods (cf. Chapter 7) by Chapman and Shearer (1989) where it was confirmed that the coupling solution was a good approximation to the complete solution (see Exercise 10.2). A fundamental matrix of system (10.2.56) is simply

$$\mathbf{\Phi}^{-1} = \begin{pmatrix} \cos\phi & \sin\phi \\ -\sin\phi & \cos\phi \end{pmatrix}, \tag{10.2.59}$$

which exactly compensates for the rotation of the infinitesimal-anisotropy polarizations with respect to the isotropic polarizations. The resultant polarization is the isotropic polarization. This solution is the degenerate behaviour. Near degenerate points, neither the degenerate nor non-degenerate behaviour apply alone, and we have to consider both terms in equation (10.2.56) and solve the equation numerically.

Before we investigate numerical solutions, let us consider equation (10.2.50) when the isotropic polarizations, $\overline{\mathbf{g}}_\nu$, are used. Then

$$\overline{\gamma} = 0 \tag{10.2.60}$$

as the isotropic polarizations satisfy equation (5.6.8). Thus equation (10.2.50) reduces to

$$\frac{\mathrm{d}\overline{\mathbf{v}}^{(0)}}{\mathrm{d}T} = -\frac{\mathrm{i}\omega}{2\beta^2}\,\overline{\mathbf{B}}\,\overline{\mathbf{v}}^{(0)}. \tag{10.2.61}$$

In the isotropic limit, the right-hand side of this equation is zero and the amplitude coefficients, $\overline{\mathbf{v}}^{(0)}$, are constant, i.e. the isotropic rays. In general, the matrix $\overline{\mathbf{B}}$ is symmetric and full, but it can be diagonalized using the eigen-matrix $\boldsymbol{\Phi}$, i.e.

$$\boxed{\frac{\mathrm{d}\overline{\mathbf{v}}^{(0)}}{\mathrm{d}T} = -\frac{\mathrm{i}\omega}{2\beta^2}\,\boldsymbol{\Phi}\mathbf{b}\,\boldsymbol{\Phi}^{-1}\overline{\mathbf{v}}^{(0)},} \tag{10.2.62}$$

an alternative to equation (10.2.56). As the GRA compensated amplitudes can be expressed with respect to the anisotropic or isotropic polarizations, i.e.

$$\check{\mathbf{v}}^{(0)} = \widetilde{\mathbf{G}}\,\widetilde{\mathbf{v}}^{(0)} = \overline{\mathbf{G}}\,\overline{\mathbf{v}}^{(0)}, \tag{10.2.63}$$

we have

$$\overline{\mathbf{v}}^{(0)} = \boldsymbol{\Phi}\,\widetilde{\mathbf{v}}^{(0)}. \tag{10.2.64}$$

Substituting in equation (10.2.62), it is straightforward to show that this differential equation is exactly equivalent to equation (10.2.56). Analytically it doesn't matter which polarizations we use, although different limits are most easily discussed using different polarizations – the degenerate and non-degenerate anisotropic limits with equation (10.2.56) and the isotropic limit with equation (10.2.62). Neither equation is ideal for direct numerical solution as the terms $\dot{\phi}$ and ω may be large.

The approximate solutions of the differential systems (10.2.56) and (10.2.62) suggest that we should write the solution as

$$\overline{\mathbf{v}}^{(0)}(T) = \boldsymbol{\Phi}(T)\boldsymbol{\Lambda}(T)\,\mathbf{r}(T). \tag{10.2.65}$$

This equation defines the *quasi-shear component vector*, $\mathbf{r}$. By including the shear-ray splitting propagation term, $\boldsymbol{\Lambda}$, in this definition, the component vector, $\mathbf{r}$, is constant in both the non-degenerate and isotropic limits. Substituting definition

(10.2.65) in the differential equation (10.2.62), the equation becomes

$$\frac{d\mathbf{r}}{dT} = \dot{\phi} \begin{pmatrix} 0 & e^{-i\omega\Delta T} \\ -e^{i\omega\Delta T} & 0 \end{pmatrix} \mathbf{r}, \tag{10.2.66}$$

where

$$\Delta T = \Delta T_1 - \Delta T_2, \tag{10.2.67}$$

is the shear-ray splitting (non-negative, by definition). It is particularly easy to solve this equation numerically. The shear-wave splitting is continuous and will not vary rapidly along the ray. For a small step along the ray, δT, it can be taken constant. Then the exact solution of (10.2.66) for the step is

$$\mathbf{r}(\omega, T + \delta T) = \begin{pmatrix} \cos\delta\phi & e^{-i\omega\Delta T}\sin\delta\phi \\ -e^{i\omega\Delta T}\sin\delta\phi & \cos\delta\phi \end{pmatrix} \mathbf{r}(\omega, T), \tag{10.2.68}$$

where $\delta\phi = \phi(T + \delta T) - \phi(T)$ is the change in angle ϕ over the step. In the time domain, the exact solution (whatever the variation $\dot{\phi}$ or however large $\delta\phi$) is

$$r_1(t, T + \delta T) = r_1(t, T)\cos\delta\phi + r_2(t + \Delta T, T)\sin\delta\phi \tag{10.2.69}$$

$$r_2(t, T + \delta T) = -r_1(t - \Delta T, T)\sin\delta\phi + r_2(t, T)\cos\delta\phi. \tag{10.2.70}$$

This solution represents coupling due to the rotation angle, $\delta\phi$. The time shifts $\pm\Delta T$ occur in the cross-coupling terms as the quasi-shear ray component functions, $r_\nu(t, T)$, propagate with the quasi-shear ray times, T_ν. Solving equations (10.2.69) and (10.2.70) with initial conditions such as $r_\nu(t, 0) = \delta_{I\nu}\delta(t)$ will lead to the frequency-dependent pulse distortion in quasi-isotropic shear-ray propagation.

Equations (10.2.69) and (10.2.70) provide an extremely simple, robust algorithm for modelling quasi-shear-ray coupling or QI ray theory. As our convention is that r_1 is the quasi-shear component of the slower ray, and r_2 is the faster ray, the support of $r_1(t)$ is $-\Delta T \le t \le 0$ and for $r_2(t)$ is $0 \le t \le \Delta T$. In practice, these time functions are approximated by finite length, discrete time series with time step Δt, say. For the numerical algorithm we round ΔT to the nearest discrete time point, so operations (10.2.69) and (10.2.70) are simply weighted additions of finite series. In Figure 10.19, we have illustrated the functions r_1 and r_2 for the region near the degeneracy in Figure 10.17. For the interval in the enlargement from $T = 0.130$ to 0.145, the shear wave splitting is only $\Delta T \simeq 1.12 \times 10^{-6}$. With a discrete time step, Δt, larger than this, the time series r_1 and r_2 are represented by one point. This value is plotted in Figure 10.19. The ray illustrated in Figure 10.17 is the *qSV* ray, which is the faster ray near the symmetry axis in the TI Green Horn shale (Figure 5.13). Thus initially $r_2 = 1$ and $r_1 = 0$. The *qSH* polarization, $\hat{\mathbf{g}}_1$, is into the page. As the degeneracy is approached these polarizations rotate

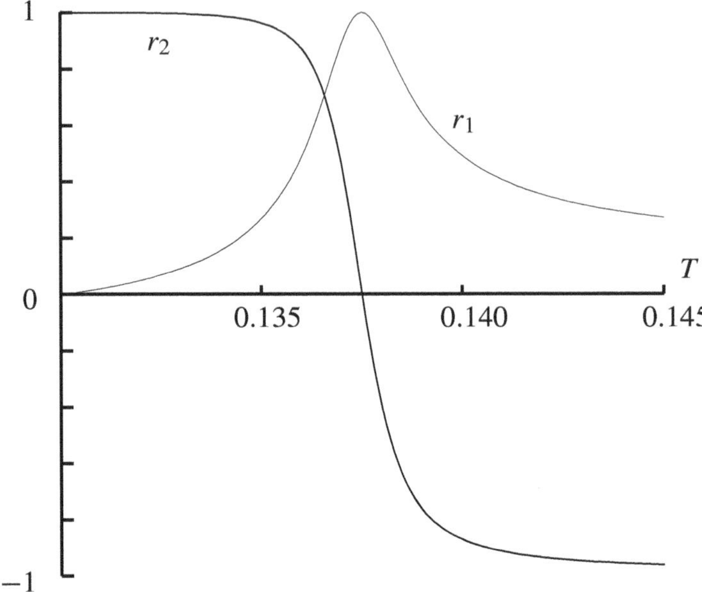

Fig. 10.19. The functions r_1 and r_2 plotted as a function of travel time, T, near the degenerate point in Figure 10.17. The initial conditions at $T = 0.130$ are $r_1 = 0$ and $r_2 = 1$. The functions have been calculated with the discrete time step larger than the shear-ray splitting, $\Delta t > \Delta T$, so the times series r_1 and r_2 are restricted to one point, $t = 0$ (for continuous time functions, $r_1(t, T)$ and $r_2(t, T)$), the support would be restricted to an interval $\mp\Delta T$, and effectively the functions' integrals or D.C. level are plotted).

about the ray until at the nearest point, $\hat{\mathbf{g}}_1$ is in the plane of the ray. The rotation angle is $\phi = \pi/2$ and so $r_2 = 0$ and $r_1 = 1$. Continuing along the ray, the polarizations continue to rotate until $\phi = \pi$ and $r_2 = -1$ and $r_1 = 0$. Although very small time steps $\delta T = 0.0001$ have been taken along the ray so that the continuous variation of the r_ν's can be illustrated, it is important to note that the algorithm is robust whatever the step size. Result (10.2.68) (and equations (10.2.69) and (10.2.70) in the time domain) are exact solutions provided ΔT can be taken constant through the step, which is a very good approximation near the degeneracy. Even one large step over the degeneracy would lead to $\delta\phi = \pi$ and the sign change in r_2. This sign change would be an example of a coupling event as described in Section 6.8.1.

In Figure 10.20, the QI polarizations (10.2.65) using the component amplitudes from Figure 10.19 are illustrated. Unlike the results in Figure 10.17 which showed a rapid rotation about the ray, now the polarization remains in the same direction in the plane orthogonal to the ray. This behaviour agrees with the isotropic behaviour

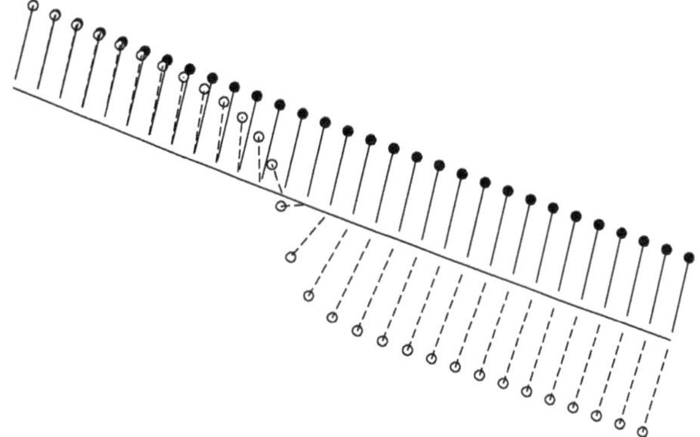

Fig. 10.20. The polarizations predicted using QI ray theory through the degeneracy in Figure 10.17. These are plotted just as in the enlargement in Figure 10.17. The polarizations predicted by anisotropic ray theory (Figure 10.17) are illustrated again with dotted lines, and the QI polarizations with solid lines.

predicted by equation (5.6.8). It applies when the shear-ray splitting, ΔT, is small compared with the pulse length of interest, i.e. $\Delta t > \Delta T$. For shorter pulses, the behaviour will be more complicated. This is illustrated in Figure 10.21 for the same degeneracy.

Figure 10.21 illustrates coupling and polarizations when the pulse length is less than the shear splitting. In the discrete time series, $\Delta t = 2 \times 10^{-8}$. The time series, $r_1(t, T)$ and $r_2(t, T)$, are plotted at various times T through the degeneracy (at intervals of 0.002 from $T = 0.130$). The shear splitting ΔT relative to $T = 0.130$ is plotted and the origin of the series $r_1(t, T)$ is aligned with it. Thus the support of $r_2(t, T)$ from $t = 0$ to $t = \Delta T$ and of $r_1(t, T)$ from $t = -\Delta T$ to $t = 0$ are coincident. Note around $T = 0.138$, the shear splitting is almost stationary and an r_1 pulse is generated which tracks the splitting. The resultant polarization depends on the pulse length and is frequency dependent.

Finally we may need the first-order additional term for shear waves. Equation (5.6.13) is modified to include the QI term

$$v_3^{(1)} = \hat{\mathbf{g}}_3^{\mathrm{T}} \left(\beta \frac{\partial v_\nu^{(0)}}{\partial g_\nu} + \frac{v_\nu^{(0)}}{\alpha^2 - \beta^2} \left((\alpha^2 + 3\beta^2) \nabla\beta + \beta^3 \nabla(\ln \rho) \right) \right) - \mathrm{i}\omega v_\nu^{(0)} \frac{B_{3\nu}}{\alpha^2 - \beta^2}. \tag{10.2.71}$$

As before, overall the QI term is independent of frequency and describes the correction to the shear polarization, in agreement with degenerate perturbation theory (Jech and Pšenčík, 1989).

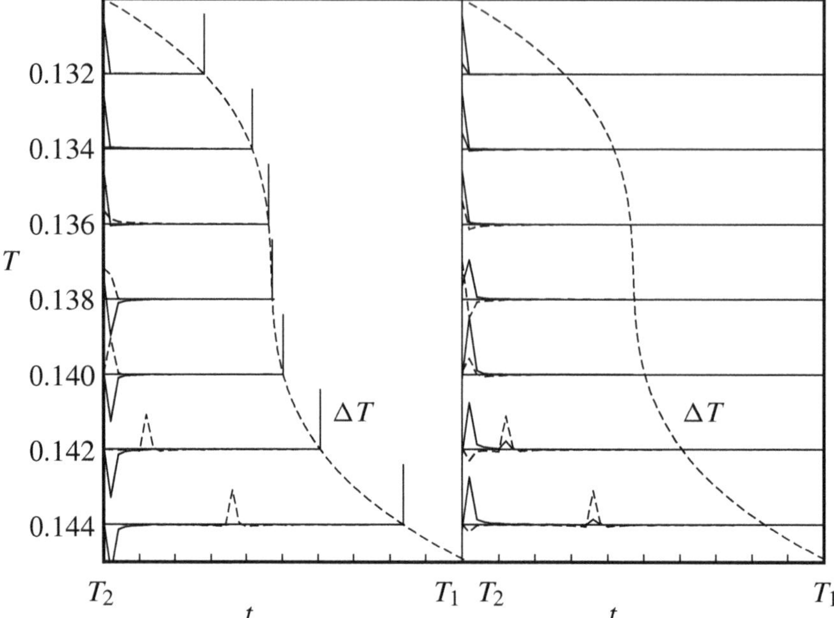

Fig. 10.21. Shear ray coupling through the degeneracy in Figure 10.17. On the left are plots of the times series $r_1(t, T)$ (dashed) and $r_2(t, T)$ (solid). The vertical axis is the travel time, T. The shear splitting ΔT is plotted (dashed) as a function of the travel time (relative to a zero at $T = 0.130$). The origin of the $r_1(t, T)$ series is aligned with the shear splitting time so the support of $r_2(t, T)$ from $t = 0$ to $t = \Delta T$ and of $r_1(t, T)$ from $t = -\Delta T$ to $t = 0$ are coincident. The vertical line at ΔT has unit magnitude. On the right are plots of the resultant polarization. The solid line is the component in the plane of the ray, and the dashed line is the component perpendicular to the plane of the ray. On the time axis, the tick marks are at intervals of 10^{-7}.

10.3 Born scattering theory

When a wave propagates through a heterogeneous medium, signals are generated by the interaction of the incident wave with heterogeneities in the medium. Most approximate modelling methods based on asymptotic ray theory (Chapter 5), e.g. geometrical ray theory, Maslov asymptotic ray theory (Section 10.1), etc., only model geometrical properties of the wavefronts and signals scattered by *discontinuities* in the model. Partial reflections generated by continuous gradients are not modelled.

The Born approximation can be used to model signals scattered by heterogeneities, and to provide a linear approximation for inverse problems. In the standard Born method, the model is separated into a *background* or *reference model*, and a *perturbation*. The scattered signal from the perturbation is obtained in terms of the solution in the reference model. Hudson and Heritage (1981), Ben-Menahem and Gibson (1990) and Gibson and Ben-Menahem (1991) have investigated the Born approximation in anisotropic, elastic media in some detail. Often the reference model is taken as homogeneous, so the reference Green functions are particularly simple (Section 4.5.5), but for realistic problems a smooth, inhomogeneous reference model is neeeded (the *distorted Born approximation*).

Using the distorted Born approximation introduces a new difficulty – how to divide the model into a reference model and a perturbation? Related to this problem, the Green function in the heterogeneous reference model will normally be approximate. If the heterogeneity included in the reference model is increased and the perturbation correspondingly decreased, then the errors in the Green function will increase but the signals scattered by the perturbation will decrease. This is clearly a fundamental, theoretical difficulty which can only be resolved by acknowledging that the Green function used in the reference model is only approximate. Coates and Chapman (1990*a*) and Chapman and Coates (1994) have introduced a generalization of Born scattering theory which solves this problem. In this section, we develop this generalized Born scattering theory for acoustic and anisotropic media. It is then straightforward to specialize the results to isotropic media. Generalized Born scattering theory uses an approximate Green function in a reference model. Errors in the Green function cause scattered signals which are combined with scattered signals from perturbations to the reference model. The scattered signals due to errors in the Green function are analogous to the WKBJ iterative solution (the Bremmer series) in one dimension (Section 7.2.6).

The Born scattering method and the Kirchhoff method (Section 10.4) are used for forward modelling (and inverse methods) in complicated Earth models. Sometimes to save expense they are applied in two-dimensional models. The basic methods remain the same so for brevity we do not explicitly give expressions in two dimensions – in our expressions, volume integrals must be interpreted as area integrals, and surface integrals as line integrals, in two dimensions. In addition, the Green functions must be the two-dimensional Green functions. The final result can be converted for three-dimensional wave propagation (2.5D wave propagation) using the usual conversion factor (8.2.66) – remember that if the dimensions of the integrals are reduced, then the Green functions in the integrals should be the two-dimensional functions not those for 2.5D wave propagation.

10.3.1 Perturbation and error Born scattering

10.3.1.1 Acoustic scattering

Born scattering theory in an acoustic medium is naturally significantly simpler than for an elastic medium so we consider it first. It is convenient to work in the frequency domain, although the results are easily transformed into simple expressions in the time domain. The equation of motion (4.5.1) is rewritten

$$-\nabla \underline{P} = -\omega^2 \rho \, \underline{\mathbf{u}} - \mathbf{I}\, \delta(\mathbf{x} - \mathbf{x}_1), \tag{10.3.1}$$

for the Green functions with a source at $\mathbf{x}_1$ with the argument $(\omega, \mathbf{x}; \mathbf{x}_1)$ understood. The constitutive relation (4.5.2) is written

$$\nabla \cdot \underline{\mathbf{u}} = -k \, \underline{P}, \tag{10.3.2}$$

or

$$\underline{P} = -\kappa \, \nabla \cdot \underline{\mathbf{u}}. \tag{10.3.3}$$

The model is divided into two parts

$$\rho(\mathbf{x}) = \rho^{\mathrm{A}}(\mathbf{x}) + \rho^{\mathrm{B}}(\mathbf{x}) \tag{10.3.4}$$
$$k(\mathbf{x}) = k^{\mathrm{A}}(\mathbf{x}) + k^{\mathrm{B}}(\mathbf{x}) \tag{10.3.5}$$
$$\kappa(\mathbf{x}) = \kappa^{\mathrm{A}}(\mathbf{x}) + \kappa^{\mathrm{B}}(\mathbf{x}). \tag{10.3.6}$$

Part A is a model in which the approximate Green function is known (the A stands for approximate). Part B is a perturbation, possibly unknown, to this model for which a linear scattering theory is obtained (the B stands for Born scattering). Note that this division of the model differs from normal Born scattering theory. The part A is not required to be a smooth background model in which the Green function is taken to be known exactly. Model A can be as complicated as desired and can be the true model, i.e. $\rho^{\mathrm{B}} = 0$ and $k^{\mathrm{B}} = 0$, but it is acknowledged that the Green function, e.g. asymptotic ray theory, is only approximate.

The approximate Green functions for the model A are denoted by $\underline{\mathbf{u}}^{\mathrm{A}}$ and $\underline{P}^{\mathrm{A}}$. These do not satisfy the equation of motion (10.3.1) nor the constitutive relations, (10.3.2) or (10.3.3), exactly, but can be used to define *error* terms

$$\underline{\mathbf{F}}^{\mathrm{A}} = \omega^2 \rho^{\mathrm{A}} \underline{\mathbf{u}}^{\mathrm{A}} - \nabla \underline{P}^{\mathrm{A}} + \mathbf{I}\, \delta(\mathbf{x} - \mathbf{x}_2), \tag{10.3.7}$$

for the force error in the equation of motion, with the argument $(\omega, \mathbf{x}; \mathbf{x}_2)$ understood. The dilatation, Θ, or pressure, Φ, errors are

$$\underline{\Theta}^{\mathrm{A}} = \nabla \cdot \underline{\mathbf{u}}^{\mathrm{A}} + k^{\mathrm{A}} \underline{P}^{\mathrm{A}} \tag{10.3.8}$$
$$\underline{\Phi}^{\mathrm{A}} = \underline{P}^{\mathrm{A}} + \kappa^{\mathrm{A}} \nabla \cdot \underline{\mathbf{u}}^{\mathrm{A}}. \tag{10.3.9}$$

If the Green functions in model A were exact, $\underline{\mathbf{F}}^{\mathrm{A}}$, $\underline{\Theta}^{\mathrm{A}}$ and $\underline{\Phi}^{\mathrm{A}}$ would be zero. As the model A and the approximate Green functions are known, the errors can be determined. For simplicity in what follows, we assume that the approximate Green functions are accurate at the source, although this is not essential.

The equations (10.3.7) and (10.3.8) can be rewritten

$$-\nabla \underline{P}^{\mathrm{A}} = -\omega^2 \rho \, \underline{\mathbf{u}}^{\mathrm{A}} - \mathbf{I}\,\delta(\mathbf{x} - \mathbf{x}_2) + \underline{\boldsymbol{\mathcal{E}}}^{\mathrm{N}} \tag{10.3.10}$$

$$\nabla \cdot \underline{\mathbf{u}}^{\mathrm{A}} = -k\, \underline{P}^{\mathrm{A}} + \underline{\mathcal{E}}^{\mathrm{H}}, \tag{10.3.11}$$

where the complete scattering terms are

$$\boxed{\underline{\boldsymbol{\mathcal{E}}}^{\mathrm{N}} = \underline{\mathbf{F}}^{\mathrm{A}} + \omega^2 \rho^{\mathrm{B}} \underline{\mathbf{u}}^{\mathrm{A}}} \tag{10.3.12}$$

$$\boxed{\underline{\mathcal{E}}^{\mathrm{H}} = \underline{\Theta}^{\mathrm{A}} + k^{\mathrm{B}} \underline{P}^{\mathrm{A}}} \tag{10.3.13}$$

(the notation N stands for Newton, the error in the equation of motion, and H for Hooke, the error in the constitutive relation). Although the equations (10.3.10) and (10.3.11) have been arranged as the differential equations (10.3.1) and (10.3.2), there is an important distinction: equations (10.3.1) and (10.3.2) are partial differential equations for the unknowns $\underline{\mathbf{u}}$ and $\underline{P}$; equations (10.3.10) and (10.3.11) are definitions for the scatterers, $\underline{\boldsymbol{\mathcal{E}}}^{\mathrm{N}}$ and $\underline{\mathcal{E}}^{\mathrm{H}}$ – the differential operators apply to the known functions, $\underline{\mathbf{u}}^{\mathrm{A}}$ and $\underline{P}^{\mathrm{A}}$.

We now combine equations (10.3.1), (10.3.2), (10.3.10) and (10.3.11). Changing the source $\mathbf{x}_1$ in equation (10.3.1) to $\mathbf{x}_{\mathrm{R}}$, we transpose the equation and post-multiply by $\underline{\mathbf{u}}^{\mathrm{A}}(\omega, \mathbf{x}; \mathbf{x}_{\mathrm{S}})$. Then pre-multiplying equation (10.3.10) by $\underline{\mathbf{u}}^{\mathrm{T}}(\omega, \mathbf{x}; \mathbf{x}_{\mathrm{R}})$ changing $\mathbf{x}_2$ to $\mathbf{x}_{\mathrm{S}}$, the two equations are subtracted and integrated over a volume V to obtain

$$\begin{aligned}\underline{\mathbf{u}}(\omega, \mathbf{x}_{\mathrm{R}}; \mathbf{x}_{\mathrm{S}}) &= \underline{\mathbf{u}}^{\mathrm{A}}(\omega, \mathbf{x}_{\mathrm{R}}; \mathbf{x}_{\mathrm{S}}) + \int_V \underline{\mathbf{u}}^{\mathrm{T}}(\omega, \mathbf{x}; \mathbf{x}_{\mathrm{R}})\, \underline{\boldsymbol{\mathcal{E}}}^{\mathrm{N}}(\omega, \mathbf{x}; \mathbf{x}_{\mathrm{S}})\, \mathrm{d}V \\ &- \int_V \Big((\nabla \underline{P})^{\mathrm{T}}(\omega, \mathbf{x}; \mathbf{x}_{\mathrm{R}})\, \underline{\mathbf{u}}^{\mathrm{A}}(\omega, \mathbf{x}; \mathbf{x}_{\mathrm{S}}) - \underline{\mathbf{u}}^{\mathrm{T}}(\omega, \mathbf{x}; \mathbf{x}_{\mathrm{R}})\, \nabla \underline{P}^{\mathrm{A}}(\omega, \mathbf{x}; \mathbf{x}_{\mathrm{S}}) \Big)\, \mathrm{d}V,\end{aligned} \tag{10.3.14}$$

where we have used the reciprocity relation $\underline{\mathbf{u}}(\omega, \mathbf{x}_{\mathrm{R}}; \mathbf{x}_{\mathrm{S}}) = \underline{\mathbf{u}}^{\mathrm{T}}(\omega, \mathbf{x}_{\mathrm{S}}; \mathbf{x}_{\mathrm{R}})$ (4.5.19) (note we have arranged the analysis so reciprocity is used on the exact solution, when it always applies, not on the approximate solution, when it may not apply). The integrands contain the products of Green functions of vectors, i.e. each term is of dimension 3×3 as is the overall expression. In these expressions, the Green functions of vectors, e.g. displacement and pressure gradient, can be treated as matrices, where the 'Green' index is the second, i.e. column, index. Products, transposes, etc. of Green functions are obtained just as for normal matrices.

Integrating the second volume integral in expression (10.3.14) by parts, the derivatives of the displacement can be replaced by strain, and using results (10.3.2) and (10.3.9) we obtain

$$\boxed{\begin{aligned}&\underline{\mathbf{u}}(\omega,\mathbf{x}_\mathrm{R};\mathbf{x}_\mathrm{S}) = \underline{\mathbf{u}}^\mathrm{A}(\omega,\mathbf{x}_\mathrm{R};\mathbf{x}_\mathrm{S})\\ &\quad -\int_S \left(\underline{P}^\mathrm{T}(\omega,\mathbf{x};\mathbf{x}_\mathrm{R})\,\underline{\mathbf{u}}^\mathrm{A}(\omega,\mathbf{x};\mathbf{x}_\mathrm{S}) - \underline{\mathbf{u}}^\mathrm{T}(\omega,\mathbf{x};\mathbf{x}_\mathrm{R})\,\underline{P}^\mathrm{A}(\omega,\mathbf{x};\mathbf{x}_\mathrm{S})\right)\mathrm{d}\mathbf{S}\\ &\quad +\int_V \left(\underline{\mathbf{u}}^\mathrm{T}(\omega,\mathbf{x};\mathbf{x}_\mathrm{R})\,\underline{\boldsymbol{\mathcal{E}}}^\mathrm{N}(\omega,\mathbf{x};\mathbf{x}_\mathrm{S}) + \underline{P}^\mathrm{T}(\omega,\mathbf{x};\mathbf{x}_\mathrm{R})\,\underline{\mathcal{E}}^\mathrm{H}(\omega,\mathbf{x};\mathbf{x}_\mathrm{S})\right)\mathrm{d}V,\end{aligned}} \tag{10.3.15}$$

where S is the surface of the volume V and $\mathrm{d}\mathbf{S}$ is the outward normal. This is our fundamental result for the scattering integral.

It is convenient to summarize result (10.3.15) as

$$\underline{\mathbf{u}} = \underline{\mathbf{u}}^\mathrm{A} + \underline{\mathbf{u}}^\mathrm{S} + \underline{\mathbf{u}}^\mathrm{E} + \underline{\mathbf{u}}^\mathrm{B}, \tag{10.3.16}$$

where the surface integral is

$$\boxed{\begin{aligned}\underline{\mathbf{u}}^\mathrm{S}(\omega,\mathbf{x}_\mathrm{R};\mathbf{x}_\mathrm{S}) = \int_S \Big(&\underline{\mathbf{u}}^\mathrm{T}(\omega,\mathbf{x};\mathbf{x}_\mathrm{R})\,\underline{P}^\mathrm{A}(\omega,\mathbf{x};\mathbf{x}_\mathrm{S})\\ &- \underline{P}^\mathrm{T}(\omega,\mathbf{x};\mathbf{x}_\mathrm{R})\,\underline{\mathbf{u}}^\mathrm{A}(\omega,\mathbf{x};\mathbf{x}_\mathrm{S})\Big)\,\mathrm{d}\mathbf{S}\end{aligned}} \tag{10.3.17}$$

(cf. the representation theorem (4.5.27)), and the volume integral is made up of two parts

$$\boxed{\begin{aligned}\underline{\mathbf{u}}^\mathrm{E}(\omega,\mathbf{x}_\mathrm{R};\mathbf{x}_\mathrm{S}) = \int_V \Big(&\underline{\mathbf{u}}^\mathrm{T}(\omega,\mathbf{x};\mathbf{x}_\mathrm{R})\,\underline{\mathbf{F}}^\mathrm{A}(\omega,\mathbf{x};\mathbf{x}_\mathrm{S})\\ &+ \underline{P}^\mathrm{T}(\omega,\mathbf{x};\mathbf{x}_\mathrm{R})\,\underline{\Theta}^\mathrm{A}(\omega,\mathbf{x};\mathbf{x}_\mathrm{S})\Big)\,\mathrm{d}V,\end{aligned}} \tag{10.3.18}$$

due to errors in the approximate Green function, and

$$\boxed{\begin{aligned}\underline{\mathbf{u}}^\mathrm{B}(\omega,\mathbf{x}_\mathrm{R};\mathbf{x}_\mathrm{S}) = \int_V \Big(&\omega^2\underline{\mathbf{u}}^\mathrm{T}(\omega,\mathbf{x};\mathbf{x}_\mathrm{R})\,\rho^\mathrm{B}(\mathbf{x})\,\underline{\mathbf{u}}^\mathrm{A}(\omega,\mathbf{x};\mathbf{x}_\mathrm{S})\\ &+ \underline{P}^\mathrm{T}(\omega,\mathbf{x};\mathbf{x}_\mathrm{R})\,k^\mathrm{B}(\mathbf{x})\,\underline{P}^\mathrm{A}(\omega,\mathbf{x};\mathbf{x}_\mathrm{S})\Big)\,\mathrm{d}V,\end{aligned}} \tag{10.3.19}$$

due to model perturbations. The dimension of the Green functions $\underline{P}$ and $\underline{\Theta}$ is 1×3, making these expressions 3×3 overall. We will discuss the two parts of the volume scattering due to the approximation (10.3.18) and perturbation (10.3.19) in the next sections (Section 10.3.2 and 10.3.3, respectively).

It is straightforward to manipulate the two integrals (10.3.18) and (10.3.19) into various forms. One of interest is

$$\boxed{\begin{aligned}\underline{\mathbf{u}}^{\mathrm{E}'}(\omega, \mathbf{x}_\mathrm{R}; \mathbf{x}_\mathrm{S}) = \int_V \Big(&\underline{\mathbf{u}}^{\mathrm{T}}(\omega, \mathbf{x}; \mathbf{x}_\mathrm{R})\, \underline{\mathbf{F}}^{\mathrm{A}}(\omega, \mathbf{x}; \mathbf{x}_\mathrm{S}) \\ &+ \Big(\nabla \cdot \underline{\mathbf{u}}(\omega, \mathbf{x}; \mathbf{x}_\mathrm{R})\Big)^{\mathrm{T}}\, \underline{\Phi}^{\mathrm{A}}(\omega, \mathbf{x}; \mathbf{x}_\mathrm{S})\Big)\, \mathrm{d}V,\end{aligned}} \tag{10.3.20}$$

and

$$\boxed{\begin{aligned}\underline{\mathbf{u}}^{\mathrm{B}'}(\omega, \mathbf{x}_\mathrm{R}; \mathbf{x}_\mathrm{S}) = \int_V \Big(&\omega^2 \underline{\mathbf{u}}^{\mathrm{T}}(\omega, \mathbf{x}; \mathbf{x}_\mathrm{R})\, \rho^{\mathrm{B}}(\mathbf{x})\, \underline{\mathbf{u}}^{\mathrm{A}}(\omega, \mathbf{x}; \mathbf{x}_\mathrm{S}) \\ &- \big(\nabla \cdot \underline{\mathbf{u}}^{\mathrm{T}}(\omega, \mathbf{x}; \mathbf{x}_\mathrm{R})\big)^{\mathrm{T}}\, \kappa^{\mathrm{B}}(\mathbf{x})\, \nabla \cdot \underline{\mathbf{u}}^{\mathrm{A}}(\omega, \mathbf{x}; \mathbf{x}_\mathrm{S})\Big)\, \mathrm{d}V,\end{aligned}} \tag{10.3.21}$$

where

$$\underline{\mathbf{u}}^{\mathrm{E}} + \underline{\mathbf{u}}^{\mathrm{B}} = \underline{\mathbf{u}}^{\mathrm{E}'} + \underline{\mathbf{u}}^{\mathrm{B}'}. \tag{10.3.22}$$

This form (10.3.21) with the stiffness perturbations, κ^{B}, corresponds to the standard Born scattering method described by Hudson and Heritage (1981), Ben-Menahem and Gibson (1990) and Gibson and Ben-Menahem (1991).

It is also straightforward to rewrite all the above integrals in the time domain. The products in the integrands become time-domain convolutions (3.1.18). For brevity, we do not write these out.

Expression (10.3.15) is an exact equation. Unfortunately, the unknown solutions $\underline{\mathbf{u}}$ and $\underline{P}$ are contained in the integrals and normally the equations can only be solved iteratively. Let us assume that the surface integral $\underline{\mathbf{u}}^{\mathrm{S}}$ can be ignored. This can normally be achieved if the surface S is sufficiently distant from the source, the receiver and the ray paths between them. Alternatively if the Green function is known exactly outside V and on the surface S, the surface integral is zero. The solution can then be obtained as a series

$$\underline{\mathbf{u}}(\omega, \mathbf{x}_\mathrm{R}; \mathbf{x}_\mathrm{S}) = \sum_{j=0}^{\infty} \underline{\mathbf{u}}^{(j)}(\omega, \mathbf{x}_\mathrm{R}; \mathbf{x}_\mathrm{S}), \tag{10.3.23}$$

where the zeroth-order term is

$$\underline{\mathbf{u}}^{(0)}(\omega, \mathbf{x}_\mathrm{R}; \mathbf{x}_\mathrm{S}) = \underline{\mathbf{u}}^{\mathrm{A}}(\omega, \mathbf{x}_\mathrm{R}; \mathbf{x}_\mathrm{S}), \tag{10.3.24}$$

i.e. the approximate solution, and the higher-order iterations are

$$\begin{aligned}&\underline{\mathbf{u}}^{(j)}(\omega, \mathbf{x}_\mathrm{R}; \mathbf{x}_\mathrm{S}) \\ &= \int_V \Big(\underline{\mathbf{u}}^{(j-1)^{\mathrm{T}}}(\omega, \mathbf{x}; \mathbf{x}_\mathrm{R})\, \underline{\boldsymbol{\mathcal{E}}}^{\mathrm{N}}(\omega, \mathbf{x}; \mathbf{x}_\mathrm{S}) + \underline{P}^{(j-1)^{\mathrm{T}}}(\omega, \mathbf{x}; \mathbf{x}_\mathrm{R})\, \underline{\mathcal{E}}^{\mathrm{H}}(\omega, \mathbf{x}; \mathbf{x}_\mathrm{S})\Big)\, \mathrm{d}V.\end{aligned} \tag{10.3.25}$$

It is worth noting that although signals propagate from the source, $\mathbf{x}_S$, to the scattering point, $\mathbf{x}$, and onwards to the receiver, $\mathbf{x}_R$, the receiver Green functions in all the volume and surface integrals are from the receiver to the scattering point, i.e. have argument $(\omega, \mathbf{x}; \mathbf{x}_R)$. By reciprocity, these can be converted to the propagation direction, but in the theoretical development are in the reversed direction. The distinction can be important in defining the signs of polarization and slowness vectors consistently.

Each iteration can be divided into parts due to errors, $\underline{\mathbf{u}}^{\mathrm{E}}$, and due to perturbations, $\underline{\mathbf{u}}^{\mathrm{B}}$. The first-order term due to perturbations, $\underline{\mathbf{u}}^{(1)\,\mathrm{B}'}$, is the standard Born scattering integral. Scattering is due to perturbations B to the model A. In our development using the first-order equation of motion and the constitutive relation, the physical origin of the scattered signals is obvious. In equation (10.3.21), scattered signals are caused by density and bulk modulus perturbations. The scattering due to the density perturbation, ρ^{B}, arises from the 'error' in the equation of motion and is caused by the mismatch in the rate of change of momentum. The force provided by the pressure field in the reference model is inadequate to provide the rate of change of momentum in the true model. This mismatch acts as an effective force source in the direction of the polarization of the incident field. The radiation pattern for a point scatterer can be described using the point source results in Section 4.5.5. The scattering due to the bulk modulus perturbation, κ^{B}, arises from 'errors' in the constitutive relation which act as a dilatation source, or in the compressibility scatterer, k^{B}, which act as a pressure source. The radiation pattern for a point pressure source has been given in Section 4.6.2.

The important generalization of Born scattering theory is the integral $\underline{\mathbf{u}}^{\mathrm{E}}$ or $\underline{\mathbf{u}}^{\mathrm{E}'}$, which represents 'scattering' due to errors in the approximate Green functions. The magnitude of this term depends on the reference model A. If only the standard Born scattering term is considered, $\underline{\mathbf{u}}^{\mathrm{B}}$ or $\underline{\mathbf{u}}^{\mathrm{B}'}$, then the reference, background model must be chosen to be smooth and simple enough that the Green functions can be taken as exact. Including the error integrals, this requirement is no longer necessary. For forward modelling, the division into models A and B is arbitrary. At one extreme, we take $\rho^{\mathrm{B}} = 0$ and $k^{\mathrm{B}} = 0$ and obtain the scattered signals from errors in the approximate Green functions. At the other, we take ρ^{A} and k^{A} to be homogeneous so the Green functions are known exactly. The integral equation (10.3.15) is still exact but usually the iterative solution converges slowly. Although the division into models A and B is arbitrary, the rate of convergence of the iterative series (10.3.23) will depend on this division (and, of course, on the approximate Green functions chosen).

To emphasize the distinction between the two volume scattering terms in expressions (10.3.16), (10.3.18) and (10.3.19), and the generalization of Born scattering theory, we refer to the terms as *error* (10.3.18) and *perturbation* (10.3.19)

Born scattering. Before we investigate these terms in more detail, we develop the equivalent elastic scattering theory.

10.3.1.2 Elastic scattering

The development of Born scattering theory for elastic, isotropic or anisotropic media follows the same procedure and we just summarize the results and the differences. The equation of motion (4.5.45) is rewritten

$$\frac{\partial \underline{\mathbf{t}}_j}{\partial x_j} = -\omega^2 \rho\, \underline{\mathbf{u}} - \mathbf{I}\,\delta(\mathbf{x} - \mathbf{x}_1), \tag{10.3.26}$$

for the Green function with source at $\mathbf{x}_1$ (the argument $(\omega, \mathbf{x}; \mathbf{x}_1)$ is understood). The form of the constitutive relation with the compliances is used, i.e. equation (4.4.41)

$$\underline{\mathbf{e}}_j = \mathbf{s}_{jk} \underline{\mathbf{t}}_k. \tag{10.3.27}$$

The model is divided into two parts with equation (10.3.4) and

$$\mathbf{s}_{jk}(\mathbf{x}) = \mathbf{s}^{\mathrm{A}}_{jk}(\mathbf{x}) + \mathbf{s}^{\mathrm{B}}_{jk}(\mathbf{x}). \tag{10.3.28}$$

The approximate Green functions for the model A are denoted by $\underline{\mathbf{u}}^{\mathrm{A}}$ and $\underline{\mathbf{t}}^{\mathrm{A}}_j$. These do not satisfy the equation of motion (10.3.26) nor the constitutive relation (10.3.27) exactly, but can be used to define *error* terms

$$\underline{\mathbf{F}}^{\mathrm{A}} = \omega^2 \rho^{\mathrm{A}} \underline{\mathbf{u}}^{\mathrm{A}} + \frac{\partial \underline{\mathbf{t}}^{\mathrm{A}}_j}{\partial x_j} + \mathbf{I}\,\delta(\mathbf{x} - \mathbf{x}_2) \tag{10.3.29}$$

$$\underline{\mathbf{E}}^{\mathrm{A}}_j = \underline{\mathbf{e}}^{\mathrm{A}}_j - \mathbf{s}^{\mathrm{A}}_{jk} \underline{\mathbf{t}}^{\mathrm{A}}_k \tag{10.3.30}$$

$$\underline{\mathbf{T}}^{\mathrm{A}}_j = \underline{\mathbf{t}}^{\mathrm{A}}_j - \mathbf{c}^{\mathrm{A}}_{jk} \underline{\mathbf{e}}^{\mathrm{A}}_k, \tag{10.3.31}$$

for a source at $\mathbf{x}_2$ (and argument $(\omega, \mathbf{x}; \mathbf{x}_2)$ is understood throughout). $\underline{\mathbf{F}}^{\mathrm{A}}$ is a *force* error, $\underline{\mathbf{E}}^{\mathrm{A}}_j$ a *strain* error and $\underline{\mathbf{T}}^{\mathrm{A}}_j = -\mathbf{c}^{\mathrm{A}}_{jk} \underline{\mathbf{E}}^{\mathrm{A}}_k$ an equivalent *stress* error. If the Green functions in model A were exact, $\underline{\mathbf{F}}^{\mathrm{A}}$, $\underline{\mathbf{E}}^{\mathrm{A}}_j$ and $\underline{\mathbf{T}}^{\mathrm{A}}_j$ would be zero.

Equations (10.3.29) and (10.3.30) can be rewritten as

$$\frac{\partial \underline{\mathbf{t}}^{\mathrm{A}}_j}{\partial x_j} = -\omega^2 \rho\, \underline{\mathbf{u}}^{\mathrm{A}} - \mathbf{I}\,\delta(\mathbf{x} - \mathbf{x}_2) + \underline{\boldsymbol{\mathcal{E}}}^{\mathrm{N}} \tag{10.3.32}$$

$$\underline{\mathbf{e}}^{\mathrm{A}}_j = \mathbf{s}_{jk} \underline{\mathbf{t}}^{\mathrm{A}}_k + \underline{\boldsymbol{\mathcal{E}}}^{\mathrm{H}}_j. \tag{10.3.33}$$

The complete scatterer terms are

$$\boxed{\underline{\boldsymbol{\mathcal{E}}}^{\mathrm{N}} = \underline{\mathbf{F}}^{\mathrm{A}} + \omega^2 \rho^{\mathrm{B}} \underline{\mathbf{u}}^{\mathrm{A}}} \tag{10.3.34}$$

$$\boxed{\underline{\boldsymbol{\mathcal{E}}}_j^{\mathrm{H}} = \underline{\mathbf{E}}_j^{\mathrm{A}} - \mathbf{s}_{jk}^{\mathrm{B}} \underline{\mathbf{t}}_k^{\mathrm{A}},} \tag{10.3.35}$$

with a similar notation as before (except $\boldsymbol{\mathcal{E}}_j^{\mathrm{H}}$ are vectors).

Again we combine equations (10.3.26), (10.3.27), (10.3.32) and (10.3.33) to obtain

$$\begin{aligned} \underline{\mathbf{u}}(\omega, \mathbf{x}_{\mathrm{R}}; \mathbf{x}_{\mathrm{S}}) &= \underline{\mathbf{u}}^{\mathrm{A}}(\omega, \mathbf{x}_{\mathrm{R}}; \mathbf{x}_{\mathrm{S}}) + \int_V \underline{\mathbf{u}}^{\mathrm{T}}(\omega, \mathbf{x}; \mathbf{x}_{\mathrm{R}}) \, \underline{\boldsymbol{\mathcal{E}}}^{\mathrm{N}}(\omega, \mathbf{x}; \mathbf{x}_{\mathrm{S}}) \, \mathrm{d}V \\ &+ \int_V \left(\frac{\partial \underline{\mathbf{t}}_j^{\mathrm{T}}(\omega, \mathbf{x}; \mathbf{x}_{\mathrm{R}})}{\partial x_j} \underline{\mathbf{u}}^{\mathrm{A}}(\omega, \mathbf{x}; \mathbf{x}_{\mathrm{S}}) - \underline{\mathbf{u}}^{\mathrm{T}}(\omega, \mathbf{x}; \mathbf{x}_{\mathrm{R}}) \frac{\partial \underline{\mathbf{t}}_j^{\mathrm{A}}(\omega, \mathbf{x}; \mathbf{x}_{\mathrm{S}})}{\partial x_j} \right) \mathrm{d}V, \end{aligned} \tag{10.3.36}$$

where we have used the reciprocity relation $\underline{\mathbf{u}}(\omega, \mathbf{x}_{\mathrm{R}}; \mathbf{x}_{\mathrm{S}}) = \underline{\mathbf{u}}^{\mathrm{T}}(\omega, \mathbf{x}_{\mathrm{S}}; \mathbf{x}_{\mathrm{R}})$ (4.5.42).

Integrating the second volume integral in expression (10.3.36) by parts, the derivatives of displacement can be replaced by strain, and using equations (10.3.27) and (10.3.33) we obtain

$$\boxed{\begin{aligned} \underline{\mathbf{u}}(\omega, \mathbf{x}_{\mathrm{R}}; \mathbf{x}_{\mathrm{S}}) &= \underline{\mathbf{u}}^{\mathrm{A}}(\omega, \mathbf{x}_{\mathrm{R}}; \mathbf{x}_{\mathrm{S}}) \\ &+ \int_S \left(\underline{\mathbf{t}}_j^{\mathrm{T}}(\omega, \mathbf{x}; \mathbf{x}_{\mathrm{R}}) \, \underline{\mathbf{u}}^{\mathrm{A}}(\omega, \mathbf{x}; \mathbf{x}_{\mathrm{S}}) - \underline{\mathbf{u}}^{\mathrm{T}}(\omega, \mathbf{x}; \mathbf{x}_{\mathrm{R}}) \, \underline{\mathbf{t}}_j^{\mathrm{A}}(\omega, \mathbf{x}; \mathbf{x}_{\mathrm{S}}) \right) \mathrm{d}S_j \\ &+ \int_V \left(\underline{\mathbf{u}}^{\mathrm{T}}(\omega, \mathbf{x}; \mathbf{x}_{\mathrm{R}}) \, \underline{\boldsymbol{\mathcal{E}}}^{\mathrm{N}}(\omega, \mathbf{x}; \mathbf{x}_{\mathrm{S}}) - \underline{\mathbf{t}}_k^{\mathrm{T}}(\omega, \mathbf{x}; \mathbf{x}_{\mathrm{R}}) \, \underline{\boldsymbol{\mathcal{E}}}_k^{\mathrm{H}}(\omega, \mathbf{x}; \mathbf{x}_{\mathrm{S}}) \right) \mathrm{d}V, \end{aligned}} \tag{10.3.37}$$

where S is the surface of the volume V and $\mathrm{d}\mathbf{S}$ is the outward normal. This is our fundamental result for the elastic scattering integral.

Again it is convenient to summarize result (10.3.37) using the expansion (10.3.16). The surface integral is

$$\boxed{\begin{aligned} \underline{\mathbf{u}}^{\mathrm{S}}(\omega, \mathbf{x}_{\mathrm{R}}; \mathbf{x}_{\mathrm{S}}) = \int_S \Big(& \underline{\mathbf{t}}_j^{\mathrm{T}}(\omega, \mathbf{x}; \mathbf{x}_{\mathrm{R}}) \, \underline{\mathbf{u}}^{\mathrm{A}}(\omega, \mathbf{x}; \mathbf{x}_{\mathrm{S}}) \\ & - \underline{\mathbf{u}}^{\mathrm{T}}(\omega, \mathbf{x}; \mathbf{x}_{\mathrm{R}}) \, \underline{\mathbf{t}}_j^{\mathrm{A}}(\omega, \mathbf{x}; \mathbf{x}_{\mathrm{S}}) \Big) \mathrm{d}S_j \end{aligned}} \tag{10.3.38}$$

(cf. the representation theorem (4.5.49)), and the volume integral is made up of two parts

$$\boxed{\begin{aligned}\underline{\mathbf{u}}^{\mathrm{E}}(\omega,\mathbf{x}_{\mathrm{R}};\mathbf{x}_{\mathrm{S}}) = \int_V \Big(&\underline{\mathbf{u}}^{\mathrm{T}}(\omega,\mathbf{x};\mathbf{x}_{\mathrm{R}})\,\underline{\mathbf{F}}^{\mathrm{A}}(\omega,\mathbf{x};\mathbf{x}_{\mathrm{S}}) \\ &-\underline{\mathbf{t}}_k^{\mathrm{T}}(\omega,\mathbf{x};\mathbf{x}_{\mathrm{R}})\,\underline{\mathbf{E}}_k^{\mathrm{A}}(\omega,\mathbf{x};\mathbf{x}_{\mathrm{S}})\Big)\,\mathrm{d}V,\end{aligned}} \tag{10.3.39}$$

due to errors in the approximate Green function, and

$$\boxed{\begin{aligned}\underline{\mathbf{u}}^{\mathrm{B}}(\omega,\mathbf{x}_{\mathrm{R}};\mathbf{x}_{\mathrm{S}}) = \int_V \Big(&\omega^2\underline{\mathbf{u}}^{\mathrm{T}}(\omega,\mathbf{x};\mathbf{x}_{\mathrm{R}})\,\rho^{\mathrm{B}}(\mathbf{x})\,\underline{\mathbf{u}}^{\mathrm{A}}(\omega,\mathbf{x};\mathbf{x}_{\mathrm{S}}) \\ &+\underline{\mathbf{t}}_k^{\mathrm{T}}(\omega,\mathbf{x};\mathbf{x}_{\mathrm{R}})\,\mathbf{s}_{kj}^{\mathrm{B}}(\mathbf{x})\,\underline{\mathbf{t}}_j^{\mathrm{A}}(\omega,\mathbf{x};\mathbf{x}_{\mathrm{S}})\Big)\,\mathrm{d}V,\end{aligned}} \tag{10.3.40}$$

due to model perturbations.

It is straightforward to manipulate the two volume integrals (10.3.39) and (10.3.40) into various forms. One of interest is

$$\boxed{\begin{aligned}\underline{\mathbf{u}}^{\mathrm{E}'}(\omega,\mathbf{x}_{\mathrm{R}};\mathbf{x}_{\mathrm{S}}) = \int_V \Big(&\underline{\mathbf{u}}^{\mathrm{T}}(\omega,\mathbf{x};\mathbf{x}_{\mathrm{R}})\,\underline{\mathbf{F}}^{\mathrm{A}}(\omega,\mathbf{x};\mathbf{x}_{\mathrm{S}}) \\ &-\underline{\mathbf{e}}_k^{\mathrm{T}}(\omega,\mathbf{x};\mathbf{x}_{\mathrm{R}})\,\underline{\mathbf{T}}_k^{\mathrm{A}}(\omega,\mathbf{x};\mathbf{x}_{\mathrm{S}})\Big)\,\mathrm{d}V,\end{aligned}} \tag{10.3.41}$$

and

$$\boxed{\begin{aligned}\underline{\mathbf{u}}^{\mathrm{B}'}(\omega,\mathbf{x}_{\mathrm{R}};\mathbf{x}_{\mathrm{S}}) = \int_V \Big(&\omega^2\underline{\mathbf{u}}^{\mathrm{T}}(\omega,\mathbf{x};\mathbf{x}_{\mathrm{R}})\,\rho^{\mathrm{B}}(\mathbf{x})\,\underline{\mathbf{u}}^{\mathrm{A}}(\omega,\mathbf{x};\mathbf{x}_{\mathrm{S}}) \\ &-\frac{\partial\underline{\mathbf{u}}^{\mathrm{T}}(\omega,\mathbf{x};\mathbf{x}_{\mathrm{R}})}{\partial x_k}\,\mathbf{c}_{kj}^{\mathrm{B}}(\mathbf{x})\,\frac{\partial\underline{\mathbf{u}}^{\mathrm{A}}(\omega,\mathbf{x};\mathbf{x}_{\mathrm{S}})}{\partial x_j}\Big)\,\mathrm{d}V,\end{aligned}} \tag{10.3.42}$$

where again expression (10.3.22) applies, and $\underline{\mathbf{T}}_k^{\mathrm{A}}$ was defined in equation (10.3.31). This form (10.3.42) with the stiffness perturbations $\mathbf{c}_{jk}^{\mathrm{B}}$ corresponds to the standard Born scattering method (Hudson and Heritage, 1981; Ben-Menahem and Gibson, 1990; and Gibson and Ben-Menahem, 1991).

Again these results are easily rewritten in the time domain with products being replaced by convolutions (3.1.18). For brevity, we do not write these out.

As in the acoustic case, the exact solution (10.3.37) is written as an iterative series (10.3.23), where the zeroth-order term is the approximation solution (10.3.24) and the first-order term is the Born approximation. Higher-order iterations are given by

$$\underline{\mathbf{u}}^{(j)}(\omega, \mathbf{x}_\mathrm{R}; \mathbf{x}_\mathrm{S}) = \int_V \Big(\underline{\mathbf{u}}^{(j-1)^\mathrm{T}}(\omega, \mathbf{x}; \mathbf{x}_\mathrm{R})\, \underline{\boldsymbol{\mathcal{E}}}^\mathrm{N}(\omega, \mathbf{x}; \mathbf{x}_\mathrm{S}) - \underline{\mathbf{t}}_k^{(j-1)^\mathrm{T}}(\omega, \mathbf{x}; \mathbf{x}_\mathrm{R})\, \underline{\boldsymbol{\mathcal{E}}}_k^\mathrm{H}(\omega, \mathbf{x}; \mathbf{x}_\mathrm{S}) \Big)\, \mathrm{d}V. \quad (10.3.43)$$

The physical interpretation of the scattered terms is similar to the acoustic case. A density perturbation, ρ^B, effectively introduces a force source due to the mismatch in the rate of change of momentum. The compliance scatterer, $\mathbf{s}_{jk}^\mathrm{B}$, arises from 'errors' in the constitutive relation which acts as a strain source, or the stiffness scatterer, $\mathbf{c}_{jk}^\mathrm{B}$, which acts as a stress source. The radiation patterns for point stress sources have been given in Section 4.6.2. These have been described in detail by Ben-Menahem and Gibson (1990) and Gibson and Ben-Menahem (1991).

10.3.2 Error Born scattering using ray theory

First we investigate the error Born scattering caused by approximations in the Green functions, e.g. expressions (10.3.18), (10.3.20), (10.3.39) or (10.3.41). Alternative expressions exist as the total response also contains the corresponding perturbation scattering term. We only consider the first iteration in the scattering series (10.3.23), i.e. $j = 1$ in expressions (10.3.25) or (10.3.43). As mentioned above, the results can be used in two or three dimensions. The three-dimensional results given can be interpreted as two-dimensional provided the volume integrals are replaced by area integrals, and the surface integrals by line integrals. The differences in the ray approximations for the Green functions are contained in the source spectral term $f^{(\ell)}(\omega)$, (5.2.64) and (5.2.78)

$$f^{(2)}(\omega) = -\frac{\mathrm{i}\omega\lambda(\omega)}{2^{1/2}\pi} \quad (10.3.44)$$

$$f^{(3)}(\omega) = -\frac{\mathrm{i}\omega}{2\pi}, \quad (10.3.45)$$

which are used throughout this section and the next (Section 10.4). Remember as well that the geometrical spreading is different in two dimensions from three, as true two-dimensional Green functions are required not 2.5D functions. If 2.5D wave propagation is required, the conversion is applied after the Born integral.

10.3.2.1 Acoustic error scattering

In order to evaluate the scattered signals due to error Born scattering, we need an approximate Green function. The results in Section 10.3.1 apply whatever approximation is used, be it analytic or numerical. An ideal approximation, as it is easily evaluated from analytic expressions, is the zeroth-order term in asymptotic ray theory, (5.4.28) and (5.4.31). For simplicity, we ignore the possibility of multi-pathing in the Green functions – it can be included by summations over the multiple rays. The error terms in equation (10.3.18) are defined by expressions (10.3.7) and (10.3.8) and lead to

$$\underline{\mathbf{F}}^{\mathrm{A}}(\omega, \mathbf{x}; \mathbf{x}_{\mathrm{S}}) = -f^{(\ell)}(\omega)\, \Lambda(\omega, \mathbf{x}, \mathcal{L}_{\mathrm{S}})\, \nabla \underline{P}^{(0)}(\mathbf{x}, \mathcal{L}_{\mathrm{S}}) \tag{10.3.46}$$

$$\underline{\Theta}^{\mathrm{A}}(\omega, \mathbf{x}; \mathbf{x}_{\mathrm{S}}) = -\frac{1}{\mathrm{i}\omega} f^{(\ell)}(\omega)\, \Lambda(\omega, \mathbf{x}, \mathcal{L}_{\mathrm{S}})\, \nabla \cdot \underline{\mathbf{v}}^{(0)}(\mathbf{x}, \mathcal{L}_{\mathrm{S}}), \tag{10.3.47}$$

where we have used equations (5.2.1) and (5.2.2) to cancel the leading terms. Substituting these expressions in integral (10.3.18) for the first iteration ($j = 1$ in integral (10.3.25)), we obtain

$$\begin{aligned}\underline{\mathbf{u}}^{\mathrm{E}}(\omega, \mathbf{x}_{\mathrm{R}}; \mathbf{x}_{\mathrm{S}}) = \frac{\left(f^{(\ell)}(\omega)\right)^2}{\mathrm{i}\omega} \int_V \Big(&\underline{\mathbf{v}}^{(0)\,\mathrm{T}}(\mathbf{x}, \mathcal{L}_{\mathrm{R}})\, \nabla \underline{P}^{(0)}(\mathbf{x}, \mathcal{L}_{\mathrm{S}}) \\ &- \underline{P}^{(0)\,\mathrm{T}}(\mathbf{x}, \mathcal{L}_{\mathrm{R}})\, \nabla \cdot \underline{\mathbf{v}}^{(0)}(\mathbf{x}, \mathcal{L}_{\mathrm{S}})\Big)\, \mathrm{e}^{\mathrm{i}\omega\,(T(\mathbf{x},\mathcal{L}_{\mathrm{R}})+T(\mathbf{x},\mathcal{L}_{\mathrm{S}}))}\mathrm{d}V.\end{aligned} \tag{10.3.48}$$

The form of this integral is extremely straightforward. The frequency dependence is contained just in the phase with the travel time to and from the scattering point. Let us denote this as

$$\widetilde{T}(\mathbf{x}, \mathcal{L}_{\mathrm{R}}, \mathcal{L}_{\mathrm{S}}) = T(\mathbf{x}, \mathcal{L}_{\mathrm{R}}) + T(\mathbf{x}, \mathcal{L}_{\mathrm{S}}). \tag{10.3.49}$$

The amplitude is independent of frequency and is due to spatial derivatives of the amplitude coefficients. Simple though it is, the result is not obviously reciprocal, although this might be expected as the ray approximation is reciprocal. However, simply integrating by parts establishes a reciprocal result. For brevity, we indicate the arguments just by a subscript so the integral in equation (10.3.48) becomes

$$\begin{aligned}&\int_V \left(\underline{\mathbf{v}}_{\mathrm{R}}^{(0)\,\mathrm{T}}\, \nabla \underline{P}_{\mathrm{S}}^{(0)} - \underline{P}_{\mathrm{R}}^{(0)\,\mathrm{T}}\, \nabla \cdot \underline{\mathbf{v}}_{\mathrm{S}}^{(0)}\right) \mathrm{e}^{\mathrm{i}\omega\widetilde{T}}\, \mathrm{d}V \\ &= \int_S \left(\underline{\mathbf{v}}_{\mathrm{R}}^{(0)\,\mathrm{T}}\, \hat{\mathbf{n}}\, \underline{P}_{\mathrm{S}}^{(0)} - \underline{P}_{\mathrm{R}}^{(0)\,\mathrm{T}}\, \hat{\mathbf{n}}^{\mathrm{T}}\, \underline{\mathbf{v}}_{\mathrm{S}}^{(0)}\right) \mathrm{e}^{\mathrm{i}\omega\widetilde{T}}\, \mathrm{d}S \\ &\quad - \mathrm{i}\omega \int_V \left(\underline{\mathbf{v}}_{\mathrm{R}}^{(0)\,\mathrm{T}}\, (\mathbf{p}_{\mathrm{R}} + \mathbf{p}_{\mathrm{S}}) \underline{P}_{\mathrm{S}}^{(0)} - \underline{P}_{\mathrm{R}}^{(0)\,\mathrm{T}}\, (\mathbf{p}_{\mathrm{R}} + \mathbf{p}_{\mathrm{S}})^{\mathrm{T}}\, \underline{\mathbf{v}}_{\mathrm{S}}^{(0)}\right) \mathrm{e}^{\mathrm{i}\omega\widetilde{T}}\, \mathrm{d}V \\ &\quad - \mathrm{i}\omega \int_V \left(\left(\nabla \cdot \underline{\mathbf{v}}_{\mathrm{R}}^{(0)}\right)^{\mathrm{T}} \underline{P}_{\mathrm{S}}^{(0)} - \left(\nabla \underline{P}_{\mathrm{R}}^{(0)}\right)^{\mathrm{T}} \underline{\mathbf{v}}_{\mathrm{S}}^{(0)}\right) \mathrm{e}^{\mathrm{i}\omega\widetilde{T}}\, \mathrm{d}V.\end{aligned} \tag{10.3.50}$$

The first integral over the surface S is combined with $\underline{\mathbf{u}}^{\mathrm{S}}$, (10.3.17), and neglected as before. If in the second integral the pressure is replaced using expression (5.2.2), it reduces to zero. Combining the final volume integral with the original, we can replace result (10.3.48) with

$$\underline{\mathbf{u}}^{\mathrm{E}}(\omega, \mathbf{x}_{\mathrm{R}}; \mathbf{x}_{\mathrm{S}}) = -\frac{\left(f^{(\ell)}(\omega)\right)^2}{\mathrm{i}\omega} \int_V \underline{\mathcal{K}}^{\mathrm{E}}(\mathbf{x}, \mathcal{L}_{\mathrm{R}}, \mathcal{L}_{\mathrm{S}})\, \mathrm{e}^{\mathrm{i}\omega \widetilde{T}(\mathbf{x}, \mathcal{L}_{\mathrm{R}}, \mathcal{L}_{\mathrm{S}})}\, \mathrm{d}V, \qquad (10.3.51)$$

where

$$\begin{aligned}
&\underline{\mathcal{K}}^{\mathrm{E}}(\mathbf{x}, \mathcal{L}_{\mathrm{R}}, \mathcal{L}_{\mathrm{S}}) \\
&\quad = -\frac{1}{2}\Big(\underline{\mathbf{v}}^{(0)\,\mathrm{T}}(\mathbf{x}, \mathcal{L}_{\mathrm{R}})\, \nabla \underline{P}^{(0)}(\mathbf{x}, \mathcal{L}_{\mathrm{S}}) - \underline{P}^{(0)\,\mathrm{T}}(\mathbf{x}, \mathcal{L}_{\mathrm{R}})\, \nabla \cdot \underline{\mathbf{v}}^{(0)}(\mathbf{x}, \mathcal{L}_{\mathrm{S}}) \\
&\qquad + \left(\nabla \underline{P}^{(0)}(\mathbf{x}, \mathcal{L}_{\mathrm{R}})\right)^{\mathrm{T}} \underline{\mathbf{v}}^{(0)}(\mathbf{x}, \mathcal{L}_{\mathrm{S}}) - \left(\nabla \cdot \underline{\mathbf{v}}^{(0)}(\mathbf{x}, \mathcal{L}_{\mathrm{R}})\right)^{\mathrm{T}} \underline{P}^{(0)}(\mathbf{x}, \mathcal{L}_{\mathrm{S}})\Big).
\end{aligned} \qquad (10.3.52)$$

This error scattering function, $\underline{\mathcal{K}}^{\mathrm{E}}$, is a 3×3 matrix, the kernel for the Born error scattering integral (10.3.51) and has a reciprocal form.

With the ray approximation (5.4.31), it is straightforward if tedious to evaluate all the terms in result (10.3.52). Spatial derivatives of four types exist: derivatives of the polarizations at the scattering point; derivatives of the impedance at the scattering point; derivatives of the geometrical propagation term; and derivatives of the polarizations at the source and receiver. The latter are normally less significant as the ray directions at the source and receiver do not vary much as the scattering point varies. For brevity, we omit these terms. However, geometrical ray theory breaks down on nodes of the source radiation pattern and the receiver conversion coefficients, and derivatives of the source and receiver polarizations would then be significant and correct these errors.

Neglecting the derivatives of the source and receiver polarizations, we can rewrite result (10.3.52) as

$$\underline{\mathcal{K}}^{\mathrm{E}}(\mathbf{x}, \mathcal{L}_{\mathrm{R}}, \mathcal{L}_{\mathrm{S}}) = \Gamma^{\mathrm{E}}(\mathbf{x}, \mathcal{L}_{\mathrm{R}}, \mathcal{L}_{\mathrm{S}})\, \underline{\mathcal{D}}^{(\ell)}(\mathbf{x}, \mathcal{L}_{\mathrm{R}}, \mathcal{L}_{\mathrm{S}}), \qquad (10.3.53)$$

where

$$\underline{\mathcal{D}}^{(\ell)}(\mathbf{x}, \mathcal{L}_{\mathrm{R}}, \mathcal{L}_{\mathrm{S}}) = -\mathbf{g}(\mathbf{x}_{\mathrm{R}}, \mathcal{L}_{\mathrm{R}})\, \mathcal{T}^{(\ell)}(\mathbf{x}, \mathcal{L}_{\mathrm{R}})\, \mathcal{T}^{(\ell)}(\mathbf{x}, \mathcal{L}_{\mathrm{S}})\, \mathbf{g}^{\mathrm{T}}(\mathbf{x}_{\mathrm{S}}, \mathcal{L}_{\mathrm{S}}), \qquad (10.3.54)$$

and

$$\begin{aligned}
\Gamma^{\mathrm{E}}(\mathbf{x}, \mathcal{L}_{\mathrm{R}}, \mathcal{L}_{\mathrm{S}}) &= \frac{1}{4}\left(\nabla\left(\ln Z(\mathbf{x})\right) - \nabla\right) \cdot \left(\hat{\mathbf{g}}(\mathbf{x}, \mathcal{L}_{\mathrm{R}}) + \hat{\mathbf{g}}(\mathbf{x}, \mathcal{L}_{\mathrm{S}})\right) \\
&\quad - \frac{1}{4}\nabla\left(\ln \frac{\mathcal{T}^{(\ell)}(\mathbf{x}, \mathcal{L}_{\mathrm{R}})}{\mathcal{T}^{(\ell)}(\mathbf{x}, \mathcal{L}_{\mathrm{S}})}\right) \cdot \left(\hat{\mathbf{g}}(\mathbf{x}, \mathcal{L}_{\mathrm{R}}) - \hat{\mathbf{g}}(\mathbf{x}, \mathcal{L}_{\mathrm{S}})\right),
\end{aligned} \qquad (10.3.55)$$

and $\mathcal{T}^{(\ell)}(\mathbf{x}, \mathcal{L})$ is the geometrical part of the ray propagation (5.4.34). We refer to $\underline{\boldsymbol{\mathcal{D}}}^{(\ell)}$ as the *scattering dyadic*, and Γ^{E} as the *scalar Born error scattering term*. Note that the first term in definition (10.3.55) includes the gradient of the impedance, ∇Z, and the second term is the divergence of the sum of the polarizations. The divergence of the acoustic polarization can be obtained from result (5.6.11). Normally the final term from the gradient of the propagation terms will be less significant than the first two terms. For zero-offset seismograms it will be exactly zero. We have introduced the negative sign in definition (10.3.54) (and the compensating sign in definition (10.3.55)) as $\mathbf{g}(\mathbf{x}_{\mathsf{R}}, \mathcal{L}_{\mathsf{R}})$ is for the reversed ray from $\mathbf{x}_{\mathsf{R}}$ to $\mathbf{x}$; $-\mathbf{g}(\mathbf{x}_{\mathsf{R}}, \mathcal{L}_{\mathsf{R}})$ will be appropriate for the ray from $\mathbf{x}$ to $\mathbf{x}_{\mathsf{R}}$ and so will be compatible with the propagation from $\mathbf{x}_{\mathsf{S}}$ to $\mathbf{x}_{\mathsf{R}}$. The scalar scattering term, Γ^{E}, due to the approximate ray Green function, represents the local scattering at the scattering point, $\mathbf{x}$. The source, receiver and propagation terms have been factored out in expression (10.3.54).

To reduce this result (10.3.51) to the time domain, we need to specify the dimensionality.

10.3.2.2 Two-dimensional error scattering

In two dimensions, $f^{(2)}(\omega)$ is given by expression (5.2.78; 10.3.44) and the volume integral (10.3.51), actually an area integral, is simply

$$\underline{\mathbf{u}}^{\mathrm{E}}(t, \mathbf{x}_{\mathsf{R}}; \mathbf{x}_{\mathsf{S}}) = \frac{1}{2\pi} \int_V \mathrm{Re}\Big(\underline{\boldsymbol{\mathcal{K}}}^{\mathrm{E}}(\mathbf{x}, \mathcal{L}_{\mathsf{R}}, \mathcal{L}_{\mathsf{S}})\, \Delta\left(t - \widetilde{T}(\mathbf{x}, \mathcal{L}_{\mathsf{R}}, \mathcal{L}_{\mathsf{S}})\right)\Big)\, \mathrm{d}V. \tag{10.3.56}$$

The physical significance of expressions (10.3.51) and (10.3.56) is obvious and is illustrated in Figure 10.22. The volume integral models signals scattered by volume elements. The arrival time of these signals (10.3.49) is given by the geometrical travel time from the source to the scatterer and from the scatterer to the receiver. The strength of the scatterer is given by the expression (10.3.53) which can be derived from the geometrical properties of the ray approximation.

For a fixed time, t, the delta function is singular on isochron lines defined by

$$\widetilde{T}(\mathbf{x}, \mathcal{L}_{\mathsf{R}}, \mathcal{L}_{\mathsf{S}}) = t. \tag{10.3.57}$$

The volume (area) integral (10.3.56) can be reduced to a surface (line) integral

$$\boxed{\underline{\mathbf{u}}^{\mathrm{E}}(t, \mathbf{x}_{\mathsf{R}}; \mathbf{x}_{\mathsf{S}}) = \frac{1}{2\pi} \mathrm{Re}\left(\Delta(t) * \int_{\widetilde{T}=t} \frac{\underline{\boldsymbol{\mathcal{K}}}^{\mathrm{E}}(\mathbf{x}, \mathcal{L}_{\mathsf{R}}, \mathcal{L}_{\mathsf{S}})}{\left|\nabla \widetilde{T}(\mathbf{x}, \mathcal{L}_{\mathsf{R}}, \mathcal{L}_{\mathsf{S}})\right|}\, \mathrm{d}S \right),} \tag{10.3.58}$$

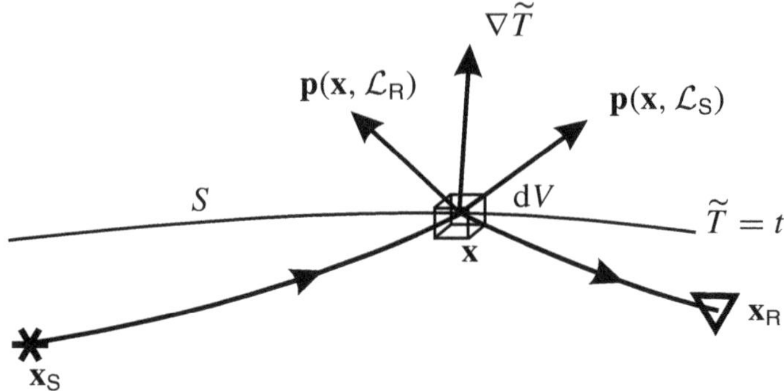

Fig. 10.22. Rays from a source, $\mathbf{x}_\mathsf{S}$, to a scattering point, $\mathbf{x}$, and to a receiver, $\mathbf{x}_\mathsf{R}$. Integral (10.3.58) is over the isochron line (in two dimensions), or integral (10.3.61) is over the isochron surface (in three dimensions), defined by equation (10.3.57). Note that $\mathbf{p}(\mathbf{x}, \mathcal{L}_\mathsf{R})$, and the acoustic polarization $\hat{\mathbf{g}}(\mathbf{x}, \mathcal{L}_\mathsf{R})$ in the same direction, is for the ray from $\mathbf{x}_\mathsf{R}$ to $\mathbf{x}$.

where

$$\nabla\widetilde{T}(\mathbf{x}, \mathcal{L}_\mathsf{R}, \mathcal{L}_\mathsf{S}) = \mathbf{p}(\mathbf{x}, \mathcal{L}_\mathsf{R}) + \mathbf{p}(\mathbf{x}, \mathcal{L}_\mathsf{S}), \tag{10.3.59}$$

which is necessarily normal to the isochron surface (Figure 10.22).

It is now obvious why in many circumstances we can neglect the surface integrals. Signals scattered near the ray path arrive with times close to the geometrical time. For a fixed time, t, signals may be scattered from volume elements on a surface (line) defined by $\widetilde{T}(\mathbf{x}, \mathcal{L}_\mathsf{R}, \mathcal{L}_\mathsf{S}) = t$ (10.3.57). As the volume expands away from the geometrical ray, the scattered signals arrive later (the only exception to this is for geometrical rays with maximum travel times – then the arrival time may initially decrease but as the scattering points tend to infinity, the time must eventually increase). Provided we are only interested in a finite time window less than the arrival time of reflections from the surface, we can neglect the surface integrals.

10.3.2.3 Three-dimensional error scattering

In three dimensions, $f^{(3)}(\omega)$ is given by expression (5.2.64; 10.3.45) and the volume integral (10.3.51) becomes

$$\underline{\mathbf{u}}^\mathrm{E}(t, \mathbf{x}_\mathsf{R}; \mathbf{x}_\mathsf{S}) = \frac{1}{4\pi^2}\frac{\mathrm{d}}{\mathrm{d}t}\int_V \mathrm{Re}\Big(\underline{\boldsymbol{\mathcal{K}}}^\mathrm{E}(\mathbf{x}, \mathcal{L}_\mathsf{R}, \mathcal{L}_\mathsf{S})\,\Delta\left(t - \widetilde{T}(\mathbf{x}, \mathcal{L}_\mathsf{R}, \mathcal{L}_\mathsf{S})\right)\Big)\,\mathrm{d}V. \tag{10.3.60}$$

Replacing result (10.3.58), we have

$$\boxed{\underline{\mathbf{u}}^{\mathrm{E}}(t, \mathbf{x}_{\mathrm{R}}; \mathbf{x}_{\mathrm{S}}) = \frac{1}{4\pi^2} \frac{\mathrm{d}}{\mathrm{d}t} \operatorname{Re}\left(\Delta(t) * \int_{\widetilde{T}=t} \frac{\underline{\mathcal{K}}^{\mathrm{E}}(\mathbf{x}, \mathcal{L}_{\mathrm{R}}, \mathcal{L}_{\mathrm{S}})}{\left|\nabla \widetilde{T}(\mathbf{x}, \mathcal{L}_{\mathrm{R}}, \mathcal{L}_{\mathrm{S}})\right|} \mathrm{d}S\right),} \tag{10.3.61}$$

where the integral is now over surfaces defined by the isochron condition (10.3.57), again illustrated by Figure 10.22.

10.3.2.4 Elastic scattering

The development for the error Born scattering of elastic waves is essentially the same as for acoustic waves, except that the final expression for the scalar scattering amplitude, Γ^{E} (10.3.55), is more complicated. The zeroth-order term in asymptotic ray theory is given by expression (5.4.29) with the amplitude coefficients (5.4.32). Again for simplicity we ignore the possibility of multi-pathing and the existence of multiple ray types in elastic media. In all expressions, a summation should be included over different ray types and possible multi-pathing. With this approximation it is straightforward to obtain expressions for the error terms (10.3.29) and (10.3.30). These are

$$\underline{\mathbf{F}}^{\mathrm{A}}(\omega, \mathbf{x}; \mathbf{x}_{\mathrm{S}}) = \mathrm{i}\omega \frac{\partial \underline{\mathbf{t}}_j^{(0)}(\mathbf{x}, \mathcal{L}_{\mathrm{S}})}{\partial x_j} \mathrm{e}^{\mathrm{i}\omega T(\mathbf{x}, \mathcal{L}_{\mathrm{S}})} \tag{10.3.62}$$

$$\underline{\mathbf{E}}^{\mathrm{A}}(\omega, \mathbf{x}; \mathbf{x}_{\mathrm{S}}) = \underline{\mathbf{e}}_j^{(0)}(\mathbf{x}, \mathcal{L}_{\mathrm{S}}) \, \mathrm{e}^{\mathrm{i}\omega T(\mathbf{x}, \mathcal{L}_{\mathrm{S}})}, \tag{10.3.63}$$

where in result (10.3.62) we have used equation (5.3.2) with $m = 0$, and in (10.3.63), (5.3.3) with $m = 0$ together with result (4.4.44). In expression (10.3.63), $\underline{\mathbf{e}}_j^{(0)}$ is the strain coefficient corresponding to $\underline{\mathbf{v}}^{(0)}$, i.e.

$$\left(\underline{\mathbf{e}}_j^{(0)}\right)_k = \frac{1}{2}\left(\frac{\partial v_j^{(0)}}{\partial x_k} + \frac{\partial v_k^{(0)}}{\partial x_j}\right). \tag{10.3.64}$$

Substituting these errors (10.3.62) and (10.3.63) in the error scattering integral (10.3.39), together with the same approximate Green function from the receiver, we obtain for the first iteration ($j = 1$ in integral (10.3.43))

$$\begin{aligned} &\underline{\mathbf{u}}^{\mathrm{E}}(\omega, \mathbf{x}_{\mathrm{R}}; \mathbf{x}_{\mathrm{S}}) \\ &= -\mathrm{i}\omega \int_V \left(\underline{\mathbf{v}}^{(0)\,\mathrm{T}}(\mathbf{x}, \mathcal{L}_{\mathrm{R}}) \frac{\partial \underline{\mathbf{t}}_j^{(0)}(\mathbf{x}, \mathcal{L}_{\mathrm{S}})}{\partial x_j} - \underline{\mathbf{t}}_j^{(0)\,\mathrm{T}}(\mathbf{x}, \mathcal{L}_{\mathrm{R}}) \frac{\partial \underline{\mathbf{v}}^{(0)}(\mathbf{x}, \mathcal{L}_{\mathrm{S}})}{\partial x_j}\right) \\ &\quad \times \mathrm{e}^{\mathrm{i}\omega\left(T(\mathbf{x}, \mathcal{L}_{\mathrm{R}}) + T(\mathbf{x}, \mathcal{L}_{\mathrm{S}})\right)} \mathrm{d}V. \end{aligned} \tag{10.3.65}$$

Again, the form of this integral is extremely straightforward with the frequency dependence contained in the phase with the total travel time (10.3.49). To make the amplitude reciprocal we integrate by parts

$$\begin{aligned}
&\int_V \left(\underline{\mathbf{v}}_{\mathsf{R}}^{(0)\,\mathrm{T}} \frac{\partial \underline{\mathbf{t}}_{\mathsf{S}\,j}^{(0)}}{\partial x_j} - \underline{\mathbf{t}}_{\mathsf{R}\,j}^{(0)\,\mathrm{T}} \frac{\partial \underline{\mathbf{v}}_{\mathsf{S}}^{(0)}}{\partial x_j} \right) \mathrm{e}^{\mathrm{i}\omega \widetilde{T}} \, \mathrm{d}V \\
&= \int_S \left(\underline{\mathbf{v}}_{\mathsf{R}}^{(0)\,\mathrm{T}} \underline{\mathbf{t}}_{\mathsf{S}\,j}^{(0)} - \underline{\mathbf{t}}_{\mathsf{R}\,j}^{(0)\,\mathrm{T}} \underline{\mathbf{v}}_{\mathsf{S}}^{(0)} \right) \mathrm{e}^{\mathrm{i}\omega \widetilde{T}} \, \mathrm{d}S_j \\
&\quad - \mathrm{i}\omega \int_V \left(\underline{\mathbf{v}}_{\mathsf{R}}^{(0)\,\mathrm{T}} \underline{\mathbf{t}}_{\mathsf{S}\,j}^{(0)} - \underline{\mathbf{t}}_{\mathsf{R}\,j}^{(0)\,\mathrm{T}} \underline{\mathbf{v}}_{\mathsf{S}}^{(0)} \right) (p_{\mathsf{R}\,j} + p_{\mathsf{S}\,j}) \, \mathrm{e}^{\mathrm{i}\omega \widetilde{T}} \, \mathrm{d}V \\
&\quad - \int_V \left(\frac{\partial \underline{\mathbf{v}}_{\mathsf{R}}^{(0)\,\mathrm{T}}}{\partial x_j} \underline{\mathbf{t}}_{\mathsf{S}\,j}^{(0)} - \frac{\partial \underline{\mathbf{t}}_{\mathsf{R}\,j}^{(0)\,\mathrm{T}}}{\partial x_j} \underline{\mathbf{v}}_{\mathsf{S}}^{(0)} \right) \mathrm{e}^{\mathrm{i}\omega \widetilde{T}} \, \mathrm{d}V. \qquad (10.3.66)
\end{aligned}$$

The first integral over the surface S is combined with $\underline{\mathbf{u}}^{\mathsf{S}}$ and neglected as before. The volume integral can be rewritten as result (10.3.51) where

$$\begin{aligned}
&\underline{\mathcal{K}}^{\mathrm{E}}(\mathbf{x}, \mathcal{L}_{\mathsf{R}}, \mathcal{L}_{\mathsf{S}}) \\
&= \frac{1}{2} \left(\underline{\mathbf{v}}^{(0)\,\mathrm{T}}(\mathbf{x}, \mathcal{L}_{\mathsf{R}}) \frac{\partial \underline{\mathbf{t}}_j^{(0)}(\mathbf{x}, \mathcal{L}_{\mathsf{S}})}{\partial x_j} - \underline{\mathbf{t}}_j^{(0)\,\mathrm{T}}(\mathbf{x}, \mathcal{L}_{\mathsf{R}}) \frac{\partial \underline{\mathbf{v}}^{(0)}(\mathbf{x}, \mathcal{L}_{\mathsf{S}})}{\partial x_j} \right. \\
&\quad \left. + \frac{\partial \underline{\mathbf{t}}_j^{(0)\,\mathrm{T}}(\mathbf{x}, \mathcal{L}_{\mathsf{R}})}{\partial x_j} \underline{\mathbf{v}}^{(0)}(\mathbf{x}, \mathcal{L}_{\mathsf{S}}) - \frac{\partial \underline{\mathbf{v}}^{(0)\,\mathrm{T}}(\mathbf{x}, \mathcal{L}_{\mathsf{R}})}{\partial x_j} \underline{\mathbf{t}}_j^{(0)}(\mathbf{x}, \mathcal{L}_{\mathsf{S}}) \right), \qquad (10.3.67)
\end{aligned}$$

which obviously has a reciprocal form. This error kernel, $\underline{\mathcal{K}}^{\mathrm{E}}$, is still a 3×3 matrix.

The acoustic time-domain results, e.g. expressions (10.3.58) and (10.3.61), still apply with the more complicated elastic scattering term (10.3.67). With the amplitude coefficients (5.4.32), it is straightforward to evaluate expression (10.3.67) but it contains many spatial derivatives. As before, we neglect derivatives of the source and receiver polarizations, and write it as expression (10.3.53). Then the scalar scattering term representing local scattering at the point $\mathbf{x}$ is

$$\begin{aligned}
&\Gamma^{\mathrm{E}}(\mathbf{x}, \mathcal{L}_{\mathsf{R}}, \mathcal{L}_{\mathsf{S}}) \\
&= \frac{1}{2} \left(\mathbf{g}^{\mathrm{T}}(\mathbf{x}, \mathcal{L}_{\mathsf{R}}) \left(\mathbf{Z}_j^{\mathrm{T}}(\mathbf{x}, \mathcal{L}_{\mathsf{R}}) - \mathbf{Z}_j(\mathbf{x}, \mathcal{L}_{\mathsf{S}}) \right) \frac{\partial \mathbf{g}(\mathbf{x}, \mathcal{L}_{\mathsf{S}})}{\partial x_j} \right. \\
&\quad - \frac{\partial \mathbf{g}^{\mathrm{T}}(\mathbf{x}, \mathcal{L}_{\mathsf{R}})}{\partial x_j} \left(\mathbf{Z}_j^{\mathrm{T}}(\mathbf{x}, \mathcal{L}_{\mathsf{R}}) - \mathbf{Z}_j(\mathbf{x}, \mathcal{L}_{\mathsf{S}}) \right) \mathbf{g}(\mathbf{x}, \mathcal{L}_{\mathsf{S}}) \\
&\quad - \mathbf{g}^{\mathrm{T}}(\mathbf{x}, \mathcal{L}_{\mathsf{R}}) \frac{\partial}{\partial x_j} \left(\mathbf{Z}_j^{\mathrm{T}}(\mathbf{x}, \mathcal{L}_{\mathsf{R}}) + \mathbf{Z}_j(\mathbf{x}, \mathcal{L}_{\mathsf{S}}) \right) \mathbf{g}(\mathbf{x}, \mathcal{L}_{\mathsf{S}}) \\
&\quad \left. - \mathbf{g}^{\mathrm{T}}(\mathbf{x}, \mathcal{L}_{\mathsf{R}}) \frac{\partial}{\partial x_j} \ln \left(\frac{\mathcal{T}^{(3)}(\mathbf{x}, \mathcal{L}_{\mathsf{R}})}{\mathcal{T}^{(3)}(\mathbf{x}, \mathcal{L}_{\mathsf{S}})} \right) \left(\mathbf{Z}_j^{\mathrm{T}}(\mathbf{x}, \mathcal{L}_{\mathsf{R}}) - \mathbf{Z}_j(\mathbf{x}, \mathcal{L}_{\mathsf{S}}) \right) \mathbf{g}(\mathbf{x}, \mathcal{L}_{\mathsf{S}}) \right) \\
&\qquad (10.3.68)
\end{aligned}$$

(in anisotropic media, it is really only sensible to consider the three-dimensional problem). Apart from the extra complications of this term, and the multiplicity of ray types, the error Born scattering signal is as easily calculated in elastic as acoustic media.

10.3.3 Perturbation Born scattering using ray theory

Now we investigate the terms (10.3.19), (10.3.21), (10.3.40) or (10.3.42) which occur in the perturbation Born scattering. Alternative expressions are possible as, as well as the perturbation scattering, there is also the corresponding error scattering. We only consider the first iteration in the scattering series (10.3.23), i.e. $j = 1$ in expressions (10.3.25) or (10.3.43).

10.3.3.1 Acoustic perturbation scattering

For acoustic waves, the perturbation Born scattering (10.3.19) with the geometric ray approximation (5.4.28) becomes

$$\underline{\mathbf{u}}^{\mathrm{B}}(\omega, \mathbf{x}_{\mathrm{R}}; \mathbf{x}_{\mathrm{S}}) = -\left(f^{(\ell)}(\omega)\right)^2 \int_V \underline{\mathcal{K}}^{\mathrm{B}}(\mathbf{x}, \mathcal{L}_{\mathrm{R}}, \mathcal{L}_{\mathrm{S}})\, \mathrm{e}^{\mathrm{i}\omega \widetilde{T}(\mathbf{x}, \mathcal{L}_{\mathrm{R}}, \mathcal{L}_{\mathrm{S}})} \mathrm{d}V, \qquad (10.3.69)$$

where

$$\begin{aligned}\underline{\mathcal{K}}^{\mathrm{B}}&(\mathbf{x}, \mathcal{L}_{\mathrm{R}}, \mathcal{L}_{\mathrm{S}}) \\ &= \underline{\mathbf{v}}^{(0)\,\mathrm{T}}(\mathbf{x}, \mathcal{L}_{\mathrm{R}})\, \rho^{\mathrm{B}}(\mathbf{x})\, \underline{\mathbf{v}}^{(0)}(\mathbf{x}, \mathcal{L}_{\mathrm{S}}) + \underline{P}^{(0)\,\mathrm{T}}(\mathbf{x}, \mathcal{L}_{\mathrm{R}})\, k^{\mathrm{B}}(\mathbf{x})\, \underline{P}^{(0)}(\mathbf{x}, \mathcal{L}_{\mathrm{S}}), \qquad (10.3.70)\end{aligned}$$

the perturbation kernel for the Born perturbation scattering integral, again a 3×3 matrix. Substituting for the ray amplitude coefficients (5.4.31), we expand it as

$$\underline{\mathcal{K}}^{\mathrm{B}}(\mathbf{x}, \mathcal{L}_{\mathrm{R}}, \mathcal{L}_{\mathrm{S}}) = \Gamma^{\mathrm{B}}(\mathbf{x}, \mathcal{L}_{\mathrm{R}}, \mathcal{L}_{\mathrm{S}})\, \underline{\mathcal{D}}^{(\ell)}(\mathbf{x}, \mathcal{L}_{\mathrm{R}}, \mathcal{L}_{\mathrm{S}}), \qquad (10.3.71)$$

where the scattering dyadic has been given in equation (10.3.54) and

$$\Gamma^{\mathrm{B}}(\mathbf{x}, \mathcal{L}_{\mathrm{R}}, \mathcal{L}_{\mathrm{S}}) = \mathbf{g}^{\mathrm{T}}(\mathbf{x}, \mathcal{L}_{\mathrm{R}})\, \rho^{\mathrm{B}}(\mathbf{x})\, \mathbf{g}(\mathbf{x}, \mathcal{L}_{\mathrm{S}}) - \frac{1}{2} Z(\mathbf{x})\, k^{\mathrm{B}}(\mathbf{x}), \qquad (10.3.72)$$

is the scalar acoustic Born perturbation scattering term. Notice that the scattered wave due to the density perturbation has a directional dependence, whereas the scattered signal due to compressibility perturbation is isotropic.

To reduce the result (10.3.69) to the time domain, we need to specify the dimensionality.

10.3.3.2 Two-dimensional perturbation scattering

In two dimensions. using expression (5.2.78; 10.3.44) for $f^{(2)}(\omega)$, the result (10.3.69) can be transformed into the time domain

$$\underline{\mathbf{u}}^{\mathrm{B}}(t, \mathbf{x}_{\mathrm{R}}; \mathbf{x}_{\mathrm{S}}) = -\frac{1}{2\pi}\frac{\mathrm{d}}{\mathrm{d}t}\int_V \mathrm{Re}\Big(\underline{\mathcal{K}}^{\mathrm{B}}(\mathbf{x}, \mathcal{L}_{\mathrm{R}}, \mathcal{L}_{\mathrm{S}})\, \Delta\left(t - \widetilde{T}(\mathbf{x}, \mathcal{L}_{\mathrm{R}}, \mathcal{L}_{\mathrm{S}})\right)\Big)\, \mathrm{d}V. \tag{10.3.73}$$

As before the volume (area) integral can be converted into a surface (line) integral on surfaces (lines) defined by the isochron (10.3.57). The result is then

$$\boxed{\underline{\mathbf{u}}^{\mathrm{B}}(t, \mathbf{x}_{\mathrm{R}}; \mathbf{x}_{\mathrm{S}}) = -\frac{1}{2\pi}\frac{\mathrm{d}}{\mathrm{d}t}\mathrm{Re}\left(\Delta(t) * \int_{\widetilde{T}=t} \frac{\underline{\mathcal{K}}^{\mathrm{B}}(\mathbf{x}, \mathcal{L}_{\mathrm{R}}, \mathcal{L}_{\mathrm{S}})}{\left|\nabla\widetilde{T}(\mathbf{x}, \mathcal{L}_{\mathrm{R}}, \mathcal{L}_{\mathrm{S}})\right|}\right)\mathrm{d}S.} \tag{10.3.74}$$

It is worth commenting on the scattered signal from a point scatterer. If $\underline{\mathcal{K}}^{\mathrm{B}}(\mathbf{x}, \mathcal{L}_{\mathrm{R}}, \mathcal{L}_{\mathrm{S}})$ is only non-zero at a point, the scattered signal has the form $\dot{\delta}(t - \widetilde{T})$ compared with $\lambda(t)$ for the geometrical rays. Thus the scattered signal is $O(\omega^{3/2})$ compared with the geometrical arrivals. It is straightforward to understand this amplification of high frequencies. Consider a point density perturbation. The scattered signal is due to the effective force source caused by the mismatch in force and acceleration. The scattering source has a factor ω^2 from the acceleration compared with the true source (10.3.12). A factor of $\omega^{-1/2}$ arises because of the two-dimensionality of the problem. Effectively, scatterers are integrated in the third dimension. Thus overall we obtain a factor of $\omega^{3/2}$.

10.3.3.3 Three-dimensional perturbation scattering

In three dimensions, using expression (5.2.64; 10.3.45) for $f^{(3)}(\omega)$, we obtain

$$\underline{\mathbf{u}}^{\mathrm{B}}(t, \mathbf{x}_{\mathrm{R}}; \mathbf{x}_{\mathrm{S}}) = -\frac{1}{4\pi^2}\frac{\mathrm{d}}{\mathrm{d}t}\int_V \mathrm{Re}\Big(\underline{\mathcal{K}}^{\mathrm{B}}(\mathbf{x}, \mathcal{L}_{\mathrm{R}}, \mathcal{L}_{\mathrm{S}})\, \Delta\left(t - \widetilde{T}(\mathbf{x}, \mathcal{L}_{\mathrm{R}}, \mathcal{L}_{\mathrm{S}})\right)\Big)\, \mathrm{d}V, \tag{10.3.75}$$

for the scattering volume integral. Reducing to a surface integral, we have

$$\boxed{\underline{\mathbf{u}}^{\mathrm{B}}(t, \mathbf{x}_{\mathrm{R}}; \mathbf{x}_{\mathrm{S}}) = -\frac{1}{4\pi^2}\frac{\mathrm{d}^2}{\mathrm{d}t^2}\mathrm{Re}\left(\Delta(t) * \int_{\widetilde{T}=t} \frac{\underline{\mathcal{K}}^{\mathrm{B}}(\mathbf{x}, \mathcal{L}_{\mathrm{R}}, \mathcal{L}_{\mathrm{S}})}{\left|\nabla\widetilde{T}(\mathbf{x}, \mathcal{L}_{\mathrm{R}}, \mathcal{L}_{\mathrm{S}})\right|}\right)\mathrm{d}S.} \tag{10.3.76}$$

For a point scatterer, the scattered signal is now $\ddot{\delta}(t)$, i.e. $O(\omega^2)$. The amplification of the high frequencies arises from the effective acceleration source (10.3.12).

The Born approximation is ideal for modelling scattering by small isolated scatterers. As the Green functions in the integral are calculated in the unperturbed model, they do not include any phase correction which might occur if waves propagate through extended perturbations. However, the integral solution (10.3.15) is complete and will be valid even for extended scatterers. If the complete, iterative series is included it must model the perturbation to the Green function caused by propagation through the perturbation, as well as scattered signals. We demonstrate below (Section 10.3.5) that the Born approximation does in fact model the travel-time correction to the original Green function, but first we generalize the perturbation Born scattering to elastic waves and describe the numerical algorithm for Born seismograms.

10.3.3.4 Elastic perturbation scattering

The results for elastic perturbation scattering, given by expressions (10.3.40) or (10.3.42), are very similar to the acoustic case. Volume integral (10.3.69) still applies with

$$\underline{\mathcal{K}}^{\mathrm{B}}(\mathbf{x}, \mathcal{L}_{\mathrm{R}}, \mathcal{L}_{\mathrm{S}}) \\ = \underline{\mathbf{v}}^{(0)\,\mathrm{T}}(\mathbf{x}, \mathcal{L}_{\mathrm{R}})\, \rho^{\mathrm{B}}(\mathbf{x})\, \underline{\mathbf{v}}^{(0)}(\mathbf{x}, \mathcal{L}_{\mathrm{S}}) - \underline{\mathbf{t}}_k^{(0)\,\mathrm{T}}(\mathbf{x}, \mathcal{L}_{\mathrm{R}})\, \mathbf{s}_{kj}^{\mathrm{B}}(\mathbf{x})\, \underline{\mathbf{t}}_j^{(0)}(\mathbf{x}, \mathcal{L}_{\mathrm{S}}), \tag{10.3.77}$$

for the scattering kernel. Writing the amplitude coefficients as in result (5.4.32), we write this as in expression (10.3.71) with

$$\Gamma^{\mathrm{B}}(\mathbf{x}, \mathcal{L}_{\mathrm{R}}, \mathcal{L}_{\mathrm{S}}) = \mathbf{g}^{\mathrm{T}}(\mathbf{x}, \mathcal{L}_{\mathrm{R}}) \left(\rho^{\mathrm{B}}(\mathbf{x}) - \mathbf{Z}_k^{\mathrm{T}}(\mathbf{x}, \mathcal{L}_{\mathrm{R}})\, \mathbf{s}_{kj}^{\mathrm{B}}(\mathbf{x})\, \mathbf{Z}_j(\mathbf{x}, \mathcal{L}_{\mathrm{S}})\right) \mathbf{g}(\mathbf{x}, \mathcal{L}_{\mathrm{S}}). \tag{10.3.78}$$

With this scattering amplitude substituted, the time domain results, (10.3.73) to (10.3.76), still apply. Apart from the change in the scattering amplitude, and summation that will be necessary over multiple ray types, it is as simple to evaluate elastic scattering as acoustic scattering.

10.3.4 Numerical band-limited Born seismograms

The volume integrals of Born scattering are difficult to evaluate efficiently at high frequencies. The integrands of the spectral results, (10.3.51) and (10.3.69), are highly oscillatory. In general the oscillations will cancel and the significant contributions will only come from stationary and end-points, and regions where the scattering amplitude, $\underline{\mathcal{K}}$, is varying rapidly. The time-domain integrals, (10.3.58), (10.3.61), (10.3.74) and (10.3.76), are more easily calculated as normally the integrand is only significant in limited regions. An efficient algorithm that avoids aliasing problems by band-limiting the response is an extension of that used in one dimension for WKBJ seismograms (Section 8.4.2) and in two dimensions for Maslov seismograms (Section 10.1.2.1).

All the time-domain results (10.3.58), (10.3.61), (10.3.74) and (10.3.76), for two and three dimensions and error and perturbation scattering, can be band-limited to

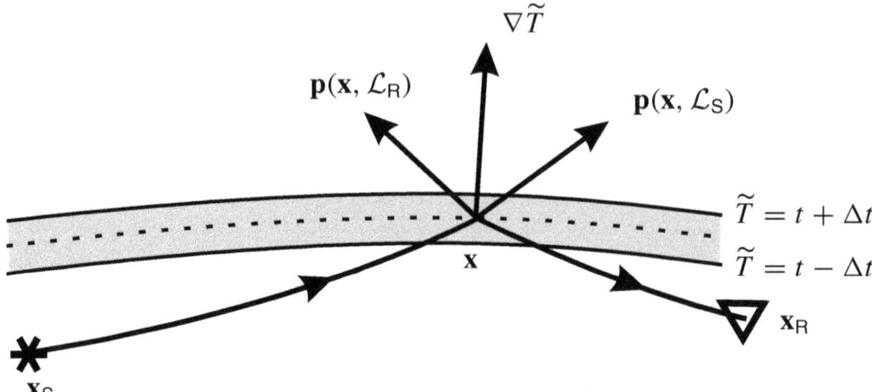

Fig. 10.23. An illustration of the volume slice contributing to the band-limited Born integral (10.3.79), where the slice is defined by $t - \Delta t < \widetilde{T} < t + \Delta t$.

obtain a robust algorithm. For instance, applying the boxcar smoothing (8.4.13) to equation (10.3.58), we obtain

$$\boxed{\begin{aligned}&\frac{1}{\Delta t} B\left(\frac{t}{\Delta t}\right) * \underline{\mathbf{u}}^\mathrm{E}(t, \mathbf{x}_\mathrm{R}; \mathbf{x}_\mathrm{S})\\ &\qquad = \frac{1}{4\pi\,\Delta t} \mathrm{Re}\left(\Delta(t) * \int_{\widetilde{T}=t\pm\Delta t} \underline{\mathcal{K}}^\mathrm{E}(\mathbf{x}, \mathcal{L}_\mathrm{R}, \mathcal{L}_\mathrm{S})\,\mathrm{d}V\right),\end{aligned}} \tag{10.3.79}$$

where the integral is over a thin slice of volume defined by

$$\widetilde{T}(\mathbf{x}, \mathcal{L}_\mathrm{R}, \mathcal{L}_\mathrm{S}) = t \pm \Delta t. \tag{10.3.80}$$

This is illustrated in Figure 10.23. Similar band-limited expressions are obtained for the other time-domian results, (10.3.61), (10.3.74) and (10.3.76).

The two-dimensional, smoothed versions of the integrals (10.3.58) and (10.3.74) can be evaluated using the same algorithm as for the Maslov seismogram in three dimensions (Section 10.1.2.1). The two-dimensional integral is just a spatial rather than a slowness integral. In three dimensions, the method is simply extended to volume elements. This can be applied to the smoothed versions of the integrals (10.3.61) and (10.3.76). These integrals are evaluated by dividing the volume into tetrahedral elements and assuming linear interpolation between the four nodes. In each tetrahedron, the volume slice is bounded by two plane, parallel surfaces (Figure 10.24), and the volume integral is a summation over tetrahedra. Within each element, the phase, $\widetilde{T}$, and amplitude, $\underline{\mathcal{K}}^\mathrm{E}$, are interpolated linearly. An efficient, numerical algorithm has been given by Spencer, Chapman and Kragh (1997) in terms of the integrals over tetrahedra. With linear interpolation, the isochron surfaces (10.3.57) are planar in each tetrahedron. Let us consider

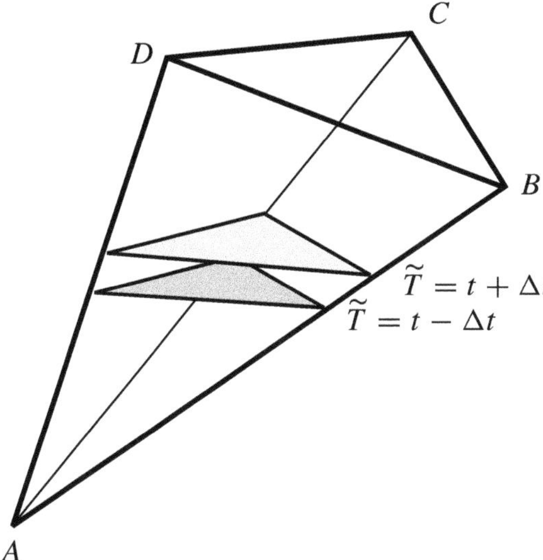

Fig. 10.24. An illustration of the volume slice in a tetrahedron contributing to the band-limited Born integral (10.3.79). The parallel, planar isochrons are defined by $\widetilde{T} = t \pm \Delta t$.

a tetrahedron $ABCD$ illustrated in Figure 10.25, where the points have been ordered by phase, $\widetilde{T}$, for a time t such that $\widetilde{T}_A < t < \widetilde{T}_B < \widetilde{T}_C < \widetilde{T}_D$. We define points C' on AC and D' on AD with the same phase as B, i.e. $BC'D'$ define the isochron surface $\widetilde{T}_B$. The isochron surface t intersects AB, AC and AD at B'', C'' and D'', respectively. The volumes of the similar tetrahedra are related by

$$\begin{aligned} V_{AB''C''D''} &= \left(\frac{t - \widetilde{T}_A}{\widetilde{T}_B - \widetilde{T}_A} \right)^3 V_{ABC'D'} \\ &= \frac{(t - \widetilde{T}_A)^3}{(\widetilde{T}_B - \widetilde{T}_A)(\widetilde{T}_C - \widetilde{T}_A)(\widetilde{T}_D - \widetilde{T}_A)} V_{ABCD}. \end{aligned} \tag{10.3.81}$$

The amplitudes are linearly interpolated between the values $\underline{\mathcal{K}}^{\mathrm{E}}_A$, $\underline{\mathcal{K}}^{\mathrm{E}}_B$, $\underline{\mathcal{K}}^{\mathrm{E}}_C$ and $\underline{\mathcal{K}}^{\mathrm{E}}_D$, to give

$$\underline{\mathcal{K}}^{\mathrm{E}}_{B''} = \underline{\mathcal{K}}^{\mathrm{E}}_A + (\underline{\mathcal{K}}^{\mathrm{E}}_B - \underline{\mathcal{K}}^{\mathrm{E}}_A) \left(\frac{t - \widetilde{T}_A}{\widetilde{T}_B - \widetilde{T}_A} \right) \tag{10.3.82}$$

$$\underline{\mathcal{K}}^{\mathrm{E}}_{C''} = \underline{\mathcal{K}}^{\mathrm{E}}_A + (\underline{\mathcal{K}}^{\mathrm{E}}_C - \underline{\mathcal{K}}^{\mathrm{E}}_A) \left(\frac{t - \widetilde{T}_A}{\widetilde{T}_C - \widetilde{T}_A} \right) \tag{10.3.83}$$

$$\underline{\mathcal{K}}^{\mathrm{E}}_{D''} = \underline{\mathcal{K}}^{\mathrm{E}}_A + (\underline{\mathcal{K}}^{\mathrm{E}}_D - \underline{\mathcal{K}}^{\mathrm{E}}_A) \left(\frac{t - \widetilde{T}_A}{\widetilde{T}_D - \widetilde{T}_A} \right). \tag{10.3.84}$$

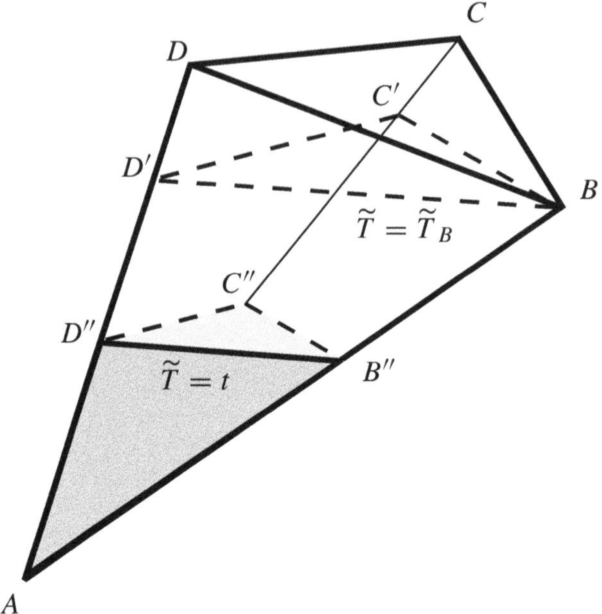

Fig. 10.25. A tetrahedral volume element $ABCD$ with a sub-tetrahedron $AB''C''D''$ defined by the plane surface $\widetilde{T} = t$.

Averaging these over the tetrahedron $AB''C''D''$, we have

$$\int_{\widetilde{T}_A}^{t} \underline{\mathcal{K}}^{\mathrm{E}} \mathrm{d}V$$
$$= \left\{ \underline{\mathcal{K}}_A^{\mathrm{E}} + \frac{1}{4} \left(\frac{\underline{\mathcal{K}}_B^{\mathrm{E}} - \underline{\mathcal{K}}_A^{\mathrm{E}}}{\widetilde{T}_B - \widetilde{T}_A} + \frac{\underline{\mathcal{K}}_C^{\mathrm{E}} - \underline{\mathcal{K}}_A^{\mathrm{E}}}{\widetilde{T}_C - \widetilde{T}_A} + \frac{\underline{\mathcal{K}}_D^{\mathrm{E}} - \underline{\mathcal{K}}_A^{\mathrm{E}}}{\widetilde{T}_D - \widetilde{T}_A} \right) (t - \widetilde{T}_A) \right\} V_{AB''C''D''}. \tag{10.3.85}$$

Together with expression (10.3.81), this gives the volume integral over sub-tetrahedra. Similar formulae can be obtained for $\widetilde{T}_B < t < \widetilde{T}_D$ by considering extensions of the tetrahedron. The integrals over slices (Figure 10.24) are obtained by subtraction – many terms in equations (10.3.81) and (10.3.85) need not be re-calculated.

10.3.4.1 Acoustic Born error and perturbation modelling

In Chapter 6, we introduced a simple three-dimensional model to illustrate three-dimensional ray tracing – the French (1974) model, Figure 6.12. We used this model to demonstrate modelling using the Born scattering method. The model parameters have been given in Section 6.8, and for simplicity the model is taken as acoustic.

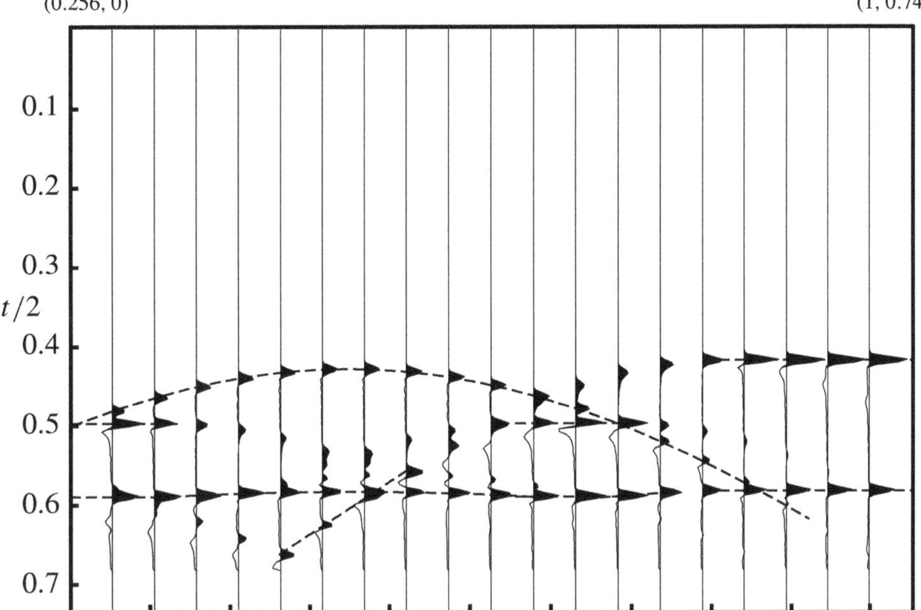

Fig. 10.26. Born error scattering zero-offset seismograms for the profile illustrated in Figure 6.12, i.e. arrivals corresponding to those in Figure 6.15. The time axis is half-time.

In Figure 10.26, band-limited, zero-offset, impulse seismograms calculated using Born error scattering theory are shown. The source is an explosion and the vertical component of displacement is plotted. The coincident source and receivers ($\mathbf{x}_S = \mathbf{x}_R$) lie on the profile shown in Figure 6.12. This corresponds to the rays plotted in Figure 6.14 and the one-way travel times in Figure 6.15. The Born error scattering seismograms are calculated using expression (10.3.55) for the scalar Born acoustic scattering term, Γ^E. The interface model is sampled on a uniform grid ($dx = dy = 0.0025$), and smoothed slightly using a Gaussian window with standard deviation equal to the sampling interval, in order to evaluate the scattering volume integral. The interfaces are smoothed by the discrete grid sampling and the Gaussian smoothing. The divergence of the polarizations can be calculated using results (5.6.11) and Exercise 5.5. Even in a homogeneous medium, this term is non-zero and corrects the near field. For simplicity it is omitted in Figure 10.26. The band-limited seismograms are calculated in the time domain using the algorithm described above (result (10.3.79) using approximation (10.3.85)). Arrivals corresponding to the travel times in Figure 6.15 are visible in Figure 10.26, together with non-geometrical arrivals extending the branches of arrivals. Because the rays have been traced in the 'correct' model (it only differs from Figure 6.12 by being sampled and smoothed on a discrete grid), the arrivals are at the correct times.

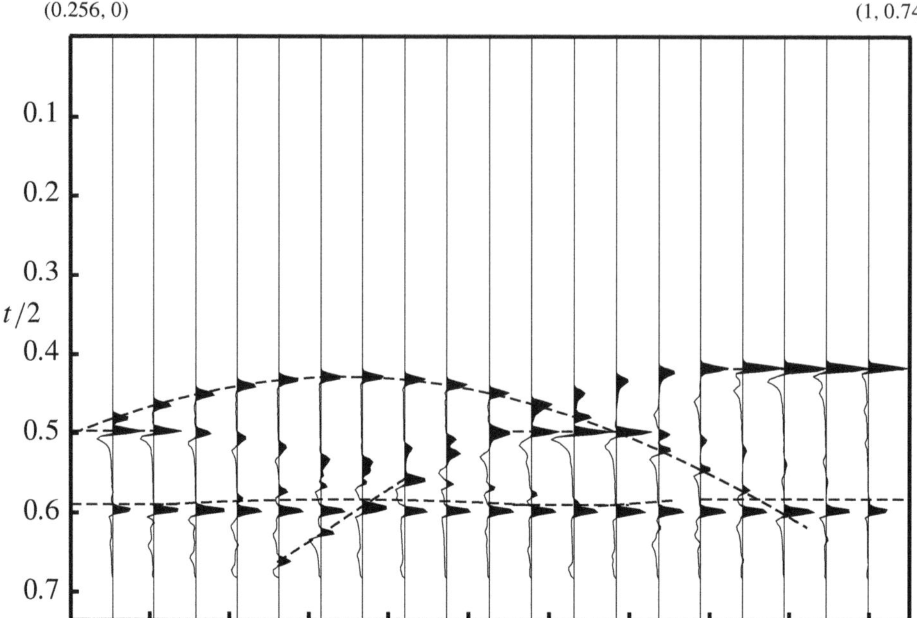

Fig. 10.27. Born perturbation scattering zero-offset seismograms for the profile illustrated in Figure 6.12, i.e. arrivals corresponding to those in Figure 6.15. The time axis is half-time.

In Figure 10.27, band-limited, zero-offset, impulse seismograms calculated using Born perturbation scattering theory are shown. Details are identical to Figure 10.26 except that Born perturbation theory is used rather than Born error theory. Therefore these seismograms correspond to the traditional Born method. To calculate these seismograms, a uniform background model with the properties of the top layer is used. Expression (10.3.72) is used for the scalar Born acoustic perturbation term, Γ^B. The same algorithm is used for the volume integral as in the error results (as equations (10.3.79) and (10.3.85)). Similar arrivals are visible in Figures 10.27 and 10.26, but important differences exist. First the arrivals from the first interface are very similar, as the reference model corresponds to the layer above this interface. Arrivals from the second interface are misplaced, because the rays are traced in the reference model with constant velocity. Because the velocity is too slow, the arrivals are late (in fact because the velocity is uniform, the arrivals are spaced according to the interfaces in the depth model, Figure 6.12), and are also misplaced laterally.

Figures 10.26 and 10.27 compared Born error and perturbation scattering theory. The results of Born error scattering theory are more accurate as the rays are traced in the true model compared with the reference model, giving arrivals at the correct times. There are also two computational advantages. Comparing expression (10.3.61) with (10.3.76), although they are very similar the important difference is the extra time differentiation in the Born error result (10.3.76) (this is, of course, compensated for by the extra spatial derivatives in the error scattering term, Γ^{E} (10.3.55), compared with the perturbation scattering term, Γ^{B} (10.3.72)). This extra differentiation makes the Born perturbation result harder to compute accurately as numerical noise due to approximations in the volume integral (caused by the discretization and linear interpolation (10.3.85)) is amplified. In addition, the error scattering integral can normally be evaluated more efficiently as the error scattering term, Γ^{E}, is only significant in a smaller volume than the perturbation scattering term, Γ^{B}. The computational disadvantage of the Born error scattering method is that the rays must be traced in the true, heterogeneous model, whereas in the Born perturbation scattering method the rays can be traced in a homogeneous (or simpler) reference model.

10.3.5 Travel-time perturbation from perturbation Born scattering

The perturbation Born scattering approximation is ideal for isolated, small scatterers where the approximate Green function at each scatterer is not unduly influenced by other scatterers. Physically, two effects of the other scatterers can be distinguished, depending whether they are isolated or extensive. First, for isolated scatterers, multiple scattering may be important which can be modelled by higher-order terms in the iterative series (10.3.25). In principle, these are straightforward to include but in practice the higher dimensionality of the multiple volume integrals, or multiplicity of the scattered ray paths, makes it expensive. Secondly, if the scatterers are extensive, propagation of the waves through the perturbation may introduce significant errors in the Green function. For instance, if an extended perturbation to velocity exists, it will introduce a significant perturbation to the travel time, and the Green function will be in error. Physically, one would not expect single or multiple scattering to be important in an extended smooth scatterer, and yet in the perturbed Born solution they exist. Using the perturbed Born approximation, i.e. the first term in the iterative solution ($j = 1$ in equation (10.3.25)), we would expect significant errors in the scattered signals if extended scatterers exist, as the travel times will be in error. Nevertheless, the iterative series models the complete solution and must include the travel-time perturbation. In this section, we outline how the travel-time perturbation is contained in the perturbed Born series, i.e. why the perturbed Born method can be used with extensive perturbations even though it appears counter-intuitive.

We consider the approximate value of the integral (10.3.69) for perturbations near the geometrical ray path. The phase function is stationary on the ray path and

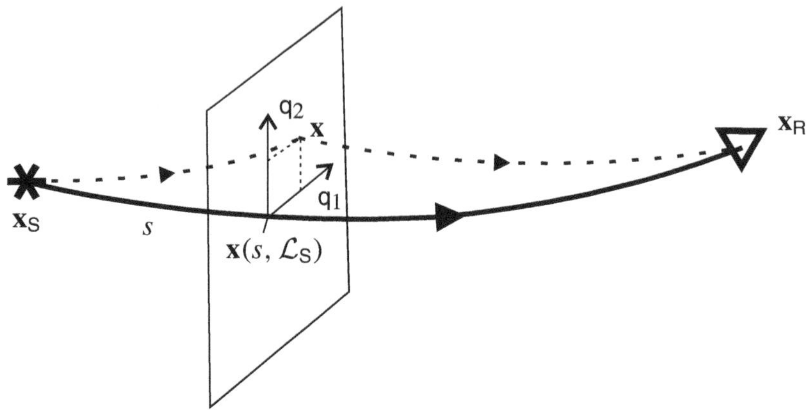

Fig. 10.28. Wavefront coordinates q, and ray length s, for perturbations at the point $\mathbf{x}$ near the geometrical ray point $\mathbf{x}(s, \mathcal{L}_S)$.

can be approximated by a quadratic expansion. At a distance, s, along the ray from the source, we have the expansion

$$\widetilde{T}(\mathbf{x}, \mathcal{L}_R, \mathcal{L}_S) \simeq T(\mathbf{x}_R, \mathcal{L}_S) + \frac{1}{2}\,\mathsf{q}^{\mathrm{T}}(s)\;\nabla_{\mathsf{q}}(\nabla_{\mathsf{q}}\widetilde{T})^{\mathrm{T}}\big|_s\,\mathsf{q}(s), \tag{10.3.86}$$

where q are the coordinates of $\mathbf{x}$ relative to the point on the ray in the wavefront, i.e. $\mathsf{q}(s) = \mathbf{x} - \mathbf{x}(s, \mathcal{L}_S)$ (Figure 10.28). Using the stationary-phase approximation (Appendix D.1), the scattering amplitude, $\underline{\boldsymbol{\mathcal{K}}}^{\mathrm{B}}$, can be treated as independent of q, with the value on the ray, i.e. when $\mathsf{q} = \mathsf{0}$. The variation along the ray, i.e. with respect to s, is significant. For simplicity, let us consider the case of acoustic scattering, results (10.3.70) and (10.3.72), and only include a velocity perturbation. To first order

$$k^{\mathrm{B}} \simeq -\frac{2\alpha^{\mathrm{B}}}{\rho\,\alpha^3}, \tag{10.3.87}$$

and the scalar scattering coefficient (10.3.72) reduces to

$$\Gamma^{\mathrm{B}}(\mathbf{x}, \mathcal{L}_R, \mathcal{L}_S) \simeq -\frac{\alpha^{\mathrm{B}}}{\alpha^2}. \tag{10.3.88}$$

Combining the approximations (10.3.86) and (10.3.88) with expression (10.3.71) in integral (10.3.69), which is rewritten as a volume integral with respect to q and s, we have

$$\underline{\mathbf{u}}^{\mathrm{B}}(\omega, \mathbf{x}_R; \mathbf{x}_S) \simeq -\frac{\omega^2}{4\pi^2}\,\mathbf{g}(\mathbf{x}_R, \mathcal{L}_R)\,\mathbf{g}^{\mathrm{T}}(\mathbf{x}_S, \mathcal{L}_S)\,\mathrm{e}^{\mathrm{i}\omega T(\mathbf{x}_R,\mathcal{L}_S)}$$
$$\times \int_0^{\mathcal{L}} \mathcal{T}^{(3)}(\mathbf{x}, \mathcal{L}_R)\,\frac{\alpha^{\mathrm{B}}(s)}{\alpha^2(s)}\,\mathcal{T}^{(3)}(\mathbf{x}, \mathcal{L}_S) \iint_{-\infty}^{\infty} \mathrm{e}^{\mathrm{i}\omega \mathsf{q}^{\mathrm{T}}\nabla_{\mathsf{q}}(\nabla_{\mathsf{q}}\widetilde{T})^{\mathrm{T}}\mathsf{q}/2}\mathrm{d}\mathsf{q}\,\mathrm{d}s, \tag{10.3.89}$$

where $\mathcal{L}$ is the length of the ray. Evaluating the wavefront integral using the second-order saddle-point method (D.1.18), we obtain

$$\underline{\mathbf{u}}^{\mathrm{B}}(\omega, \mathbf{x}_{\mathrm{R}}; \mathbf{x}_{\mathrm{S}}) \simeq -\frac{\omega}{2\pi}\, \mathbf{g}(\mathbf{x}_{\mathrm{R}}, \mathcal{L}_{\mathrm{R}})\, \mathbf{g}^{\mathrm{T}}(\mathbf{x}_{\mathrm{S}}, \mathcal{L}_{\mathrm{S}})\, \mathrm{e}^{\mathrm{i}\omega\, T(\mathbf{x}_{\mathrm{R}}, \mathcal{L}_{\mathrm{S}})}$$
$$\times \int_0^{\mathcal{L}} \frac{\mathcal{T}^{(3)}(\mathbf{x}, \mathcal{L}_{\mathrm{R}})\, \mathcal{T}^{(3)}(\mathbf{x}, \mathcal{L}_{\mathrm{S}})}{\left\| \nabla_{\mathrm{q}}(\nabla_{\mathrm{q}}\widetilde{T})^{\mathrm{T}} \right\|^{1/2}} \, \frac{\alpha^{\mathrm{B}}(s)}{\alpha^2(s)}\, \mathrm{e}^{\mathrm{i}\omega\, \mathrm{sgn}(\omega \nabla_{\mathrm{q}}(\nabla_{\mathrm{q}}\widetilde{T})^{\mathrm{T}})/4} \mathrm{d}s\, . \tag{10.3.90}$$

Using the chain rule (5.2.56) with results (5.2.68) and (5.4.34), we have

$$\mathcal{T}^{(3)}(\mathbf{x}_{\mathrm{R}}, \mathcal{L}_{\mathrm{S}}) = \mathcal{T}^{(3)}(\mathbf{x}, \mathcal{L}_{\mathrm{R}})\, \left\| \nabla_{\mathrm{q}}(\nabla_{\mathrm{q}}\widetilde{T})^{\mathrm{T}} \right\|^{-1/2} \mathcal{T}^{(3)}(\mathbf{x}, \mathcal{L}_{\mathrm{S}}). \tag{10.3.91}$$

The wavefront matrices, M, in expression (5.2.56) for a simple ray are given by

$$\mathsf{M}(T, T_0) = \nabla_{\mathrm{q}}(\nabla_{\mathrm{q}} T)^{\mathrm{T}}. \tag{10.3.92}$$

The combination $\mathsf{M}(T, T_1) - \mathsf{M}(T, T_0)$ reduces to $-\nabla_{\mathrm{q}}(\nabla_{\mathrm{q}}\widetilde{T})^{\mathrm{T}}$ as in expression (5.2.56) all the propagator matrices are solutions of the dynamic ray equations (5.2.19), solved in the forward direction. For the reversed ray from the receiver, the slowness changes direction and, as M is an odd function of slowness, $\mathsf{M}(T, T_1)$ is *minus* the wavefront matrix for the ray traced in the reversed direction (see the discussion between equations (5.2.36) and (5.2.37)).

Using result (10.3.91), the first factor in the line integral (10.3.90) reduces to $\mathcal{T}^{(3)}(\mathbf{x}_{\mathrm{R}}, \mathcal{L}_{\mathrm{S}})$ for all positions, s. Hence, expression (10.3.90) reduces to

$$\underline{\mathbf{u}}^{\mathrm{B}}(\omega, \mathbf{x}_{\mathrm{R}}; \mathbf{x}_{\mathrm{S}}) \simeq \mathrm{i}\omega\, \delta T^{\mathrm{B}}\, \underline{\mathbf{u}}^{\mathrm{A}}(\omega, \mathbf{x}_{\mathrm{R}}; \mathbf{x}_{\mathrm{S}}), \tag{10.3.93}$$

using expression (5.4.32), where

$$\delta T^{\mathrm{B}} = -\int_0^{\mathcal{L}} \frac{\alpha^{\mathrm{B}}(s)}{\alpha^2(s)}\, \mathrm{d}s. \tag{10.3.94}$$

The KMAH indices are connected by

$$\sigma(\mathbf{x}_{\mathrm{R}}, \mathcal{L}_{\mathrm{S}}) = \sigma(\mathbf{x}, \mathcal{L}_{\mathrm{R}}) + \sigma(\mathbf{x}, \mathcal{L}_{\mathrm{S}}) + 1 - \frac{1}{2}\,\mathrm{sgn}\left(\nabla_{\mathrm{q}}\left(\nabla_{\mathrm{q}}\widetilde{T}\right)^{\mathrm{T}}\right) \tag{10.3.95}$$

(normally $\mathrm{sgn}\left(\nabla_{\mathrm{q}}\left(\nabla_{\mathrm{q}}\widetilde{T}\right)^{\mathrm{T}}\right) = 2$, and the KMAH indices for the two ray segments just add).

Expression (10.3.94) is, to first order, the perturbed travel time according to Fermat's principle, i.e. the perturbed travel time can be obtained, to first order, by integrating the slowness along the unperturbed ray path. This follows as although the ray path is in error, Fermat's principle states that the error in the travel time is a second-order error. Thus expression (10.3.93) is the change in the waveform, again to first order, due to a time shift δT^{B}. More accurately we would require

$$\underline{\mathbf{u}}(\omega, \mathbf{x}_{\mathrm{R}}; \mathbf{x}_{\mathrm{S}}) = \underline{\mathbf{u}}^{\mathrm{A}}(\omega, \mathbf{x}_{\mathrm{R}}; \mathbf{x}_{\mathrm{S}}) + \delta\underline{\mathbf{u}}(\omega, \mathbf{x}_{\mathrm{R}}; \mathbf{x}_{\mathrm{S}}), \tag{10.3.96}$$

where

$$\begin{aligned} \delta\underline{\mathbf{u}}(\omega, \mathbf{x}_{\mathrm{R}}; \mathbf{x}_{\mathrm{S}}) &= (\mathrm{e}^{\mathrm{i}\omega\,\delta T^{\mathrm{B}}} - 1)\,\underline{\mathbf{u}}^{\mathrm{A}}(\omega, \mathbf{x}_{\mathrm{R}}; \mathbf{x}_{\mathrm{S}}) \\ &\simeq \mathrm{i}\omega\,\delta T^{\mathrm{B}}\underline{\mathbf{u}}^{\mathrm{A}}(\omega, \mathbf{x}_{\mathrm{R}}; \mathbf{x}_{\mathrm{S}}) \\ &= \underline{\mathbf{u}}^{\mathrm{B}}(\omega, \mathbf{x}_{\mathrm{R}}; \mathbf{x}_{\mathrm{S}}). \end{aligned} \tag{10.3.97}$$

Thus the perturbation Born approximation does model the perturbed travel time, but only to first order in the Taylor expansion of the exponential, $\exp(\mathrm{i}\omega\,\delta T^{\mathrm{B}})$. Higher-order terms in the iterative series will model higher-order terms in the travel-time perturbation expansion (and other perturbations to ray theory – Coates and Chapman, 1990*a*, and Chapman and Coates, 1994 – see Exercise 10.4), but will be a very inefficient way to model these effects. Thus in principle, perturbation Born theory works for extended scatterers, but in practice it is better to include the extended scatterers in the reference model, so the travel times are correct in the approximate Green function $\underline{\mathbf{u}}^{\mathrm{A}}$, and to model scattered signals using the error Green theory.

10.4 Kirchhoff surface integral method

The ray method describes the high-frequency behaviour of reflections from interfaces. It takes into account the amplitude changes due to the reflection/transmission coefficients, and the spreading changes that may occur due to the change in ray type and the curvature of the interface. However, diffracted signals that occur due to roughness and discontinuities in the interface are not modelled. The Kirchhoff surface integral method is a useful generalization of ray theory that models, at least approximately, such signals by integrating over non-specular reflections on the interface. Rays are traced from the source to the interface, and from the interface to the receiver (or by reciprocity, from the receiver to the interface) without satisfying Snell's law at the interface. The basic method has been described by Clay and Medwin (1977, pp. 505–508) and Bleistein (1984, p. 282). Frazer and Sen (1985), and references therein to earlier papers, have described its application to seismology for acoustic or isotropic media. Recent papers, e.g. de Hoop and Bleistein (1997) and Ursin and Tygel (1997), have extended the method to anisotropic media.

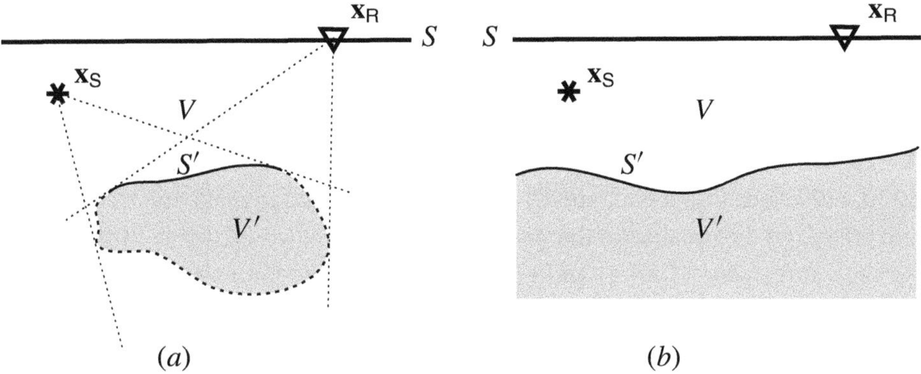

Fig. 10.29. The scattering volume V' with boundary S' embedded in the volume V with boundary S. The source and receiver are in V not V'. Two cases are illustrated: (*a*) when the scattering object is finite; and (*b*) when the surface S' is infinite and the volume V' semi-infinite. In the former case, the dashed portions of the surface S' are not illuminated from the source and receiver.

10.4.1 Acoustic Kirchhoff surface integral

Consider the situations illustrated in Figure 10.29. The source and receiver are contained in a volume V surrounded by a surface S. Within this volume we have a scatterer contained in a volume V' surrounded by a surface S', but the sources and receivers are outside volume V'. Normally S' is taken coincident with the boundary of the scatterer, i.e. at a discontiuity. For simplicity, let us define V as the volume bounded by surfaces S and S' so it excludes the volume V'. Two cases are of particular interest: the volume V' is finite, a scattering body, and the surface S' is closed; or the volume V' is semi-infinite with the surface S' consisting of an infinite interface of interest, plus an infinite hemisphere, S'_∞. Normally, the surface S is also taken as an infinite sphere. Applying the representation (Betti's) theorem (4.5.27) to the volume V, we can write the total solution in the frequency domain as

$$\underline{\mathbf{u}}(\omega, \mathbf{x}_\mathrm{R}; \mathbf{x}_\mathrm{S}) = \underline{\mathbf{u}}^\mathrm{D}(\omega, \mathbf{x}_\mathrm{R}; \mathbf{x}_\mathrm{S}) + \underline{\mathbf{u}}^\mathrm{K}(\omega, \mathbf{x}_\mathrm{R}; \mathbf{x}_\mathrm{S}), \tag{10.4.1}$$

where

$$\underline{\mathbf{u}}^\mathrm{D}(\omega, \mathbf{x}_\mathrm{R}; \mathbf{x}_\mathrm{S}) = \int_V \underline{\mathbf{u}}^\mathrm{T}(\omega, \mathbf{x}; \mathbf{x}_\mathrm{R})\, \mathbf{I}\, \delta(\mathbf{x} - \mathbf{x}_\mathrm{S})\, \mathrm{d}V, \tag{10.4.2}$$

and

$$\underline{\mathbf{u}}^\mathrm{K}(\omega, \mathbf{x}_\mathrm{R}; \mathbf{x}_\mathrm{S}) = \int_{S'} \Big(\underline{P}^\mathrm{T}(\omega, \mathbf{x}; \mathbf{x}_\mathrm{R})\, \hat{\mathbf{n}}^\mathrm{T}(\mathbf{x})\, \underline{\mathbf{u}}(\omega, \mathbf{x}; \mathbf{x}_\mathrm{S}) - \underline{\mathbf{u}}^\mathrm{T}(\omega, \mathbf{x}; \mathbf{x}_\mathrm{R})\, \hat{\mathbf{n}}(\mathbf{x})\, \underline{P}(\omega, \mathbf{x}; \mathbf{x}_\mathrm{S}) \Big)\, \mathrm{d}S', \tag{10.4.3}$$

where the point $\mathbf{x}$ lies on the surface S' of the integral and the *outward normal* is $\hat{\mathbf{n}}(\mathbf{x})$ (outward from the volume V containing the source and receiver, inward to the volume V'). We have generalized equation (4.5.27) to the Green function, $\underline{\mathbf{u}}(\omega, \mathbf{x}_\mathrm{R}; \mathbf{x}_\mathrm{S})$, by including the unit component sources, $\mathbf{I}\,\delta(\mathbf{x} - \mathbf{x}_\mathrm{S})$, in equation (10.4.2). We will assume that the integrals over surfaces at infinity can be neglected, either because some (small) attenuation is added to make the frequency integral converge, or because in the time domain, signals would arrive infinitely late.

In the expressions (10.4.2) and (10.4.3), we take the Green functions from the receiver – i.e. with argument $(\omega, \mathbf{x}; \mathbf{x}_\mathrm{R})$ – as the Green functions in a medium without the scatterer, so that expression (10.4.2), $\underline{\mathbf{u}}^\mathrm{D}$, represents the direct waves from the source that do not interact with the scatterer. Only if the volume V is homogeneous, the case usually discussed in textbooks but of limited practical use, is it entirely obvious how to define this reference medium. If homogeneous, the medium is continued through the volume V' to make a homogeneous whole space and the Green functions are known exactly (Section 4.5.5). In this case, the Green functions are sometimes called the *free-space Green functions* (Bleistein, 1984, p. 282). In most problems, however, the volume V is inhomogeneous and the Green functions are only known approximately. The Green function must include propagation through heterogeneities and interfaces in volume V, apart from the scatterer V', so the name free-space Green functions is hardly adequate. Usually we use the ray approximation for the Green function. The medium in V is continued smoothly through V' so, within the ray approximation, no scattered signal is generated in V'. Usually with the ray approximation, the rays of interest, incident on the surface S', have only propagated through the volume V. Thus, in practice, the question of how to continue the model V through V' does not arise. The question does arise if the entire surface S' is not visible from the receiver, $\mathbf{x}_\mathrm{R}$, through the volume V, i.e. if the scatterer shadows part of the surface S', e.g. the backside of the scatterer in Figure 10.29. Although rays from this part of the surface would pass through the continued volume V', it is usually assumed that this part of the surface makes no contribution. In general, the techniques of the previous section (Section 10.3) must be used to study errors in the ray approximation. In this section we only consider signals scattered from the volume V', and ignore errors in the Green function.

Thus we approximate the Green functions from the receiver – argument $(\omega, \mathbf{x}; \mathbf{x}_\mathrm{R})$ – by ray theory (Section 5.4.2) using expressions (5.4.28) and (5.4.31), i.e.

$$
\begin{aligned}
&\begin{pmatrix} \underline{\mathbf{v}} \\ -\underline{P} \end{pmatrix}(\omega, \mathbf{x}; \mathbf{x}_\mathrm{R}) \\
&\quad \simeq f^{(\ell)}(\omega)\, \mathcal{T}^{(\ell)}(\mathbf{x}, \mathcal{L}_\mathrm{R})\, \Phi(\omega, \mathbf{x}, \mathcal{L}_\mathrm{R}) \begin{pmatrix} \hat{\mathbf{g}} \\ -(Z/2)^{1/2} \end{pmatrix}(\mathbf{x}, \mathcal{L}_\mathrm{R})\, \mathbf{g}^\mathrm{T}(\mathbf{x}_\mathrm{R}, \mathcal{L}_\mathrm{R})\,.
\end{aligned}
\tag{10.4.4}
$$

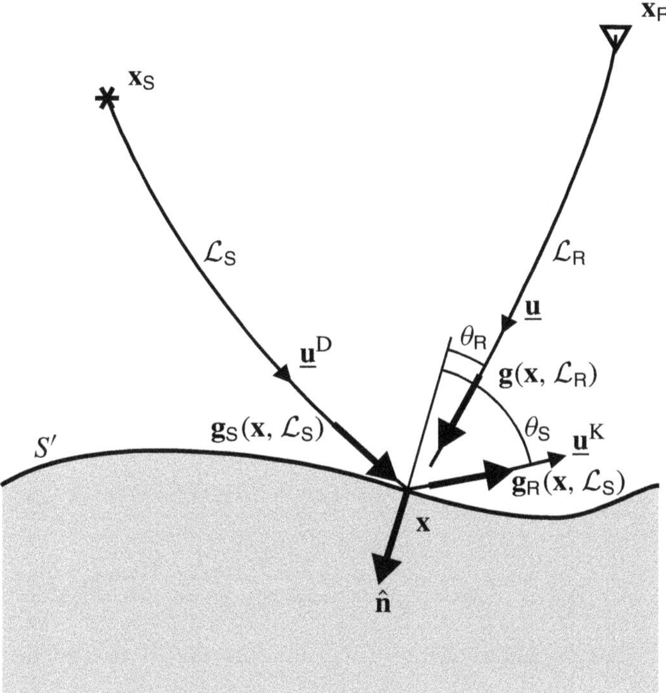

Fig. 10.30. The incident ray and reflection (10.4.5) on the surface in the Kirchhoff integral (10.4.3), and the ray from the receiver to the surface (10.4.4). The illustrated 'reflection' point is not the spectral point, so the direction of the reflected ray, $\underline{\mathbf{u}}^K$, differs from the receiver ray, $\underline{\mathbf{u}}$.

We use the shorthand $\mathcal{L}_R$ to indicate the ray path from $\mathbf{x}_R$ to $\mathbf{x}$, and for simplicity, omit a summation over multi-pathing and ray type. The product of coefficients in $\mathcal{T}^{(\ell)}(\mathbf{x}, \mathcal{L}_R)$ is for interfaces between $\mathbf{x}_R$ and $\mathbf{x}$, but not the interface at $\mathbf{x}$. In this expression, the polarization $\mathbf{g}(\mathbf{x}, \mathcal{L}_R)$ is for the ray traced from the receiver $\mathbf{x}_R$ to the interface, $\mathbf{x}$ (Figure 10.30).

In the surface integral (10.4.3), the Green functions from the source, $\mathbf{x}_S$, i.e. $\underline{\mathbf{u}}(\omega, \mathbf{x}; \mathbf{x}_S)$ and $\underline{P}(\omega, \mathbf{x}; \mathbf{x}_S)$, which include the scattered field (10.4.1), are unknown. Various approximations are made so that the integral can be evaluated and an approximate solution for the scattered wave obtained. These approximations are referred to as the *Kirchhoff approximation*. Let us list the various approximations:

- Except in homogeneous media, the Green function is rarely known exactly and we approximate it by the ray-theory Green function (Chapters 5 and 6).
- The ray-theory Green function on the interface consists of the incident plus reflected rays. The surface integral of the incident ray is generally ignored. This point is discussed further below.

- The reflected ray-theory Green function is taken as the incident ray times the reflection coefficient for an infinite plane wave reflected from an infinite plane interface (Chapter 6). This is consistent with the ray approximation.
- Sometimes the reflection coefficient is assumed to be constant and taken outside the surface integral. Except for asymptotic analysis, this approximation is not necessary and numerically we can include the variation within the integral.
- As the Green function is approximated by the incident and reflected rays, multiple scattering is ignored.

The solution on the surface of the scatterer, S', is given by the sum of the direct and scattered waves (10.4.1). On the illuminated portions of the surface, we take the incident wave as the direct ray (10.4.2). Using the ray approximation, this is given by

$$\begin{pmatrix} \underline{\mathbf{v}}^{\mathrm{D}} \\ -\underline{P}^{\mathrm{D}} \end{pmatrix}(\omega, \mathbf{x}; \mathbf{x}_{\mathrm{S}}) \simeq f^{(\ell)}(\omega)\, \mathcal{T}^{(\ell)}(\mathbf{x}, \mathcal{L}_{\mathrm{S}})\, \Phi(\omega, \mathbf{x}, \mathcal{L}_{\mathrm{S}}) \times \begin{pmatrix} \mathbf{g}_{\mathrm{S}}(\mathbf{x}, \mathcal{L}_{\mathrm{S}}) \\ -(Z(\mathbf{x})/2)^{1/2} \end{pmatrix} \mathbf{g}^{\mathrm{T}}(\mathbf{x}_{\mathrm{S}}, \mathcal{L}_{\mathrm{S}}), \tag{10.4.5}$$

where $\mathbf{g}_{\mathrm{S}}(\mathbf{x}, \mathcal{L}_{\mathrm{S}})$ is the polarization of the incident ray (in this section, as in Chapter 6, subscripts are used on the polarization and slowness vectors, when the position and path argument, $(\mathbf{x}, \mathcal{L}_{\mathrm{S}})$, is insufficient to distinguish the various rays that coexist at the reflection point). The surface integral of this term is normally ignored. This is based on the application of Betti's theorem (4.5.12) to the volume V'. As the volume contains neither the source nor receiver, the surface integral is zero, i.e.

$$\int_{S'} \Big(\underline{P}^{\mathrm{T}}(\omega, \mathbf{x}; \mathbf{x}_{\mathrm{R}})\, \hat{\mathbf{n}}^{\mathrm{T}}(\mathbf{x})\, \underline{\mathbf{u}}^{\mathrm{D}}(\omega, \mathbf{x}; \mathbf{x}_{\mathrm{S}}) - \underline{\mathbf{u}}^{\mathrm{T}}(\omega, \mathbf{x}; \mathbf{x}_{\mathrm{R}})\, \hat{\mathbf{n}}(\mathbf{x})\, \underline{P}^{\mathrm{D}}(\omega, \mathbf{x}; \mathbf{x}_{\mathrm{S}}) \Big)\, \mathrm{d}S' = \mathbf{0}. \tag{10.4.6}$$

Thus in the scattering integral (10.4.3), we need only include the scattered field

$$\underline{\mathbf{u}}^{\mathrm{K}}(\omega, \mathbf{x}_{\mathrm{R}}; \mathbf{x}_{\mathrm{S}}) = \int_{S'} \Big(\underline{P}^{\mathrm{T}}(\omega, \mathbf{x}; \mathbf{x}_{\mathrm{R}})\, \hat{\mathbf{n}}^{\mathrm{T}}(\mathbf{x})\, \underline{\mathbf{u}}^{\mathrm{K}}(\omega, \mathbf{x}; \mathbf{x}_{\mathrm{S}}) - \underline{\mathbf{u}}^{\mathrm{T}}(\omega, \mathbf{x}; \mathbf{x}_{\mathrm{R}})\, \hat{\mathbf{n}}(\mathbf{x})\, \underline{P}^{\mathrm{K}}(\omega, \mathbf{x}; \mathbf{x}_{\mathrm{S}}) \Big)\, \mathrm{d}S'. \tag{10.4.7}$$

Within the integral, we approximate the scattered field by the ray approximation which is the direct ray (10.4.5) times the appropriate reflection coefficient, i.e.

$$\begin{pmatrix} \underline{\mathbf{v}}^{\mathrm{K}} \\ -\underline{P}^{\mathrm{K}} \end{pmatrix}(\omega, \mathbf{x}; \mathbf{x}_{\mathrm{S}}) \simeq f^{(\ell)}(\omega)\, \mathcal{T}^{(\ell)}(\mathbf{x}, \mathcal{L}_{\mathrm{S}})\, \Phi(\omega, \mathbf{x}, \mathcal{L}_{\mathrm{S}}) \times \mathcal{T}_{\mathrm{RS}} \begin{pmatrix} \mathbf{g}_{\mathrm{R}}(\mathbf{x}, \mathcal{L}_{\mathrm{S}}) \\ -(Z(\mathbf{x})/2)^{1/2} \end{pmatrix} \mathbf{g}^{\mathrm{T}}(\mathbf{x}_{\mathrm{S}}, \mathcal{L}_{\mathrm{S}}), \tag{10.4.8}$$

where now $\mathbf{g}_\mathsf{R}(\mathbf{x}, \mathcal{L}_\mathsf{S})$ is the polarization of the reflected ray at the interface as illustrated in Figure 10.30. The reflection coefficient from the interface, $\mathcal{T}_{\mathsf{RS}}$, is given by expression (6.3.7). The reflected polarization, $\mathbf{g}_\mathsf{R}(\mathbf{x}, \mathcal{L}_\mathsf{S})$, and the reflection coefficient, $\mathcal{T}_{\mathsf{RS}}$, are calculated assuming the normal reflection laws (e.g. Snell's law) for the incident ray direction. The product of coefficients in $\mathcal{T}^{(\ell)}(\mathbf{x}, \mathcal{L}_\mathsf{S})$ is for interfaces between $\mathbf{x}_\mathsf{S}$ and $\mathbf{x}$ (as in expression (10.4.4)), but does not, of course, include the coefficient at $\mathbf{x}$, i.e. $\mathcal{T}_{\mathsf{RS}}$.

Reducing the surface integral to the integral of the *reflected* ray seems intuitively obvious, and is generally not considered further. However, we should comment that although the surface integral of the direct Green function is zero (10.4.6), the same is not necessarily true for the approximate ray Green function (10.4.5). It is easily shown that the contribution from the direct ray is asymptotically lower order in frequency than the reflected ray (see further comments below), but in general is non-zero. Because it is lower order, for compatibility it is necessary to include higher-order terms in the ray series (5.1.1) for the reflected ray. By studying the situation in a homogeneous case, it can be seen that the contribution from the far-field term (the ray approximation) cancels with the contribution from the near-field term. It is necessary to use the exact Green function (4.5.71) rather than the far-field, ray theory approximation (4.5.72). In addition, result (10.4.6) applies when the direct field is integrated over the complete surface. In the Kirchhoff approximation, the surface integral is normally only performed over the illuminated portions of the surface. It is assumed but unproven that it is better to ignore the surface integral of the direct field completely even though, with the ray approximation and the incomplete surface integral, the result is non-zero.

Combining the ray-theory approximation for the reflected wave from the source at the interface (10.4.8) with the receiver Green function (10.4.4) in the surface integral (10.4.7), we have

$$\underline{\mathbf{u}}^{\mathrm{K}}(\omega, \mathbf{x}_\mathsf{R}; \mathbf{x}_\mathsf{S}) = -\frac{\left(f^{(\ell)}(\omega)\right)^2}{\mathrm{i}\omega} \int_{S'} \underline{\mathcal{K}}^{\mathrm{K}}(\mathbf{x}, \mathcal{L}_\mathsf{R}, \mathcal{L}_\mathsf{S})\, \mathrm{e}^{\mathrm{i}\omega \widetilde{T}(\mathbf{x}, \mathcal{L}_\mathsf{R}, \mathcal{L}_\mathsf{S})}\, \mathrm{d}S', \qquad (10.4.9)$$

where

$$\widetilde{T}(\mathbf{x}, \mathcal{L}_\mathsf{R}, \mathcal{L}_\mathsf{S}) = T(\mathbf{x}, \mathcal{L}_\mathsf{R}) + T(\mathbf{x}, \mathcal{L}_\mathsf{S}), \qquad (10.4.10)$$

is the travel time from the source to the receiver via the interface point. The Kirchhoff scattering kernel, $\underline{\mathcal{K}}^{\mathrm{K}}$, is a 3×3 matrix and can be written

$$\underline{\mathcal{K}}^{\mathrm{K}}(\mathbf{x}, \mathcal{L}_\mathsf{R}, \mathcal{L}_\mathsf{S}) = \Gamma^{\mathrm{K}}(\mathbf{x}, \mathcal{L}_\mathsf{R}, \mathcal{L}_\mathsf{S})\, \underline{\mathcal{D}}^{(\ell)}(\mathbf{x}, \mathcal{L}_\mathsf{R}, \mathcal{L}_\mathsf{S}), \qquad (10.4.11)$$

where the scattering dyadic has been given in equation (10.3.54) and the scalar Kirchhoff scattering term is $\Gamma^{\mathrm{K}}(\mathbf{x}, \mathcal{L}_\mathsf{R}, \mathcal{L}_\mathsf{S})$. As in the Born expressions, the ray path in Green function (10.4.4) is from the receiver, $\mathbf{x}_\mathsf{R}$, to the reflection point, $\mathbf{x}$, not in the actual propagation direction. It is important to define the polarization and slowness vectors consistently in all the terms in the kernel (10.4.11).

After some manipulation, the scalar Kirchhoff term reduces to

$$\Gamma^{\mathrm{K}}(\mathbf{x}, \mathcal{L}_{\mathrm{R}}, \mathcal{L}_{\mathrm{S}}) = \frac{1}{2}\mathcal{T}_{\mathrm{RS}}(\cos\theta_{\mathrm{R}} + \cos\theta_{\mathrm{S}}), \tag{10.4.12}$$

where

$$\cos\theta_{\mathrm{S}} = -\hat{\mathbf{g}}_{\mathrm{R}}(\mathbf{x}, \mathcal{L}_{\mathrm{S}}) \cdot \hat{\mathbf{n}} \tag{10.4.13}$$

$$\cos\theta_{\mathrm{R}} = \hat{\mathbf{g}}(\mathbf{x}, \mathcal{L}_{\mathrm{R}}) \cdot \hat{\mathbf{n}}, \tag{10.4.14}$$

and θ is the angle between the polarization, $\hat{\mathbf{g}}$ (or ray slowness $\hat{\mathbf{p}} = \hat{\mathbf{g}}$, as it is acoustic) and the normal to the surface, $\hat{\mathbf{n}}$. The polarizations and angles are indicated in Figure 10.30 – expressions (10.4.13) and (10.4.14) are positive and the angles acute. To evaluate the surface integral (10.4.9) either numerically or asymptotically, we consider the two and three-dimensional results independently.

10.4.1.1 Two-dimensional Kirchhoff integral

In two dimensions, the external frequency factor in integral (10.4.9) reduces to

$$(-\mathrm{i}\omega)^{-1}\left(f^{(2)}(\omega)\right)^2 = \frac{1}{2\pi}. \tag{10.4.15}$$

Inverting the frequency transform, we obtain the impulse response

$$\underline{\mathbf{u}}^{\mathrm{K}}(t, \mathbf{x}_{\mathrm{R}}; \mathbf{x}_{\mathrm{S}}) = \frac{1}{2\pi}\,\mathrm{Re}\left(\Delta(t) * \int_{S'} \delta\left(t - \widetilde{T}(\mathbf{x}, \mathcal{L}_{\mathrm{R}}, \mathcal{L}_{\mathrm{S}})\right) \underline{\mathcal{K}}^{\mathrm{K}}(\mathbf{x}, \mathcal{L}_{\mathrm{R}}, \mathcal{L}_{\mathrm{S}})\,\mathrm{d}S'\right), \tag{10.4.16}$$

where, in two dimensions, the surface integral is a line integral, i.e. the surface variable, S', is the length along the interface. This integral is easy to evaluate as the delta function only contributes at points where

$$\widetilde{T}(\mathbf{x}, \mathcal{L}_{\mathrm{R}}, \mathcal{L}_{\mathrm{S}}) = t, \tag{10.4.17}$$

on the interface, i.e. where an isochron surface defined by equation (10.4.17) interfaces the interface (Figure 10.31). The result reduces to

$$\boxed{\underline{\mathbf{u}}^{\mathrm{K}}(t, \mathbf{x}_{\mathrm{R}}; \mathbf{x}_{\mathrm{S}}) = \frac{1}{2\pi}\,\mathrm{Re}\left(\Delta(t) * \sum_{\widetilde{T}=t} \frac{\underline{\mathcal{K}}^{\mathrm{K}}(\mathbf{x}, \mathcal{L}_{\mathrm{R}}, \mathcal{L}_{\mathrm{S}})}{|\partial\widetilde{T}/\partial S'|}\right),} \tag{10.4.18}$$

where the summation is just over those points where equation (10.4.17) is satisfied, and the partial derivative of the total travel time, $\widetilde{T}$, is along the interface. It can be

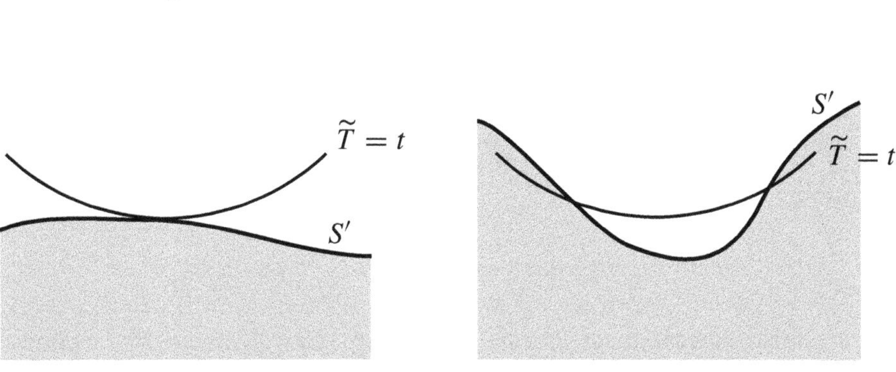

Fig. 10.31. The isochron surface $\widetilde{T}(\mathbf{x}, \mathcal{L}_R, \mathcal{L}_S) = t$ and the interface S' near the spectral reflection point, when the KMAH index (10.4.32) is incremented (right) or not (left). For simplicity, we have illustrated the isochron for coincident source and receiver – zero offset.

written

$$\left|\partial\widetilde{T}/\partial S'\right| = \left|\hat{\mathbf{n}} \times \nabla\widetilde{T}\right| \tag{10.4.19}$$

$$= \left|\hat{\mathbf{n}} \times \left(\mathbf{p}_S(\mathbf{x}, \mathcal{L}_S) + \mathbf{p}(\mathbf{x}, \mathcal{L}_R)\right)\right| \tag{10.4.20}$$

$$= |\mathbf{p}_\perp(\mathbf{x}, \mathcal{L}_S) + \mathbf{p}_\perp(\mathbf{x}, \mathcal{L}_R)| \tag{10.4.21}$$

$$= |\sin\theta_S - \sin\theta_R| \big/ \alpha(\mathbf{x}), \tag{10.4.22}$$

where θ_S and θ_R are both positive, acute angles with the normal (remember that the slowness of the receiver ray is reversed, Figure 10.30). Equation (10.4.21) uses the notation of equation (6.2.1).

The algorithm used to obtain the result (10.4.18) is very similar to that used for the Maslov asymptotic seismogram in two dimensions (cf. equation (10.1.14)), except that the integration parameter is the position on the interface rather than the ray parameter. Geometrical arrivals occur when the total travel time (10.4.10), $\widetilde{T}$, is stationary, i.e.

$$\partial\widetilde{T}/\partial S' = 0. \tag{10.4.23}$$

Thus the gradient is perpendicular to the interface, i.e.

$$\nabla\widetilde{T} = \mathbf{p}_S(\mathbf{x}, \mathcal{L}_S) + \mathbf{p}(\mathbf{x}, \mathcal{L}_R), \tag{10.4.24}$$

is parallel to the normal, $\hat{\mathbf{n}}$, from equation (10.4.20). The components of slownesses $\mathbf{p}_S(\mathbf{x}, \mathcal{L}_S)$ and $\mathbf{p}(\mathbf{x}, \mathcal{L}_R)$ parallel to the interface must be equal and opposite (using expression (10.4.21) in equation (10.4.23)). This corresponds to the reflection condition (Snell's law, Section 6.2.1)

$$\theta_{\mathrm{S}} = \theta_{\mathrm{R}}. \tag{10.4.25}$$

We refer to the positions on the interface that satisfy (10.4.23) as the *spectral reflection points*, where Snell's reflection law is satisfied. Denoting these points by $S' = S'_{\mathrm{ray}}$, or $\mathbf{x} = \mathbf{x}_{\mathrm{ray}}$, we can approximate the total travel time by a quadratic expansion, i.e.

$$\widetilde{T}(\mathbf{x}, \mathcal{L}_{\mathrm{R}}, \mathcal{L}_{\mathrm{S}}) \simeq T(\mathbf{x}_{\mathrm{R}}, \mathbf{x}_{\mathrm{S}}) + \frac{1}{2}\frac{\partial^2 \widetilde{T}}{\partial S'^2}(S' - S'_{\mathrm{ray}})^2. \tag{10.4.26}$$

We can either approximate the spectral result (10.4.9) using the second-order saddle-point method (Appendix D), or use approximation (10.4.26) directly in result (10.4.18) to obtain the first-motion approximation

$$\underline{\mathbf{u}}^{\mathrm{K}}(t, \mathbf{x}_{\mathrm{R}}; \mathbf{x}_{\mathrm{S}}) \simeq \frac{1}{2^{1/2}\pi} \operatorname{Re}\left(\frac{\underline{\mathcal{K}}^{\mathrm{K}}(\mathbf{x}, \mathcal{L}_{\mathrm{R}}, \mathcal{L}_{\mathrm{S}})\, \mathrm{e}^{-\mathrm{i}\pi\left(1-\operatorname{sgn}(\partial\widetilde{T}^2/\partial S'^2)\right)/4}\, \Lambda\left(t - T(\mathbf{x}_{\mathrm{R}}, \mathbf{x}_{\mathrm{S}})\right)}{\left|\partial\widetilde{T}^2/\partial S'^2\right|^{1/2}} \right). \tag{10.4.27}$$

At the saddle point, the scalar reflection term (10.4.12) reduces to

$$\Gamma^{\mathrm{K}}(\mathbf{x}, \mathcal{L}_{\mathrm{R}}, \mathcal{L}_{\mathrm{S}}) = \cos\theta_{\mathrm{S}}\, \mathcal{T}_{\mathrm{RS}}, \tag{10.4.28}$$

as condition (10.4.25) is satisfied. The perturbation on the interface is related to a wavefront perturbation by

$$\delta S' = S' - S'_{\mathrm{ray}} = \frac{\delta \mathsf{q}}{\cos\theta_{\mathrm{S}}}, \tag{10.4.29}$$

and so we can relate the second derivative of $\widetilde{T}$ on the interface to the wavefront curvature (5.2.47)

$$\frac{\partial^2 \widetilde{T}}{\partial S'^2} = \cos^2\theta_{\mathrm{S}}\left(\mathsf{M}(\mathbf{x}, \mathbf{x}_{\mathrm{R}}) + \mathsf{M}(\mathbf{x}, \mathbf{x}_{\mathrm{S}})\right) \tag{10.4.30}$$

(in two dimensions, M is a scalar). Thus using the chain rule (5.2.57), factors in expression (10.4.27) with definition (10.4.11) simplify as

$$\frac{\mathcal{T}^{(2)}(\mathbf{x}_{\mathrm{ray}}, \mathcal{L}_{\mathrm{R}})\, \mathcal{T}^{(2)}(\mathbf{x}_{\mathrm{ray}}, \mathcal{L}_{\mathrm{S}})\, \mathcal{T}_{\mathrm{RS}}}{\left|\partial\widetilde{T}^2/\partial S'^2\right|^{1/2}} = \frac{\mathcal{T}^{(2)}(\mathbf{x}_{\mathrm{R}}, \mathcal{L}_{\mathrm{S}})}{\cos\theta_{\mathrm{S}}}, \tag{10.4.31}$$

where $\mathcal{T}^{(2)}(\mathbf{x}_{\mathrm{R}}, \mathcal{L}_{\mathrm{S}})$ is for the complete ray (6.8.2) and includes the product of all the reflection/transmission coefficients in the two segments, $\mathcal{T}^{(2)}(\mathbf{x}_{\text{ray}}, \mathcal{L}_{\text{R}})$ and $\mathcal{T}^{(2)}(\mathbf{x}_{\text{ray}}, \mathcal{L}_{\text{S}})$, and the interface coefficient $\mathcal{T}_{\text{RS}}$ (6.3.7). Combining results (10.4.28) and (10.4.31) in approximation (10.4.27) we obtain result (5.4.37) for the geometrical reflection in two dimensions. The KMAH indices combine as

$$\sigma(\mathbf{x}_{\text{R}}, \mathcal{L}_{\text{S}}) = \sigma(\mathbf{x}, \mathcal{L}_{\text{R}}) + \sigma(\mathbf{x}, \mathcal{L}_{\text{S}}) + \frac{1}{2}\left(1 - \text{sgn}\left(\frac{\partial^2 \widetilde{T}}{\partial S'^2}\right)\right). \tag{10.4.32}$$

Normally, the total travel-time function (10.4.10), $\widetilde{T}(\mathbf{x}, \mathcal{L}_{\text{R}}, \mathcal{L}_{\text{S}})$, is minimum at the spectral reflection point, $\mathbf{x} = \mathbf{x}_{\text{ray}}$, so the final term in expression (10.4.32) is zero. If the KMAH indices for the individual segments are zero, and all the reflection/transmission coefficients are real, then the first-motion approximation (10.4.27) has the form $\lambda\left(t - T(\mathbf{x}_{\text{R}}, \mathbf{x}_{\text{S}})\right)$. Occasionally, if the curvature of the interface is greater than the curvature of the isochron $\widetilde{T}(\mathbf{x}, \mathcal{L}_{\text{R}}, \mathcal{L}_{\text{S}}) = t$ (Figure 10.31), the second derivative (10.4.30) is negative, the final term in expression (10.4.32) increments the KMAH index, and the first-motion approximation (10.4.27) has the form of the Hilbert transform $\bar{\lambda}\left(t - T(\mathbf{x}_{\text{R}}, \mathbf{x}_{\text{S}})\right)$.

It is interesting to consider what happens if the direct ray is included in the scattering integral (10.4.9). The total field (10.4.1), with approximations (10.4.5) and (10.4.8), is used in the integral so the Kirchhoff scatterer (10.4.11), $\underline{\mathcal{K}}^{\text{K}}$, is replaced by the total scatterer, $\underline{\mathcal{K}} = \underline{\mathcal{K}}^{\text{D}} + \underline{\mathcal{K}}^{\text{K}}$, and the scalar scattering amplitude (10.4.12), Γ^{K}, by $\Gamma = \Gamma^{\text{D}} + \Gamma^{\text{K}}$ where

$$\Gamma^{\text{D}}(\mathbf{x}, \mathcal{L}_{\text{R}}, \mathcal{L}_{\text{S}}) = \frac{1}{2}(\cos\theta_{\text{R}} - \cos\theta_{\text{S}}). \tag{10.4.33}$$

As already mentioned, this term makes zero contribution to the asymptotc result as condition (10.4.25) is satisfied at the spectral point. However, if $\underline{\mathcal{K}}$ were substituted in expression (10.4.18), the contribution from Γ^{D} is not zero but is lower order than the geometrical arrival (10.4.27), i.e. the waveform is of the form $\mu(t)$ (9.2.30), half the integral of $\lambda(t)$. If the exact Green function were known for $\underline{\mathbf{u}}^{\text{D}}$, its contribution in the scattering integral must be exactly zero (10.4.6), but the contribution from the ray approximation is non-zero. As its contribution is lower order, it is necessary to include higher-order terms in the ray series (5.1.1) in order to obtain cancellation. This can be demonstrated explicitly for a homogeneous medium, when the exact Green function (4.5.71) is known. In general, however, higher-order terms are not included in the ray approximation.

Finally, we comment that numerical evaluation of result (10.4.18) is unstable as the expression contains singularities. It can be made numerically robust by band-limiting the result as in the WKBJ seismogram (8.4.14). Thus expression (10.4.18) is replaced by

$$\boxed{\begin{aligned}&\frac{1}{\Delta t}B\left(\frac{t}{\Delta t}\right)*\underline{\mathbf{u}}^{\mathrm{K}}(t,\mathbf{x}_{\mathrm{R}};\mathbf{x}_{\mathrm{S}})\\&\quad=\frac{1}{4\pi\,\Delta t}\,\mathrm{Re}\left(\Delta(t)*\int_{\widetilde{T}=t\pm\Delta t}\underline{\mathcal{K}}^{\mathrm{K}}(\mathbf{x},\mathcal{L}_{\mathrm{R}},\mathcal{L}_{\mathrm{S}})\,\mathrm{d}S'\right).\end{aligned}} \tag{10.4.34}$$

Exactly the same numerical algorithm can be used as for the WKBJ or two-dimensional Maslov seismograms.

10.4.1.2 Three-dimensional Kirchhoff integral

In three dimensions, the Kirchhoff surface integral method is essentially the same as in two dimensions except that the integral is over a two dimensional surface. Equations (10.4.3) through (10.4.14) still all apply except that the external frequency factor in expression (10.4.9) is

$$(-\mathrm{i}\omega)^{-1}\left(f^{(3)}(\omega)\right)^2=-\frac{\mathrm{i}\omega}{4\pi^2}. \tag{10.4.35}$$

Inverting the Fourier transform, equation (10.4.9) becomes

$$\begin{aligned}&\underline{\mathbf{u}}^{\mathrm{K}}(t,\mathbf{x}_{\mathrm{R}};\mathbf{x}_{\mathrm{S}})\\&\quad=\frac{1}{4\pi^2}\frac{\mathrm{d}}{\mathrm{d}t}\mathrm{Re}\left(\Delta(t)*\int_{S'}\delta\left(t-\widetilde{T}(\mathbf{x},\mathcal{L}_{\mathrm{R}},\mathcal{L}_{\mathrm{S}})\right)\underline{\mathcal{K}}^{\mathrm{K}}(\mathbf{x},\mathcal{L}_{\mathrm{R}},\mathcal{L}_{\mathrm{S}})\,\mathrm{d}S'\right),\end{aligned} \tag{10.4.36}$$

where $\mathrm{d}S'$ is an area element. The isochron surface defined by equation (10.4.17) now intersects the interface in a line. The surface integral can be reduced to a line integral

$$\boxed{\underline{\mathbf{u}}^{\mathrm{K}}(t,\mathbf{x}_{\mathrm{R}};\mathbf{x}_{\mathrm{S}})=\frac{1}{4\pi^2}\frac{\mathrm{d}}{\mathrm{d}t}\mathrm{Re}\left(\int_{\widetilde{T}=t}\frac{\underline{\mathcal{K}}^{\mathrm{K}}(\mathbf{x},\mathcal{L}_{\mathrm{R}},\mathcal{L}_{\mathrm{S}})}{\left|\nabla_{\mathrm{s}}\widetilde{T}(\mathbf{x},\mathcal{L}_{\mathrm{R}},\mathcal{L}_{\mathrm{S}})\right|}\,\mathrm{d}s'\right),} \tag{10.4.37}$$

along lines where equation (10.4.17) is satisfied, and the gradient, $\nabla_{\mathrm{s}}\widetilde{T}$, is in the interface. The algorithm used to obtain this result is similar to the three-dimensional Maslov result (Section 10.1.2), except that the integration parameter is the position on the interface rather than the slowness. This is illustrated in Figure 10.32*a*.

Again geometrical arrivals occur when the total travel time (10.4.17), $\widetilde{T}$, is stationary

$$\nabla_{\mathrm{s}}\widetilde{T}=\mathbf{0}. \tag{10.4.38}$$

Again this implies that the gradient (10.4.24) is perpendicular to the interface or parallel to the normal, $\hat{\mathbf{n}}$. This reduces to Snell's reflection condition (10.4.25),

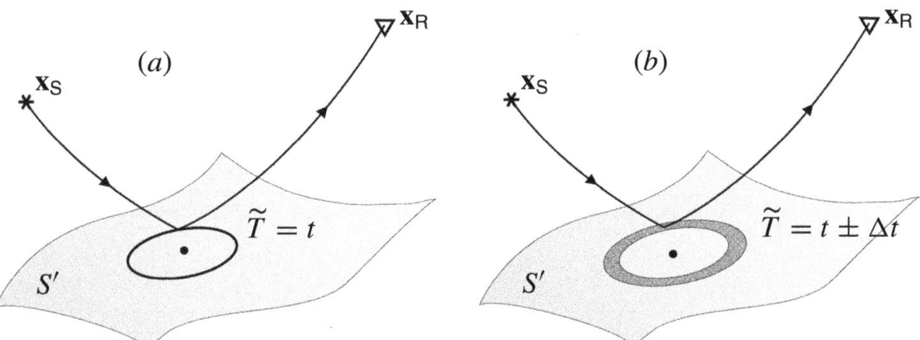

Fig. 10.32. The (*a*) isochron line (10.4.37) and (*b*) strip (10.4.45) integrals on the surface in the Kirchhoff surface integral in the time domain.

where, in addition, the slownesses must be coplanar. Expanding about the spectral point, $\mathsf{s}' = \mathsf{s}'_{\text{ray}}$, the total travel time is

$$\widetilde{T}(\mathbf{x}, \mathcal{L}_{\mathsf{R}}, \mathcal{L}_{\mathsf{S}}) \simeq T(\mathbf{x}_{\mathsf{R}}, \mathbf{x}_{\mathsf{S}}) + \frac{1}{2}\,\delta\mathsf{s}'^{\mathsf{T}}\nabla_{\mathsf{s}}(\nabla_{\mathsf{s}}\widetilde{T})^{\mathsf{T}}\,\delta\mathsf{s}', \tag{10.4.39}$$

where

$$\delta\mathsf{s}' = \mathsf{s}' - \mathsf{s}'_{\text{ray}}. \tag{10.4.40}$$

Evaluating the saddle-point approximation of integral (10.4.9) using result (D.1.18), we obtain

$$\underline{\mathbf{u}}^{\mathrm{K}}(\omega, \mathbf{x}_{\mathsf{R}}; \mathbf{x}_{\mathsf{S}}) \simeq -\frac{\mathrm{i}}{2\pi}\,\frac{\underline{\mathcal{K}}^{\mathrm{K}}(\mathbf{x}, \mathcal{L}_{\mathsf{R}}, \mathcal{L}_{\mathsf{S}})}{\left\|\nabla_{\mathsf{s}}(\nabla_{\mathsf{s}}\widetilde{T})^{\mathsf{T}}\right\|^{1/2}}\,\mathrm{e}^{\mathrm{i}\omega T + \mathrm{i}\pi\,\mathrm{sgn}\left(\omega\nabla_{\mathsf{s}}(\nabla_{\mathsf{s}}\widetilde{T})^{\mathsf{T}}\right)/4}. \tag{10.4.41}$$

Equation (10.4.30) is replaced by

$$\nabla_{\mathsf{s}}\left(\nabla_{\mathsf{s}}\widetilde{T}\right)^{\mathsf{T}} = \cos^2\theta_{\mathsf{S}}\left(\mathsf{M}(\mathbf{x}, \mathbf{x}_{\mathsf{R}}) + \mathsf{M}(\mathbf{x}, \mathbf{x}_{\mathsf{S}})\right) \tag{10.4.42}$$

where the wavefront matrices are of dimension 2×2. Thus equation (10.4.31) is replaced by

$$\frac{\mathcal{T}^{(3)}(\mathbf{x}_{\text{ray}}, \mathcal{L}_{\mathsf{R}})\,\mathcal{T}^{(3)}(\mathbf{x}_{\text{ray}}, \mathcal{L}_{\mathsf{S}})\,\mathcal{T}_{\mathsf{RS}}}{\left\|\nabla_{\mathsf{s}}(\nabla_{\mathsf{s}}\widetilde{T})^{\mathsf{T}}\right\|^{1/2}} = \frac{\mathcal{T}^{(3)}(\mathbf{x}, \mathcal{L}_{\mathsf{S}})}{\cos\theta_{\mathsf{S}}}. \tag{10.4.43}$$

Equation (10.4.28) still applies, and combining these results we obtain the geometrical result, (5.4.28) and (5.4.31), in three dimensions. The KMAH indices combine as

$$\sigma(\mathbf{x}_{\mathsf{R}}, \mathcal{L}_{\mathsf{S}}) = \sigma(\mathbf{x}, \mathcal{L}_{\mathsf{R}}) + \sigma(\mathbf{x}, \mathcal{L}_{\mathsf{S}}) + 1 - \frac{1}{2}\,\mathrm{sgn}\left(\nabla_{\mathsf{s}}(\nabla_{\mathsf{s}}\widetilde{T})^{\mathsf{T}}\right). \tag{10.4.44}$$

Normally, the total travel-time function (10.4.17), $\widetilde{T}(\mathbf{x}, \mathcal{L}_{\mathrm{R}}, \mathcal{L}_{\mathrm{S}})$, is minimum at the spectral point, and the total KMAH index is just the sum of the segment values. Depending on the curvature of the interface compared with the isochron, the reflection may introduce a Hilbert transform or a sign change.

Finally, the evaluation of the integral (10.4.37) is numerically robust if the result is smoothed to give the band-limited response. Thus expression (10.4.37) is replaced by

$$\boxed{\begin{aligned}&\frac{1}{\Delta t} B\left(\frac{t}{\Delta t}\right) * \underline{\mathbf{u}}^{\mathrm{K}}(t, \mathbf{x}_{\mathrm{R}}; \mathbf{x}_{\mathrm{S}}) \\ &\quad = \frac{1}{8\pi^2 \Delta t} \frac{\mathrm{d}}{\mathrm{d}t} \operatorname{Re}\left(\Delta(t) * \int_{\widetilde{T}=t\pm\Delta t} \underline{\mathcal{K}}^{\mathrm{K}}(\mathbf{x}, \mathcal{L}_{\mathrm{R}}, \mathcal{L}_{\mathrm{S}})\, \mathrm{d}s'\right),\end{aligned}} \tag{10.4.45}$$

where the integral is over strips on the surface. The band-limited result, (10.4.45), which is in a form suitable for numerical evaluation, has surface integrals over strips defined by $\widetilde{T}(\mathbf{x}, \mathcal{L}_{\mathrm{R}}, \mathcal{L}_{\mathrm{S}}) = t \pm \Delta t$. If the surface is divided into triangular elements, where the ray results are known at the apexes, then exactly the same algorithm used for three-dimensional Maslov seismograms (Section 10.1.2.1) can be used to evaluate these surface integrals efficiently (Spencer, Chapman and Kragh, 1997 – see Section 10.1). The two-dimensional slowness integral is replaced by a surface integral. These surface integrals are illustrated in Figure 10.32*b*.

As an example of the Kirchhoff surface integral method, we have computed seismograms for the French (1974) model (Figure 6.12) used to illustrate Born scattering theory. The model details are the same as before. Rays are traced to the two interfaces in the French model (Figure 6.12) from positions on the profile used in Figures 6.15, 10.26 and 10.27. Zero-offset seismograms calculated using result (10.4.45) are shown in Figure 10.33. Signals are similar to those in Figure 10.26. The computational cost is less for the Kirchhoff surface integral method as the three-dimensional volume integral is replaced by two-dimensional surface integrals.

10.4.2 Anisotropic Kirchhoff surface integral

The elastic, isotropic or anisotropic, Kirchhoff surface integral method is developed in the same manner as the acoustic case, with the extra complication that more than one reflected ray may exist. The representation (Betti's) theorem (4.5.49) gives the reflected signal in terms of a surface integral. Thus, following the same procedure as for acoustics, we have for the scattered signal in the frequency

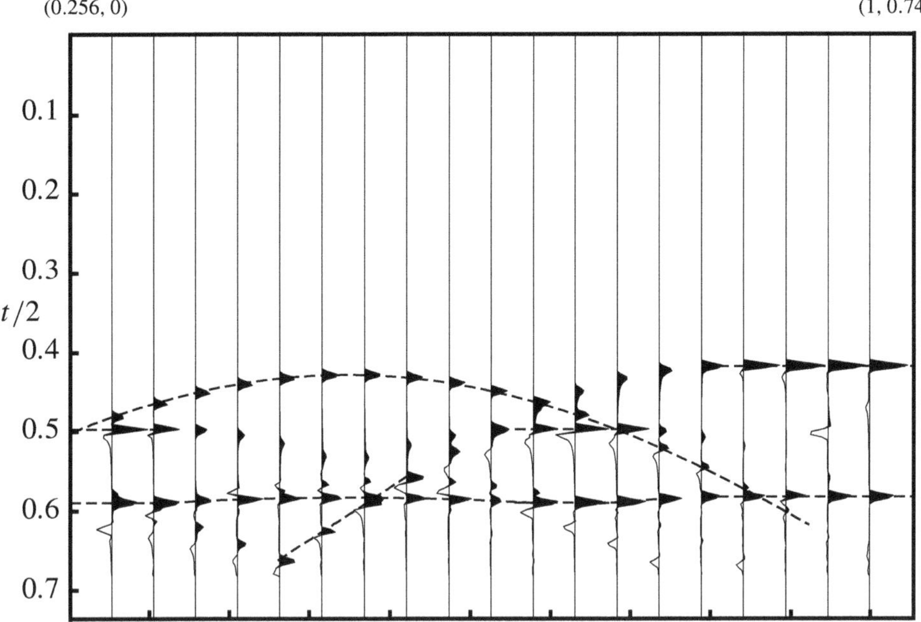

Fig. 10.33. Kirchhoff surface integral zero-offset seismograms for the profile illustrated in Figure 6.12, i.e. arrivals corresponding to those in Figure 6.15.

domain

$$\underline{\mathbf{u}}^{\mathrm{K}}(\omega, \mathbf{x}_{\mathrm{R}}; \mathbf{x}_{\mathrm{S}}) = \int_{S'} \Big(\underline{\mathbf{u}}^{\mathrm{T}}(\omega, \mathbf{x}; \mathbf{x}_{\mathrm{R}})\, \underline{\mathbf{t}}_j(\omega, \mathbf{x}; \mathbf{x}_{\mathrm{S}}) - \underline{\mathbf{t}}_j^{\mathrm{T}}(\omega, \mathbf{x}; \mathbf{x}_{\mathrm{R}})\, \underline{\mathbf{u}}(\omega, \mathbf{x}; \mathbf{x}_{\mathrm{S}}) \Big)\, \hat{n}_j(\mathbf{x})\, \mathrm{d}S', \qquad (10.4.46)$$

where the point $\mathbf{x}$ lies on the surface S' of the integral, and the normal $\hat{\mathbf{n}}(\mathbf{x})$ is out of the volume V (into the volume V').

The Green functions from the receiver – argument $(\omega, \mathbf{x}; \mathbf{x}_{\mathrm{R}})$ – are again approximated by the Green functions in a medium without the scatterer, i.e. when the medium V is continued through V'. They are approximated by ray theory (Section 5.4.2) using expression (5.4.32), i.e.

$$\begin{pmatrix} \underline{\mathbf{v}} \\ \underline{\mathbf{t}}_j \end{pmatrix}(\omega, \mathbf{x}; \mathbf{x}_{\mathrm{R}}) \simeq f^{(3)}(\omega)\, \mathcal{T}^{(3)}(\mathbf{x}, \mathcal{L}_{\mathrm{R}})\, \Phi(\omega, \mathbf{x}, \mathcal{L}_{\mathrm{R}}) \times \begin{pmatrix} \mathbf{g} \\ -\mathbf{Z}_j \mathbf{g} \end{pmatrix}(\mathbf{x}, \mathcal{L}_{\mathrm{R}})\, \mathbf{g}^{\mathrm{T}}(\mathbf{x}_{\mathrm{R}}, \mathcal{L}_{\mathrm{R}}) \qquad (10.4.47)$$

(cf. equation (10.4.4) – for brevity we only consider the result in three dimensions). For simplicity, we only include one ray type in the Green function. Others can be solved and combined independently. We use the shorthand $\mathcal{L}_{\mathsf{R}}$ to indicate the ray path from $\mathbf{x}_{\mathsf{R}}$ to $\mathbf{x}$, and for simplicity, omit a summation over multi-pathing and ray type. The product of coefficients in $\mathcal{T}^{(3)}(\mathbf{x}, \mathcal{L}_{\mathsf{R}})$ is for interfaces between $\mathbf{x}_{\mathsf{R}}$ and $\mathbf{x}$, but not the interface at $\mathbf{x}$.

As in the acoustic case, the solution on the surface S' can be divided into the direct and scattered part (10.4.1). Again the representation theorem (4.5.49) can be used to prove that the contribution of the direct part to the surface integral is zero. Thus the surface integral (10.4.46) reduces to

$$\underline{\mathbf{u}}^{\mathsf{K}}(\omega, \mathbf{x}_{\mathsf{R}}; \mathbf{x}_{\mathsf{S}}) = \int_{S'} \left(\underline{\mathbf{u}}^{\mathsf{T}}(\omega, \mathbf{x}; \mathbf{x}_{\mathsf{R}})\, \underline{\mathbf{t}}_j^{\mathsf{K}}(\omega, \mathbf{x}; \mathbf{x}_{\mathsf{S}}) - \underline{\mathbf{t}}_j^{\mathsf{T}}(\omega, \mathbf{x}; \mathbf{x}_{\mathsf{R}})\, \underline{\mathbf{u}}^{\mathsf{K}}(\omega, \mathbf{x}; \mathbf{x}_{\mathsf{S}}) \right) \hat{n}_j(\mathbf{x})\, \mathrm{d}S'. \tag{10.4.48}$$

However, as the direct term is only zero for the exact Green function, for the moment we retain it.

The solution on the surface S' from the source – Green functions with arguments $(\omega, \mathbf{x}; \mathbf{x}_{\mathsf{S}})$ in (10.4.46) – is again approximated by the ray approximation. Thus generalizing expression (5.4.32) so it applies on the surface (Section 6.6), we have

$$\begin{pmatrix} \underline{\mathbf{v}} \\ \underline{\mathbf{t}}_j \end{pmatrix}(\omega, \mathbf{x}; \mathbf{x}_{\mathsf{S}}) \simeq f^{(3)}(\omega)\, \mathcal{T}^{(3)}(\mathbf{x}, \mathcal{L}_{\mathsf{S}})\, \Phi(\omega, \mathbf{x}, \mathcal{L}_{\mathsf{S}})\, \mathbf{w}(\mathbf{x}, \mathcal{L}_{\mathsf{S}})\, \mathbf{g}^{\mathsf{T}}(\mathbf{x}_{\mathsf{S}}, \mathcal{L}_{\mathsf{S}}), \tag{10.4.49}$$

with

$$\mathbf{w}(\mathbf{x}, \mathcal{L}_{\mathsf{S}}) = \begin{pmatrix} \mathbf{h} \\ \mathbf{s}_j \end{pmatrix}(\mathbf{x}, \mathcal{L}_{\mathsf{S}}), \tag{10.4.50}$$

where $\mathbf{h}$ is the interface polarization conversion (Section 6.6) and $\mathbf{s}_j$ is the corresponding traction. Expressing these in terms of the incident and reflected rays (recalling equation (6.8.5)), they are

$$\mathbf{h} = \mathbf{g}_{\mathsf{S}}(\mathbf{x}, \mathcal{L}_{\mathsf{S}}) + \sum_{l=1}^{3} \mathcal{T}_{l\mathsf{S}}\mathbf{g}_l(\mathbf{x}, \mathcal{L}_{\mathsf{S}}) = \sum_{l=0}^{3} \mathcal{T}_{l\mathsf{S}}\mathbf{g}_l(\mathbf{x}, \mathcal{L}_{\mathsf{S}}) \tag{10.4.51}$$

$$\mathbf{s}_j = -\mathbf{Z}_{\mathsf{S}j}\mathbf{g}_{\mathsf{S}}(\mathbf{x}, \mathcal{L}_{\mathsf{S}}) - \sum_{l=1}^{3} \mathcal{T}_{l\mathsf{S}}\mathbf{Z}_{lj}\mathbf{g}_l(\mathbf{x}, \mathcal{L}_{\mathsf{S}}) = -\sum_{l=0}^{3} \mathcal{T}_{l\mathsf{S}}\mathbf{Z}_{lj}\mathbf{g}_l(\mathbf{x}, \mathcal{L}_{\mathsf{S}}). \tag{10.4.52}$$

In these expressions we have included a subscript to indicate the ray type, S, for the incident ray and l for the reflected waves. Then $\mathcal{T}_{l\mathsf{S}}$ is the reflection coefficient for the incident ray, S, converting into the receiver ray, l. We have simplified the expressions including $l = 0$ for the incident ray, i.e. defining $\mathcal{T}_{0\mathsf{S}} = 1$. Note that the

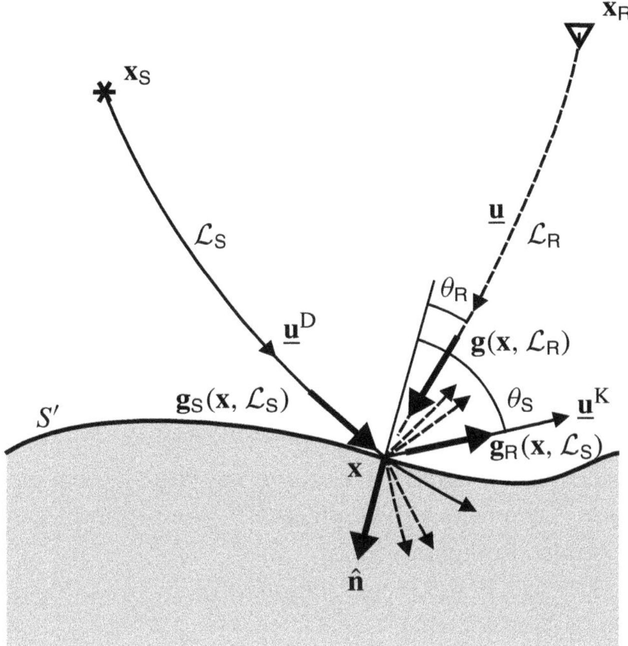

Fig. 10.34. As Figure 10.30 except that up to three reflected rays are generated at the reflection point. For illustrative purposes, the quasi-shear rays are shown dashed.

matrix impedances (5.3.22), for each term in the tractions, differ as the slowness vectors, $\mathbf{p}_S$ and $\mathbf{p}_{1l}$, differ (as do the polarizations, $\mathbf{g}_l$, of course).

Combining the ray-theory approximations (10.4.49) for the wave from the source to the interface with the Green functions to the receiver (10.4.47), in the surface integral (10.4.46), we have

$$\underline{\mathbf{u}}^K(\omega, \mathbf{x}_R; \mathbf{x}_S) = -\frac{\left(f^{(3)}(\omega)\right)^2}{\mathrm{i}\omega} \int_{S'} \mathrm{e}^{\mathrm{i}\omega \widetilde{T}(\mathbf{x}, \mathcal{L}_R, \mathcal{L}_S)} \underline{\mathcal{K}}^K(\mathbf{x}, \mathcal{L}_R, \mathcal{L}_S) \, \mathrm{d}S', \quad (10.4.53)$$

where the total travel time has been defined in equation (10.4.10). The scattering kernel, $\underline{\mathcal{K}}^K$, is a 3×3 matrix and can be written

$$\underline{\mathcal{K}}^K(\mathbf{x}, \mathcal{L}_R, \mathcal{L}_S) = \Gamma^K(\mathbf{x}, \mathcal{L}_R, \mathcal{L}_S) \, \underline{\mathcal{D}}^{(\ell)}(\mathbf{x}, \mathcal{L}_R, \mathcal{L}_S) \,, \quad (10.4.54)$$

where the scattering dyadic has been given in equation (10.3.54) and

$$\Gamma^K(\mathbf{x}, \mathcal{L}_R, \mathcal{L}_S) = -\mathbf{g}^{\mathrm{T}}(\mathbf{x}, \mathcal{L}_R) \left\{ \mathbf{Z}_j^{\mathrm{T}}(\mathbf{x}, \mathcal{L}_R) \, \mathbf{h}(\mathbf{x}, \mathcal{L}_S) + \mathbf{s}_j(\mathbf{x}, \mathcal{L}_S) \right\} \hat{n}_j, \quad (10.4.55)$$

is a scalar Kirchhoff scattering term (cf. definitions (10.4.11)). The physical significance of the terms in Γ^K (10.4.55) is illustrated in Figure 10.34.

For numerical purposes, it is straightforward to transform expression (10.4.53) with result (10.4.35) into the time-domain

$$\mathbf{u}^{\mathrm{K}}(t, \mathbf{x}_{\mathrm{R}}; \mathbf{x}_{\mathrm{S}}) = \frac{1}{4\pi^2}\frac{\mathrm{d}}{\mathrm{d}t}\,\mathrm{Re}\left(\Delta(t) * \int_{\widetilde{T}=t} \frac{\underline{\mathcal{K}}^{\mathrm{K}}(\mathbf{x}, \mathcal{L}_{\mathrm{R}}, \mathcal{L}_{\mathrm{S}})}{\left|\nabla_{\mathrm{s}}\widetilde{T}(\mathbf{x}, \mathcal{L}_{\mathrm{R}}, \mathcal{L}_{\mathrm{S}})\right|}\,\mathrm{d}s'\right) \tag{10.4.56}$$

$$\simeq \frac{1}{8\pi^2\Delta t}\frac{\mathrm{d}}{\mathrm{d}t}\,\mathrm{Re}\left(\Delta(t) * \int_{\widetilde{T}=t\pm\Delta t} \underline{\mathcal{K}}^{\mathrm{K}}(\mathbf{x}, \mathcal{L}_{\mathrm{R}}, \mathcal{L}_{\mathrm{S}})\,\mathrm{d}S'\right). \tag{10.4.57}$$

The integral in the impulse response, (10.4.56), is evaluated along lines on the surface where $\widetilde{T}(\mathbf{x}, \mathcal{L}_{\mathrm{R}}, \mathcal{L}_{\mathrm{S}}) = t$, illustrated in Figure 10.32*a*. The band-limited result, (10.4.57), in a form suitable for numerical evaluation, has surface integrals over strips defined by $\widetilde{T}(\mathbf{x}, \mathcal{L}_{\mathrm{R}}, \mathcal{L}_{\mathrm{S}}) = t \pm \Delta t$, and can be evaluated as in the acoustic case using the same algorithm as the three-dimensional Maslov seismograms (Spencer, Chapman and Kragh, 1997 – see Section 10.1). These surface integrals are illustrated in Figure 10.32*b*.

The surface integral (10.4.53) has stationary points (and result (10.4.56) has corresponding singularities) when

$$\nabla_{\mathrm{s}}\widetilde{T}(\mathbf{x}, \mathcal{L}_{\mathrm{R}}, \mathcal{L}_{\mathrm{S}}) = \nabla_{\perp}T(\mathbf{x}, \mathcal{L}_{\mathrm{R}}) + \nabla_{\perp}T(\mathbf{x}, \mathcal{L}_{\mathrm{S}}) = \mathbf{p}_{\perp}(\mathbf{x}, \mathcal{L}_{\mathrm{R}}) + \mathbf{p}_{\perp}(\mathbf{x}, \mathcal{L}_{\mathrm{S}}) = \mathbf{0}, \tag{10.4.58}$$

which corresponds to the Snell's law condition for a specular reflection (cf. equation (6.2.2) – remember $\mathbf{p}(\mathbf{x}, \mathcal{L}_{\mathrm{R}})$ is reversed compared with the combined ray). The appropriate reflected ray in expressions (10.4.51) and (10.4.52) will match the receiver ray at this point. Usually only this term is retained in the Kirchhoff integral and the other parts of the wavefield on the interface in expressions (10.4.51) and (10.4.52) are dropped. However, if they are retained, to first order they make no contribution to the integral. Let us consider the scattering term (10.4.55), Γ^{K}, substituting the summation (10.4.51) and (10.4.52)

$$\begin{aligned}\Gamma^{\mathrm{K}}(\mathbf{x}, \mathcal{L}_{\mathrm{R}}, \mathcal{L}_{\mathrm{S}}) &= \sum_{l=0}^{3} \mathcal{T}_{l\mathrm{S}}\,\hat{n}_j\,\mathbf{g}^{\mathrm{T}}(\mathbf{x}, \mathcal{L}_{\mathrm{R}})\left(\mathbf{Z}_{lj}(\mathbf{x}, \mathcal{L}_{\mathrm{S}}) - \mathbf{Z}_j^{\mathrm{T}}(\mathbf{x}, \mathcal{L}_{\mathrm{R}})\right)\mathbf{g}_l(\mathbf{x}, \mathcal{L}_{\mathrm{S}})\\ &= \sum_{l=0}^{3} \mathcal{T}_{l\mathrm{S}}\left(\hat{n}_k\,p_{l\,j}(\mathbf{x}, \mathcal{L}_{\mathrm{S}}) - \hat{n}_j\,p_k(\mathbf{x}, \mathcal{L}_{\mathrm{R}})\right)\\ &\quad\times \mathbf{g}^{\mathrm{T}}(\mathbf{x}, \mathcal{L}_{\mathrm{R}})\,\mathbf{c}_{kj}(\mathbf{x})\,\mathbf{g}_l(\mathbf{x}, \mathcal{L}_{\mathrm{S}}),\end{aligned} \tag{10.4.59}$$

using results (5.3.22), (10.4.51) and (10.4.52). As before terms from the free-space ray from the receiver are indicated by the argument $(\mathbf{x}, \mathcal{L}_{\mathrm{R}})$, whereas for the ray from the source and reflections, the argument $(\mathbf{x}, \mathcal{L}_{\mathrm{S}})$ is supplemented with the extra subscript $l = 0$ to 3. The directionality or *obliquity factor* in this expression is mainly through the products of slowness and normal components, though there is a complicated interaction with the anisotropy through the product with matrices $\mathbf{c}_{jk}$. At the saddle point defined by equation (10.4.58), we will have

$$\mathbf{p}(\mathbf{x}, \mathcal{L}_\mathsf{R}) + \mathbf{p}_\mathsf{R}(\mathbf{x}, \mathcal{L}_\mathsf{S}) = \mathbf{0}, \tag{10.4.60}$$

for the reflection that matches the receiver ray (which we denote with $l = \mathsf{R}$). Only equation (10.4.58) will apply for all the other reflections and incident ray, i.e. $l \neq \mathsf{R}$, as they are defined to satisfy Snell's law. It is clear that if we return to the original Kirchhoff surface integral (10.4.48) and use the orthogonality relationship (6.3.33), then at the saddle point the contribution to expression (10.4.59) from these three terms (incident ray and two reflections) is zero. Therefore, at the saddle point, expression (10.4.59) reduces to

$$\Gamma^\mathrm{K}(\mathbf{x}, \mathcal{L}_\mathsf{R}, \mathcal{L}_\mathsf{S}) = -\mathcal{T}_\mathsf{RS}\, \widehat{\mathbf{V}}_\mathsf{R}(\mathbf{x}, \mathcal{L}_\mathsf{S}) \cdot \hat{\mathbf{n}}(\mathbf{x}), \tag{10.4.61}$$

where $\mathcal{T}_\mathsf{RS}$ is the reflection coefficient from the interface at $\mathbf{x}$. Expression (10.4.59) has been reduced using the group velocity of the receiver ray (5.3.20) and energy-flux normalized polarizations (5.4.33). Thus although we have included all four rays in expression (10.4.59) to give the total field on the interface, to first order only the matching ray, $l = \mathsf{R}$, contributes to the saddle point (10.4.61). The Kirchhoff integral method consists of evaluating the surface integral, (10.4.56) or (10.4.57), without the saddle-point approximation. In most publications, only the matching term, $l = \mathsf{R}$, is retained in this integral, even though the other terms are non-zero except at the saddle point. Whether these terms are important has not been investigated yet.

The saddle-point contribution can be evaluated using the stationary phase method (Bleistein, 1984, p. 88) and reduces to the ray-theory result (5.4.35) with definition (6.8.2), where the saddle point values give

$$T(\mathbf{x}_\mathsf{R}, \mathcal{L}_\mathsf{S}) = \widetilde{T}(\mathbf{x}, \mathcal{L}_\mathsf{R}, \mathcal{L}_\mathsf{S}) = T(\mathbf{x}, \mathcal{L}_\mathsf{R}) + T(\mathbf{x}, \mathcal{L}_\mathsf{S}) \tag{10.4.62}$$

$$\sigma(\mathbf{x}_\mathsf{R}, \mathcal{L}_\mathsf{S}) = \sigma(\mathbf{x}, \mathcal{L}_\mathsf{R}) + \sigma(\mathbf{x}, \mathcal{L}_\mathsf{S}) + 1 - \frac{1}{2}\,\mathrm{sgn}\left(\nabla_\mathsf{S}\left(\nabla_\mathsf{S}\widetilde{T}\right)^\mathsf{T}\right) \tag{10.4.63}$$

(Ursin and Tygel, 1997). Using the symmetries of the dynamic propagator (5.2.57), the factor $|\mathcal{T}^{(3)}(\mathbf{x}, \mathcal{L}_\mathsf{R})\, \mathcal{T}^{(3)}(\mathbf{x}, \mathcal{L}_\mathsf{S})|$ can be related to $|\mathcal{T}^{(3)}(\mathbf{x}_\mathsf{R}, \mathcal{L}_\mathsf{S})|$ for the complete ray using $\|\nabla_\mathsf{S}\left(\nabla_\mathsf{S}\widetilde{T}\right)^\mathsf{T}\|$ using the same result as for the acoustic integral (Coates and Chapman, 1990*a*; Ursin and Tygel, 1997).

This completes the theoretical development of the Kirchhoff surface integral method. In this final chapter, four extensions of ray theory – Maslov asymptotic ray theory, quasi-isotropic ray theory, Born error and perturbation scattering theory, and the Kirchhoff surface integral method – have been developed. Although these methods have been demonstrated to be useful, research problems remain. An incomplete list includes end-point errors and pseudo-caustics in Maslov theory, off-ray effects in quasi-isotropic ray theory, multiple-scattering in Born theory and robust implementations in very heterogeneous models, and questions concerning which terms to include in Kirchhoff surface integrals (see also Exercise 10.6). Hybrid methods combined with numerical solutions of the wave equations will probably be necessary in realistic heterogeneous media. Considering the heterogeneity that exists in the Earth on all scales, the success of the methods developed in this book – ray theory, transform methods and extensions of ray theory – is a pleasant surprise which is sometimes difficult to justify theoretically.

Exercises

10.1 In Section 9.2.7, we have investigated the waveforms at Airy caustics in some detail. In three dimensions, more general caustics are possible, e.g. the Pearcey (1946) caustic. Using Maslov asymptotic theory, investigate the waveforms at caustics possible in three dimensions.

10.2 Using transform methods, show that coupling between quasi-shear plane waves exists (using methods similar to Section 7.2.6) and that the form of coupling is similar to the ray result (10.2.56) (see Chapman and Shearer, 1989).

10.3 Using the stationary-phase method to evaluate the Born scattering integral, linearized reflection coefficients can be obtained for a small-contrast interface (Shaw and Sen, 2004). By considering a perturbation to a half-space, show that the coefficients obtained in this matter agree with those in Section 6.7.

10.4 *Further reading:* In Section 10.3.5, we have shown that Born perturbation scattering theory predicts to lowest order the travel-time perturbation. Investigate how Born scattering theory predicts other corrections to ray theory (Coates and Chapman, 1990*a*; Chapman and Coates, 1994).

10.5 Confirm the result (10.3.55) for the acoustic scalar Born error scattering term, Γ^{E}.

Investigate expressions for the spatial derivatives needed in the anisotropic scalar Born error scattering term, Γ^{E} (10.3.68), and how they might be calculated.

Investigate the simplifications that occur in the Born scattering terms, Γ^{E} (10.3.68) and Γ^{B} (10.3.78), in isotropic media.

10.6 For a free acoustic surface, show that the Kirchhoff surface integral method is robust to the numerical specification of the shape of the interface, i.e. the surface can be represented as a smooth curve or a staircase and provided the steps are small compared with the wavelength, approximately the same results are obtained. This result depends on the reflection coefficient being independent of angle, and a similar result is not available for general interfaces.

Appendices

A
Useful integrals

The following integrals are used in this book and can be found in many reference books, or are easily proved by differentiation

$$\int \frac{\mathrm{d}x}{(x^2-1)^{1/2}} = \cosh^{-1} x \tag{A.0.1}$$

(Abramowitz and Stegun, 1965, §4.6.38),

$$\int \frac{\mathrm{d}x}{(x^2+1)^{1/2}} = \sinh^{-1} x \tag{A.0.2}$$

(Abramowitz and Stegun, 1965, §4.6.37),

$$\int \sec x \, \mathrm{d}x = \tanh^{-1}(\sin x) \tag{A.0.3}$$

(Abramowitz and Stegun, 1965, §4.3.117 and §4.6.22),

$$\int \frac{\mathrm{d}x}{(1-x^2)^{1/2}} = \sin^{-1} x \tag{A.0.4}$$

(Abramowitz and Stegun, 1965, §4.4.52), and

$$\int \frac{\mathrm{d}x}{x(x^2-1)^{1/2}} = \sec^{-1} x \tag{A.0.5}$$

(Abramowitz and Stegun, 1965, §4.4.56).

A.1 Multiple triangular integrals

In discussing the convergence of the WKBJ iterative solution for a 'thin' interface (Section 9.1.2.1), a simple, multi-dimensional volume integral occurs which is related to the volume of a multi-dimensional simplex. These integrals can be defined as

$$b_1 = \int_0^1 \mathrm{d}x_1 \tag{A.1.1}$$

$$b_2 = -\int_0^1 \mathrm{d}x_1 \int_{x_1}^1 \mathrm{d}x_2 \tag{A.1.2}$$

$$b_3 = -\int_0^1 \mathrm{d}x_1 \int_{x_1}^1 \mathrm{d}x_2 \int_0^{x_2} \mathrm{d}x_3 \tag{A.1.3}$$

$$b_4 = \int_0^1 \mathrm{d}x_1 \int_{x_1}^1 \mathrm{d}x_2 \int_0^{x_2} \mathrm{d}x_3 \int_{x_3}^1 \mathrm{d}x_4 \tag{A.1.4}$$

$$b_5 = \int_0^1 \mathrm{d}x_1 \int_{x_1}^1 \mathrm{d}x_2 \int_0^{x_2} \mathrm{d}x_3 \int_{x_3}^1 \mathrm{d}x_4 \int_0^{x_4} \mathrm{d}x_5, \tag{A.1.5}$$

etc. (note the alternating limits on the integrals, and the changes of sign which are introduced to simplify later results).

These integrals appear to be so simple that one would expect to find them in a classic textbook. Similar integrals

$$c_n = \int_0^1 \mathrm{d}x_1 \int_0^{x_1} \mathrm{d}x_2 \ldots \int_0^{x_{n-1}} \mathrm{d}x_n = \frac{1}{n!}, \tag{A.1.6}$$

are well known and trivial, being related to the volume of an n-dimensional simplex, e.g. Sommerville (1929, p. 124). However, I have not found a text for the integrals, b_n, and am unable to give an appropriate reference, so it is shown in this appendix how the b_n's reduce to well-known numbers: for n even, they are the coefficients in the series expansion for $\operatorname{sech} x$; and for n odd, coefficients in the series expansion for $\tanh x$ (which in turn are related to the Euler and Bernoulli numbers, respectively). In our application (Section 9.1.2.1), even-order integrals are needed for transmissions and odd-order for reflections.

The integrals are straightforward to evaluate although as n increases it gets tedious. The lowest-order values of b_n are tabulated in Table A.1. The odd and even values can be recognized as coefficients in the series expansions of $\operatorname{sech} x$ and $\tanh x$, respectively (Abramowitz and Stegun, 1965, §4.5.66 and §4.5.64), hinting at the general result.

Note that as each integral is evaluated, the indefinite integrals are naturally written as polynomials in x_i for n odd, or $(1 - x_i)$ for n even, and that the same coefficients appear in sequence for each b_i. Let us define an array of coefficients a_{nk} for the powers in the n-th polynomial. Then for the final integrals we have for n odd

$$b_n = \sum_{k=1}^{n} a_{nk} x_1^k \Big|_0^1 = \sum_{k=1}^{n} a_{nk} \tag{A.1.7}$$

Table A.1. *The polynomial coefficients a_{nk}, the resultant integrals b_n, and the Bernoulli and Euler numbers, B_n and E_n.*

	a_{nk}									
$n\backslash k$	1	2	3	4	5	6	7	b_n	B_{n+1}	E_n
0	1							1		1
1	1							1	$\frac{1}{6}$	
2	0	$-\frac{1}{2}$						$-\frac{1}{2}$		-1
3	$-\frac{1}{2}$	0	$\frac{1}{6}$					$-\frac{1}{3}$	$-\frac{1}{30}$	
4	0	$\frac{1}{4}$	0	$-\frac{1}{24}$				$\frac{5}{24}$		5
5	$\frac{5}{24}$	0	$-\frac{1}{12}$	0	$\frac{1}{120}$			$\frac{2}{15}$	$\frac{1}{42}$	
6	0	$-\frac{5}{48}$	0	$\frac{1}{48}$	0	$-\frac{1}{720}$		$-\frac{61}{720}$		-61
7	$-\frac{61}{720}$	0	$\frac{5}{144}$	0	$-\frac{1}{240}$	0	$\frac{1}{5040}$	$-\frac{17}{315}$	$-\frac{1}{30}$	

(where the terms with k even are zero), and for n even

$$b_n = -\sum_{k=1}^{n} a_{nk}(1-x_1)^k \Big|_0^1 = \sum_{k=1}^{n} a_{nk} \tag{A.1.8}$$

(with the terms with k odd zero). The coefficients a_{nk} are included in Table A.1.

Considering the penultimate integral, we can obtain a recurrence relationship between these coefficients. For n odd we have

$$\begin{aligned} b_n &= \int_0^1 \mathrm{d}x_1 \sum_{k=1}^{n-1} a_{n-1\,k} x_2^k \Big|_{x_1}^1 \\ &= \int_0^1 \mathrm{d}x_1 \left(\sum_{k=1}^{n-1} a_{n-1\,k} - \sum_{k=1}^{n-1} a_{n-1\,k} x_1^k \right) \\ &= \left(\sum_{k=1}^{n-1} a_{n-1\,k} x_1 - \sum_{k=1}^{n-1} \frac{a_{n-1\,k}}{k+1} x_1^{k+1} \right) \Bigg|_0^1, \end{aligned} \tag{A.1.9}$$

which we compare with equation (A.1.7). Hence for n odd, we have

$$a_{n1} = b_{n-1} \qquad \text{and} \qquad a_{n\,k+1} = -\frac{a_{n-1\,k}}{k+1}, \tag{A.1.10}$$

using also equation (A.1.8). For n even

$$
\begin{aligned}
b_n &= \int_0^1 \mathrm{d}x_1 \sum_{k=1}^{n-1} a_{n-1\,k}(1-x_2)^k \bigg|_{x_1}^{1} \\
&= -\int_0^1 \mathrm{d}x_1 \sum_{k=1}^{n-1} a_{n-1\,k}(1-x_1)^k \\
&= \sum_{k=1}^{n-1} \frac{a_{n-1\,k}(1-x_1)^{k+1}}{k+1} \bigg|_0^1, \qquad \text{(A.1.11)}
\end{aligned}
$$

to compare with equation (A.1.8). Hence for n even, we have

$$
a_{n\,k+1} = -\frac{a_{n-1\,k}}{k+1}. \qquad \text{(A.1.12)}
$$

The numbers in Table A.1 can be constructed using these formulae, (A.1.10) and (A.1.12). The numbers are moved down the diagonal, divided by $k+1$ (the new k) with a sign change. The new number which enters in a_{n1}, every odd row, is b_{n-1} from the previous row. The number b_n is the sum of the numbers in the row. The $n=0$ term – no integral – has been included for completeness. From this construction, we see that

$$
\begin{aligned}
a_{nk} &= (-1)^{k-1}\frac{b_{n-k}}{k!} \qquad \text{for } n-k \text{ even}, \qquad \text{(A.1.13)} \\
&= 0 \qquad \text{for } n-k \text{ odd}. \qquad \text{(A.1.14)}
\end{aligned}
$$

It is then obvious that b_n (n even) are the coefficients in the power series expansion of $\operatorname{sech} x$. Assuming they are, we have

$$
(\cosh x)\left(\sum_{n=0}^{\infty} b_{2n}x^{2n}\right) = 1, \qquad \text{(A.1.15)}
$$

so

$$
\left(1+\frac{x^2}{2!}+\frac{x^4}{4!}+\cdots\right)\left(b_0+b_2x^2+b_4x^4+\cdots\right) = 1. \qquad \text{(A.1.16)}
$$

Hence

$$
b_0 = 1, \qquad \text{(A.1.17)}
$$

and

$$
\sum_{k=0}^{n} \frac{b_{2k}}{(2n-2k)!} = 0, \qquad \text{(A.1.18)}
$$

which with result (A.1.13) is equivalent to result (A.1.8). The standard expression for the coefficients in the series expansion of $\operatorname{sech} x$ is (Abramowitz and Stegun, 1965, §4.5.66)

$$b_n = \frac{E_n}{n!}, \tag{A.1.19}$$

where E_n are the Euler numbers, which are included in Table A.1.

Result (A.1.18) can be rewritten

$$b_{2n} = -\sum_{k=0}^{n-1} b_{2k} c_{2n-2k} \tag{A.1.20}$$

(c_n is defined in equation (A.1.6)). This result can be understood geometrically. The alternating integrals in (A.1.2), (A.1.4), etc. have an upper limit of unity, rather than a lower limit of zero compared with the simplex integral (A.1.6). This corresponds to subtracting the simplex from the corresponding 'square' with volume unity. This 'unity' is then integrated through two less dimensions, giving rise to the sequence of terms in c_{2n-2k} in equation (A.1.20).

For b_n with n odd, we consider the series expansion for $\tanh x$ together with the expansion for $\operatorname{sech} x$, i.e.

$$(\sinh x)(\operatorname{sech} x) = \sum_{n=0}^{\infty} b_{2n+1} x^{2n+1}. \tag{A.1.21}$$

Then

$$\left(x + \frac{x^3}{3!} + \frac{x^5}{5!} + \cdots\right)\left(b_0 + b_2 x^2 + b_4 x^4 + \cdots\right) = \left(b_1 x + b_3 x^3 + b_5 x^5 + \cdots\right), \tag{A.1.22}$$

and

$$b_{2n+1} = \sum_{k=0}^{n} \frac{b_{2k}}{(2n+1-2k)!} = \sum_{k=0}^{n} b_{2k} c_{2n+1-2k}, \tag{A.1.23}$$

which is equivalent to result (A.1.7) with result (A.1.13). Again, as in equation (A.1.20), this result is understood geometrically from the sequence of subtracting the simplex from a unit square. The standard expression for the coefficients in the series expansion of $\tanh x$ is (Abramowitz and Stegun, 1965, §4.5.64)

$$b_{n-1} = \frac{2^n(2^n-1)B_n}{n!}, \tag{A.1.24}$$

where B_n are the Bernoulli numbers, which are included in Table A.1.

B

Useful Fourier transforms

B.1 Exponentials

The simple exponential in the frequency domain corresponds to a time shift, i.e.

$$e^{i\omega c} \quad \longleftrightarrow \quad \delta(t-c), \tag{B.1.1}$$

when c is real. When multiplied by another spectrum, this can be written as a convolution or simply a time shift

$$f(\omega)\, e^{i\omega c} \quad \longleftrightarrow \quad \delta(t-c) * f(t) = f(t-c). \tag{B.1.2}$$

These results can be generalized when c and f are complex constants, provided the condition (3.1.9) applies. Suppose

$$c = a + i\, \mathrm{sgn}(\omega)\, b, \tag{B.1.3}$$

where a and b are positive constants. Then with $f = 1$ we have

$$e^{i\omega c} = e^{i\omega a - |\omega| b} \quad \longleftrightarrow \quad \frac{1}{\pi} \frac{b}{(t-a)^2 + b^2} = \frac{1}{\pi} \mathrm{Im}\left(\frac{1}{t-c}\right), \tag{B.1.4}$$

a result that is well known for attenuation and evanescent waves. As $b \to 0$, the right-hand side tends to the Dirac delta function, $\delta(t-a)$. The Hilbert transform of this gives

$$-i\, \mathrm{sgn}(\omega)\, e^{i\omega c} \quad \longleftrightarrow \quad \frac{1}{\pi} \frac{t-a}{(t-a)^2 + b^2} = -\frac{1}{\pi} \mathrm{Re}\left(\frac{1}{t-c}\right). \tag{B.1.5}$$

Again as $b \to 0$,

$$-\frac{1}{\pi} \mathrm{Re}\left(\frac{1}{t-c}\right) \to -\frac{1}{\pi} \frac{1}{t-a} = \bar{\delta}(t-a). \tag{B.1.6}$$

Thus the analytic Dirac delta function (3.1.21)

$$\Delta(t) = \delta(t) + \mathrm{i}\,\bar{\delta}(t) = \delta(t) - \frac{\mathrm{i}}{\pi t}, \tag{B.1.7}$$

can be generalized for complex argument as

$$\Delta(t - c) = -\frac{1}{\pi}\frac{\mathrm{i}}{t - c}, \tag{B.1.8}$$

where for c real, we include a small imaginary part $\mathrm{i}\,\mathrm{sgn}(\omega)\epsilon$ and take the limit $\epsilon \to 0$.

Then for a complex constant, f, we have the general inverse transform

$$f\mathrm{e}^{\mathrm{i}\omega c} \longleftrightarrow \mathrm{Re}\,(f\,\Delta(t - c)) = \frac{1}{\pi}\mathrm{Im}\left(\frac{f}{t - c}\right). \tag{B.1.9}$$

B.2 Inverse square roots

The inverse square root frequently occurs in (two-dimensional) wave propagation problems. It is convenient to define a special function

$$\boxed{\lambda(t) = H(t)\,t^{-1/2}.} \tag{B.2.1}$$

Its Fourier transform is

$$\lambda(\omega) = \left(\frac{\pi}{|\omega|}\right)^{1/2} \mathrm{e}^{\mathrm{i}\,\mathrm{sgn}(\omega)\,\pi/4}. \tag{B.2.2}$$

The Hilbert transform is

$$\boxed{\bar{\lambda}(t) = H(-t)(-t)^{-1/2},} \tag{B.2.3}$$

with the Fourier transform

$$\bar{\lambda}(\omega) = \left(\frac{\pi}{|\omega|}\right)^{1/2} \mathrm{e}^{-\mathrm{i}\,\mathrm{sgn}(\omega)\,\pi/4}. \tag{B.2.4}$$

The analytic function is sometimes useful:

$$\Lambda(t) = \lambda(t) + \mathrm{i}\,\bar{\lambda}(t) = t^{-1/2}. \tag{B.2.5}$$

These Fourier transforms are special cases of the general result

$$\Gamma(k)(-\mathrm{i}\omega)^{-k} \longleftrightarrow H(t)\,t^{k-1} \tag{B.2.6}$$

($k > 0$, Abramowitz and Stegun, 1965, §29.3.7). The time-reversed pulse has the conjugate spectrum.

B.3 Exponentials and inverse square roots

The results of the previous two sections are sometimes needed together. Conveniently an inverse square root and an exponential can be written as a modified Bessel function of order 1/2 (Abramowitz and Stegun, 1965, §10.2.17)

$$K_{1/2}(b\omega) = \left(\frac{\pi}{2b\omega}\right)^{1/2} \mathrm{e}^{-b\omega}, \tag{B.3.1}$$

where for simplicity we assume $b > 0$ and $\omega > 0$. The inverse Fourier transforms of the modified Bessel functions are known (Erdélyi, Magnus, Oberhettinger and Tricomi, 1954, §1.3(22) and §2.3(20)) and we find the Fourier transforms

$$\frac{\left((t^2+b^2)^{1/2} \pm t\right)^{1/2}}{(t^2+b^2)^{1/2}} \longleftrightarrow \left(\frac{2\pi}{\omega}\right)^{1/2} \mathrm{e}^{\pm \mathrm{i}\pi/4} \mathrm{e}^{-b\omega}. \tag{B.3.2}$$

The product in the frequency domain is equivalent to a convolution in the time domain, so the time functions are equivalent to

$$\frac{\left((t^2+b^2)^{1/2} + t\right)^{1/2}}{(t^2+b^2)^{1/2}} = 2^{1/2}\lambda(t) * \frac{b}{\pi(t^2+b^2)} \tag{B.3.3}$$

$$\frac{\left((t^2+b^2)^{1/2} - t\right)^{1/2}}{(t^2+b^2)^{1/2}} = 2^{1/2}\bar{\lambda}(t) * \frac{b}{\pi(t^2+b^2)}. \tag{B.3.4}$$

B.4 Bessel and Hankel functions

Various inverse Fourier transforms of Bessel functions are given in the standard reference books, but not exactly in the form we need. For instance,

$$\int_{-\infty}^{\infty} \mathrm{e}^{-\mathrm{i}\omega t} J_n(t)\, \mathrm{d}t = \frac{2(-\mathrm{i})^n T_n(\omega)}{(1-\omega^2)^{1/2}} \tag{B.4.1}$$

(Abramowitz and Stegun, 1965, §11.4.21) where $T_n(x)$ is the Chebyshev polynomial

$$T_n(x) = \cos\left(n \cos^{-1} x\right) \tag{B.4.2}$$

(Abramowitz and Stegun, 1965, §22.3.15), so

$$T_0(x) = 1 \tag{B.4.3}$$

$$T_1(x) = x \tag{B.4.4}$$

$$T_2(x) = 2x^2 - 1 \tag{B.4.5}$$

$$T_3(x) = 4x^3 - 3x \text{ , etc.} \tag{B.4.6}$$

Alternatively, we have

$$\int_0^\infty e^{ibt} J_0(at)\,dt = \frac{1}{(a^2-b^2)^{1/2}} \quad 0 \le b < a \tag{B.4.7}$$

$$= \frac{i}{(b^2-a^2)^{1/2}} \quad 0 < a < b \tag{B.4.8}$$

$$\int_0^\infty e^{ibt} Y_0(at)\,dt = \frac{2i}{\pi(a^2-b^2)^{1/2}} \sin^{-1}\frac{b}{a} \quad 0 \le b < a \tag{B.4.9}$$

$$= -\frac{1}{(b^2-a^2)^{1/2}} + \frac{2i}{(b^2-a^2)^{1/2}} \ln\frac{b-(b^2-a^2)^{1/2}}{a} \quad 0 < a < b \tag{B.4.10}$$

(Abramowitz and Stegun, 1965, §11.4.39 and §11.4.40). The result we require (by analogy with the spatial Fourier transform – see equation (3.3.7)) is

$$\frac{1}{2\pi}\int_{-\infty}^{\infty} e^{-i\omega t} i H_0^{(1)}(a\omega)\,d\omega = \frac{1}{\pi}\mathrm{Re}\int_0^\infty e^{-i\omega t} i H_0^{(1)}(a\omega)\,d\omega \tag{B.4.11}$$

$$= \frac{1}{\pi}\int_0^\infty \Big(\sin(\omega t) J_0(a\omega) - \cos(\omega t) Y_0(a\omega)\Big)d\omega \tag{B.4.12}$$

$$= \frac{2H(t-a)}{\pi(t^2-a^2)^{1/2}}, \tag{B.4.13}$$

where we have substituted $t \to \omega$ and $b \to t$ in equations (B.4.7)–(B.4.10). This result is also equivalent to result (B.4.1) with $n = 0$, $t \to a\omega$ and $\omega \to t/a$ substituted.

The general result we require is

$$\boxed{\frac{1}{2\pi}\int_{-\infty}^{\infty} e^{-i\omega t} H_n^{(1)}(a\omega)\,d\omega = \frac{2(-i)^{n+1} T_n(t/a)}{\pi(t^2-a^2)^{1/2}}.} \tag{B.4.14}$$

Again this is equivalent to result (B.4.1) with $t \to a\omega$ and $\omega \to t/a$ substituted. It can also be obtained iteratively from result (B.4.13). For each iteration, we integrate with respect to time (which is equivalent to dividing the left-hand side by $-i\omega$) and differentiate with respect to a. The integral with the higher-order Hankel function is then obtained using

$$2H_n^{(1)\prime}(z) = H_{n-1}^{(1)}(z) - H_{n+1}^{(1)}(z) \tag{B.4.15}$$

(Abramowitz and Stegun, 1965 §9.1.27) (and $H_0^{(1)\prime}(z) = -H_1^{(1)}(z)$).

C
Ordinary differential equations

Ordinary differential equations arise in the ray equations (Chapter 5), in the transformed wave equations (Chapter 7) and in quasi-isotropic ray theory (Chapter 10). In this appendix we review some of the standard terminology and results used to describe the solutions of systems of ordinary differential equations.

C.1 Propagator matrices

Differential systems of the form

$$\frac{\mathrm{d}\mathbf{y}(x)}{\mathrm{d}x} = \mathbf{A}(x)\mathbf{y}(x) + \mathbf{w}(x), \tag{C.1.1}$$

of various orders, occur in wave propagation. The vector $\mathbf{y}$ is $n \times 1$ and the matrix $\mathbf{A}$ is $n \times n$. We call the vector $\mathbf{w}$ a *source term* (avoiding the ambiguous term, *inhomogeneous*). We discuss some general concepts, notation and terminology introduced by Gilbert and Backus (1966) to seismology, which describe the solutions of such a system (see also Gantmacher, 1959).

A matrix $\mathbf{F}(x)$ is called an *integral matrix* if it satisfies

$$\frac{\mathrm{d}\mathbf{F}(x)}{\mathrm{d}x} = \mathbf{A}(x)\mathbf{F}(x), \tag{C.1.2}$$

i.e. each column is a solution of the equation without source term (the homogeneous equation). An integral matrix is called a *fundamental matrix* if it is non-singular at all x, i.e. equation (C.1.2) and $|\mathbf{F}(x)| \neq 0$ for all x. In other words, $\mathbf{F}$ is $n \times n$ and the n solutions are independent. We shall see later that non-singularity at one x implies it is non-singular at all x. An integral matrix is called a *propagator matrix from* x_0, if $\mathbf{F}(x_0) = \mathbf{I}$, the identity matrix. We denote it by $\mathbf{P}(x, x_0)$ so

$$\frac{\mathrm{d}\mathbf{P}(x, x_0)}{\mathrm{d}x} = \mathbf{A}(x)\mathbf{P}(x, x_0), \tag{C.1.3}$$

and $\mathbf{P}(x_0, x_0) = \mathbf{I}$.

We can always form a propagator matrix from a fundamental matrix

$$\mathbf{P}(x, x_0) = \mathbf{F}(x)\mathbf{F}^{-1}(x_0), \tag{C.1.4}$$

just by post-multiplying equation (C.1.2) by $\mathbf{F}^{-1}(x_0)$. Obviously, $\mathbf{F}^{-1}(x_0)$ exists and $\mathbf{P}(x_0, x_0) = \mathbf{I}$.

A *chain rule* applies for propagator matrices

$$\mathbf{P}(x, x_0) = \mathbf{P}(x, x_1)\mathbf{P}(x_1, x_0), \tag{C.1.5}$$

just by post-multiplying equation (C.1.3) applied from x_1, by $\mathbf{P}(x_1, x_0)$, and noting equality at $x = x_1$.

Substituting $x = x_0$ in equation (C.1.5), we obtain

$$\mathbf{P}(x_0, x_0) = \mathbf{I} = \mathbf{P}(x_0, x_1)\mathbf{P}(x_1, x_0), \tag{C.1.6}$$

so

$$\mathbf{P}(x_0, x_1) = \mathbf{P}^{-1}(x_1, x_0). \tag{C.1.7}$$

Thus the *inverse propagator* is easily obtained by reversing the arguments. We must be careful with this result as it depends on $\mathbf{A}(x)$ being independent of the direction of integration. In some systems, terms in $\mathbf{A}(x)$ depend on the direction, e.g. ray tracing (5.2.20) depends on the slowness vectors and polarizations.

We note that

$$\mathbf{y}(x) = \mathbf{P}(x, x_0)\mathbf{y}(x_0) \tag{C.1.8}$$

is the solution of equation (C.1.1) without the source term, with initial condition $\mathbf{y}(x_0)$ (just post-multiply equation (C.1.3) by $\mathbf{y}(x_0)$). Equation (C.1.8) justifies the name propagator – it *propagates* the solution from x_0 to x.

For the differential equation (C.1.1) with the source term, $\mathbf{w}$, it is straightforward to show by substitution that the solution is

$$\mathbf{y}(x) = \mathbf{P}(x, x_0)\mathbf{y}(x_0) + \int_{x_0}^{x} \mathbf{P}(x, \zeta)\mathbf{w}(\zeta)\mathrm{d}\zeta, \tag{C.1.9}$$

or using any fundamental matrix

$$\mathbf{y}(x) = \mathbf{F}(x)\left(\mathbf{F}^{-1}(x_0)\mathbf{y}(x_0) + \int_{x_0}^{x} \mathbf{F}^{-1}(\zeta)\mathbf{w}(\zeta)\mathrm{d}\zeta\right). \tag{C.1.10}$$

Finally we note that the determinant of a fundamental matrix satisfies

$$\frac{\mathrm{d}}{\mathrm{d}x}|\mathbf{F}(x)| = \mathrm{tr}\left(\mathbf{A}(x)\right)|\mathbf{F}(x)|. \tag{C.1.11}$$

The proof proceeds

$$\frac{\mathrm{d}}{\mathrm{d}x}|\mathbf{F}(x)| = \begin{vmatrix} F'_{11} & \dots & F'_{1n} \\ F_{21} & \dots & F_{2n} \\ . & \dots & . \\ F_{n1} & \dots & F_{nn} \end{vmatrix} + \begin{vmatrix} F_{11} & \dots & F_{1n} \\ F'_{21} & \dots & F'_{2n} \\ . & \dots & . \\ F_{n1} & \dots & F_{nn} \end{vmatrix} + \cdots \tag{C.1.12}$$

$$\begin{aligned} &= \begin{vmatrix} A_{11}F_{11} + A_{12}F_{21} + \cdots & \dots & A_{11}F_{1n} + A_{12}F_{2n} + \cdots \\ F_{21} & \dots & F_{2n} \\ . & \dots & . \\ F_{n1} & \dots & F_{nn} \end{vmatrix} \\ &+ \begin{vmatrix} F_{11} & \dots & F_{1n} \\ A_{21}F_{11} + A_{22}F_{21} + \cdots & \dots & A_{21}F_{1n} + A_{22}F_{2n} + \cdots \\ . & \dots & . \\ F_{n1} & \dots & F_{1n} \end{vmatrix} \\ &+ \cdots \end{aligned} \tag{C.1.13}$$

$$\begin{aligned} &= A_{11}\begin{vmatrix} F_{11} & \dots & F_{1n} \\ F_{21} & \dots & F_{2n} \\ . & \dots & . \\ F_{n1} & \dots & F_{nn} \end{vmatrix} + A_{12}\begin{vmatrix} F_{21} & \dots & F_{2n} \\ F_{21} & \dots & F_{2n} \\ . & \dots & . \\ F_{n1} & \dots & F_{nn} \end{vmatrix} + \cdots \\ &+ A_{21}\begin{vmatrix} F_{11} & \dots & F_{1n} \\ F_{11} & \dots & F_{1n} \\ . & \dots & . \\ F_{n1} & \dots & F_{1n} \end{vmatrix} + A_{22}\begin{vmatrix} F_{11} & \dots & F_{1n} \\ F_{21} & \dots & F_{2n} \\ . & \dots & . \\ F_{n1} & \dots & F_{1n} \end{vmatrix} + \cdots \\ &+ \cdots \end{aligned} \tag{C.1.14}$$

$$= (A_{11} + A_{22} + \cdots)\begin{vmatrix} F_{11} & \dots & F_{1n} \\ F_{21} & \dots & F_{2n} \\ . & \dots & . \\ F_{n1} & \dots & F_{nn} \end{vmatrix}. \tag{C.1.15}$$

In equation (C.1.12) we have differentiated by row using a prime for $\mathrm{d}/\mathrm{d}x$, and substituted using equation (C.1.2) to get equation (C.1.13). Expanding equation (C.1.13), the determinants with two equal rows in equation (C.1.14) are zero, so it reduces to equation (C.1.15) which is equal to equation (C.1.11). From equation (C.1.11), we obtain

$$|\mathbf{F}(x)| = |\mathbf{F}(x_0)| \exp\left(\int_{x_0}^{x} \mathrm{tr}(\mathbf{A}(\zeta))\mathrm{d}\zeta\right). \tag{C.1.16}$$

This result is known as the *Jacobi identity* (Gantmacher, 1959, Vol. 2, p. 114), and establishes that non-singularity of an integral matrix at one x, e.g. x_0, implies non-singularity at all x, i.e. independence of the solutions.

C.2 Smirnov's lemma

Smirnov's lemma (Smirnov, 1964, p. 442–3) provides a result for the Jacobian of the solution of a system of ordinary differential equations such as equation (C.1.1). Consider a more general system

$$\frac{\mathrm{d}\mathbf{y}}{\mathrm{d}x} = \mathbf{B}(\mathbf{y}). \tag{C.2.1}$$

The solution will depend on the initial condition, e.g. $\mathbf{y} = \mathbf{Y}$ at $x = 0$, say, so in general $\mathbf{y} = \mathbf{y}(x, \mathbf{Y})$ and $\mathbf{Y} = \mathbf{y}(0, \mathbf{Y})$. The Jacobian

$$D = \frac{\partial\,(y_1, y_2, \ldots)}{\partial\,(Y_1, Y_2, \ldots)} \tag{C.2.2}$$

$$= \begin{vmatrix} \frac{\partial y_1}{\partial Y_1} & \frac{\partial y_1}{\partial Y_2} & \cdots \\ \frac{\partial y_2}{\partial Y_1} & \frac{\partial y_2}{\partial Y_2} & \cdots \\ \cdots & \cdots & \cdots \end{vmatrix}, \tag{C.2.3}$$

describes the mapping from $\mathbf{Y}$ to $\mathbf{y}$ of a volume element in the $\mathbf{y}$ space, where x is kept constant in the partial derivatives. Then

$$\frac{\mathrm{d}D}{\mathrm{d}x} = \begin{vmatrix} \frac{\partial y_1'}{\partial Y_1} & \frac{\partial y_1'}{\partial Y_2} & \cdots \\ \frac{\partial y_2}{\partial Y_1} & \frac{\partial y_2}{\partial Y_2} & \cdots \\ \cdots & \cdots & \cdots \end{vmatrix} + \begin{vmatrix} \frac{\partial y_1}{\partial Y_1} & \frac{\partial y_1}{\partial Y_2} & \cdots \\ \frac{\partial y_2'}{\partial Y_1} & \frac{\partial y_2'}{\partial Y_2} & \cdots \\ \cdots & \cdots & \cdots \end{vmatrix} + \cdots \tag{C.2.4}$$

$$= \begin{vmatrix} \frac{\partial B_1}{\partial Y_1} & \frac{\partial B_1}{\partial Y_2} & \cdots \\ \frac{\partial y_2}{\partial Y_1} & \frac{\partial y_2}{\partial Y_2} & \cdots \\ \cdots & \cdots & \cdots \end{vmatrix} + \begin{vmatrix} \frac{\partial y_1}{\partial Y_1} & \frac{\partial y_1}{\partial Y_2} & \cdots \\ \frac{\partial B_2}{\partial Y_1} & \frac{\partial B_2}{\partial Y_2} & \cdots \\ \cdots & \cdots & \cdots \end{vmatrix} + \cdots \tag{C.2.5}$$

$$= \frac{\partial B_1}{\partial y_1}\begin{vmatrix} \frac{\partial y_1}{\partial Y_1} & \frac{\partial y_1}{\partial Y_2} & \cdots \\ \frac{\partial y_2}{\partial Y_1} & \frac{\partial y_2}{\partial Y_2} & \cdots \\ \cdots & \cdots & \cdots \end{vmatrix} + \frac{\partial B_1}{\partial y_2}\begin{vmatrix} \frac{\partial y_2}{\partial Y_1} & \frac{\partial y_2}{\partial Y_2} & \cdots \\ \frac{\partial y_2}{\partial Y_1} & \frac{\partial y_2}{\partial Y_2} & \cdots \\ \cdots & \cdots & \cdots \end{vmatrix} + \cdots$$

$$+ \frac{\partial B_2}{\partial y_1}\begin{vmatrix} \frac{\partial y_1}{\partial Y_1} & \frac{\partial y_1}{\partial Y_2} & \cdots \\ \frac{\partial y_1}{\partial Y_1} & \frac{\partial y_1}{\partial Y_2} & \cdots \\ \cdots & \cdots & \cdots \end{vmatrix} + \frac{\partial B_2}{\partial y_2}\begin{vmatrix} \frac{\partial y_1}{\partial Y_1} & \frac{\partial y_1}{\partial Y_2} & \cdots \\ \frac{\partial y_2}{\partial Y_1} & \frac{\partial y_2}{\partial Y_2} & \cdots \\ \cdots & \cdots & \cdots \end{vmatrix} + \cdots + \cdots \tag{C.2.6}$$

$$= \left(\frac{\partial B_1}{\partial y_1} + \frac{\partial B_2}{\partial y_2} \cdots\right) D. \tag{C.2.7}$$

where in equation (C.2.4) we differentiated the Jacobian determinant by rows, in equation (C.2.5) we substituted using equation (C.2.1), in equation (C.2.6) we expanded using the chain rule for partial derivatives, and in equation (C.2.7) we eliminated the zero determinants with equal rows. Hence if $\mathbf{y}$ satisfies equation (C.2.1), Smirnov's lemma states that the Jacobian (C.2.2) satisfies

$$\frac{\mathrm{d}}{\mathrm{d}x} \ln D = \nabla \cdot \mathbf{B}, \tag{C.2.8}$$

equivalent to result (C.2.7).

D

Saddle-point methods

Frequently in wave propagation problems we have integrals of the form

$$I = \int f(z) e^{i\omega g(z)} \, dz, \tag{D.0.1}$$

where ω is a large parameter, e.g. frequency. In general, such integrals are difficult to evaluate numerically, as even if $f(z)$ and $g(z)$ are simple functions, the integrand is highly oscillatory. To evaluate the integral numerically, we use the Filon method (Section 8.5.2.3). In each cycle, the positive and negative contributions to the integral almost cancel. The overall value of the integral is due to changes and special features of the functions $f(z)$ and $g(z)$. A large body of mathematics exists concerning the analytic and asymptotic evaluation of integrals such as equation (D.0.1). In this appendix, we summarize some of the simplest, basic results: the second-order saddle-point method, and the third-order and incomplete saddle-point methods. The latter two serve as definitions of the Airy and Fresnel functions. We do not consider the details of these functions nor higher-order asymptotic results as there are many excellent textbooks analysing these methods and tabulating the functions. However, we do discuss their inverse Fourier transforms.

D.1 Second-order saddle points

If ω is large and $g'(z)$ non-zero, the contribution to the integral (D.0.1) is small. Only when $g'(z)$ is small or zero, is there a significant contribution. Points where

$$g'(z) = 0, \tag{D.1.1}$$

are known as *stationary* or *saddle points*. Taking Taylor expansions about such a point, $z = z_0$, we have

$$I = \int \left(f(z_0) + f'(z_0)(z - z_0) + \cdots \right) e^{i\omega \left(g(z_0) + \frac{1}{2} g''(z_0)(z - z_0)^2 + \ldots \right)} \, dz \tag{D.1.2}$$

$$\simeq f(z_0) \, e^{i\omega g(z_0)} \int e^{i\omega \frac{1}{2} g''(z_0)(z - z_0)^2} dz, \tag{D.1.3}$$

retaining the leading terms in the Taylor expansions. If higher-order terms are included we can investigate the asymptotic expansion. By a simple change of variable, we can reduce this integral (D.1.3) to a standard form.

Let us first consider the integral

$$I = \int_{-\infty}^{\infty} \mathrm{e}^{-ax^2}\, \mathrm{d}x, \tag{D.1.4}$$

where a is real and positive. The real axis is known as the *steepest descent path* as the integrand decreases most rapidly from the saddle point at the origin $x = 0$. To evaluate this integral we use a trick taught in most elementary mathematics courses. The integral does not depend on the symbol used for the dummy variable, x, so we consider the same integral with dummy variable y and multiply the two together

$$I^2 = \iint_{-\infty}^{\infty} \mathrm{e}^{-a(x^2+y^2)}\, \mathrm{d}x\, \mathrm{d}y. \tag{D.1.5}$$

This can be considered as an area integral over the entire cartesian x–y plane, and can be evaluated by changing the variables of integration to the polar coordinates r and θ. Thus

$$I^2 = \int_0^{2\pi} \int_0^{\infty} \mathrm{e}^{-ar^2} r\, \mathrm{d}r\, \mathrm{d}\theta = \left(-\frac{\mathrm{e}^{-ar^2}}{2a} \right)\Bigg|_0^{\infty} \theta \Bigg|_0^{2\pi} = \frac{\pi}{a}. \tag{D.1.6}$$

Thus we obtain the result

$$I = \int_{-\infty}^{\infty} \mathrm{e}^{-az^2} \mathrm{d}z = \sqrt{\frac{\pi}{a}}, \tag{D.1.7}$$

probably one of the most important mathematic results for wave propagation!

A variant on this result is for the integral

$$I = \int_{-\infty}^{\infty} \mathrm{e}^{\mathrm{i}az^2}\, \mathrm{d}z, \tag{D.1.8}$$

with a real (either sign). The integrand has a constant amplitude but oscillates more and more rapidly away from the saddle point at the origin $z = 0$. Evaluating the integral along the real axis is known as the *stationary phase method*. It can be simply evaluated by distorting the contour so it lies in the valleys of the saddle point and the integral reduces to integral (D.1.7).

When a is positive, the saddle point is at $\pi/4$ to the real axis and with a change of variable $y = z \exp(-\mathrm{i}\pi/4)$, we distort the contour to the real y axis

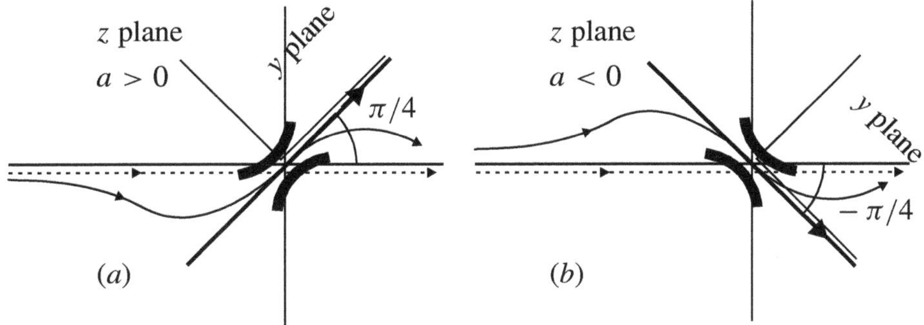

Fig. D.1. The saddle points when a is (a) positive and (b) negative.

(Figure D.1a). Then

$$I = \int_{-\infty}^{\infty} \mathrm{e}^{-ay^2} \mathrm{e}^{\mathrm{i}\pi/4}\, \mathrm{d}y = \mathrm{e}^{\mathrm{i}\pi/4}\sqrt{\frac{\pi}{a}}, \tag{D.1.9}$$

using integral (D.1.7). Similarly, when a is negative the saddle point is at $-\pi/4$ to the real axis and with a change of variable $y = z\exp(\mathrm{i}\pi/4)$, we distort the contour to the real y axis (Figure D.1b). Then

$$I = \int_{-\infty}^{\infty} \mathrm{e}^{-ay^2} \mathrm{e}^{-\mathrm{i}\pi/4}\, \mathrm{d}y = \mathrm{e}^{-\mathrm{i}\pi/4}\sqrt{\frac{\pi}{|a|}}. \tag{D.1.10}$$

Thus the general result for integral (D.1.8) is

$$I = \int_{-\infty}^{\infty} \mathrm{e}^{\mathrm{i}az^2}\, \mathrm{d}z = \sqrt{\frac{\pi}{|a|}}\, \mathrm{e}^{\mathrm{i}\,\mathrm{sgn}(a)\pi/4}. \tag{D.1.11}$$

D.1.1 Multi-dimensional saddle point

The above results are easily extended to multi-dimensional saddle points. Analogous to result (D.1.11) we consider the integral

$$I = \int_{-\infty}^{\infty} \mathbf{B}\, \mathrm{e}^{\mathrm{i}a\mathbf{q}^{\mathrm{T}}\mathbf{C}\mathbf{q}}\, \mathrm{d}\mathbf{q}, \tag{D.1.12}$$

where $\mathbf{q}$ is an m-dimensional vector and $\mathbf{C}$ is a real, $m \times m$ matrix. We can assume that the matrix $\mathbf{C}$ is symmetric, so it can be diagonalized by a rotation matrix $\mathbf{R}$ where $\mathbf{q}' = \mathbf{R}\mathbf{q}$. Thus

$$I = \int_{-\infty}^{\infty} \mathbf{B}\, \mathrm{e}^{\mathrm{i}a\mathbf{q}'^{\mathrm{T}}\mathbf{R}\mathbf{C}\mathbf{R}^{\mathrm{T}}\mathbf{q}'}\, \mathrm{d}\mathbf{q}' \tag{D.1.13}$$

$$= \int_{-\infty}^{\infty} \mathbf{B}\, \mathrm{e}^{\mathrm{i}a\sum_{i=1}^{m} c_i q_i'^2}\, \mathrm{d}\mathbf{q}', \tag{D.1.14}$$

as the matrix $\mathbf{RCR}^{\mathrm{T}}$ is diagonal. The eigenvalues of matrix $\mathbf{C}$ are c_i. Note that as $\mathbf{R}$ is an orthonormal rotation matrix, the Jacobian mapping area elements $\mathrm{d}\mathbf{q}$ to $\mathrm{d}\mathbf{q}'$ is unity. Integral (D.1.14) can now be evaluated as m examples of integral (D.1.11). Thus

$$I = \mathbf{B} \prod_{i=1}^{m} \sqrt{\frac{\pi}{|ac_i|}}\, \mathrm{e}^{\mathrm{i}\pi\, \mathrm{sgn}(ac_i)/4} = \mathbf{B} \left(\frac{\pi}{|a|}\right)^{m/2} \frac{\mathrm{e}^{\mathrm{i}\pi\, \mathrm{sgn}(a\mathbf{C})/4}}{\|\mathbf{C}\|^{1/2}}, \tag{D.1.15}$$

where

$$|\mathbf{C}| = \prod_{i=1}^{m} c_i, \tag{D.1.16}$$

is the determinant, and

$$\sum_{i=1}^{m} \mathrm{sgn}(c_i) = \mathrm{sgn}(\mathbf{C}), \tag{D.1.17}$$

the signature of the matrix, defined as the number of positive eigenvalues minus the number of negative eigenvalues. This result

$$\int_{-\infty}^{\infty} \mathbf{B}\mathrm{e}^{\mathrm{i}a\mathbf{q}^{\mathrm{T}}\mathbf{C}\mathbf{q}}\, \mathrm{d}\mathbf{q} = \mathbf{B} \left(\frac{\pi}{|a|}\right)^{m/2} \frac{\mathrm{e}^{\mathrm{i}\, \mathrm{sgn}(a\mathbf{C})\pi/4}}{\|\mathbf{C}\|^{1/2}}, \tag{D.1.18}$$

is given in Bender and Orzag (1978). It is needed most frequently with $m = 2$.

D.2 Third-order saddle points – Airy functions

If two saddle points exist close together, the second-order saddle-point method fails as the second derivative of the exponent varies significantly near the saddle points. Instead, we must use the third-order saddle-point method. If we expand about the point where the second derivative is zero, the required integral can be written

$$I = \int_{-\infty}^{\infty} \mathrm{e}^{\pm\mathrm{i}(az+bz^3)}\, \mathrm{d}z, \tag{D.2.1}$$

replacing integral (D.1.8).

The special function with the required form is the Airy function

$$Ai(x) = \frac{1}{2\pi} \int_{-\infty}^{\infty} \mathrm{e}^{\pm\mathrm{i}(z^3/3+xz)}\, \mathrm{d}z \tag{D.2.2}$$

(from Abramowitz and Stegun, 1965, §10.4.32). Terms arising from higher-terms in the expansion of the integrand can be converted into the Airy function or its derivative (Chester, Friedman and Ursell, 1957). The Airy function is described

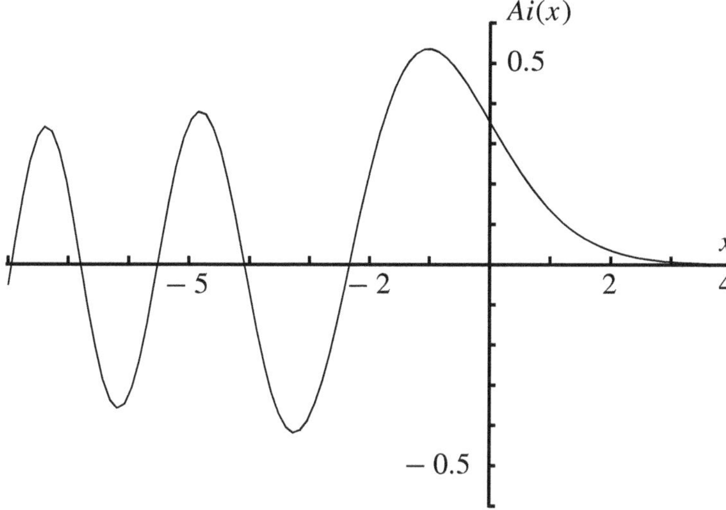

Fig. D.2. The Airy function $Ai(x)$.

in detail and tabulated in Abramowitz and Stegun (1965, §10.4) and is plotted in Figure D.2. By a simple change of variable we can convert integral (D.2.1) into definition (D.2.2)

$$I = \int_{-\infty}^{\infty} e^{\pm i(az+bz^3)}\, dz = 2\pi (3b)^{-1/3} Ai\left((3b)^{-1/3} a\right), \tag{D.2.3}$$

where for simplicity we have assumed $b > 0$.

D.2.1 Asymptotic forms

When the argument of the Airy function is negative, two stationary points exist in the integral (D.2.3). Approximating these by the second-order saddle-point method, we obtain the leading term in the asymptotic expansion for the Airy function. For completeness, we give the results for both types of Airy functions (Abramowitz and Stegun, 1965, §10.4.60 and §10.4.64)

$$Ai(-\xi) \simeq \frac{1}{\pi^{1/2}\xi^{1/4}} \sin(\zeta + \pi/4) \tag{D.2.4}$$

$$Bi(-\xi) \simeq \frac{1}{\pi^{1/2}\xi^{1/4}} \cos(\zeta + \pi/4), \tag{D.2.5}$$

where

$$\zeta = \frac{2}{3}\xi^{3/2}. \tag{D.2.6}$$

For positive argument, the saddle points move off the real axis and only one contributes to the Airy function. The asymptotic forms are (Abramowitz and Stegun,

1965, §10.4.59 and §10.4.63)

$$Ai(\xi) \simeq \frac{1}{2\pi^{1/2}\xi^{1/4}}\,\mathrm{e}^{-\zeta} \tag{D.2.7}$$

$$Bi(\xi) \simeq \frac{1}{\pi^{1/2}\xi^{1/4}}\,\mathrm{e}^{\zeta}. \tag{D.2.8}$$

Higher-order terms in the asymptotic expansions, and the power series expansion when $|\xi|$ is small, are given in many handbooks (for example, Abramowitz and Stegun, 1965, §10.4).

D.2.2 An inverse Fourier transform

Frequently the Airy function arises in the spectral domain. We modify expression (D.2.3) to

$$\int_{-\infty}^{\infty} \mathrm{e}^{\pm \mathrm{i}\omega(az+bz^3)}\,\mathrm{d}z = 2\pi(3\omega b)^{-1/3} Ai\left(\omega^{2/3}(3b)^{-1/3}a\right) \tag{D.2.9}$$

(in this section, we write the spectra for $\omega > 0$ – results for negative frequencies can be obtained using relationship (3.1.9)). Multiplying this by $\lambda(\omega)$, we can take the inverse Fourier transform (3.1.2) of the left-hand side, reverse the order of integration and obtain the transform pair

$$2\pi^{3/2}\mathrm{e}^{\mathrm{i}\pi/4}\omega^{-5/6}(3b)^{-1/3}Ai\left(\omega^{2/3}(3b)^{-1/3}a\right) \longleftrightarrow \int_{(\,)>0} \frac{\mathrm{d}z}{\left(t \mp (az+bz^3)\right)^{1/2}}. \tag{D.2.10}$$

The integral on the right-hand side is over values of z for which the bracket in the denominator is positive, i.e. its square root is real. This integral has been investigated by Burridge (1963*a,b*) who defined the function

$$C(t,y) = \int_{(\,)>0} \frac{\mathrm{d}x}{(t - 3yx + 4x^3)^{1/2}}, \tag{D.2.11}$$

equivalent to result (D.2.10) with the positive sign and $a = -3y$, $b = 4$ (the factors of 3 and 4 are introduced to facilitate finding roots of the cubic, and to simplify the numerical factors in the asymptotic results – see below). With these substitutions, result (D.2.10) reduces to the transform pair

$$\left(\frac{2}{3}\right)^{1/3} \pi^{3/2}\mathrm{e}^{\mathrm{i}\pi/4}\omega^{-5/6}Ai\left(-y\left(\frac{3\omega}{2}\right)^{2/3}\right) \longleftrightarrow C(t,y). \tag{D.2.12}$$

Although the function $C(t,y)$ has two arguments, only three cases need be considered, $C(t,0)$ and $C(t,\ \pm 1)$, for with a simple change of variable $x = |y|^{1/2}x'$

in definition (D.2.11), we have

$$C(t,y) = |y|^{-1/4} \int_{(\,)>0} \frac{\mathrm{d}x'}{(t|y|^{-3/2} \mp 3x' + 4x'^3)^{1/2}}$$
$$= |y|^{-1/4} C\left(t|y|^{-3/2},\ \mathrm{sgn}(y)\right). \tag{D.2.13}$$

The function $C(t, y)$ can be expressed in terms of the complete elliptic integral of the first kind, $K(m)$. Two 'standard' results are needed to handle the cubic in the denominator. If the cubic has three real roots, $a > b > c$, say, we have

$$\int_a^\infty \frac{\mathrm{d}x}{\left((x-a)(x-b)(x-c)\right)^{1/2}} = \int_c^b \frac{\mathrm{d}x}{\left((x-a)(x-b)(x-c)\right)^{1/2}}$$
$$= \frac{2}{(a-c)^{1/2}} K\left(\frac{b-c}{a-c}\right) \tag{D.2.14}$$

(see, for instance, Magnus, Oberhettinger and Soni, 1966, pp. 366–367 or Gradshteyn and Ryzhik, 1980, §3.131(4) and §3.131(8)). The integral can be reduced to the Legendre standard form letting $x - a = \xi^2$. If the cubic only has one real root, we have

$$\int_a^\infty \frac{\mathrm{d}x}{\left((x-a)((x-b)^2+c^2)\right)^{1/2}} = \frac{2}{p^{1/2}} K\left(\frac{p+b-a}{2p}\right), \tag{D.2.15}$$

where $p^2 = (a-b)^2 + c^2$ (Gradshteyn and Ryzhik, 1980, §3.138(7)).

D.2.2.1 At the caustic – $C(t, 0)$ ($a = y = 0$)

Exactly at the caustic, the function reduces to

$$C(t,0) = \int_{(\,)>0} \frac{\mathrm{d}x}{(t+4x^3)^{1/2}} \tag{D.2.16}$$
$$= \frac{1}{2^{2/3}|t|^{1/6}} \int_{\mp 1}^\infty \frac{\mathrm{d}x}{(x^3 \pm 1)^{1/2}} \tag{D.2.17}$$
$$= \frac{1}{|t|^{1/6}} \frac{2^{1/3}}{3^{1/4}} K\left(\frac{1}{2} \pm \frac{3^{1/2}}{4}\right). \tag{D.2.18}$$

The sign depends on whether $t > 0$ or $t < 0$. When $t > 0$, we have $a = -1$, $b = 1/2$, $c = \sqrt{3}/2$ and $p = \sqrt{3}$ in (D.2.15). Similarly when $t < 0$, $a = 1$, $b = -1/2$, $c = \sqrt{3}/2$ and $p = \sqrt{3}$. The argument of the elliptic integral in result (D.2.18) is $\sin^2(5\pi/12)$ or $\sin^2(\pi/12)$. This function $C(t, 0)$ is illustrated in Figure D.3.

Alternatively, we can consider the spectrum (D.2.12) with

$$Ai(0) = \frac{1}{3^{2/3}\Gamma(2/3)} \tag{D.2.19}$$

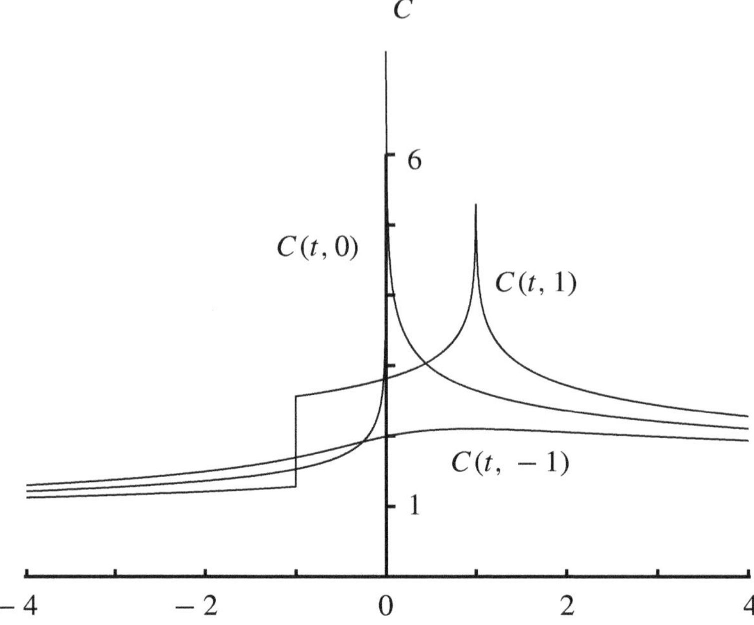

Fig. D.3. The functions $C(t, 0)$ and $C(t, \pm 1)$, defined in equation (D.2.13).

(Abramowitz and Stegun, 1965 §10.4.4), i.e.

$$\frac{2^{1/3}\pi^{3/2}}{3\Gamma(2/3)} \frac{e^{i\pi/4}}{\omega^{5/6}} \longleftrightarrow C(t, 0). \tag{D.2.20}$$

The inverse Fourier transform is obtained using result (B.2.6) and we obtain

$$C(t, 0) = \frac{2^{1/3}\pi^{3/2}}{3\Gamma(5/6)\Gamma(2/3)} \frac{3^{1/2}H(t) + H(-t)}{|t|^{1/6}}. \tag{D.2.21}$$

Expressions (D.2.18) and (D.2.21) are equal and are approximately

$$C(t, 0) \simeq \frac{2.65H(t) + 1.53H(-t)}{|t|^{1/6}}. \tag{D.2.22}$$

D.2.2.2 Illuminated region – $C(t, 1)$ ($a < 0, y = 1$)

The function $C(t, 1)$ is more complicated so first we consider the asymptotic form of the spectrum. Using the asymptotic result (D.2.4), the spectrum in result (D.2.12) becomes

$$\sim -\frac{1}{i\omega} \frac{\pi}{6^{1/2}} \left(e^{i\omega} - i\, e^{-i\omega}\right) \longleftrightarrow C(t, 1). \tag{D.2.23}$$

Thus the singularities of the function are

$$C(t,1) \sim \frac{\pi}{6^{1/2}}\left(H(t+1) - \frac{1}{\pi}\ln|t-1|\right). \tag{D.2.24}$$

Now we express the function $C(t, 1)$ in terms of the complete elliptic integral of the first kind to obtain an exact result.

Consider first the case when $|t| < 1$. The cubic $4x^3 - 3x + t = 0$ has three real roots. Let $x = \cos\theta$ so $4x^3 - 3x = \cos 3\theta$. Thus the roots, $a > b > c$, are

$$x = \cos\left(\frac{1}{3}\cos^{-1}(-t)\right) \quad \text{and} \quad \cos\left(\frac{1}{3}\cos^{-1}(-t) \pm \frac{2\pi}{3}\right), \tag{D.2.25}$$

and using result (D.2.14), we obtain

$$C(t,1) = \frac{2}{(a-c)^{1/2}} K\left(\frac{b-c}{a-c}\right). \tag{D.2.26}$$

For $t > 1$ only one real root exists. Letting $x = -\cosh\theta$, $3x - 4x^3 = \cosh 3\theta$, the root is

$$x = a = -\cosh\left(\frac{1}{3}\cosh^{-1} t\right). \tag{D.2.27}$$

Comparing definition (D.2.11) with result (D.2.15), the constants are $b = -a/2$ and $c^2 = 3(a^2 - 1)/4$ so

$$p^2 = 3a^2 - 3/4, \tag{D.2.28}$$

and we obtain

$$C(t,1) = \frac{1}{p^{1/2}} K\left(\frac{p - 3a/2}{2p}\right). \tag{D.2.29}$$

For $t < -1$, again one root exists. Letting $x = \cosh\theta$, $4x^3 - 3x = \cosh 3\theta$, the root is

$$x = a = \cosh\left(\frac{1}{3}\cosh^{-1}(-t)\right), \tag{D.2.30}$$

and result (D.2.29) can be used again.

The complete function $C(t, 1)$ is illustrated in Figure D.3. It is interesting to confirm the singularity (D.2.24). At $t = -1 - \epsilon$, $a = 1$ from equation (D.2.30) and $p = 3/2$ from equation (D.2.28). Hence (D.2.29) gives

$$C(-1-\epsilon, 1) = \left(\frac{2}{3}\right)^{1/2} K(0) = \frac{\pi}{6^{1/2}}, \tag{D.2.31}$$

as $K(0) = \pi/2$ (Abramowitz and Stegun, 1965, §17.3.11). At $t = -1 + \epsilon, a = 1$, $b = c = -1/2$ in equation (D.2.14), and result (D.2.29) gives

$$C(-1 + \epsilon, 1) = \left(\frac{8}{3}\right)^{1/2} K(0) = \frac{2\pi}{6^{1/2}}. \quad \text{(D.2.32)}$$

The difference between values (D.2.32) and (D.2.31) agrees with discontinuity (D.2.24).

D.2.2.3 Shadow region – $C(t, -1)$ ($a > 0, y = -1$)

Again we can investigate the approximate behaviour from the asymptotic form (D.2.7) of the spectrum (D.2.12)

$$\sim -\frac{e^{-i\pi/4}}{i\omega} \frac{\pi}{6^{1/2}} e^{-\omega} \longleftrightarrow C(t, -1). \quad \text{(D.2.33)}$$

Thus the leading term of the function is

$$C(t, -1) \sim \frac{1}{2.3^{1/2}} \left(\tan^{-1} t + \ln(t^2 + 1)^{1/2}\right), \quad \text{(D.2.34)}$$

from the integrals of results (B.1.4) and (B.1.5).

The exact function (D.2.11) can be reduced to a standard form. For all t, the cubic has one root and with $x = \sinh\theta$, $4x^3 + 3x = -\sinh 3\theta$. Thus the root is

$$a = -\sinh\left(\frac{1}{3}\sinh^{-1} t\right), \quad \text{(D.2.35)}$$

and result (D.2.29) applies again, except

$$p^2 = 3a^2 + 3/4. \quad \text{(D.2.36)}$$

The function $C(t, -1)$ is illustrated in Figure D.3.

The Airy function arises at caustics in wave propagation. The attractive feature of the results in this appendix is that the waveforms can be described by just three 'standard' functions, $C(t, \pm 1)$ and $C(t, 0)$. These functions have a relatively simple behaviour (Figure D.3) and are straightforward to compute. No knowledge or computations of the Airy functions for the spectrum are needed.

D.2.3 A Fourier transform

The Airy transform also arises in the time domain (Section 9.3.7). Consider the spectrum

$$Sh^{(2)}(\omega) = (3\omega)^{-2/3} \exp\left(-(3\omega)^{1/3} e^{-i\pi/6} - 2i\pi/3\right), \quad \text{(D.2.37)}$$

for $\omega > 0$ (and defined with relationship (3.1.9) for $\omega < 0$ so that the inverse Fourier transform is real – the notation is used as the spectrum applies in two dimensions in a shadow). The inverse Fourier transform of $Sh^{(2)}(\omega)$ is given by

$$Sh^{(2)}(t) = \frac{1}{\pi}\,\mathrm{Re}\int_0^\infty (3\omega)^{-2/3}\mathrm{e}^{-(3\omega)^{1/3}\mathrm{e}^{-\mathrm{i}\pi/6}-2\mathrm{i}\pi/3-\mathrm{i}\omega t}\,\mathrm{d}\omega \qquad \text{(D.2.38)}$$

$$= \frac{t^{-1/3}}{\pi}\,\mathrm{Re}\int_0^\infty \mathrm{e}^{-\mathrm{i}t^{-1/3}\xi-\mathrm{i}\xi^3/3}\,\mathrm{d}\xi, \qquad \text{(D.2.39)}$$

where we have made the substitution

$$\omega^{1/3} = \mathrm{e}^{2\mathrm{i}\pi/3}(3t)^{-1/3}\xi. \qquad \text{(D.2.40)}$$

This integral (D.2.39) can be recognized as the Airy integral (D.2.2) so the result is

$$Sh^{(2)}(t) = t^{-1/3}Ai(t^{-1/3}). \qquad \text{(D.2.41)}$$

Because of the singular term $t^{-1/3}$, the behaviour of the function $Sh^{(2)}(t)$ is interesting and worth investigating. As $t \to \infty$, the function has a long, slowly decaying tail

$$Sh^{(2)}(t) \to Ai(0)\,t^{-1/3}. \qquad \text{(D.2.42)}$$

Near the origin the behaviour is more interesting. Using the asymptotic form (D.2.7), we have

$$Sh^{(2)}(t) = \xi Ai(\xi) \simeq \frac{1}{2\pi^{1/2}}\,\xi^{3/4}\mathrm{e}^{-2\xi^{3/2}/3} \to 0, \qquad \text{(D.2.43)}$$

as $t \to 0$ ($\xi = t^{-1/3}$). All derivatives are zero at $t = 0$ so the function is emergent. The function is stationary when

$$\frac{\mathrm{d}\,Sh^{(2)}(t)}{\mathrm{d}\xi} = Ai(\xi) + \xi Ai'(\xi) = 0, \qquad \text{(D.2.44)}$$

which, from the asymptotic form (D.2.7), has a root when $\xi \simeq 1$ (more accurately at $\xi \simeq 0.88$ or $t \simeq 1.47$). Between the origin and this maximum the function must have an inflexion point. This occurs when

$$\frac{\mathrm{d}^2 Sh^{(2)}(t)}{\mathrm{d}t\,\mathrm{d}\xi} = -\frac{\xi^3}{3}\left(\xi^3 Ai(\xi) + 6\,\xi Ai'(\xi) + 4Ai(\xi)\right) = 0. \qquad \text{(D.2.45)}$$

Using the asymptotic form (D.2.7), this reduces to the quadratic in $\xi^{3/2}$

$$\xi^3 - 6\,\xi^{3/2} + 4 = 0. \qquad \text{(D.2.46)}$$

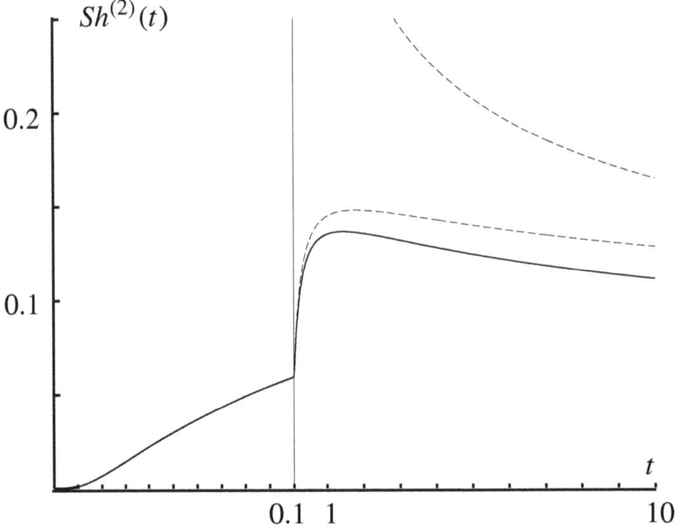

Fig. D.4. The function $Sh^{(2)}(t)$ (D.2.41) with the asymptotic behaviour near $t = 0$ using (D.2.43) and for $t \to \infty$ (D.2.42) shown with dashed lines. Note that the ordinate scale is divided at $t = 0.1$ – each part of the axis is linear, and for $t < 0.1$ the difference between the exact and approximate expressions, (D.2.41) and (D.2.43), is not visible. The inflexion point at $t \simeq 0.036$ and the maximum at $t \simeq 1.47$ are visible.

The required root $\xi^{3/2} = 3 + \sqrt{5}$ gives $t \simeq 0.03647$. The complete function $Sh^{(2)}(t)$ (D.2.41) is illustrated in Figure D.4. Included in dashed lines are the asymptotic behaviour near $t = 0$ using (D.2.43) and for $t \to \infty$ (D.2.42).

In three dimensions, we require the function

$$Sh^{(3)}(t) = \frac{\mathrm{d}}{\mathrm{d}t}\left(\lambda(t) * Sh^{(2)}(t)\right) \tag{D.2.47}$$

$$= \int_0^t \frac{Sh^{(2)\prime}(t - t')}{t'^{1/2}}\,\mathrm{d}t' \tag{D.2.48}$$

$$= 2\int_0^{\sqrt{t}} Sh^{(2)\prime}(t - \xi^2)\,\mathrm{d}\xi. \tag{D.2.49}$$

The function $Sh^{(2)\prime}(t)$ is well behaved

$$Sh^{(2)\prime}(t) = -\frac{1}{3}t^{-4/3}\left(Ai(t^{-1/3}) + t^{-1/3}Ai'(t^{-1/3})\right), \tag{D.2.50}$$

and illustrated in Figure D.5. The integral (D.2.49) is easily evaluated numerically to give the function $Sh^{(3)}(t)$ illustrated in Figure D.6.

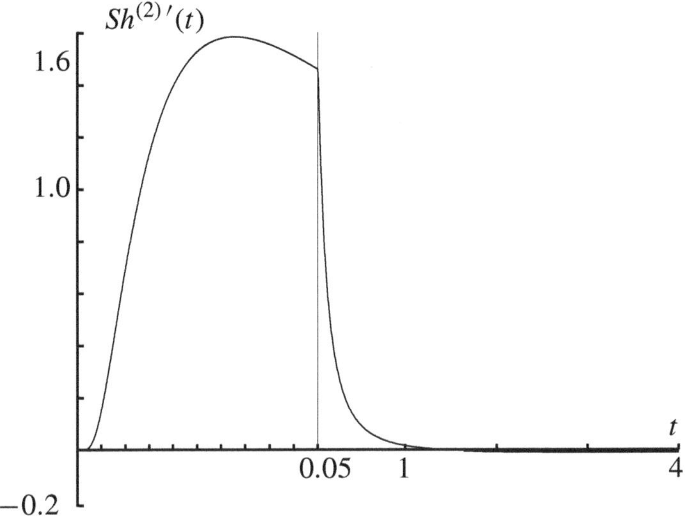

Fig. D.5. The function $Sh^{(2)\prime}(t)$ (D.2.50). Note that the ordinate scale is divided at $t = 0.05$ – each part of the axis is linear.

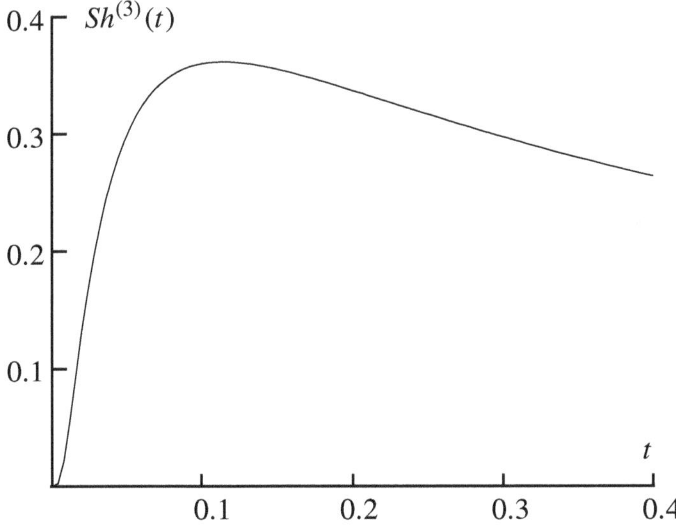

Fig. D.6. The function $Sh^{(3)}(t)$ (D.2.49).

D.3 Incomplete saddle points – Fresnel functions

If the function $f(z)$ in integral (D.0.1) is discontinuous, we need the incomplete, second-order saddle-point result, i.e.

$$I = \int_b^\infty \mathrm{e}^{\mathrm{i}ax^2}\,\mathrm{d}x, \tag{D.3.1}$$

instead of integral (D.1.8).

The standard functions are the cosine and sine Fresnel integrals

$$C(z) = \int_0^z \cos\left(\frac{\pi}{2}t^2\right) \mathrm{d}t \tag{D.3.2}$$

$$S(z) = \int_0^z \sin\left(\frac{\pi}{2}t^2\right) \mathrm{d}t, \tag{D.3.3}$$

defined in Abramowitz and Stegun (1965, §7.3.1 and §7.3.2). Combining these we define (the non-standard function)

$$Fr(z) = \frac{1}{2} - \frac{\mathrm{e}^{-\mathrm{i}\pi/4}}{2^{1/2}}\left(C(z) + \mathrm{i}S(z)\right). \tag{D.3.4}$$

Combining definitions (D.3.2) and (D.3.3) in expression (D.3.4) with result (D.1.11), we obtain

$$Fr(z) = \frac{\mathrm{e}^{-\mathrm{i}\pi/4}}{2^{1/2}} \int_z^\infty \mathrm{e}^{\mathrm{i}\pi t^2/2}\, \mathrm{d}t. \tag{D.3.5}$$

By a simple change of variable, the integral (D.3.1) can be mapped into this function

$$I = \int_b^\infty \mathrm{e}^{\mathrm{i}ax^2}\, \mathrm{d}x = \mathrm{e}^{\mathrm{i}\pi/4}\left(\frac{\pi}{a}\right)^{1/2} Fr\left(\left(\frac{2a}{\pi}\right)^{1/2} b\right), \tag{D.3.6}$$

where for simplicity we have assumed that $a > 0$. The non-standard function $Fr(z)$ can also be written in terms of the complementary error function

$$Fr(z) = \frac{1}{2}\,\mathrm{erfc}\left(\left(\frac{\pi}{2}\right)^{1/2} \mathrm{e}^{-\mathrm{i}\pi/4} z\right), \tag{D.3.7}$$

using Abramowitz and Stegun (1965, §7.3.22). The real part of this function is plotted in Figure D.7. Note that $Fr(-\infty) = 1$, $Fr(0) = 1/2$ and $Fr(\infty) = 0$.

D.3.1 Asymptotic forms

For $z \gg 1$, we can use the asymptotic form of the function (D.3.4). We have

$$C(z) \simeq \frac{1}{2} + \frac{1}{\pi z}\sin\left(\frac{\pi z^2}{2}\right) \tag{D.3.8}$$

$$S(z) \simeq \frac{1}{2} - \frac{1}{\pi z}\cos\left(\frac{\pi z^2}{2}\right), \tag{D.3.9}$$

from Abramowitz and Stegun (1965, §7.3.9, §7.2.10 and §7.3.27). Hence

$$Fr(z) \simeq \frac{1}{2^{1/2}\pi z}\mathrm{e}^{\mathrm{i}\pi z^2/2 + \mathrm{i}\pi/4}. \tag{D.3.10}$$

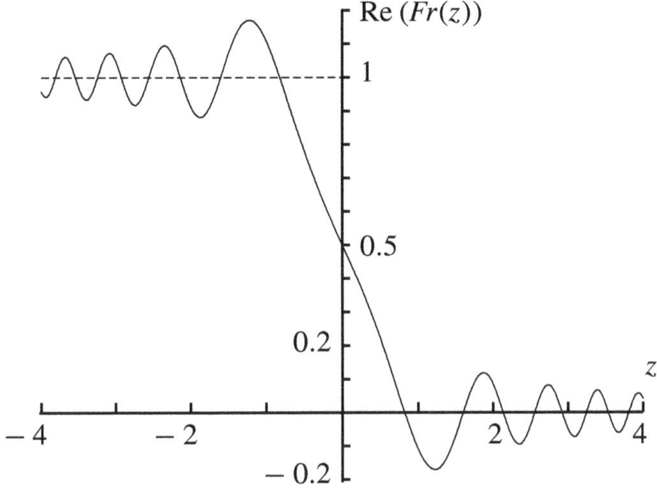

Fig. D.7. The real part of the function (D.3.7), $Fr(z)$.

For $z \ll -1$, we can use

$$Fr(-z) = 1 - Fr(z), \tag{D.3.11}$$

which follows from Abramowitz and Stegun (1965, §7.3.17).

D.3.2 Inverse Fourier transforms

Frequently the Fresnel function arises in the spectral domain. We modify expression (D.3.6) to

$$\int_b^\infty e^{i\omega a x^2}\, dx = e^{i\pi/4} \left(\frac{\pi}{a\omega}\right)^{1/2} Fr\left(\left(\frac{2a\omega}{\pi}\right)^{1/2} b\right). \tag{D.3.12}$$

Although the integral has two parameters, a and b, it can always be reduced to a single parameter as

$$\int_b^\infty e^{i\omega a x^2}\, dx = b \int_1^\infty e^{i\omega(ab^2)x^2}\, dx, \tag{D.3.13}$$

where we have assumed $b > 0$. For $b < 0$ we can use result (D.3.11) to reduce it to the positive case. We therefore need only consider the standard form

$$\int_1^\infty e^{i\omega a x^2}\, dx = e^{i\pi/4} \left(\frac{\pi}{a\omega}\right)^{1/2} Fr\left(\left(\frac{2a\omega}{\pi}\right)^{1/2}\right) \tag{D.3.14}$$

(remember that for simplicity we assume $a > 0$ and use the conjugate when $a < 0$).

Using the asymptotic result (D.3.10), the asymptotic form for the Fresnel function is

$$Fr\left(\left(\frac{2a\omega}{\pi}\right)^{1/2}\right) \sim \frac{1}{2(\pi\omega a)^{1/2}}\,\mathrm{e}^{\mathrm{i}a\omega+\mathrm{i}\pi/4}. \tag{D.3.15}$$

Thus the singularity of its inverse Fourier transform is

$$Fr\left(\left(\frac{2a\omega}{\pi}\right)^{1/2}\right) \to \sim \frac{1}{2\pi a^{1/2}}\frac{H(t-a)}{(t-a)^{1/2}}. \tag{D.3.16}$$

The spectrum can also be written in terms of the complementary error function (D.3.7)

$$Fr\left(\left(\frac{2a\omega}{\pi}\right)^{1/2}\right) = \frac{1}{2}\,\mathrm{erfc}\left((-\mathrm{i}\omega a)^{1/2}\right). \tag{D.3.17}$$

The inverse Fourier transform of this is given in many handbooks

$$Fr\left(\left(\frac{2a\omega}{\pi}\right)^{1/2}\right) \longleftrightarrow \frac{a^{1/2}\,H(t-a)}{2\pi t(t-a)^{1/2}} \tag{D.3.18}$$

(e.g. Abramowitz and Stegun, 1965, §29.3.114). Clearly when $t \simeq a$, this agrees with the approximation (D.3.16).

Returning now to the incomplete saddle-point integral (D.3.14), it is convenient to multiply it by $\lambda(\omega)$

$$\lambda(\omega)\int_1^\infty \mathrm{e}^{\mathrm{i}\omega a x^2}\,\mathrm{d}x = -\frac{1}{\mathrm{i}\omega}\frac{\pi}{a^{1/2}}\,Fr\left(\left(\frac{2a\omega}{\pi}\right)^{1/2}\right), \tag{D.3.19}$$

to obtain the integral of the above time series (D.3.18). Thus

$$\lambda(\omega)\int_1^\infty \mathrm{e}^{\mathrm{i}\omega a x^2}\,\mathrm{d}x \longleftrightarrow \int_a^t \frac{\mathrm{d}t}{2t(t-a)^{1/2}} = \frac{1}{a^{1/2}}\cos^{-1}\left(\frac{a}{t}\right)^{1/2}. \tag{D.3.20}$$

This result can also be obtained without knowledge of the Fresnel function or its inverse Fourier transform by taking the inverse transform of the incomplete saddle-point integral, i.e.

$$\begin{aligned}\lambda(\omega)\int_1^\infty \mathrm{e}^{\mathrm{i}\omega a x^2}\,\mathrm{d}x &\longleftrightarrow \int_1^{(t/a)^{1/2}} \frac{\mathrm{d}x}{(t-ax^2)^{1/2}}\\ &= \frac{1}{a^{1/2}}\cos^{-1}\left(\frac{a}{t}\right)^{1/2} = a^{-1/2}Fi(t/a),\end{aligned} \tag{D.3.21}$$

say. The function $Fi(t)$ is illustrated in Figure D.8 ($Fi(\infty) = \pi/2$).

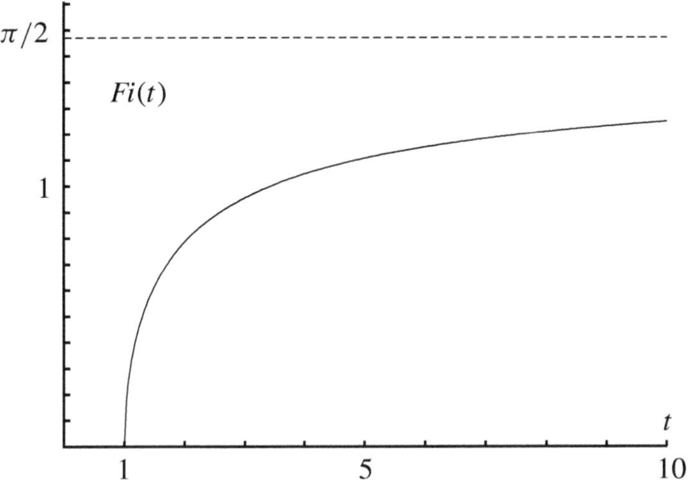

Fig. D.8. The function $Fi(t)$ (D.3.21).

In general, results for an incomplete saddle point can be characterized by three cases

$$\lambda(\omega)\int_{-1}^{\infty} e^{i\omega x^2} dx = -\frac{\pi}{i\omega} Fr\left(-\left(\frac{2\omega}{\pi}\right)^{1/2}\right) \longleftrightarrow \pi H(t) - Fi(t) \quad (D.3.22)$$

$$\lambda(\omega)\int_{0}^{\infty} e^{i\omega x^2} dx = -\frac{\pi}{2i\omega} \longleftrightarrow \frac{\pi}{2} H(t) \quad (D.3.23)$$

$$\lambda(\omega)\int_{1}^{\infty} e^{i\omega x^2} dx = -\frac{\pi}{i\omega} Fr\left(\left(\frac{2\omega}{\pi}\right)^{1/2}\right) \longleftrightarrow Fi(t), \quad (D.3.24)$$

describing the signal in the illuminated region (D.3.22), at the shadow edge (D.3.23), and in the shadow (D.3.24).

To combine these results, it is convenient to define a function $F(t, y)$ which is the inverse Fourier transform of $\pi^{-1}\lambda(\omega)\int_y^{\infty} \exp(i\omega x^2)\, dx$. We only need the results for $y = \pm 1$ and $y = 0$

$$F(t, +1) = Fi(t)/\pi \quad (D.3.25)$$

$$F(t, 0) = H(t)/2 \quad (D.3.26)$$

$$F(t, -1) = H(t) - Fi(t)/\pi, \quad (D.3.27)$$

as

$$F(t, y) = F\left(\frac{t}{y^2}, \text{sgn}(y)\right). \quad (D.3.28)$$

This function is illustrated in Figure D.9.

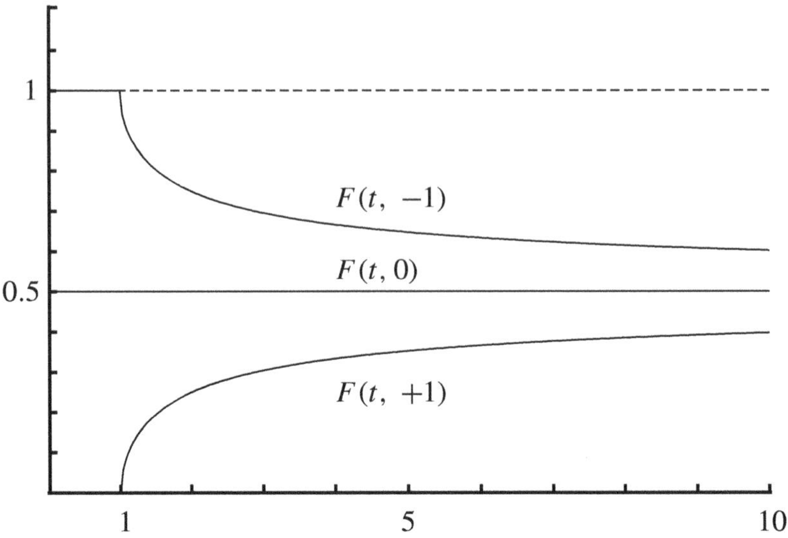

Fig. D.9. The functions $F(t, 0)$ and $F(t, \pm 1)$ defined in equations (D.3.25)–(D.3.27).

The Fresnel function arises at shadows in wave propagation. The attractive feature of the results in this appendix is that the waveforms can be described by the single 'standard' function $F(t, y)$ or $Fi(t)$. This function has a simple behaviour (Figure D.8) and is straightforward to compute. No knowledge or computations of the Fresnel function for the spectrum are needed.

Bibliography

Abo-Zena, A., 1979. Dispersion function computations for unlimited frequency values, *Geophys. J.R. Astr. Soc.*, **58**, 91–105.

Abramovici, F., 1970. Numerical seismograms for a layered elastic solid, *Bull. Seism. Soc. Amer.*, **60**, 1861–76.

1978. A generalization of the Cagniard method, *J. Comp. Phys.*, **29**, 328–43.

1984. The exact solution to the problem of an *SH* pulse in a layered elastic half-space, *Bull. Seism. Soc. Amer.*, **74**, 377–93.

Abramovici, F. and Gal-Ezer, J., 1978. Numerical seismograms for a vertical point-force in a single layer over a half-space, *Bull. Seism. Soc. Amer.*, **68**, 81–101.

Abramovici, F., Kanasewich, E. R. and Kelamis, P. G., 1982. Seismic waves from a horizontal stress discontinuity in a layered solid, *Bull. Seism. Soc. Amer.*, **72**, 1483–98.

Abramowitz, M. and Stegun, I. A., 1965. *Handbook of Mathematical Functions*, New York: Dover.

Aki, K. and Richards, P. G., 1980. *Quantitative Seismology: Theory and Methods*, San Francisco: W. H. Freeman and Company.

2002. *Quantitative Seismology: Theory and Methods*, 2nd edn, Sausalito: University Science Books.

Alsop, L. E., 1968. An orthnormality relation for elastic body waves, *Bull. Seism. Soc. Amer.*, **58**, 1949–54.

Ansell, J. H., 1978. On the scattering of *SH* waves from a point source by a sphere, *Geophys. J. R. Astr. Soc.*, **54**, 349–87.

Arnol'd, V. I., 1967. Characteristic classes entering in quantization conditions, *Funct. Anal. Appl.*, **1**, 1–13.

Auld, B. A., 1973. *Acoustic Fields and Waves in Solids*, vol. I, New York: John Wiley and Sons.

Backus, G. E., 1970. A geometrical picture of anisotropic elastic tensors, *Rev. Geophys. Space Phys.*, **8**, 633–71.

Backus, G. E. and Mulcahy, M., 1976*a*. Moment tensors and other phenomenological descriptions of seismic sources – I. Continuous displacements, *Geophys. J. R. Astr. Soc.*, **46**, 341–61.

1976*b*. Moment tensors and other phenomenological descriptions of seismic sources – II. Discontinuous displacements, *Geophys. J.R. Astr. Soc.*, **47**, 301–29.

Baerheim, R., 1993. Harmonic decomposition of the anisotropic tensor, *Q. Jl Mech. Appl. Math.*, **46**, 391–418.

Bakker, P. M., 1998. Phase shift at caustics along rays in anisotropic media, *Geophys. J. Int.*, **134**, 515–8.

Båth, M., 1969. *Mathematical Aspects of Seismology*, Amsterdam: Elsevier Publishing Company.

Bechmann, R., 1958. Elastic and piezoelectric constants of alpha-quartz, *Phys. Rev.*, **110**, 1060–1.

Bender, C. M. and Orzag, S. A., 1978. *Advanced Mathematical Methods for Scientists and Engineers*, New York: McGraw-Hill.

Ben-Menahem, A. and Gibson, R. L., 1990. Scattering of elastic waves by localized anisotropic inclusions, *J. Acoust. Soc. Amer.*, **87**, 2300–9.

Ben-Menahem, A. and Singh, S. J., 1981. *Seismic Waves and Sources*, New York: Springer Verlag.

Bennett, H. F., 1968. An investigation into velocity anisotropy through measurements of ultrasonic wave velocities in snow and ice cores from Greenland and Antartica, Ph.D. thesis, University of Wisconsin.

Berry, M. V., 1982. Quantal phase factors accompanying adiabatic changes, *Proc. Roy. Soc.*, **A392**, 45–57.

1988. The geometric phase, *Sci. Am.*, **259**, 6, 26–32.

Biot, M. A., 1957. General theorems on the equivalence of group velocity and energy transport, *Phys. Rev.*, **105**, 1129–37.

Bleistein, N., 1984. *Mathematical Methods for Wave Phenomena*, Orlando, FL: Academic Press.

Bond, W., 1943. The mathematics of the physical properties of crystals, *Bell Sys. Tech. J.*, **22**, 1–72.

Bortfield, R., 1961. Approximation of the reflection and transmissioon coefficients of plane longitudinal and transverse waves, *Geophys. Prosp.*, **9**, 485–503.

Brandsberg-Dahl, S., Hoop, M. V. de and Ursin, B., 2003. Focusing in dip and AVA compensation on scattering-angle/azimuth common image gathers, *Geophysics*, **68**, 232–54.

Brekhovskikh, L. M., 1960. *Waves in Layered Media*, New York: Academic Press.

Bremmer, H., 1939. Geometrisch-optische benadering van de golfvergelijking, in *Handelingen Natuur- en Geneeskundig Congres, Nijmegen*, 88–91, Haarlem, the Netherlands: Ruygrok (in Dutch).

1949*a*. *Terrestrial Radio Waves*, Amsterdam: Elsevier Publishing Company.

1949*b*. The propagation of electromagnetic waves through a stratified medium and its W.K.B. approximation for oblique incidence, *Physica*, **15**, 593–608.

1951.The W.K.B. approximation as the first term in a geometric-optical series, *Commun. Pure Appl. Math.*, **4**, 105–15.

Brillouin, L., 1926. Remarques sur la méchanique ondulatoire, *J. Phys. Radium*, **7**, 353–68.

Brokešova, J., 1992. High frequency 2.5D modelling of seismic wavefield applied to a finite extent source simulation, in *Proceedings of XXIII General Assembly of the European Seismological Commission*, Prague, Czechoslovak Academy of Sciences.

Buchen, P. W. and Ben-Hador, R., 1996. Free-mode surface-wave computations, *Geophys. J. Intl.*, **124**, 869–87.

Buchwald, V. T., 1959. Elastic waves in anisotropic media, *Proc. R. Soc. Lond., A*, **253**, 563–80.

Budden, K. G., 1961. *The Wave-Guide Mode Theory of Wave Propagation*, London: Logos Press.

1966. *Radio Waves in the Ionosphere*, Cambridge: Cambridge University Press.

Bullen, K. E., 1963. *An Introduction to the Theory of Seismology*, 3rd edn, Cambridge: Cambridge University Press.

Burridge, R., 1963*a*. The reflexion of a disturbance within an elastic sphere, Ph.D. thesis, Cambridge University.

1963*b*. The reflexion of a disturbance within a solid sphere, *Proc. Roy. Soc.*, **A276**, 367–400.

1967. The singularity on the plane lids of the wave surface of elastic media with cubic symmetry, *Q. J. Mech. Appl. Math.*, **20**, 41–56.

1968. A new look at Lamb's problem, *Jour. Phys. Earth*, **16**, 168–71.

Burridge, R. and Knopoff, L., 1964. Body force equivalents for seismic dislocations, *Bull. Seism. Soc. Amer.*, **54**, 1875–88.

Burridge, R., Hoop, M. V. de, Miller, D. and Spencer, C., 1998. Multiparameter inversion in anisotropic elastic media, *Geophys. J. Intl.*, **134**, 757–77.

Cagniard, L., 1939. *Réflexion et réfraction des ondes séismiques progressives*, Paris: Gauthier-Villars.

1962. *Reflexion and refraction of progressive seismic waves*, trans. E. A. Flinn and C. H. Dix, New York: McGraw-Hill.

Červený, V., 1972. Seismic rays and ray intensities in inhomogeneous anisotropic media, *Geophys. J.R. Astron. Soc.*, **29**, 1–13.

1982. Direct and inverse kinematic problems for inhomogeneous anisotropic media – linearization approach. *Contr. Geophys. Inst. Slov. Acad. Sci.*, **13**, 127–33.

1985. The application of ray tracing to the numerical modeling of seismic wavefield in complex structures, in *Handbook of Geophysical Exploration, Section I: Seismic Exploration*, ed. K. Helbig and S. Treitel, vol. 15A, Seismic Shear Waves (ed. G. Dohr), London: Geophysical Press, 1–124.

2001. *Seismic Ray Theory*, Cambridge; Cambridge University Press.

2002. Fermat's variational principle for anisotropic inhomogeneous media, *Studia Geophy. et Geod.*, **46**, 567–88.

Červený, V. and Hron, F., 1980. The ray series and dynamic ray tracing system for three-dimensional inhomogeneous media, *Bull. Seism. Soc. Amer.*, **70**, 47–77.

Červený, V. and Jech, J., 1982. Linearized solutions of kinematic problems of seismic body waves in inhomogeneous slightly anisotropic media, *J. Geophys.*, **51**, 96–104.

Červený, V., Langer, J. and Pšenčík, I., 1974. Computations of geometrical spreading of seismic body waves in laterally inhomogeneous media with curved interfaces, *Geophys. J.R. Astr. Soc.*, **38**, 9–19.

Červený, V. and Ravindra, R., 1971. *Theory of Seismic Head Waves*, Toronto: University of Toronto Press.

Chapman, C. H., 1972. Lamb's problem and comments on the paper 'On Leaking Modes' by Usha Gupta, *Pageoph.*, **94**, 233–47.

1973. The Earth flattening transformation in body wave theory, *Geophys. J. R. Astr. Soc.*, **35**, 55–70.

1974*a*. Generalized ray theory for an inhomogeneous medium, *Geophys. J.R. Astr. Soc.*, **36**, 673–704.

1974*b*. The turning point of elastodynamic waves, *Geophys. J.R. astr. Soc.*, **39**, 613–21.

1976. Exact and approximate generalized ray theory in vertically inhomogeneous media, *Geophys. J. R. astr. Soc.*, **46**, 201–33.

1977. The computation of synthetic body-wave seismograms, *Appl. Mech. Div. of ASME*, **25**, 42–77.

1978. A new method for computing synthetic seismograms, *Geophys. J.R. Astr. Soc.*, **54**, 481–518.
1979. On impulsive wave propagation in a spherically symmetric model, *Geophys. J. R. astr. Soc.*, **58**, 229–34.
1981. Long-period corrections to body-waves: theory, *Geophys. J. R. astr. Soc.*, **64**, 321–72.
1994. Reflection/transmission coefficient reciprocities in anisotropic media, *Geophys. J. Int.*, **116**, 498–501.
2003. Yet another layer matrix algorithm, *Geophys. J. Intl.*, **154**, 212–23.
Chapman, C. H., Chu Jen-Yi and Lyness, D. G., 1988. The WKBJ Seismogram Algorithm, in *Seismological Algorithms*, ed. D. J. Doornbos, 47–74, London: Academic Press.
Chapman, C. H. and Coates, R. T., 1994. Generalized Born scattering in anisotropic media, *Wave Motion*, **19**, 309–41.
Chapman, C. H. and Drummond, R., 1982. Body-wave seismograms in inhomogeneous media using Maslov asymptotic theory, *Bull. Seism. Soc. Amer.*, **72**, S277–S317.
Chapman, C. H. and Keers, H., 2002. Application of the Maslov seismogram method in three dimensions, *Studia Geophy. et Geod.*, **46**, 615–49.
Chapman, C. H. and Phinney, R. A., 1970. Diffraction of *P* waves by the core and an inhomogeneous mantle, *Geophys. J. R. Astr. Soc.*, **21**, 185–205.
1972. Diffracted seismic signals, *Meth. Comp. Phys.*, **12**, 166–230.
Chapman, C. H. and Shearer, P. M., 1989. Ray tracing in azimuthally anisotropic media – II. Quasi-shear wave coupling, *Geophys. J.*, **86**, 65–83.
Chester, C., Friedman, B. and Ursell, F., 1957. An extension of the method of steepest descent, *Proc. Camb. Phil. Soc.* **53**, 599–611.
Choy, G. L., 1977. Theoretical seismograms of core phases calculated using a frequency-dependent full wave theory, and their interpretation, *Geophys. J.R. Astr. Soc.*, **51**, 275–311.
Cisternas, A., Betancourt, O. and Leiva, A., 1973. Body waves in a 'Real Earth', Part I, *Bull. Seism. Soc. Amer.*, **63**, 145–56.
Clay, C. S. and Medwin, H., 1977. *Acoustical Oceanography: Principles and Applications*, New York: John Wiley.
Clemmow, P. C. and Heading, J., 1954. Coupled forms of the differential equations governing radio propagation in the ionosphere, *Proc. Camb. Phil. Soc.*, **50**, 319–33.
Coates, R. T. and Chapman, C. H., 1990*a*. Ray perturbation theory and the Born approximation, *Geophys. J. Int.*, **100**, 379–92.
1990*b*. Quasi-shear wave coupling in weakly anisotropic 3–D media, *Geophys. J. Int.*, **103**, 301–20.
Coddington, E. A. and Levinson, N., 1955. *Theory of Ordinary Differential Equations*, New York: McGraw-Hill.
Cormier, V. and Richards, P. G., 1977. Full-wave theory applied to a discontinuous velocity increase: the inner core boundary, *J. Geophys. Res.*, **43**, 3–31.
Courant, R. and Hilbert, D., 1966. *Methods of Mathematical Physics*, New York: John Wiley.
Cowin, S. C., 1989. Properties of the anisotropic elasticity tensor, *Q. Jl Mech. Appl. Math.*, **42**, 249–66.
1993. Corrigendum: Properties of the anisotropic elasticity tensor, *Q. Jl Mech. Appl. Math.*, **46**, 541–2.
Cowin, S. C., Mehrabadi, M. M. and Sadegh, A. M., 1991. Kelvin's formulation of the anisotropic Hooke's law, in *Modern Theory of Anisotropic Elasticity and Applications*, ed. J. J. Wu, T. C. T. Ting and D. M. Barnett, ch. 24, SIAM, Philadelphia.

Dahlen, F. A. and Tromp, J., 1998. *Theoretical Global Seismology*, Princeton, NJ: Princeton University Press.
Daley, P. F. and Hron, F., 1983. High-frequency approximation of the nongeometrical S^* arrival, *Bull. Seism. Soc. Amer.*, **73**, 109–23.
Drijkoningen, G. G. and Chapman, C. H., 1988. Tunneling rays using the Cagniard–de Hoop method, *Bull. Seism. Soc. Amer.*, **78**, 898–907.
Drijkoningen, G. G. and Fokkema, J. T., 1987. The exact response of an ocean and a N–layer configuration, *Geophys. Prosp.*, **35**, 33–61.
Duff, G. F. D., 1960. The Cauchy problem for elastic waves in an anisotropic media, *Phil. Trans. Roy. Soc. Lond.*, **A252**, 249–73.
Duistermaat, J. J., 1995. *Fourier Integral Operators*, Boston: Birkhäuser.
Dunkin, J.W., 1965. Computation of modal solutions in layered, elastic media at high frequencies, *Bull. Seism. Soc. Amer.*, **55**, 335–58.
Duwalo, G. and Jacobs, J. A., 1959. Effects of a liquid core on propagation of seismic waves, *Can. J. Phys.*, **37**, 109–28.
Dziewonski, A. M. and Anderson, D. L., 1981. Preliminary reference Earth model, *Phys. Earth Plant. Int.*, **25**, 297–356.
Eisner, L. and Pšenčík, I., 1996. Computation of additional components of the first-order ray approximation in isotropic media, *Pageoph*, **148**, 227–53.
Erdélyi, A., Magnus, W., Oberhettinger, F. and Tricomi, F. G., 1954. *Tables of Integral Transforms*, vol. 1, New York: McGraw-Hill.
Ewing, M., Jardetsky, W. and Press, F., 1957. *Elastic Waves in Layered Media*, New York: McGraw-Hill.
Farra, V., 1989. Ray perturbation theory for heterogeneous hexagonal anisotropic medium, *Geophys. J. Intl.*, **99**, 723–38.
Farra, V. and Le Bégat, S., 1995. Sensitivity of qP-wave traveltimes and polarization vectors to heterogeneity, anisotropy and interfaces, *Geophys. J. Intl.*, **121**, 371–84.
Farra, V. and Madariaga, R., 1987. Seismic waveform modeling in heterogeneous media by ray perturbation theory, *J. Geophys. Res.*, **92**, 2697–712.
Fatti, J. L., Vail, P. J., Smith, G. C., Strauss, P. J. and Levitt, P. R., 1994. Detection of gas in sandstone reservoirs using AVO analysis: a 3-D seismic case history using the Geostack technique, *Geophysics*, **59**, 1362–76.
Fedorov, F. I., 1968. *Theory of Elastic Waves in Crystals*, New York: Plenum Press.
Filon, L. N. G., 1929. On a quadrature method for trigonometric integrals, *Proc. Roy. Soc. Edinburgh*, **49**, 38–47.
Frazer, L. N. and Fryer, G. J., 1989. Useful properties of the system matrix for a homogeneous anisotropic visco-elastic solid, *Geophys. J.*, **97**, 173–7.
Frazer, L. N. and Gettrust, J. F., 1984. On a generalization of Filon's method and the computation of the oscillatory integrals of seismology, *Geophys. J. R. Astr. Soc.*, **76**, 461–81.
Frazer, L. N. and Phinney, R. A., 1980. The theory of finite frequency body wave synthetic seismograms in inhomogeneous elastic media, *Geophys. J. R. Astr. Soc.*, **63**, 691–717.
Frazer, L. N. and Sen, M. K., 1985. Kirchhoff–Helmholtz reflection seismograms in a laterally inhomogeneous multi-layered elastic medium – I. Theory, *Geophys. J. R. Astr. Soc.*, **80**, 121–47.
French, W. S., 1974. Two-dimensional and three-dimensional migration of model-experiment reflection profiles, *Geophysics*, **39**, 265–77.
Fryer, G. J. and Frazer, L. N., 1984. Seismic waves in stratified media, *Geophys. J. R. Astr. Soc.*, **78**, 691–710.

Fuchs, K. and Müller, G., 1971. Computation of synthetic seismograms with the reflectivity method and comparison with observations, *Geophys. J. R. Astr. Soc.*, **23**, 417–33.

Fuchs, K. and Schulz, K., 1976. Tunneling of low-frequency waves through the subscrustal lithosphere, *J. Geophys.*, **42**, 175–90.

Fung, Y. C., 1965. *Foundations of Solid Mechanics*, Englewood Cliffs, NJ: Prentice Hall.

Gajewski, D. and Pšenčík, I., 1987. Computation of high-frequency seismic wavefields in 3–D laterally inhomogeneous anisotropic media, *Geophys. J.R. Astr. Soc.*, **91**, 383–411.

1990. Vertical seismic profile synthetics by dynamic ray tracing in laterally varying anisotropic structures, *J. Geophys. Res.*, **95**, 11301–15.

Gantmacher, F. R., 1959. *Matrix Theory*, New York: Chelsea Publ. Co.

Garmany, J., 1988*a*. Seismograms in stratified anisotropic media – I. WKBJ theory, *Geophys. J.*, **92**, 365–77.

1988*b*. Seismograms in stratified anisotropic media – II. Uniformly asymptotic approximations, *Geophys. J.*, **92**, 379–89.

2000. Phase shifts at caustics in anisotropic media, in *Proceedings of the 9th International Workshop on Anisotropy*, at: http://www.seg.org/9iwsa/proceedings/26_Garmany.pdf.

Garvin, W. W., 1956. Exact transient solution of the buried line source problem, *Proc. Roy. Soc. Lon.*, **A234**, 528–41.

Gebrande, H., 1976. A seismic ray tracing method for two-dimensional inhomogeneous media, in *Exploration Seismology in Central Europe: Data and Results*, ed. P. Giese, C. Prodehl and A. Stein, Berlin: Springer, 162–7.

Gibson, R. L. and Ben-Menahem, A., 1991. Elastic wave scattering by anisotropic obstacles: application to fractured volumes, *J. Geophys. Res.*, **96**, 19905–24.

Gilbert, F., 1960. Scattering of impulsive elastic waves by a smooth convex cylinder, *J. Acous. Soc. Am.*, **32**, 841–56.

1967. Gravitationally perturbed elastic waves, *Bull. Seism. Soc. Amer.*, **57**, 783–94.

1998. Elastic waves in transitional solid with arbitrarily small rigidity, *Geophys. J. Int.*, **133**, 230–2.

Gilbert, F. and Backus, G. E., 1966. Propagator matrices in elastic wave and vibration problems, *Geophysics*, **31**, 326–32.

Gilbert, F. and Dziewonski, A. M., 1975. An application of normal mode theory to the retrieval of structural parameters and source mechanisms from seismic spectra, *Phil. Trans. Roy. Soc. Lond.*, **A278**, 187–269.

Gilbert, F. and Knopoff, L., 1961. The directivity problem for a buried line source, *Geophysics*, **26**, 626–34.

Gilbert, F. and Laster, S. J., 1962. Excitation and propagation of pulses on an interface, *Bull. Seism. Soc. Amer.*, **52**, 299–319.

Golub, G. H. and Loan, C. F. van, 1996. *Matrix Computations*, 3rd edn, Baltimore: John Hopkins University Press.

Gradshteyn, I. S. and Ryzhik, I. M., 1980. *Table of Integrals, Series and Products*, New York: Academic Press.

Gray, S. H., 1982. A geometric-optic series and a WKB paradox, *Q. Appl. Math.*, **40**, 73–81.

1983. On the convergence of the time-domain Bremmer series, *Wave Motion*, **5**, 249–55.

Green, R., 1962. The hidden layer problem, *Geophys. Prosp.*, **10**, 166–70.

Guest, W. S., Thomson, C. J. and Kendall, J.-M., 1992. Displacement and rays for splitting *S*-waves: perturbation theory and the failure of the ray method, *Geophys. J. Int.*, **108**, 372–8.
Gutowski, P. R., Hron, F., Wagner, D. E. and Treitel, S., 1984. *S**, *Bull. Seism. Soc. Amer.*, **74**, 61–78.
Hanyga, A., 1982*a*. Dynamic ray tracing in anisotropic media, *Tectonophysics*, **90**, 243–51.
1982*b*. The kinematic inverse problem for weakly laterally inhomogeneous anisotropic media, *Tectonophysics*, **90**, 253–62.
Haskell, N. A., 1953. The dispersion of surface waves on multilayered media, *Bull. Seism. Soc. Amer.*, **43**, 17–43.
Helbig, K. and Schoenberg, M., 1987. Anomalous polarization of elastic waves in transversely isotropic media, *J. Acoust. Soc. Am.*, **81**, 1235–45.
Helmberger, D. V., 1968. The crust–mantle transition in the Bering Sea, *Bull. Seism. Soc. Amer.*, **58**, 179–214.
Helmberger, D. V. and Hadley, D. M., 1981. Seismic source functions and attenuation from local and teleseismic observations of the NTS events Jorum and Handley, *Bull. Seism. Soc. Amer.*, **71**, 51–67.
Herrera, I., 1964. On a method to obtain a Green's function for a multi-layered half-space, *Bull. Seism. Soc. Amer.*, **54**, 1087–96.
Hijden, J. H. M. T. van der, 1987. *Propagation of Transient Elastic Waves in Stratified Anisotropic Media*, Amsterdam: Elsevier Science Publishers.
Hong, T.-L. and Helmberger, D. V., 1977. Generalized ray theory for dipping structure, *Bull. Seism. Soc. Amer.*, **67**, 995–1008.
Hoop, A. T. de, 1960. A modification of Cagniard's method for solving seismic pulse problems, *Appl. Scient. Res.*, **B8**, 349–56.
1990. Acoustic radiation from an impulsive point source in a continuously layered fluid – an analysis based on the Cagniard method, *J. Acoust. Soc. Am.*, **88**, 2376–88.
Hoop, M. V. de and Bleistein, N., 1997. Generalized Radon transform inversions for reflectivity in anisotropic elastic media, *Inverse Problems*, **13**, 669–90.
Hörmander, L., 1971. Fourier integral operators, *Acta Math.*, **127**, 79–183.
Hron, F., 1971. Criteria for selection of phases in synthetic seismogram for layered media, *Bull. Seism. Soc. Amer.*, **61**, 765–79.
1972. Numerical methods of ray generation in multilayered media, *Methods in Computational Physics*, **12**, 1–34.
Hron, F. and Mikhailenko, B. G., 1981. Numerical modeling of nongeometrical effects by the Alekseev–Mikhailenko method, *Bull. Seism. Soc. Amer.*, **71**, 1011–29.
Hron, M. and Chapman, C. H., 1974*a*. Seismograms for Epstein transition zones, *Geophys. J.R. Astr. Soc.*, **37**, 305–22.
1974*b*. The 'shadow' from a velocity reversal, *Bull. Seism. Soc. Amer.*, **64**, 25–32.
Hudson, J. A., 1980. *The Excitation and Propagation of Elastic Waves*, Cambridge: Cambridge University Press.
Hudson, J. A. and Heritage, J. R., 1981. Use of the Born approximation in seismic scattering problems, *Geophys. J. R. Astr. Soc.*, **66**, 221–40.
Jech, J. and Pšenčík, I., 1989. First-order perturbation method for anisotropic media, *Geophys. J. Int.*, **99**, 369–76.
Jeffreys, H., 1924. On certain approximate solutions of linear differential equations of the second order, *Proc. London Math. Soc.*, **23**, 428–36.
1931. *Cartesian Tensors*, Cambridge: Cambridge University Press.
1939. The time of core waves, *Mon. Not. R. Astr. Soc.*, **4**, 548–61.

Jilek, P., 2002. Converted *PS*-wave coefficients in weakly anisotropic media, *Pure Appl. Geophys.*, **159**, 1527–62.
Jones, L. E. A. and Wang, H. F., 1981. Cretaceous shales from the Williston Basin, *Geophysics*, **46**, 288–97.
Keller, J. B., 1958. Corrected Bohr–Sommerfeld quantum conditions for non-separable systems, *Ann. Phys.*, **4**, 180–8.
Kendall, J.-M., Guest, W. S. and Thomson, C. J., 1992. Ray-theory Green's function reciprocity and ray-centred coordinates in anisotropic media, *Geophys. J. Int.*, **108**, 364–71.
Kendall, J.-M. and Thomson, C. J., 1993. Maslov ray summation, pseudo-caustics, Lagrangian equivalence and transient seismic waveforms, *Geophys. J. Int.*, **113**, 186–214.
Kennett, B. L. N., 1974. Reflections, rays and reverberations, *Bull. Seism. Soc. Amer.*, **64**, 1685–96.
1983. *Seismic Wave Propagation in a Stratified Medium*, Cambridge: Cambridge University Press.
2001. *The Seismic Wavefield*, vols. I and II, Cambridge: Cambridge University Press.
Kennett, B. L. N. and Illingworth, M. R., 1981. Seismic waves in a stratified half-space – III. Piecewise smooth models, *Geophys. J. R. Astr. Soc.*, **66**, 633–75.
Kennett, B. L. N., Kerry, N. J. and Woodhouse, J. H., 1978. Symmetries in the reflection and transmission of elastic waves, *Geophys. J. R. Astr. Soc.*, **52**, 215–30.
Kim, J.Y. and Behrens, J., 1986. Experimental evidence for the S^* wave, *Geophys. Props.*, **34**, 100–8.
Klimeš, L., 1997. Phase shift of the Green function due to caustics in anisotropic media, SEG expanded abstract, **ST 14.2**, 1834–7.
Knopoff, L., 1958. Love waves from a line *SH* source, *J. Geophys. Res.*, **63**, 619–30.
1964. A matrix method for elastic wave problems, *Bull. Seism. Soc. Amer.*, **454**, 431–8.
Knopoff, L. and Gangi, A. F., 1959. Seismic reciprocity, *Geophysics*, **24**, 681–91.
Knopoff, L. and Gilbert, F., 1961. Diffraction of elastic waves in the core of the Earth, *Bull. Seism. Soc. Amer.*, **51**, 35–49.
Knott, C. G., 1899. On the reflexion and refraction of elastic waves, with seismological applications, *Phil. Mag.*, 5th ser., **48**, 64–97.
Koefoed, O., 1955. On the effect of Poisson's ratio of rock strata on the reflection coefficients of plane waves, *Geophys. Prosp.*, **3**, 381–7.
Kramers, H. A., 1926. Wellenmechanik und halbzahige Quantisierung, *Z. Physik*, **39**, 828–40.
Kravtsov, Yu. A. and Orlov, Yu. I., 1990. *Geometrical Optics in Inhomogeneous Media*, Berlin: Springer-Verlag.
Lamb, H., 1904. On the propagation of tremors over the surface of an elastic solid, *Phil. Trans. Roy. Soc. Lond.*, **A203**, 1–42.
Lamé, M. G., 1852. *Leçons sur théorie mathématique de l'élasticité des corps solides*, Paris.
Langer, R. E., 1937. On the connection formulas and the solution of the wave equation, *Physical Reviews*, Ser. 2, **51**, 669–76.
Lapwood, E. R., 1949. The disturbance due to a line source in a semi-infinite elastic medium, *Phil. Trans. Roy. Soc. Lond.*, **A242**, 63–100.
Lewis, R. M., 1965. Asymptotic theory of wave-propagation, *Arch. Rat. Mech. Anal.*, **20**, 191–250.
Lighthill, M. J., 1960. Studies on magneto-hydrodynamic waves and other anisotropic wave motions, *Phil. Trans. Roy. Soc. Lond.*, **A252**, 397–430.

Love, A. E. H., 1944. *A Treatise on the Mathematical Theory of Elasticity*, New York: Dover.

Magnus, W., Oberhettinger, F. and Soni, R. P., 1966 *Formulas and Theorems for the Special Functions of Mathematical Physics*, New York: Springer-Verlag.

Mallick, S. and Frazer, L. N., 1987. Practical aspects of reflectivity modeling, *Geophysics*, **52**, 1355–64.

1991. Reflection/transmission coefficients and azimuthal anisotropy in marine seismic studies, *Geophys. J. Intl.*, **105**, 241–52.

Maslov, V. P., 1965 [1972]. *Perturbation Theory and Asymptotic Methods*, Moskov. Gos. Univ., Moscow (in Russian); trans. into French by J. Lascoux and R. Sénéor, Paris: Dunod.

Muir, F., 2003. Translating the tensors of elasticity into matrices, *personal communication and unpublished notes*.

Muir, F. and Dellinger, J., 1985. A practical anisotropic system, *Stanford Exploration Project Reports*, **44**, 55–8.

Musgrave, M. J. P., 1970. *Crystal Acoustics*, San Francisco: Holden-Day.

Nussenzveig, H. M., 1965. High frequency scattering by an impenetrable sphere, *Ann. Phys. New York*, **34**, 23–95.

1969*a*. High frequency scattering by a transparent sphere, I. Direct reflection and transmission, *J. Math. Phys.*, **10**, 82–124.

1969*b*. High frequency scattering by a transparent sphere, II. Theory of the rainbow and the glory, *J. Math. Phys.*, **10**, 125–76.

Nye, J. F., 1957. *Physical Properties of Crystals*, Oxford: Oxford University Press.

Ott, H., 1943. Die Sattelpunktsmethode in der Umgebung eines Poles mit Anwendungen auf die Wellenoptik und Akustik, *Ann. Physik*, **43**, 393–403.

Pearcey, T., 1946. The structure of an electromagnetic field in the neighbourhood of a caustic, *Phil. Mag. Ser. 7*, **37**, 311–7.

Pekeris, C. L., 1948. Theory of propagation of explosive sound in shallow water, *Geol. Soc. Amer. Memoir*, **27**.

1955*a*. The seismic surface pulse, *Proc. Nat. Acad. Sci. USA*, **41**, 469–80.

1955*b*. The seismic buried pulse, *Proc. Nat. Acad. Sci. USA*, **41**, 629–39.

Pekeris, C. L., Alterman, Z. and Abramovici, F., 1963. Propagation of an *SH*–torque pulse in a layered solid, *Bull. Seism. Soc. Amer.*, **53**, 39–57.

Pekeris, C. L., Alterman, Z., Abramovici, F. and Jarosh, H., 1963. Propagation of compressional pulse in a layered solid, *Rev. Geophys.*, **3**, 25–47.

Petrashen, G. I. and Kashtan, B. M., 1984. Theory of body wave propagation in inhomogeneous anisotropic media, *Geophys. J.R. Astr. Soc.*, **76**, 29–39.

Phinney, R. A., 1961. Propagation of leaking interface waves, *Bull. Seism. Soc. Amer.*, **51**, 527–55.

Phinney, R. A. and Alexander, S. S., 1966. *P* wave diffraction theory and the structure of the core-mantle boundary, *J. Geophys. Res.*, **71**, 5959–75.

Pitteway, M. L. V., 1965. The numerical calculation of wave-fields, reflexion coefficients and polarisations for long radio waves in the lower ionosphere, *Phil. Trans. Roy. Soc. Lond.*, **A257**, 219–41.

Pol, B. van der and Bremmer, H., 1937*a*. The diffraction of electromagnetic waves from an electric point source round a finitely conducting sphere, with applications to radiotelegraphy and the theory of the rainbow, Part I, *Phil. Mag.*, **24**, 141–76.

1937*b*. The diffraction of electromagnetic waves from an electric point source round a finitely conducting sphere, with applications to radiotelegraphy and the theory of the rainbow, Part II, *Phil. Mag.*, **24**, 825–64.

1964. *Operational Calculus*, Cambridge: Cambridge University Press.

Popov, M. M. and Pšenčík, I., 1978. Computation of ray amplitudes in inhomogeneous media with curved interfaces, *Studia Geophy. et Geod.*, **22**, 248–58.

Press, W. H., Flannery, B. P., Teulolsky, S. A. and Vetterling, W. T., 1986. *Numerical Recipes, The Art of Scientific Computing*, Cambridge: Cambridge University Press.

Pšenčík, I., 1979. Ray amplitudes of compressional, shear, and converted seismic body waves in 3D laterally inhomoegeneous media with curved interfaces, *J. Geophys.*, **45**, 381–90.

1998. Green functions for inhomogeneous weakly anisotropic media, *Geophys. J. Int.*, **135**, 279–88.

Pšenčík, I. and Gajewski, D., 1998. Polarization, phase velocity and NMO velocity of *qP* waves in arbitrary weakly anisotropic media, *Geophysics*, **63**, 1754–66.

Pšenčík, I. and Vavrycuk, V., 1998. Weak contrast *PP* wave displacement R/T coefficients in weakly anisotropic elastic media, *Pure Appl. Geophys.*, **151**, 699–718.

Pujol, J., 2003. *Elastic Wave Propagation and Generation in Seismology*, Cambridge: Cambridge University Press.

Rayleigh, J. W. S., 1885. On waves propagating along the plane surface of an elastic solid, *Proc. Lond. Math. Soc.*, **17**, 4–11.

Richards, P. G., 1971. Waves in stratified media, *Geophysics*, **36**, 798–809.

Richards, P. G. and Frasier, C. W., 1976. Scattering of elastic waves from depth-dependent inhomogeneities, *Geophysics*, **41**, 441–58.

Riley, K. F., Hobson, M. P. and Bence, S. J., 2002. *Mathematical Methods for Physics and Engineering*, 2nd edn, Cambridge: Cambridge University Press.

Roever, W. L. and Vining, T. F., 1959*a*. Propagation of elastic wave motion from an impulsive source along a fluid/solid interface, I. Experimental pressure response, *Phil. Trans. Roy. Soc.*, **A251**, 455–65.

1959*b*. Propagation of elastic wave motion from an impulsive source along a fluid/solid interface, II. Theoretical pressure response, *Phil. Trans. Roy. Soc.*, **A251**, 465–88.

Roth, M. and Holliger, K., 2000. The non-geometric $\bar{P}S$ wave in high-resolution seismic data: observations and modelling, *Geophys. J. Intl.*, **140**, F5–F11.

Ruger, A., 1997. *P*-wave reflection coefficients for transversely isotropic models with vertical and horizontal axis of symmetry, *Geophysics*, **62**, 713–22.

1998. Variation of *P*-wave reflectivity with offset and azimuth in anisotropic media, *Geophysics*, **63**, 935–47.

Rutherford, S. R. and Williams, R. H., 1989. Amplitude-versus-offset variations in gas sands, *Geophysics*, **54**, 680–8.

Schoenberg, M. and Protazio, J., 1992. 'Zoeppritz' rationalized and generalized to anisotropy, *J. Seism. Expl.*, **1**, 125–44.

Scholte, J. G. J., 1947. The range of existence of Rayleigh and Stoneley waves, *Roy. Astr. Soc. Monthly Notices Geophys. Supp.*, **5**, 120–6.

1956. On seismic waves in spherical Earth, *Meded. Verh. K. Ned. Met. Inst.*, **65**, 1–55.

1958. Rayleigh waves in isotropic and anisotropic elastic media, *Meded. Verh. K. Ned. Met. Inst.*, **72**, 1–44.

1962. Oblique propagation of waves in inhomogeneous media, *Geophys. J.R. Astr. Soc.*, **7**, 244–61.

Schwab, F. and Knopoff, L., 1972. Fast surface wave and free-mode computations, *Meths. Comp. Phys.*, **11**, 87–180.

Shaw, R. K. and Sen, M. K., 2004. Born integral, stationary phase and linearized reflection coefficients in weak anisotropic media, *Geophys. J. Intl.*, in press.

Shearer, P. M., 1999. *Introduction to Seismology*, Cambridge: Cambridge University Press.

Shearer, P. M. and Chapman, C. H., 1988. Ray tracing in anisotropic media with a linear gradient, *Geophys. J.*, **94**, 575–80.
Shuey, R. T., 1985. A simplification of the Zoeppritz equations, *Geophysics*, **50**, 609–14.
Smirnov, V. I., 1964. *A Course in Higher Mathematics*, vol. 4, trans. D. E. Brown and I. N. Sneddon, Oxford: Pergamon Press.
Smith, G. C. and Gidlow, P. M., 1987. Weighted stacking for rock property estimation and detection of gas, *Geophys. Prosp.*, **35**, 993–1014.
Sommerville, D. M. Y., 1929. *An Introduction to the Geometry of N Dimensions*, London: Methuen.
Spencer, T. W., 1960. The method of generalized reflection and transmission coefficients, *Geophysics*, **25**, 625–41.
1965. Refraction along a layer, *Geophysics*, **30**, 369–88.
Spencer, C. P., Chapman, C. H. and Kragh, J. E., 1997. A fast, accurate, integration method for Kirchhoff, Born and Maslov synthetic seismogram generation, SEG expanded abstract, **ST 14.3**, 1838–41.
Stephen, R. A. and Bolmer, S. T., 1985. The direct wave root in marine seismology, *Bull. Seism. Soc. Amer.*, **75**, 57–67.
Stokes, G. G., 1851. On the dynamical theory of diffraction, *Trans. Camb. Phil. Soc.*, **9**, 1–62.
Stoneley, R., 1924. Elastic waves at the surface of separation of two solids, *Proc. Roy. Soc. Lond., Ser.*, **A106**, 416–28.
Strick, E., 1959. Propagation of elastic wave motion from an impulsive source along a fluid/solid interface, III. The pseudo-Rayleigh wave, *Phil. Trans. Roy. Soc.*, **A251**, 488–523.
Takeuchi, H. and Saito, M., 1972. Seismic surface waves, *Meth. Comp. Phys.*, **11**, 217–95.
Telford, W. M., Geldart, L. P., Sheriff, R. E. and Keys, D. A., 1976. *Applied Geophysics*, Cambridge: Cambridge University Press.
Teng, T., 1970. Inversion of the spherical layer-matrix, *Bull. Seism. Soc. Amer.*, **60**, 317–20.
Thomsen, L., 1986. Weak elastic anisotropy, *Geophysics*, **51**, 1954–66.
1990. Poisson was *not* a geophysicist!, *The Leading Edge*, **9**, 12, 27–9.
Thomson, W. (Lord Kelvin), 1856. Elements of a mathematical theory of elasticity, *Phil. Trans. Roy. Soc.*, **116**, 481–98.
1877. Mathematical theory of elasticity, in *Encyclopaedia Britannica*, 9th edn, vol. 7, Edinburgh: A. and C. Black, 819–25.
Thomson, W. T., 1950. Transmission of elastic waves through a stratified solid medium, *J. Appl. Phys.*, **21**, 89–93.
Thomson, C. J. and Chapman, C. H., 1984. On approximate solutions for reflection of waves in a stratified medium, *Geophys. J. R. Astr. Soc.*, **79**, 244–61.
1986. End-point contributions to synthetic seismograms, *Geophys. J.R. Astr. Soc.*, **87**, 285–94.
Thrower, E. N., 1965. The computation of dispersion of elastic waves in layered media, *J. Sound Vib.*, **2**, 210–26.
Towne, D. H., 1967. *Wave Phenomena*, New York: Dover.
Ursin, B., 1990. Offset dependent geometrical spreading in a layered medium, *Geophysics*, **55**, 492–6.
Ursin, B. and Tygel, M., 1997. Reciprocal volume and surface scattering integrals for anisotropic elastic media, *Wave Motion*, **26**, 31–42.
Vavrycuk, V., 1999. Weak-contrast reflection/transmission coefficients in weakly anisotropic elastic media: *P*-wave incidence, *Geophys. J. Int.*, **138**, 553–62.

Vavrycuk, V. and Pšenčík, I., 1998. *PP*-wave reflection coefficients in weakly anisotropic elastic media, *Geophysics*, **63**, 2129–41.

Verweij, M. D. and Hoop, A. T. de, 1990. Determination of seismic wavefields in arbitrarily continuously layered media using the modified Cagniard method, *Geophys. J. Int.*, **103**, 731–54.

Virieux, J., Farra, V. and Madariaga, R., 1988. Ray tracing for earthquake location in laterally heterogeneous media, *J. Geophys. Res.*, **93**, 6585–99.

Voigt, W., 1910. *Lehrbuch der Kristallphysik*, Leipzig: B. G. Teubner.

Waerden, B. L. van der, 1950. On the method of saddle points, *Appl. Sci. Res. B*, **2**, 33–45.

Wang, R., 1999. A simple orthonormalization for stable and efficient computations of Green's functions, *Bull. Seism. Soc. Amer.*, **89**, 733–41.

Wasow, W., 1965. *Asymptotic Expansions of Ordinary Differential Equations*, vol. 14, of *Pure and Applied Mathematics*, Interscience Publishers, New York: John Wiley.

Watson, G. N., 1918. The diffraction of electric waves by the Earth, *Proc. Roy. Soc. Lon.*, **A95**, 83–99.

Watson, T. H., 1970. A note on fast computation of Rayleigh wave dispersion in the multilayered half-space, *Bull. Seism. Soc. Amer.*, **60**, 161–6.

Wentzel, G., 1926. Eine Verallgemeinerung des Quantenbedingungen für die Zwecke der Wellenmechanik, *Z. Physik*, **38**, 518–29.

Wiggins, R. A. and Helmberger, D. V., 1974. Synthetic seismogram computation by expansion in generalized rays, *Geophys. J. R. Astr. Soc.*, **37**, 73–90.

Woodhouse, J. H., 1974. Surface waves in laterally varying media, *Geophys. J. R. Astr. Soc.*, **37**, 29–51.

1978. Asymptotic results for elastodynamic propagator matrices in plane stratified and spherically stratified Earth models, *Geophys. J. R. Astr. Soc.*, **54**, 263–80.

1980. Efficient and stable methods for performing seismic calculations in stratified media, in *Proceedings of the International School of Physics 'Enrico Fermi', Course LXXVIII. Physics of the Earth's Interior*, ed. A. M. Dziewonski and E. Boschi, Amsterdam: North Holland Publishing Company, 127–51.

Wright, J., 1986. Reflection coefficients at pore-fuild contacts as a function of offset, *Geophysics*, **51**, 1858–60.

Young, R. A. and LoPiccolo, R. D., 2003. A comprehensive AVO classification, *The Leading Edge*, **22**, 10, 1030–7.

Zillmer, M., Gajewski, D. and Kashtan, B. M., 1997. Reflection coefficients for weak anisotropic media, *Geophys. J. Intl.*, **129**, 389–98.

1998. Anisotropic reflection coefficients for a weak-contrast interface, *Geophys. J. Intl.*, **132**, 159–66.

Zillmer, M., Kashtan, B. M. and Gajewski, D., 1998. Quasi-isotropic approximation of ray theory for anisotropic media, *Geophys. J. Intl.*, **132**, 643–53.

Zoeppritz, K., 1919. Über Erdbebenwellen VII B. Über Reflexion und Durchgang Seismischer Wellen durch Unstetigkeitsflächen, *Nöchr. der Königlichen Gesell. Wiss., Göttingen, Math.-Phys. KI.*, 66–84.

Author index

Abo-Zena, A., 301, 587
Abramovici, F., 330, 587, 595
Abramowitz, M., xxii, 23, 42, 50, 53, 69, 287, 289, 290, 325, 339, 450, 453, 555, 556, 559, 561–3, 572–4, 576, 578, 582–4, 587
Aki, K., ix, 45, 113, 132, 231, 237, 357, 366, 587
Alexander, S. S., 309, 458, 595
Alsop, L. E., 213, 587
Alterman, Z., 330, 595
Anderson, D. L., 54, 591
Ansell, J. H., 456, 587
Arnol'd, V. I., xvii, 159, 587
Auld, B. A., 92, 95, 587

Backus, G. E., 119, 121, 131, 564, 587, 592
Baerheim, R., 131, 587
Bakker, P. M., 176, 197, 588
Båth, M., 310, 588
Bechmann, R., 168, 588
Behrens, J., 404, 594
Ben-Hador, R., 301, 588
Ben-Menahem, A., 113, 505, 509, 513, 514, 588, 592
Bence, S. J., x, 58, 196, 596
Bender, C. M., 572, 588
Bennett, H. F., 191, 588
Berry, M. V., 180, 588
Betancourt, O., 272, 590
Biot, M. A., 213, 588
Bleistein, N., 532, 534, 549, 588, 593
Bolmer, S. T., 404, 597
Bond, W., 93, 588
Bortfeld, R., 224, 232, 588
Brandsberg-Dahl, S., 145, 478, 588
Brekhovskikh, L. M., 588
Bremmer, H., 3, 60, 286, 377, 588, 595
Brillouin, L., xvii, 588
Brokešova, J., 185, 588
Buchen, P. W., 301, 588
Buchwald, V. T., 132, 174, 588
Budden, K. G., 286, 438, 588
Bullen, K. E., 25, 589
Burridge, R., 121, 133, 138, 152, 174, 376, 428, 446, 574, 589

Cagniard, L., 3, 310, 313, 589
Červený, V., 57, 143, 163, 165, 169, 181, 194, 196, 204, 458, 488, 491, 589
Chapman, C. H., 34, 42, 168, 183, 190, 193, 218, 237, 274, 278, 281, 286, 290, 292, 298–301, 309, 346, 348, 349, 356, 357, 376, 377, 384, 401, 406, 424, 456, 458, 460, 472, 477, 479, 497–9, 505, 524, 532, 544, 548–50, 589–91, 593, 597
Chester, C., 446, 572, 590
Choy, G. L., 357, 590
Chu Jen-Yi, 218, 356, 424, 479, 590
Cisternas, A., 272, 590
Clay, C. S., 532, 590
Clemmow, P. C., 286, 590
Coates, R. T., 497–9, 505, 532, 549, 550, 590
Coddington, E. A., 277, 590
Cormier, V., 357, 590
Courant, R., 69, 141, 590
Cowin, S. C., 92, 131, 590

Dahlen, F. A., ix, xv, 78, 81, 591
Daley, P. F., 404, 591
Dellinger, J., 52, 595
Drijkoningen, G. G., 330, 406, 591
Drummond, R., 460, 590
Duff, G. F. D., 174, 591
Duistermaat, J. J., 460, 463, 591
Dunkin, J. W., 300, 302, 591
Duwalo, G., 449, 456, 458, 591
Dziewonski, A. M., 54, 56, 356, 591, 592

Eisner, L., 182, 591
Erdélyi, A., 443, 562, 591
Ewing, M., 310, 591

Farra, V., 52, 204, 206, 491, 591, 598
Fatti, J. L., 232, 591
Fedorov, F. I., 92, 591
Filon, L. N. G., 359, 591

Flannery, B. P., 364, 596
Fokkema, J. T., 330, 591
Frasier, C. W., 286, 596
Frazer, L. N., 209, 226, 281, 359, 363, 366, 369, 467, 470, 472, 532, 591, 595
French, W. S, 240–3, 526, 544, 591
Friedman, B., 446, 572, 590
Fryer, G. J., 209, 226, 281, 591
Fuchs, K., 356, 404, 592
Fung, Y. C., 78, 133, 313, 393, 592

Gajewski, D., 204, 232, 488, 496, 497, 592, 596, 598
Gal-Ezer, J., 330, 587
Gangi, A. F., 103, 594
Gantmacher, F. R., 303, 564, 567, 592
Garmany, J., 176, 197, 278, 292, 592
Garvin, W. W., 310, 592
Gebrande, H., 50, 592
Geldart, L. P., 50, 597
Gettrust, J. F., 359, 366, 591
Gibson, R. L., 505, 509, 513, 514, 588, 592
Gidlow, P. M., 232, 597
Gilbert, F., 54, 56, 308, 356, 400, 449, 458, 564, 592, 594
Golub, G. H., 196, 592
Gradshteyn, I. S., 575, 592
Gray, S. H., 282, 286, 341, 592
Green, R., 40, 592
Guest, W. S., 133, 158, 174, 593, 594
Gutowski, P. R., 404, 593

Hadley, D. M., 404, 593
Hanyga, A., 488, 491, 593
Haskell, N. A., 255, 300, 593
Heading, J., 286, 590
Helbig, K., 130, 593
Helmberger, D. V., 138, 337, 404, 406, 593, 598
Heritage, J. R., 504, 509, 513, 593
Herrera, I., 213, 593
Hijden, J. H. M. T. van der, 314, 330, 593
Hilbert, D., 69, 141, 590
Hobson, M. P., x, 58, 196, 596
Holliger, K., 401, 596
Hong, T.-L., 404, 406, 593
Hoop, A. T. de, 3, 284, 310, 313, 593, 598
Hoop, M. V. de, 138, 145, 478, 532, 588, 589, 593
Hörmander, L., xvii, 159, 593
Hron, F., 181, 194, 273, 274, 404, 589, 591, 593
Hron, M., 34, 42, 593
Hudson, J. A., 113, 117, 132, 504, 509, 513, 593

Illingworth, M. R., 292, 298, 594

Jacobs, J. A., 449, 456, 458, 591
Jardetsky, W., 310, 591
Jarosh, H., 330, 595
Jech, J., 488, 491, 492, 496, 503, 589, 593
Jeffreys, H., xvii, 97, 132, 426, 446, 593
Jilek, P., 232, 594
Jones, L. E. A., 189, 489, 594

Kanasewich, E. R., 330, 587
Kashtan, B. M., 181, 232, 488, 497, 595, 598
Keers, H., 460, 472, 590
Kelamis, P. G., 330, 587
Keller, J. B., xvii, 159, 594
Kendall, J.-M., 133, 158, 174, 462, 593, 594
Kennett, B. L. N., ix, 263, 264, 292, 298, 300, 301, 594
Kerry, N. J., 301, 594
Keys, D. A., 50, 597
Kim, J. Y., 404, 594
Klimeš, L., 176, 197, 594
Knopoff, L., 103, 121, 301, 375, 449, 458, 589, 592, 594, 596
Knott, C. G., 216, 245, 301, 594
Koetoed, O., 224, 594
Kragh, J. E., 477, 524, 544, 548, 597
Kramers, H., xvii, 594
Kravtsov, Yu. A., 57, 180, 488, 594

Lamb, H., 3, 310, 594
Lamé, M. G., 3, 97, 594
Langer, J., 204, 589
Langer, R. E., 287, 594
Lapwood, E. R., 3, 310, 438, 594
Laster, S. J., 400, 592
Le Bégat, S., 204, 206, 491, 591
Leiva, A., 272, 590
Levinson, N., 277, 590
Levitt, P. R., 232, 591
Lewis, R. M., 197, 594
Lighthill, M. J., 132, 174, 595
Loan, C. F. van, 196, 592
LoPiccolo, R. D., 244, 598
Love, A. E. H., 3, 133, 595
Lyness, D. G., 218, 356, 424, 479, 590

Madariaga, R., 52, 491, 591, 598
Magnus, W., 443, 562, 575, 591, 595
Mallick, S., 226, 359, 363, 369, 595
Maslov, V. P., xvii, 159, 460, 595
Medwin, H., 532, 590
Mehrabadi, M. M., 92, 590
Mikhailenko, B. G., 404, 593
Miller, D., 138, 589
Muir, F., 52, 133, 595
Mulcahy, M., 119, 121, 587
Müller, G., 356, 592
Musgrave, M. J. P., 92, 95, 97, 99, 168, 185, 193, 195, 595

Nussenzveig, H. M., 376, 456, 595
Nye, J. F., 92, 95, 595

Oberhettinger, F., 443, 562, 575, 591, 595
Orlov, Yu. I., 57, 180, 488, 594
Orzag, S. A., 572, 588
Ott, H., 438, 595

Pearcey, T., 485, 550, 595
Pekeris, C. L., 3, 310, 313, 330, 376, 595

Petrashen, G. I., 181, 595
Phinney, R. A., 300, 309, 357, 377, 401, 456, 458, 467, 470, 472, 590, 591, 595
Pitteway, M. L. V., 300, 595
Pol, B. van der, 3, 60, 377, 595
Popov, M. M., 181, 596
Press, F., 310, 591
Press, W. H., 364, 596
Protazio, J., 301, 596
Pšenčík, I., 181, 182, 204, 232, 488, 492, 496, 497, 503, 589, 591–3, 596, 597
Pujol, J., ix, 113, 132, 596

Ravindra, R., 458, 589
Rayleigh, J. W. S., 225, 596
Richards, P. G., ix, 45, 113, 132, 231, 237, 278, 286, 357, 366, 587, 590, 596
Riley, K. F., x, 58, 196, 596
Roever, W. L., 337, 401, 596
Roth, M., 401, 596
Ruger, A., 232, 596
Rutherford, S. R., 224, 596
Ryzhik, I. M., 575, 592

Sadegh, A. M., 92, 590
Saito, M., 133, 309, 597
Schoenberg, M., 130, 301, 593, 596
Scholte, J. G. J., 228, 231, 237, 286, 377, 401, 456, 596
Schulz, K., 404, 592
Schwab, F., 301, 596
Sen, M. K., 232, 532, 550, 591, 596
Shaw, R. K., 232, 550, 596
Shearer, P. M., ix, 168, 183, 190, 193, 281, 499, 550, 590, 596, 597
Sheriff, R. E., 50, 597
Shuey, R. T., 224, 231, 597
Singh, S. J., 113, 588
Smirnov, V. I., 567, 597
Smith, G. C., 232, 591, 597
Sommerville, D. M. Y., 556, 597
Soni, R. P., 575, 595
Spencer, C. P., 138, 477, 524, 544, 548, 589, 597
Spencer, T. W., 272, 330, 375, 597
Stegun, I. A., xxii, 23, 42, 50, 53, 69, 287, 289, 290, 325, 339, 450, 453, 555, 556, 559, 561–3, 572–4, 576, 578, 582–4, 587
Stephen, R. A., 404, 597
Stokes, G. G., 113, 597
Stoneley, R., 3, 219, 401, 597
Strauss, P. J., 232, 591
Strick, E., 337, 401, 597

Takeuchi, H., 133, 309, 597
Telford, W. M., 50, 597
Teng, T., 309, 597
Teulolsky, S. A., 364, 596
Thomsen, L., 130, 131, 223, 232, 597
Thomson, C. J., 133, 158, 174, 299, 349, 384, 462, 593, 594, 597
Thomson, W. (Lord Kelvin), 92, 597
Thomson, W. T., 300, 597
Thrower, E. N., 300, 302, 597
Towne, D. H., 40, 597
Treitel, S., 404, 593
Tricomi, F. G., 443, 562, 591
Tromp, J., ix, xv, 78, 81, 591
Tygel, M., 532, 549, 597

Ursell, F., 446, 572, 590
Ursin, B., 48, 145, 478, 532, 549, 588, 597

Vail, P. J., 232, 591
Vavrycuk, V., 232, 596, 597
Verweij, M. D., 284, 598
Vetterling, W. T., 364, 596
Vining, T. F., 337, 401, 596
Virieux, J., 52, 598
Voigt, W., 91, 598

Waerden, B. L. van der, 438, 598
Wagner, D. E., 404, 593
Wang, H. F., 189, 489, 594
Wang, R., 300, 598
Wasow, W., 277, 290, 292, 298, 598
Watson, G. N., 376, 598
Watson, T. H., 300, 598
Wentzel, G., xvii, 598
Wiggins, R. A., 337, 598
Williams, R. H., 224, 596
Woodhouse, J. H., 95, 292, 298, 301, 594, 598
Wright, J., 232, 598

Young, R. A., 224, 598

Zillmer, M., 232, 488, 497, 598
Zoeppritz, K., 216, 245, 301, 598

Subject index

1066B, 54

2.5D wave propagation, 337, 464, 505, 514

acausal signals, 362, 370–1, 418, 420
acoustic medium, 89–90
acoustic waves, 100–7
action, 144
additional components, 162, 171
admittance, xxi, 384
Airy caustic, xxiii, 33, 34, 425–30, 446–8, 575–8
  illuminated region, 426, 447, 576–8
  shadow region, 426, 447, 578
  time function, xxii
Airy function, 61, 287, 288, 342, 572–80
  asymptotic, 573–4
  Fourier transform, 578–80
  generalized, xxii
  inverse Fourier transform, 574–8
Airy phase, 375
aliasing, 351, 365
amplitude coefficients, 136, 163, 164
amplitude versus offset, 224, 232, 237
analytic continuation, 61
analytic Dirac delta function, xxii, 397
analytic lambda function, xxii
analytic time series, 61–2, 64
anisotropic elastic waves, 107–10
anisotropic medium, 90–6
anisotropic perturbation, 488
anti-Stokes line, 61
ART, *see* asymptotic ray theory
asymptotic ray theory, 134, 459
  anisotropic operators, xx, 164
  ansatz
    acoustic, 135–8
    anisotropic, 163
  dynamic, 134
    acoustic, 145–63
    anisotropic, 170–8
    isotropic, 180–2
    Jacobian volume mapping, xx
  dynamic fundamental matrix, xx
  dynamic propagator, xx, 173
  dynamic ray discontinuity, 198, 204–6
  dynamic ray equations, xx, 149
  eikonal equation
    acoustic, 138–40
    anisotropic, 164–6
  geometrical Green dyadic, 157–9, 174–8, 237–45
    with interfaces, 244–5
  high-order terms, 161–3
  higher-order amplitude coefficients
    isotropic, 181–2
  kinematic, 134
    acoustic, 134–45
    anisotropic, 163–70
    isotropic, 178–9
  Maslov, 460–87, 504
    three dimensions, 472–87
    two dimensions, 461–72
  Maslov ansatz, 460
  one and two-dimensional media, 182–93
  polarization, 134
  quasi-isotropic, 174, 459, 487–503
    eikonal equation, 495
    transport equation, 495–503
  ray series
    acoustic, 138
    anisotropic, 163–4
  transport equation
    acoustic, 146–9
    anisotropic, 170–4
  two-dimensional
    acoustic, 159–61
AVO, *see* amplitude versus offset

background model, 504
Bernoulli numbers, 557, 559
Berry's topological phase, 180
Bessel function, 61, 69, 74, 325, 334, 376
Betti's theorem, 101–2, 107–8, 533, 544
Binet–Cauchy formula, 303
body force source, xiii, xix
Bond transformation, 93
Born scattering theory, 384, 459, 504–32

acoustic, 505–11
acoustic error, 514–9
acoustic perturbation, 521–3
band-limited, 523–9
elastic, 511–4
elastic error, 519–21
elastic perturbation, 523
error scattering term, 516, 520
error terms, xxi, 505–23
force error, 511
Hooke error, 507
Newton error, 507
perturbation, 505–23
ray theory error, 514–21
ray theory perturbation, 521–3
scalar error term, 517, 520
scalar scattering term, 517
scattering dyadic, 517
strain error, 511
stress error, 511
three dimensions, 518–9
three-dimensional perturbation, 522–3
travel-time perturbation, 529–32
two dimensions, 517–8
two-dimensional perturbation, 522
boundary conditions, 76, 86–9, 200–201
acoustic, 201
dynamic, 87
elastic, 201
fluid–fluid, 88–9
fluid–solid, 88–9
free, 89
kinematic, 87
welded interface, 87–8
boxcar function, xxii, 351, 352
branch cut, 66, 67, 69, 318, 353, 401
branch point, 65, 69, 380, 386, 387, 409
Bremmer series, *see* WKBJ iterative solution
Bromwich contour, 59–61
bulk modulus, xix, 90, 128
Bullen's ray parameter, 25

Cagniard contour, 314–8, 325, 342, 343, 380, 386, 387, 393, 397, 401, 406
asymptote, 316, 317
general waves, 316–7
saddle point, 316
uni-velocity waves, 314–6
Cagniard method, 28, 40, 284, 310, 311, 313, 376, 378, 420, 424, 436, 458
first-motion approximation, 379–415
generalized rays, sources and receivers, 330–5
inverse transforms, 317–20, 323–35
stratified media, 340–6
three dimensions, 323–40
two dimensions, 313–23
without Bessel or Hankel functions, 326–30
Cagniard–de Hoop–Pekeris method, *see* Cagniard method
Cauchy formula, 80
causality, 59–61
caustic, 173, 176, 349, 378, 417, 418, 420–2, 425, 460, 465; *see also* Airy caustic; Pearcey caustic
chain-dimension rule, xiii
characteristic equation, 196
characteristic function, 195
characteristic matrix, 195
characteristic polynomial, 195
Chebyshev polynomials, 74, 334, 562
Christoffel equation, 129, 164, 175, 185, 488
Christoffel matrix, 165, 195
circular frequency, xviii, 7, 58
circular ray, 48
commutator, xxii, 234, 278
complementary error function, 582
complete elliptic integral, 575
complex spectrum, 58
compliance tensor, 96
component vector, 255
compressibility, xix, 90
conical wave, *see* head wave
constant gradient media, 189–93
constitutive relations, 89–100
contact transform, *see* Legendre transform
continuum mechanics, 76–127
contravariant tensor, 78
convolution, 62–4
convolution model, 383
coupling coefficient, 381, 384
coupling equation, 283
coupling event, 238
coupling matrix, xxi
covariant tensor, 78
critical angle, 10, 15
critical point, 460
cross impedance, 296
cross-sectional area function, xx, 147, 150, 194
curvature matrix, 155
cusp, 188

Debye expansion, 377
deep shadow, xxii, xxiii, 35, 378, 433, 449–56
deformation tensor, 81
degenerate directions, 174, 487
degenerate perturbation theory, 492
degenerate polarizations, 496
delay time, *see* intercept-time function
Δ-matrix method, 300
density, xix, 76
depth, xv
derivative, 64
Eulerian, 83
Lagrangian, 83
material, 83
differential coefficients, xxi, 198
differential systems, xxi
fourth-order $P-SV$, 253
one-dimensional, 247–53
second-order SH, 253
three dimensions
acoustic, 251

differential systems (*cont.*)
anisotropic, 251–2
isotropic, 252–3
two dimensions
acoustic, 248–51
diffracted signal, 432, 433
dilatation, xix, 86
dilatational modulus tensor, 131
dimensions, xii
dipole source, 124–5
Dirac delta function, xxii, 63–4, 72, 73, 137
integrals, xxii
direct ray, xxiii, 15, 26, 27, 29, 31–3, 41, 255–8, 320–3, 338–40, 378–80
direct wave root, 404
directivity source function, 178
distorted Born approximation, 505
diving ray, *see* turning ray
double-couple source, 125–7, 128
dyadic Green function, 113, 175, 177
dynamic analogues, 274
dynamic propagator, 159

Earth flattening transformation, 24–5
effective medium theory, 77
effective ray length, xviii, 159, 175, 176
eigenvalue matrix, 254, 257
eigenvector matrix, 210, 254
inverse, 212–3
Einstein summation convention, xiii, 80
elastic compliances, xix, 96, 98–9
elastic stiffnesses, xix, 90–100, *see also* Lamé elastic parameters
energy flux conservation, 215
energy flux vector, xx, 147, 173
Epstein layer, 34, 42
equation of motion, 76, 83–4
equivalent body-force density source, 120
equivalent surface-force density source, 120
Euler numbers, 557, 559
Euler–Lagrange equations, 144
Eulerian coordinates, 78, 81
evanescent wave, 11, 204
explosion source, 124, 338–40

far-field approximation, 113, 335–7, 348, 379
far-field term, 322, 340
fast Fourier transform, *see* Fourier transforms, fast
fault source, 120, 121
slip-discontinuity, 121–3
stress-glut surface density, 121
Fermat's principle, 144, 490, 531
FFT, *see* Fourier transform, fast
Filon method, 347, 356, 359, 363–4, 569
first-motion approximation, 322, 346, 356, 420–21
fluid, 394
focusing layer, 42
Fourier integral operator, 463
Fourier series, 68, 360
Fourier transforms, 560–3
Bessel and Hankel, 562–3
exponential, 560–1
exponentials and inverse square roots, 562
fast, 360
inverse square roots, 561
Fourier–Bessel transform, 68–9, 74, 325, 335, 358
free surface, xv, xvi, 198
French model, 240–4, 526–9, 544
frequency, xviii, 7
Fresnel cosine and sine integrals, 582
Fresnel function, 581–6
Fresnel integrals
asymptotic, 582–3
inverse Fourier transform, 583–6
Fresnel shadow, xxiii, 30, 35, 378, 430–3, 448–9, 585
spectral function, xxii, 583
time function, xxii, 584
frustrated total reflection, 40
full-wave theory, 357
fundamental matrix, 254, 564

general curvilinear coordinates, 78
generalized ray response, 272, 311–3
generalized rays, 272
generalized reflection and transmission coefficients, 272
generalized time function, xviii
generating function, 21
geometrical Green dyadic, *see* asymptotic ray theory
geometrical ray approximation, 138, 159, 244, 497
GRA *see* geometrical ray approximation
GRA compensated amplitude coefficients, 497
gradient operator, xiv–xv
gravity, xv, 308
grazing ray, 28
Greek subscript, xiii, 67
Green function, 76, 100–18
free space, 534
isotropic, homogeneous, 112–8
line force dyadic, 116–8
notation, xiii
point force dyadic, 114–6
units, xix
Green Horn shale, 189, 489, 501
group velocity, xviii, 166, 167, 175

half-space, xv, xvi, 10
Hamilton equations, 143
Hamilton's principle, 144
Hamilton–Jacobi equations, 141, 165
Hamiltonian, xx, 141, 143, 165, 169, 186, 190
Hankel function, 69, 325, 334
Haskell matrix, 255, 300, 302
head wave, xxiii, 15, 27, 31, 32, 41, 310, 353, 386–90, 409, 421–5, 436–8, 457
general, 388–90
length, xxi
Heaviside step function, xxii, 137
Helmholtz equation, 6
Herglotz–Wiechert–Bateman method, 45
hexagonal anisotropy, *see* transversely isotropic medium

hidden layer problem, 40
high-velocity layer, 38
Hilbert transform, 63–4, 73, 74, 159, 245, 291, 347, 412, 417, 418, 420, 424, 447, 465, 469, 471, 483, 544, 561
homogeneous, xi, *see also* inhomogeneous
  boundary conditions, xi
  half-space, 8, 14
  media, 254–5
  model, 6, 8, 13
Hooke's law, 90
horizontal, xv
hydrostatic pressure, 83

identity matrix, xiv
imaginary part, xxii
impedance, xx
  scalar, 177
  tensor, 166, 177
incomplete saddle, 448
incompressible solid, 394
infinitesimal deformation, 81
infinitesimal layer, 274–7
infinitesimal rotation, 85
inhomogeneous, xi, *see* homogeneous
  medium
  wave equation, xi
  waves, xi, 11
integral matrix, 564
integrals, 555
integration, 64
intercept-time function, xviii, 19, 20
interface, xv, xvi, 7, 10, 14, 198
  fluid-solid, 215
interface basis vectors, xxi, 199
interface polarization conversion, xxi, 228–30
  acoustic, 228–9
  anisotropic, 229–30
  isotropic, 230
interface wave, xxiii, 378, 392–402, 439–40
invariant Gibbs notation, xv
inverse transforms, 311–2
isochron, 343–6
  line, 518
  surface, 518, 525, 539, 542
isotropic elastic waves, 110–18
isotropic medium, 96–9

Jacobi identity, 156, 255, 567
Jacobian, 71, 567, 572

Kennett's ray expansion, 263–71, 292, 308
  receiver rays, 268–71
  source rays, 268–71
kinematic analogues, 273
Kirchhoff approximation, 535, 537
Kirchhoff surface integral method, 459, 532–50
  acoustic, 533–44
  anisotropic, 544–50
  band-limiting, 542, 544
  scalar scattering term, 547
  scattering kernel, 547
  three dimensions, 542–4
  two dimensions, 538–42
KMAH index, xx, 159, 176, 195, 291, 417, 420–2, 531, 539, 541, 543, 544
Knopoff method, 301
Kronecker delta, xiv

Lagrangian, 143, 144, 165, 169
Lagrangian brackets, 154
Lagrangian coordinates, 78, 81
Lamb's problem, 310, 313
lambda function, xxii, 117, 160, 321, 336, 348, 380
Lamé elastic parameters, xix, 97, 235
Langer asymptotic expansion, 247, 284, 286–308, 342, 449
  differential matrix, xxi
  elastic waves, 292–308
  propagator matrix, xxi
  transformation matrix, xxi
Langer decomposition, 302
Langer iterative solution, 299
Laplace transform, 60
layer, xv, xvi, 16
  phase propagator, xxi
layer stack, 260
layered model, xv, xvi
leaking Rayleigh wave, 399–402
leaky modes, 376
Legendre standard form, 575
Legendre transform, 19, 69–72, 74, 143, 144, 165, 169, 309
line explosion source, 320–3
linear anisotropic velocity, 190–3
linear squared-slowness medium, 189–90
linear squared-slowness model, 52
linear velocity model, 48
Liouville's theorem, 156, 472
locked modes, 375
longitudinal waves, 111
Love wave, 375
low-velocity shadow, 34

Maslov asymptotic ray, 240
Maslov method, 459
Maslov seismogram, 28, 30, 34, 35, 464, 523, 539
  band-limited, 464–5, 475–8
  geometrical arrivals, 465–6, 478–84
  three dimensions, 542, 544, 548
  two dimensions, 542
Matlab, 306
matrix, xii–xiv, *see also* vector-matrix algebra
  column, xii
  compound, 302
  identity, xii
  notation, xii–xv
  row, xii
  signature, xiii
  signature function, xxii
  sub-matrices, xiv

min-max phase, 418, 419
modified Bessel function, 562
Mohorovičić velocity function, 56, 57
moment tensor source, xix, 120, 123–7, 313
mu function, xxii, 423
multiple arrivals, 425
multiple triangular integrals, 555–9
∇, *see* gradient operator

Navier wave equation, 76, 100–18
near-field term, 113, 323, 340
Newton's second law of motion, 83
NMO, *see* normal moveout
normal moveout, 45
  correction, 46
  stacking, 46
normal rays, 242
Nyquist frequency, 351, 360, 386

obliquity factor, 549
order, xii
ordinary differential equations, 564–8

P quasi-isotropic rays, 495–6
P waves, 111
parabolic squared-slowness model, 57
paraxial ray equations, 149–56
paraxial rays, 176, 488
particle displacement, xix
particle velocity, xix, 83
  amplitude coefficients, xx
Pearcey caustic, 485
period, 7
perturbation coefficients, xxi, 198
perturbation matrix, 233, 278
perturbation model, 504
perturbed eigenvectors, 233
phase, 7
phase space, 143
phase velocity, xviii, 7, 175
physical tensor components, 78
PKIKP, 54, 355
PKiKP, 54, 355
PKP, 54, 355
plane waves, 6–11
plane-wave approximation, 383, 384
point source, 6, 12–5, 313, 330
point stress-glut source, 331
point symmetry, 167
point, pressure source, 324–30
Poisson brackets, 154
Poisson solid, 98, 394
Poisson's ratio, 129, 393, 395, 399, 458
polar anisotropy, *see* transversely isotropic medium
polar reciprocal, 169
polarization, 146, 167, 175, 195, 516, 520, 535
  degenerate, 492–4
  dyadic, xx
  energy-flux normalized, xx, 178
  generalized energy-flux normalized, 257
  infinitesimal-anisotropy, 493
  normalized, xx
  perturbations, 491–4
  shear ray, 180–1
polynomial depth-slowness model, 52
position vector, xviii, 7
Poynting vector, *see* energy flux vector
Preliminary Reference Earth Model, 54
pressure, xix
  amplitude coefficients, xx
pressure line source, 249–51
principal components, 162, 171
principal radii of curvature, 194
propagator matrix, 254, 309, 564–7
  chain rule, 258, 565
  inverse, 309, 565
pseudo-caustic, 466–72, 484–7
pseudo-differential operator, 463
pseudo-Rayleigh wave, 401

QI, *see* asymptotic ray theory, quasi-isotropic
qP wave, 167, 219
qS_1 wave, 167, 200
qS_2 wave, 167, 200
qSH wave, 186, 200, 219
qSV wave, 200, 219
α-quartz, 168
quasi-isotropic ray theory, *see* asymptotic ray theory, quasi-isotropic
quasi-shear component vector, 500
quasi-shear ray coupling, 499
quasi-shear rays, 167, 487
quasi-shear slowness surface, 168, 169
quasi-*P* rays, 167

radius, xv
Radon transform, 72–5, 138, 346
rainbow expansion, 377
range function, xviii, 17
ray code, 201, 237
ray cone, 141
ray descriptor, 135, 240
ray event, 237
ray expansion, 40, 136, 258, 260–77
ray history, 135, 201
ray integrals, 18, 183
ray parameters, 18, 142
ray path, 167, 176
ray perturbation theory, 488–94
ray phase term, xx
ray propagation term, xx
ray response, 271–4
ray scalar amplitude, xx
ray segments, 201
ray shooting, 140
ray signature, 198, 201, 237–40
ray speed, 169
ray spreading function, xx
ray table, 198, 237–40
ray theory, *see* asymptotic ray theory

ray tube, 147, 148, 172
ray velocity, 140, 166, 195
Rayleigh pole, 401
Rayleigh wave, 310, 375, 393–402, 458
  velocity, xxi, 395
real part, xxii
receiver conversion coefficients, xx, 198, 258, 516
receivers
  notation, xiii
reciprocity, 102–5, 108–9, 176, 195, 515, 519
reduced Δ-matrix method, 300
reduced travel time, xviii, 56
reference model, 504
reflection, 26, 258–60, 419–20
  partial, xxiii, 11, 15, 27, 29, 31, 32, 36, 41, 378–86
  partial in stratified media, 380–86
  total, xxiii, 10, 11, 15, 27, 31, 32, 41, 173, 378, 386, 390–2
reflection/transmission coefficients, xxi, 198, 207–24, 537, 541
  acoustic, 207–9
  anisotropic, 209–215
  denominator, xxi
  differential, 285
  fluid–fluid, 263
  fluid–solid, 225–8, 263
    anisotropic, 227
    isotropic, 227–8
  free surface, 224–25
    acoustic, 224
    anisotropic, 225
    isotropic, 225
  isotropic, 216–9
  linearized, 231–7
  reciprocity, 209, 214–5, 246
  small contrast
    acoustic, 235–6
    isotropic, 236–7
  transversely isotropic, 219–20
reflectivity method, 356
refraction, 30, 36
representation theorem, 105–7, 109–110, 508, 512, 533, 544
reverberation matrix, 265, 268
reverberations, 38, 415
  thin layer, 441–5
Ricatti differential equation, 194
Riemann sheet, 66, 318, 392, 393, 400, 401, 437
right-handed system, xv
root-mean-square velocity, 46
Rytov's field-vectors rotation law, 180

S quasi-isotropic rays, 496–503
S waves, 111
saddle point, 379, 380, 462, 569
saddle-point method, 569–86
  incomplete, 581–6
  multi-dimensional, 571–2
  second-order, 460, 464, 531, 540, 543, 569–72
  third-order, 572–80
saltus function, xxii
sans serif font, *see* vector, subspace
scalar amplitude function, 147
scalar Born error kernel, xxi
scalar Born perturbation kernel, xxi
scalar impedance, 147
scalar Kirchhoff kernel, xxi
scalar propagation, 178
scalar seismic moment source, 121
scaling rule, 64
scattering dyadic, xxi
Scholte wave, 401
second-order discontinuity, 35, 240, 281–2
second-order minors, 300–8
  notation, xxii
seismic potentials, 116
shadow, *see* deep shadow; Fresnel shadow; low-velocity shadow
shear wave degeneracy, 173, 487, 496
shear wave splitting, 200
shear-ray splitting, 499, 500
shift rule, 64
sign function, xiii, xxii
sign of the imaginary root, 10
signature, 176, 572
signature of matrix, *see* sign function
sinc function, 351
slant slack, 73
sloth, *see* linear squared-slowness model
slowness, xviii, 7, 65
  vector, 7, 167, 195
slowness integral
  avoiding singularities, 364–5
  improved convergence, 365–74
  numerical, 358–74
  three dimensions, 358
  two dimensions, 357
  weighting functions, 368–74
slowness method, real, 346–54
slowness surface, 165–70, 191
slowness vector, 9, 10, 13, 16, 138
Smirnov's lemma, 148, 172, 567–8
Snell's law, 8–11, 198, 201–4, 245, 532, 537, 540, 542, 548
source excitation, xx, 258
source radiation pattern, 113–15, 510, 514, 516
sources
  notation, xiii
spatial Fourier transform, 65–9, 72, 248, 311, 362–74
spectral method, 347, 356–74, 378, 425, 433–56
spectral reflection point, 539, 540
spherical Bessel function, 309
spherical spreading, 113
spherical system, xv, 376
spherical wave, 12, 13
spreading function, 158, 159, 175, 337
spreading matrix, 150
  string rule, 155–56
stacking velocity, 46, 73
static equilibrium, 80–81
stationary phase method, 434–6, 570

stationary phase method (*cont.*)
three dimensions, 434–5
two dimensions, 434
stationary point, 569
stiffness tensor, 90
Stokes equation, 287, 288
Stokes phenomena, 61
Stoneley wave, 365, 401
strain energy function, 94
strain tensor, xix, 76, 84–6
normal components, 86
shear components, 86
stress glut, 76
stress glut tensor source, 118–27
stress tensor, xix, 76, 78–84
Cauchy, 81
normal components, 82
Piola–Kirchhoff, 81
shear components, 82
symmetry, 81–3
superposition principle, 106
surface unit normal vector, xviii, 79, 81
symplectic symmetry, 153–4, 204, 309
symplectic transform, 153, 309

tangent transform, *see* Legendre transform
target surface, 461
tau-p curve, *see also* intercept-time function, 19
tau-p transform, 69–75
telescopic point, *see* pseudo-caustic
temporal Fourier transform, 58–64, 72–5, 248, 311, 360–2
tensor
dimensions, xii
order, xii
units, xii
thin interface limit, 383–6
TI, *see* transversely isotropic medium
time, xviii
traction vector, xix, 76, 78–84
amplitude coefficients, xx
transformed source matrix, xxi, 249, 251, 252
transforms, 58–75
transmission, 39, 258–60, 378
transverse waves, 111
transversely isotropic medium, 99–100, 129, 130, 185–8, 195, 200, 223, 314, 330
tranversely isotropic medium, 232
travel time, 136
travel-time curve, 18, 20
travel-time function, xviii, 13, 16, 17
travelling wave, 11
triplication, 30, 349
tunnelling wave, xxiii, 38, 39, 378, 403–15, 440–5
spectral function, xxii
time function, xxii
turning branch
reversed branch, 416
turning point, 21–4, 341, 343–6, 421
turning ray, 21, 22, 28, 31–3, 36, 37, 43, 44, 349
forward branch, xxiii, 34, 416
reversed branch, xxiii, 34, 417
two turning points, 417–9
two to three dimensions, 336–7
stationary phase method, 435–6
two-dimensional Green function, xxii

unit cartesian vector, 79
unit function, xxii
unit surface normal vector, 122, 198, 199, 534
units, xii
upwards direction, xv

Valhall gas-cloud model, 145, 478
vector, xii–xiv
normalization function, xxii
normalized, xii
notation, xii–xv
sub-space, xiii, 68
unit, xii
vector-matrix algebra, xiii, xiv
velocity maximum, 42
vertical, xv–xvii
Voigt notation, 91
Voigt tensor, 131
volume injection source, 101

Watson transform, 376
wave vector, 7
wavefront coordinates, xviii
wavefront curvature, xx, 154–5, 181, 194
wavefront propagator, 152
wavefronts, 8–10, 139, 166–70
wavelength, xviii, 7, 8
wavenumber, xviii, 7, 65
whispering gallery mode, 30
WKB paradox, 282, 341
WKBJ asymptotic expansion, 38, 247, 277–82, 286, 299, 313, 341, 347
WKBJ iterative solution, 34, 38, 247, 283–6, 299, 310, 313, 342–6, 380, 383, 505, 555
WKBJ seismogram, 28, 30, 34, 35, 310, 347–54, 376, 378, 436, 458–60, 463, 477, 523, 541, 542
band-limited, 351–5
first-motion approximation, 415–33
wrapped signals, 361

$X^2 - T^2$ method, 47

Young's modulus, 129

zero-offset rays, 242

For EU product safety concerns, contact us at Calle de José Abascal, 56–1°,
28003 Madrid, Spain or eugpsr@cambridge.org.

www.ingramcontent.com/pod-product-compliance
Ingram Content Group UK Ltd.
Pitfield, Milton Keynes, MK11 3LW, UK
UKHW051127260726
13967UKWH00010B/2913

* 9 7 8 0 5 2 1 8 9 4 5 4 8 *